ADVANCES AND TRENDS IN ENGINEERING SCIENCES AND TECHNOLOGIES III

PROCEEDINGS OF THE 3RD INTERNATIONAL CONFERENCE ON ENGINEERING SCIENCES AND TECHNOLOGIES (ESAT 2018), HIGH TATRAS MOUNTAINS, TATRANSKÉ MATLIARE, SLOVAK REPUBLIC, 12–14 SEPTEMBER 2018

Advances and Trends in Engineering Sciences and Technologies III

Mohamad Al Ali & Peter Platko
Institute of Structural Engineering, Faculty of Civil Engineering, Technical University of Košice, Slovak Republic

CRC Press
Taylor & Francis Group
Boca Raton London New York

CRC Press is an imprint of the
Taylor & Francis Group, an **informa** business

A BALKEMA BOOK

Library of Congress Cataloging-in-Publication Data
Applied for

Published by:
CRC Press/Balkema
P.O. Box 447, 2300 AK Leiden, The Netherlands
e-mail: Pub.NL@taylorandfrancis.com
www.crcpress.com – www.taylorandfrancis.com

First issued in paperback 2020

Typeset by V Publishing Solutions Pvt Ltd., Chennai, India

ISBN 13: 978-0-367-73127-4 (pbk)
ISBN 13: 978-0-367-07509-5 (hbk)

Visit the Taylor & Francis Web site at
http://www.taylorandfrancis.com

and the CRC Press Web site at
http://www.crcpress.com

Advances and Trends in Engineering Sciences and Technologies III – Al Ali & Platko (Eds)
© 2019 Taylor & Francis Group, London, ISBN 978-0-367-07509-5

Table of contents

Part B
Buildings and structures
Construction technology and management
Environmental engineering
Heating, ventilation and air condition
Materials and technologies
Water supply and drainage

Part C
Geodesy, surveying and mapping
Roads, bridges and geotechnics

Advances and Trends in Engineering Sciences and Technologies III – Al Ali & Platko (Eds)
© *2019 Taylor & Francis Group, London, ISBN 978-0-367-07509-5*

Preface

The International Conference on Engineering Sciences and Technologies (ESaT 2018) was organized under the auspices of the Faculty of Civil Engineering, Technical University in Košice – Slovak Republic and Peter the Great St. Petersburg Polytechnic University, Institute of Civil Engineering – Russia.

The 3rd annual Conference of ESaT was held on 12th–14th September 2018 in the scenic High Tatras Mountains, Tatranské Matliare, Slovak Republic.

Main purpose of the Conference was to organize a meeting focused on various fields of Civil engineering sciences that would stimulate valuable discussions on novel and fundamental advances in the areas of engineering sciences and technologies for scientists, researchers and professionals around the world. Conference participants from different universities, research institutes and companies all around the world had the opportunity to present their ongoing research activities and exchange their experiences with other colleagues.

These proceedings provide original developments and trends in various fields of engineering science and technology. Academics, scientists, researchers and professionals from universities and practice have contributed to the book that contains different topics and disciplines.

Conference Topics:

- Buildings and structures
- Computer simulation and modeling
- Construction technology and management
- Environmental engineering
- Geodesy, surveying and mapping
- Heating, ventilation and air condition
- Materials and technologies
- Mechanics and dynamics
- Reliability and durability of structures
- Roads, bridges and geotechnics
- Water supply and drainage

The editors would like to express their great thanks to the members of Organizing, Scientific and Reviewers Committees. The editors would also like to express special thanks to all reviewers, sponsors and conference participants for intensive cooperation to make this conference successful. Many thanks to the publisher for valuable advice and cooperation during the preparation of the Proceedings.

Mohamad Al Ali   Peter Platko

Advances in Engineering Sciences and Technologies III – AIAA & Piatkovskej (eds)
© 2020 Taylor & Francis Group, London ISBN 978-0-367-07300-9

Preface

The International Conference on Engineering Science and Technologies (ESaT 2015) was organized under the auspices of the Faculty of Civil Engineering, Technical University in Košice – Slovak Republic and Peter the Great St. Petersburg Polytechnic University hosting at Civil Engineering, Russia.

The 2... annual Conference of ESaT was held on 17...19 September 2015 in the scenic High Tatras Mountains, Tatranské Matliare, Slovak Republic.

Main purpose of the Conference was to organize a forum focused on various fields of Civil Engineering sciences that would stimulate valuable insights on novel and innovative advances in the areas of engineering sciences and technologies for scientists, researchers and professionals around the world. Conference participants from different universities, research institutes and companies all around the world had the opportunity to present their ongoing research activities and exchange their experiences with other colleagues.

The meeting time provide original ideas, concepts and novel in various related engineering science and technology. Academics, scientists, researchers and professionals from universities and practice have contributed to the pool that contains different topics and disciplines.

Conference Topics:

- Buildings and structures
- Computer simulation and modelling
- Construction technology and management
- Environmental engineering
- Geodesy, surveying and mapping
- Heating, ventilation and air condition
- Materials and technologies
- Mechanics and dynamics
- Reliability and durability of structures
- Roads, bridges and geotechnics
- Water supply and drainage

The editors would like to express their great thanks to the members of Organizing, Scientific and Reviewers Committees. Therefore, we would like to express great thanks to all supervisors, sponsors and conference participants for their continuous cooperation to make this conference success. With many thanks to the publisher for valuable advice and cooperation during the preparation of the Proceedings.

Michael ALAli   Peter Piatko

Acknowledgements

These proceedings are prepared within the research project VEGA No. 1/0188/16 "Static and Fatigue Resistance of Joints and Members of Steel and Composite Structures", supported by the Scientific Grant Agency of the Ministry of Education of Slovak Republic and by the Slovak Academy of Sciences.

Committees of the ESaT 2018

SCIENTIFIC COMMITTEE

Prof. Stanislav Kmeť
Rector of the University, Technical University of Košice, Slovak Republic

Prof. Vincent Kvočák
Dean of the Faculty of Civil Engineering, Technical University of Košice, Slovak Republic

Prof. Vitaliy V. Sergeev
Vice-rector of the University, Peter the Great St. Petersburg Polytechnic University, Russia

Prof. Nikolai I. Vatin
Institute of Civil Engineering, Peter the Great St. Petersburg Polytechnic University, Russia

Prof. RNDr. Magdaléna Bálintová
Director of the Institute of Environmental Engineering, Technical University of Košice, Slovak Republic

Prof. Mária Kozlovská
Director of the Institute of Construction Technology and Management, Technical University of Košice, Slovak Republic

Prof. Zuzana Vranayová
Director of the Institute of Architectural Engineering, Technical University of Košice, Slovak Republic

Assoc. Prof. Ján Mandula
Director of the Institute of Structural Engineering, Technical University of Košice, Slovak Republic

Prof. Abdel Hakim Abdel Khalik Khalil
Tanta University, Egypt

Prof. Mohamad A.E. Al Ali
University of Aleppo, Syria

Prof. Safar Al Hilal
Zirve University, Turkey

Prof. Ivan Baláž
Slovak University of Technology in Bratislava, Slovak Republic

Prof. Abdelhamid Bouchair
Blaise Pascal University, France

Prof. Ján Brodniansky
Slovak University of Technology in Bratislava, Slovak Republic

Prof. Radim Čajka
Technical University of Ostrava, Czech Republic

Prof. Ján Čelko
University of Žilina, Slovak Republic

Assoc. Prof. Jana Peráčková
Slovak University of Technology in Bratislava, Slovak Republic

Assoc. Prof. Elżbieta Radziszewska-Zielina
Cracow University of Technology, Poland

Assoc. Prof. Sergej Priganc
Technical University of Košice, Slovak Republic

Assoc. Prof. Miloslav Řezáč
Technical University of Ostrava, Czech Republic

Assoc. Prof. Brigita Salaiová
Technical University of Košice, Slovak Republic

Assoc. Prof. Anna Sedláková
Technical University of Košice, Slovak Republic

Assoc. Prof. Vojtěch Václavík
Technical University of Ostrava, Czech Republic

Assoc. Prof. Silvia Vilčeková
Technical University of Košice, Slovak Republic

Assoc. Prof. Roman Vodička
Technical University of Košice, Slovak Republic

Assoc. Prof. Martina Zeleňáková
Technical University of Košice, Slovak Republic

Dr. Mohamad Al Ali
Technical University of Košice, Slovak Republic

Dr. Ali M. Almagar
Environmental Energy Engineering Consultants, Saudi Arabia

Dr. Enayat Danishjoo
THK Rhythm Automotive GmbH, Germany

Dr. François Hanus
ArcelorMittal Global R&D, Luxembourg

Dr. Bernhard Hauke
Bauforumstahl e. V., Germany

Dr. Peter Platko
Technical University of Košice, Slovak Republic

Dr. Jaroslav Vojtuš
Technical University of Košice, Slovak Republic

ORGANIZING COMMITTEE

Dr. Mohamad Al Ali & Dr. Peter Platko
Technical University of Košice, Slovak Republic, esat2018@esat-conference.com

Dr. Štefan Kušnír, Dr. Marcela Spišáková, Eng. Katarína Čákyová, Eng. Zdenka Kováčová, Eng. Adam Repel & Eng. Martina Wolfová
Technical University of Košice, Slovak Republic

MSc. Daria Zaborova
Peter the Great St. Petersburg Polytechnic University, Russia

REVIEWERS COMMITTEE

Prof. Vincent Kvočák
Institute of Structural Engineering, Technical University of Košice, Slovak Republic
Bearing Structures

Prof. Magdaléna Bálintová
Institute of Environmental Engineering, Technical University of Košice, Slovak Republic
Material and Environmental Engineering

Prof. Mária Kozlovská
Institute of Construction Technology and Management, Technical University of Košice, Slovak Republic
Construction Technology and Management

Prof. Zuzana Vranayová
Institute of Architectural Engineering, Technical University of Košice, Slovak Republic
Building Physics and Facilities

Assoc. Prof. Peter Blišťan
Institute of Geodesy, Cartography and Geographical Information Systems, Košice, Slovak Republic
Geodesy, Surveying and Mapping

Assoc. Prof. Ján Mandula
Institute of Structural Engineering, Technical University of Košice, Slovak Republic
Geotechnics and Traffic engineering

Prof. Nikolai I. Vatin
Institute of Civil Engineering, Peter the Great St. Petersburg Polytechnic University, Russia
Buildings and Architectural Engineering

LIST OF REVIEWERS

Abdelgaied M., Afonso A., Al Ali M., Aroch R., Arsić D., Bajzecerová V., Baláž I., Balintova M., Bašková R., Bilek V., Blišťan P., Brodniansky J., Brožovský J., Cachim P., Cajka R., Čabala J., Čelko J., Čop G., Demcak S., Demjan I., Djokovic J., Ďurčanská D., Dvorsky T., Ellingerová H., Estokova A., Fazekašová D., Fillo L., Gašparík J., Gravit M., Halvonik J., Harabinová S., Höger A., Holub M., Hulínová Z., Hybská H., Chodasova Z., Jankovichová E., Jármai K., Junák J., Kabeel A. E., Kanócz J., Kanuchova M., Karmazínová M., Katunský D., Keršner Z., Kisilewicz T., Kmeť S., Konasova S., Kormaníková E., Korsun V., Košičanová D., Kotrasová K., Kozik R., Kozlovska M., Krawiec S., Kryžanowski A., Kušnír M., Kušnír Š., Kvočák V., Lazarev J., Lazarová E., Lazić V., Lee M. C., Lesniak A., Lopusniak M., Luptakova A., Mačingová E., Mandičák T., Mandula J., Markovič G., Mečiarová Ľ., Melcher J., Mesároš P., Miladinovic S., Molčíková S., Nagy R., Nemova D., Nikolić R., Novotný M., Ondova M., Panulinová E., Pauliková A., Petrichenko M., Pipiska M., Plášek J., Platko P., Poórová Z., Radziszewska-Zielina E., Reiterman P., Rovňák M., Řezáč M., Sabol P., Salaiová B., Samešová D., Sandanus J., Sedláková A., Schnabl S., Sičáková A., Spisakova M., Stec A., Strigáč J., Struková Z., Šoltýs R., Švajlenka J., Talian J., Tažiková A., Terpakova E., Tomko M., Vatin N., Veličković S., Venkrbec V., Vernársky P., Vičan J., Vilcekova S., Voznyak O., Vranayová Z., Zach J., Zeleňáková M., Zima K., Zinicovscaia I., Zöld A., Zozulák M.

Part A
Computer simulation and modeling
Mechanics and dynamics
Reliability and durability of structures

Advances and Trends in Engineering Sciences and Technologies III – Al Ali & Platko (Eds)
© 2019 Taylor & Francis Group, London, ISBN 978-0-367-07509-5

Behavior of the 2D frames for different approach to soil modeling

V. Akmadzic & A. Vrdoljak
Faculty of Civil Engineering, University of Mostar, Mostar, Bosnia & Herzegovina

ABSTRACT: The paper presents the influence of the soil reaction coefficient modeling on the 2D frame behavior. The foundations are square and the ground is granular. Different authors perform different expressions for values of the soil reaction coefficient. So, for each of these coefficients, a 2D frame was calculated using the Tower 3D Model Builder program package. In the next step the average value of the soil reaction coefficient is obtained by the SE_Calc software. Then the dimensions of the foundation were corrected in the way to obtain the same response value of the reaction coefficient for each author. The consequences of the both approaches on the behavior of the 2D frame are shown.

1 INTRODUCTION

Generally, any constructive system is commonly based on the concrete foundation. In the case of shallow foundations, Winkler springs modeling is used to model the connection between the soil and the foundation. This implies that it is necessary to determine the soil reaction coefficient, which in its essence represents the relationship between the stress under the foundation and its deflection. Numerous authors have dealt with this issue and have given their solutions. That is why the end user remains in a dilemma over which solution to choose and what consequences this solution would have on the behavior of the static system. In the following text the answer to this dilemma will be presented on a simple 2D frame example.

2 SOIL REACTION COEFFICIENT

2.1 *Software SE_Calc—the programmatically solution*

As explained with more details in references (Akmadzic 2018), a software SE_Calc—the programmatic solution—was developed for the purpose of determining the soil reaction coefficients by different authors' expressions. It has been tested on various examples from literature. Therefore, only the expressions and the table values of the coefficients are mentioned in this paper.

The expressions are (Nonveiller 1990, Selimović 2000):

a. by Vesic:

$$k_s = \frac{0.65 \cdot E_s}{B \cdot (1 - v^2)} \cdot \sqrt[12]{\frac{E_s \cdot B^4}{E_b \cdot I}}$$

(1)

where Es = modulus of foundation elasticity; I = moment of inertia; and B = foundation width.

b. by Biot:

$$k_s = \frac{0.95 \cdot E_s}{B \cdot (1 - v^2)} \cdot \left[\frac{E_s \cdot B^4}{(1 - v^2) \cdot E_b \cdot I} \right]^{0.108} \tag{2}$$

c. by Meyerhof & Baike:

$$k_s = \frac{E_s}{B \cdot (1 - v_s^2)} \tag{3}$$

d. by Kloppe & Glock:

$$k_s = \frac{2 \cdot E_s}{B \cdot (1 + v_s)} \tag{4}$$

Table 1. Comparison of different coefficient of subgrade reaction.

Author	Area m²	Coefficient kN/m³	Average kN/m³	Deviation %
Vesic	1.00	104107.95	137955.16	−24.53
Biot	1.00	162490.76	137955.16	+17.79
Meyerhof & Baike	1.00	130208.33	137955.16	−5.62
Kloppe & Glock	1.00	208333.33	137955.16	+51.02
Selvadurai	1.00	84635.42	137955.16	−38.65

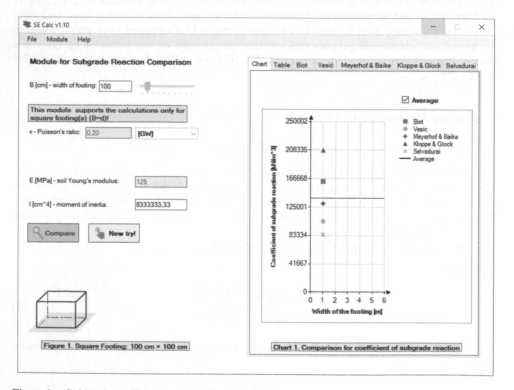

Figure 1. Subgrade coefficient reaction from SE_Calc.

4

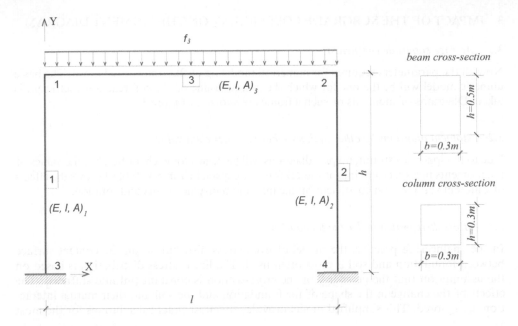

Figure 2.   2D frame.

Table 2.   Different foundation dimensions B for the same subgrade coefficient.

Author	Area m²	Side B m'	Average subgrade coefficient kN/m³
Vesic	0.56	0.75	137955.16
Biot	1.39	1.18	137955.16
Meyerhof & Baike	0.90	0.95	137955.16
Kloppe & Glock	2.25	1.50	137955.16
Selvadurai	0.40	0.63	137955.16

e. by Selvadurai:

$$k_s = \frac{0.65 \cdot E_s}{B \cdot (1 - v_s^2)} \tag{5}$$

In the development of the latest version for programmatic solution SE_Calc, we have gone a step further. Now the program can calculate the mean value of the ground reaction coefficient for all considered authors. This average value will be the starting point for tracking the behavior of the 2D frame. Therefore, the ground reaction coefficients for a base dimension of the square 1.00 m × 1.00 m foundation are shown in Table 1 and graphically in Figure 1.

Looking back, it is interesting to see in Table 2 what would happened with the dimensions B and the area of the foundation if the coefficient of soil reaction by each author would have a value as the average one.

2.2   2D frame

For the purpose of demonstrating the structural response to different values of the soil reaction coefficient, a simple frame was taken. Its range is $l = 6.00$ m and height $h = 4.00$ m. It is made of concrete C25/30. The foundations of the frame are dimensions 1.00 m × 1.00 m and the thickness is 0.60 m. It is loaded with uniformly distributed load in the amount of 30 kN/m². The load also includes the own weight of each construction element.

3 IMPACT OF THE SUBGRADE COEFFICIENT ON THE MOMENT DIAGRAM

3.1 *Average subgrade coefficient*

Now all the parameters required for numerical modeling of 2D frames are known. The basic numeric model will be the one for which the average value of the soil reaction coefficient is taken. Diagrams of moments of such a frame are shown in Figure 3.

3.2 *Moment diagrams by other authors subgrade coefficient values*

Due to the space constraint, no plot diagrams will be drawn for each author, but the values of the moments in a special points of the 2D frame are given in Table 3. Table 4 shows the effects of different soil reaction coefficient on the maximum displacements and rotation.

3.3 *Stress distribution under the foundation*

In the engineering practice, the model of even stress distribution on the contact surface between foundation and soil is most often used. The linear stress distribution is based on the assumption that the geometry of the cross-section is constant (infinite stiffness). The effects of the change in the shape of the foundation and the soil and their mutual interaction are ignored. This simplified model provides the most practical solutions for the most

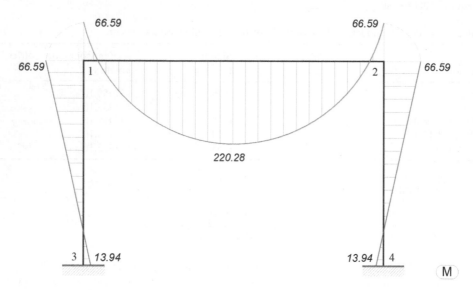

Figure 3.   Diagram of moments.

Table 3.   Comparison of the moments on 2D frame.

Author	Point 1 kNm	Point 3 kNm	Maximum kNm	Deviation %
Vesic	65.69	11.52	221.18	+0.41
Biot	67.13	15.40	219.75	−0.24
Meyerhof & Baike	66.40	13.43	220.47	+0.09
Kloppe & Glock	67.95	17.68	218.93	−0.61
Selvadurai	65.06	9.94	221.81	+0.69
Average	66.59	13.94	220.28	0.00

Table 4. Comparison of the displacements.

Author	max X mm	max Y mm	max Zr rad/1000	Deviation %
Vesic	2.71	−10.32	3.90	+5.52
Biot	2.60	−9.52	3.86	−2.66
Meyerhof & Baike	2.65	−9.88	3.88	+1.02
Kloppe & Glock	2.54	−9.19	3.84	−6.03
Selvadurai	2.76	−10.83	3.91	+10.73
Average	2.64	−9.78	3.87	0.00

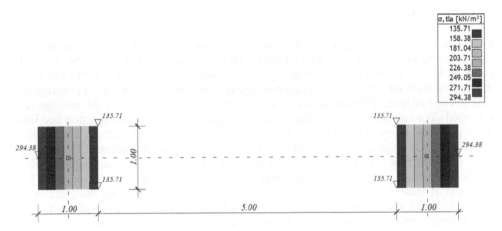

Figure 4. Stress distribution.

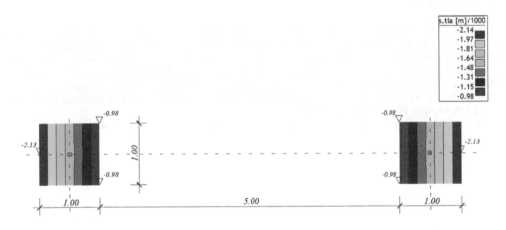

Figure 5. Vertical displacement of the foundation.

engineering problems. Diagram of stress distribution for the average subgrade coefficient is shown in Figure 4, while in Figure 5 the diagram of the vertical displacement of the foundation is shown.

Due to the space constraint, the values of the maximum and minimum stresses and displacements of the edge points in the contact area are given in Table 5. Comparing the results from the Table 5 and Table 1 it is obviously that the higher subgrade coefficient gives the higher strain.

Table 5. Comparison of the stress and vertical foundation displacement.

Author	σ (max) kN/m^2	σ (min) kN/m^2	s (max) mm	s (min) mm
Vesic	280.86	149.33	−2.70	−1.43
Biot	302.67	127.36	−1.86	−0.78
Meyerhof & Baike	291.52	138.60	−2.24	−1.06
Kloppe & Glock	315.58	114.32	−1.51	−0.55
Selvadurai	271.70	158.54	−3.21	−1.87
Average	294.38	135.71	−2.13	−0.98

4 CONCLUSION

For the numerical modeling of the structure it is necessary to know the soil characteristics so that it can be simulated by the Winkler spring model (Prskalo 2010, Caselunghe 2012). The paper deals with the simplest 2D frame supported on shallow square foundations. The rigidity of the ground is set by the soil reaction coefficient. The software SE_Calc—a programmatic solution—that gives different values of subgrade coefficient, as well as the average value, was used. The calculation results show that the higher value of the soil reaction coefficient gives less peak moments and fewer vertical displacements of the structure. The deviation of the midpoint moment value is within 1.00% and the vertical displacement to about 10%. Also, from the point of view in regard the stress under the foundation, the higher subgrade coefficient gives the higher stress and lower vertical displacement of the foundation. From the point of view in regard the departure of the results, the approach that using the average subgrade coefficient seems to be correct. The results from the new analyses of multiple and spatial structures will be considered for further development of the software.

REFERENCES

Akmadzic, Vlaho & Vrdoljak, Anton 2018. Determination of the soil coefficient reaction value—software solution, *e-Zbonik: Electronic collection of papers of the Faculty of Civil Engineering*, Issue 15.
Caselunghe, Aron & Eriksson, Jonas 2012. *Structural Element Approaches for Soil-Structure Interaction*, Göteborg: Department of Civil and Environmental Engineering, Chalmers University of Technology.
Nonveiller, Ervin (ed. 3) 1990. *Mehanika tla i temeljenje građevina*, Zagreb: Školska knjiga.
Prskalo, Maja, Akmadzic, Vlaho & Colak, Ivo 2010. Numerical modeling of raft foundations, Annals of DAAAM & Proceedings: p. 851–852.
Selimović, Mustafa (ed. 2) 2000. *Mehanika tla i temeljenje*, Mostar: Građevinski fakultet.

Advances and Trends in Engineering Sciences and Technologies III – Al Ali & Platko (Eds)
© 2019 Taylor & Francis Group, London, ISBN 978-0-367-07509-5

The change of the NVH characteristics of composite vehicle components as a result of visible and not visible damages

S.J. Alsarayefi & K. Jalics
University of Miskolc, Miskolc, Hungary

ABSTRACT: The paper presents the investigation of the Noise, Vibration and Harshness (NVH) behavior of damaged and not damaged fiber reinforced polymer FRP test probes with simple geometry. For that purpose, artificial defect/failure generation procedure for the test probes was developed and applied. Afterwards some NVH measurements, e.g. natural frequencies, modal damping and mode shapes were evaluated. The results show on one hand the sensitivity of the measurement method for damages and on the other the change of the NVH characteristics of the probes through damages.

1 MOTIVATION

Composite materials, e.g. fiber reinforced polymers (FRP), are increasingly utilized in the vehicle and machine industry. FRP components can show damages that are not visible after an impact or a crash. A component that is damaged (even not visibly) loses its load capacity, its original energy absorption capacity and its NVH behavior can change significantly.

2 GOAL OF THE INVESTIGATIONS

In order to understand the changes of the NVH behavior of the FRP vehicle components which arise due to certain damages (e.g. fiber or matrix cracks), the investigation of simple test specimen is recommended first. For that purpose, overall 20 specimen were prepared, which makes the repetition of the planned investigations possible, respectively allows some statistics. The generation procedure of the damages on the specimen should be also considered. The high-speed impact of balls, bullets, etc. by means of drop tests, or gun seems to be simply realized.

One of the test methods that can describe the NVH characteristics of a structure, is the modal analysis. This method delivers the modal frequencies, mode shapes and modal damping where the parameters can be more or less responsible for the NVH behavior of a vehicle. If the excitation meets a modal frequency, and the mode shape at that modal frequency has a good radiation efficiency, or the path of vibration transfer from that part to another in the vehicle is sufficient, this method is applied for the investigations. After getting the results of the modal test, they were analyzed, and the change of the modal behavior between undamaged and damaged specimen was explained. At the end the eligibility, respectively the sensitivity of the modal analysis method for crack detection at FRP were verified.

3 TYPES OF THE FRP MATERIAL

The general composition of the FRP consists of fibers (carbon, glass, etc.) and resin (polymer). The common types of fibers are armed, glass, carbon and basalt.

Figure 1.   The rectangular FRP test specimen (with the marks of the measurement points).

The resin can be of two categories:

– Thermoset Resin: It is common in structural uses; it cannot be reformed.
– Thermoplastic Resin: It has the property to be reformed.

Common Types of Thermoset Resin are polyester, vinyl ester, polyurethane and epoxy (Amin 2017).

4   THE TEST SPECIMEN

For the further investigations, a glass reinforced plastic plate with the material type of MF GC 201 (melamin resin laminate) was selected due to the simple accessibility and similar types are often used by vehicles. The specimen had a simple rectangular shape with the dimensions of $500 \times 200 \times 3$ mm (Figure 1).

5   POSSIBLE DAMAGES IN COMPOSITES

The mechanism of damages in composite materials is not easily predicted and understood due to the nature of the material. Defects and fracture generally may occur during the manufacturing process or service life of the structure or parts. The damage mechanisms in a fibrous composite are broadly categorized as:

A. Micro-level damage: This can be classified into fiber level damage and matrix level damage mechanisms. Regarding both levels, many damages can occur such as fiber breaking, fiber buckling, fiber bending, fiber splitting, and matrix cracking.
B. Macro-level damage: The macro-level mechanisms are laminate level mechanisms. It is seen that the adjacent layers are bonded together by a thin layer of resin between them. This interface layer transfers the displacement and force from one layer to another layer. When this interface layer weakens or damages completely, it causes the adjacent layers to separate. This mode of failure is called delamination.
C. Coupled Micro-Macro Level Failure Mechanisms: The through thickness transverse crack may propagate to neighboring lamina causing it to break (Ever 2010).

6   THE GENERATION OF ARTIFICIAL DAMAGES IN THE COMPOSITE STRUCTURES

In order to generate any type of damages (e.g. matrix cracks) in the material of the test specimen as mentioned in section 5, several experiments were performed. For the first considerations, a simple bearing ball drop test was considered. Within the test a roughly 8.5 m drop height in the building, which was proved insufficient to generate any damage. In this case the maximum kinetic energy could be achieved of the falling ball with m = 20 g weight was approx. 1.6 J. The ball weight was increased to m = 40, so the kinetic energy achieved was 3.2 J, without success. The point of the ball-specimen contact could not even be found.

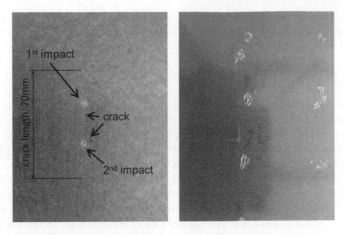

Figure 2.   Damage of the specimen (left: side of the impact; right: back side).

Thus, an air gun, type Diana 300R cal.177 was useful. It has the maximum kinetic energy of E = 7.5 J at the muzzle. The gun "fires" lead pellets of the weight m = 0.53 g of a calculated speed at the muzzle of 170 m/s. After firing with this air gun from 10 m distance, the contact place could be found as a small, flat buckle. The hit could be detected also on the back side of the plate. After the repetition of the shots from 7,5 m, 5 m, and 2,5 m, small flat buckles were detected without any cracks of the specimen.

At the next firing tests, the distance to the specimen was reduced to 1.5 m and two shots were fired. At the first shot only a small buckle could be seen (Figure 2 left: 1st impact), but by the 2nd shot the specimen has cracked, producing a crack of a length of 70 mm (Figure 2 left and right). So, that cracked specimen was used for the further investigations.

7   THE METHOD OF INVESTIGATION

7.1   *The modal analysis in general*

The experimental modal analysis is a techniques that has been used to study dynamic characteristics of mechanical components. furthermore, in recent years the modal analysis has been used to measure the performance and efficiency loss of working structures as a result of eventual degradation of individual components. The progression of such factors leads gradually to change the material performance resulting in changes the variation characteristics. The characteristics meant in such cases are modal frequencies, mode shapes and damping properties of structural components. These parameters can be delivered by modal analysis testing. the experimental modal analysis are proven to be able to test in site structures and with even large scales.

There are two methods used for the excitation of the structure for modal measurements. Namely, these two methods are the input-output (active excitation) and output only (operational excitation). In the Input-output methods of excitation, the procedure involved consists of an exciting function introduced to initiate vibration of the structure. Typically, the forms of excitation entail are impact hammers, drop weights, shakers or displacement-release. The waveforms used in modal analysis can be of various natures, including harmonic and random input, as well as impulsive excitation. In the Output only method, excitation is present if the structure is in service and under some form of external excitation, e.g. traffic or wind loads. In field testing, dynamic properties are extracted by placing several motion sensors (commonly accelerometers) at predetermined locations along the structure. To suit the need for full-motion recording, triaxial accelerometers are commonly given preference. The objective of placing sensors in multiple locations is to attain a sufficient amount of frequency response functions (FRF), such that individual modes can be identified from the modal test. Herein, the highest measurable mode depends largely on the optimal placement of accelerometers; i.e., the extraction of higher modes demands a higher number of accelerometers (Karbhari et al. 2005).

7.2 *Performing the test*

Before performing the tests, overall of 30 measurement points on the plate were defined, to be able to represent also the mode shapes, which tend to have more local displacement, than global. This amount of points should be sufficient enough for that. The distance between the points was 50 mm along the long side and 75 mm along the short side of the plate (visible in Figure 1). Measurement point Nr. 30 was also the excitation point.

For the measurements, the test equipment B&K Pulse frontend, B&K Pulse Labshop, B&K 4397 uniaxial accelerometer and Endevco 2202-10 impact hammer were used. It was decided to perform the fixed hammer excitation method. The schematic representation of the measurement setup is described in Figure 3 (Ward Heylen et al, 1998). the plate was laid on elastic foam supports on both ends and one accelerometer was placed on the 1st measurement point. After the hit was done on the excitation point 10 times, the recorded FRFs were averaged. The measurement also were repeated for the remaining measurement points. During the test, the quality of the FRFs, the coherences, the spectrum and time signal of the excitation were checked (e.g. to avoid double hits). A typical FRF, coherence and autopower spectrum of excitation is shown in Figure 4. The quality of the measurements seems to be logical; the excitation shows no significant drop of level over the frequency and the coherence, beside a few anti-resonances, is high, reaches the value of nearly 1. Before the test on the probe which was cracked by the shots, recording of a few FRFs on 6 probes were performed with the same material and dimension (incl. the probe which was cracked later). The goal was to see the scatter of the resonance frequencies on similar/same probes. The FRFs showed no significant differences between the resonance frequencies.

7.3 *Results of the test*

At first sight on the average of the transfer functions (Figure 5), no significant difference can be seen in the overall characteristics of the FRFs. From 550 Hz there is some difference, but this is rather due to the poor excitation level and the high damping of the material. Beyond that frequency no more distinct resonances can be found in the FRFs.

Considering the individual resonances (roughly 20 pieces up to 500 Hz) in the range from 110 to 190 Hz (Figure 6), differences in level of the peaks, and missing new peaks can be found in the averaged FRF of the cracked plate compared to the uncracked one. It can be stated that an existing crack in the material causes appearance and disappearance of certain resonances. Presumably, the modes are not appearing/disappearing, they are always there,

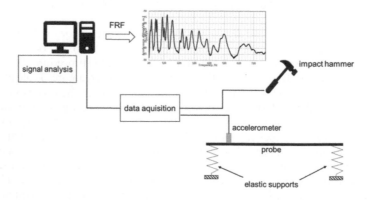

Figure 3.   Schematic representation of our test setup.

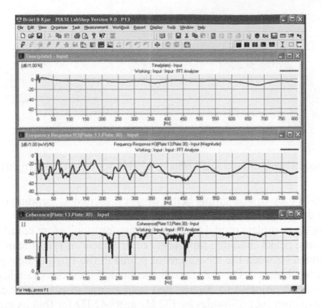

Figure 4. Typical excitation spectrum, FRF and coherence during the measurements (upper: autopower of excitation; middle: FRF; lower: coherence).

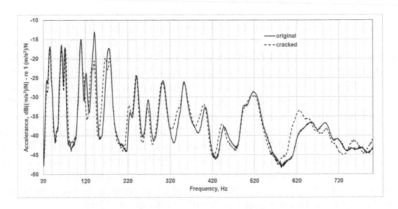

Figure 5. The average of the measured FRFs for the uncracked and for the cracked specimen.

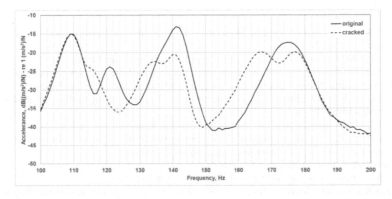

Figure 6. The average of the measured FRFs for the uncracked and for the cracked specimen from 100 to 200 Hz.

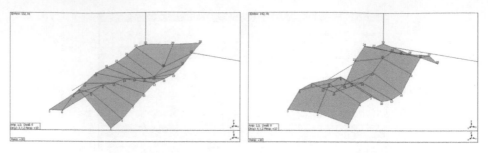

Figure 7.   Two mode shapes of the cracked plate by 132 Hz (left) and 142 Hz (right).

but due to the crack they will be simply better or worse excited, depending on the frequency, mode shape and damping.

By performing the complete modal analysis, the assumption before could be proved. That means the resonance peaks changing (appearing or disappearing) significantly, where a large relative displacement between the measurement points in the surrounding of the crack of a mode shape can be observed. On the cracked specimen the crack is ranging approximately from the measurement point 14 to 20, and here also a large displacement can be observed. Figure 7 shows two examples of modes by 132 Hz and 142 Hz for that as evidence.

8   CONCLUSION, FUTURE WORK

Typically, The modal analysis can help to detect a crack in the FRP material. In this investigation, The method can only work in the low frequency range due to the poor excitation by the impact hammer over 800 Hz. With an excitation by a light shaker, this limit can be raised, but due to the gradually growing modal density, the investigation of resonances and modes will be difficult. In this case as it is known that there is a crack, so some changes in the modal behavior of the specimen were expected. In real cases by a relatively complex shaped vehicle component, other effects have to be considered, even a not optimal support of the component by the test can cause some change in the modal behavior. The modal analysis still requires a further investigation for the purpose of crack detection in the FRP, but also in general for other types of materials. At the time of writing this paper only a single test is existing, so the test has to be repeated several times, in order to obtain the sensitivity of the method for external influences and for the crack generation. More promising acoustic method for crack detection is e.g. the acoustic emission which will be also utilized later.

REFERENCES

Amin, Ghobadi 2017. Common Type of Damages in Composites and Their Inspections. *World Journal of Mechanics* (7): 24–33
Cremer, L. & Heckl, M. 2005. *Structure—Born Sound. Berlin*: Springer Verlag.
Ever, Barbero 2010. *Introduction to Composite Materials Design*. London, New York: CRC Press.
Ewins, D.J 2000. *Modal Testing: Theory, Practice and Application*. Philadelphia: Research Studies Press.
Hajel, P. & Soeiro, F.J. 1990. Structural damage detection based on static and modal analysis. *AAIAA Journal* 28(6): 1110–1115.
Randall, R.B. 1987. *Frequency Analysis*. Denmark: Brüel & Kjaer.
Ramesh, Talreja 2012. Damage Mechanics of Composite Materials, *Department of Aerospace Engineering, Advanced & Composite & Materials*. Texas A&M University.
Karbhari, V.M, & Kaiser, Henning 2005. *Methods for Detecting Defects in Composite Rehabilitated Concrete Structures*. Washington: Federal Highway Administration.
Ward, Heylen & Lammens, Stefan & Sas, Paul 1998. *Modal analysis theory and testing*. Belguim: Katholieke University Leuven.

Advances and Trends in Engineering Sciences and Technologies III – Al Ali & Platko (Eds)
© 2019 Taylor & Francis Group, London, ISBN 978-0-367-07509-5

Long-term bending test of adhesively bonded timber-concrete composite slabs

V. Bajzecerová
Faculty of Civil Engineering, Technical University of Košice, Slovakia

J. Kanócz
Faculty of Art, Technical University of Košice, Slovakia

ABSTRACT: Long-term behavior of timber-concrete composite members is significantly influenced by the stiffness of the connection system. The highest stiffness can be achieved by adhesive. In this paper, results of long-term bending test of timber-lightweight concrete composite floor elements with adhesive shear connection are presented. The composite connection of two materials by the adhesive for bonding wet concrete to timber was performed. Two types of composite slabs were used. From the measured values of deflection, the global creep coefficient of both investigated composite beams was estimated. After 3 years of testing, the measured deflection of the specimens overestimates the limit values for service limit state of beams for ceiling structures. These results show that the creep and shrinkage of materials, as well as the influence of the environmental changes, can not be neglected in the design process. Comparison of the test with results of simplified calculation model is finally presented.

1 INTRODUCTION

Timber-concrete composite members such as ceiling structures or bridges have many advantages, so their use in construction increases (Dias et al. 2016). The joining of timber and concrete can be achieved in many ways, but in principle, it is possible with flexible mechanical connections or with an adhesive rigid connection.

Long-term behavior of timber-concrete composite members includes phenomena such as creep of materials, mechano-sorptive creep of timber, concrete shrinkage and the different reactions of connected materials to temperature and humidity changes of environment. The impact of these phenomena on the behavior of the timber-concrete elements is significantly influenced by the stiffness of the connection system. For example, the higher is the stiffness of the connection, the more the concrete shrinkage influences the deflection and the internal stresses of the composite element. With lower stiffness of the composite connections, these phenomena are partly eliminated by the slip of the mechanical fasteners. Even this high stiffness of the connection, which is advantageous in terms of short-term bearing capacity, is a disadvantage in terms of long-term action. Therefore, the long-term behavior of timber-concrete composite elements with rigid connection must be carefully analyzed.

Bonding the wood and concrete can be made in two ways: by bonding precast cured concrete slab and the wood element, or by application of fresh concrete in the wet adhesive applied to the wood element—wet on wet (Brunner et al. 2007, Negrão et al. 2010, Eisenhut et al. 2016b, Schmid et al. 2016). Both methods have their advantages and disadvantages. Bonding of the precast concrete slab significantly eliminates the effect of the concrete shrinkage and creep. In this case, however, it is necessary to adhere to strict bonding conditions, to provide for pressing pressure and reinforcement to concrete. In the second method, the production is less time-consuming and there is no need for concrete reinforcement, but conversely, the shrinkage of the concrete causes the increase of deformation and stresses in the composite cross-section in time.

Few bending tests of the adhesively bonded timber-concrete composite beams under long-term loading were published. In (Eisenhut et al. 2016a), long-term bending test of composite beams with precast concrete slabs is presented. Duration of the test was 2.0 years. In (Tannert et al. 2017), 4.5 years long bending test is presented. Within this experiment, the cast of concrete was performed on site, into the wet adhesive applied on timber surface.

In this paper, results of long-term bending test of the timber-lightweight concrete composite slabs with adhesive connection are presented. The adhesive connection was performed using the method wet on wet. Two slabs were subjected to a long-term load for 3 years. The specimens were produced in the previous research (Kanócz et al. 2014). In this study, members with the same mechanical properties were subjected to the short-term bending test. Results show that using the adhesive connection is effective and the connection is rigid. The mode of the failure was in all cases brittle.

Determination of the short-term load carrying capacity of the composite timber-concrete element can be performed by the γ-model presented in Additions B of EN 1995-1-1, Design of Timber Structures (EN 1995-1-1, 2008). This model allows calculation of effective bending stiffness considering the flexure of connection. In case of adhesive connection, fully composite action could be considered and consequently, γ-factor is equal to 1.0. Calculation model of the long-term behavior, which considers all important rheological parameters, is published in (Kanócz et al. 2013). In this paper, calculated values are established using this model.

2 PARAMETERS OF SPECIMENS

2.1 Geometrical parameters

Length of the beam specimens signed as TC1_4,5 and TC1_6,0 were 4.5 m and 6.0 m, respectively. The length value resulted from the aim to apply investigated timber-concrete composite members as building slabs. Vertically glued laminated timber slab was used. The width of the slab was 600.0 mm (Fig. 1). The depths of the timber part of beams TC1_4,5 and TC1_6,0 were 80.0 mm and 120.0 mm, respectively. The depth of light-weight concrete slab was 50.0 mm. The composite action between the timber and concrete part was performed by adhesive Sikadur T35 LVP. No reinforcement in concrete part was used.

2.2 Material parameters

The concrete layer was produced by lightweight aggregates—ceramic Liapor Expanded Clay Spheres. Material parameters were evaluated according to the relevant standard. Mean value of concrete cylinder compressive strength and middle value of the modulus of elasticity at 28 days were 25.6 MPa and 17.7 GPa, respectively. Average value of the density of concrete was 1792.0 kg/m³. Expected strength class of the concrete was LC25/28, but the obtained concrete parameters correspond with the lover strength class.

Strength class of the glued laminated timber was GL24. Material parameters were evaluated according to (EN 408, 2013). Values of the modulus of elasticity, bending strength and tensile strength along the grain were 13.9 GPa, 54.8 MPa and 32.9 MPa, respectively. Average value of the density of timber was 425.0 kg/m³.

The composite connection of timber and concrete was performed by 2-component epoxy resin adhesive Sikadur T35 LVP. Wet concrete on the wet adhesive was applied. Shear strength

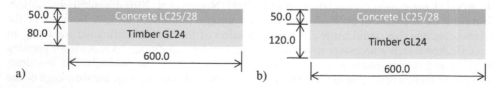

Figure 1.   Geometrical parameters of specimens: TC1_4,5 (a), TC1_6,0 (b).

of the glued connection between timber and concrete was evaluated according to (EN 392, 1998). Mean value of shear strength was 5.27 MPa.

3 LONG TERM BENDING TEST

From August 2013 to September 2016, two beams with the length of 4.5 m and 6.0 m were subjected to four-point bending test with long-term constant static load (Fig. 2). The loading history is summarized in Table 1. Beam specimen denoted as TC1_4,5 with a theoretical span of 4.4 m was loaded by 2×3.0 kN at the distance of 1.5 m from supports. Beam specimen denoted as TC1_6,0 with a theoretical span of 5.8 m was loaded by 2×4.0 kN at the distance of 2.0 m from supports. The value of the long-term load resulted from the supposed quasi-permanent uniform load of 2.2 kN/m² on building slabs.

Using the acquisition device Spider 8 middle span deflection was continuously gauged. The experiment was carried out inside. Humidity and temperature of the internal environment were continuously recorded (Fig. 3). The temperature was between 6°C and 32°C, the values of relative humidity changed between 23% and 68%.

Figure 2.   Set up of the bending test.

Table 1.   History of loading.

	TC1_4,5	TC1_6,0	
Beginning of drying shrinkage*	3	3	days
Removal of temporary support*	69	75	days
Application of permanent load*	82	82	days
Duration of long-term test	3.13	3.11	years
Permanent load	2×3.0	2×4.0	kN

*Time measured from concrete casting.

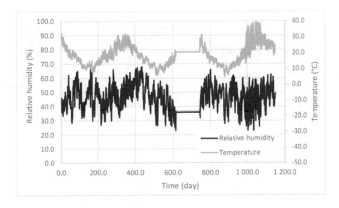

Figure 3.   Environmental conditions during the long-term test.

4 RESULTS AND DISCUSSION

4.1 *Measured values of mid-span deflection*

The measured values of mid-span deflection are illustrated in Figure 4 and Figure 5. After placing the beams on the supports, the initial mid-span deflection of 10 mm of both specimens was measured. This value includes the self-weight and the initial part of the concrete shrinkage. Thereafter, continually measuring of deflection using gauges placed in the middle of the span started. After the application of the permanent load on beams, initial deflections with a value of 6.5 mm and 10.2 mm were measured for beams TC1_4,5 and TC1_6,0, respectively. These values are similar to that calculated using the calculation model according to (Kanócz et al. 2013). The overall initial deflections caused by the self-weight and the applied permanent load are equal to 1/267 and 1/287 of the span of beams TC1_4,5 and TC1_6,0, respectively.

The deformation increased rapidly at the beginning of the experiment (Figs. 4–5). After about half year of loading, deformation increased at a slower rate. In the behavior, obvious annual fluctuations due to thermal and humidity changes of the environment can be observed. The differences between the maximum and minimum measured deflection in the 2nd year of the experiment were 6.3 mm (1/704 of span) and 8.4 mm (1/687 of span) for beams TC1_4,5

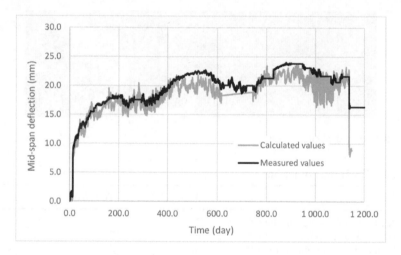

Figure 4. Measured and calculated mid-span deflection of beam TC1_4,5.

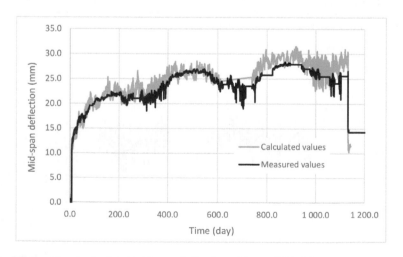

Figure 5. Measured and calculated mid-span deflection of beam TC1_6,0.

18

and TC1_6,0, respectively. In the 3rd year, the differences were 5.4 mm (1/819 of span) and 9.3 mm (1/621 of span) for beams TC1_4,5 and TC1_6,0, respectively.

In the comparison of the measured and calculated values (Figs. 4–5), we can see more significant differences in the daily changes of the deflection due to the differences in humidity and temperature of the environment within a day. The calculation model takes into the account the assumption that the entire cross-section will adapt to the external environment immediately, which in fact does not occur. However, the overall trend of the measured and calculated values over time is similar. The comparison shows that the calculation model is sufficient to estimate the long-term behavior of timber-concrete composite beams with the adhesive connection.

Maximum measured mid-span deflections during the test were 24.0 mm and 28.4 mm for the beams TC1_4,5 and TC1_6,0, respectively. By adding the initials deflection measured before the start of continually measuring using deflection gauges, the real maximum deflections were 34.0 and 38.4 mm, with a ratio of 1/129 and 1/150 of the span for beams TC1_4,5 and TC1_6,0, respectively. These values overestimate the limit values for service limit state of beams for ceiling structures. These results show that the creep and shrinkage of materials, as well as the influence of the environmental changes, can not be neglected.

During the bending test, no degradation of the adhesive occurred. The residual load-bearing capacity of the beams has not yet been determined.

4.2 Global creep coefficient of the timber-concrete composite beams

The creep coefficient of materials is defined as the ratio of the creep strain and the current elastic strain. For the timber-concrete composite members, the global creep coefficient can be defined as the ratio of the creep and the elastic deformation. Before the estimation of the global creep coefficient of investigated beams, it is necessary to subtract the influence of thermal and humidity changes of environment from the measured values of deflection. The mid-span deflection caused by the changes of environment can be calculated according to (Bajzecerová & Kanócz 2016). However, the modified measured deflection includes the deflection caused by the mechano-sorptive creep and can not be considered as a deflection under constant environmental conditions. The resulting creep coefficient values are valid only for the investigated elements and can be used to predict the deflection of beams at the end of their service life.

In Figure 6, the calculated values and logarithm approximation of the global creep coefficient of both investigated composite beams are presented. According to the approximations, at 50 years the global creep coefficients reach values of 1.4 and 1.1 for beams TC1_4,5 and TC1_6,0, respectively. It can be assumed that the final deformations of investigated beams after 50 years of loading, without any fluctuations due to the thermal and humidity environmental changes, can be 39,2 mm a 41,9 mm for beams TC1_4,5 and TC1_6,0, respectively.

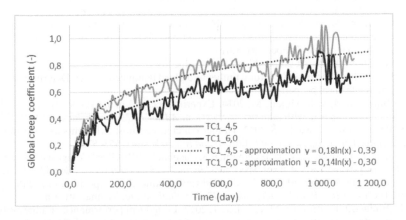

Figure 6.   Creep coefficient curve of the specimens.

5 CONCLUSION

The adhesive connection of timber-concrete composite members could be an effective solution to achieve higher bending stiffness. The advantage of gluing the wet concrete is also cutting down on reinforcement in the concrete section, which in the prefabricated concrete slab is necessary because of their handling. On the other hand, the adhesive connection increases the effect of concrete shrinkage on the deflection of composite beams.

Experimental investigation of two types of adhesively bonded timber-concrete composite members under long-term loading was carried out. After 3 years of the test, the measured deflection of the specimens overestimates the limit values for service limit state of beams for ceiling structures. These results show that the creep and shrinkage of materials, as well as the influence of the environmental changes, can not be neglected in the design process.

From the measured values of deflection, the global creep coefficient of both investigated composite beams was estimated. According to the approximations, the expected values of the global creep coefficient are 1.4 and 1.1 for beams TC1_4,5 and TC1_6,0, respectively. The resulting creep coefficient values are valid only for the investigated elements and can be used to predict the deflection of beams at the end of their service life.

The measured values were compared to calculated values of deflection according to the simplified theoretical model. The overall trend of the deflections in time shows that the model is applicable to estimation of the real long-term behavior of timber-concrete composite beams.

ACKNOWLEDGMENT

Paper is the result of the Project implementation: University Science Park TECHNICOM for Innovation Applications Supported by Knowledge Technology, ITMS: 26220220182, supported by the Research & Development Operational Programme funded by the ERDF. This paper was prepared with supporting of the grant VEGA Project No. 1/0538/16.

REFERENCES

Bajzecerová, V. & Kanócz, J. 2016. The Effect of Environment on Timber-concrete Composite Bridge Deck. *Procedia Engineering* 156: 32–39.

Brunner, M., Romer, M. & Schnüriger, M. 2007. Timber-concrete-composite with an adhesive connector (wet on wet process). *Materials and Structures/Materiaux et Constructions* 40 (1): 119–126.

Dias, A., Skinner, J., Crews, K. & Tannert, T. 2016. Timber-concrete-composites increasing the use of timber in construction. *European Journal of Wood and Wood Products* 74 (3): 443–451.

Eisenhut, L. & Seim, W. 2016a. Long term-behavior of glued full-scale specimens made from wood and high performance concrete at natural climate conditions (Langzeitverhalten geklebter Bauteile aus Holz und hochfestem Beton bei natürlichem Klima). *Bautechnik* 93 (11): 807–816.

Eisenhut, L., Seim, W. & Kühlborn, S. 2016b. Adhesive-bonded timber-concrete composites—Experimental and numerical investigation of hygrothermal effects. *Engineering Structures* 125: 167–178.

EN 1995-1-1, 2008: Design of Timber Structures.

EN 392, 1998: Glued laminated timber. Shear test of glue lines.

EN 408, 2013: Timber structures. Structural timber and glued laminated timber. Determination of some physical and mechanical properties.

Kanócz, J., Bajzecerová, V. & Mojdis, M. 2014. Experimental and numerical analysis of timber-lightweight concrete composite with adhesive connection. *Advanced Materials Research* 969:155–160.

Kanócz, J., Bajzecerová V. & Šteller, Š. 2013. Timber-concrete composite elements with various composite connections. Part 1: Screwed connection. *Wood research* 58(4): 555–570.

Negrão, J., Leitão de Oliveira, C., Maia de Oliveira, F. & Cachim, P. 2010. Glued Composite Timber-Concrete Beams. II: Analysis and Tests of Beam Specimens. *Journal of Structural Engineering*, 136(10): 1246–1254.

Schmid, V., Zauft, D. & Polak, M.A. 2016. Bonded timber-concrete composite floors with lightweight concrete. *WCTE 2016—World Conference on Timber Engineering*. Vienna, Austria.

Tannert, T., Endacott, B., Brunner, M. & Vallée, T. 2017. Long-term performance of adhesively bonded timber-concrete composites. *International Journal of Adhesion and Adhesives* 72: 51–61.

Battened built-up member under compression and bending with one end fixed and the other end free

I.J. Baláž
Department of Metal and Timber Structures, Faculty of Civil Engineering, STU in Bratislava, Slovakia

Y.P. Koleková & L.O. Moroczová
Department of Structural Mechanics, Faculty of Civil Engineering, STU in Bratislava, Slovakia

ABSTRACT: Eurocodes EN 1993 Design of steel structures and EN 1999 Design of aluminium structures give guidance on how to design and verify uniform built-up member under compression with hinged ends laterally supported. The built-up members in Eurocodes are considered as the columns with a local bow imperfection $e_0 = L/500$ (Ramm, W. & Uhlmann, W.). This paper investigates the more general case. The second order analysis is used for analysis of the battened built-up member with two IPE 240 sections under combined compression and bending. The bottom member end is fixed and the upper end is free. The global initial sway imperfection of the analyzed built-up member is taken into account instead of the local bow imperfection.

1 INTRODUCTION

1.1 *Task and input values*

The analysis of the 2nd order with imperfection is used in calculations. The geometrical equivalent global initial sway imperfection is taken according to EN 1993-1-1. The column is made of steel S355. The yield strength $f_y = 355$ MPa, the safety factors $\gamma_{M0} = 1.0$, $\gamma_{M1} = 1.0$. The design values of the external actions applied at the column top (Figure 2d) are:

$$F_{Ed} = 860\,kN, \quad H_{Ed} = 22\,kN \tag{1}$$

The cross-section properties of 2 IPE 240 sections (Figure 1) are as follows:

$$A = 78.2\,10^2\,mm^2, \quad I_y = 77.8\,10^6\,mm^4, \quad I_z = 318.5\,10^6\,mm^4 \tag{2}$$

$$i_y = 99.7\,mm, \quad i_z = 201.8\,mm \tag{3}$$

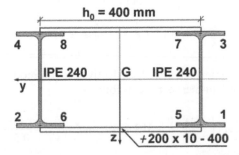

Figure 1. The section 2 IPE 240 – DIN 1025-5: 1994 of the battened built-up column.

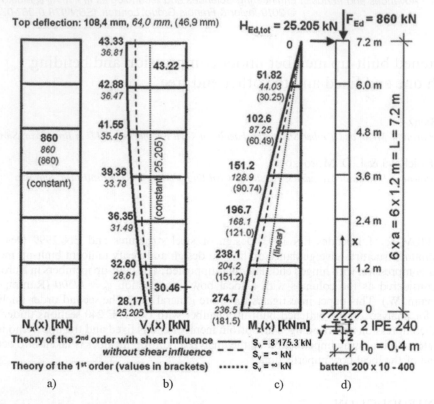

Figure 2. Column geometry, actions and distributions of the internal forces $N_x(x)$, $V_y(x)$, $M_z(x)$.

The cross-section properties of the one chord. IPE 240 is Class 2 cross-section:

$$A_{ch} = 39.1\,10^2\,mm^2, \quad I_{ch,y} = 38.9\,10^6\,mm^4, \quad I_{ch,z} = 2.84\,10^6\,mm^4 \tag{4}$$

$$W_{ch,el,y} = 324.7\,10^3\,mm^3, \quad W_{ch,el,z} = 47.3\,10^3\,mm^3, \quad i_{ch,y} = 99.7\,mm \tag{5}$$

$$W_{ch,pl,y} = 366\,10^3\,mm^3, \quad W_{ch,pl,z} = 73.92\,10^3\,mm^3, \quad i_{ch,z} = 26.9\,mm \tag{6}$$

The geometry of the column and the actions are shown in Figure 2d. $L_{cr,z} = 2L$, $L_{cr,y} = L$. The bending stiffness and shear stiffness of the battened built-up column defining stiffnesses of an equivalent member:

a. The bending stiffness EI_{eff}

$$I_1 = 2\left[I_{ch,z} + A_{ch}\left(h_0/2 \right)^2 \right] = 318.48\,10^6\,mm^4, \quad i_0 = \sqrt{I_1/\left(2A_{ch} \right)} = 201.8mm \tag{7}$$

$$L_{cr.z} = 2L = 2\,7.2m = 14.4m, \quad \lambda = \frac{L_{cr.z}}{i_0} = 71.35 < 75 \rightarrow \mu = 1 \tag{8}$$

The effective second moment of area I_{eff}, effective section modulus W_{eff} and the bending stiffness EI_{eff} of the equivalent member are as follows

$$I_{eff} = 2\left[\mu I_{ch,z} + A_{ch}\left(\frac{h_0}{2} \right)^2 \right] = 318.48\,10^6\,mm^4, \quad W_{eff} = \frac{I_{eff}}{0.5h_0} = 1592\,10^3\,mm^3 \tag{9}$$

$$EI_{eff} = 66\,880\,kNm^2 \tag{10}$$

22

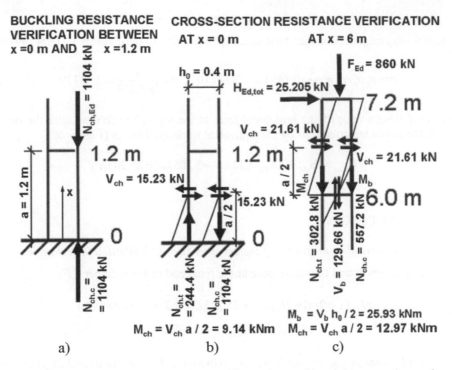

Figure 3.   Verification of the chord: a) in-plane buckling resistance, b) and c) cross-sections resistance.

b. The shear stiffness S_v of the battened panel defining shear stiffness S_v of equivalent member. The shear stiffness S_v is limited by the value $S_{v,max}$, which takes into account the possible failure of a battened panel by shear. The dimensions of the batten are shown in Figure 1 and Figure 2d.

$$n_b = 2, b_b = 200\,mm,\ t_b = 10\,mm,\ I_b = b_b t_b/12 = 6.667 10^6\,mm^4 \tag{11}$$

$$S_v = \frac{24 EI_{ch,z}}{a^2 \left[1 + \dfrac{2I_{ch,z}}{n_b I_b}\dfrac{h_0}{a}\right]} = 8704\,kN,\ S_{v,max} = \frac{\pi^2 EI_{ch,z}}{a^2} = 8175.3\,kN \tag{12}$$

$$S_v = S_{v,max} = 8175.3\,kN \tag{13}$$

The parameter of the member ε used in the analysis of the 2nd order with taking into account the influence of the shear deformations

$$\varepsilon = L\sqrt{\frac{\gamma N_{Ed}}{EI_{eff}}} = 7.2\,m\sqrt{\frac{1.118\ 860\,kN}{66\ 880\,kNm^2}} = 0.863 \tag{14}$$

where the parameter γ takes into account the shear deformations

$$\gamma = \frac{1}{1 - N_{Ed}/S_v} = \frac{1}{1 - 860\,kN/8175.3\,kN} = 1.118 \tag{15}$$

The parameter of the member $\varepsilon = 0.816$ was used in the analysis of the 2nd order without taking into account the influence of the shear deformations ($S_v = \infty\,kN$, $\gamma = 1.0$).

1.2 Geometrical equivalent global sway initial imperfection

The initial sway imperfection may be determined according to the clause 5.3.2 (EN 1993-1-1):

$$\phi = \phi_0\, \alpha_h\, \alpha_m = \frac{1}{200}\,0.745\,1.0 = 3.727\,10^{-3}, \quad \phi = \frac{1}{268.328} = 3.727\,10^{-3} \tag{16}$$

The total design value of the horizontal force at the top of the column after the replacement of the initial imperfections by the equivalent horizontal forces (Figure 2d):

$$H_{Ed,tot} = H_{Ed} + \phi N_{Ed} = 22\,kN + 3.205\,kN = 25.205\,kN \tag{17}$$

2 INTERNAL FORCES

2.1 Calculation based on the theory of the 1st order (ε = 0) with initial imperfection

The bending moment and the shear force at the fixed end of the member

$$M_{Ed}(x=0\,m) = H_{Ed.tot}L = 25.205\,kN\ 7.2\,m = 181.48\,kNm \tag{18}$$

$$V_{Ed}(x) = H_{Ed,tot} = 25.205\,kN \tag{19}$$

The distributions of the internal forces N_{Ed} (constant), V_{Ed} (constant) and M_{Ed} (linear) are drawn in Figure 2a, b, c with dotted lines. The values are given in the brackets. If these values will be multiplied by the ratio 22/25.205 one obtains the results valid for a case without imperfection.

2.2 Calculation based on the theory of the 2nd order with initial imperfection and without the influence of the shear deformations $(S_v = \infty\ kN, \gamma = 1.0, \varepsilon = 0.816, \xi_H = 1.0)$

The distributions of the internal forces N_{Ed} (constant), V_{Ed} (non-linear) and M_{Ed} (non-linear) are drawn in Figure 2a, b, c with dashed lines. Their values are written in italics. The values were calculated in the similar way as it was done in the next paragraph 2.3. The values were obtained with the same formulae but with using parameters $\gamma = 1.0$ and $\varepsilon = 0.816$.

2.3 Calculation based on the theory of the 2nd order with the sway initial imperfection and with the influence of the shear deformations $(S_v = 8175.3\ kN, \gamma = 1.118, \varepsilon = 0.863, \xi_H = 1.0)$

The distributions of the internal forces N_{Ed} (constant), V_{Ed} (non-linear) and M_{Ed} (non-linear) are drawn in Figure 2a, b, c with solid lines. Their values are written in bold. The values were calculated from the following formulae (Stahlbau Handbuch, 1982), which are valid for any point of horizontal force application x_H from the interval $(0\ m \le x_H \le L)$ or $(0 \le \xi_H = x_H / L \le 1.0)$.

The bending moment due to horizontal force $H_{Ed,tot}$ for $\xi \le \xi_H$.

$$M_{Ed}(\xi) = \left\{ \frac{\gamma \sin\left[(1-\xi_H)\varepsilon\right]\sin(\xi\varepsilon)}{\varepsilon \sin(\varepsilon)} - \frac{\gamma \sin\left[(1-\xi)\varepsilon\right]}{\sin(\varepsilon)}\,\frac{\sin(\varepsilon)-\sin\left[(1-\xi_H)\varepsilon\right]}{\varepsilon\cos(\varepsilon)} \right\} LH_{Ed,tot} \tag{20}$$

The bending moment due to horizontal force $H_{Ed,tot}$ for $\xi > \xi_H$

$$M_{Ed}(\xi) = \left\{ \begin{array}{l} \dfrac{\gamma\sin(\xi_H\varepsilon)\sin\left[(1-\xi)\varepsilon\right]}{\varepsilon\sin(\varepsilon)} - \\ -\dfrac{\gamma\sin\left[(1-\xi)\varepsilon\right]}{\sin(\varepsilon)}\dfrac{\sin(\varepsilon)-\sin\left[(1-\xi_M)\varepsilon\right]}{\varepsilon\cos(\varepsilon)} \end{array} \right\} LH_{Ed,tot} \tag{21}$$

The shear force

$$V_{Ed}(\xi) = [\mathrm{d}M_{Ed}(\xi)/\mathrm{d}\xi]/L. \tag{22}$$

3 VERIFICATION OF THE IN-PLANE BUCKLING RESISTANCE—ABOUT AXIS Z

3.1 Verification of a chord buckling resistance about weak axis z

The axial forces in the chords $N_{ch,Ed}$ are determined at the bottom fixed end (Figure 3a)

$$N_{ch,Ed} = \frac{N_{Ed}}{2} \pm \frac{M_{Ed}(x=0)}{W_{eff}} A_{ch} = 1104.4\,kN, \ -244.4\,kN \ \text{(tension)} \tag{23}$$

The in-plane buckling resistance of a chord (buckling curve = "b", $\alpha_{ch,z} = 0.34$)

$$N_{cr,ch,z} = \pi^2 EI_{ch,z}/a^2 = 4087.7\,kN, \ \lambda_{ch,z} = a/i_{ch,z} = 44.6 \tag{24}$$

$$\overline{\lambda}_{ch,z} = \sqrt{\frac{A_{ch}f_y}{N_{cr,ch,z}}} = 0.583, \ \phi_{ch,z} = 0.5\left[1+\alpha_{ch,z}\left(\overline{\lambda}_{ch,z} - 0.2\right) + \overline{\lambda}_{ch,z}\right] = 0.735 \tag{25}$$

$$\chi_{ch,z} = \frac{1}{\phi_{ch,z}+\sqrt{\phi_{ch,z}^2 - \overline{\lambda}_{ch,z}^2}} = 0.846, \ N_{ch,b,z,Rd} = \chi_{ch,z}A_{ch}\frac{f_y}{\gamma_{M1}} = 1173.8\,kN \tag{26}$$

Verification criterion, utility factor $U_{ch,b,z}$

$$U_{ch,b,z} = \frac{N_{ch,Ed}}{N_{ch,b,z,Rd}} = \frac{1104.4}{1173.8} = 0.941 < 1.0 \tag{27}$$

If the linear moment distribution would be taken into account according to the formula (6.62) and Tables B1 and B3 in EN 1993-1-1: $\psi_z = -1$, $C_{m,z} = 0.6 + 0.4\ \psi_z = 0.2 < 0.4 \rightarrow C_{m,z} = 0.4$, $2\ \overline{\lambda}_{ch,z} - 0.6 = 2\ 0.735 - 0.6 = 0.565 < 1.4 \rightarrow 0.565$, $k_{zz} = 0.4\ (1+0.565\ 0.941) = 0.613$.

$$U_{ch,b,z} = \frac{N_{ch,Ed}}{N_{ch,b,z,Rd}} + k_{zz}\frac{\gamma_{M1}M_{ch,Ed}}{W_{ch,pl,z}f_y} = 0.941 + 0.613\frac{9.138}{26.243} = 0.941 + 0.213 = 1.154 \tag{28}$$

$1.154 > 1.0$ and the battened built-up member is not safe

3.2 Verification of chord cross-section resistance in the bottom end panel in section $x = 0\ m$

Internal forces are calculated in Figure 3b. The influence of the shear force is negligible:

$$\frac{15.23\,kN}{2bt_f\dfrac{f_y}{\gamma_{M0}\sqrt{3}}} = 0.032, \ n = \frac{1104.4\,kN}{A_{ch}\dfrac{f_y}{\gamma_{M0}}} = 0.796, \ \overline{a} = \frac{A_{ch} - 2bt_f}{A_{ch}} = 0.398 \tag{29}$$

$$n \geq a \rightarrow M_{ch,N,pl,z,Rd} = W_{ch,pl,z,Rd}\frac{f_y}{\gamma_{M0}}\left[1 - \left(\frac{n-\overline{a}}{1-\overline{a}}\right)\right] = 14.80\,kNm \tag{30}$$

25

Utility factor for the section x = 0 m (Figure 3b) is 0.617. For section x = 6 m (Figure 3c) is 0.494.

$$U_{ch.c.z} = \frac{M_{ch}}{M_{ch,N,pl,z,Rd}} = \frac{9.14\,kNm}{14.80\,kNm} = 0.617 < 1.0 \tag{31}$$

4   VERIFICATION OF OUT-OF-PLANE BUCKLING RESISTANCE—ABOUT AXIS Y

4.1   *Influence of the flexural buckling about axis y*

$$L_{cr.y} = L = 7.2m, \ N_{cr.y} = \frac{\pi^2 EI_y}{L_{cr.y}^2} = 3111kN, \ \overline{\lambda}_y = \sqrt{\frac{Af_y}{N_{cr.y}}} = 0.945 \tag{32}$$

$$b.c. = \text{``a''}, \ \alpha_y = 0.21, \ \phi_y = 0.5\left[1 + \alpha_y\left(\overline{\lambda}_y - 0.2\right) + \overline{\lambda}_y\right] = 1.024 \tag{33}$$

$$\chi_y = \frac{1}{\phi_y + \sqrt{\phi_y^2 - \overline{\lambda}_y^2}} = 0.704, \ N_{b,y,Rd} = \chi_y A \frac{f_y}{\gamma_{M1}} = 1954\,kN \tag{34}$$

Utility factor $U_{b,y}$ if only the flexural buckling about axis y is taken into account

$$U_{b,y} = \frac{N_{Ed}}{N_{b,y,Rd}} = \frac{860}{1954} = 0.44 \tag{35}$$

4.2   *Influence of the bending moment about axis y due to the horizontal force H_{Ed}*

The bending moment in the section x = 3.6 m due to the horizontal force H_{Ed} = 22 kN (Fig. 2d):

$$M_{z,Ed,H} = M_{z,Ed,H,tot}\,(x=3.6m)\frac{H_{Ed}}{H_{Ed,tot}} = 151.2\,kNm\frac{22\,kN}{25.205\,kN} = 131.97\,kNm \tag{36}$$

Utility factor $U_{M,z}$ if only the bending moment $M_{z,Ed,H}$ is taken into account

$$U_{M,z} = \frac{\dfrac{M_{z,Ed,H}\,(x=3.6m)}{W_{eff}}A_{ch}}{A_{ch}f_y / \gamma_{M0}} = \frac{324\,kN}{1388\,kN} = 0.233 \tag{37}$$

Verification criterion, the total utility factor

$$U_y = U_{b,y} + U_{M,z} = 0.44 + 0.233 = 0673 < 1.0 \tag{38}$$

5   CONCLUSION

Eurocodes EN 1993-1-1 and EN 1999-1-1 give guidance on how to verify the basic case: the simply supported built-up member under compression. No rules are given there concerning general boundary conditions and different types of actions.

The paper shows how it is possible to verify the battened built-up members with any boundary conditions under any combination of actions. The distributions of internal forces drawn in Figure 2 show influence of the theory of the 2nd order with and without shear deformations comparing with results of the 1st order analysis. The influence of the shear deformation is at battened built-up members very important. Figure 3 gives details of the structure which were in the paper verified.

Both the in plane and also the out-of-plane buckling resistances of the battened built-up column are verified in detail. In the out-of-plane buckling verification also the action acting in the plane is taken into account. As a tool was used the theory of the 2nd order with geometrical equivalent global sway initial imperfection on the model of the column with real boundary conditions ($L = 7.2$ m) and not on the model of equivalent member ($L_{cr,z} = 2L = 14.4$ m).

The analysis of the battened, laced and closely spaced built-up members according to the former Slovak standard STN 73 1401 is in (Baláž & Agócs, 1994).

ACKNOWLEDGEMENT

Project No. 1/0603/17 was supported by the Slovak Grant Agency VEGA.

REFERENCES

Baláž, I.J. & Agócs, Z. 1994. Metal Structures. I. part. (in Slovak). ES STU Bratislava, 3rd edition, 1–496.
EN 1993-1-1: March 2005 and Corrigendum AC: February 2006. Eurocode 3   Design of steel structures, Part 1-1: General rules and rules for buildings. CEN Brussels.
EN 1999-1-1: May 2007 + A1: July 2009 + A2: December 2013. Design of Aluminium Structures. Part 1-1 General Rules and Rules for Buildings. CEN Brussels.
Ramm, W. & Uhlmann, W. 1981. Zur Anpassung des Stabilitätsnachweises für mehrteilige Druckstäbe and das Nachweiskonzept. Der Stahlbau. 50. Jahrgang, Heft 6, pp. 161–172.
Stahlbau Handbuch Für Studium und Praxis. Band 1. 1982. Stahlbau-Verlags-GmbH. Köln.

Advances and Trends in Engineering Sciences and Technologies III – Al Ali & Platko (Eds)
© *2019 Taylor & Francis Group, London, ISBN 978-0-367-07509-5*

Behaviour of steel laced built-up columns

I.J. Baláž
Department of Metal and Timber Structures, Faculty of Civil Engineering, STU in Bratislava, Slovakia

Y.P. Koleková & L.O. Moroczová
Department of Structural Mechanics, Faculty of Civil Engineering, STU in Bratislava, Slovakia

ABSTRACT: Eurocodes EN 1993 Design of steel structures gives guidance on how to design and verify uniform built-up members under compression with hinged ends laterally supported only. The built-up members in Eurocodes are considered as members with equivalent bending and shear stiffnesses with a local bow imperfection $e_0 = L/500$. This paper investigates a more general case. The second order analysis is used for analysis of the laced built-up member with two IPE 240 sections under combined compression and bending action. The bottom member end is fixed and the upper end is free for in-plane buckling. The member is restrained against out-of-plane buckling at both ends. The global initial sway imperfection Φ is taken into account instead of the local bow imperfection with an amplitude e_0.

1 INTRODUCTION

1.1 *Introduction*

The analysis of the 2nd order with imperfection is used in calculations. The geometrical equivalent global initial sway imperfection is taken according to EN 1993-1-1. The column is made of steel S355. The yield strength $f_y = 355$ MPa, Young modulus $E = 210$ GPa, the safety factors $\gamma_{M0} = 1.0$, $\gamma_{M1} = 1.0$. The design values of the external actions applied at the column top (Figure 2d) are:

$$F_{Ed} = 860\,kN, \quad H_{Ed} = 22\,kN, \quad M_{Ed,o} = 350\,kNm \tag{1}$$

The height of the column $L = 7.2$ m. The boundary conditions are defined by the buckling lengths $L_{cr,y} = L$, $L_{cr,z} = 2L$. The distance between the posts of the lacing $a = 1.2$ m. The initial sway imperfection is determined according to the clause 5.3.2 in EN 1993-1-1:

$$\phi = \phi_0\, \alpha_h\, \alpha_m = \frac{1}{200} 0.745\, 1.0 = 3.727\,10^{-3}, \quad \phi = \frac{1}{268.328} = 3.727\,10^{-3} \tag{2}$$

The total design value of the horizontal force at the top of the column after the replacement of the initial imperfections by the equivalent horizontal forces:

$$H_{Ed,tot} = H_{Ed} + \phi N_{Ed} = 22\,kN + 3.205\,kN = 25.205\,kN \tag{3}$$

The cross-section properties of 2 IPE 240 sections (Figure 1) are as follows:

$$A = 78.2\,10^2\,mm^2, \quad I_y = 77.8\,10^6\,mm^4, \quad I_z = 318.5\,10^6\,mm^4 \tag{4}$$

$$i_y = 99.7\ mm, \quad i_z = 201.8\ mm \tag{5}$$

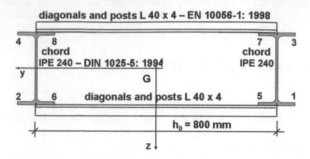

Figure 1. The components of the laced built-up column.

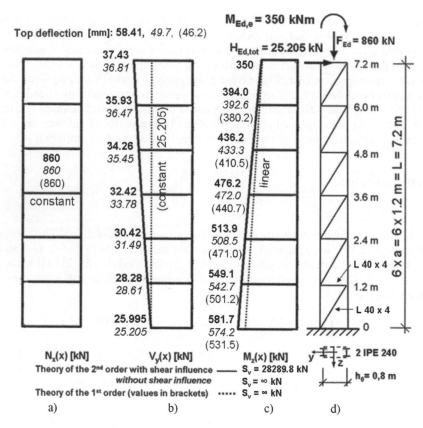

Figure 2. Column geometry, actions and distributions of the internal forces $N_x(x)$, $V_y(x)$, $M_z(x)$.

The cross-section properties of the one chord. IPE 240 is Class 2 cross-section:

$$A_{ch} = 39.1\,10^2\,mm^2, \quad I_{ch,y} = 38.9\,10^6\,mm^4, \quad I_{ch,z} = 2.84\,10^6\,mm^4 \tag{6}$$

$$W_{ch,el,y} = 324.7\,10^3\,mm^3, \quad W_{ch,el,z} = 47.3\,10^3\,mm^3, \quad i_{ch,y} = 99.7\,mm \tag{7}$$

$$W_{ch,pl,y} = 366\,10^3\,mm^3, \quad W_{ch,pl,z} = 73.92\,10^3\,mm^3, \quad i_{ch,z} = 26.9\,mm \tag{8}$$

The bending stiffness and shear stiffness of the laced built-up column defining stiffnesses of an equivalent member:

a. The bending stiffness EI_{eff}. The effective second moment of area I_{eff}, effective section modulus W_{eff} and the bending stiffness EI_{eff} of the equivalent member are for $\mu = 0$ as follows

$$I_{eff} = 2\left[\mu I_{ch,z} + A_{ch}\left(h_0/2 \right)^2 \right] = 1251\,10^6\,mm^4, \quad W_{eff} = \frac{I_{eff}}{0.5h_0} = 3128\,10^3\,mm^3 \tag{9}$$

$$EI_{eff} = 262\,752\,kNm^2 \tag{10}$$

b. The shear stiffness Sv of the laced panel defining shear stiffness Sv of the equivalent member. The properties of the diagonals and the posts have L 40×4 Class 4 cross-section. See Figure 1 and Figure 2d.

$$n = 2, \quad A_d = 15.5\,cm^2, \quad A_v = 12.3\,cm^2, \quad d = \sqrt{a^2 + h_0^2} = 1.442m \tag{11}$$

$$S_v = \frac{nEA_d a h_0}{d^3\left[1 + \dfrac{A_d h_0^3}{A_v d^3} \right]} = 28\,289.8\,kN \tag{12}$$

The parameter of the member ε used in the analysis of the 2nd order with taking into account the influence of the shear deformations

$$\varepsilon = L\sqrt{\frac{\gamma N_{Ed}}{EI_{eff}}} = 7.2m\sqrt{\frac{1.031\,860\,kN}{262752\,kNm^2}} = 0.418 \tag{13}$$

where the parameter γ takes into account the shear deformations

$$\gamma = \frac{1}{1 - N_{Ed}/S_v} = \frac{1}{1 - 860kN/28289.8\,kN} = 1.031 \tag{14}$$

The geometry of the column and the actions are shown in Figure 2d. $L_{cr,z} = 2L$, $L_{cr,y} = L$.

2 INTERNAL FORCES

2.1 Calculation based on the theory of the 1st order ($\varepsilon = 0$) with initial imperfection and without influence of the shear deformation ($S_v = \infty$ kN, $\gamma = 1.0$)

The bending moment and the shear force at the fixed end of the member

$$M_{Ed}(x = 0\,m) = H_{Ed,tot}L = 25.205kN\ 7.2m = 181.48kNm \tag{15}$$

$$V_{Ed}(x) = H_{Ed,tot} = 25.205kN \tag{16}$$

The distributions of the internal forces N_{Ed} (constant), V_{Ed} (constant) and M_{Ed} (linear) are drawn in Figure 2a, b, c with dotted lines. The values are given in the brackets. If these values will be multiplied by the ratio 22/25.205 one obtains the results valid for a case without imperfection.

2.2 Calculation based on the theory of the 2nd order with initial imperfection and without the influence of the shear deformations ($S_v = \infty$ kN, $\gamma = 1.0$, $\varepsilon = 0.412$, $\xi_H = 1.0$, $\xi_M = 1.0$)

The distributions of the internal forces N_{Ed} (constant), V_{Ed} (non-linear) and M_{Ed} (non-linear) are drawn in Figure 2. Their values are written in italics. The values were calculated in the similar way as it was done in the next paragraph 2.3. The values were obtained with the same formulae but with parameters $\gamma = 1.0$ and $\varepsilon = 0.417$.

2.3 Calculation based on the theory of the 2nd order with the sway initial imperfection and with the influence of the shear deformations ($S_v = 28289.8$ kN, $\gamma = 1.031$, $\varepsilon = 0.418$, $\xi_H = 1$, $\xi_H = 1$)

The distributions of the internal forces N_{Ed} (constant), V_{Ed} (non-linear) and M_{Ed} (non-linear) are drawn in Figure 2a, b, c with solid lines. Their values are written in bold. The values were calculated from the following formulae (Stahlbau Handbuch, 1982), which are valid for any point of horizontal force application x_H from the interval (0 m $\leq x_H \leq L$) or (0 $\leq \xi_H = x_H/L \leq 1.0$) and for any point of bending moment $M_{Ed,e}$ application x_M from the interval (0 m $\leq x_M \leq L$) or (0 $\leq \xi_M = x_M/L \leq 1.0$).

The bending moment due to horizontal force $H_{Ed,tot}$ at the top of column for $\xi \leq \xi_H$

$$M_{Ed}^H(\xi) = \left\{ \frac{\gamma \sin\left[(1-\xi_H)\varepsilon\right]\sin(\xi\varepsilon)}{\varepsilon \sin(\varepsilon)} - \frac{\gamma \sin\left[(1-\xi)\varepsilon\right]}{\sin(\varepsilon)} \frac{\sin(\varepsilon) - \sin\left[(1-\xi_H)\varepsilon\right]}{\varepsilon \cos(\varepsilon)} \right\} LH_{Ed,tot} \tag{17}$$

The bending moment due to horizontal force $H_{Ed,tot}$ for $\xi > \xi_H$:

$$M_{Ed}^H(\xi) = \left\{ \frac{\gamma \sin\left[(1-\xi_H)\varepsilon\right]\sin(\xi\varepsilon)}{\varepsilon \sin(\varepsilon)} - \frac{\gamma \sin\left[(1-\xi)\varepsilon\right]}{\sin(\varepsilon)} \frac{\sin(\varepsilon) - \sin\left[(1-\xi_H)\varepsilon\right]}{\varepsilon \cos(\varepsilon)} \right\} LH_{Ed,tot} \tag{18}$$

The bending moment due to external bending moment $M_{Ed,e}$ at the top of column for $\xi \leq \xi_M$:

$$M_{Ed}^M(\xi) = \left\{ -\frac{\cos\left[(1-\xi_M)\varepsilon\right]\sin(\xi\varepsilon)}{\sin(\varepsilon)} - \frac{\sin\left[(1-\xi)\varepsilon\right]}{\sin(\varepsilon)} \frac{\cos\left[(1-\xi_M)\varepsilon\right]}{\cos(\varepsilon)} \right\} M_{Ed,e} \tag{19}$$

The bending moment due to external bending moment $M_{Ed,e}$ at the top of column for $\xi > \xi_M$:

$$M_{Ed}^M(\xi) = \left\{ \frac{\cos\left[(\xi_M\varepsilon)\right]\sin\left[(1-\xi)\varepsilon\right]}{\sin(\varepsilon)} - \frac{\sin\left[(1-\xi)\varepsilon\right]}{\sin(\varepsilon)} \frac{\cos\left[(1-\xi_M)\varepsilon\right]}{\cos(\varepsilon)} \right\} M_{Ed,e} \tag{20}$$

The total bending moment:

$$M_{Ed}(\xi) = M_{Ed}^H(\xi) + M_{Ed}^M(\xi) \tag{21}$$

The shear force:

$$V_{Ed}(\xi) = [dM_{Ed}(\xi)/d\xi]/L. \tag{22}$$

3 VERIFICATION OF THE IN-PLANE BUCKLING RESISTANCE—ABOUT AXIS Z

3.1 Verification of a chord buckling resistance about weak axis z

The axial forces in the chords $N_{ch,Ed}$ used for the verification of the in-buckling resistance of the chord determined at the bottom fixed end (Figure 3a)

$$N_{ch,c,Ed} = \frac{N_{Ed}}{2} + \frac{M_{Ed}(x=0)}{W_{eff}} A_{ch} = 1153.86 kN \text{ (right chord in compression)} \qquad (23)$$

$$N_{ch,c,Ed} = \frac{N_{Ed}}{2} - \frac{M_{Ed}(x=0)}{W_{eff}} A_{ch} = -293.86 kN \text{ (left chord in tension)} \qquad (24)$$

The in-plane buckling resistance of the chord in compression at the bottom fixed end (Figure 2a), buckling curve = "b", $\alpha_{ch,z} = 0.34$

$$N_{cr,ch,z} = \frac{\pi^2 EI_{ch,z}}{a^2} = 4087.7\, kN, \lambda_{ch,z} = \frac{a}{i_{ch,z}} = 44.6 \qquad (25)$$

$$\overline{\lambda}_{ch,z} = \sqrt{\frac{A_{ch}f_y}{N_{cr,ch,z}}} = 0.583, \phi_{ch,z} = 0.5\left[1 + \alpha_{ch,z}\left(\overline{\lambda}_{ch,z} - 0.2\right) + \overline{\lambda}_{ch,z}\right] = 0.735 \qquad (26)$$

$$\chi_{ch,z} = \frac{1}{\phi_{ch,z} + \sqrt{\phi_{ch,z}^2 - \overline{\lambda}_{ch,z}^2}} = 0.846 \qquad (27a)$$

$$N_{ch,b,z,Rd} = \chi_{ch,z} A_{ch} \frac{f_y}{\gamma_{M1}} = 1173.8 kN \qquad (27b)$$

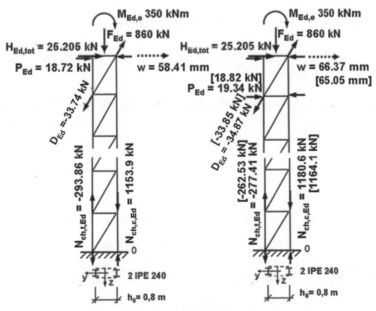

Calculation models:
a) Member with equivalent stiffnesses,   b) Frame structure

INTERNAL FORCES PER 1 CHORD, 1 DIAGONAL, 1 POST

Figure 3.   Comparisons of results of three calculation models.

Verification criterion, utility factor $U_{ch,b,z}$

$$U_{ch,b,z} = \frac{N_{ch,Ed}}{N_{ch,b,z,Rd}} = \frac{1153.86}{1173.8} = 0.983 < 1.0 \tag{28}$$

Figure 3 shows the comparisons of the axial forces in the chords, diagonals and posts calculated on the model of: a) the continuum with effective bending and smeared shear stiffnesses according to EN 1993-1-1, b) the simple frame structures with discrete components, in which the joints are not required to resist moments. The b) model was calculated twice: b1) the influence of axial tension forces in the left chord and in the diagonals was neglected, b2) the influence of tension axial forces was taken into account to obtain more exact values given in the brackets [ ]. The internal forces and deformations of both b) models were calculated with the program IQ 100 (Rubin, H, Aminbaghai, M. and Weier, H.). The differences between b1 and b2 in the values of the axial forces in the components of laced built-up member are not important. The good agreement was achieved between the results of the continuum with effective bending and smeared shear stiffnesses and the simple frame structures with discrete components.

4 VERIFICATION OF OUT-OF-PLANE BUCKLING RESISTANCE—ABOUT AXIS Y

4.1 Influence of the flexural buckling about axis y

Buckling curve = "a", $\alpha_y = 0.21$

$$L_{cr.y} = L = 7.2\,m, \; N_{cr.y} = \frac{\pi^2 EI_y}{L_{cr.y}^2} = 3111kN, \; \bar{\lambda}_y = \sqrt{\frac{Af_y}{N_{cr.y}}} = 0.945 \tag{29}$$

$$\phi_y = 0.5\left[1 + \alpha_y\left(\bar{\lambda}_y - 0.2\right) + \bar{\lambda}_y\right] = 1.024 \tag{30}$$

$$\chi_y = \frac{1}{\phi_y + \sqrt{\phi_y^2 - \bar{\lambda}_y^2}} = 0.704, \; N_{b,y,Rd} = \chi_y A \frac{f_y}{\gamma_{M1}} = 1954\,kN \tag{31}$$

Utility factor $U_{b,y}$ if only the flexural buckling about axis y is taken into account

$$U_{b,y} = \frac{N_{Ed}}{N_{b,y,Rd}} = \frac{860}{1954} = 0.44 \tag{32}$$

4.2 Influence of the bending moment about axis y due to the horizontal force H_{Ed}

The bending moment in the section $x = 3.6$ m due to the horizontal force $H_{Ed} = 22$ kN:

$$M_{z,Ed,H}\left(x = 3.6\,m\right) = 460.29\,kNm \tag{33}$$

Utility factor $U_{M,z}$ if only the bending moment $M_{z,Ed,H}$ is taken into account

$$U_{M,z} = \frac{\dfrac{M_{z,Ed,H}\left(x = 3.6\,m\right)}{W_{eff}} A_{ch}}{A_{ch}\dfrac{f_y}{\gamma_{M0}}} = \frac{572.4\,kN}{1388\,kN} = 0.413 \tag{34}$$

Verification criterion, the total utility factor

$$U_y = U_{b,y} + U_{M,z} = 0.44 + 0.413 = 0.853 < 1.0 \qquad (35)$$

5 CONCLUSION

The paper shows how it is possible to verify the laced built-up members with any boundary conditions under any combination of actions. The distributions of internal forces drawn in Figure 2 show influence of the theory of the 2nd order with and without shear deformations comparing with results of the 1st order analysis. The influence of the shear deformation is at laced built-up members not very important. Figure 3 gives details of the structure which were in the paper verified. In the out-of-plane buckling verification also the action acting in the plane is taken into account. As a tool was used the theory of the 2nd order with geometrical equivalent global sway initial imperfection on the model of the column with real boundary conditions ($L = 7.2$ m) and not on the model of equivalent member with $L_{cr,z} = 2L = 14.4$ m.

ACKNOWLEDGEMENT

Project No. 1/0603/17 was supported by the Slovak Grant Agency VEGA.

REFERENCES

Rubin, H, Aminbaghai, M. and Weier, H. IQ 100, Vollversion, Feb. 2010, TU Wien. (2010).
Stahlbau Handbuch Für Studium und Praxis. Band 1. 1982. Stahlbau-Verlags-GmbH. Köln.

Piezoresistive behavior of mortars loaded with graphene and carbon fibers for the development of self-sensing composites

A. Belli, A. Mobili, T. Bellezze & F. Tittarelli
Department of Materials, Environmental Sciences and Urban Planning (SIMAU),
Università Politecnica delle Marche, INSTM Research Unit, Ancona, Italy

P.B. Cachim
Department of Civil Engineering (DECIVIL), Universidade de Aveiro, RISCO Research Unit,
Aveiro, Portugal

ABSTRACT: Structural monitoring systems are gaining increasing interest in the field of civil engineering research, due to the recent commitment for the preservation of building heritage, for the saving of resources and for an eco-friendly construction industry. Recent researches show that the addition of conductive fillers and fibers within cement materials could originate cement-composites able to diagnose their own state of strain and tension, measuring the variation of their electrical characteristics (resistance). In this work, resistivity and piezoresistivity of mortars complemented with Graphene Nanoplatelets (GNP), and Carbon Fibers (CF) were evaluated. The variations in electrical resistivity as a function of strain were analyzed under cyclic uniaxial compression of the mortars samples. The results showed a high piezoresistivity behavior of the mortars with an optimal dispersion of GNP and CF, with a quite reversible relation between Fractional Change in Resistivity (FCR) and compressive strain.

1   INTRODUCTION

In the last decades, infrastructures have acquired increasing importance within modern society. These regulate resources supply and the proper functioning of communication technologies and modes of transport. Their efficiency and integrity is an essential factor for industry and for the economic development. In the field of reinforced concrete structures, civil engineering is increasingly focusing on innovative control systems that guarantee safety in their use and an efficient operation (Brownjohn, 2007). This aims to introduce non-destructive systems able to investigate the health of concrete through the detection of parameters such as pH, humidity, corrosion rate of the rebars or through sensors for the detection of cracks and damages.

Strain gauges are a widely used instrument for these types of applications. Strain gauges, applied to the surface of the analyzed material, allow an accurate measurement of its deformation thanks to piezoresistivity effect. Piezoresistivity is a physical characteristic of electrically conductive materials which leads to a variation in the electrical resistivity of the material subjected to strain stress (Zhu et al., 2007). Since the 1990s, studies have been carried out to implement cementitious materials through piezoresistive properties, to obtain an effective reading of the state of health of a concrete structure (Chung et al., 1993).

Material engineering contributes to this field of study, thanks to the development of innovative, multifunctional and high-performance materials, such as graphene. This material, once described in a purely theoretical way, was successively isolated in 2004 and its use has spread to many research fields because of its peculiar properties. In particular, its electrical properties make it an interesting material (as a filler) to increase the electrical conductivity of cement-based materials and to develop their piezoresistive behavior. Graphene, dispersed in a cement-based mixture, forms an effective network of electrically conductive particles. When the hardened composite is subjected to compressive deformation, the particles move closer, changing the conductive paths and giving the material piezoresistive properties (Chung et al., 1993). In presence of an elastic deformation, during the unloading phase the material returns to its original conditions, with a perfectly repeatable effect.

In recent years, materials such as carbon micro-fibers and carbon nanotubes have been incorporated into cement matrix to increase the conductivity of the composites and their piezoresistive behaviors under different types of mechanical loads (Jeevanagoudar et al., 2017). Graphene, thanks to its physical, chemical and morphological characteristics, could have an interesting effect in this type of application. However, there are few studies performed on piezoresistive cements with graphene addition. In this work, the electrical behavior of cementitious composites (structural mortars) with graphene nanoplatelets (GNP) and carbon fibers (CF) addition was investigated. Furthermore, the combined effect of these two materials on both mechanical properties and piezoresistive behavior of mixtures was investigated.

2 MATERIALS AND METHODS

2.1 *Mixtures*

In order to study the effects of GNP and CF (technical data in Table 1) in cement-based materials, mortars made of CEM II 32.5 N cement and silica sand (diameter <1 mm) with an aggregate/cement ratio equal to 3 and a water/cement ratio equal to 0.5 were realized. GNP (Pentagraf, Pentachem S.r.l.) were added in two different quantities: 4% and 7% by cement weight; as well as CF (Apply carbon): 0.05% and 0.2% by mortar volume.

For a better comparison of performances, a reference mixture without the addition of fillers and fibers was also produced. In this type of application, an effective dispersion of the fillers in the matrix is required, in order to guarantee a uniform distribution of the conductive particles and eliminate the segregation effect. In this experimentation, the dispersion of GNP in the composites was facilitated mixing the filler in a solution of water and a polycarboxylate ether agent (Melflux 4930F, Basf S.E.) as superplasticizer (SP) and thorough the sonication of the solution for 30 minutes in an ultrasonic bath. SP is also used to improve the workability of the mixtures, since all mortars show an approximate flow value of 180 mm at the slump test (UNI EN 1015-3: 2007). In Table 2 the composition of the seven mixtures realized and the relative slump test flow values.

The mixtures were sampled in prismatic specimens of 40 × 40 × 160 mm dimensions for both mechanical resistance tests and resistivity tests. Four stainless steel meshes (as electrodes) were placed inside the specimens used for the resistivity tests, with an immersion area of 30 × 30 mm. The electrodes are placed at a distance of 40 mm from each other. The samples were cured in an air-conditioned cell at 20°C with a humidity level of > 95% for 7 days, and with a humidity level of > 65% for another 21 days (UNI EN 1015-11: 2007).

2.2 *Mechanical characterization*

In order to verify the influence of the additions on the mechanical properties of the composites, 3-point flexural strength tests and compressive strength tests were carried out on the 40 × 40 × 160 mm samples at 2, 7 and 28 days of curing (UNI EN 1015-11:2007).

Table 1. Technical data of GNP and CF used in this research.

Properties	GNP	CF
Density (g/cm³)	2.0	1.8
Bulk density (g/cm³)	0.03–0.10	0.55
Particle diameter (μm)	< 5	–
Length (mm)	–	5.5–6.5
Specific surface area (m²/g)	120–150	–

Table 2. Mix proportions and slump flow values of tested mortars.

Mixtures	Cement (g/L)	Water (g/L)	Sand (g/L)	GNP (g/L)	CF (g/L)	SP (g/L)	Flow value (mm)
REF	512	256	1535	–	–	–	185
GNP4	512	256	1535	20	–	1	173
GNP7	512	256	1535	36	–	2	200
0.05CF	512	256	1535	–	2	–	177
0.2CF	512	256	1535	–	7	–	170
GNP4 + 0.05CF	512	256	1535	20	2	1	200
GNP4 + 0.2CF	512	256	1535	20	7	2	177

2.3 Electrical characterization

The electrical conductivity of the mortars was investigated through the measurement of the electrical resistivity of the samples at 7, 14, 21 and 28 days of curing and in the specimens in dried conditions. The electrical resistivity ρ is defined as the inverse of the conductivity σ ($\rho = \sigma^{-1}$), therefore the lower the resistivity, the greater the electrical conductivity of the sample.

For the measurements, a power supply (Protek) and a data acquisition device (Data Taker DT80) connected to a computer were used.

A potential difference was applied in the external electrodes of the samples, in order to apply well defined values of current $I_{(t)}$ (from 0.05 to 10 mA) to the material. The voltage $U_{(t)}$ between the two inner electrodes was detected by the data acquisition device, and the resistivity ρ_s is thus calculated through the first and the second Ohm's law, From which (Equation 1):

$$\rho_S = \frac{U_S}{I_S} \frac{A}{l} \tag{1}$$

where U_s = voltage across the inner electrodes: I_s = current measured between the outer electrodes; A = contact area between the electrodes and the material; l = spacing between the inner electrodes. Scheme in Figure 1 shows the test setup.

2.4 Piezoresistivity tests

For piezoresistivity tests, a continuous acquisition of the resistivity of the material subjected to compression load cycles was carried out. After 28 days of curing, samples were dried at 60°C until a constant mass was reached, thus eliminate the influence of the samples humidity during the tests. For the measurement of compression deformation, a 15 mm, 120 Ω strain gauge connected to the DAQ was applied in the middle of the sample.

A potential difference of $\approx$ 20V was applied to the outer electrodes until the stabilization of the resistivity value. A Shimadzu AG-IC press applied a pressure increase of 250 N·s^{-1} to the sample, up to a maximum of 25 kN (15.6 MPa) and a related decrease until 2 kN. For each test, 10 sequential cycles were applied, with a total time of 180 s each. The piezoresistive properties of the materials were evaluated through the calculation of the fractional change in resistivity (FCR) of the sample according to equation (Equation 2):

$$FCR = \frac{\rho_{(t)} - \rho_0}{\rho_0} \tag{2}$$

where $\rho_{(t)}$ = maximum load sample resistivity; ρ_0 = initial sample resistivity. The correlation between FCR and strain, defined as sensitivity, was evaluated as (Equation 3):

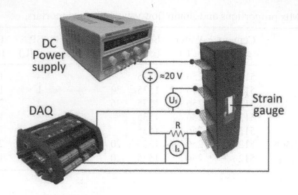

Figure 1.   Setup of electrical measurements.

$$sensitivity = \frac{FCR}{\mu\varepsilon} \tag{3}$$

where $\mu\varepsilon$ = axial deformation of the sample.

3   RESULTS AND DISCUSSIONS

3.1   *Mechanical properties*

The mechanical strength tests demonstrated that the additions do not significantly improve the mechanical strength of the mixtures (Figure 2). On the contrary, high quantity of GNP, lead to a decrease in mechanical compressive strength (22% less for GNP7, compared to REF). This is due to the hydrophobicity of graphene and its physical characteristics (specific surface area and bulk density) and to the consequent difficulty in mixing during the samples preparation (Li et al., 2017). This leads to the segregation of the composites and to an increase in porosity in the hardened mortar (and a decrease in mechanical strength).

Best results in terms of mechanical performance were obtained on samples containing CF (with increments of 10% compared to REF). This is related to bridging action of fibers (Mastali et al., 2016). This effect offers more resistance to crack opening in cement-based materials, with transferring stress from fibers to the cement matrix regard to interfacial shear strength.

3.2   *Electrical properties*

The tests on the wet samples (28 days of curing) show that the resistivity of the mortars with GNP and CF addition is slightly lower compared to REF (Figure 3). However, the tests carried out on the dried samples show a high increase in the resistivity of the mortars (two orders of magnitude), and the samples with GNP show higher values compared to the reference (260% of the REF value for the mortar GNP4-0.05CF). This is due to the greater porosity of the mixtures containing GNP (as seen in the mechanical strength tests). The increased porosity of the cement matrix caused by GNP leads to a higher water content in the material, and, therefore, to a higher electrical conductivity. However, when the amount of water decreases, the material has a greater amount of voids, and therefore a higher resistivity.

On the contrary, two of the realized mixtures show very low resistivity values, even after the samples drying. 0.2CF and GNP4-0.2CF mortars show a resistivity of 328 Ω·m (87% less than REF) and 5.3 Ω·m (three orders of magnitude less than REF) respectively. The high conductivity of these mixtures is probably due to an optimal distribution of the additions within the matrix, which form an extremely effective conductive network within the material (Han et al., 2011). The very high conductivity of GNP4-0.2CF could be related to the reaching of the "percolation threshold", because of the combined addition of GNP and a high dosage of CF (Han et al., 2015).

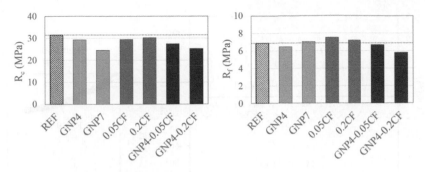

Figure 2.   Results of compressive (left) and flexural (right) strength tests at 28 days of curing.

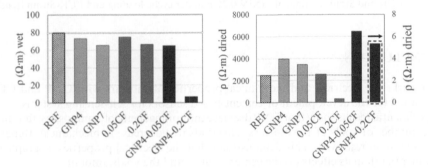

Figure 3.   Resistivity of wet samples at 28 days of curing (left) and dried samples (right).

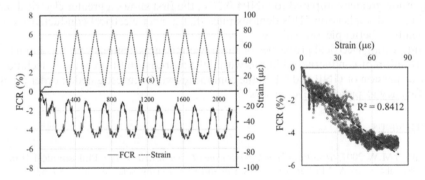

Figure 4.   FCR and Strain vs. time of 0.2CF under cyclic loading and FCR-Strain trendline.

3.3 *Piezoresistivity behavior*

Most composites, due to their low conductivity, showed poor electrical sensitivity when subjected to load cycles. However, 0.2CF and GNP4-0.2CF mixtures, thanks to their high conductivity, showed interesting piezoresistive properties. The correlation between FCR and Strain during load application shows a decrease in the resistivity of the two mixtures as load and deformation increase (Monteiro et al., 2017). 0.2CF shows a very high sensitivity (613.5 MPa^{-1} and 5% of FCR), and the change in resistivity is constant and repeatable. However, the reading has some noise, as can be seen from the irregularity of the FCR-Time curve (Figure 4).

On the contrary, GNP4-0.2CF shows a lower sensitivity (295.4 MPa^{-1} and 2.2% of FCR with similar strain values), but a very high regularity of reading, as seen by the trendline of the FCR-Strain points (Figure 5). The high repeatability is due to the high conductivity of the material, which shows a piezoresistive behavior more similar to a traditional strain gauge.

41

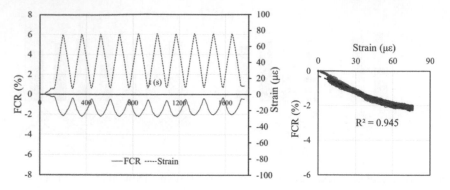

Figure 5.   FCR and Strain vs. time of GNP4-0.2CF under cyclic loading and FCR-Strain trendline.

4   CONCLUSIONS

In this work, the electrical and mechanical properties of mortars with GNP and CF have been studied for the development of cement-based sensors for self-sensing systems.

Experimental results show that GNP decrease the homogeneity of composites, thus increasing the number of voids of the mortars. This leads to a decrease in mechanical strength and to an increase in resistivity. This demonstrates that the electrical properties of graphene are less influential than its effects on the cement matrix and the voids amount.

On the contrary, 0.2CF and GNP4-0.2CF mixtures show a very high electrical conductivity, and clear piezoresistive properties. Although the resistivity of the 0.2CF was two orders of magnitude greater compared to GNP4-0.2CF, the first shows a greater electrical sensitivity during load application. This demonstrates that greater electrical conductivity does not always leads to better piezoresistivity.

The interesting electrical properties of these two mixtures demonstrate the importance of the distribution of conductive additions within the mortars. In future studies, techniques for a better dispersion of GNP and CF in the composites will be investigated in order to improve their efficiency as piezoresistive sensors.

REFERENCES

Brownjohn, J.M.W. 2007. *Structural health monitoring of civil infrastructure.* Philosophical Transactions of the Royal Society A: Mathematical, Physical and Engineering Sciences, 365, 589–622.

Chung D.D.L., Pu-Woei C. 1993. *Carbon fiber reinforced concrete for smart structures capable of non-destructive flaw detection.* Smart Materials and Structures, 2 (1), p. 22.

Han B., Yu X., Ou J. 2011. *Multifunctional and smart nanotube reinforced cement-based materials.* Nanotechnology in Civil Infrastructure. A Paradigm shift. Editors—Springer, 1–48.

Han B., Ding S., Yu X. 2015. *Intrinsic self-sensing concrete and structures: A review.* Measurement, 59, 110–128.

Jeevanagoudar Y.V., Hari Krishna R., Gowda R., Preetham R., Prabhakara R. 2017. *Improved mechanical properties and piezoresistive sensitivity evaluation of MWCNTs reinforced cement mortars.* Construction and Building Materials, 144, 188–194.

Li X., Liu Y.M., Li W.G., Li C.Y., Sanjayan J.G., Duan W.H., Li Z. 2017. *Effects of graphene oxide agglomerates on workability, hydration, microstructure and compressive strength of cement paste.* Construction and Building Materials, 145, 402–410.

Mastali M, Dalvand A. 2016. *The impact resistance and mechanical properties of self-compacting concrete reinforced with recycled CFRP pieces.* Composites Part B, 92. 360–376.

Monteiro A.O., Cachim P.B., Costa P.M.F.J. 2017. *Self-sensing piezoresistive cement composite loaded with carbon black particles.* Cement and Concrete Composites, 81, 59–65.

Zhu S., Chung D.D.L. 2007. *Analytical model of piezoresistivity for strain sensing in carbon fiber polymer-ematrix structural composite under flexure.* Carbon, 45 (8), 1606–1613.

Advances and Trends in Engineering Sciences and Technologies III – Al Ali & Platko (Eds)
© *2019 Taylor & Francis Group, London, ISBN 978-0-367-07509-5*

Bending behavior of steel and GFRP reinforced concrete girders

V. Borzovič, R. Švachula, D. Lániová & S. Šarvaicová
Faculty of Civil Engineering, Slovak University of Technology in Bratislava, Slovakia

ABSTRACT: The goal of this experimental work was to compare the behavior of the girders reinforced with classical steel and GFRP (Glass Fiber Reinforced Polymer) reinforcement. The flexural performance of the girders was verified through the four point load test. Two series of girders with different reinforcement ratio were examined. Together 14 beams with length of 1.5 m were tested. The actual material properties of the concrete and steel and GFRP reinforcement were tested. Both types of the girders were compared within the ultimate and serviceability limit states.

1 INTRODUCTION

1.1 *Glass fiber reinforced polymer*

In the previous work, the research in this area was focused on strengthening the existing structures where FRP reinforcement was applied to the surface of structural members. Nowadays, FRP reinforcement application is verified to replace steel reinforcement for the use in areas of increased environmental load, with the need to avoid steel corrosion or to ensure the electromagnetic neutrality of individual members of the structures. These disadvantages of steel reinforcements can be eliminated by substituting with a corrosion resistive material such as reinforcement made of composite materials: glass, (GFRP – glass fiber reinforced polymer), carbon (CFRP – carbon fiber reinforced polymer) or aramid fibers (AFRP – aramid fiber reinforced polymer). The present experimental study focuses on assessment of the flexural resistance of steel and GFRP reinforced girders (*fib* Bulletin no. 40, 2007; Benko, 2015).

1.2 *Research significance*

The application of GFRP is meant as a replacement of ordinary steel reinforcement (Vijay, 2001; Li, 2012). In many cases this replacement is based on the equivalence of stiffness of the reinforcement types. According to this presumption, approximately 4 times lower stiffness of GFRP requires designing of bigger diameters of GFRP compared to steel reinforcement. This is followed for instance in a prediction model with the same assumptions for shear resistance of members without shear reinforcement. The presented experimental program was inspired by the idea of replacing a steel reinforcement with a GFRP, which was recommended by the GFRP producer. By comparing short-term ultimate tensile force of chosen steel and GFRP bars, the equivalence between steel and GFRP bars was not achieved in our case. The compared reinforcements are shown in Table 1. In this case of comparison, the steel bar is replaced with a GFRP bar of a lower diameter. This comparison does not take into account the influence of alkali environment on a long-term GFRP strength. At first sight the ultimate tensile force based equivalence seems to be correct for calculation of bending resistance. The approach to determine the bending resistance however differs considering the absence of plastic behavior of GFRP compared to steel reinforcement. Hence these differences and the behavior of girders with different types of reinforcement are discussed next in this paper.

Table 1. The properties of reinforcing bars assumed in the comparison.

Steel reinforcement			GFRP reinforcement		
Mean yield strength (MPa) 550			Mean tensile strength (MPa) 1200		
Diameter (mm)	Cross-section area (mm²)	Ultimate tensile force (kN)	Diameter (mm)	Cross-section area (mm²)	Ultimate tensile force (kN)
6	28.27	15.55	4	8.62	10.34
12	113.1	62.21	10	69.07	82.88
14	153.94	84.67	12	106.76	128.11

2 EXPERIMENTAL PROGRAM

2.1 *Material properties*

The concrete strength was determined as an average value measured on three concrete cubes and cylinders. The cubes where of dimensions $150 \times 150 \times 150$ mm, the diameter of cylinders was 150 mm and the height 300 mm. The modulus of elasticity was measured on prismatic specimens with dimensions $100 \times 100 \times 400$ mm. Table 2 summarizes results of these measurements. The tensile strength test showed the ultimate tensile force of steel and GFRP bars. The failure mode for both type of reinforcement are presented in the Figure 1 and Figure 2. The Figure 3 presents stress-strain diagram for both type of reinforcements.

2.2 *The girders 1st series*

The design of the reinforced girders for the purposes of a comparative study was based on the replacing of steel by GFRP bars recommended by the producer of GFRP (Table 1). Together 6 beams with length of 1.5 m and cross-sectional dimensions 100×100 mm were tested in the 1st series. The concrete cover was 15 mm for all specimens. Steel reinforced girders had 2 bars of diameter 6 mm (reinforcement ratio 0.57%), GFRP reinforced girders had 2 bars of diameter 4 mm (reinforcement ratio 0.21%). The girders were subjected to the four-point load test (Figure 4). Results of the loading test are presented in the Figure 5. The failure of the Steel reinforced girders occurred under the acting force approximately 11.5 kN whereas the GFRP reinforced girders collapsed already nearly by 4.7 kN what is 41% of the resistance of steel reinforced specimens. The ultimate tensile force determined from tensile strength test for steel bar of diameter 6 mm was 15.6 kN and for GFRP reinforcement of diameter 4 mm it was 10.3 kN what is 66% of a steel reinforcement. The first loading steps of GFRP reinforced beams induced a large crack which developed rapidly along the height of the concrete section and subsequently the collapse of the beam occurred due to the bar split. The reinforcement ratio of the beams was near to the so called balanced reinforcement ratio which expects the beam failure in bending by collapse of the reinforcement. Considering the lever arm of reinforced cross-section on the level of 90% of effective height, the stress in the GFRP bar reached value of about 800 to 900 MPa at the moment of failure. Characteristic effect for failure of the girders with GFRP reinforcement in the first series was that the reinforcement bar rupture appeared not in the section where the crack was. By collapse the girders broke into two pieces and the GFRP reinforcement was sticking out approximately 50 mm from the crack, from the point of the beam failure (Figure 7). This mode of failure suggests bond problems resulting from a large stiffness difference between the cracked and non-cracked section.

2.3 *The girders 2nd series*

In the next series 8 girders were tested. The specimens were 1.5 m long with cross-section dimensions 175×75 mm. The concrete cover was 15 mm. Two steel reinforced girders were

Table 2. Material properties of the concrete.

Series of girders	1st	2nd
Mean cube strength (MPa)	78	59
Mean cylinder strength (MPa)	58	41
Mean modulus of elasticity (GPa)	42	39

Figure 1. The failure mode in tensile strength test for the steel reinforcement.

Figure 2. The GFRP reinforcement and its failure mode in tensile strength test.

reinforced with two bars of diameter 12 mm (reinforcement ratio 2.4%) and another two girders were reinforced with two bars of diameter 14 mm (reinforcement ratio 3.3%). Then two girders with GFRP of diameter 10 mm (reinforcement ratio 1.5%) and two girders with GFRP of diameter 12 mm (reinforcement ratio 2.3%) were prepared. The results of experimental measurements are presented in the Figure 6. The girders with GFRP reinforcement had bigger deflections even loaded with lower force than steel reinforced girders.

Steel reinforced girders with lower reinforcement ratio (steel ø 12 mm, GFRP ø 10 mm) collapsed at acting force approximately 23.7 kN, whereas the girders reinforced with GFRP reinforcement collapsed already at a force nearly 13.7 kN, which is 59% of the resistance of the girders with steel reinforcement. Based on the tensile strength test, the ultimate tensile force of the steel reinforcement with diameter 12 mm was 62.2 kN and 82.9 kN for the GFRP reinforcement of diameter 10 mm which is 133% of steel reinforcement. Steel reinforced girders with higher reinforcement ratio (steel ø 14 mm, GFRP ø 12 mm) collapsed at acting

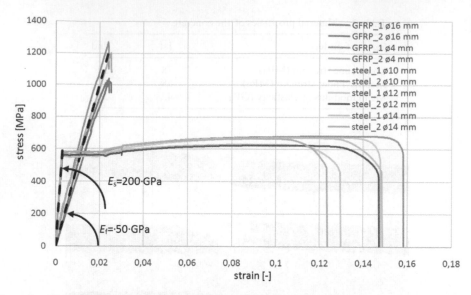

Figure 3.　The stress-strain diagram for the steel and GFRP reinforcement.

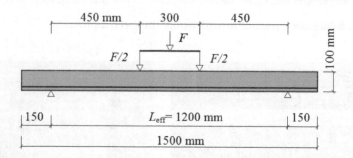

Figure 4.　The four-point loading test arrangement.

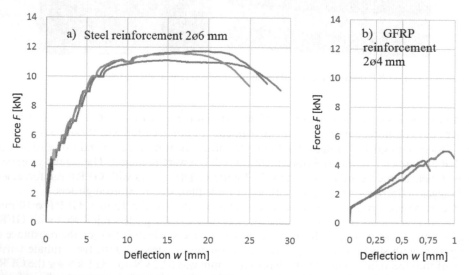

Figure 5.　Results of the loading tests of the girders 1st series with a) steel reinforcement ø 6 mm, b) GFRP reinforcement ø 4 mm.

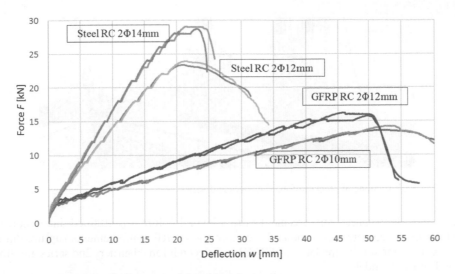

Figure 6.   Results of the loading tests of the girders 2nd series.

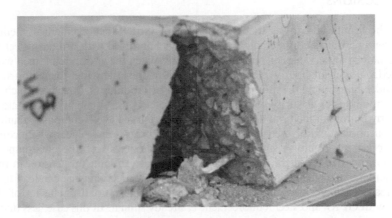

Figure 7.   Failure mode of GFRP reinforced girders 1st series.

Figure 8.   Failure mode of GFRP reinforced girders 2nd series.

force approximately 28.9 kN and the GFRP reinforced girders collapsed already at loading level 15.9 kN what is 55% of resistance of the steel reinforced girders. Based on the tensile strength test, the ultimate tensile force of the steel reinforcement with diameter 14 mm was 84.7 kN and 128.1 kN for the GFRP reinforcement of diameter 12 mm which is 151% of steel reinforcement. The reinforcement ratio of GFRP reinforced girders of 2nd series was

Figure 9. Failure mode of steel reinforced girders 2nd series.

higher than the balanced reinforcement ratio. Flexural failure was attained due to concrete crushing in the compression zone and therefore the GFRP reinforcements rupture did not occur. Characteristic failure detail of GFRP and steel reinforced girders 2nd series are shown in the Figures 8 and 9.

3 CONCLUSIONS

Based on the results, we can say that the behavior of GFRP reinforcement in members subjected to bending differs considerably from members reinforced with ordinary steel reinforcement especially at lower values of reinforcement ratio. When the girders reinforced with GFRP were loaded, a dominant crack was observed which developed right after formation and quickly attained width higher than 0.3 mm. This was caused by modulus of elasticity which is 4 times lower than modulus of elasticity of the steel reinforcement. Serviceability limit state shows as a particularly important factor in design of members reinforced with composite reinforcement. GFRP application could be used in parking lots, road bridges and places where conventional reinforced concrete structures suffer from aggressive properties of the environment. When designing, in order to avoid fragile collapse it is necessary to limit the tension in the GFRP reinforcement and also ensure maximum crack width and deflection what can lead to a considerably increased reinforcement ratio. By higher reinforcement ratio, the behavior of the beams reinforced with GFRP and steel reinforcement is already similar. In this case, less ductile mode of failure based on the collapse of concrete in compression zone becomes the main deficiency. During experimental verification, failures were sudden without any prior notice.

ACKNOWLEDGEMENT

This work was supported by the Slovak Research and Development Agency under the contract No. APVV-15-0658. This work was supported by the University Science Park (USP) of the Slovak University of Technology in Bratislava (ITMS: 26240220084).

The authors appreciate the suggestion and the support of experimental work by company STRABAG.

REFERENCES

Benko, V., Bilčík, J. et al. 2015.*Manuálnanavrhovanie GFRP výstuže do betónových konštrukcii*. Bratislava: The Slovak Chamber of Civil Engineers.
fib Bulletin no. 40, 2007.*FRP reinforcement in RC structures*. Lausanne: Fédération internationale du béton (fib).
Li, X., Lv, H., Zhou, S. 2012. *Flexural Behavior of Innovative Hybrid GFRP-Reinforced Concrete Beams*. Key Engineering Materials. Vol. 517.
Vijay, P. V., Ganga Rao, H. V. S. 2001. *Bending behavior and deformability of glass fiber-reinforced polymer reinforced concrete members*. ACI Structural Journal, 98(6), 834–842.

Advances and Trends in Engineering Sciences and Technologies III – Al Ali & Platko (Eds)
© 2019 Taylor & Francis Group, London, ISBN 978-0-367-07509-5

FEM modelling of gas pipeline components

J. Brodniansky Jr., J. Brodniansky & M. Magura
Faculty of Civil Engineering, Slovak University of Technology in Bratislava, Bratislava, Slovakia

ABSTRACT: Department of Steel and Timber Structures has a long-term partnership with EUSTREAM (transit gas pipeline administrator in Slovakia) to ensure uninterrupted supply of natural gas to western Europe. Since 1998 a regular diagnostic inspections on bridgings and other important gas pipeline components have been performed by Department of Steel and Timber Structures. According to the long-term partnership EUSTREAM and Department of Steel and Timber structures, Faculty of Civil Engineering, Slovak University of Technology was asked to carry out a static assessment of selected structures. Two of the most important gas pipeline structural components are described in the paper. A friction clamp ring on the newly built anchoring block near Sikenica bridging on 1st transit line (pipe diameter—DN1200). The whole bridging Sikenica and interaction bridging- 1st and 2nd transit line according to planned rectification works is described.

1 INTRODUCTION

Department of Steel and Timber Structures (hereafter as—KKDK) has a long-term partnership with EUSTREAM (transit gas pipeline administrator in Slovakia) to ensure uninterrupted supply of natural gas to Western Europe. Since 1998 regular diagnostic inspections on bridgings and other important gas pipeline components (compressor stations, road and rail underpasses, anchoring blocks) have been performed by KKDK. This paper is focused on two of them:

- The whole bridging Sikenica and interaction bridging – 1st and 2nd transit line according to planned rectification works
- Friction clamp ring on the newly built anchoring block near Sikenica bridging on the 1st transit line in Slovakia (pipe diameter—DN1200)

For both models program (RFEM) was used. For further information about other solved examples of transit gas pipeline components, the authors recommend previous papers (Magura M., Brodniansky J.).

2 MODEL OF BRIDGING SIKENICA

Transit gas pipeline line no.1 TP-I. DN1200 and no.2 TP-II. DN1200 are overpassing river Sikenica on three hinge truss arch (Fig. 1). Pipeline bridge has span of 57318 mm and camber of 7678 mm. The structure is made of typified structural parts which were made for bridge overpasses for transit gas pipelines (structural assembly PM-10-B).

Pipelines tend to get closer in the middle of the arch span, this phenomenon is occurring on all gas pipeline bridgings in Slovakia. Spring supports (Fig. 2) under the pipelines are pressed to their maximum and most of them are dysfunctional. It is proposed to perform a complex reconstruction and rectification of pipeline on bridging. All actual spring supports under the gas pipelines need to be changed with new special designed telescopic spring supports (Fig. 2) (new telescopic spring supports were experimentally designed by KKDK). Transit line no.1

Figure 1.   Bridging Sikenca.

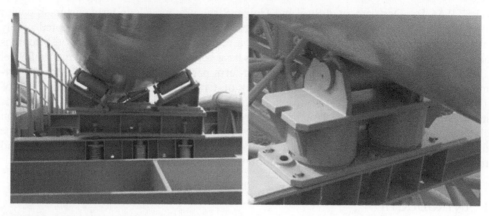

Figure 2.   Old (left) and new (right) telescope.

is proposed to be rectified to a new position (horizontal displacement approx. 250 mm from transit line no. 2).

Four separate models were done, each of them represents a different stage of rectification procedure. For modelling there were used only beam elements, but with different characteristics (only tension, only compression, sprig beam with definition of spring constant). Soil interaction was considered as Winkler spring constant with a value estimated from (Dický J., Mistríková Z., Sumec J.) in perpendicular horizontal and vertical directions to the pipeline.

Modelled construction phases:

– Phase I – Original position of pipelines on bridging (without proposing rectification). The following loads were considered: self-weight, other dead loads (inspection path, cable path and old contact spring seats on each pipeline bridge cross-beam), inside gas pressure, water, inspection vehicle, wind, temperature +, temperature –.
– Phase II – Horizontal rectification. The following loads were considered: self-weight, other dead loads (inspection path, cable path and old contact spring seats on each pipeline bridge cross-beam), inside gas pressure and forces introduced to push the pipeline in to the new position.
– Phase III – Vertical rectification (newly mounted telescopic spring seats and newly mounted spacing beams between pipelines and on the edges of bridging). The following loads were considered: self-weight, other dead loads (inspection path, cable path and new contact spring seats on each other pipeline bridge cross-beam), inside gas pressure, compression prestress of spring saddle beams.
– Phase IV – New position of pipeline. The following loads were considered: self-weight, other dead loads (inspection path, cable path and new contact spring seats on each other pipeline bridge cross-beam), inside gas pressure, inspection vehicle, wind, temperature +, temperature –.

Each of the presented loads were combined, according to (EN 1990).

Figure 3. Bridging Sikenca—Model of unrectified position of pipelines on bridging (Phase I).

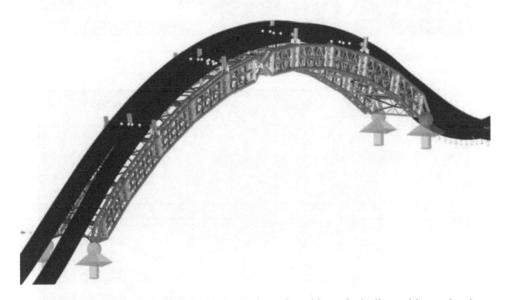

Figure 4. Bridging Sikenca—Model of new horizontal position of pipelines with spacing beams (Phase III and Phase IV).

3 FRICTION CLAMP RING

Horizontal axial force from pipeline is transferred to the anchoring block through friction clamp ring (next in text FCR) consisting of four parts. FCR is leaning on the front of the anchoring block and it is necessary to assemble FCR right after sandblast cleaning of pipeline surface. Minimal prescribed inside pressure of gas is 4,0 MPa (standard operating pressure is between 6–7 MPa). After full activation of newly build anchoring block, it is possible to start restoration works on part of the pipeline affected by the old block structure. Design of FCR is made to enable the transfer of horizontal axial force by friction area FCR-pipe DN 1220.18.9 (diameter 1220 mm). All four parts of FCR are connected by 18pic M30

Figure 5. Friction Clamp Ring (FCR).

Figure 6. Friction Clamp Ring (FCR) – detail of bolt connection between segments.

(10.9) bolts. Material of pipe is steel XC60 and material of FCR is S355. Contact plates of FCR are from P20 (steel plate thickness 20 mm), triangular stiffeners are P20 and plate of bolt connection is P30. Contact plate width is 990 mm rounded with radius R = 610 mm. Friction coefficient used of by the value 0,3 (in real interaction it will be higher). Main aim

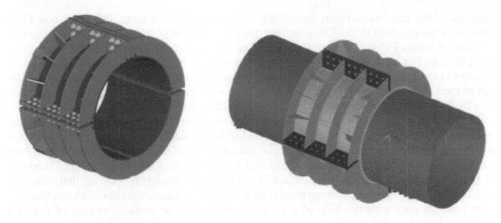

Figure 7.   Friction Clamp Ring (FCR) CAD and FEM model.

of the whole FEM analysis was to predict FCR behavior in interaction of inside pressure—bolt prestress—friction. Besides the of FEM model there were performed simple analytical calculations which could prove correct approach by FEM modelling. After obtaining working FEM model the simple procedure was enhanced and corrected with results from the final FEM model. Calculated internal forces in bolts allowed design of tightening torque according to (STN 73 1495).

In presented FEM model there were used the following finite elements:

– shell elements for pipe, FCR stiffeners, FCR frontal and central plates, plates with bolt connection,
– volume elements for pipe in contact area, contact plate and contact zone between pipe and contact plate,
– beam elements for bolts.

In a large parametric study each of the load cases was calculated separately and some of the relevant load combinations were introduced (inside pressure increase + bolt prestress + horizontal axial force). Linear material model for calculation was used, but because of contact volumes and tension bolt elements the whole calculation procedure had to be solved by means of Newton-Rapson nonlinear calculation procedure. To speed up the calculation three desktop PCs each for a different set of load cases and combinations were used.

4   CONCLUSION

4.1   Rectification works on bridging Sikenica

Assessment of all cross-sections on the structure was done by the means of STN EN standards, both bridging (material S235) and pipelines (material XC60) are satisfactory. For the control of the real-situation tensometric values were modelled for phase II and phase III (phases during rectification works) and it was proved that stresses in pipelines do not overcome 20% of material strength of XC60 steel by vertical rectification and 5% of material strength of XC60 steel by horizontal rectification. Calculated middle span deformation from temperature −35°C is 55 mm (in extreme situation this value may reach 80 mm). There is no limit deformation for such structures, main limiting factor is spring support capacity. Designed shifting force in spring seats is 60 kN and this value produced deformation of the spring support of 100 mm.

4.2   Friction Clamp Ring (FCR)

Results prove correct approach cross checked by simplified calculation of FCR. The most important results from the modelling were internal forces in the bolts due to the internal prestress and prestress of the bolt beams (bolts themselves) and combination of them. Design of new FCR structure is more robust then FCR structures designed in the past, mainly due to better transfer of prestress forces. Deformations of FCR bolts and stiffening plates are unwanted. Full prestress was prescribed in bolts near the friction plate and stiffeners (these bolt positions are the most effective in load transfer). As an example the FCR mounted by internal gas pressure of 4 MPa can be used, minimum of 9 bolts on each connection is prestressed by force 150 kN, gas pressure rises by 3 MPa—this load combination can ensure friction force of 2694 kN.

The lifetime of steel structures can be extended and operational capability and safety can be ensured by regular maintenance. It is advised to keep a detailed record of the diagnosis in written records from all maintenance works, where all the identified defects and repairs are marked. Large FEM modelling was introduced to ensure safety and real structural behavior of presented structural components. Each of the presented models is proving; that relatively difficult problems from engineering praxis can be solved. Not long time ago such calculations were possible only with the help of a supercomputer and the implementation of detailed FEM modelling.

ACKNOWLEDGEMENT

The authors would like to express thanks to the Grant agency of the Ministry of Education, Science, Research and Sports of the Slovak Republic for providing a grant from the research program VEGA Nr.1/0747/16 – Safety and reliability of modern load-bearing members and structures from metal, glass and membranes.

REFERENCES

Magura, M. – Brodniansky, J.: Structural analysis and of braking block sleeves on transit gaspipeline. In Procedia Engineering: Steel Structures and Bridge 2012. Czech and Slovak International Conference. Podbanské, SR, 26.–28. 9. 2012. Vol. 40 (2012), s.257–261. ISSN 1877-7058.
Magura, M. – Brodniansky, J.: Repair Works, Experimental Resarch and Monitoring on Pipeline and Pipeline Bridges. In IASS—APCS 2012: From spatial structures to space structures. Seoul, Korea, 21.–24.5.2012. Seoul: Korean Association for Spatial Structures, 2012.
Dlubal RFEM 5.07: Structural Engineering Software for Analysis and Design—www.dlubal.com.
Dický J., Mistríková Z., Sumec J.: Pružnosť a plasticita v stavebníctve 2, Slovenská technická univerzita v Bratislave, 2006. ISBN 80-227-2515-3.
STN EN 1990: 2009 Eurocode. Basis of structural design.
STN EN 1993-1-1: 2005 Eurocode 3: Design of steel structures—Part 1–1: General rules and rules for buildings.

Advances and Trends in Engineering Sciences and Technologies III – Al Ali & Platko (Eds)
© 2019 Taylor & Francis Group, London, ISBN 978-0-367-07509-5

Slip joint connection of conical towers subjected to bending and torsion

J. Brodniansky Jr., J. Brodniansky, J. Recký, M. Magura & T. Klas
Faculty of Civil Engineering, Slovak University of Technology in Bratislava, Bratislava, Slovakia

M. Botló
Ingsteel s.r.o, Tomášikova 17, Bratislava, Slovakia

ABSTRACT: Slip joint connection research began in Department of Steel and Timber Structures at the Slovak University of Technology in year 2007 by Jozef Recký. Michal Botló continued the research and his achievements were published in his dissertation thesis. Both research works give us good base of knowledge and show us where to continue with the slip joint connection experimental research. In the recent time appeared practical requirement to describe slip joint connection exposed to bending and torsional loads. Slovak company ELV (Elektrovod Slovakia) prepared the specimens for experimental research for us. Proposed research is practically based and achieved results will be used in actual mast and tower structures.

1 INTRODUCTION

Slip-joint connection offers an alternative to the traditionally used flanged one, the most common solution in structures of this kind. Despite many advantages, slip-joint connections are rarely used in engineering practice. The reasons for this are various, but it is certain that the lack of design methods is an important one. There is no mention of these connections in the Eurocodes describing the design of steel structures and their connections. Only the standard for overhead electrical lines (STN EN50341-1) gives us some notations of these connections. The most important parameter, the overlapping length, is not mentioned in any standard and can therefore only be considered through conservative ratios – 1.5 multiple of the diameter. Design procedures of slip-joint connection are affected by standards (STN EN50341-1), (STN EN 1993-3-1) and (DNV-OS-J101).

In the recent time there has been a huge demand from praxis to assess such structures. Slovak company ELV (Elektrovod Slovakia) will provide the specimens for experimental research foe us and the obtained results will be used for safer and economically reliable design of slip join connections mainly for electrical transmission towers (masts) and cellphone (mobile) provider towers.

This paper is divided in to three parts, in first part is described research supervised by Jozef Recký (Recký J.), in second part is described research done by Michal Botló during his PhD. study (Botló M.) and in third part is description of proposed new continuation of research based on obtained knowledge.

2 ELECTRICAL TRANSMISSION LINE TOWER

The geometry, loads and other parameters, was extracted from (Recký J.). To ensure the analysis is accurate, it is necessary to simulate real structure under real conditions, i.e. to sustain the ratio between its parameters and the parameters of loads it withstands. Material is steel S355, it has hexadecagonal cross-sectional shape and is 41.40 m high.

Figure 1. Electrical transmission tower in testing facility and occurred plastic deformation near anchoring.

Figure 2. Tower structure in supporting frame and deformation during experiment.

The structure was fixed to the anchoring platform. The steel cables were used for inducing the loads through steel cables and system of pulleys to hinges on the other end. The hinges were formed by connecting flat articles which hung from a pair of perpendicular portals. The unit construction form of the hinge enabled insertion of pulleys between the flat articles, making it easier to pilot and attach the cables in the horizontal direction. From the hinges,

Figure 3.   Assembly of slip joint deformation measuring device and nowadays functional transmission line near Ružomberok.

cables stretched down to the groove of the fixtures and attached to the hook of the manual pulley block. Afterwards, it is possible to further increase the loads to the ultimate state of the structure which causes a permanent plastic deformation in the critical spot.

The methodology of the experiment with the selected tower was elaborated in accordance with the international standard IEC 60652, Ed. 2.0 – Loading tests on overhead line structures. Ten articulated hinges were used to induce the horizontal forces in the direction of the transmission line. All of them attached to the truss portals on one end and to the fixtures on the other one. During the experimental testing of the conical tower, the structure was gradually exposed to the load states determined by the documentation and its deformations during this process were measured.

3   EXPERIMENTAL ANALYSIS OF SLIP-JOINT CONNECTION

Due to the spatial limits of the laboratory premises, it was necessary to test a smaller sample. The dimensions (the slenderness, in particular) were calculated based on the real structure and the load was adjusted to match the actual load levels. Shown are the results for a sample with the overlapping length equal to 1,5-multiple of the sample's diameter.

The sample was fixed into the loading frame and afterwards loaded by axial force, through a hydraulic press, and the bending moment, by moving the upper base. The sample and the final loading state are demonstrated in the Figure 4.

The loading phases:

– Centric compression, the axial force of previously determined value (13 kN) was applied.
– Bending, the upper base was deflected; the deflection was calculated based on the intended value of the bending moment
– Combined load, where the axial force (13 kN again) was applied on the structure bent in the previous phase

The most important parameter for measuring was the tangential stress in the web of the structure by means of HBM 1-LY6/120 strain gauges (the arrangement is shown in the Figure 5). The extremes of the tangential stress were measured by two devices—one attached to the upper base of the lower segment on the compressed side (i.e. inside the segment) and attached to the lower base of the upper segment, on the side of the highest tension.

Deflection measuring devices were applied in four locations—by the top base (to measure its deflection), at the slip joint (to measure the possible further insertion of the lower segment into the upper one), and two more at the fixation point of the sample's lower base, to ensure that the possible semi-rigid reaction of the fixation is well documented.

Figure 4. The experimental sample; left – before the loading, right – after the loading.

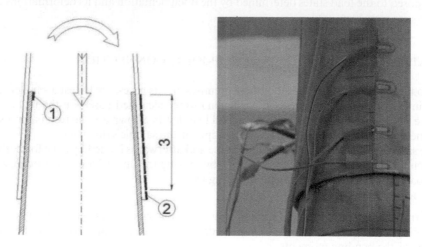

Figure 5. The strain gauges—the arrangement.

3.1 *Phase 1 – Centric compression*

The only acting load was axial force introduced through the hydraulic press. The intended overlapping length was the 1,5-multiple of the sample's diameter (in this case 195 mm). The friction coefficient was measured in the laboratory, its value equal to 0,342.

3.2 *Phase 2 – Bending*

The force from the phase 1 was lowered to zero in order to enable the application of the bending moment. The deflection of the upper base was amounted to 60 mm and was determined by calculation.

Table 1. Comparison of results.

Phase	Tangential stress [MPa]		
	Experiment	Numerical method	Discrepancy
Phase 1 – compression	8.35	10.47	20.2%
Phase 2 – bending	45.58	45.55	0.1%
Phase 3 – combined	49.03	56.03	12.5%

Table 2. Samples.

Sample	Pieces	Length (m)	Maximal diameter (mm)	Minimal diameter (mm)	Web thickness (mm)	End plate thickness (mm)
Round tube	5	3.828	200	147	3	14
12 edge	5	3.828	200	147	3	14
16 edge	5	3.828	200	147	3	14

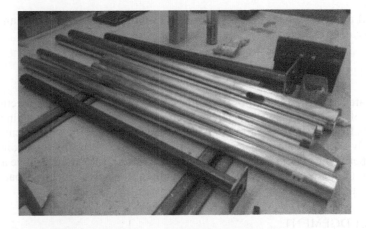

Figure 6. The test samples in preparation stage for experimental research.

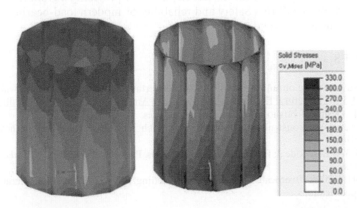

Figure 7. Outer (left) and inner (right) part of slip-joint connection with equivalent stresses.

Due to the semi-rigid fixation of the upper base, it was necessary to calculate the stiffness of the support and take it into the software model.

3.3 *Phase 3 – Combined loading*

Combined load, where the axial force was applied on the bent structure in the previous phase.

4   CONTINUATION OF SLIP-JONT CONNECTION RESEARCH

In recent time there is huge demand from praxis to assess structures which are assembled from more than two conical parts slipped together. With cooperation with company ELV, there were procured first three test samples for the calibration of test devices. Two of them were formerly street lights with small damage (it will not affect purpose of these calibration mesurements) and one is calibration sample for proposed test under dynamic low-cycle loading. Zinc coated round street light samples will be used only for compression tests (because of their shape without anchoring plates) (Figure 6). Raw steel sample (octagonal cross-sectional shape) is for mentioned low cycle dynamic testing (Figure 6).

Be size of calibration samples there will be delivered and tested samples (material S355) under compression, bending and torsional loads. Mainly torsional loads will be more investigated.

First FEM models (RFEM) were done for better calibration of test samples and for reason what results can be expected.

5   CONCLUSION

It can be stated that the reduced overlapping length of the slip-joint connection equal to 1.2 multiple of the tower diameter safely withstands the least favourable loading conditions due to results from previous research. FEM-software results and the experiment proved that the commonly accepted empirical length (1.5 to 1.8 multiple of the tower) is conservative.

A new experimental research is scheduled and obtained results should prove and enhance the results in (Botĺo M.). Incorporation of torsion and low-cycle dynamic load is step forward in our outgoing research.

ACKNOWLEDGEMENT

The authors would like to express thanks to the Grant agency of the Ministry of Education, Science, Research and Sports of the Slovak Republic for providing a grant from the research program VEGA Nr.1/0747/16 – Safety and reliability of modern load-bearing members and structures from metal, glass and membranes.

REFERENCES

Botló M., Nasúvané spoje ohraňovaných stožiarov. Dissertation thesis, Bratislava, Slovakia, 2017.
Dlubal RFEM 5.07: Structural Engineering Software for Analysis and Design—www.dlubal.com
DNV-OS-J101: Design of offshore wind turbine structures.
Recký J., Teoretické a konštrukčné problémy oceľových stožiarov. Dissertation thesis, Bratislava, Slovakia, 2013.
STN EN 1993-3-1: Eurocode 3: Design of Steel Structures. Part 3–1: Towers, masts and chimneys. Towers and masts.
STN EN 50341-1-2006: Overhead electrical lines exceeding AC 45 kV. Part 1: General requirements. Common specifications.

Advances and Trends in Engineering Sciences and Technologies III – Al Ali & Platko (Eds)
© *2019 Taylor & Francis Group, London, ISBN 978-0-367-07509-5*

Minimum mass container production for ships and airplanes, a review

A.A. Deli, K. Jármai & G. Kovacs
Faculty of Mechanical Engineering and Informatics, University of Miskolc, Hungary

ABSTRACT: The research summarizes the latest published works in the most modern applications, theoretical developments and new designs of the sandwich structures consisting of metal and/or core (honeycomb, foam or corrugated plates), where the advantages are combined because there is a great need to reduce the mass and costs. Analytical models and methods of analysis of sandwich structures are shown by representative problems. The most important applications of the sandwich structures included in this review are centered around aerospace and marine engineering including shipping and Air Containers. Lightweight cargo containers provide a huge savings compared to conventional containers. According to the IATA calculations, the weight of fuel required to carry 1 kg additional weight per hour 0,04 Kg. The researchers suggest further research into the global container industry.

1 INTRODUCTION

Cargo shipping containers can withstand all kinds of weather and environments without deteriorating, depending on the mechanical properties of the structure to be manufactured. Conventional containers are made of a type of corrugated steel known as corten. The Corten is a compound based on the fusion of steel and copper with additives of chromium, manganese, and nickel. New cargo containers have a lighter weight than traditional freight containers due to the new designs of sandwich structures and the type of material they are made of. Air cargo containers and shipping containers are used to load baggage, freight and mail by linking a large quantity of goods in one unit. The aim of this paper is to review the relevant research on the possibility of replacing the walls, floor and roof of conventional containers with lightweight composite materials structure panels. The result of these savings in weight is the annual reduction in costs or increased aircraft turnover or movement of ships, trains and trucks.

2 LITERATURE REVIEW

Review relevant research on the possibility of replacing the walls, floor and roof of conventional air and shipping cargo containers with lightweight composites materials as well as the researches that deals with minimum weight optimization design of honeycomb, foam and corrugated cores sandwich panels. The result of these savings in weight is the annual reduction in cost or increase aircraft turnover or movement of ships, trains and trucks.

Borvik & et al. 2008a,b. Developed of a new, cost-effective and lightweight protection concept for a 20 ft container to be used as shelter in international operations.

Crupi, V. Epasto, G. & Guglielmino E. 2013. Studied the parameters which influence on the static and response of low-velocity impact for two aluminum sandwich structures: foam and honeycomb. The collapse of honeycomb cells is strongly influenced by cell size.

Aimmanee, S. & Vinson, J.R. 2002. Analyzed and optimized of simply supported sandwich plates with a foam-reinforced web core subjected to in-plane compressive loads.

The sandwich plate is optimized for the minimum weight by subjected to a uniaxial compressive load.

Kevin Giriunas and et al. 2012. Investigated the structural limitations of shipping containers to develop them according to design requirements through the use of computer modeling of finite elements where computer simulations show the effectiveness of container walls and roof to resist loads. The main objective of the research is to develop structural guidelines for shipping containers of the International Organization for Standardization (ISO).

Demsetz, L.A. & Gibson, L.J. 1986. Identified the core density and face thicknesses to minimize the weight of a sandwich plate for a given bending stiffness. The bending test results is confirm the analysis results for simply supported circular sandwich plates with polyurethane foam cores and skins of aluminum.

Banhart, J. Schmoll C. Neumann U. 1998. Characterized the corrosion behavior of lightweight aluminum foam samples in salt water. Sandwich structures consisting of aluminum foam cores and aluminum face panels are manufactured by adhesive bonding.

Triantafillou, T.C. & Gibson, L.J. 1987. Designed the minimum weight of the foam beam or sandwich panel using the developed failure equations by restricting the face and substance of the failure simultaneously. The core density relationship was used as one of the beam design parameters to be found in the optimization analysis. The results gave the of thickness core and density as well as the face thickness that reduce the weight of a foam core sandwich beam or plate. Minimum Weight Design of Foam Core Sandwich Panels for a Given Strength.

Bode, W. 2016. Investigated in the replacement of the current aluminum floor of Nordisk container 14.1 kg with lighter composite 40% to savings cost reductions. Analytical and finite element calculations was performed. The results showed the composites materials do not have sufficient stiffness so, caused deflections in the full scale roller tests. The wear on the aluminum plate after 13000 cycles compare with composite plates after 800 cycles.

Vinson, J.R. & Sidney, Shore. 1971. Developed the analytical methods to designed web-core sandwich panels with minimum weight for a given length and width of the panel, load index and specified materials of face and web. The numerical results showed that by using the second generation composite materials such as boron epoxy & graphite epoxy the weight will be saved.

William, G. & et al. Developing innovative of lightweight design concepts that would allow weight reduction in the air cargo container through the applications of lightweight composites. Where, built a scaled model prototype of a typical air cargo container to assess the technical feasibility and economic viability of creating such a container from fiber-reinforced polymer (FRP) composite materials.

DSM, & Samskip company. 2012. Developed the new High Q™ container as shown in the Figure 1. The aim of HighQ™ container is 20% lightweight than its steel counterparts. The mass reduction of High Q™ container is achieved thanks by replace the commonly used corrugated steel with composite panels. The main ad-vantage of new High Q™ container is efficiency of cost that could lead to considerable cost savings and increased efficiency in the transport sector. The decrease in the weight of tare container as well as with an aerodynamic design leads to significant fuel savings during transport when compared with a steel container.

Bode. 2016. Technical discussions between DSM & Nordisk 2012 to design composite bottom plate to reduction weight of at least 40% compared to an aluminium bottom plate. Nordisk Aviation Products has designed, developed and manufactured Nordisk UltraLite® 55 kg AKE container as shown in the Figure 2 which provides the lowest tare weight and highest strength-to-weight performance. Nordisk's proven lightweight design combined with high quality materials allows a weight reduction of 15 kg to 40 kg per ULD or between 300–600 kg per loaded aircraft. Nordisk Ultralite® is the world's lightest AKE. While carrying the same cargo volume, it will save you at least 25% weight compared to traditional aluminium containers. Savings on fuel and increased payload, low maintenance costs, less damage to bags and aircraft and fewer injuries to loaders. This Nordisk container has a 2.5 mm thick aluminium floor measuring 1440 mm by 1412 mm. The current weight of the floor panel is 14.1 kg.

Figure 1. HighQ™ container developed by DSM, & Samskip company 2012.

Figure 2. UltraLite® 55 kg AKE container developed by DSM, & Nordisk company 2012.

Li, X. & et al. 2011a,b. Developed a method of minimum weight optimization for sandwich structure subjected to torsion with and without bending load. The results showed the core weight value is 66.7% of the whole sandwich structure at optimum design. Derived the optimum solutions for three cases to requirement design constraints of torsion and bending stiffness. Numerical examples are presented to illustrate the developed optimum design solutions for sandwich structures made of either isotropic face skins or orthotropic composite face skins.

Vinson, J.R. 1986. Analytical solutions of Closed-form are presented to analysis and design of minimum weight composite honeycomb sandwich panels with different types of cell core subjected to in-plane uniaxial compressive loads such as hex-cell and square cell. In this method included monocell buckling, overall buckling, face wrinkling, core shear instability and over-stressing. The Analytically solution are found the optimum design for cell wall thickness, core depth, cell size and face thickness. In this methods provided compare between various material such as polymer, metal, and ceramic matrix and honeycomb sandwich construction with other panel architectures.

Birman, V. & Kardomateas, G.A. 2018. Discussion on the analytical models and analysis methods of new designs in the elements of sandwich structures focus on the type of core and introduction of nanotubes and smart materials.

Zaid, N.Z.M. & et al. 2016. Presented and discussed a review of sandwich structure dependent on with different cores, such as honeycomb, foam core and corrugated core for future research development.

Salem, A.I. & Donaldson, S.L. 2017. Presented optimization for a combined weight and cost for sandwich plates with thin hybrid composite face sheets which consisted of carbon/epoxy and E-glass/epoxy fiber reinforced polymer and foam core based on the bending and torsional stiffness's using the Active Set Algorithm.

Japan Airlines (JAL). 2014. Started using new light-weight cargo containers weighs 58 kg, which weighs 41 kg less than the current aluminum container decreased by 40%. The structure of the new cargo container is made of synthetic resin in honeycomb. The Boeing 777-300ER can be loaded by 44 containers with weighing 1,804 kg less compared with current cargo contains loaded so, to reduce the fuel consumption 800 liters.

3 THE PROBLEM STATEMENT

U.S. Department of Energy. 2018. Since the year 2000, the kerosene price went up. In the year 2000, 15% of the ticket price of a plane was made up by jet fuel costs. Now, this percentage has risen up to 40%. So, for this reason, airliners are looking for lightweight air cargo containers. To solve this problem the companies development the composite sandwich structures plates which are used in the industry of container to satisfies the requirements. Designed a composite floor for air cargo containers with comparable performance to aluminium floors to lower lightweight container tare weight.

4 RESULTS AND DISCUSSION

The walls, roof, and floor of containers can be designed as follows (Department of DSP 2017).

- Wall panels are made of corrugated or flat sheet steel, a riveted or bonded aluminum sheet and wall post assembly, fiber reinforcement plastic FRP, foam and beams, aluminum, or honeycomb material that forms the side wall or end wall as shown in the Figures 3 and 4.
- Roof panels are made of corrugated or flat sheet steel, sheet aluminum, fiber reinforcement plastic FRP, or foam and beam and aluminum honeycomb panel that forms the top closure of the container.
- Flooring is made of material that is supported by the cross members and bottom rails to form a load bearing surface of the cargo. The flooring is usually constructed of laminated wood planks, plywood sheets, or other composite material and is screwed or bolted to the

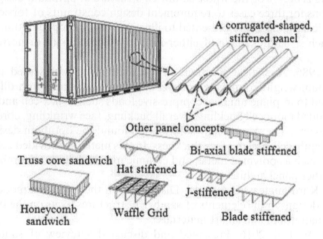

Figure 3.   Shipping container design wall 2012.

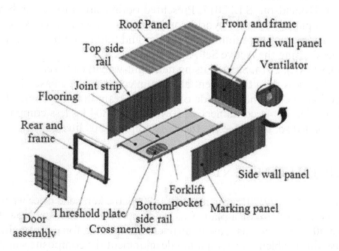

Figure 4.   Typical steel container, exploded view (Department of DSP 2017).

64

cross members. Some containers have welded steel or aluminum flooring, sandwich panels or a combination of metal and wood.

All types of shipping containers meet the following test requirements as set forth in ISO 1496/1. 1990. For loading strength of side walls, front walls, door and roof as shown in Table 1. The Table 2 shows the deflection results of the modified composites with an improved stiffness due to the addition of composite material deformation to match the aluminum sample distortion used in the manufacture of the air cargo container floor by increasing thickness and weight. The material that offers the same stiffness performance at the lowest weight is the concept of full carbon while the concept of full glass requires higher weight than aluminum to get the same stiffness. The felt and aramid concepts achieved weight savings of 32% and 35%, respectively (Bode 2016).

Table 1. Loading strength of side walls, front walls, door and roof (ISO 1496/1 1990).

Construction element	Test load
Side walls	0.6 times the weight of the max. payload
Front wall and doors	0.4 times the weight of the max. payload
Roof	300 kg at surface of 60 cm × 30 cm

Table 2. Deflection results of the improved composite (Bode 2016).

Thin plate	ω_{max} [mm]	t [mm]	Weight [kg]
Al7021-T6	0.70	2.5	14.1
Aramid/felt/glass	0.76	3.7	9.6
Aramid	0.67	3.7	9.2
Glass	0.67	3.8	14.5
Carbon	0.74	2.7	8.0

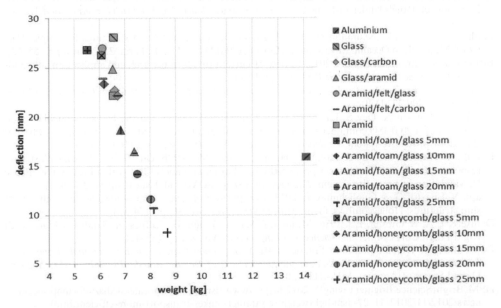

Figure 5. Overview of deflection vs. weight for different composite concepts and alminium as calculated with FEA (Bode 2016).

Figure 6. Halfway consolidated sheet of Tegris.

Table 3. Deflection results of a Tegris composite (Bode 2016).

Thin plate	ω_{max} [mm]	t [mm]	Weight [kg]
Tegris consolidated sheet	0.73	6.0	9.5

5 ALTERNATIVE SOLUTIONS

Milliken & Company, 2015. Tegris is a polypropylene thermoplastic composite fabric woven from specially modified ribbon yarns, which are made of three layers of polymer in the ABA structure. Melt the outer layers, or "A" at a temperature lower than the base layer "B". For consolidation, multiple layers of cloth are stacked together and heat and pressure are applied to form a rigid, impact resistant part see Figure 6. Tegris composite material has low mechanical properties, whereas the Tegris consolidated sheet has a tensile stiffness of 5–6 GPa, but the density of Tegris is very low at 780 kg/m³. See Table 3.

REFERENCES

Aimmanee, S. & Vinson, J.R. 2002. Analysis and Optimization of Foam-Reinforced Web Core Composite Sandwich Panels Under In-Plane Compressive Loads. *Journal of Sandwich Structures and Materials*. Vol. 4.

Banhart, J. Schmoll C. Neumann U. 1998. Light-weight aluminium foam structures for ships. Proc. Conf. Materials in Oceanic Environment (Euromat '98) Vol. 1, Ed.: L. Faria; 22–24 July. pp. (55–63).

Birman, V. & Kardomateas, G.A. 2018. Review of current trends in research and applications of sandwich structures. *Composites Part B* 142 pp. (221–240).

Bode W. 2016. *Evaluation of a Lightweight Composite Bottom Plate for Air Cargo Containers.* Master Thesis.1311409 TU Delft.

Borvik, T. Hanssen A. Dey S. Langberg H. & Langseth M. 2008a. On the ballistic and blast load response of a 20 ft ISO container protected with aluminum panels filled with a local mass—Phase I: Design of protective system. *Engineering Structures*. Vol. 30, Issue 6, pp. (1605–1620).

Borvik, T. Hanssen A. Dey S. Langberg H. & Langseth M. 2008b. On the ballistic and blast load response of a 20 ft ISO container protected with aluminium panels filled with a local mass—Phase II: Validation of protective system. *Engineering Structures*. Vol. 30, Issue 6, pp. (1621–1631).

Crupi, V. Epasto, G. & Guglielmino, E. 2013. Comparison of aluminium sandwiches for lightweight ship structures: Honeycomb vs. foam. *Marine Structures*. Vol. 30, pp. (74–96).

Demsetz, L.A. & Gibson, L.J. 1986. Minimum Weight Design for Stiffness in Sandwich Plates with Rigid Foam Cores. *Materials Science and Engineering*. Vol. 85, pp. (33–42).

Department of Defense Standard Practice. (2017). *Inspection Criteria for International Organization for Standardization (ISO) Containers and Department of Defense Standard Family of ISO Shelters.* United States of America. MIL-STD–3037.

DSM, BrightScience, BrighterLiving™. 2012. https://www.dsm.com/corporate/media/informationcenter-news/2012/11/2012–11–27-dsm-lightweight-container-makes-transport-more-efficient.html.

ISO 1496/1.1990. Series 1 freight containers—Specification and testing Part 1: General cargo containers for general purposes URL: https://law.resource.org/pub/us/cfr/ibr/004/iso.1496-1.1990.pdf.

Japan Airlines (JAL). 2014. Introduces New Light-weight Cargo Container on its International Routes URL: http://press.jal.co.jp/en/bw_uploads/20140507-JAL%20Introduces%20New%20light-weight%20 Cargo%20Container%20on%20its%20International%20Routes.pdf.

Kevin Giriunas, Halil Sezen & Rebecca B. Dupaix. 2012. Evaluation, modeling, and analysis of shipping container building structures. *Engineering Structures*. Vol. 43, pp. (48–57).

Li, X. Li, G. Wang, C.H. & You M. 2011a. Minimum Weight Sandwich Structure Optimum Design Subjected to Torsional Loading. *Springer Science+Business Media*. Applied Composite Materials. Vol. 19, Issue 2, pp. (117–126).

Li, X. Li, G. Wang, C.H. & You M. 2011b. Optimum Design of Composite Sandwich Structures Subjected to Combined Torsion and Bending Loads. *Springer Science+Business Media*. Applied Composite Materials. Vol. 19, Issue 3–4, pp. (315–331).

Milliken & Company. 2015. *Milliken Tegris Thermoplastic Composites*. Online brochure. URL: http://tegris.milliken.com/en-us/technology/Documents/Tegris%20Overview%202015.pdf.

Nordisk Aviation Products. 2012. Nordisk UltraLite® AKE. URL: http://http://www.nordisk-aviation.com/en/ld-containers/nordisk-lite-family/nordisk-ultralite-ake/.

Salem, A.I. & Donaldson, S.L. 2017. Weight and Cost Multi-objective Optimization of Hybrid Composite Sandwich Structures. *International Journal of Computational Methods and Experimental Measurements*. Vol. 5, issue 2, pp. (200–210).

Triantafillou, T.C. & Gibson, L.J. 1987. Minimum Weight Design of Foam Core Sandwich Panels for a Given Strength. *Materials Science and Engineering*. Volume 95, pp. (55–62).

U.S. Department of Energy. 2018. U.S. Gulf Coast Kerosene-Type Jet Fuel Spot Price FOB. URL: http://www.eia.gov/dnav/pet/hist/LeafHandler.ashx?n=pet&s=eerepjkpf4rgcdpg&f=d.

Vinson, J.R. & Shore, S. 1971. Minimum weight web-core sandwich panels subjected to uniaxial compression. *Journal of Aircraft*, Vol. 8, Issue 11, pp. (843–847).

Vinson, J.R. 1986. Optimum Design of Composite Honeycomb Sandwich Panels Subjected to Uniaxial Compression. *AIAA Journal*, Vol. 24, No. 10, pp. (1690–1696).

William, G., Shoukry, S., Prucz, J., & William, M. 2016. Lightweight Composite Air Cargo Containers. *SAE International Journal of Aerospace*. Vol. 9, No. 1, pp. (185–189).

Zaid, N.Z.M. Rejab, M.R.M. & Mohamed, N.A.N. 2016. Sandwich Structure Based On Corrugated-Core: A Review. *MATEC Web of Conferences*. Vol.74, 00029.

Advances and Trends in Engineering Sciences and Technologies III – Al Ali & Platko (Eds)
© 2019 Taylor & Francis Group, London, ISBN 978-0-367-07509-5

Temperature dependence of the yield strength of heat-resistant steels

A. Erdős & K. Jármai
University of Miskolc, Hungary

ABSTRACT: The paper describes the applicable heat resistant steels for pressure vessels, working on high temperature. Such a high temperature apparatus is, for example, an ammonia synthesizer. The wall thicknesses of a converter are planned to be determined with three different internal pressures and temperature values. The investigated heat-resistant steels are classified in three groups: ferritic, austenitic and duplex (ferrite + austenite). There is a great variety of heat resistant steels in material properties, heat resistance, weldability and price. For the curve fitting calculations, the Table Curve program was used.

1 INTRODUCTION

Heat resistant steels are those steels that retain their mechanical properties even over a large number of repetitive stresses above 550°C. At this high temperature they also resist aggressive impacts, such as hot gases, molten metals and salts and combustion products. However, it is worth noting that this resilience is highly dependent on the characteristics of the stress, so the comparison and characterization of each type is not a simple task. At a temperature above 550°C, various reactions occur between the steel surface and the organ. The result of these reactions is the formation of an oxide layer on the surface of the steel. The formation of rust is a process of oxidation, the beginning of which is the affinity between the reactants and the diffusion is decisive at the end of the process. Therefore, these steels must have corrosion resistance. This effect can be achieved by alloying. The main alloy of heat and corrosion-resistant steels is chromium, which is above 18%. Its primary task is to prevent the surface oxidation process. Reveal alloys also include silicon and aluminium. Typical heat treatment of heat resistant steels is a high temperature tempering around 1050–1150°C. This is followed by cooling that may occur in air or water. This heat treatment process provides a 223 HB hardness. Their application area is very wide (Böhler 2017). It is used in heat treatment plants as tank material, for construction of furnaces and stream boilers, as a material for grids, fittings and cladding. They also occur in the glass and ceramic industry. They can also be found in the manufacture of machinery primarily as protective tubes for thermocouples and in the gas and petroleum industry as pipeline materials.

2 HEAT RESISTING STEELS

Heat-resistant steels should be classified into three groups according to the basic element of the base steel. Thus, we can distinguish ferritic, austenitic and duplex (ferrite + austenite) heat resistant steels. Typically, small carbon and high alloying content, besides chromium, typically contain manganese and nickel or titan and molybdenum (Weld-technology 2018).

Among the three types mentioned above, the ferritic group possesses the least degree of toughness. The main alloy of these steels is chromium, but it also can contain a small amount of silicon and aluminium to help the formation of the ferrite, and they also increase the corrosion resistance rate of the steel. One of the great advantages of this type is that it is highly resistant to sulphur-containing materials. It is susceptible to rupture around 475–500°C.

Nickel is an important alloy of duplex heat resistant steels, which is always below 9%, because if it goes over 9% the basic texture of the steel will be purely austenitic. This type combines the properties of ferritic and austenitic steels. They have better toughness characteristic than pure ferrite, and also have better weldability, cold forming and heat strength. Sharpening due to particulate suppression occurs at higher temperatures at ferrite.

The purely austenitic structure is also available, but only 9% or above the Ni content. Their warmth and toughness are excellent. Their debilitating tendency is considerably lower than the previous two types and it is only between 650 and 900°C after a long time and is not at all above a certain temperature. Their reversible properties are very high in the oxidizing atmosphere. They have a good cold forming and a great advantage that they can be welded by almost all procedures. It maintains its heat resistance in air up to 1150°C. When cutting these steels pay attention to cold form hardening. Their disadvantage compared to the other two is that their resistance to sulphur-containing materials is considerably lower.

The chemical composition of heat-resistant steels is very similar to corrosion-resistant steels, so their weldability and the problems during the process are the same. These general problems are as follows:

– the secession of the σ-phase, which problem dangerous between 500 and 900°C,
– the growth of the grains, this problem steps up only above 1150°C,
– the grain boundary corrosion,
– the hot cracking above 1250°C
– the cold cracking between 400 and 475°C.

3   HEAT RESISTING STEELS AND THEIR WELDING

Before the welding of the ferritic type, the raw material must be heated to a temperature of 200–300°C and then at the end of the process a high temperature tempering (around 650–700°C) is applied to increase the toughness of the joint. Their hot cracking sensitivity is also higher. Currently, the increasing temperature causes the growth of the grains and this means a constant toughness decreasing. The structure of the ferrite steels does not undergo transformation; therefore, their toughness cannot be improved by heat treatment. It is not recommended to choose a high heat input method for welding. The grain boundary corrosion takes place relatively fast in ferritic steels. This cannot be suppressed by rapid cooling in water. For this reason, the heat treatment of ferritic steels consists of a 950°C solvent dewatering and a high temperature 750°C two hours heat retention.

Duplex is called heat resistant steels with structure consisting almost 50% to 50% of ferrite and austenite. Accordingly, it is between its mechanical characteristic of the two types. There are also differences in composition, duplex steels contain much more chromium up to 20–31%, and 5% manganese, but less nickel and nitrogen, up to 9% and 0,5% respectively. By decreasing the nickel content, strength is increased and in addition, its cost is greatly reduced. Their primary use is oil and gas pipelines and tanks. These steels are therefore more resistant to corrosion and mechanical stress but cannot be used at temperatures such as ferritic or austenitic counterparts. The duplex steels can be classified into three additional groups according to alloying content:

– Low-duplex steels, which generally contain 22% chromium,
– Standard duplex steels with chromium content up to 25%,
– The chromium content of hyper duplex steels is much higher, up to 31%.

During the welding, the intensive heat input should be used as fast as possible to minimize the amount of ferrite produced. This can also be reduced by reducing the heat input. However, the heat input should not be less than 0,5 kJ/mm. The nickel content of the filler materials used during their welding must be higher than the base material. This is important because nickel is helping in the formation of austenite over ferrite. Duplex steels are generally not required to be preheated prior to welding unless the ambient temperature is below 5°C or if there is a risk of condensation on the welded surface. In such cases, it is advisable to heat the duplex steel to 50–75°C and then complete the welding. But with some kinds of arc welding this temperature can be up to 100°C. Intermediate temperature has a significant effect on the structure of the steel and the heat-effect zone. Its value is recorded in standards and, some at 200°C, but at 150°C for other more stringent specifications (Stainless steels 2018).

This low temperature greatly influences the time of welding. In the preparation of the welding, all excess chips of the surfaces must be removed, the grease-free surface is important and then dried. Failure to do so or improperly impairs the bond quality and corrosion resistance. Hydrogen fracture is a rare but unimaginable phenomenon when welding duplex heat-resistant steels. The defence is the same as for conventional steels.

After welding of duplex steels, it is rarely used for subsequent heat treatment to avoid the formation of σ-phases. However, if heat treatment is required, it must be carried out at high temperatures (1000–1100°C) for the entire structure and then cooled in water. In duplex steels, temperature changes above 300°C cause significant changes in mechanical properties. For this reason, thermal smoothing is not recommended.

Austenitic heat-resistant steels can be welded by all conventional methods, and there is a standard selection of filler materials. The austenitic steels have a thermal conductivity coefficient and a coefficient of thermal expansion different from the ferrites, the former being about one third of the ferrite, while the latter is 30% below. This results in a higher thermal expansion of the heat affected zone, resulting in greater residual stress, leading to greater warping (for stitching, set a gap of 1 to 1.5 mm higher than it would be needed). This is especially a problem when welding plates. Here, the warp can be compensated by accelerating the cooling, so that the required tolerances can be met. The toothed austenitic Cr-Ni steels are very tough, well-formed, highly elongated ($A_5 > 40\%$). Due to the high elongation, they can be welded without any risk of cracking, without preheating and after treatment. There is no γ-α transformation, and hardening does not occur during welding.

So that the connection can be written between the yield strength and the temperature, a series of tensile tests shall be carried out at specified temperatures. The tensile strength of the structural steels increases with the temperature up to about 300°C, followed by a steep decline, and then the difference between the lower and upper yields disappears. However, their yield strength is constantly decreasing as temperature rises. The deformation characteristics, however, increase with the temperature. The modulus of elasticity of the ferrite steels decreases linearly as the temperature rises up to 500°C, from which it begins to decline steeply. Compared to the austenitic, their modulus starts to fall steeply above 700°C. In the case of manganese and molybdenum alloys, the temperature in which a sharp change is found at about 400°C. For chrome-molybdenum alloy steels this temperature is around 450°C. 9–12% pure chromium alloys also have similar values to the previous case.

4 DIFFERENT STEELS AND THEIR APPLICABILITY

The tests were performed on 20 different heat-resistant steels, one duplex, four ferrites and 15 austenitic. The data are summarized in Table 1.

The following steels were examined:

– The duplex steel: X2 CrNiMoN 22-5-3
– The ferritic steels: X2CrTi 12, X2CrTiNb18, X3CrTi 17, S433
– The austenitic steels: X12CrMnNiN 17-7-5, X10CrNi18-8 (annealed), X9CrNi18-9, X5CrNi18-10, X2CrNi 18-9, X4CrNi1812, 309S, X12CrNi 23-13, X8CrNi 25-21, 310S, X2CrNiMo 18-14-3, X2CrNiMo 1815-4, X6 CrNiTi 18-10, Alloy 610, Alloy 611.

The process was based on the data of the online catalogue of American ATI metals, which contained the yield strength of the steel at various temperatures in tabular form. The processing was carried out with the temperature in Fahrenheit, while the voltage values were in kilovolt/square inch, so some changes were needed in some places. In addition, the American steel marking, which we have used, is the one I've used to go through the marking system we use in Hungary. The properties and uses of each steel are as follows.

Table 1. Different steels.

Temperature	Type of the steel			
	X2CrTi 12	X2CrTiNb18	X3CrTi 17	S433
20°C	237,869 MPa	301,301 MPa	277,859 MPa	291,648 MPa
93,33°C	206,153 MPa	277,859 MPa	255,106 MPa	258,554 MPa
204,44°C	160,64 MPa	250,969 MPa	230,969 MPa	233,732 MPa
315,56°C	146,169 MPa	245,453 MPa	189,606 MPa	210,98 MPa
426,67°C	139,964 MPa	224,08 MPa	184,09 MPa	190,295 MPa
537,78°C	122,37 MPa	165,474 MPa	157,2 MPa	194,432 MPa
648,89°C	114,453 MPa	131 MPa	82,737 MPa	137,895 MPa
760°C	55,158 MPa	67,569 MPa	34,474 MPa	56,537 MPa
871,11°C	20,684 MPa	35,511 MPa	–	33,095 MPa
Melting point	1450°C	1440°C	1510°C	1450°C

309S and 310S. This high alloy steels are very resistant to creep caused by cracking and environmental influences. It is therefore widely used in the heat treatment industry such as furnaces, conveyor belts, cylinders, burners, refractory supports, tube clamps, baskets and trays for small parts. This alloy is also used by chemical process, especially where they work with hot concentrated acids, ammonia or sulphur oxide. It is also used in the food industry, especially where hot acetic acid or citric acid process. This type is typically used in structures and installations operating at 1100°C.

S433. This type contains 20% chromium, which is also good for oxidation and chlorine atmospheres at high temperatures. It contains 0,5% copper, which further improves corrosion resistance. It also contains 0,5% niobium, which improves its strength and properties at an elevated temperature. Its welding is similar to other stabilized ferrite alloys.

X2 CrNiMoN 22-5-3. It is a nitrogen-reinforced duplex steel alloy. Nitrogen significantly improves the corrosion resistance of the steel, especially in welded state. Contrary to other duplex steel, they were moderately resistant to general and stress corrosion mainly in welded state, which was improved by the addition of nitrogen. Furthermore, it contains high chromium, molybdenum and nickel. ASME classifies the boiler and pressure vessel code in the VIII group, which can be used up to 316°C. It is also suitable for nuclear construction.

X2 CrTiNb 18, X3 CrTi 17. These types have good oxidation and corrosion properties and are therefore used as a vehicle for the exhaust system motor vehicles. Ferritic steels are not typically used at high temperature, but niobium greatly improves creep resistance. Also applied with titanium improves weldability and grain boundary corrosion in the heat-affected zone. Good resistance to stress corrosion.

X2 CrTi 12. This alloy has good corrosion resistance and structural properties. Widely used as a material for ignition pipes, catalysts, exhaust systems, silencers and tail pipes. It is also used in heating systems, thermostats, heat exchangers, turbines and airplanes. Titanium and small niobium content improve resistance to particulate corrosion. This type is also suitable for brazing.

Alloy 610, Alloy 611. They are typically used as base material for components where substance is permanently in contact with high concentration of nitric acid. Such as vessels, heat exchangers, piping, valves, pumps. As regards their composition, low carbon and high silicon-containing arachinite stainless steels. Thanks to their high silicon content, they have excellent resistance to environmental corrosion.

X12CrMnNiN 17-7-5. It has excellent mechanical properties, it can be perfectly moulded in softened condition. For this reason, it is suitable, for example as a base material for washing machines. This alloy is resistant to a wide range of mildly to moderately corrosive materials. It maintains its resistance to oxidation up to 840°C, therefore its maximum operating temperature is around 815°C. This type can be welded by any conventional method.

X10CrNi18-8 (annealed). An austenitic stainless-steel type, which is particularly resistant to atmospheric corrosion, has a steel surface that is bright and is suitable for use as a decorative structural element. Used in automobiles, conveyors, kitchen machines. It also resists many corrosive media. It is not recommended to use above 871°C.

X5CrNi18-10. The main field of use of the material is food pots, as its corrosion resistance does not deteriorate during long-term contact with foodstuffs. Easy to clean, good oxidation resistance and good heat distribution factor and durability. It can also be used as an induction heater. It exhibits excellent corrosion resistance to slightly corrosive chemical media, oxidation and drinking water, as it has a highly adhesive corrosion-resistant oxide film on its surface. In durable 650°C operation, corrosion resistance may decrease due to chromium secession.

X2CrNiMo 18-14-3, X2CrNiMo 18-15-4. These are molybdenum-alloy stainless steels that are resistant to general corrosion and have good resilience to pitting and slit corrosion. These alloys have higher resistance to creep and maintain good mechanical properties at high temperatures. They also have good resistance to sulphur-containing acids and solutions.

X6 CrNiTi 18-10. Stabilized stainless-steel alloy with excellent resistance to granular corrosion. It has a long service life of between 427 and 816°C, with excellent mechanical properties. Compared to other stainless-steel rivals, shear and tension breakage are better. Its general corrosion resistance is the same as the non-stabilized chrome-nickel steels. During the welding process, the two determining aspects are to maintain corrosion resistance and to avoid cracks. There may also be a problem during the welding process of the stabilizing element, loss of titanium. It is also necessary to protect the bond from the absorption of nitrogen. Due to stabilization with titanium, there is a risk of hot rupture.

X9CrNi18-9, X2CrNi18-9, X4CrNi18-12. They are widely used when important for corrosion resistance, impurities, resistance to oxidation, good deformability, lightweight manufacturing, easy cleaning, high strength and low weight, good strength and toughness at cryogenic temperatures. They are used in the food industry and in healthcare, but can also be a source of pressure in a cryogenic temperature pressure vessel. Due to 18–19% chromium, these alloys have good resistance to oxidizing or slightly acidic media such as diluted nitric acid. Aggressive acidic media, such as acetic acid or phosphoric acid, is substantially less resistant. This can be improved by increasing the nickel content.

X8CrNi 25-21, X15CrNiSi 20-10. These austenitic corrosion-resistant steels are typically used at high temperatures. High chromium and nickel content have good corrosion and oxidation resistance. They have good mechanical properties at high temperatures, as well as cracking and environmental influences. It is widely used as part of heat treatment furnace. Its good corrosion resistance is utilized as a container for the storage of aggressive acidic atmosphere.

5 THE DEPENDENCE OF THE YIELD STRENGTH ON THE TEMPERATURE

After setting up the data set, it was possible to prescribe the dependence of the yield strength temperature using some mathematical function. To do this, Version 5.01 of TableCurve 2D provided assistance. In the program, you can specify which values are to be present on the "x" and "y" axes (in my case the temperature is the latter, the latter means the yield strength values). It is also possible to weigh the figures entered. Then, you add different functions to the points. They are also listed and can be displayed individually. Then a function corresponding to the actual behaviour should be selected (the yield strength should be reduced, but at least not increased with the increase in temperature).

Based on the data, the set functions are as follows:

– X2CrTi 12:
$$f_y^{-1} = 0{,}0042 + 7{,}289 \cdot 10^{-6} \cdot T - 7{,}314 \cdot 10^{-9} \cdot T^2 + 1{,}605 \cdot 10^{-10} \cdot T^3 - 5{,}327 \cdot 10^{-13} \cdot T^4 + 4{,}686 \cdot 10^{-16} \cdot T^5$$

– X12CrMnNiN 17-7-5:
$$f_y^{-1} = 0{,}0027 + 1{,}25 \; 10^{-5} \; T - 1{,}652 \; 10^{-8} \; T^2 + 2{,}586 \; 10^{-12} \; T^3 + 1{,}299 \; 10^{-14} \; T^4$$

– X10CrNi18-8 (annealed):
$$Inf_y = 5{,}62 - 0{,}0045 \; T + 9{,}399 \; 10^{-6} \; T^2 - 6{,}998 \; 10^{-9} \; T^3$$

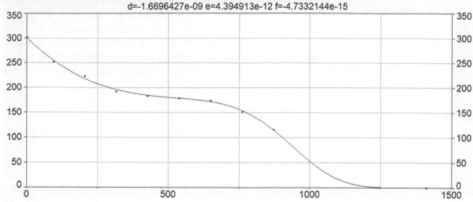

Figure 1.   Curve fitting using TableCurve2D.

– X2CrNi 18-9:
   $f_y = 240{,}191 + 4{,}446\ 10^{-6}\ T^{2{,}5} - 1{,}238\ 10^{-7}\ T^3 - 5{,}741\ T^{0{,}5}$
– 309S:
   $lnf_y = 5{,}858 - 0{,}0011\ T - 3{,}675\ 10^{-6}\ T^2 + 1{,}08\ 10^{-8}\ T^3 - 7{,}947\ 10^{-12}\ T^4$
– X2CrNiMo 18-14-3:
   $Lnf_y = 5{,}708 - 0{,}002\ T + 2{,}284\ 10^{-6}\ T^2 - 1{,}67\ 10^{-9}\ T^3 + 4{,}395\ 10^{-12}\ T^4 - 4{,}733\ 10^{-15}\ T^5$

As can be seen these relationships are quarter of five-pole polynomials or exponential functions polynomial exponents. The one of the curves drawn by the "TableCurve 2D" is shown on Figure 1. The functions can be used to determine the yield strength of the tested steels at any temperature, which is to be substituted in the formulas at position T in °C.

ACKNOWLEDGEMENT

The described article was carried out as part of the EFOP-3.6.1-16-2016-00011 "Younger and Renewing University—Innovative Knowledge City—institutional development of the University of Miskolc aiming at intelligent specialisation" project implemented in the framework of the Széchenyi 2020 program. The realization of this project is supported by the European Union, co-financed by the European Social Fund.

REFERENCES

Böhler 2017. Tool Steel, Stainless Steel Catalog (in Hungarian).
Stainless steels 2018. https://www.atimetals.com/Products/stainless-steel (accessed June 15, 2018).
Weld-technology 2018. Highly alloyed cracking and heat-resistant steels, http://www.weld-technology.com/anyagok/erossen-oetvoezoett-kuszas-es-hoallo-acelok (accessed June 15, 2018).

Advances and Trends in Engineering Sciences and Technologies III – Al Ali & Platko (Eds)
© 2019 Taylor & Francis Group, London, ISBN 978-0-367-07509-5

Solid glass bricks masonry—material properties of bricks and mortar

J. Fíla, M. Eliášová & Z. Sokol
Faculty of Civil Engineering, Czech Technical University in Prague, Prague, Czech Republic

ABSTRACT: The use of flat glass on load-bearing structures, such as glass beams, columns, staircases or railings is quite common in contemporary architecture. There is also strong effort to use other glass products such as tubes, hollow blocks or solid bricks in addition to flat glass. The paper is focused on experimental verification of solid glass bricks usability for glass masonry. Key factor is the material properties of the components: glass bricks and mortar. Small size tests were used to determine flexural strength and modulus of elasticity of glass bricks and compressive strength, flexural strength and modulus of elasticity of mortar. Critical evaluation of the experimental results was used to select the appropriate mortar for subsequent testing of masonry pillars in scale 1:1.

1 INTRODUCTION

The current architectural requirements for better interior lighting and the emphasis on transparency are increasing the use of flat glass in the building industry. Glazing of openings is hundreds of years old, but in the last decades, glass is also used on the supporting structural elements. Load bearing glass beams, ribs, plates or columns are nowadays quite common. At the beginning of the twentieth century, hollow glass blocks began to be used, but they did not fulfill the load bearing function. Still, they are still used today, often on the facades of administrative or representative buildings in city centers. Due to the light translucence and aesthetic appearance, solid glass bricks are also used in the contemporary architecture. Unlike common masonry, glass solid bricks are glued or bonded with special mortars whose properties are close to the adhesives. An advantage is the high compressive strength of glass.

One of the first structures where full glass bricks were used was Atocha Memorial in Madrid (Christoph & Knut 2008). This structure is completely made of glass. The walls are made of solid glass bricks and the roof consists of glass plates supported by glass beams. The bricks are connected in the bed joint by UV cured adhesive and in the head joint there is a lock formed by a bricks shape. The second structure is Crystal House in Amsterdam (Oikonomopoulou et al. 2017) with a unique facade made of solid glass bricks also held together by UV cured adhesive. These two examples show that use of masonry made of solid glass bricks is not only possible but also that the resulting construction is functional and very interesting for both professionals and the general public.

Experimental research performed in the laboratories of the Faculty of Civil Engineering, CTU in Prague was focused on the verification of the material properties of individual components for brick masonry using special mortars suitable for joining glass. The aim is to create a similar masonry with traditional one using solid glass bricks. The results of the small-scale tests are part of the long-term research in cooperation with Vitrablok Company. The influence of the mortar type on resistance of the bed joint under shear, surface treatment of glass bricks and bed joint thickness have been studied at (Fíla et al. 2017).

For all tests, the bricks supplied by Vitrablok were used, the dimensions and shape of which are shown in Figure 1 and Figure 2. Their shape is tapered to allow easy removal of the finished brick from the mould.

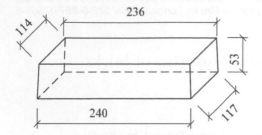

Figure 1. Geometrical dimensions of the brick. Figure 2. Glass brick.

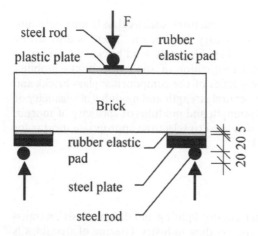

Figure 3. Schema of the test. Figure 4. Specimen in testing machine.

2 SOLID GLASS BLOCKS TESTS

Tests conducted at the FCE CTU was focused on the determination of flexural tensile strength and modulus of elasticity of solid glass bricks. The compressive strength of the bricks was taken from the research report (IKATES 2015).

2.1 Three-point bending flexural test

The test specimen was placed in the testing machine in horizontal position as shown in Figure 3 and Figure 4. Rubber elastic pads of 5 mm thickness was inserted between the glass brick and the steel supporting plates to reduce the risk of local peak stresses generation and the subsequent fracture of the glass bricks in the contact areas. The steel plates were supported by steel rods with diameter of 20 mm. The load was transmitted by steel rod placed in the middle of the plastic plate with rubber pad placed between the glass brick and the plastic plate.

Two linear displacement transducers placed over the supports were used to measure the deformation of the rubber pads and one laser extensometer was used to measure the displacement in the middle of the span. Normal stress was measured indirectly by one strain gauge LY 11-10/120 attached to the glass surface at the mid-span on the tensile side. Test specimens were loaded with loading rate 0.5 mm/min. Totally six specimens were tested.

2.2 Evaluation of the experiments

According to the assumption, in all cases, the failure occurred suddenly due to the exceeding of the tensile strength. No crushing of the glass was observed near the supports neither load introduction point, see Figure 5. In all cases, the crack was formed in the middle of the span

Figure 5.   Test specimens after the test.

Table 1.   Solid glass blocks tests.

Specimen number	Section modulus W mm³	Failure load F_{max} kN	Failure stress σ_{max} MPa	Modulus of elasticity E GPa
H-01	54354	74.67	65.01	74.49
H-02	49157	81.27	79.31	82.61
H-03	47249	71.85	70.44	89.49
H-04	64250	82.19	60.62	63.05
H-05	52049	59.96	56.20	76.02
H-06	56023	64.28	55.01	69.14
		Average values 64.43		75.80

on the bottom tensile surface of the brick. The section modulus was calculated after test from the dimensions measured in the rupture place. The modulus of elasticity was calculated based on the values of the relative strains from the strain gauge and the deformation in the middle of the span was obtained from the values of the laser extensometer and the linear displacement transducers. The results are summarized in Table 1.

3   MORTAR TESTS

As a part of the experimental program, tests in accordance with (EN 1015-11:1999) and (EN 13412:2006) were carried out in order to determine flexural and compressive strength and modulus of elasticity of the used mortar Vetromalta with latex admixture. Vetromalta is mortar intended for traditional installation of walls from hollow glass. This mortar was chosen according to the preliminary small-scale tests (Fíla et al. 2017). Material properties of the selected mixture will be used subsequently as an input parameters to the numerical model.

3.1   Flexural strength test of hardened mortar

Flexural strength tests were performed to determine the tensile strength of the mortar. Six test specimens were prepared. Test specimens sized 40 × 40 × 160 mm were made using a triplet form, see Figure 8, and were tested after 28 days hardening period. Test set-up corresponds to three-point bending test, see Figure 6. One laser extensometer was used to measure the displacement in the middle of the span. The specimens were loaded by rate 100 N/s up to the failure. A summary of all test specimens and test results is shown in Table 2.

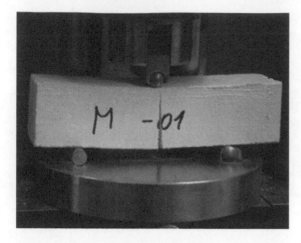

Figure 6.   Specimen after flexural test.

Figure 7. Specimen after compressive strength test.

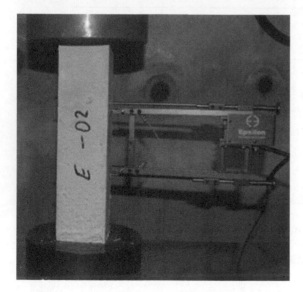

Figure 8.   Determination of modulus of elasticity.

Figure 9.   Triplet form.

3.2   *Compressive strength test of hardened mortar*

Half prisms from the flexure tests were used for the compressive test. Totally, twelve specimens have been tested. The specimen was placed into the testing device and continuously loaded by rate 100 N/s by a centric compression up to the failure, see Figure 7. One laser extensometer was used to measure the displacement. A summary of all test specimens and their results with maximal force at failure and maximal stress is resumed in Table 3.

3.3   *Determination of modulus of elasticity in compression*

To determine the modulus of elasticity there were used specimens with same dimensions as in case of flexural test, i.e. $40 \times 40 \times 160$ mm. The specimen was placed in the testing machine vertically and it was continuously loaded by rate 0.5 mm/min by a centric compression up to the failure, see Figure 8. One axial extensometer was used to measure the deformation of the specimen. A summary of all specimens and their results with maximal force at failure and calculated modulus of elasticity is encapsulated in Table 4.

Table 2. A summary of flexural strength tests.

Specimen number	Section modulus W mm³	Failure load F_{max} kN	Failure stress σ_{max} MPa
M-01	10933	4.60	10.51
M-02	10933	4.58	10.47
M-03	10933	4.77	10.90
M-04	11767	4.54	9.65
M-05	11767	4.16	8.83
M-06	12642	4.44	8.77
		Average stress	9.86

Table 3. A summary of compressive strength tests.

Specimen number	Section area A mm²	Failure load F_{max} kN	Failure stress σ_{max} MPa
M-01-01	1600	47.23	29.52
M-01-02	1600	43.48	27.17
M-02-01	1600	44.74	27.96
M-02-02	1600	43.84	27.40
M-03-01	1600	35.10	21.94
M-03-02	1600	45.96	28.73
M-04-01	1600	35.89	22.43
M-04-02	1600	39.99	24.99
M-05-01	1600	39.17	24.48
M-05-02	1600	41.90	26.19
M-06-01	1600	40.77	25.48
M-06-02	1600	40.51	25.32
		Average stress	25.97

Table 4. A summary of modulus of elasticity tests.

Specimen number	Modulus of elasticity E GPa
E-01	7.73
E-02	7.26
E-03	8.45
E-04	11.81
E-05	13.83
E-06	9.78
Average modulus	9.81

4 DISCUSSION

The modulus of elasticity of conventional float glass is generally reported in the literature (Fanderlík 1996), (Wurm 2007) at approximately 70.00 GPa. The average value of the elastic modulus based on the performed experiments is 75.80 GPa, see Table 1. It can therefore be stated that the glass bricks have a modulus of elasticity approximately the same as the elasticity of floated soda-lime-silica glass. The failure of glass bricks occurred always in the middle of the span at the bottom tensile surface. In one case the failure point was in the middle of the cross-section width, in all other cases, the failure point was at the edge, see

Figure 5. The failure always occurred suddenly without prior warning and the brittle failure was accompanied by a pronounced loud sound.

The modulus of elasticity of the mortars depends on many factors, the most important are these: the composition of the mixture, the water-cement ratio and the hardening time. Typically, the modulus of elasticity of commonly used mortars is in the order of GPa units, (Nagarajan et al. 2014), (Narayanan & Sirajuddin 2013), (Binda et al. 1988). The average value of the elastic modulus of the tested mortar is 9.81 GPa (see Table 4). The tested mortar therefore does not differ fundamentally from the commonly used mortars. Similarly, the average values of compressive strength (25.97 MPa, see Table 3) and flexural strength (9.86 MPa, see Table 2) correspond to commonly used mortars. The location and failure mode of the prisms in bending and compression also coincides with conventional mortars.

5 CONCLUSIONS

The aim of this research was to determine the mechanical properties of glass bricks and masonry mortar. For joints between glass bricks, ordinary mortars are not usable due to the smooth surface of the glass. Applicable are mortars with a plasticizer additive whose material properties do not differ from traditional mortars. Obtained data will be used in the follow-up research as an input to the numerical model of glass brick masonry. It has been shown that glass bricks have properties very similar to those of float glass.

ACKNOWLEDGEMENTS

This research was supported by grant CTU No. SGS18/168/OHK1/3T/11 and by grant No. GA16-17461S of the Czech Science Foundation. In addition, the authors are grateful to company Vitrablok, s.r.o. for the co-operation and providing the glass bricks for the experiments.

REFERENCES

Binda, L., Fontana, A., Frigerio, G. 1988. Mechanical behaviour of brick masonries derived from unit and mortar characteristics. Brick and block masonry. 1, 205–216.

Christoph, P., Knut, G. 2008. Innovative Glass Joints – The 11 March Memorial in Madrid. In: Bos, F., Louter, C., Veer, F. (eds.) *Challenging Glass: Conference on Architectural and Structural Applications of Glass*, Delft, the Netherlands p. 113/668. IOS Press, The Netherlands.

EN 1015-11:1999, Methods of test for mortar for masonry—Part 11: Determination of flexural and compressive strength of hardened mortar, European Committee for Standardization, Brussels.

EN 13412:2006, Products and systems for the protection and repair of concrete structures. Test methods. Determination of modulus of elasticity in compression, European Committee for Standardization, Brussels.

Fanderlík, I. 1996. Vlastnostiskel. Informatorium, Praha.

Fíla, J., Eliášová, M., Sokol, Z. 2017. Glass Masonry – Experimental Verification of Bed Joint under Shear. In: *IOP Conference Series: Materials Science and Engineering* doi: 10.1088/1757-899X/251/1/012097.

IKATES, s.r.o. 2015. Protokol o zkoušce č. 74/2015.

Nagarajan, T., Viswanathan, S., Ravi, S., Srinivas, V., Narayanan, P. 2014. Experimental Approach to Investigate the Behaviour of Brick Masonry for Different Mortar Ratios. In: *International Conference on Advances in Engineering and Technology (ICAET'2014)* March 29–30, 2014. Singapore. p. 7. International Institute of Engineers.

Narayanan, S.P., Sirajuddin, M. 2013. Properties of Brick Masonry for FE modeling. In: *American Journal of Engineering Research (AJER)*. 1, 6–11.

Oikonomopoulou, F., Bristogianni, T., Veer, F., Nijsse, R. 2017. The construction of the Crystal Houses façade: challenges and innovations, In: *Glass Struct Eng, Springer International Publishing*, doi: 10.1007/s40940-017-0039-4.

Wurm, J. 2007. Glass structures. Birkhäuser, Basel – Boston – Berlin.

Advances and Trends in Engineering Sciences and Technologies III – Al Ali & Platko (Eds)
© 2019 Taylor & Francis Group, London, ISBN 978-0-367-07509-5

Optimum dynamic analysis of a robot arm using flower pollination algorithm

H.N. Ghafil
University of Miskolc, Miskolc, Hungary
University of Kufa, Najaf, Iraq

K. Jármai
University of Miskolc, Miskolc, Hungary

ABSTRACT: Dynamic analysis of the robot manipulator is a crucial subject in robot arm design. Dynamic equations are very important for estimating the specifications of the robot actuators. In this research, Flower Pollination Algorithm (FPA) was used to find the minimum required torques for the robot manipulator actuation based on optimum trajectories. FPA was applied in two stages. First, the optimum path was planned by this algorithm, and then the algorithm was applied to establish the kinematic equations of the robot to map points from the Cartesian space to the joint space, where the trajectory planning occurred. The robot considered in this paper is the two link planar revolute joint robot RR. Euler-Lagrange formula was used to derive the equations of motion of the robot arm.

1   INTRODUCTION

Calculating the dynamic equations for the robot arm is necessary for the mechanical design of the robot. By deriving Euler-Lagrange equations, one can estimate the stress-strain distribution along the robot's physical links. This is necessary to design the arm, according to the fatigue or theories of failure limits. Also, by calculating Euler-Lagrange equations, it could be easy to do vibration analysis, which is very important in position tracking and control.

In this work, a complete example to design robot arm was done starting from trajectory planning based on the flower pollination algorithm FPA (Yang 2012) and (Alyasseri et al. 2018) to calculate velocity and acceleration trajectories needed for dynamical analysis. FPA is used for global optimization by a wide range of researchers in different topics. The example is built from scratch where the flower pollination algorithm was used for the path planning process to calculate the optimum via points for the trajectory planning process. The same algorithm was used to optimize the position of the robot tip or say for the inverse kinematic which is necessary to transfer all points; start, end and via points from the Cartesian space to the joint space where the trajectory occurs (Spong et al. 2006). Two links planar robot manipulator was used as an example during this work because of its simple Kinematics, which is shown in the Figure 1 with its Denavit-Hartonberg parameters which are illustrated in Table 1. Each link was assumed to be 750 mm long. Also q_1 and q_2 are the rotational angles of the link 1 and link 2 respectively with z-axis in the direction of the reader.

This paper presents an overall description of what is needed for optimal dynamic analysis, from the viewpoint of optimum trajectory planning. As a result, all numerical data on the quantities of the dynamic equation will be left to the reader.

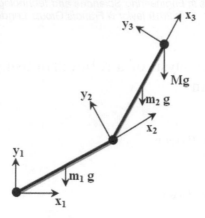

Figure 1.　Two link planar robot manipulator.

Table 1.　DH parameters of the robot.

Link	a_i mm	$\propto_i$	d_i mm	q_i
1	750	0	0	q_1
2	750	0	0	q_2

2 FPA FOR TRAJECTORIES

It is assumed that there are obstacles in the configuration space of the robot such that they constrained the motion of the robot arm while moving from start to goal point. Three points usually used in cubic spline curve for the path planning process, these three points will be later the via points for the trajectory planning. Flower pollination algorithm is used for the optimization process by handling iterative three points which are responsible for forming the cubic spline. FPA main parameters are explained in Table 2 while workspace parameters which are handled to the algorithm are; minimum x-value = 0, minimum y-value = 0, maximum x-value = 1620, maximum y-value = 1630 with four circular obstacles; arranged as (x-axis, y-axis, radius): (1402, 353, 110), (1080, 530, 50), (1062, 950, 120) and (510, 1100, 185). The cubic spline curve, which is governed by the three points consists of 100 sub-points and start with (1470,33) and end with (174,1350) and the fitness function is taken to be the total distances between each of these points.

$$fitness = \sum \sqrt{(x_{i+1} - x_i)^2 + (y_{i+1} - y_i)^2} + penalty \qquad (1)$$

where $i =1....99$ which is the order of the point in the curve and penalty taken to be 1000 whenever the cubic spline curve touch one of the obstacles. It is necessary to move all points in the working space; start, goal and via points to the joint space by using inverse kinematics for the sake of the trajectory planning. FPA also has used to optimize the forward kinematic equations (Ghafil et al. 2015) of the two link planar robot with the same parameters mentioned in Table 2. The fitness function of the inverse problem is the distance between the task point and the end-effector frame of the robot (Gilbert and Johnson 1985). The variable size of the flower pollination algorithm was taken three and each variable is a structure of two explicit variables represent the x and y-axis for each point. Inverse kinematic side of the problem is a straightforward problem and the fitness function which has to be minimized is the position vector of the end-effector:

$$f = \sqrt{(x_d - x_t)^2 + (y_d - y_t)^2} \qquad (2)$$

82

Table 2. Flower pollination parameters.

Parameter	Amount
Population size	25
Probability switch	0.5
No. of iterations	2000
No. of variables	3

where the subscription denotes the desired coordinate while t denotes task point coordinate and they can be extracted from the first two numbers of the fourth column of the product of equations (9) and (10) which they explained in the following matrix form:

$$T_1 * T_2 = \begin{bmatrix} r11 & r12 & r13 & x \\ r21 & r22 & r23 & y \\ r31 & r32 & r33 & 0 \\ 0 & 0 & 0 & 1 \end{bmatrix} \quad (3)$$

R11 to *r33* are rotation matrix elements and shall not consider all of them in our calculations. The variable size of the inverse kinematics in FPA are two variables; each variable represents the angular displacement for each of the two joints of the robot. For the start and end points of the trajectory, there are three boundary conditions in addition to the three conditions to the via points, so there should be nine boundary conditions for the problem. Accordingly, a polynomial equation of order 8 with nine coefficients shall be used to develop the overall trajectory equation. The equation takes the following form:

$$q(t) = a_1 + a_2 t + a_3 t^2 + a_4 t^3 + a_5 t^4 + a_6 t^5 + a_7 t^6 + a_8 t^7 + a_9 t^8 \quad (4)$$

where t is the time variable and a_1 .... a_9 are constants. By assuming that the actuation time starts from 0 and ends at 1 second. This equation can be solved by using the following boundary conditions which are considered using the data of the inverse Kinematics:

– For joint variable 1:

$q(0) = 12.7$, $\dot{q}(0) = 0$, $\ddot{q}(0) = 0$, $q(1) = 107.5$, $\dot{q}(1) = 0$, $\ddot{q}(1) = 0$, $q(0.25) = 48.2$, $q(0.5) = 91.2$, $q(0.75) = 109$

– For joint variable 2:

$q(0) = 22.82$, $\dot{q}(0) = 0$, $\ddot{q}(0) = 0$, $q(1) = -49.7$, $\dot{q}(1) = 0$, $\ddot{q}(1) = 0$, $q(0.25) = -71.1$, $q(0.5) = -101.7$, $q(0.75) = -111.7$

where $q(0.25)$, $q(0.5)$, $q(0.75)$ are the optimized via points at time 0.25, 0.5 and 0.75 respectively, which they are the result of the inverse kinematic problem. Solving equation (4) in MATLAB is a common work and the result is:

$$q_1(t) = 12.7 + 8383.4 t^3 - 41147 t^4 + 88190 t^6 - 98866 t^7 + 56593 t^8 - 13058 t^9 \quad (5)$$

$$q_2(t) = 22.82 - 33651 t^3 + 2.0385 * 10^5 t^4 - 5.1 * 10^5 t^6 + 6.41 * 10^5 t^7 - 4 * 10^5 t^8 + 98851 t^9 \quad (6)$$

Driving equations (5) and (6) lead to velocity and acceleration profile for each joint at any time during the time interval [0,1]. The angular velocity and acceleration are inputs to the dynamic equations which can be driven as follows:

3 DRIVING DYNAMIC EQUATIONS

By using the Denavit-Hartenberg convention, it is possible to calculate the homogeneous transformation matrices at the center of each link as well as at the end-effector where the robot is holding a weight of mass (M) with a moment of inertia (I_m). Tc_1, Tc_2, T_1 and T_2 represent the homogeneous transformation matricies (HTM) at the center of link 1, the center of link 2, the tip of the link 1 and the end-effector respectively.

$$Tc_1 = \begin{bmatrix} \cos(q_1) & -\sin(q_1) & 0 & lc_1 \cdot \cos(q_1) \\ \sin(q_1) & \cos(q_1) & 0 & lc_1 \cdot \sin(q_1) \\ 0 & 0 & 1 & 0 \\ 0 & 0 & 0 & 1 \end{bmatrix} \tag{7}$$

$$Tc_2 = \begin{bmatrix} \cos(q_2) & -\sin(q_2) & 0 & lc_2 \cdot \cos(q_2) \\ \sin(q_2) & \cos(q_2) & 0 & lc_2 \cdot \sin(q_2) \\ 0 & 0 & 1 & 0 \\ 0 & 0 & 0 & 1 \end{bmatrix} \tag{8}$$

$$T_1 = \begin{bmatrix} \cos(q_1) & -\sin(q_1) & 0 & l_1 \cdot \cos(q_1) \\ \sin(q_1) & \cos(q_1) & 0 & l_1 \cdot \sin(q_1) \\ 0 & 0 & 1 & 0 \\ 0 & 0 & 0 & 1 \end{bmatrix} \tag{9}$$

$$T_2 = \begin{bmatrix} \cos(q_2) & -\sin(q_2) & 0 & l_2 \cdot \cos(q_2) \\ \sin(q_2) & \cos(q_2) & 0 & l_2 \cdot \sin(q_2) \\ 0 & 0 & 1 & 0 \\ 0 & 0 & 0 & 1 \end{bmatrix} \tag{10}$$

The position vector o_n^1 expressed at the base frame which is the first three elements of the fourth column of the HTM will be used to drive linear Jacobian J_v by the following equation:

$$J_v = \partial o_n^1 / \partial q_i \text{ where } i = 1,2 \tag{11}$$

The linear Jacobian at the center of the mass of the link 1 is as follow:

$$J_{vc1} = \partial o_{c1}^1 \Big/ \partial q_i = \begin{bmatrix} -l_{c1} * \sin(q_1) & 0 \\ l_{c1} * \cos(q_1) & 0 \\ 0 & 0 \end{bmatrix} \tag{12}$$

where $i = 1,2$ and o_{c1}^1 is the position vector of Tc_1. Also, the angular Jacobian J_w is just the z-vector at the rotation matrix in the HTM, so it is necessary to drive angular Jacobians at the center of mass of link 1, link 2 and end-effector J_{wc1}, J_{wc2} and J_{w2} respectively

$$J_{wc1} = \begin{bmatrix} z_1 & 0 \end{bmatrix} \tag{13}$$

$$J_{vc2} = \partial o_{c2}^1 \Big/ \partial q_i = \begin{bmatrix} -l_1 * \sin(q_1) - l_{c2} * \sin(q_1 + q_2) & -l_{c2} * \sin(q_1 + q_2) \\ l_1 * \cos(q_1) + l_{c2} * \cos(q_1 + q_2) & l_{c2} * \cos(q_1 + q_2) \\ 0 & 0 \end{bmatrix} \tag{14}$$

where o_{c2}^1 is the position vector of $T_1 \times Tc_2$

84

$$J_{wc2} = \begin{bmatrix} z_1 & z_2 \end{bmatrix} = \begin{bmatrix} 0 & 0 \\ 0 & 0 \\ 1 & 1 \end{bmatrix} \quad \text{where } z_2 \text{ is } z - \text{vector of } T_1 x T c_2 \tag{15}$$

$$J_{v2} = \frac{\partial o_2^1}{\partial q_i} = \begin{bmatrix} -l_1 * \sin(q_1) - l_2 * \sin(q_1 + q_2) & -l_2 * \sin(q_1 + q_2) \\ l_1 * \cos(q_1) + l_2 * \cos(q_1 + q_2) & l_2 * \cos(q_1 + q_2) \\ 0 & 0 \end{bmatrix} \tag{16}$$

o_2^1 is the position vector of T_2

$$J_{w2} = \begin{bmatrix} z_1 & z_2 \end{bmatrix} = \begin{bmatrix} 0 & 0 \\ 0 & 0 \\ 1 & 1 \end{bmatrix} \tag{17}$$

Translational part of the kinetic energy:

$$k_v = \frac{1}{2} \dot{q}^T \{ m_1 J_{vc1}^T J_{vc1} + m_2 J_{vc2}^T J_{vc2} + M J_{v2}^T J_{v2} \} \dot{q} \tag{18}$$

where m_1, m_2, denotes a mass of link 1 and 2 respectively. After doing the mathematical operation, the translational part can be written in matrix form:

$$k_v = \frac{1}{2} \dot{q}^T \begin{bmatrix} k_{v11} & k_{v12} \\ k_{v21} & k_{v22} \end{bmatrix} \dot{q} \tag{19}$$

where (Equations (20), (21), (22)):

$$k_{v11} = M(l_1^2 + l_2^2) + 2l_1 \left(\cos(q_1) \cos(q_1 + q_1) + \sin(q_1) \sin(q_1 + q_1) \right) (Ml_2 + m_2 l_{c2}) \\ + m_2 (l_1^2 + l_{c2}^2) + m_1 l_{c1}^2 \tag{20}$$

$$k_{v12} = k_{v21} = Ml_2^2 + m_2 l_{c2}^2 + l_1 \left(\cos(q_1) \cos(q_1 + q_1) + \sin(q_1) \sin(q_1 + q_1) \right)(Ml_2 + m_2 l_{c2}) \tag{21}$$

$$k_{v22} = Ml_2^2 + m_2 l_{c2}^2 \tag{22}$$

The rotational part of the kinetic energy is:

$$\begin{aligned} k_w &= \frac{1}{2} \dot{q}^T \{ J_w^T R_i I_i R_i^T J_w \} \dot{q} \\ &= \frac{1}{2} \dot{q}^T \{ J_{wc1}^T R_1 I_1 R_1^T J_{wc1} + J_{wc2}^T R_2 I_2 R_2^T J_{wc2} + J_{w2}^T R_2 I_m R_2^T J_{w2} \} \dot{q} \\ &= \frac{1}{2} \dot{q}^T \left\{ \begin{bmatrix} I_1 + I_2 + I_m & I_2 + I_m \\ I_2 + I_m & I_2 + I_m \end{bmatrix} \right\} \dot{q} \end{aligned} \tag{23}$$

The first, second and third term of equation (23) are the rotational kinetic energy for link 1, link 2 and the mass carried by the tooltip respectively. The overall kinetic energy can be got by adding (23) to (29), and it will be written in the following matrix form:

$$K = \frac{1}{2} \dot{q}^T \{ D(q) \} \dot{q} \tag{24}$$

where $D(q)$ is a 2×2 matrix, where (Equations (25), (26), (27)):

85

$$d_{11} = M(l_1^2 + l_2^2) + 2l_1 \left(\cos(q_1)\cos(q_1 + q_1) + \sin(q_1)\sin(q_1 + q_1) \right)(Ml_2 + m_2 l_{c2})$$
$$+ m_2 (l_1^2 + l_{c2}^2) + m_1 l_{c1}^2 + I_1 + I_2 + I_m \tag{25}$$

$$d_{12} = d_{21} = Ml_2^2 + m_2 l_{c2}^2 + l_1 \left(\cos(q_1)\cos(q_1 + q_1) + \sin(q_1)\sin(q_1 + q_1) \right)(Ml_2 + m_2 l_{c2})$$
$$+ I_2 + I_m \tag{26}$$

$$d_{22} = Ml_2 + m_2 l_{c2}^2 + I_2 + I_m \tag{27}$$

Total potential energy is (Equation (28)):

$$P = m_1 g l_{c1} \sin(q_1) + m_2 g (l_1 \sin(q_1) + l_{c2}\sin(q_1 + q_2)) + Mg(l_1 \sin(q_1) + l_2\sin(q_1 + q_2)) \tag{28}$$

Now we can estimate the Lagrangian of the robot arm $\mathcal{L}$ which is the difference between kinetic and potential energies:

$$\mathcal{L} = K - P \tag{29}$$

We are ready to drive the dynamic equations for the robot using the following equation

$$\frac{d}{dt}\left(\frac{\partial \mathcal{L}}{\partial \dot{q}_i} \right) - \frac{\partial \mathcal{L}}{\partial q_i} = \tau_i \tag{30}$$

and its results:

$$a_1 \ddot{q}_1 + b_1 \ddot{q}_2 + h_1 \dot{q}_1 \dot{q}_2 + k_1 \dot{q}_2^2 + p_1 = \tau_1 \tag{31}$$
$$a_2 \ddot{q}_1 + b_2 \ddot{q}_2 + k_2 \dot{q}_1^2 + p_2 = \tau_2 \tag{32}$$

where:

$$a_1 = m_1 l_{c1}^2 + m_2 \left(l_1^2 + l_{c2}^2 + 2l_1 l_{c2} \cos(q_2) \right) + M\left(l_1^2 + l_2^2 + 2l_1 l_2 \cos(q_2) \right) + I_1 + I_2 + I_m \tag{33}$$

$$b_1 = a_2 = m_2 \left(l_{c2}^2 + l_1 l_{c2} \cos(q_2) \right) + M\left(l_2^2 + l_1 l_2 \cos(q_2) \right) + I_2 + I_m \tag{34}$$

$$b_1 = a_2 = m_2 \left(l_{c2}^2 + l_1 l_{c2} \cos(q_2) \right) + M\left(l_2^2 + l_1 l_2 \cos(q_2) \right) + I_2 + I_m \tag{35}$$

$$k_1 = 2h_1 \tag{36}$$

$$p_1 = g(l_{c1} m_1 + l_1 m_2 + l_1 M)\cos(q_1) + g(l_{c2} m_2 + l_2 M)\cos(q_1 + q_2) \tag{37}$$

$$b_2 = l_{c2}^2 m_2 + l_2^2 M + I_2 + I_m \tag{38}$$

$$k_2 = (m_2 l_{c2} + Ml_2)l_1 \sin(q_2) \tag{39}$$

$$p_2 = g(m_2 l_{c2} + Ml_2)\cos(q_1 + q_2) \tag{40}$$

So far, we have driven all the equations that describe the relationship between robot arm parameters from mass, a moment of inertia to joint trajectories and the corresponding torques that should be at the joints of the robotic arm. These parameters can be measured at works or assumed for simulation, and for a specific task or multi-task space, we can calculate angular positions, velocities and accelerations based on the optimality of the chosen via points which are calculated based on flower pollination algorithm. Moments of inertia for the two links and for the mass carried by the end-effector are left to the readers to use their own assumptions or calculations.

4 CONCLUSION

The dynamic loads during runtime of the robot manipulator should be considered in the mechanical design of the robot. The acceleration of the actuators of the robot is a crucial parameter that affects the output of the dynamic equations. This work presents a full example of designing a robot arm from task planning to the estimation of the generalized forces on the robot. Axiomatically, calculating the optimum value of the robot acceleration should lead to the best evaluation of the forces. Flower pollination algorithm was presented as an optimization tool to the optimum trajectories during robot work time. We recommend using the procedure which is described in work to design any robot manipulators especially when the desired torques parameter of the motors of the robot are needed. Also, the presented analysis forms the first step for a further advanced study like fatigue or modal analysis.

ACKNOWLEDGEMENT

The described article was carried out as part of the EFOP-3.6.1-16-2016-00011 "Younger and Renewing University—Innovative Knowledge City—institutional development of the University of Miskolc aiming at intelligent specialisation" project implemented in the framework of the Széchenyi 2020 program. The realization of this project is supported by the European Union, co-financed by the European Social Fund.

REFERENCES

Alyasseri Z.A.A., Khader A.T., Al-Betar M.A., Awadallah M.A., Yang X.-S. (2018). Variants of the flower pollination algorithm: a review. In: Nature-Inspired Algorithms and Applied Optimization. Springer, pp. 91–118.

Ghafil H.N., Mohammed A.H., Hadi N.H. (2015). A virtual reality environment for 5-DOF robot manipulator based on XNA framework. International Journal of Computer Applications 113 (3).

Gilbert E., Johnson D. (1985). Distance functions and their application to robot path planning in the presence of obstacles. IEEE Journal on Robotics and Automation 1 (1):21–30.

Spong M.W., Hutchinson S., Vidyasagar M. (2006). Robot modeling and control, vol 3. Wiley New York.

Yang X.-S. Flower pollination algorithm for global optimization. In: International conference on unconventional computing and natural computation, 2012. Springer, pp. 240–249.

Advances and Trends in Engineering Sciences and Technologies III – Al Ali & Platko (Eds)
© 2019 Taylor & Francis Group, London, ISBN 978-0-367-07509-5

Four-point bending tests of double laminated glass panels with PVB interlayer in different loading rates

T. Hána, M. Vokáč & K.V. Machalická
Klokner Institute, Czech Technical University, Prague, Czech Republic

Z. Sokol & M. Eliášová
Faculty of Civil Engineering, Czech Technical University, Prague, Czech Republic

ABSTRACT: Looking at the current architecture, we may notice various examples of glass load bearing structures such as panels, beams, stairs or even columns. Most of these members are made of laminated safety glass with polymeric interlayer between the individual glass panes. Polymeric interlayer is able to provide the shear coupling of glass panes due to its shear stiffness, but this stiffness is temperature and load duration dependent. Therefore, the exact laminated glass panel analysis becomes difficult. This paper is focused on double laminated heat toughened glass panels in four-point bending tests. Panels are laminated with PVB interlayer and they are loaded in two different loading rates. In particular, normal stress distribution along the critical cross section and maximal vertical deflections depending on the loading rate are highlighted. Further, the comparison of the experimental data with simplified analytical calculation is elaborated. All bending tests were performed at CTU in Prague.

1 INTRODUCTION

Laminated glass is becoming more extensively used in a civil engineering practice as a load bearing structural element. It consists of two or more glass panes bonded by transparent polymeric interlayer. Bonding process is usually performed in an autoclave at the temperature between 120–140°C and ambient pressure 0.8 MPa depending on type of the interlayer. The use of laminated glass instead of monolithic glass becomes necessary because of its behavior in the post breakage phase. When one glass pane in a laminated panel breaks, the remaining panes are still able to carry part of the load and enable the structure users to leave the endangered area. Moreover, glass fragments stay adhered to the interlayer, preventing harmful injuries. Laminated glass can therefore be used above the utility zones. The stress state analysis becomes rather difficult in case of perpendicularly loaded laminated glass panels, because the shear stiffness of the interlayer changes with temperature and load duration. Civil engineers thus neglect the shear coupling of glass panes, because the shear stiffness of most interlayers is not available. This approach leads to a safe but simultaneously uneconomic design of glass structures. Polymeric interlayers used in practice are of different chemical composition (polyvinylbutyral, ethylenevinylacetate, ionoplast, etc.), therefore they have different shear stiffness. Even though this stiffness is known, there is no official European code taking glass panes shear coupling into account, therefore enhanced modelling and analysis are necessary to be performed, Ivanov (2006). This paper introduces the experimental data acquired from four-point bending tests of double laminated heat toughened glass panels with PVB (polyvinylbutyral) interlayer loaded in two different loading rates, describes the experimental program and compares the experimental data with those analytically calculated. All tests were performed at CTU in Prague.

2 EXPERIMENTAL PROGRAM

2.1 *Materials and equipment*

Six double laminated glass panels with PVB (Trosifol BG-R-20®) were tested in four-point bending tests. All panels were made of heat toughened glass. Thickness of one glass pane was 10 mm and thickness of PVB interlayer was 0.76 mm. Nominal dimensions of glass panels were 1100 × 360 mm. The tests were performed in MTS loading device with the maximum load capacity 100 kN. Static schema of four-point bending test and specimen's position in MTS device are shown in Figure 1. To measure the normal stresses, there were totally 6 strain gauges LY 11-10/120 attached to the glass surface at the mid-span, particularly three strain gauges in tension on the lower glass surface and three strain gauges in compression on the upper glass surface. This is displayed in Figure 2. Vertical deflection at the mid-span was measured by two displacement sensors. The cross section of laminated panel and the position of displacement sensors I and II are shown in Figure 3.

2.2 *Test set-up*

The loading was displacement controlled with the appropriate MTS cross-head speed. Three panels were loaded with the speed of 0.5 mm/min and three panels with the speed of 0.125 mm/min to find the influence of the loading rate on laminated glass panel performance. The loading rate was constant during the entire loading phase. Every test specimen was loaded in two loading phases. The first phase resulted into the lower glass pane breakage

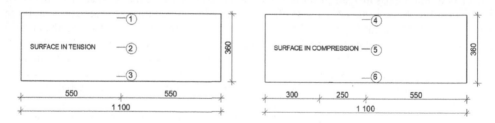

Figure 1.   Static schema of four-point bending test and position of the test specimen in MTS device.

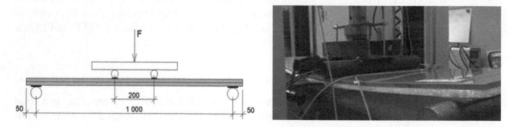

Figure 2.   Strain gauges position on the test specimen.

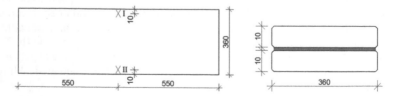

Figure 3.   Displacement sensors I and II and their position on the test specimen; cross section of laminated panel.

when its tensile strength was exceeded. The specimen was then unloaded and loaded again in the second loading phase to find its residual load bearing capacity. As soon as it was achieved, the whole laminated panel collapsed. Values of all strain gauges and displacement sensors were offset before each loading step. Temperature during the experiments was measured in the range of 19–22°C. Composition of all testing specimens, their numbering in the text, and the loading rate during the test are shown in Table 1.

2.3 Analytical calculation

The experimental data are compared to the simplified analytical calculation executed under the following assumptions—PVB interlayer provides full shear and no shear coupling of the glass panes. In case of full shear coupling ($G_{PVB} = \infty$), the thickness of the panel is considered as 20 mm and the calculation of normal stress and vertical deflection is performed for this reported thickness and experimentally measured loading force. When the shear transfer between the glass panes is neglected ($G_{PVB} = 0$), the applied load is uniformly distributed between both glass panes in terms of their bending stiffness. In this case, every glass pane carries one half of measured force, therefore the normal stress and deflections are calculated for one glass pane 10 mm thick. This approach is in correlation with DIN 18008-2 (2010) and it can be interpreted according to Figure 4, where the pane bending stiffness is substituted by the spring stiffness k_i. Normal stress on the glass surface σ is calculated according to the Eq. 1.

$$\sigma = M / W \tag{1}$$

where M = bending moment at the mid-span based on the level of shear coupling (full or no glass panes shear transfer); and W = cross section modulus (whole composite or one pane only). Vertical deflections are determined for four-point bending test schema according to the Eq. 2.

$$w = \frac{0,5\,F}{24\,E\,I}(3a\,l^2 - 4a^3) \tag{2}$$

where F = applied MTS load based on the considered level of glass panes shear coupling; E = Young's modulus of glass 70 GPa, DIN18008-1 (2011); I = moment of inertia (full glass

Table 1. Numbering, loading rate and composition of testing specimens.

Specimen's number	Type of glass*	Glass panel thickness [mm]	Number of panes	Panel thickness [mm]	Type of interlayer	Interlayer thickness	Loading rate [mm/min]
1	HTG	10	2	20.76	PVB	0.76	0.5
2	HTG	10	2	20.76	PVB	0.76	0.5
3	HTG	10	2	20.76	PVB	0.76	0.5
4	HTG	10	2	20.76	PVB	0.76	0.125
5	HTG	10	2	20.76	PVB	0.76	0.125
6	HTG	10	2	20.76	PVB	0.76	0.125

*HTG = heat toughened glass.

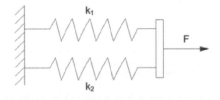

Figure 4. Model of load distribution between the glass panes when the shear transfer is neglected.

panes shear coupling with $G_{PVB} = \infty$ then I corresponds to the whole panel cross section or no glass panes shear coupling with $G_{PVB} = 0$ then I corresponds to one pane only); $a = 0.4$ m; and $l = 1.0$ m. Both equations hold for the stress and deflection analysis of simply supported beam vertically loaded.

3 RESULTS AND DISCUSSION

3.1 *First loading phase*

The following paragraph describes the experimental relationships of two representative specimens and compares them to those analytically calculated assuming full ($G_{PVB} = \infty$) or no glass panes shear coupling ($G_{PVB} = 0$). Figure 5 shows force-stress relationships of the specimen number 2 loaded with the cross-head speed of 0.5 mm/min and the specimen number 5 loaded with the cross-head speed of 0.125 mm/min. Normal stress is measured by the decisive strain gauge number 1stuck on the lower surface of the lower glass pane. Figure 5 also shows the analytical relationships of two shear transfer border-line cases referring to the state $G_{PVB} = \infty$ and $G_{PVB} = 0$. When comparing the experimental force-stress relationships, it becomes noticeable that the specimen number 2 loaded with a higher loading rate achieves more favorable normal stress values in the whole loading phase than the specimen number 5. It can be explained by different PVB shear stiffness modulus which is loading rate dependent as documented in Hána et al. (2017). Higher the rate of the load, the stiffer response of polymeric interlayers is because their viscosity effect is stifled, Ferry (1980). If one neglected the glass panes shear coupling, too conservative results would be obtained. Figure 5 shows that the rate of conservativeness increases with the increasing force.

Force-stress relationships are further accompanied by force-deflection relationships of the same testing specimens as displayed in Figure 6. Experimental curves in this figure refer to the average vertical deflection calculated from both displacement sensors (sensor I and II). Specimen number 2 attains lower vertical deflections for a certain force value in the entire loading phase than the specimen number 5. Higher loading rate thus results in higher bending stiffness of the laminated panel. In case of the glass panes shear coupling neglection, too conservative vertical deflections are calculated.

To compare the presented results, the numerical values based on the experiment and on the analytical calculation for the value of the applied load 10 kN are shown in Table 2. The experimental stresses are stated for the decisive strain gauge. These numbers confirm the aforementioned relationships. If one neglects PVB shear stiffness ($G_{PVB} = 0$), the difference between calculated and experimental normal stresses would be 38 MPa in average. In case of deflections, this difference would be approximately 11 mm.

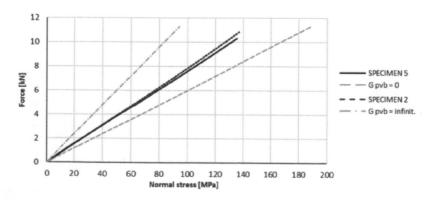

Figure 5. Experimental force-stress relationships measured by strain gauge 1 of specimen number 2 and 5, border line relationships analytically calculated.

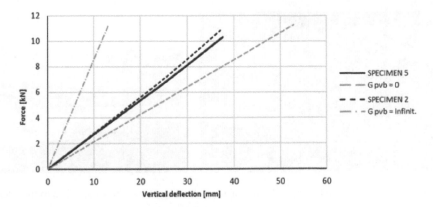

Figure 6.   Experimental force-vertical deflection relationships of four specimens, border line relationships analytically calculated.

Table 2.   Comparison of experimental data and analytical results for the value of the applied force 10 kN.

Specimen's number	Loading ratio [mm/min]	$\sigma_{exp.}$ [MPa]	Strain gauge number	$\sigma_{cal.}$ ($G_{pvb}=0$) [MPa]	$\sigma_{cal.}$ ($G_{pvb}=\infty$) [MPa]	$w_{exp.}$ [mm]	$w_{cal.}$ ($G_{pvb}=0$) [mm]	$w_{cal.}$ ($G_{pvb}=\infty$) [mm]
1	0.5	128.7	1	166.6	83.3	—	46.8	11.7
2	0.5	126.6	1	166.6	83.3	34.5	46.8	11.7
3	0.5	126.5	1	166.6	83.3	34.4	46.8	11.7
4	0.125	130.6	3	166.6	83.3	36.5	46.8	11.7
5	0.125	131.2	1	166.6	83.3	36.6	46.8	11.7
6	0.125	129.3	1	166.6	83.3	36.1	46.8	11.7

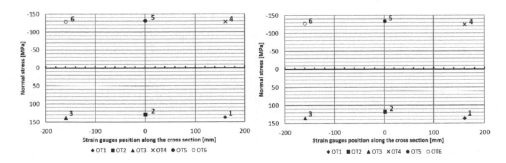

Figure 7.   Normal stress distribution along the cross section in case of the lower glass pane breakage; left chart-specimen number 2 (0.5 mm/min), right chart-specimen number 5 (0.125 mm/min).

Figure 7 shows, how the normal stress was distributed along the cross section in case of the lower glass pane breakage (the end of 1st loading phase) for the representative specimens 2 and 5. When observing these, one can find out that normal stress is not uniformly distributed along the cross section width as Navier bending hypotheses for slender beams assumes. For wide beams, Poisson's effect generates nonnegligible normal stresses in the longitudinal direction thus they cannot be uniformly distributed along the cross section width, ČSN EN (1288-1).

To summarize the experimental extremes of all specimens, maximum force 14.0 kN evoked maximum measured tensile stress 177 MPa (SG1, specimen number 1). Minimum force 10.3 kN evoked minimum measured tensile stress 116 MPa (SG2, specimen number 5). When the glass tensile strength of the lower glass pane was exceeded, this pane abruptly collapsed with no previous warning. The failure mode was typical for a heat toughened glass. Lower

Figure 8.   Lower glass pane failure at the and of the 1ˢᵗ loading phase, shards attached to the interlayer.

Figure 9.   Vertical deflections and failure mode of the glass panel during the 2ⁿᵈ loading phase.

pane fell apart into small glass shards in its entire volume, the upper pane remained undamaged as shown in Figure 8. Shards stayed adhered to the interlayer. Failure modes were in the same manner for all testing specimens.

3.2   Second loading phase

The purpose of the second loading phase was to find the residual load bearing capacity of the upper glass pane since this was still able to carry the load. Maximum residual load bearing capacity for tested specimens was 6.5 kN, minimum residual capacity was 4.8 kN, and average residual capacity was 5.3 kN. Every laminated panel performed vertical deflections exceeding 45 mm, which was the maximum range of displacement sensors. Deflections are shown in Figure 9. When the glass tensile strength was exceeded, every glass panel abruptly collapsed with no additional bending stiffness as shown in Figure 9.

4   CONCLUSIONS

In this paper, important experimental results concerning displacement controlled four-point bending tests of double laminated heat toughened glass panels with PVB interlayer (Trosifol BG-R-20®) performed at CTU in Prague were elaborated. Panels loaded with the cross-head speed of 0.5 mm/min performed stiffer response, more favorable vertical deflections, and tensile stresses at the mid-span cross section than those loaded with the cross-head speed of 0.125 mm/min. The shear stiffness of PVB, which is loading rate dependent, therefore has a practical impact on laminated glass panel performance. Experimental results were further compared to the simplified analytical calculation considering full or no glass panes shear coupling according to DIN 18008-2 (2010). The analytical calculation neglecting PVB shear stiffness ($G_{PVB} = 0$) provided too conservative results. The experimental data also showed that the normal stress was not uniformly distributed along the cross section width at the moment of the lower glass pane breakage as Navier bending hypotheses for slender beams assumes. The type of panel failure was brittle, every loaded panel suddenly collapsed, and glass shards stayed adhered to the interlayer. Experimentally verified behavior of perpendicularly loaded glass panels is the way to include the interlayer's shear stiffness into the design of laminated glass structures and to extend the use of these structures in practice.

ACKNOWLEDGEMENT

This project was supported by project MPO TRIO FV10295.

REFERENCES

ČSN EN 1288-1, 2001. Sklo ve stavebnictví-Stanovení pevnosti skla v ohybu, Část 1: Podstata zkoušení skla, *Český normalizační institut*. Praha.
DIN 18008-1, 2011. Glas im Bauwesen – Bemessungs- und Konstruktionsregeln – Teil 1: Begriffe und allgemeine Grundlagen, *Deutsches Institut für Normung e.V.*, Berlin.
DIN 18008-2, 2010. Glas im Bauwesen – Bemessungs- und Konstruktionsregeln – Teil 2: Linienförmig gelagerte Verglasungen, *Deutsches Institut für Normung e.V.*, Berlin.
Ferry, J.D. 1980. Viscoelastic properties of polymers, *John Wiley & Sons*. New York.
Hána, T. & Eliášová, M. & Machalická, K. & Vokáč, M. 2017. Determination of PVB interlayer's shear modulus and its effect on normal stress distribution in laminated glass panels, *IOP Conf. Ser.: Mater. Sci. Eng.251 012076*. Riga.
Ivanov, I.V. 2006. Analysis, modelling and optimization of laminated glasses as a plane beam, *International Journal of Solids and Structures 43*: 6887–6907. doi: 10.1016/j.ijsolstr.2006.02.014.

Advances and Trends in Engineering Sciences and Technologies III – Al Ali & Platko (Eds)
© *2019 Taylor & Francis Group, London, ISBN 978-0-367-07509-5*

The effect of the subsoil simulation on the punching resistance of footings

J. Hanzel, A. Bartók & J. Halvonik
Slovak University of Technology, Bratislava, Slovakia

ABSTRACT: Within the last three decades, many experiments were carried out on testing of the punching capacity of footings. The tests differ manly by subsoil simulation. Most of the tests were carried out in a way where soil pressure was modelled by springs and later by a set of hydraulic jacks. In order to know the differences in behaviour of the footings resting on the more realistic soil, several tests were carried out at the university laboratory. Four footings with different shear slenderness resting on 2 m thick gravel cushion were tested for punching. The paper presents obtained results and comparison with the assessed punching resistances using three different models for the prediction of the punching capacity. Using results of further 55 tests that were carried out on different types of the subsoil the model's safety was assessed for each of the three models.

1   INTRODUCTION

Many experiments were carried out for punching of the footings up to now, Ricker (2009); Siburg (2014); Simões (2018). They differ mainly by the test setup, where six types of subsoil modelling can be distinguished, see Figure 1.

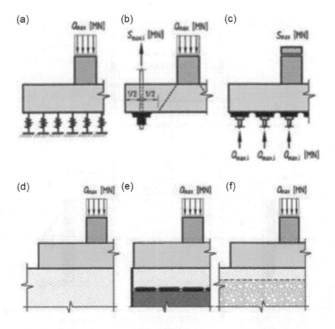

Figure 1.   Subsoil types, part 1, a) set of springs, b) line support, c) point support or point load. Subsoil types, part 2, d) thick sand layer; e) thin sand layer; f) deep gravel cushion with a thin sand layer.

Figure 1a) shows a set of springs with footing loaded through the column stub. Springs represents perfectly elastic subsoil. Their use is outdated. In Figure 1b) the footing rests on the line support that used to be located in half distance between the end of the footing and the section where critical shear crack crosses bending reinforcement. Accuracy of the results is highly affected by the precision of the critical control perimeter assessment, Hallgren (1998). In Figure 1c) subsoil is modelled by a set of hydraulic jacks, 16 or 25, depending on footing's dimensions. Higher number of jacks is able to create a perfectly uniform ground pressure. This way of testing is currently used, Diterle (1987); Siburg (2014). In Figure 1d) the subsoil is created by a thick layer of sand which is more realistic configuration of the tests. However, non-uniform ground pressure can be developed here depending on the shear slenderness of the footing, greater slenderness lower stiffness and thus higher concentration of ground pressure under the column area. In Figure 1e) is the latest way of subsoil simulation, Simões (2018). The layer of sand, 300 mm deep, is confined in a stiff box. This setup creates conditions for uniform ground pressure development under the footing, similar to a set of jacks. In Figure 1f) is the last type of subsoil simulation, consisting of a thick layer of gravel with thin layer of sand. In this case non-uniform soil pressure can be expected.

2 EXPERIMENTAL CAMPAIGN

2.1 *The test setup*

Type of subsoil in Figure 1e) was used for our experimental program. The footings rested on 2 m deep layer of compacted gravel (E_{def} = 50 MPa) with thin, 100 mm deep layer of sand spread under the footing, see Figure 2. The test setup is composed of three parts: tank, 4 × 4 m, with side stiffening, main and secondary beams anchored to the heavy concrete floor and bracing and stabilizing elements. The punching force was generated by hydraulic jack located above the footing. Hydraulic jack was supported by massive steel beam and trough the load cell pushed to the column stub. Measuring devices were attached to the independent

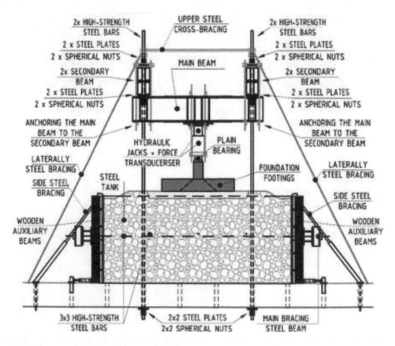

Figure 2.   The test setup for the experimental program—the footings resting on the gravel.

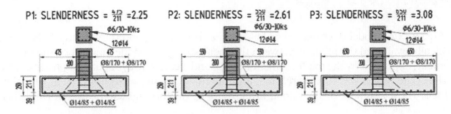

Figure 3.   Test specimens geometry and reinforcement layout.

Table 1.   Main properties of tested footings and obtained maximum forces Q_{max} [MN].

Specimen	Dimensions $(a.b)$ [mm]	Shear slenderness (a/d)	Concrete strength (f_{cm}) [MPa]	Reiforcement ratio (ρ_l) [%]	Yield strength (f_{ym}) [MPa]	Reinforcement layout $(A_{sx}; A_{sy})$	Column dimen. $(c_1.c_2)$ [mm]	Jack force Q_{max} [MN]
P1	1.15×1.15	2.25	41.33	0.833		13Ø14; 13Ø14	200 × 200	1.625[1]
P2.1	1.30×1.30	2.61	40.30		544.8	15Ø14; 15Ø14		1.608
P2.2	1.30×1.30	2.61	43.20	0.84		15Ø14; 15Ø14		1.714
P3	1.50×1.50	3.08	40.40			17Ø14; 17Ø14		1.550

[1]Test load test was terminated due to the failure of the subsoil.

beam supported by the columns located out of the tank. The measuring devices included 4 LVDT gauges, 8 displacement gauges and 4 inclinometers, all located on the top surface of the footings.

2.2   *The experimental footings*

Three types of footings without transverse reinforcement were proposed for the tests, see Figure 3. As main variable was chosen shear slenderness. Shear slenderness is defined as the ratio a/d, where a is distance between the face of a column stub and edge of the footing, while d is an effective depth of the footing. Following three different shear slenderness were used: 2.25, 2.61 and 3.08. Together four footings (P1, P2.1, P2.2 and P3) were cast and tested. Important properties and parameters are introduced in Table 1.

2.3   *Results of the tests*

Following main variables were monitored during the tests: settlement and deformations (displacement and rotation) of the footings and jack forces. The footings with higher shear slenderness failed due to the punching, while specimen P1 with the slenderness of 2.25 by failure of the subsoil when ground pressure reached a value of 1.289 MPa. The footings, that failed by punching, were able to carry remaining 86–88% of the punching force after the failure.

3   MODELS FOR THE ASSESSMENT OF THE PUNCHING RESISTANCE

Three design models were used for the safety verification. They are: current EC2 (2004) model and two models based on Critical Shear Crack Theory: Model Code 2010, LoAIII and CSCT model expressed in closed form. The fundamental difference between the empirical EC2 (2004) model and models based on the CSCT is in determining of the position of critical control perimeter a_{crit}. In the case of the EC2 (2004) model it is iterative process, while in the case of the CSCT models the critical control perimeter is fixed at distance $d/2$ from the face of a column.

3.1 The EC2 (2004) model

The EC2 (2004) model developed was based on a formula proposed by Zsutty (1968). Zsutty statistically evaluated the relation between the shear strength and the amount of the main reinforcement, expressed by the reinforcement ratio ρ, and a concrete cylinder compressive strength f_{ck}[MPa]. The model was further refined, and the final formula was published in Model Code 1990. For the assessment of the punching capacity of the footings can be used formula (1)

$$V_{Rd,c} = \frac{2 \times d}{a_{cr}} \frac{C_{Rk,c}}{\gamma_c} k \left(100 \rho f_{ck}\right)^{1/3} u_{cr} d \tag{1}$$

where:

$C_{Rk,c}$ is the empirical factor $C_{Rk,c} = 0.18$ MPa

ρ reinforcement ratio, $\rho = (\rho_x \rho_y)^{0.5}$

$\rho_{x(y)}$ reinforcement ratios of the main reinforcement in orthogonal directions $\rho_{x(y)} = A_{sx(y)}/(d_{x(y)} b)$

d effective depth, the average value of the effective depths in two orthogonal directions d_x and d_y

f_{ck} concrete cylinder compressive strength [MPa]

k factor, which takes into account the effect of size $k = 1 + (200\,[\text{mm}]/d)^{0.5}$

γ_c partial safety factor, $\gamma_c = 1.5$, for evaluation of the model safety $\gamma_c = 1.0$

u_{cr} length of the critical control perimeter measured at a distance a_{cr} from the face of a column, where $V_{Rd,c}/V_{Ed,red}$ is minimum

$V_{Ed,red}$ shear force at assumed control perimeter, $V_{Ed,red} = Q_{max} - \sigma_{gd} A_f(a)$

Q_{max} axial force in a column, see Figure 1

σ_{gd} upward soil pressure

$A_f(a)$ contact area of the footing surrounded by the control perimeter considered at distance a from the face of a column

3.2 The model code 2010 model

The Model Code 2010 model is based on the Critical Shear Crack Theory. The punching capacity depends on the slab rotation in the vicinity of a column. The relation between the punching capacity $V_{Rd,c}$ and slab rotation ψ is expressed by the so-called failure criterion. Failure criterion (3) was proposed for the Model Code 2010 model

$$V_{Rd,c} = k_{\psi} \frac{\sqrt{f_{ck}}}{\gamma_c} b_0 d_v \tag{2}$$

$$k_{\psi} = \frac{1}{1.5 + 0.9 k_{dg} \psi d} \leq 0.6 \tag{3}$$

where:

d_v is the effective depth of the slab for shear, usually $d_v = d$

b_0 length of the control perimeter at a distance $d_v/2$ from the face of a column

k_{dg} factor depending on the maximum aggregate size d_g[mm], $k_{dg} = 32/(16 + d_g)$

ψ slab rotation [rad]

The Model Code 2010 also provides methods for the assessment of ψ, depending on the load. The assessment isbased on four levels of approximation (LoA), depending on the degree of accuracy required. Except for LoAIV, the rotation ψ can be calculated using formula (4), which is derived from a quadri-linear model.

$$\psi = k_m \frac{r_s}{d} \frac{f_{yd}}{E_s} \left(\frac{m_{Sd}}{m_{Rd}}\right)^{1.5} \tag{4}$$

where:

r_s is the distance measured from the column axis to the line of the contra flexure of the radial bending moments, for footings $r_s = B/2$ or $D/2$

B, D dimension of the square, circular footing

k_m factor, $k_m = 1.5$ for LoAI, LoAII and 1.2 for LoAIII

f_{yd} yield strength of the main reinforcement

m_{Rd} average design bending capacity of a slab per unit length

m_{Sd} average design bending moments per unit length
– square footing: $m_{Sd} = (1 - b_c/B)Q_{max}/8$; circular footing: $m_{Sd} = (1 - d_c/D)Q_{max}/(2\pi)$

b_c, d_c cross-section dimension of the square, circular column.

3.3 The closed form CSCT model

The Closed form of the CSCT model "CF CSCT", is based on the CSCT. In order to simplify the design for punching, the basic formula (2) has been changed and expressed in a closed form (6); as such, it looks very similar to the EC2 (2004) formula now. Muttoni and Ruiz (2017) assumed the yielding of the main reinforcement and, instead of r_s, they used shear span a_v. The failure criterion (3) has been replaced by a power-law criterion (5):

$$k_\psi = 0.55 \left(\frac{1}{25} \frac{d_{dg}}{\psi_d} \right)^{2/3} \leq 0.55 \tag{5}$$

$$V_{Rd,c} = \frac{k_b}{\gamma_c} \left(100\rho f_{ck} \frac{d_{dg}}{a_v} \right)^{1/3} b_0 d_v \leq 0.6 \frac{\sqrt{f_{ck}}}{\gamma_c} b_0 d_v \tag{6}$$

$$k_b = \sqrt{8\mu \frac{d}{b_0}} \tag{7}$$

where:

d_{dg} is a coefficient, that takes account of the type of concrete and its aggregate properties, i.e. 32 [mm] for a concrete of a normal weight

a_v shear span, geometric average of the shear spans in both orthogonal directions and not less than $2.5d$

μ parameter accounting for the shear force and bending moment in the region of the shear, for an internal column without an unbalanced moment $\mu = 8/(1 - b_c/B)$ or $2\pi/(1 - d_c/D)$

4 EVALUATION OF THE MODEL'S SAFETY

For the sake of a comparison of the suitability of the models, the ratio $P_i = (V_{Rtest}/V_{Rd,c})_i$ as the statistical variable was assigned. Here "i" is the number of the test; V_{Rtest} is the resistance obtained from the test, i.e. the so-called *experimental* or *test value*; and $V_{Rd,c}$ is the punching resistance obtained from the particular normative formula of one of the three theoretical models. "P_i greater than 1.0" means that the model is on the safe side; if it is not greater than 1.0, then it is unsafe. The punching capacities $V_{Rd,c}$ were assessed with the partial safety factor $\gamma_c = 1.0$. For strength f_{ck} an actual cylinder strength of concrete (the mean value) introduced by the authors has been used. The control perimeters have been assumed at a distance of a_{crit} from the face of a column stub in the case of the EC2 (2004) model and $d/2$ for the other models.

Table Together, the results of 58 tests for specimens without transverse reinforcement tested under axis-symmetric conditions were included in the analysis. They were adapted from databases published by Ricker (2009); Siburg (2014) and by Simões (2018). The value of V_{Rtest} was determined using formula (1), where $\Delta V_{Ed,red}$ is upward force within control

Table 2. Relation shear slenderness and ratio $V_{Rtest}/V_{Rd,c}$ for different types of subsoil.

Type of subsoil	Shear slenderness [a/d]	Number of specimens	EC2 (2004) $V_{Rtest}/V_{Rd,c}$	MC 2010 $V_{Rtest}/V_{Rd,c}$	CF CSCT $V_{Rtest}/V_{Rd,c}$
Line support	1.20–1.24	5	0.610	0.884	0.769
	1.24–1.47	4	0.705	1.068	0.846
Point load	1.00–1.49	11	0.770	1.005	0.818
	1.50–1.99	6	0.802	1.101	0.867
	2.00–2.33	17	0.822	1.068	0.815
Sand (thick)	1.27–1.53	4	0.845	1.243	0.993
	2.00–2.50	4	0.873	1.278	1.021
Sand (thin)	1.26–1.28	2	0.968	1.765	1.118
	1.63–1.79	2	0.967	1.914	1.115
Gravel (deep)	2.61	2	1.263	2.365	1.257
	3.08	1	1.284	2.395	1.263

perimeter considered due to soil pressure σ_{gd} and Q_{max} is jack force when punching has occurred. The analysis was carried out under the assumption of the uniform ground pressure $\sigma_{gd} = Q_{max}/A_{foot}$ where A_{foot} is contact area of the footing.

$$V_{Rtest} = Q_{max} - \Delta V_{Ed,red} \qquad (8)$$

Evaluation was carried out independently for each type of the subsoil and with shear slenderness divided into several ranges, see Table 2.

5 CONCLUSIONS

The lowest safety was obtained for line support. The highest safety was assessed for the footings resting on deep layer of gravel. This result was expected due to the concentration of the ground pressure under the area that is surrounded byte critical control perimeter and the specimens had the greatest shear slenderness. In the case of footings resting on the thick layer of sand where non-linear distribution of ground pressure can develop, the assessed model safety was significantly lower than in the case of deep gravel cushion. The safety was even lower than in the case of subsoil consisting on thin layer of sand confined by box. The effect of shear slenderness on the assessed model safety was neglectable for each type of subsoil. The safest design provides the MC2010, LoAIII model and the lowest safety was registered for the EC2 (2004) model.

ACKNOWLEDGEMENT

This work was supported Scientific Grant Agency of the Ministry of Education, science, research and sport of the Slovak Republic and the Slovak Academy of Sciences No 1/0810/16 and by the Slovak Research and Development Agency under the contract No. APVV-15-0204.

REFERENCES

Dieterle, H. & Rostásy, F.S. 1987. Tragverhalten quadratischer Einzelfundamenteaus Stahlbeton. *Schriftenreihe des DAfStb*, Heft 387, Berlin: Ernst & Sohn: 5–91.
Hallgren, M. & Kinnunen, S. & Nylander, B. 1998. Punching Shear Tests on Column Footings. *Nordic Concrete Research 21*, 1998, Nb. 3: 1–22.

Muttoni, A. & Fernández Ruiz, M. 2017. The critical shear crack theory for punching design: From mechanical model to closed-form design expressions, Punching shear of structural concrete slabs. *Proceeding of:Honoring Neil M. Hawkins ACI-fib International Symposium,* April 2017: 237–252.

Ricker, M. 2009. Zur Zuverlässigkeit der Bemessung gegen Durchstanzen bei Einzelfundamenten (Reliabilityofthepunchingshear design offootings). *Dissertation RWTH Aachen, Lehrstuhl und Institut für Massivbau (IMB),* 2009 (in German).

Siburg, C. 2014. Zur einheitlichen Bemessung gegen Durchstanzen in Flachdecken und Fundamenten. *Dissertation RWTH Aachen, Lehrstuhl und Institut für Massivbau (IMB),* 2014 (in German).

Siburg, C. & Ricker. M. & Hegger, J. 2014. Punching shear design of footings: critical review of different code provisions. *Structural Concrete, Ernst & Sohn, Germany,* Vol. 3, No. 3, 2014: 1–27.

Simões, J. & Bujnak, J. & Ruiz, F.M. & Muttoni, A. 2016. Punching shear tests on compact footings with uniform soil pressure. *Structural Concrete, Ernst & Sohn, Germany,* 2016, No. 4: 603–617.

Simões, J. 2018. The mechanics of punching in reinforced concrete slabs and footings without shear reinforcement. *THÈSE NO 8387 (2018), ÉCOLE POLYTECHNIQUE FÉDÉRALE DE LAUSANNE.*

Zsutty, T. 1968. Beam shear strength prediction by analysis of existing data. *ACI Journal* V. 65(11): 1968: 943–951.

Advances and Trends in Engineering Sciences and Technologies III – Al Ali & Platko (Eds)
© *2019 Taylor & Francis Group, London, ISBN 978-0-367-07509-5*

Implementation of the viscous Drucker-Prager and viscous Lee-Fenves nonlinear material models and finite element benchmarks

F. Hokes & J. Kala
Faculty of Civil Engineering, Brno University of Technology, Brno, Czech Republic

M. Trcala & I. Nemec
FEM Consulting, s.r.o., Brno, Czech Republic

ABSTRACT: The subjected paper is focused on the description of implementation of the Drucker-Prager and Lee-Fenves nonlinear material models with viscous enhancement in the finite element solver. An advantage of the implementation is based on possibility to enter the stress-strain curve as an input for the calculation instead of an array of material model parameters, which is common in commercial material libraries. The paper contains theoretical background describing theory of interested models and the description of used Duvaut-Lions viscous regularization. The presented results show different responses of the numerical model to dynamic loading with various strain rates. The aim of the paper is to extend the interest of engineers in the field of viscous nonlinear material models and in the modelling of mechanical response of concrete structures to rapid loading.

1 INTRODUCTION

1.1 *The state of the art*

Nowadays, an advanced analysis of structures is enabled by utilization of the material nonlinearity within a numerical simulation. The material nonlinearity is introduced in the simulation by specific constitutive law. The stress-strain relationship defined by the constitutive law can have different forms based on the theory. Utilization of specific theoretical approach often arises from the type of the material whose static or dynamic response is numerically simulated. This fact led to the current state when there is a relatively wide range of existing nonlinear material models. With respect to the above mentioned fact, the proposed contribution focuses on a concrete where the situation is very similar and a variety of material models based on different assumptions can be found.

The construction of a correct constitutive law for concrete which is able to express nonlinear response to various types of loading appears to be problematic (Cicekli et al. 2007). Basic problem that arises when formulating a material model for concrete is the different response in the compression and in the tension (Hokes 2015). This reason caused situation where several theoretical approaches are used for mathematical description of behavior of concrete (Hokes 2014). One of these approach is based on the plasticity theory, which is documented in the following work of the authors (Willam & Warnke 1974), (Chen et al. 1975), (Bazant 1978), (Dragon & Mroz 1979), (Schreyer 1983), (Chen and Byukozturk 1985), (Onate et al. 1988), (Pramono and Willam 1989), (Etse & Willam 1994), (Menetrey & Willam 1994) and (Grassl 2002). The material models described in the above mentioned publications use standard theory of plasticity, which is not sufficient due to gradual decrease of concrete in stiffness and due to occurrence of cracks (Cicekli et al. 2007). This problem can be resolved with utilization of the damage theory (Mazars 1986). However, independent damage models are not also sufficient. The problems arise when the description of irreversible deformations and inelastic volumetric expansion is required. Theoretical assumptions and practical applications of various damage material models for concrete can be found in publications by the authors (Loland 1980), (Ortiz & Popov 1982), (Krajcinovic 1996), (Resende & Martin 1984), (Simo & Ju 1987), and (Lubarda 1994). Despite the above-mentioned limitations of both approaches, there are advantages to using both of them in mutual combination. Combined models can be divided into two branches, where the first branch is based on stress-based plasticity formulated in the effective stress space. This group of models is represented by the Lee-Fenves material model (Lee & Fenves 1998), which was implemented in the RFEM computational system. The second branch of combined models is based on the stress-based plasticity formulated in the nominal (damaged) stress space. The second mentioned material models are represented by the plastic-damage model published by (Lubliner et al. 1989). A modern way of numerical simulation of concrete response to various types of loading is represented by Extended Finite Element Method (XFEM) where the need for re-meshing is removed. Theoretical details and application of this quite new approach are described in work of (Belytschko 1999). The list of approaches suitable for numerical simulation of concrete can be closed by the mention of Discrete Element Method (DEM), which, however, moves away from the continuum theory. Further information about this branch of mechanics can be found in the works of (Cundall 1988) and (Ghaboussi & Barbosa 1990).

1.2 Structure of the contribution

The submitted contribution represents the demonstration of utilization of the viscous Lee-Fenves and viscous Drucker-Prager material models in the finite element solver. In order to show both static and dynamic response to the loading of both mentioned material models, only the behavior in compression is studied. The article is divided into three main sections, where the first part deals with the current state of the art in the field of nonlinear material models of concrete. The second part of the paper briefly describes theoretical background of used material models and also describes the form of the performed benchmark calculations in RFEM computational system. The third main part of the contribution is dedicated to presentation of the achieved results and the final conclusion.

2 DEFINITION OF THE PROBLEM

2.1 Description of benchmark

The demonstration of options for modelling response to static and various dynamic loading was performed as a 3D single-element benchmark. The geometry of the computational model was considered as one cube with edge length l of 1.0 m. The geometry was meshed with only one 8-node finite element with 6 degrees of freedom at each node. The boundary conditions were prescribed in the way that all rotational degrees of freedom were fixed and transverse expansion during compression was allowed. The appearance of the element benchmark is depicted in Figure 1.

Figure 1.   The appearance of the computational model.

Table 1. Basic mechanico-physical properties of used material models.

Property		Unit	Value
Young's modulus of elasticity	E	[Pa]	$34 \cdot 10^9$
Poisson's ratio	v	[–]	0.2
Shear modulus	G	[Pa]	$14.167 \cdot 10^9$
Uniaxial strength in compression	f_c	[Pa]	$43 \cdot 10^6$
Uniaxial strength in tension	f_t	[Pa]	$3.2 \cdot 10^6$
Specific weight	γ_c	[N/m³]	$25 \cdot 10^3$
Dynamic viscosity	η	[–]	$2 \cdot 10^8$

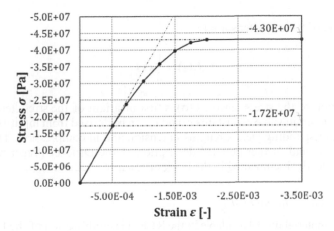

Figure 2. 1D stress-strain diagram in compression domain.

2.2 Input data

Proposed nonlinear numerical study of response of the finite element to static and dynamic loading was solved as a time analysis in RFEM computational system with use of Newmark's implicit algorithm. Load was modelled as nodal displacement in vertical direction with final value of 2.0 mm. The solution was obtained for four different deformation rates: 0.1 mm·s^{-1}, 1.0 mm·s^{-1}, 10.0 mm·s^{-1} and 100.0 mm·s^{-1}. The study, in the above mentioned configuration, was performed for Drucker-Prager and Lee-Fenves material models whose basic mechanico-physical properties are summarized in Table 1. The nonlinear behavior of used models was set up by 1D stress-strain diagram in according to present standards (EC2 2004) and (DIN 1045-1 2001). The form of the input stress-strain diagram is shown in Figure 2.

3 THEORETICAL BACKGROUND

The calculation of the above mentioned single-element benchmark was performed for two different material models in order to validate the implementation in the finite element solver. The first one, Drucker-Prager constitutive law (Drucker & Prager 1952), belongs to basic yield conditions and can be found in other commercial computational systems. On the other hand the Lee-Fenves (label by authors) is quite new and occurrence in commercial systems is not so common.

3.1 The Drucker-Prager material model

The Drucker-Prager yield criterion belongs to basic criterions within general theory of plasticity and it was designed by authors Drucker and Prager in order to approximate the

Mohr-Coulomb's yield criterion by smooth function. This approximation that extended Mohr-Coulomb's criterion for different behavior in tension and compression can be written in the following form (Nemec et al. 2018)

$$\Phi\left(\boldsymbol{\sigma},\overline{\varepsilon}^{p}\right)=\sqrt{J_{2}(\mathbf{s})}+c_{1}\frac{1}{3}I_{1}(\boldsymbol{\sigma})-c_{2}\mathrm{coh}\left(\overline{\varepsilon}^{p}\right) \tag{1}$$

where

$$\mathbf{s}=\mathbf{s}(\boldsymbol{\sigma})=\boldsymbol{\sigma}-\frac{1}{3}I_{1}(\boldsymbol{\sigma})\mathbf{I} \tag{2}$$

is deviatoric part of the stress tensor and

$$J_{2}(\mathbf{s})=\frac{1}{2}\mathbf{s}:\mathbf{s} \tag{3}$$

is second invariant of the stress tensor; $I_{1}(\boldsymbol{\sigma})$ is first invariant of the stress tensor and $\overline{\varepsilon}^{p}$ is cumulative plastic strain value. Constants c_{1} and c_{2} can be derived from input 1D stress-strain diagram and corresponding yield strengths in tension and in compression. The expressions for constants c_{1}, c_{2} and further detailed description of the Drucker-Prager material model from RFEM is clearly documented in the book of (Nemec et al. 2018).

3.2 The Lee-Fenves material model

The Lee-Fenves material model available in the RFEM is modification of the Lubliner's yield criterion (Lubliner et al. 1989). This criterion was developed in order to improve Drucker-Prager's criterion and was designed in the way that the need for the J_{3} invariant is removed. The expression of the yield criterion can be written as follows

$$\Phi\left(\boldsymbol{\sigma},\overline{\varepsilon}^{p}\right)=\sqrt{3J_{2}}+\alpha I_{1}+\beta\left(\overline{\varepsilon}_{t}^{p},\overline{\varepsilon}_{c}^{p}\right)\langle\hat{\sigma}_{\max}\rangle-\chi\langle-\hat{\sigma}_{\max}\rangle-\left(1-\alpha\right)\sigma_{c}\left(\overline{\varepsilon}_{c}^{p}\right) \tag{4}$$

where

$$\beta\left(\overline{\varepsilon}_{t}^{p},\overline{\varepsilon}_{c}^{p}\right)=\frac{\sigma_{c}\left(\overline{\varepsilon}_{c}^{p}\right)}{\sigma_{t}\left(\overline{\varepsilon}_{t}^{p}\right)}(1-\alpha)-(1+\alpha) \tag{5}$$

$$\alpha=\frac{f_{b0}-f_{c0}}{2f_{b0}-f_{c0}} \tag{6}$$

where $\alpha\in\langle1.10;1.16\rangle$ because ratio $f_{b0}/f_{c0}\in\langle0.08;0.12\rangle$. The coefficient χ is occuring only in case of three dimensional stress state and average value of the coefficient χ for concrete is about 3.0 (Nemec et al. 2018). The great advantage of the Lee-Fenves material model, against Drucker-Prager model, is based on existence of two separated plastic cumulative strains for tension and compression.

3.3 Viscous enhancement

The contribution deals also with demonstration of response of studied models to various dynamic loading. Modelling response of the structure in time can be enabled by introduction of the viscous formulation (viscous enhancement). The viscous enhancement of the above mentioned models were performed with use of Duvaut-Lions regularization (Duvaut & Lions 1972). The Duvaut-Lions formulation is based on calculation of viscoplastic strain ε_n^{vp} from previously calculated plastic strain $\mathbf{\varepsilon}^p$

$$\dot{\mathbf{\varepsilon}}^{vp} = \frac{1}{\eta}\mathbf{C}^{-1} : \left(\mathbf{\sigma} - \mathbf{C} : \left(\mathbf{\varepsilon} - \mathbf{\varepsilon}^p \right) \right) \tag{7}$$

where

$$\mathbf{\sigma} = \mathbf{C} : \left(\mathbf{\varepsilon} - \mathbf{\varepsilon}^{vp} \right) \tag{8}$$

and η is viscous (dimensionless) parameter. The expression for viscoplastic strain can be then rewrite as follows

$$\dot{\mathbf{\varepsilon}}^{vp} = \frac{1}{\eta}\left(\mathbf{\varepsilon} - \mathbf{\varepsilon}^{vp} - \mathbf{\varepsilon} + \mathbf{\varepsilon}^p \right) = \frac{1}{\eta}\left(\mathbf{\varepsilon}^p - \mathbf{\varepsilon}^{vp} \right) \tag{9}$$

Viscous enhancement with use of Duvaut-Lions formulation is also accompanied by increase of both yield strength in tension and compression, in the same way as it was realized for LS-Dyna material model 159 (Murray 2007).

4 RESULTS

Results of all performed time analyses in the RFEM computational system are graphically illustrated in Figure 3. The form of all final load-displacement curves were in compliance with theoretical assumptions. Load-displacement curves for load applied with speed of 1.0 mm/s were close to solution achieved with nonviscous model. On the opposite side, load-displacement curves for rapid load with speed of 100 mm/s were closed to linear solution. However, final values of reactions at maximum loading were not realistic. This problem was caused by unknown value of viscous parameter that was chosen for calculation on the empirical basis.

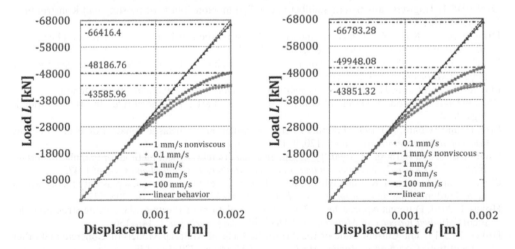

Figure 3.   Resultant load-displacement diagrams: Drucker-Prager (left); Lee-Fenves (right).

5 CONCLUSION

Although the final reaction values were burdened by a certain degree of uncertainty, the overall behavior of the element corresponded to the used theory. The usability of both models for numerical simulation of response to static and dynamic loading was proved in this way. It is necessary to note that in order to achieve the same results with both models, the 1D input stress-strain diagram for compression and tension was the same in the case of the Drucker-Prager material model, however, different 1D stress-strain diagrams for tension and compression were used for Lee-Fenves material model. This modification was caused by the existence of only one cumulative strain in Drucker-Prager material model. The Lee-Fenves does not contain this defect and thus it is more suitable for modelling nonlinear behavior of concrete structures for dynamic loading.

ACKNOWLEDGEMENT

This outcome has been achieved with the financial support of project GACR 17–23578S "Damage assessment identification for reinforced concrete subjected to extreme loading" provided by the Czech Science Foundation.

REFERENCES

Bazant, Z.P. 1978. Endochronic inelasticity and incremental plasticity. *International Journal of Solids and Structures.* 14, 9, 691–714.

Belytschko, T. and Black, T. 1999. Elastic crack growth in finite elements with minimal remeshing. *International Journal for Numerical Methods in Engineering.* 45, 5, 601–620.

Chen and Buyukozturk, O. 1985. Constitutive Model for Concrete in Cyclic Compression. *Journal of Engineering Mechanics.* 111, 6, 797–814.

Chen, A.C.T. and Chen, W.F. 1975. Constitutive relations for concrete. Journal of the Engineering Mechanical Division. 101, 465–481.

Cicekli, U., Voyiadjis, G.Z. and Abu Al-Rub, R.K. 2007. A plasticity and anisotropic damage model for plain concrete: the origin, evolution, and impact of doi moi. *International Journal of Plasticity.* 23, 10–11, 1874–1900.

Cundall, P. 1988. Formulation of a three-dimensional distinct element model—Part I: A scheme to detect and represent contacts in a system composed of many polyhedral blocks. *International Journal of Rock Mechanics & Mining Sciences & Geomechanics Abstracts.* 25, 3, 107–116.

DIN 1045-1. Tragwerke aus Beton, Stahlbeton und Spannbeton, Teil 1: Bemessung und Konstruktion, Beuth-Verlag. 2001.

Dragon, A. and Mroz, Z. 1979. A continuum model for plastic-brittle behaviour of rock and concrete. *International Journal of Engineering Science.* 17, 2, 121–137.

Drucker, D.C. and Prager, W. 1952. Soil mechanics and plastic analysis for limit design. *Quarterly of Applied Mathematics.* 10, 157–165.

Duvaut, G. and Lions, J.L. 1972. *Les Inequations en Mecanique et en Physique.* Paris: Dunod.

EN 1992-1-1: Eurocode 2: Design of concrete structures—Part 1-1: General rules and rules for buildings. 2004.

Etse, G. and Willam, K.J. 1994. Fracture Energy Formulation for Inelastic Behavior of Plain Concrete. *Journal of Engineering Mechanics.* 120, 9, 1983–2011.

Ghaboussi, J. and Barbosa, R. 1990. Three-dimensional discrete element method for granular materials. *International Journal for Numerical and Analytical Methods in Geomechanics.* 14, 451–472.

Grassl, P. and Lundgren, K. and Gylltoft, K. 2002. Concrete in compression: a plasticity theory with a novel hardening law. *International Journal of Solids and Structures.* 39, 20, 5205–5223.

Hokes, F. 2014. Different Approaches to Numerical Simulations of Prestressed Concrete Structural Elements. *Applied Research in Materials and Mechanics Engineering.* 621, 148–156.

Hokes, F. 2015. The Current State-of-the-Art in the Field of Material Models of Concrete and other Cementitious Composites. *Applied Mechanics and Materials.* 729, 134–139.

Krajcinovic, D. 1996. *Damage mechanics.* North-Holland Series in applied mathematics and mechanics. Vol. 41, Amsterdam: North-Holland, 761 p.

Lee, J. and Fenves, G. 1998. Plastic-Damage Model for Cyclic Loading of Concrete Structures. *Journal of Engineering Mechanics*. 124, 8, 892–900.

Løland, K. 1980. Continuous damage model for load-response estimation of concrete. *Cement and Concrete Research*. 10, 3, 395–402.

Lubarda, V., Krajcinovic, D. and Mastilovic, S. 1994. Damage model for brittle elastic solids with unequal tensile and compressive strengths. *Engineering Fracture Mechanics*. 49, 5, 681–697.

Lubliner, J., Oliver, J., Oller, S. and Oñate, E. 1989. A plastic-damage model for concrete. *International Journal of Solids and Structures*. 25, 3, 299–326.

Mazars, J. 1986. A description of micro- and macroscale damage of concrete structures. *Engineering Fracture Mechanics*. 25, 729–737.

Murray, Y.D. 2007. User's manual for LS-DYNA concrete material model 159. Report No. FHWA-HRT-05-063, Federal Highway Administration.

Nemec, I., Trcala, M. and Rek, V. 2018. *Nelinearni mechanika*. Brno: VUTIUM.

Menetrey, P. and Willam, K.J. 1995. Triaxial failure criterion for concrete and its generalization. *ACI Structural Journal*. 92, 3, 311–318.

Onate, E., Oller, S., Oliver, J. and Lubliner, J. 1988. A constitutive model for cracking of concrete based on the incremental theory of plasticity. *Engineering Computations*. 5, 4, 309–319.

Ortiz, M. and Popov, E. 1982. Plain concrete as a composite material. *Mechanics of Materials*. 1, 2, 139–150.

Pramono, E. and Willam, K.J. 1989. Fracture Energy – Based Plasticity Formulation of Plain Concrete. *Journal of Engineering Mechanics*. 115, 6, 1183–1204.

Resende, L. and Martin, J. 1984. A progressive damage "continuum" model for granular materials. *Computer Methods in Applied Mechanics and Engineering*. 42, 1, 1–18.

Schreyer, H. 1983. A Third-Invariant Plasticity Theory for Frictional Materials. *Journal of Structural Mechanics*. 11, 2, 177–196.

Simo, J. and Ju, J. 1987. Strain- and stress-based continuum damage models – I. Formulation. *International Journal of Solids and Structures*. 23, 7, 821 840.

Simo, J. and Ju, J. 1987. Strain- and stress-based continuum damage models – II. Computational aspects. *International Journal of Solids and Structures*. 23, 7, 841–869.

William, K.J. and Warnke, E.P. 1974. Constitutive model for the triaxial behavior of concrete. *International Association of Bridge and Structural Engineers*. 19, 1–30.

Advances and Trends in Engineering Sciences and Technologies III – Al Ali & Platko (Eds)
© 2019 Taylor & Francis Group, London, ISBN 978-0-367-07509-5

Experimental verification of properties of concrete mixtures for concrete prefabricated elements

M. Janda
S.O.K. stavební, s.r.o., Třebíč, Czech Republic

M. Zich
Faculty of Civil Engineering, Brno University of Technology, Brno, Czech Republic

ABSTRACT: Manipulation with concrete prefabricate usually takes place soon after the concreting. Concrete mixtures for the production of concrete prefabricated elements shall, in addition to requirements for the resulting concrete strength, also meet the requirements for concrete handling strength. The required handling strength of the concrete shall normally be achieved at 18 hours after concreting. Concrete compressive strength, tensile strength and modulus of elasticity at the time of 18 hours, 42 hours, 66 hours, 14 days, and 28 days, were measured during the experiment. Furthermore, the shrinkage of the concrete was measured over time of 1 day, 3 days, 7 days, 36 days, 68 days, and 91 days in 4 blocks of 75 mm, 125 mm, 150 mm, and 200 mm thickness.

1 INTRODUCTION

Requirements for concrete mixtures for the production of prefabricated structures differ considerably from the requirements for conventional monolithic structures. Prefabricated structures, unlike monolithic structures, undergo a number of handling stages. For this reason, a very important property of the concrete mixture is the handling strength that shall be achieved before handling the prefabricated elements. Manipulation normally takes place 18 hours after concreting. Achieving handling strength is generally inspected by non-destructive methods, such as a concrete strength test using a Schmidt hardness tester.

Another important property of the concrete mixture for the production of prefabricated structures is its workability. The current trend in manufacturing processes in the construction sector is to reduce the proportion of human labor, which increasingly leads to the demand for the concrete mixture to be self-compacting.

Requirements for early achievement of handling strength and good workability of concrete commonly lead to the design of a mixture that contains a relatively large amount of cement. The undesirable effect of high cement content in the concrete mix is large volumetric changes during maturing of the concrete accompanied by shortening of the prefabricate or formation of shrinkage cracks. The advantage of prefabricated structures is that the installation of the elements does not take place immediately after the concreting and with a sufficiently long maturation period, a large part of the volume changes can take place on an unassembled statically specific prefabricated panel.

The minimum required handling strength is defined in the manufacturer's instructions for transport handles, e.g. on www.halfen.com. The concrete strength test using the Schmidt hardness tester is described in the standard (Czech Office for Standards, Metrology and Testing).

2 PERFORMED EXPERIMENTS

2.1 *Preparation of test samples*

For experiments verifying the properties of the concrete mixture for the production of prefabricated structures, the following test samples were produced:

- 30 test cubes of standard sizes for measuring mechanical properties of concrete,
- 30 test beams of standard sizes for measuring mechanical properties of concrete,
- 4 test blocks of different thicknesses to measure shrinkage of concrete.

The geometry of the test cubes and beams is evident from Figure 1, the geometry of the test blocks is shown in Figure 2. The test blocks were cast on steel washers provided with a deformation oil and a layer of plastic foil. Strain gauges with built-in thermistors were installed in the test blocks. The number and position of the strain gauges in the test blocks is shown in Figure 3. For the production of test samples, self-compacting concrete C 30/37, XC2, Dmax16 and consistency F6 (cement CEM I 52,5 R – 360 kg.m⁻³, bulk limestone VJM7/V 115 kg.m⁻³, washed aggregates KD fractions 0/4–840 kg.m⁻³, rough aggregates PKD 8/16–895 kg.m⁻³, plasticizer Isoflow 7850–3,2 kg.m⁻³).

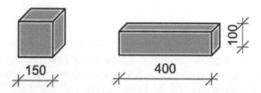

Figure 1.   Geometry of the test cubes and beams.

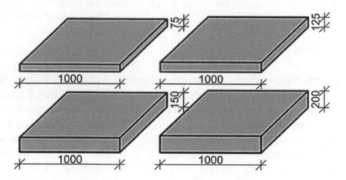

Figure 2.   Geometry of the test blocks.

LOCATION OF STRAIN GAUGES IN TEST BLOCKS

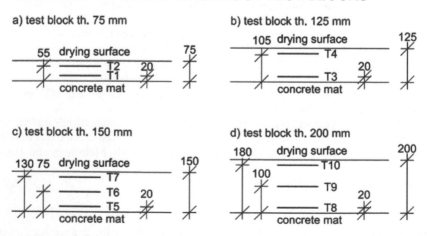

Figure 3.   Location of strain gauges in the test blocks.

114

The following tests were performed to verify the properties of the concrete mixture for the production of prefabricated structures:

- compression test of concrete on cubes of standard sizes at time 18 hours, 42 hours, 14 days, and 28 days after concreting,
- tensile splitting strength test on test cubes of standard sizes at 18 hours, 42 hours, 14 days, and 28 days after concreting,
- flexural tensile strength test of concrete on test beams of standard sizes at 18 hours, 42 hours, 14 days, and 28 days after concreting,
- test of the elastic modulus of concrete on the beams of standard sizes at 18 hours, 42 hours, 14 days, and 28 days after concreting,
- verification of concrete strength on standard size cubes using the Schmidt type R hardness tester at 18 hours after concreting,
- verification of concrete strength on standard size cubes using Schmidt type N hardness tester at 18 hours, 42 hours, 14 days, and 28 days after concreting,
- measuring shrinking on test blocks of different thicknesses between the 1st and the 36th day after concreting and between the 68th and the 91st day after concreting,
- measuring the ambient temperature of the test samples between 1st and 91st day after concreting.

The mechanical properties of concrete have been tested according to the standards (Czech Office for Standards, Metrology and Testing). Tests were performed by the S.O.K. company in cooperation with Ing. Petr Daněk, Ph.D. from the Faculty of Civil Engineering, Brno University of Technology. The compressive strength test is shown in Figure 4, the flexural tensile strength test is shown in Figure 5, the test blocks for measuring the shrinkage are shown in Figure 6, and the photo of strain gauges is shown in Figure 7.

2.2 Experiment results

The mechanical properties of the concrete mixture at 18 hours, 42 hours, 66 hours, 14 days, and 28 days after concreting are shown in Table 1. The shrinkage recorded by the individual strain gauges is shown in Table 2. The shrinkage of the individual test blocks determined as the average shrinking measured by the built-in strain gauges can be found in Table 3. The average ambient temperature of the test samples and test blocks was 23°C, the average relative humidity of the environment was 58%. The concrete compressive strength measured on the cubes at 28 days after concreting $f_{cm,cube}$ was 65 MPa. The corresponding value of characteristic compressive strength is $f_{ck,cube} = 65 - 8 = 57$ MPa. The value is significantly higher than characteristic cube compressive strength 37 MPa defined for the strength class C 30/37 (strength class of used concrete). The development of the concrete compressive strength, tensile strength, and shrinkage of concrete is shown in Figures 8, 9, and 10.

Figure 4.   Compressive strength test.

Figure 5.   Flexural tensile strength test.

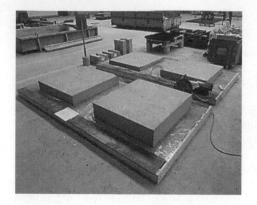

Figure 6.   Test blocks after detachement.

Figure 7.   Strain gauges in test blocks.

Table 1.   Mechanical properties of the concrete mixture.

Age of concrete	18 hours	42 hours	66 hours	14 days	28 days
Cube compressive strength [MPa]*	38.3	49.2	54.4	61.7	65.0
Comp. strength—Schmidt—type R [MPa]*	35.0	–	–	–	–
Comp. strength—Schmidt—type N [MPa]*	38.3	41.5	46.0	49.2	51.0
tensile splitting strength [MPa]*	2.80	3.41	3.63	3.90	4.06
flexural tensile strength [MPa]*	3.39	4.08	4.31	4.89	5.01
Flexural tensile elastic modulus [GPa]*	30.0	38.0	40.0	41.0	42.0

*Individual values were determined as the average of 3 test samples.

Table 2.   Shrinkage recorded by individual strain gauges.

	1 day	3 days	7 days	36 days	68 days	91 days
Age of concrete	shrinkage [‰]					
Strain gauge T1	0.42	0.43	0.44	0.59	0.66	0.70
Strain gauge T2	0.62	0.72	0.80	1.07	1.08	1.09
Strain gauge T3	0.35	0.36	0.37	0.45	0.50	0.54
Strain gauge T4	0.61	0.65	0.70	0.95	0.97	0.96
Strain gauge T5	0.32	0.32	0.35	0.37	0.42	0.44
Strain gauge T6	0.39	0.41	0.43	0.59	0.63	0.65
Strain gauge T7	0.48	0.49	0.53	0.76	0.80	0.81
Strain gauge T8	0.20	0.22	0.24	0.26	0.28	0.30
Strain gauge T9	0.40	0.41	0.42	0.55	0.58	0.60
Strain gauge T10	0.52	0.55	0.61	0.83	0.87	0.89

Table 3.   Average shrinkage in individual test blocks.

	1 day	3 days	7 days	36 days	68 days	91 days
Age of concrete	shrinkage [‰]					
Test block th. 75 mm	0.52	0.58	0.62	0.83	0.87	0.90
Test block th. 125 mm	0.48	0.51	0.54	0.70	0.74	0.75
Test block th. 150 mm	0.40	0.41	0.44	0.57	0.62	0.63
Test block th. 200 mm	0.37	0.39	0.42	0.55	0.58	0.60

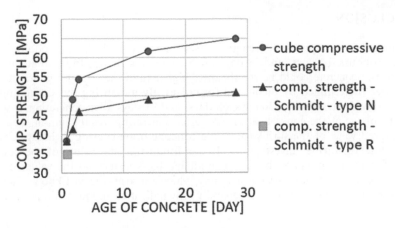

Figure 8. Development of the concrete compressive strength.

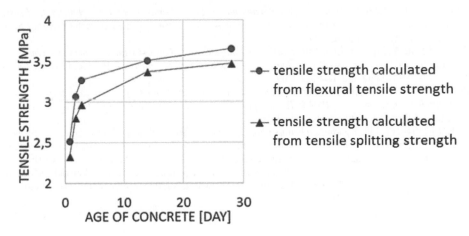

Figure 9. Development of the concrete tensile strength.

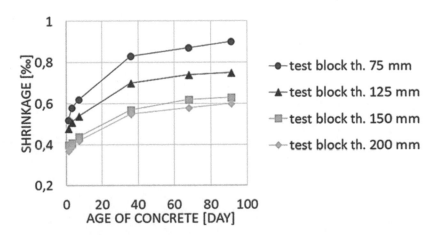

Figure 10. Development of the shrinkage.

3 CONCLUSION

By measuring the compressive strength, tensile strength, and elastic modulus 18 hours, 42 hours, 66 hours, 14 days, and 28 days after concreting, it was found that the resulting strength of the concrete could be significantly higher than the strength guaranteed by the manufacturer (determined by theoretical formulae). By measuring the shortening the test blocks at 1 day, 3 days, 7 days, 36 days, 68 days, and 91 days after concreting, it was found that the shrinkage of the concrete at 91 days after concreting was 0.9 mm/m. The relatively large measured value of the shrinkage of the concrete often causing shrinkage cracks, and this issue deserves further investigation. It was found that the values of the compressive strength of concrete measured by the Schmidt hardness tester and the values determined by the compression test may vary considerably, which confirms the well-known fact, that the test of Schmidt hammer properties is generally only indicative and must always be calibrated on the cubes.

REFERENCES

ČSN EN 12390-3 (731302). *Zkoušení ztvrdlého betonu—Část 3: Pevnost v tlaku zkušebních těles [Testing hardened concrete—Part 3: Compressive strength of test specimens]*. Prague: Czech Office for Standards, Metrology and Testing, 2009.

ČSN EN 12390-5 (731302). *Zkoušení ztvrdlého betonu—Část 5: Pevnost v tahu ohybem zkušebních těles [Testing hardened concrete—Part 5: Flexural strength of test specimense]*. Prague: Czech Office for Standards, Metrology and Testing, 2009.

ČSN EN 12390-6 (731302). *Zkoušení ztvrdlého betonu—Část 6: Pevnost v příčném tahu zkušebních těles [Testing hardened concrete—Part 6: Tensile splitting strength of test specimens]*. Prague: Czech Office for Standards, Metrology and Testing, 2010.

ČSN EN 12504-2 (731303). *Zkoušení betonu v konstrukcích—Část 2: Nedestruktivní zkoušení—Stanovení tvrdosti odrazovým tvrdoměrem [Testing concrete in structures—Part 2: Non-destructive testing—Determination of rebound number]*. Prague: Czech Office for Standards, Metrology and Testing, 2013.

HD-Socket Lifting System: Technical Product Information [online]. HALFEN Vertriebsgesellschaft mbH. 2016 [cit. 2017-01-27]. Available from: http://downloads.halfen.com/catalogues/de/media/catalogues/liftingsystems/HD_16-E.pdf.

Advances and Trends in Engineering Sciences and Technologies III – Al Ali & Platko (Eds)
© 2019 Taylor & Francis Group, London, ISBN 978-0-367-07509-5

Minimum mass design of compressed I-section columns with different design rules

K. Jármai & M. Petrik
University of Miskolc, Miskolc, Hungary

ABSTRACT: The aim of the present study is to show the minimum mass design procedure for welded steel I-section columns loaded by compression force. The normal stresses and overall stability are calculated for pinned columns. The dimensions of the I-columns are optimized by using constraints on overall stability, and local buckling of webs and flanges. The different design rules and standards, like Eurocode 3, Japan Railroad Association (JRA), American Petroleum Institute (API), and American Institute of Steel Construction (AISC) have been compared for this predesign. The calculations are made for different loadings, column length and steel grades. The yield stress varies between 235 and 690 MPa. The optimization is made using the Generalized Reduced Gradient method in the Excel Solver. Comparisons show the most economic structure.

1   INTRODUCTION

Stability is one of the most important problems in the design of welded metal structures since the instability causes in many cases failure or collapse of the structures.

The normal stresses, overall and local stability are calculated at compressed columns. These are the constraints of the optimization. The unknowns of the I-cross section to be optimized are the dimensions of the web and flanges. There can be several design rules or standards to follow, such as the Eurocode 3, the Japan Railroad Association (JRA), the American Petroleum Institute (API), and the American Institute of Steel Construction (AISC), which approach is the same, but there are differences in details. Changing the compression force, the column length and steel grades, one can get a number of results to compare them and to get a better view of the problem solving. There are several optimization techniques are available. In this case we have used the Generalized Reduced Gradient (GRG2) method, which is built in the Excel Solver. This method was found more reliable, and easy to use tool for those, who are not familiar with programming.

2   DESIGN RULES ACCORDING TO DIFFERENT STANDARDS

The strut should be checked for overall buckling in compression, like Euler (1776) has established rule for that and Ayrton-Perry (1886) modified it. The general rule is the following

$$N \le \frac{\chi A f_y}{\gamma_{M1}} \tag{1}$$

Buckling parameter χ is calculated according to Eurocode 3 (2005) on the following way

$$\chi = \frac{\phi - \sqrt{\phi^2 - \overline{\lambda}^2}}{\overline{\lambda}^2} = \frac{1}{\phi + \sqrt{\phi^2 - \overline{\lambda}^2}} \tag{2}$$

$$\phi = 0.5\left(1 + \eta_b + \bar{\lambda}^2\right) \text{ and } \eta_b = \alpha\left(\bar{\lambda} - 0.2\right) \tag{3}$$

$$\bar{\lambda} = \lambda / \lambda_E; \lambda_E = \pi\sqrt{E / f_y} \tag{4}$$

There exist other column curves used in other countries which can be applied instead of EC3 curves. Their approach is the same as EC3 for compression, but there are differences in details. The Japan Railroad Association (JRA) (2012) curve is described by the following formulae

$$\chi = 1 \text{ for } \bar{\lambda} \leq 0.2$$

$$\chi = 1.109 - 0.545\bar{\lambda} \text{ for } 0.2 \leq \bar{\lambda} \leq 1 \tag{5}$$

$$\chi = 1/\left(0.773 + \bar{\lambda}^2\right) \text{ for } \bar{\lambda} \geq 1$$

The curve of the American Petroleum Institute (API) (2004) is defined by

$$\chi = 1 - 0.25\bar{\lambda}^2 \text{ for } 0 \leq \bar{\lambda} \leq 1.41 \tag{6}$$

$$\chi = 1/\bar{\lambda}^2 \text{ for } \bar{\lambda} \geq 1.41$$

The curve of the American Institute of Steel Construction (AISC) (2013) mainly for round tubes is given by

$$\chi = 1 - 0.091\bar{\lambda} - 0.22\bar{\lambda}^2 \text{ for } \bar{\lambda} \leq 1.41 \tag{7}$$

$$\chi = 0.015 + 0.834/\bar{\lambda}^2 \text{ for } \bar{\lambda} \geq 1.41$$

In this case we used these rules to make comparisons between the optima calculated by them.

2.1 Constraints

Constraint of overall buckling is according to Equation (1). The buckling parameters are in Equations (2–7). Constraint on local buckling of web for the I-column

$$\frac{h}{t_w} \leq \frac{1}{\beta}; \text{ or } t_w \geq \beta h \tag{8}$$

where

$$1/\beta = 42\varepsilon; \varepsilon = \sqrt{\frac{235}{f_y}} \tag{9}$$

Constraint for local buckling of compressed flange of I-column

$$\frac{b}{t_f} \leq \frac{1}{\delta} = 28\varepsilon \text{ or } t_f \geq \delta l \tag{10}$$

2.2 Objective function

The objective function is the cross-section area, it is proportional to the mass of the column, when we know the length and the density.

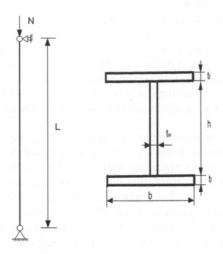

Figure 1.   The welded box and I-column.

$$A = h\, t_w + 2\, b\, t_f \tag{11}$$

2.3   *Unknowns to be optimized*

The four sizes of the cross section are the unknowns

$$h, t_w, b, t_f \tag{12}$$

There are lower and upper limits for the unknowns. The minimum thickness is 5 mm, the maximum is 20 mm. The minimum height and width are 80 mm, the maximum is 350 mm. The height is greater or equal to the width. At the I-column the overall buckling around both axes is considered.

2.4   *The optimum design procedure*

The Microsoft Excel Solver uses the Generalized Reduced Gradient (GRG2) algorithm for optimization in case of nonlinear problems. The algorithm is developed by Leon Lasdon (University of Texas at Austin) and Allen Warren (Cleveland State University). The basic concept of the method is, that it searches the solution with the expansion of Taylor-series besides non-linear criteria. The reduced gradient method separates two specified subsets of the variable, a fundamental and a non-fundamental part. The effective method searches the unconditional NLP problems with approximation. The process is repeated, until the optimization criterion is fulfilled (Hong-Tau et al. 2004).

2.5   *Design data*

Compression force $N = 20$–150 [kN], column length $L = 2$–10 [m], Yield stress $f_y = 235, 355, 460, 690$ [MPa]. Design rule EC3 = Eurocode 3, JRA = Japan Road Association, API = American Petroleum Institute, AISC = American Institute for Steel Construction. Figure 1 shows the compressed column and the cross-section.

3   OPTIMUM RESULTS, COMPARISONS OF THE WELDED I-COLUMN

A great amount of calculations has been made. The effective length factor $K = 1$, if the end conditions are pinned-pinned, $K = 0.7$, if the end conditions are pinned-fixed, $K = 0.5$, if the end conditions are fixed-fixed. Figure 2 shows the optimum cross-sectional areas [mm²] in the function of the compression force [N] for the welded I-column: $K = 0,7$; $L = 4$ m; $f_y = 460$ MPa; loading is changing. It shows that the API is the most liberal; EC3 and JRA are the most conservative rules. AISC is somewhere between them.

Figure 3 shows the optimum cross-sectional areas [mm²] in the function of the compression force [N] for the welded I-column: $K = 0,7$; $L = 6$ m; $f_y = 460$ MPa; loading is changing. The length is larger. There is a jump in the cross section at around 8000 N. The reason is the rounding effect.

Figure 4 shows the optimum cross-sectional areas [mm²] in the function of the yield stress [MPa] for the welded I-column: $K = 0,7$; $L = 10$ m; $N = 1500000$ N. This is a surprising result. It shows the applicability of the higher strength steels. Due to the local buckling effect, one cannot get smaller cross section, increasing the yield stress up to 690 MPa. The order of the different rules is similar to the Figure 6, where the compression force changed.

Figure 5 shows the optimum cross-sectional areas [mm²] in the function of the length [m] for the welded I-column: $K = 0,7$; $N = 85000$ N; $f_y = 235$ MPa. The results are similar to Figure 3, where there are some differences at 9 m length. The reason is also the rounding effect.

When we consider not only the mass, the material cost, but the welding, plate cutting and painting costs, using CO_2 welding, propane gas cutting and two layers painting, then we get additional information (Farkas & Jármai 2015). We have calculated the differences in the cost [$] in the function of the length and design rules, when the compression force is $N = 85$ kN.

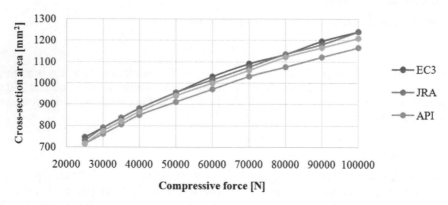

Figure 2. Optimum cross section areas [mm²] in the function of the compression force [N] for I-column, $L = 4$ m.

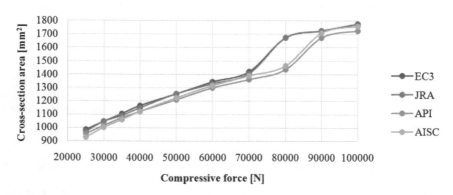

Figure 3. Optimum cross section areas [mm²] in the function of the compression force [N] for I-column, $L = 6$ m.

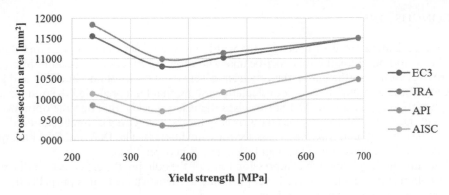

Figure 4. Optimum cross section areas [mm²] in the function of the yield stress [MPa] for I-column, $L = 10$ m; $N = 1500000$ N.

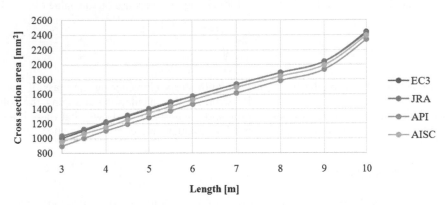

Figure 5. Optimum cross section areas [mm²] in the function of the length [m] for I-column, $N = 85000$ N; $f_y = 235$ MPa.

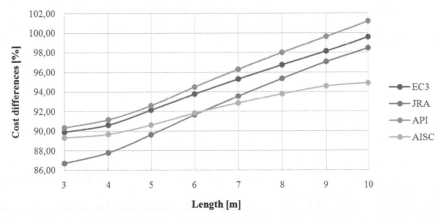

Figure 6. The differences in the costs [$] in the function of the length and design rules, when the compression force is $N = 85$ kN.

The value is calculated considering the cross sections for the largest and smallest steel grade and how much percent is the difference, when the compression force is given to be 85 kN $(A_{S690} - A_{S235}) \times 100\%$. Figure 6 shows the differences for the four design rules. The AISC has the smallest difference and the JRA has the largest.

4 CONCUSIONS

In the optimization process the height, the width and the thickness of the web and flanges have been optimized for the welded I-beams. The objective function to be minimized was the cross-section area (the mass), the constraints were the overall column buckling and local buckling of the web and flanges. The calculations show, that for a predesign the optimization is applicable and very useful. The Solver gives the result quickly. Building the optimization system in the Solver environment is relatively easy for the user. Using different standards and design rules the optimum sizes and cross sections are different. The JRA and EC3 look more conservative, the API is more liberal and the AISC is between them.

Changing the length and the compression force of the column, the cross section (the mass of the column) it is less than linearly proportional to them. The reason is probably that the local buckling does not depend on them.

The yield stress has a special effect. When increasing the yield stress from 235 MPa up to 690 MPa, first the cross sections decrease at 355 MPa, but later increase. The reason is also the local buckling constraint and the overall buckling one, because of the value of

$$\varepsilon = \sqrt{\frac{235}{f_y}},$$

the limit slenderness is smaller for higher strength steels. Considering the mass of the steel one can say, that the best is the 355 MPa steel. At the calculations changing the column length, the compression force, the steel grade and the design rule shows, that there is a complex behaviour of the structure and it is worth to evaluate and compare the alternatives.

ACKNOWLEDGEMENTS

The described article was carried out as part of the EFOP-3.6.1-16-2016-00011 "Younger and Renewing University—Innovative Knowledge City—institutional development of the University of Miskolc aiming at intelligent specialisation" project implemented in the framework of the Széchenyi 2020 program. The realization of this project is supported by the European Union, co-financed by the European Social Fund.

With this article we remember to Prof. József Farkas, who has passed away on the 15th of October 2016 in his age 89.

REFERENCES

American Institute of Steel Construction (2013), AISC, Design Guide 28: Stability Design of Steel Buildings.
American Petroleum Institute (2004), API Bulletin 2V, Design of Flat Plate Structures, Third Edition.
Ayrton, W.E., Perry, J. (1886), On struts. The Engineer, 62, p. 464.
Euler, L. (1776), Determinatio onerum, quae columnae gestare valent. Acta Acad. Sci. Petrop. 2.
Eurocode 3. (2005) Part 1.1. Design of steel structures. General rules and rules for buildings. European Committee for Standardization. Brussels.
Farkas, J., Jármai, K. (2013), Optimum design of steel structures. Springer Verlag, Berlin, Heidelberg
Farkas, J., Jármai, K. (2015), Design of innovative metal structures, Gazdász-Elasztik Kiadó és Nyomda, p. 592. (in Hungarian).
Hong-Tau Lee, Sheu-Hua Chen, He-Yau Kang (2004), A Study of Generalized Reduced Gradient Method with Different Search Directions, Measurement Management Journal, Vol. 1, No. 1, pp. 25–38.
Japan Road Association (2012), JRA, Specifications for Highway Bridges, Part I_V.

Advances and Trends in Engineering Sciences and Technologies III – Al Ali & Platko (Eds)
© 2019 Taylor & Francis Group, London, ISBN 978-0-367-07509-5

Tensfoamity—a new concept of foam integrated beams

K. Jeleniewicz
Warsaw University of Life Sciences, Warsaw, Poland

W. Gilewski
Warsaw University of Technology, Warsaw, Poland

ABSTRACT: The present paper is dedicated to the new concept of extra light beams named Tens-Foam-Ity. The idea originates within the well-known concepts of Tensegrity and Tensairity. Bending stiffness of the light timber beams is extended and stabilized via the veneer box filled with foam and covered by cables. Medium scale experimental results, made at the Warsaw University of Life Sciences, are presented and discussed. The simple computational model, proposed in the paper, is based on the Timoshenko beam theory with using the finite element method. The stiffness, geometric stiffness and mass matrices are based on the so-called physical shape functions to obtain high precision efficient formulation. Numerical results are calibrated and compared to experiments.

1 INTRODUCTION

The idea of pneumatic-cable beams is to reinforce inflated beams with cables and struts in order to improve the load bearing capacity of the system (Luchsinger et al. (2004), Pedretti (2004), Luchsinger and Crettol (2006) and de Laet et al. (2008)). The concept was named Tensairity as an acronym for tension, air and integrity. The first Tensairity construction was built in 2002 and it was a demonstration car bridge with a span of 20 m (Biernacka and Gilewski (2018)). In the coming years, the technology was continuously improved. The most famous construction is the roof of the parking garage in Montreux, which was built in 2004 and shows the full potential of the Tensairity technology (Biernacka and Gilewski (2018)). Another example is the skier's bridge in France with its 52 m span (Biernacka and Gilewski (2018)). Basic form of Tensairity beam consists of: an air-beam, a compression element, which is tightly connected with the air-beam, two cables running in helical form around the air-beam, both ends of which are connected to the compression element. In a typical beam the top is in compression and the bottom—intension. In order to reduce the weight of the beam in the

Figure 1. TensFoamIty beam.

case of Tensairity technology an inflated tube, covered by cables, is proposed. The air tube stabilizes the compression element and transports the load between compression and tension elements. The mechanics of a Tensairity beam is a mix between the beam theory and the membrane theory.

The use of a pneumatic balloon is structurally cumbersome and their construction requires the use of special textile materials. In the literature on the subject, we can find examples of using polyurethane foam as a filling for sandwich constructions (Linul and Marsavina (2015), Rizov and Mladensky (2007) and Styles et al. (2008)). The foam has good mechanical properties in compression but does not exhibit shape stability. The attempt to stabilize the shape with thin veneer sheets (Jeleniewicz and Gilewski (2018)) was promising, however, local destruction of beams in bending was observed and the full bearing capacity of the structure was not used.

The present paper is dedicated to analysis of alight timber beam resting on a foam foundation stabilized by two sheets of veneer (Figure 1). To ensure the cooperation between the wooden beam and the foam foundation, a band with a cable is used, as in the Tensairity concept.

The timber-foam/veneer-cable beam is a combination of the concept of Tensairity with good mechanical properties of the polyurethane foam. The proposed name of the concept is TensFoamIty. A simple moderately thick beam finite element model is proposed. The concept of the so-called physical shape functions is applied (see. Gilewski (2013) for details). The parameters of the model are calibrated with the use of experimental analysis in a medium laboratory scale.

2 EXPERIMENTAL INVESTIGATION

The basis for the analysis in this work are experimental studies carried out at the Water Center Laboratory at Warsaw University of Life Sciences. Two types of beams were tested—a rectangular wooden cross-section (Figure 2a) and a wooden cross-section based on a veneer thin-walled profile filled with polyurethane foam (Figure 2b).

Figure 3 shows the section of the beam profile, and its top and side views. The wooden beam rests on the foam rather than on the veneer profile (Figure 3a). Plastic strips at the ends of the beams are designed to prevent the foam profile from detaching from the wooden

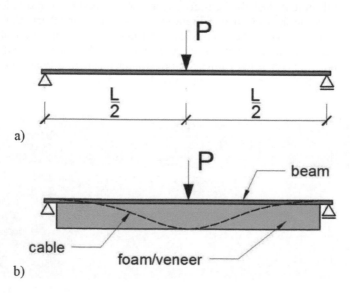

Figure 2. Beam models: Simple timber beam (a) and TensAirIty beam (b).

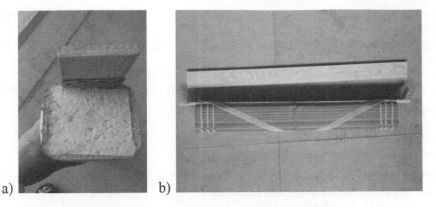

a)　　　　　　　　b)

Figure 3.　Cross-section (a) and top/side view (b) of the TensFoamIty beam.

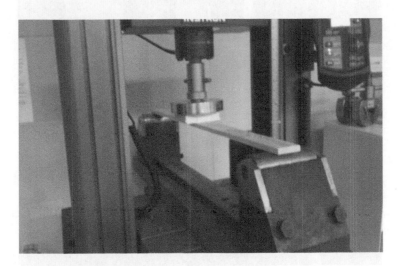

Figure 4.　Simple timber beam experiment.

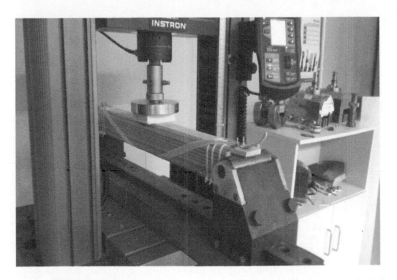

Figure 5.　TensFoamIty beam experiment.

beam and ensuring the cooperation of wooden beams and foam. The stabilizing cable is a webbing tape.

Figure 4 shows the simple beam in the Instron testing machine, and in Figure 5 the analogue TensFoamIty beam.

The damage of the proposed beam is illustrated in Figure 6.

A number of experimental bending tests has been performed with the following data:

– Length of the beam between supports – 0.78 m,
– Height of a timber beam – 0.01 m,
– Width of a timber beam – 0.042 m,
– Two glued veneer sheets of the thickness 0.0012 m.

Typical average results are presented in Figure 7. The bottom line is for the simple beam and the upper line for the TensFoamIty beam.

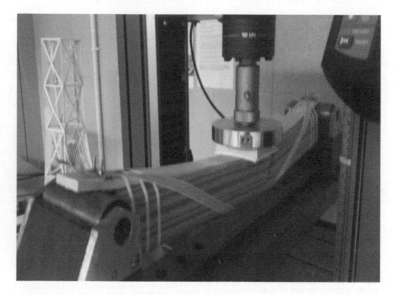

Figure 6.   Damage of TensFoamIty beam.

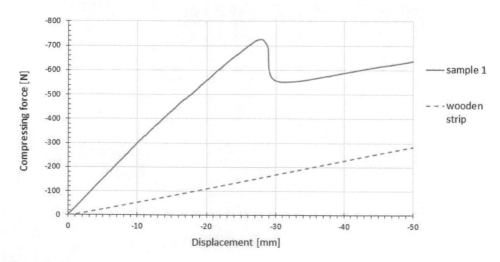

Figure 7.   The average force-displacement curves.

128

The dependence between the displacement and the level of force in the simple beam is linear in the considered range. For TensFoamIty beam, in the force range up to more than 700 N, is almost linear. Then, there is a breakdown and loss of load capacity, after which the reinforcement of the beam is observed. The load-bearing capacity of the TensFoamIty beam is more than 4 times greater than that of a simple timber beam.

3 NUMERICAL MODEL

A simple two-noded beam finite element model with high precision physical shape functions is proposed for the analysis of the pneumatic-cable structure. Physical shape functions are obtained from the static homogeneous displacement equations of the Timoshenko beam theory and are exact for static analysis (for detailed description see Gilewski (2013), Gilewski and Obara (2007)). The details of the finite element are the following: length – $2a$, cross-section – A, moment of inertia – J, Young's modulus – E, Poisson's ratio – v, elastic foundation coefficient – k_0, axial force – S and $H = 5EA/12(1-v)$, $\gamma = EJ/Ha^2$, $\mu = 3\gamma/(1+3\gamma)$.

Stiffness, geometric stiffness, as well as elastic foundation matrices and load vector respectively, are used for the analysis on the finite element level in the following form:

$$\mathbf{K}^e = \frac{3EJ}{2a^3}\begin{bmatrix} 1-\mu & a(1-\mu) & -1+\mu & a(1-\mu) \\ a(1-\mu) & a^2\left(\frac{4}{3}-\mu\right) & -a(1-\mu) & a^2\left(\frac{2}{3}-\mu\right) \\ -1+\mu & -a(1-\mu) & 1-\mu & -a(1-\mu) \\ a(1-\mu) & a^2\left(\frac{2}{3}-\mu\right) & -a(1-\mu) & a^2\left(\frac{4}{3}-\mu\right) \end{bmatrix} \tag{1}$$

$$\mathbf{K}_g^e = \frac{S}{10a}\begin{bmatrix} 5+(1-\mu)^2 & a(1-\mu)^2 & -5-(1-\mu)^2 & a(1-\mu)^2 \\ a(1-\mu)^2 & a^2\left[\frac{5}{3}+(1-\mu)^2\right] & -a(1-\mu)^2 & a^2\left[-\frac{5}{3}+(1-\mu)^2\right] \\ -5-(1-\mu)^2 & -a(1-\mu)^2 & 5+(1-\mu)^2 & -a(1-\mu)^2 \\ a(1-\mu)^2 & a^2\left[-\frac{5}{3}+(1-\mu)^2\right] & -a(1-\mu)^2 & a^2\left[\frac{5}{3}+(1-\mu)^2\right] \end{bmatrix} \tag{2}$$

$$\mathbf{K}_s^e = \frac{ak_0}{210}\begin{bmatrix} 156-18\mu+2\mu^2 & a\left(44-11\mu+2\mu^2\right) & 54+18\mu-2\mu^2 & a\left(-26-11\mu+2\mu^2\right) \\ a\left(44-11\mu+2\mu^2\right) & a^2\left(14+2(1-\mu)^2\right) & a\left(26+11\mu-2\mu^2\right) & a^2\left(-14+2(1-\mu)^2\right) \\ 54+18\mu-2\mu^2 & a\left(26+11\mu-2\mu^2\right) & 156-18\mu+2\mu^2 & a\left(-44+11\mu-2\mu^2\right) \\ a\left(-26-11\mu+2\mu^2\right) & a^2\left(-14+2(1-\mu)^2\right) & a\left(-44+11\mu-2\mu^2\right) & a^2\left(14+2(1-\mu)^2\right) \end{bmatrix} \tag{3}$$

$$\mathbf{Q}^e = \left[qa \quad \frac{1}{3}qa^2 \quad qa \quad -\frac{1}{3}qa^2\right]^T \tag{4}$$

Standard finite element procedures are applied to the global analysis. In the literature of pneumatic-cable structures one can find much more complicated finite element models, usually with the use of commercial FE systems (see Pedretti et al. (2004), Luchsinger and Cretol (2006)).

The finite element model of the beams presented in Figure 2 is very simple and consist of 2 finite elements and 4 degrees of freedom The global matrix equilibrium equation is the following

$$(\mathbf{K}-\mathbf{K}_g+\mathbf{K}_s)\mathbf{q}=\mathbf{Q}. \tag{5}$$

For a simple beam only the stiffness matrix **K** is considered. Comparison of the experimental results and computational model provide for the calibration of the Young modulus on the level of 15800 MPa.

If TensFoamIty beam is considered the axial force is calculated after Tensairity theory (see Biernacka and Gilewski (2018) for details) with the formulae $S = PL/4g$, where g is the height of the foam/veneer beam, which is 0.07 m in the experimental tests. The second test allow to calibrate the stiffness k_0 of the elastic foam/veneer foundation on the level of 93 kPa.

The finite element results with the calibrated Young modulus and stiffness of foundation, provide exactly for the results obtained from the experimental tests. This calibration allows to estimate the load capacity of the TensFoamIty beams with various boundary conditions and dimensions. More precise analysis needs more experimental tests, but the estimated results are promising for future applications in medium size timber plates with foam beams.

4 CONCLUSIONS

A new concept of the light-weight timber beams, stiffened by a foam/veneer beam is proposed. The preliminary results are promising. Medium scale experimental tests were done and a simple finite element beam bending model was introduced to calibrate the computational parameters. The load-bearing capacity of the TensFoamIty beam is more than 4 times greater than that of a simple beam. Computational estimation of the load capacity of TensFoamIty beams with various boundary conditions in a bigger range of dimensions lead to the conclusion that they can be applied in real civil engineering structures.

REFERENCES

Biernacka, M. & Gilewski, W. 2018. An overview of pneumatic-cable structures for civil engineering applications. *Theoretical Foundations of Civil Engineering VIII*, Warsaw University of Technology Publishing House, Warsaw, Poland, 131–150.

De Laet, L. et al. 2008. Deployable Tensairity Structures. *Proceedings of the International Symposium on New Materials and Technologies, New Designs and Innovations (IASS)*, Acapulco, Mexico.

Gilewski, W. 2013. *Physical shape functions in the finite element method*, Polish Academy of Sciences, Committee of Civil Engineering, Studies in Civil Engineering, Warsaw (in Polish).

Gilewski, W. & Obara, P. 2007. Exact and high precision stiffness matrices for Timoshenko beam resting on elastic foundation. *6th International Conference on New Trends in Statics and Dynamics of Buildings*, Bratislava, 69–72 (in Polish).

Jeleniewicz, K. & Gilewski, W. 2018. Medium scale experimental investigation of bending for thin-walled veneer beams with foam filling. *World Multidisciplinary Civil Engineering-Architecture-Urban Planning Symposium*, 18–22 June, Prague, Czech Republic.

Linul, E. & Marsavina, L. 2015. Assessment of sandwich beams with rigid polyurethane foam core using failure-mode maps. *Proceedings of the Romanian Academy, Series A*, 16: 522–530.

Luchsinger, R.H. & Crettol, R. 2006. Adaptable Tensairity. *International Conference On Adaptable Building Structures*, Eindhoven, Netherlands.

Luchsinger, R.H. Pedretti, A., Pedretti, M. & Steingruber, P. 2004. The new structural concept Tensairity: Basic Principles. *Progress in Structural Engineering, Mechanics and Computation*, Taylor & Francis Group, 1–5.

Pedretti, A. Steingruber, P., Pedretti, M. & Luchsinger, R.H. 2004. The new structural concept Tensairity: FE-modeling and applications. *Proceedings of the Second International Conference on Structural Engineering, Mechanics and Computation*, Lisse, Netherlands.

Pedretti, M. 2004. Tensairity. ECCOMAS 2004. *European Congress on Computational Methods in Applied Sciences and Engineering*, Jyväskylä, Finland.

Rizov, V. & Mladensky, A. 2007. Influence of the foam core material on the indentation behavior of sandwich composite panels. *Cellular Polymers*, 26: 117–131.

Styles, M., Compston, P. & Kalyanasundaram, S. 2008. Finite element modelling of core thickness effects in aluminium foam/composite sandwich structures under flexural loading. *Composite structures*, 86: 227–232.

Local buckling of thin-walled steel tubes

R. Kanishchev & V. Kvocak
Faculty of Civil Engineering, Institute of Structural Engineering, Technical University of Košice, Košice, Slovakia

ABSTRACT: The article deals with theoretical and experimental analysis of the local stability of short axially compressed rectangular thin-walled steel tubes with imperfections. This analysis is one part of the research focusing on local stability of rectangular concrete-filled thin-walled steel tubes. The work presents a theory of elastic critical stress for local buckling of right-angled wall elements and acting under the influence of a uniform compression load. It was also mentioned the preparation and procedure of experimental tests of the researched samples. The data of this research supplement the results of the analysis of local buckling of the axially compressed rectangular steel tubes filled with concrete.

1 INTRODUCTION

Requirements concerning economy and effectiveness of designed structures force designers to use progressive load-bearing structures. Composite columns made of rectangular concrete-filled hollow sections are definitely regarded as very cost-effective as they enable very fast construction and offer all the advantages of both materials—concrete and steel. These elements have distinct advantages over hollow steel tubes as described in the research works of Storozhenko et al. (2009, 2014), Duvanova & Salmanov (2014), Kanishchev (2016). However, one of the main structural advantages is its significant resistance to loss of local and global stability, which allows designers to reduce the steel part of the composite cross-section from compact class to slender class (use the thin-walled cold-formed sections).

Theory and design development of steel thin-walled cold-formed members and profiles creates a certain knowledge base for their practical application in civil engineering. However, this fact does not mean that all complex and challenging processes of their behaviour, during strain, transformation and failure are sufficiently investigated. From the material and geometric point of view, the thin-walled cold-formed profiles have specific specialties, which must take into account in their design. From this aspect, the local stability requirements related to unfavorable buckling effects of their compressed parts are very significant. Favorable effects, related to membrane stresses and post-critical behavior are also important. Different calculation procedures with different results in relevant standard (European standard EN 1993-1-3 2006) and their confrontation with experimental results (Al Ali et al. 2012) indicate the need for further investigation of post-critical and elastic-plastic behavior of these members.

In above mentioned research the thin-walled cold formed welded rectangular hollow sections RHS 200/100/3 (according to EN 10219-1 2006) were used. Although these profiles were cold-formed, EN 1993-1-3 (2006) does not apply to the design of such cross-sections. This standard specifies that EN 1993-1-1 (2005) should be used for this purpose even though this cross-section can be classified into a group of thin-walled elements.

2 STABILITY OF RECTANGULAR WALLS

The subject matter of local buckling of slender compressed walls was intensively researched by Timoshenko (1971), where he represented a differential equation for a slender wall with a length a, and width b (Figure 1), which is simply supported around its perimeter:

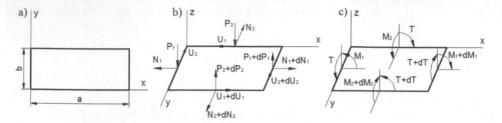

Figure 1.   Mathematical model of slender wall: a) rectangular wall; b) components force in a unit element of the wall; c) components of the bending moment in a unit element of the wall.

$$C\left(\frac{\partial^4 w}{\partial x^4} + 2\frac{\partial^4 w}{\partial x^2 \partial y^2} + \frac{\partial^4 w}{\partial y^4}\right) + N\frac{\partial^2 w}{\partial x^2} = 0 \qquad (1)$$

where w = deflection of slender walls; P = compression force; and C = cylindrical wall stiffness. A particular solution of the differential equation:

$$w = A\sin\frac{m\pi x}{a}\sin\frac{n\pi y}{b} \qquad (2)$$

The given solution satisfies the boundary conditions (Figure 1a, b): for $x = 0$ and $x = a \rightarrow$ $w = 0$ and $M_1 = 0$; for $y = 0$ and $y = b \rightarrow w = 0$ and $M_2 = 0$. The following conditions are fulfilled for stress: $N_1 = -N$; $U_1 = U_2 = N_2 = 0$. Thus, the value of elastic critical stress of the wall:

$$\sigma_{cr} = k_\sigma \sigma_E = \left(m\frac{b}{a} + \frac{1}{m}\frac{a}{b}\right)^2 \frac{\pi^2 E t^2}{12(1 - v^2)b^2} \qquad (3)$$

where k_σ = coefficient of critical stress; E = modulus of elasticity of the steel; t = the wall thickness; v = Poisson ratio.

The basic principles of designing class 4 cross-sections were stipulated by Bryan (1891), which offers a critical analysis of the elastic stress σ_{cr} for local buckling of long right-angled wall elements. The term of this stress includes various boundary conditions with the aid of the coefficient of critical stress k_σ:

$$\sigma_{cr} = k_\sigma \frac{\pi^2 E}{12(1 - v^2)}\left(\frac{t}{b}\right)^2 \qquad (4)$$

The minimum values of the coefficient of critical stress k_σ are stipulated in EN 1993-1-5 (2006). This coefficient can be used for hollow rectangular tubes. When the tube is filled with concrete, the standard does not provide a k_σ value.

3   EXPERIMENTAL TESTS OF THE RESEARCHED SAMPLES

3.1   *Material characteristics*

For determine the material characteristics of the steel used in the experimental elements, tensile tests were performed. From the rectangular thin-walled profiles supplied by the manufacturer, elements for the production of samples according to EN 10219-2 (2006) were cut. From these elements, 15 pieces of steel samples were cut, which were used in the tensile test (see Figure 2) and tested according to the standard requirements of ISO 6892-1 (2016). The average tensile test results are shown in Table 1.

3.2 *Preparation of the research samples*

For the purpose of experimental research, 1 types of short samples with and without welded front plates with dimensions of 220/120/6 mm (see Figure 3), were designed. The total number of samples is 6 pieces. Each sample was milled to align the faces. The experimental specimens without end plates were named as ST-1, ST-2, ST-3; and with welded front plates – STp-1, STp-2, STp-3. For better clarity of deformation and for measuring of geometric imperfections on each side of the elements a grid was drawn (Figure 3c). The measurement of geometric imperfections is showed in the Figures 4a, 5.

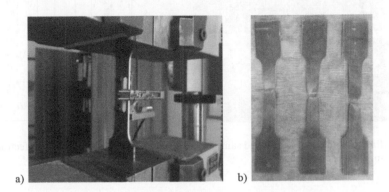

a)   b)

Figure 2. Material test of steel: a) test machine; b) samples after the tensile test.

Table 1. Steel tensile test results.

F_m [kN]	R_E [MPa]	$R_{p0.2}$ [MPa]	R_m [MPa]	E [GPa]
23.71	241.86	374.89	425.56	198.31

Note: F_m – maximum force; R_E – limit of elasticity; $R_{p0.2}$ – proof yield strength; R_m – tensile strength; E – modulus of elasticity.

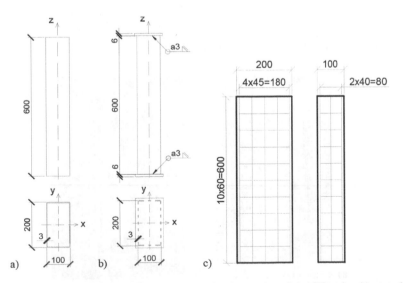

Figure 3. Types of samples for experimental research: a) samples of the ST series; b) samples of the STp series; c) grid of investigated elements.

3.3 Experimental tests of the researched samples

The stress of the walls of steel tubes specimens was measured using strain gauges placed in the middle of the broad sides along their width (Figure 4b). Step of load was 20 kN to collapse of the samples. Before the test specimens were painted in white and on the wall of cross-section was drawn grid with a mesh size of 45×60 mm. The results of the experimental tests are given in chapter 4.

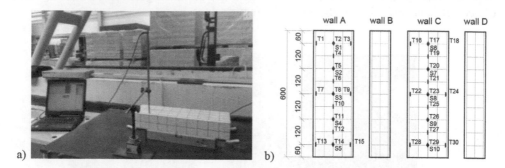

Figure 4. Preparation of the researched samples: a) measurement of geometric imperfections; b) location of strain gauges and sensors.

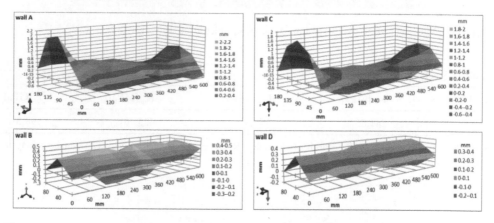

Figure 5. Geometric imperfections of the sample ST-1.

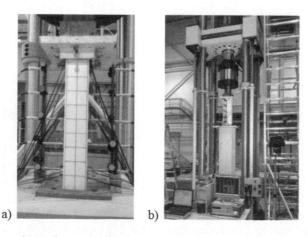

Figure 6. Experimental test of researched samples.

4 RESULTS OF THE RESEARCH

The results of the above mentioned experiments are shown in Figures 7–9. The minimum theoretical value for the coefficient of critical stress k_σ equals 4 for hollow sections (see Figure 7).

This value is given in EN 1993-1-5 (2006). The coefficients of critical stress, which were obtained from the experiment, were expressed by relationship (4). The minimum difference between theoretical and experimental values is 1.5% (specimens ST-2 and STp-1). As shown on the Figures 7–9, the welding of the end plates in samples STp group does not have a significant influence on the research results

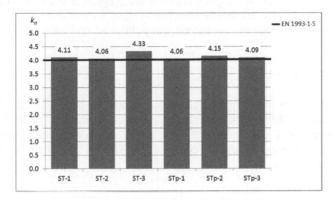

Figure 7. Comparison of the coefficient of critical stress.

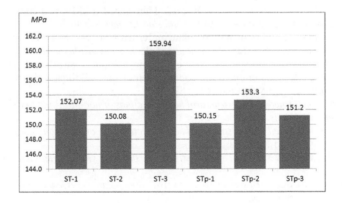

Figure 8. Comparison of the critical stress.

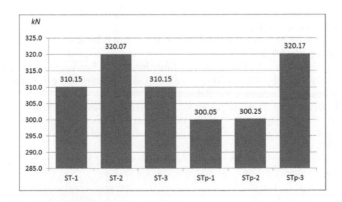

Figure 9. Comparison of the force under critical stress.

135

5 CONCLUSIONS

The paper mentioned above focuses on one part of the research, which was carried out at the Institute of Structural Engineering of the Civil Engineering Faculty of the Technical University in Kosice. The program of the research concerns the theoretical and experimental analysis of local and global stability of composite columns in the form of rectangular concrete-filled steel tubes with slender steel part of the composite cross-section. For comparison of behavior of hollow thin-walled cold formed rectangular tubes, were carried out the above mentioned experimental research. After the completion of all experiments, the members will be modelled using the numerical computational-graphics software **ABAQUS** whose results will be compared with the experimentally attained values. Based on such extensive findings, recommendations for design of these structures for the general public and professionals will be made.

ACKNOWLEDGEMENT

The paper presented was supported by the projects: VEGA 1/0188/16 "Static and Fatigue Resistance of Joints and Members of Steel and Composite Structures" of the Scientific Grant Agency of the Ministry of Education, science, research and sport of the Slovak Republic and the Slovak Academy of Sciences and the contract No. APVV-15-0486 of the Slovak Research and Development Agency.

REFERENCES

Al Ali, M., Tomko, M., Demian, I. Kvočák, V. 2012. Thin-walled cold-formed compressed steel members and the problem of initial imperfections. *Procedia Engineering* 40: 8–13.

Bryan, G.H. 1891. On the stability of a plane plate under trusts in its own plane, with applications to the "bukcling" of the sides of a ship. *Proceedings of the London Mathematical Society* 22: 54–67.

Duvanova, I.A. & Salmanov, I.D. 2014. Trubobetonnie kolonni v stroitelstve visotnych zdanii i sooruzhenii [Pipe-concrete columns in the construction of tall buildings and structures]. *Construction of Unique Buildings and Structures* 6(21): 89–103.

EN 10219-1. 2006. *Cold formed welded structural hollow sections of non-alloy and fine grain steels. Part 1: Technical delivery conditions*. Brussels: CEN.

EN 10219-2. 2006. *Cold formed welded structural hollow sections of non-alloy and fine grain steels. Part 2: Technical delivery conditions*. Brussels: CEN.

EN 1993-1-1. 2005. Eurocode 3. *Design of steel structures. Part 1-1: General rules and rules for buildings*. Brussels: CEN.

EN 1993-1-3. 2006. Eurocode 3. *Design of concrete structures—Part 1-3: General rules—Supplementary rules for cold-formed members and sheeting*. Brussels: CEN.

EN 1993-1-5. 2006. Eurocode 3. *Design of steel structures. Part 1-5: General rules—Plated structural elements*. Brussels: CEN.

ISO 6892-1. 2016. *Metallic materials—Tensile testing—Part 1: Method of test at room temperature*. The International Organization for Standardization.

Kanishchev, R.A. 2016. Analysis of local stability of the rectangular tubes filled with concrete. *Magazine of Civil Engineering* 64(4): 59–68.

Storozhenko, L., Onishchenko, P., Pichugin, S., Onishchenko, V., Semko, O., Emelanova, I., Landar, O. 2009. *Visokoefektivni tehnologii ta komplexni konstrukcii v budivnictvi [High-efficient technologies and complex structures in construction]*. Poltava: PF "Formika".

Storozhenko, L.I., Ermolenko, D.A., Demchenko, O.V. 2014. Rabota pod nagruzkoi szhatych trubobetonnych elementov s usilennymi jadrami [The work under load of compressed pipe-concrete elements with strengthened cores]. *Efficiency of resource energy of technology in the construction industry of region* 4: 288–292.

Timoshenko, S.P. 1971. *Ustojchivost sterghnej, plastin i obolochek [Stability of rods, plates and shells]* Moskva: Nauka.

Advances and Trends in Engineering Sciences and Technologies III – Al Ali & Platko (Eds)
© 2019 Taylor & Francis Group, London, ISBN 978-0-367-07509-5

Optimisation while designing portal frames

G. Kászonyi & K. Jármai
University of Miskolc, Hungary

ABSTRACT: The optimisation in this study is shown on a sway frame structure made of welded I-section members. Structural stress, stability constraints, frame strength, and load-bearing capacities were all considered. The load-bearing capacity of the structure was maximised—using a FEM (Final Element Method) AXIS package for the simulation. Further we carried out test calculations using MathCAD, where our results examined stresses on a welded beam-to-column connection. It was found that significant material could be saved this way. Further development to extend calculations will be using different steel grades and different semi-rigid beam-to-column connections.

1 INTRODUCTION

Our objective is an in-situ built steel frame optimisation. The connecting point is welded, the structure consists of welded I-section members. The exact measures and positions are defined in the design specifications. We intended to reach final parameters defined in the specs.

The task is an optimisation of a frame by means of changing its bolted connections to welded joins; This is partly why welded I-sections have been used instead of rolled sections. The hall frame was designed using the FEM software. After that we checked results using hand calculation. Finally, we discuss the financial advantages of this structural solution. In the calculation, the load-model and the load-combinations are carried outwith AXIS VM 14 software. We checked the calculations using MathCad Pro 14 software. The connected point-measuring calculations at two points (column-beam and beam-beam) were done with IDEA STATICA software.

The details of the designed frame are as follows: single-storey frame (not including built-in crane), non-compound structure, welded I-section members—the final solution uses non-series profiles. The frame is an angle-line girder, and the ground connection is non-rigid. The other connections are a sequence of rigid joins.

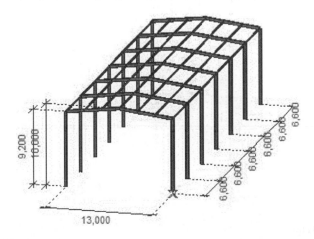

Figure 1. The hall structure.

2 GEOMETRICAL STRUCTURE—ON THE COMPLETE BUILDING

Hall ground area: 39.6 m × 13.00 metres ≈ 515 m², eave height is 9.20 metres, ridge height is ≈ 10 metres. Axis distance of frames 6.60 metres each. Axis-distance of spans 13 metres. There is no bracing system employed at the frame, the reason is we do not need (visually) any bracing system, because it is a theoretical two-dimension construction model (frame). We did not describe either lattices or stiffening elements, but, when we built our frame model with software, we took these effects into account.

2.1 Applied materials, used connection elements

Pillars, beams: S 235 base material—made from single individual beams after optimisation. Connections: 1st class corner welding joining 3 or 4 sides (Halász, Platthy 1989).

During optimisation, we reduced costs and materials by allowing the ground-pillar fixing point to be partly flexible. Naturally we experienced some remaining rigidity, but our final hall has semi-rigid footings. Uniformly distributed forces have been considered in vertical and horizontal directions. The our model has 48 different load combinations and we took into account (we count with dead load, permanent loads, technical building system loads, snow and wind loads)—we use as result the relevant combination.

2.2 Frame design—solution

At the design we have considered bending and compression and overall and local stability. The lateral torsional buckling was the first effect for calculating. After optimisation our frame members are: the pillars are individual welded 280 I-section members (length: ≈ 9.2 metres), the frame-beams made of individual welded 320 I-section girders (length: ≈ 6.5 metres). Columns and beams joined by welded connections (Németh 2006).

3 PROCESS OF OPTIMISATION

Our aim for the optimisation was to transform a previous traditional plan design. That plan was a detailed plan with rigid footings.

Our task for optimisation was to minimise structural mass, while we maximised load-bearing capacity—we defined our objective function (Farkas, Jármai 2008).

We simplified our previous plan. First part of our calculation was: rebuilding the "base version", we handled as a starting point rigidity at the fixing points (Farkas, Jármai 2008). As in the first optimisation step we did not consider the built-in crane (Seregi 2011). We

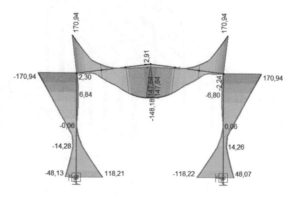

Figure 2. Rigid frame.

138

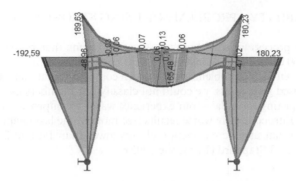

Figure 3. Non-rigid frame.

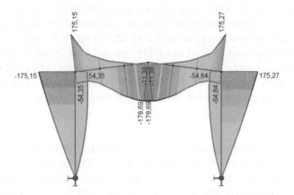

Figure 4. Semi-rigid frame I.

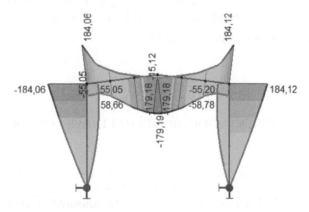

Figure 5. Semi-rigid frame II. – Solution.

transformed the footing from rigid to semi-rigid connection. We illustrated the first model obviously reducing load-bearing—for that we designed further semi-rigid footings.

In further steps for this optimisation (4th version) we built an FEM model with I-section members easily available from all suppliers. We tried to get load as close as possible to 100% of the maximum permitted load.

When the model was ready, we used the special steel-designing algorithm of the AXIS-software.

In the checking calculation we followed the catalogue form-given area-numbers (MathCad-process). Finally—the construction has use level: 91.6% – column; and 79.3% – beams.

4 LOSING STABILITY—PROBLEMS OF USING STRONGER MATERIAL

During the design process we tried to increase use by raising material yield stress—we tried to use S355 or S460 class materials instead of S235 base material (these were the 2nd and 3rd models). We had been facing problems with modern compound materials (Katula et al. 2007).

When we increased yield stress, we could not classify our members as 1st class on account of "yield stress adjusting formula"—our experience was if we improved steel parameters, the material standard Eurocode gave worse results (see more in the last chapter of Seregi 2011).

It is well known that using cross-section classes lower than 1st and 2nd steel material is limited by EuroCode 3 Standard (Eurocode 1993).

5 THE INDIVIDUAL PROFILE—ITERATION

Our aim was to achieve the 100% use of profiles. Our theory was: if we change shape-parameters and decrease these, our construction total-mass will be decreased as well. For that we followed our previous results and created a special product for the frame.

During the design stage we replaced available "ready" profiles with specially created members. We started to build individual welded I-section elements. These elements were 10 mm depth steel plates and we created I-section beams by welding these elements.

We achieved our final goal with this "product-creating" procedure.

5.1 *The results are*

1. The cross-sectional area of the welded beam (individual 320) was increased by 15%, which resulted in mass increasing from 49.1 kg/metre to 56.4 kg/metre.
2. On the other hand, the cross-sectional area of the welded column (individual 280) decreased—compared to the HEA280 the previous model – 26%, which resulted in a very large weight reserve (76.4 kg/metre instead of 57.3 kg/metre).
3. In summary, with growth and reduction, the weight of the frame has decreased 8% in total. In addition to maximising use, the available optimum-mass is fulfilled.

At the end of optimisations, the structure use is:

– Beam: 90.2%
– Column: 98.3%
– Achieved weight reduction: 8%.

6 THE CHARACTERISTIC NODE OF THE HALL (PART OF SOLUTION)

The connecting points:

– Footing-column connection. We analysed this fixing – according to the non-rigid model – as a semi-rigid connection, after that we designed this node with screws.
– "Column-beam" and "Beam-beam" connection. We measured these nodes for maximum shear and bending with welding.

The process of welding:

– MAG welding process, according the technological standard – a = 4 mm welding "root"

We use on the 2nd node fixing point a "short steel-pinch". On the 3rd node we designed 10 mm thickness front plate – from the same material, as the beam (Seregi 2011).

We designed the welds with 135 processing, in-situ version on the construction site, especially with "overmatching" measurement. We employed a tight 1.2 mm wire (type: G 46 3 M21 G3Si1), the protecting gas type is CORGON 18. We prescribe U = 18 V voltage and I = 120 A current.

IDEA STATICA – modelling with design software. Below we introduce theoretical preparatory welding on the 2nd and 3rd connecting point. The graphs are realistic axis, planes and angle—illustration from the drawing software.

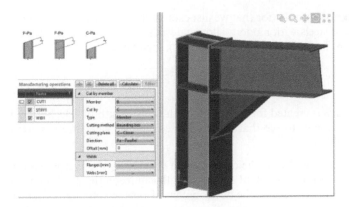

Figure 6.   2nd node.

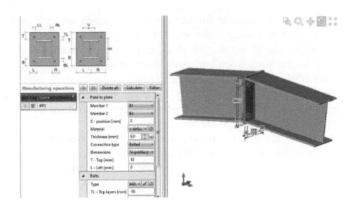

Figure 7.   3rd node.

- COLUMN-BEAM JOINING – 2nd node
- BEAM-BEAM connecting – 3rd node – junction plate with screwed join, 8.8 quality class M27 type self-holding screw.

7   THE FUTURE OF STEEL STRUCTURES IN INDUSTRIAL ARCHITECTURE

In planning practice, the basic case for steel structures is not rigid, and therefore requires support in several axes and planes. While planning these structures, besides loss of load-bearing strength, loss of stability and interactive effects must be added into the calculable failure.

Steel is one of the most capable materials for standardisation and lightness.

In many cases, such as large span roofs, bridges, work on monuments, only steel is sufficiently strong and light. Our base material was a non-compound steel member with less than 1.65% alloying elements.

The Eurocode 3 standard defines the lateral torsional buckling as the dominant calculable event leading to loss of stability—except for slim and very tall structures: where buckling is the first problem to be considered. (Halász, O., Iványi, M., 2001)

Damage caused by superposition complicates the planning of steel-member structures, especially when the structure reaches the load-bearing boundary and a "plastic hinge" deformation appears and collapses the structure.

In this project we found that many special topics in physics are not thoroughly enough understood. In detail there are two problems:

We do not know exactly (for that we just estimated) the change behaviour of the materials if we compound steels with modern alloy materials. For that our Eurocode standard gives a classification in the cross-section category using formula 1:

$$\varepsilon = \sqrt{235/f_y} \qquad (1)$$

This formula was created for a material with the strength of 235 N/mm² (Halász, Iványi 2001). It isn't difficult to see that if we choose a higher yield-stress material, the solution space would tighten.

1. The cross-sectional class is the one in which we classify those cross-sections in which the calculated stress on the element of ideal length can reach the yield boundary, but the local buckling of a plate prevents the development of plastic bending-resistance.
2. Cross-sectional division comprises those cross-sections in which, when the pressure/torque resistance is determined, the impact of the local denting must be expressly given (Halász, Iványi 2001).

We believe the following will guide further research: finding an approximating method to measure advanced materials some other way. In this case the Eurocode doesn't cover the scientifically tested approach to measuring this composite structure, and this will result in engineers having more freedom when choosing from materials during the design process. Reduction in the material mass reduces costs of course.

The second area to research: we have to try to define very precisely the remaining stress after the welding process. With sufficient data here, we could understand the behaviour of our conjugate joined materials (Farkas, Jármai 2008).

Researching any area of science and making use of the results at some stage runs up against questions of costs, the financial side of the project, and this is a separate category of problem.

What is the main lesson for engineers or for physicists? Patience, and understanding: investing in scientific research is an extremely long-term process.

ACKNOWLEDGEMENTS

The described article was carried out as part of the EFOP-3.6.1-16-2016-00011. "Younger and Renewing University—Innovative Knowledge City—institutional development of the University of Miskolc aiming at intelligent specialisation" project implemented in the framework of the Széchenyi 2020 programme. The realisation of this project is supported by the European Union, co-financed by the European Social Fund.

REFERENCES

AXIS VM 14 Software – EC 0-1 MSZ EN (1993 0/1) Loads and Effects part.

EUROCODE 3 – Steel structures design (MSZ EN 1993) Hungarian version.

Farkas, J. & Jármai, K. (2008): Design and optimisation of metal structures, Horvood Publishing, Chichester, UK, 2008., pp. 58–65.

Halász, O., Platthy, P. (1989): Steel structures. Tankönyvkiadó Budapest, pp. 197–204. (in Hungarian).

Halász, O., Iványi, M. (2001): Theory of stability. Akadémiai Publisher, pp. 54–98. and pp. 1097–1120. (in Hungarian).

Katula L., Horváth L., Strobl A. (2007): Structural technology. HEFOP, University—schoolbook, 2007, pp. 32–48. (in Hungarian).

Németh L. (2006): Frame structures Chapter III. HEFOP—University—schoolbook, pp. 7–10. (in Hungarian).

Seregi, Gy. (2011): Halls made from steel. TERC, pp. 13–22. and pp. 103–118. (in Hungarian).

Advances and Trends in Engineering Sciences and Technologies III – Al Ali & Platko (Eds)
© *2019 Taylor & Francis Group, London, ISBN 978-0-367-07509-5*

Methods for determining elastic modulus in natural plant stems

B. Kawecki & J. Podgórski
Faculty of Civil Engineering and Architecture, Lublin University of Technology, Lublin, Poland

A. Głowacka
Faculty of Agro Bioengineering, University of Life Sciences in Lublin, Lublin, Poland

ABSTRACT: The paper presents comparison of three methods for determining elastic modulus in natural plant materials. The subject arose in need of providing the simplest and the most accurate methods for determining mechanical parameters in energetic plants— sida hermaphrodita and miscanthus giganteus. Experimental tests should allow to indicate parameters in a short period of time, especially when influence of humidity level or harvesting time is tested. The subject is a preliminary approach to testing the unknown material. Mechanical parameters are going to be used for numerical modelling and strength prediction of natural plant composites in the near future. There following tests are presented: three point bending test, cantilever beam own vibration frequency test and cantilever beam deflection test. Advantages and disadvantages of each presented method are described. Quality of the results is discussed based on laboratory tests.

1 PLANTS DESCRIPTION

1.1 *Sida hermaphrodita*

Perennial plant belonging to the group of non-food renewable energy sources (Lewandowski 2000, Borkowska 2013). Perennial crop native to North America with strongly developed root system that grows intensively from the second year of vegetation. It reproduces both generatively from seeds and vegetatively from root cuttings obtained from the carp division. The height of several year old plants is 250–300 cm and the diameter of the stem ranges from 5 to 40 mm (Szczukowski 2012).

1.2 *Miscanthus giganteus*

Perennial plant belonging to the group of non-food renewable energy sources (Lewandowski 2000, Borkowska 2013). Sterile hybrid plant farmed in the 1980s in Denmark. It is characterized by rapid growth, high productivity of above-ground biomass and relatively low sensitivity to low temperatures, especially in the first year of vegetation. Miscanthus Giganteus does not produce seeds and is propagated vegetatively, most often through rhizomes cuttings obtained from the division of root carps of mother plants (Cichorz 2018).

2 LABORATORY TESTING METHODS

2.1 *Three point bending test*

There were prepared 15 samples for sida hermaphrodita (5,2% humidity level) and 15 samples for miscanthus giganteus (5,9% humidity level). Samples were presented in the Figure 1.
 Laboratory tests were done on Zwick/Roell Z2.5 testing machine basing on ASTM D790 standard. Test stand and results diagrams were presented in the Figure 2.

Figure 1.   Samples for three point bending test.

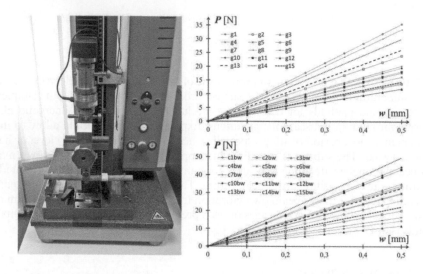

Figure 2.   Laboratory test stand for three point bending test and results diagrams.

Samples were loaded only in elastic range to prevent from plastic strains and cross-section deformations. During bending, force value and deflection in the middle of the beam was measured and force-displacement relation was obtained. Transformed formula for calculating elastic modulus might be written as:

$$E_I = \frac{L_r^{3}}{48J_x} \cdot \frac{P}{w}$$
(1)

where: L_r = distance between supports; P = load value; w = deflection in the middle of the beam; J_x = bending moment of inertia regard to natural set of the sample—x axis.

2.2   Cantilever beam vibrations test

There were prepared 10 samples for sida hermaphrodita (5,2% humidity level) and 10 samples for miscanthus giganteus (5,9% humidity level). Samples were presented in the Figure 3.

Measuring beam vibrations was based on Digital Image Correlation method (Siebert 2010 and Mat Tahir 2015). In experiment there was used fast camera Vision Research Phantom v12.1 with set of 1000 frames per second (1000 fps). Samples were successively mounted in vice. Next, measurement points were glued on free end, sample was induced and camera took 8000 pictures in 8 seconds. Using GOM Correlate software increments of free end deflection in each picture were determined. It allowed to prepare displacement-time relation diagrams. In order to determine own vibrations frequency there was performed FFT analysis of the data and pick values was read from diagram. Laboratory test stand and example results were presented in the Figure 4.

Because of samples length, cross-section was considered to be non-uniform. For that kind of sample analytical solution was prepared. It resulted in practical diagram presented in the Figure 5. Transformed formula (Chmielewski 1998) for calculating elastic modulus might be written as:

$$E_{II} = \left(\frac{2\pi L^2 \frac{f}{\eta}}{3.5151} \right)^2 \frac{\mu}{J_W}$$ (2)

where L = length of the cantilever; f = frequency from DIC method; η = non-uniform beam coefficient from diagram; J_W = moment of inertia for averaged circular cross-section in fix point; μ = evenly distributed mass of the sample.

2.3 Cantilever beam deflection test

There were used the same samples as in cantilever beam vibrations test. Experiment was based on similar test found in (Tymiński 2007). Sample was loaded successively in three steps with load increasing and deflection on millimeter paper was measured. Laboratory test stand and course of proceedings were presented in the Figure 6. Transformed formula for calculating elastic modulus might be written as:

$$E_{III} = \frac{PL_{cor}^3}{3wJ_W} \cdot \frac{D_W}{D_F}$$ (3)

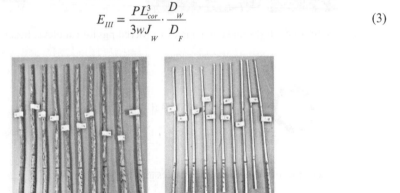

Figure 3. Samples for cantilever beam vibrations and cantilever beam deflection test.

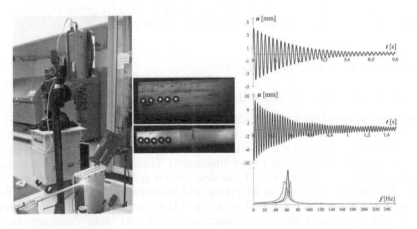

Figure 4. Laboratory test stand for cantilever beam vibrations test with example results.

145

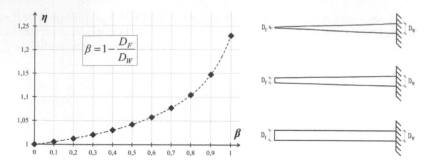

Figure 5. Vibrations coefficient for sample with non-uniform cross-section—analytical solution diagram.

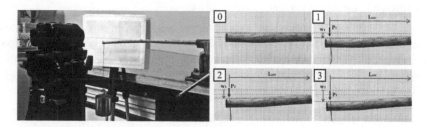

Figure 6. Laboratory test stand and course of proceedings for cantilever beam deflection test.

Figure 7. Examples of post-processed pictures and vectorial contours.

where L_{cor} = corrected length of the cantilever; P = load value; w = free end deflection; J_W = moment of inertia for averaged circular cross-section in fix point; D_W = average outer diameter in fix point; D_F = average outer diameter on free end.

3 CROSS-SECTIONAL PROPERTIES

3.1 *Natural cross-section and vectorial contours*

After laboratory tests slices for every sample were prepared. Each slice was drilled from easy removable core and coloured with black ink. Then it was photographed in high resolution next to millimeter paper and only black colour was isolated. Next, every taken picture was scaled with precision of 0.01 mm to real dimensions. Three point bending test required one slice near the middle per sample and cantilever beam tests required two slices—one near fix end and one near free end. After taking and post-processing pictures there were prepared vectorial contours using Autodesk AutoCAD software, which were presented in the Figure 7. These contours enabled to determine outer and inner diameters and then— moments of inertia.

4 RESULTS AND DISCUSSION

4.1 *Results statistical analysis*

Elastic modulus values for each kind of laboratory test were calculated basing on formulas presented in previous chapter. Data for each sample and average values were collected in the Table 1 and Table 2. Additionally there was done statistical analysis of results basing on fundamental formulas from (Taylor 1982 and Brunarski 2007):

$$E_{avg} = \frac{\sum\limits_{i=1}^{n} E_i}{n} \tag{4}$$

Table 1.   Three point bending test results (TPB) – elastic modulus [GPa].

Sample	Sida hermaphrodita E_I (TPB)	Miscanthus giganteus E_I (TPB)
1	7.85	10.75
2	10.92	15.03
3	11.67	16.34
4	10.23	15.10
5	9.43	12.38
6	7.21	17.61
7	7.45	15.66
8	8.30	14.41
9	11.37	12.07
10	8.37	13.73
11	12.39	15.78
12	11.95	11.93
13	8.59	15.08
14	6.27	13.41
15	7.85	16.01
Average value (E_{avg})	9.32	14.35
Average standard deviation (σ_x)	1.69	1.58
Average error (E_{error})	18.2%	11.0%

Table 2.   Cantilever beam vibrations (CBV) and deflection (CBD) test results—elastic modulus [GPa].

Sample	Sida hermaphrodita E_{II} (CBV)	E_{III} (CBD)	Miscanthus giganteus E_{II} (CBV)	E_{III} (CBD)
1	11.85	10.59	18.81	17.22
2	6.37	6.75	14.67	15.16
3	7.89	6.53	15.67	15.39
4	10.42	10.55	13.26	12.76
5	11.02	11.91	14.87	14.00
6	9.36	8.39	14.43	13.68
7	6.98	5.98	16.33	17.07
8	8.22	8.65	16.43	16.79
9	6.12	6.73	12.90	13.52
10	9.83	10.27	13.41	13.76
Average value (E_{avg})	8.81	8.63	15.08	14.93
Average standard deviation (σ_x)	1.69	1.76	1.39	1.39
Average error (E_{error})	19.2%	20.4%	9.2%	9.3%

$$\sigma_x = \frac{\sqrt{\sum\limits_{i=1}^{n}\left(E_i - E_{avg}\right)^2}}{n} \qquad (5)$$

$$E_{error} = \frac{\sigma_x}{E_{avg}} \cdot 100\% \qquad (6)$$

where: E_{avg} = average elastic modulus; σ_x = average standard deviation; E_{error} = average error

4.2   Results discussion and conclusions

Results obtained using three different methods gave similar average results. For natural material, which was tested, average deviation of elastic modulus was not too large. In every presented method preparation of samples was relatively easy. Some problems have occured with reliable assemble of the sample in vice without crushing fibers in case of the cantilever beam, while in case of the three point bending test there were no difficulties. Because for the cantilever experiments, samples had to be long enough to observe end vibrations and deflection, the shape was much more non-linear than in three point bending test and some cross-section approximations were done. For the three point bending test, the samples were quite short and cross-sectional parameters were taken directly from the middle slice of the sample with minimal error. Cantilever vibrations measurement using the DIC method required a high-performance computer to process all high-resolution images and was very time-consuming. The three point bending was done on a professional testing machine with certified measurement system. Cantilever beam methods were proposed mostly to make the statement that results obtained from the three point bending were reliable. Summarizing, the three point bending test is the recommended method for a relatively easy, fast and precise method for determining elastic modulus in presented natural stems.

REFERENCES

ASTM D790-17, Standard Test Methods for Flexural Properties of Unreinforced and Reinforced Plastics and Electrical Insulating Materials, ASTM International, West Conshohocken, PA, 2017.

Borkowska H., Molas R. 2012. Two extremely different crops, Salix and Sida, as sources of renewable bioenergy. Biomass Bioenergy 36, 234–240.

Borkowska H., Molas R. 2013. Yield comparison of four lignocellulosic perennial energy crop species. Biomass Bioenergy 51, 145–153.

Brunarski L., 2007. Estimation of uncertainty of building materials strength test results, Building Research Institute 141.

Chmielewski T., Zembaty Z., 1998. Fundamentals of building dynamics. Warsaw, Arkady. (in Polish).

Cichorz S., Gośka D., Mańkowski R. 2018. Regeneration system with assessment of genetic and epigenetic stability in long-term in vitro culture Industrial Crops and Products 116, 150–161.

Lewandowski I., Clifton-Brown J.C., Scurlock J.M.O., Huisman W. 2000. Miscanthus: European experience with a novel energy crop. Biomass Bioenergy 19, 209–227.

Mat Tahir M.F., Walsh S.J. and O'Boy D.J., 2015. Evaluation of the digital image correlation method for the measurement of vibration mode shapes, Inter-Noise and Noise-Con Congress and Conference Proceedings, 1986–1995.

Siebert T., Splitthof K. 2010. Vibration Analysis using 3D Image Correlation Technique. EPJ Web of Conferences 6, 11004.

Szczukowski S., Tworkowski J., Stolarski M., Kwiatkowski J., Krzyżaniak M., Lajszner W., Graban Ł. 2012. Perennial energy crops. Mulico Publishing Mouse, Warsaw, Poland. (in Polish).

Taylor J.R., 1982. An Introduction to Error Analysis—The Study of Uncertainties in Physical Measurements, Oxford University Press, Oxford.

Tymiński T., 2007. Analysis of the elastic plants influence on hydraulic conditions of flow in overgrown beds. Monograph, Wroclaw University of Environmental and Life Sciences, Wroclaw, Poland. (in Polish).

On task of thick-walled aluminum pipe deformation under dynamic superplasticity conditions

D.A. Kitaeva & G.E. Kodzhaspirov
Peter the Great St. Petersburg Polytechnic University, St. Petersburg, Russia

Ya.I. Rudaev & Yu.A. Khokhlova
Kyrgyz-Russian Slavic University, Bishkek, Kyrgyz Republic

ABSTRACT: The objective of the presented research is the mathematical formulation of the problem of thick-walled pipe loaded by the external pressure and by the stretching force in the thermal superplasticity range: differential balance equations; the kinematic ratios establishing the relationship between strain rates and displacements; an incompressibility condition in rates; the defining ratios in the form of the equations of the theory of elasto-plastic processes of small curvature, and the state equation which is a consequence of the dynamic model associated to isothermal process. Joint consideration of these equations leads to definition of one function which allows to establish stress fields, strain rates and displacements rates.

1 INTRODUCTION

Processes of the isothermal bulk metal forming under superplasticity conditions belong to one of the most perspective technological operations directed to improvement of the advanced production and which are of a certain interest to development of the bulk metal forming processes theory (Langdon 2016). As a result of the mentioned above processes it is possible to increase plastic properties of material, reduce deformation force and to realize a large strain degree.

Two basic methods for providing the superplasticity effect are known (Rudaev 1990). The first one consists in the preliminary preparation of fine-grained structure in the alloys (structural superplasticity) (Bochvar & Sviderskaya 1945, Padmanabhan et al. 2001, Sakai et al. 2014, Valiev & Semenova 2016, Ganieva et al. 2017). The other method called "dynamic superplasticity" (Rudskoy & Rudaev 2009, Kitaeva et al. 2017) consists in the combination of deformation processes and phase transformations. In the last case, the initial structure of the processed material does not matter.

The utilization of the dynamic superplasticity effect is one of the most perspective technological techniques of metal processing (Kuneev et al. 2002, Kitaeva et al. 2014). The purpose of such processes is creation of metal forming processes to produce metal billets with predetermined grain structure and mechanical properties (Kaibyshev & Valiev 1987, Metlov et al. 2012, Kaibyshev & Malopheyev 2014, Kitaeva et al. 2016).

The technological tasks certainly belong to physically and geometrically nonlinear ones (Rudskoy et al. 2015, Hirkovskis et al. 2015, Kukhar et al. 2018, Rudskoy et al. 2018). It is possible to claim that from the theoretical point of view it is necessary to solve the non-stationary tasks of mechanics investigated in two and three-dimensional statement with the complex changing of the boundary conditions. Correspondingly, it is necessary to attract the equations of state of inelastic continuous media (Kitaeva et al. 2014, Rudskoy et al. 2015, Rudskoy et al. 2018) reflecting the real properties of materials.

2 FORMULATION OF THE PROBLEM

The cylindrical pipe with length $\bar{l}$ stretches by axial force $\bar{F}$ and exposed to external pressure intensity $\bar{q}$ at a given rate of radial movement V_0 (Figure 1), where: R = the outer radius of the pipe, R_0 = the inner radius of the pipe, r = current radius.

All geometric sizes are assumed to be divided by the outer radius of the pipe (Figure 1).

$$\rho = \frac{r}{R}; \quad l = \frac{\bar{l}}{R}; \quad \rho\big|_R = 1; \quad \rho\big|_{R_0} = \rho_0; \quad z = \frac{\bar{z}}{R}. \tag{1}$$

By introducing a cylindrical coordinate system $\rho\alpha z$, we consider the problem of determining the power and kinematic parameters of the process of forming.

It is assumed that the velocity of the horizontal displacement V_z depends linearly on the coordinate and is determined by the expression:

$$V_z = b\big[z - \psi(\rho)\big], \tag{2}$$

where b = const, $\psi(\rho)$ = is an unknown function to be determined.

The representation of the velocity V_z in the Equation 2 was used to solve the problem of the reverse extrusion of a cylindrical product with a bottom in the superplasticity modes (Rudskoy et al. 2015) was borrowed in (Rudskoy & Rudaev 2009).

The mathematical formulation of the problem in terms of the theory of elastoplastic small-curvature processes includes the following:

– differential equations of equilibrium

$$\frac{\partial \sigma_\rho}{\partial \rho} + \frac{\partial \tau_{\rho z}}{\partial z} + \frac{\sigma_\rho - \sigma_\alpha}{\rho} = 0; \quad \frac{\partial \tau_{\rho z}}{\partial \rho} + \frac{\partial \sigma_z}{\partial z} + \frac{\tau_{\rho z}}{\rho} = 0; \tag{3}$$

– kinematic relations, establishing connection between strain rates and displacements

$$\dot{\varepsilon}_\rho = \frac{\partial V_\rho}{\partial \rho}; \quad \dot{\varepsilon}_\alpha = \frac{V_\rho}{\rho}; \quad \dot{\varepsilon}_z = \frac{\partial V_z}{\partial z}; \quad \dot{\gamma}_{\rho z} = \frac{\partial V_z}{\partial \rho} + \frac{\partial V_\rho}{\partial z}; \tag{4}$$

– the condition of incompressibility in strain rates

$$\dot{\varepsilon}_\rho + \dot{\varepsilon}_\alpha + \dot{\varepsilon}_z = \frac{\partial V_\rho}{\partial \rho} + \frac{V_\rho}{\rho} + \frac{\partial V_z}{\partial z} = 0; \tag{5}$$

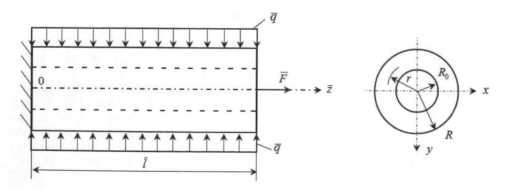

Figure 1. Scheme of the pipe loading.

- defining relations

$$\sigma_\rho - \sigma_0 = \frac{2}{3}\frac{\sigma_u}{\dot{\varepsilon}_u}\dot{\varepsilon}_\rho; \quad \sigma_\alpha - \sigma_0 = \frac{2}{3}\frac{\sigma_u}{\dot{\varepsilon}_u}\dot{\varepsilon}_\alpha; \quad \sigma_z - \sigma_0 = \frac{2}{3}\frac{\sigma_u}{\dot{\varepsilon}_u}\dot{\varepsilon}_z; \quad \tau_{\rho z} = \frac{\sigma_u}{3\dot{\varepsilon}_u}\dot{\gamma}_{\rho z}; \quad (6)$$

- equation of state (Rudskoy & Rudaev 2009) in the form of the dependence of stress intensity σ_u on strain rate intensity $\dot{\varepsilon}_u$

$$\sigma_u = 1 - m_0 - \beta + (3m_0 + \beta)\dot{\varepsilon}_u - 3m_0\dot{\varepsilon}_u^2 + m_0\dot{\varepsilon}_u^3. \quad (7)$$

Here, $\sigma_{ij}, \dot{\varepsilon}_{ij}$ = are the stress and the strain-rate tensors components respectively, v_i = are the components of the velocity vector, σ_0 = is the average stress, m_0 = is the material constant, β = is the temperature-dependent control parameter (it is constant under isothermal conditions and $\beta < 0$ under superplasticity).

All values at Equations 1 – 7 are accepted dimensionless. The stress and strain-rate components are attributed to alternative internal state parameters $\sigma^*, \dot{\varepsilon}^*$ (Kitaeva et al. 2017), and motion velocity to $R\dot{\varepsilon}^*$. Besides, external power factors are defined so:

$$q = \frac{\bar{q}}{\sigma^*}; \quad F = \frac{\bar{F}}{A\sigma^*}, \quad (8)$$

where A = pipe cross-sectional area. Boundary conditions will be formulated in solving the problem.

3 RESOLVING FUNCTION

Using Equation 6, we can write the incompressibility condition (Equation 5) in the form:

$$\frac{dV_\rho}{d\rho} + \frac{V_\rho}{\rho} + b = 0, \quad (9)$$

The solution of the received linear non-uniform equation of the first order taking into account a boundary condition $V_\rho\big|_{\rho=1} = -V_0$:

$$V_\rho = \frac{b}{2\rho}(C - \rho^2), \quad (10)$$

where constant value C is:

$$C = 1 - \frac{2V_0}{b}. \quad (11)$$

At the known the horizontal (Equation 2) and the radial (Equation 10) motion velocity vector the strain-rate tensors components will be equal:

$$\dot{\varepsilon}_\rho = -\frac{b}{2\rho^2}(C + \rho^2); \quad \dot{\varepsilon}_\alpha = \frac{b}{2\rho^2}(C - \rho^2); \quad \dot{\varepsilon}_z = -b; \quad \dot{\gamma}_{\rho z} = -b\psi'(\rho). \quad (12)$$

We will present the strain rate intensity in the form:

$$\dot{\varepsilon}_u = \frac{b}{\sqrt{3}}L^{\frac{1}{2}}(\rho), \quad (13)$$

151

where

$$L(\rho) = \frac{C^2}{\rho^4} + 3 + \psi'^2(\rho).$$ (14)

The relation $\sigma_u / \dot{\varepsilon}_u = T(\rho)$ entering the defining relations (Equation 6) taking into account Equation 13 can be written down so:

$$T(\rho) = \frac{b}{\sqrt{3}}(1 - m_0 - \beta)L^{-\frac{1}{2}} + 3m_0 + \beta - \sqrt{3}\beta m_0 L^{\frac{1}{2}} + \frac{bm_0}{3}L,$$ (15)

where $L = L(\rho)$.

For the stress tensor components we have:

$$\sigma_\rho - \sigma_0 = -\frac{1}{3}bT(\rho)\left(1 + \frac{C}{\rho^2}\right);$$ (16)

$$\sigma_\alpha - \sigma_0 = -\frac{1}{3}bT(\rho)\left(1 - \frac{C}{\rho^2}\right);$$ (17)

$$\sigma_z - \sigma_0 = -\frac{2}{3}bT(\rho);$$ (18)

$$\tau_{\rho z} = -\frac{1}{3}bT(\rho)\psi'(\rho).$$ (19)

We will substitute Equation 19 in the second equations of equilibrium from Equation 3. The for $\partial\sigma_z / \partial z$ we have:

$$\frac{\partial\sigma_z}{\partial z} = \frac{1}{3}b\left\{T'(\rho)\psi'(\rho) + T(\rho)\left[\psi''(\rho) + \frac{\psi''(\rho)}{\rho}\right]\right\}.$$ (20)

Integrating Equation 20 under a boundary condition $\sigma_z|_{z=0} = F$, we obtain:

$$\sigma_z = \frac{1}{3}b\left\{T'(\rho)\psi'(\rho) + T(\rho)\left[\psi''(\rho) + \frac{\psi''(\rho)}{\rho}\right]\right\}z + F.$$ (21)

The average stress from Equation 18 is equal:

$$\sigma_0 = \sigma_z - \frac{2}{3}bT(\rho).$$ (22)

After substituting Equation 21 into Equation 22 we obtain

$$\sigma_0 = \frac{1}{3}b\left\{T'(\rho)\psi'(\rho) + T(\rho)\left[\psi''(\rho) + \frac{\psi''(\rho)}{\rho}\right]\right\}z + F - \frac{2}{3}bT(\rho).$$ (23)

An analysis of Equations 12, 14–19, 21 shows that the stress and strain rates components can be found if an explicit form of function $\psi(\rho)$, which is called resolving, is determined. To search for function $\psi(\rho)$, we substitute Equations 16–19 into first equilibrium equation (Equation 3). Thus, we obtain

$$\frac{\partial\sigma_0}{\partial\rho} = \frac{1}{3}bT'(\rho)(1 + \frac{C}{\rho^2}).$$ (24)

152

We differentiate Equation 23 with respect to ρ and equate the obtained result to Equation 24. Therefore, we can write

$$\left\{ T''(\rho)\psi'(\rho) + T'(\rho)\psi''(\rho) + T'(\rho)\left[\psi''(\rho) + \frac{\psi'(\rho)}{\rho} \right] + T(\rho)\left[\psi'''(\rho) + \right.\right.$$
$$\left.\left. + \frac{\psi''(\rho)}{\rho} - \frac{\psi'(\rho)}{\rho^2} \right] \right\} z - T'(\rho)\left(3 + \frac{C}{\rho^2} \right) = 0. \tag{25}$$

The differential Equation 25 has to be satisfied at all values z. Therefore Equation 25 is representable as two differential equations:

$$T''(\rho)\psi'(\rho) + T'(\rho)\psi''(\rho) + T'(\rho)\left[\psi''(\rho) + \frac{\psi'(\rho)}{\rho} \right] + T(\rho)\left[\psi'''(\rho) + \right.$$
$$\left. + \frac{\psi''(\rho)}{\rho} - \frac{\psi'(\rho)}{\rho^2} \right] = 0; \tag{26}$$

$$T'(\rho) = 0. \tag{27}$$

When performing Equation 27 for definition of function $\psi(\rho)$ we obtain the equation

$$\psi'''(\rho) + \frac{\psi''(\rho)}{\rho} - \frac{\psi'(\rho)}{\rho^2} = 0. \tag{28}$$

At the solution of Equation 28 we use decrease in an order of a derivative, accepting $\psi'(\rho) = y(\rho)$. We can write

$$y'' + f(\rho)y' + g(\rho)y = 0. \tag{29}$$

If

$$g \neq 0, \frac{1}{|g|}\frac{d}{d\rho}\sqrt{|g|} + \frac{f}{\sqrt{q}} = a = const,$$

then

$$\rho^2 \frac{d}{d\rho}\left(\frac{1}{\rho} \right) + \frac{1}{\rho}\rho = \rho^2\left(-\frac{1}{\rho^2} \right) + 1 = 1 - 1 = 0.$$

Therefore for Equation 29 it is possible to accept

$$y(\rho) = \eta(\xi); \quad \xi = \int\sqrt{|g|}d\rho = \ln\rho.$$

Then at $a = 0$ we can write the equation

$$\eta'' - \eta = 0. \tag{30}$$

We can write the solution to Equation 30 in the form

$$\eta = C_1 e^{\ln\rho} + C_2 e^{\ln\rho} = C_1\rho + C_2/\rho = y. \tag{31}$$

Thus, for the resolving function $\psi(\rho)$ entering in Equations 12–17 we obtain:

153

$$\psi(\rho) = \frac{C_1}{2}\rho^2 + C_2 \ln\rho + C_3. \tag{32}$$

The components of stresses, strain rates and displacements will be established after the determination of the five constants C, C_1, C_2, C_3, b.

4 CONCLUSIONS

A boundary value problem of loading a thick-walled pipe by external pressure and tensile force in the thermal range of superplasticity is proposed. The problem is solved in the framework of the theory of elastic-plastic processes of small curvature using the equation of state of the nonlinear type, suitable not only for the intervals of superplasticity, but also for the boundary areas of thermoplastic and high-temperature creep. Stress and strain rate fields are established.

The problem is the basis for the possible development of theories of technological processes such as compression and distribution of pipes, drawing and autofreting in superplasticity.

REFERENCES

Bochvar, A.A. & Sviderskaya, Z.A. 1945. The phenomenon of superplasticity in alloys of zinc with aluminum, *Izv. Akad. Nauk SSSR, Otd. Tekh. Nauk* 9: 821–827.

Ganieva, V.R., Tulupova, O.P., Enikeev, F.U. & Kruglov, A.A. 2017. Modeling of Superplastic Structural Materials. *Russian Engineering Research* 37(5): 401–407.

Hirkovskis, A., Serdjuks, D., Goremikins, V., Pakrastins, L. and Vatin, N.I. 2015. Behaviour analysis of load-bearing aluminium members. *Magazine of Civil Engineering* 57(5): 86–96 and 116–117.

Kaibyshev, O.A. & Valiev, R.Z. 1987. *Grain Boundaries and Properties of Metals*, Moscow: Metallurgiya.

Kaibyshev, R. & Malopheyev, S. 2014. Mechanisms of dynamic recrystallization in aluminum alloys. *Materials Science Forum* 794–796: 784–789.

Kitaeva, D.A., Kodzhaspirov, G.E. & Rudaev, Ya.I. 2017. On the dynamic superplasticity. *Materials Science Forum* 879: 960–965.

Kitaeva, D.A., Pazylov, Sh.T. & Rudaev, Ya.I. 2016. Temperature-strain rate deformation conditions of aluminum alloys. *Journal of Applied Mechanics and Technical Physics* 57(2): 352–358.

Kitaeva, D.A., Rudaev, Ya.I. & Subbotina, E.A. 2014. About the volume forming of aluminium details in superplasticity conditions. In METAL 2014, *Proc. 23rd International Conference on Metallurgy and Materials, Brno, 21–23 May 2014*. Ostrava: TANGER.

Kukhar, V., Artiukh, V., Butyrin A. & Prysiazhnyi, A. 2018. Stress-Strain State and Plasticity Reserve Depletion on the Lateral Surface of Workpiece at Various Contact Conditions During Upsetting. *Advances in Intelligent Systems and Computing* 692: 201–211.

Kuneev, V.I., Pazylov, Sh. T., Rudaev, Ya.I. & Chashnikov, D.I. 2002. Dynamic superplasticity technologies. *Journal of Machinery Manufacture and Reliability* 6: 55–62.

Langdon, T.G. 2016. Forty-five Years of Superplastic Research: Recent Developments and Future Prospects. *Materials Science Forum* 838–839: 3–12.

Metlov, L.S., Myshlyaev, M.M., Khomenko, A.V. & Lyashenko, I.A. 2012. A model of grain boundary sliding during deformation. *Technical Physics Letters* 38(11): 972–974.

Padmanabhan, K.A., Vasin, R.A. & Enikeev, F.U. 2001. *Superplastic Flow: Phenomenology and Mechanics*, Berlin-Heidelberg: Springer-Verlag.

Rudaev, Ya.I. 1990. Phase transitions in superplasticity. *Strength of Materials* 22(10): 1445–1451.

Rudskoy, A.I., Kodzhaspirov, G.E., Kitaeva, D.A., Rudaev, Ya.I. & Kliber, J. 2015. To the problem of the cylindrical product with a bottom forming process optimization under superplasticity condition. *Materials Physics and Mechanics* 22(2): 191–199.

Rudskoy, A.I., Kodzhaspirov, G.E., Kitaeva, D.A., Rudaev, Ya.I. & Subbotina, E.A. 2018. On the theory of isothermal hot rolling of an aluminum alloy strip. *Russian Metallurgy (Metally)* 2018(4): 334–340.

Rudskoy, A.I. & Rudaev, Ya.I. 2009. *Mechanics of dynamic superplasticity of aluminum alloys*, St.Petersburg: Nauka.

Sakai, T., Belyakov, A., Kaibyshev, R., Miura, H. & Jonas, J.J. 2014. Dynamic and post-dynamic recrystallization under hot, cold and severe plastic deformation conditions. *Progress in Materials Science* 60: 130–207.

Valiev, R.Z. & Semenova, I.P. 2016. Advances in Superplasticity of Ultrafine-Grained Alloys: Recent Research and Development, *Materials Science Forum* 838–839: 23–33.

Advances and Trends in Engineering Sciences and Technologies III – Al Ali & Platko (Eds)
© 2019 Taylor & Francis Group, London, ISBN 978-0-367-07509-5

Application of laser vibrometry in dam health monitoring

M. Klun, D. Zupan & A. Kryžanowski
Faculty of Civil and Geodetic Engineering, University of Ljubljana, Ljubljana, Slovenia

ABSTRACT: This paper presents the monitoring of the dynamic properties of a concrete gravity dam. The dam under investigation is the new Brežice dam on the Sava River in Slovenia. In order to capture the reference state of this structure, its investigation was initiated already during the construction. During the start-up tests of the generating units, we had a unique opportunity to capture all the operational maneuvers that can occur unplanned during regular operation. Structural behavior is measured at several discrete locations on the dam structure as well as on the turbine. Various measurement equipment is used, including a state-of-the-art Laser Doppler Vibrometer. Surface velocities are captured in the time-domain; in order to identify dominant frequencies of the system, the data are then transformed to the frequency-domain.

1 INTRODUCTION

This paper reports on vibration measurements at the Brežice dam. The aim of the research is to obtain the built-in dynamic properties of the dam and to identify the effect of operative loads on the response of the dam. The investigation was initiated already during the construction and captured the reference state of the dam to be used as a baseline for long-term monitoring. The need for this investigation emerged from the recognition of the lack of knowledge on the aging phenomenon in concrete gravity dams. The design and maintenance of existing dams represent a challenge for the entire engineering community. We are increasingly faced with the problem of aging dam structures and, at the same time, with changes in the environment, especially variability in time-dependent loads and new patterns of operation with turbines.

New trends in the energy market and the trend to operate on turbines to cover peak loads have the consequence that turbines operate more often in transient and unsteady modes. In order to provide reliable operation, the effect of start/stop cycles on the turbines and the interaction with the bearing structure must be investigated. There are numerous investigations, which take the turbines, the generator and the interaction between the rotating and the stationary components of the turbine during stationary, non-stationary, transient and off-design loads (Deschênes, Fraser and Fau, 2002; Trivedi, Gandhi and Michel, 2013; Trivedi *et al.*, 2015; Fu *et al.*, 2016; Goyal and Gandhi, 2018), while the interaction with surrounding structures has not been fully investigated yet.

With this research we are introducing a new methodology for structural health monitoring of dams. The Laser Doppler Vibrometer (LDV) is used as an experimental tool to identify the effects of various operational patterns on the structural response of a concrete gravity dam. LDV is used to capture surface velocities in the time domain, while further transformation in the frequency domain allows detection of dominant eigenfrequencies and improved insight into the behavior of the system (Castellini, Martarelli & Tomasini, 2006; Goyal & Gandhi, 2018).

2 VIBRATION MONITORING AT THE BREŽICE DAM

2.1 *Layout of the experiment*

The Brežice dam is a newly built dam on the Sava River in Slovenia. Its construction started in April 2014 and concluded in 2017. It is a run-of-river dam with limited storage capacity. The dam is located in the flat Krško-Brežice Basin. To dam the Sava river, a combined dam with a total crest length of over 14 km was built. The dam is structurally divided into three parts: the left embankment, the right embankment, and the central concrete gravity dam (Figure 1).

The construction of the Brežice dam provided a unique opportunity for our investigation, as we were able to begin with the investigation already in the early phase of the dam's lifecycle. Our investigation is focused on the concrete gravity dam. This section consists of a power house and a spillway with 5 overflow sections. Each overflow section is 15 m wide and designed to a maximum discharge capacity of 1000 m³/s (Klun, Zupan and Kryžanowski, 2017). Sections are divided by pillars 2.7 m wide and 51 m long with varying height (Governmental Decree on Detailed Plan of National Importance for HPP Brežice Area, 2012). The pillars under investigation are the pillar between the first and the second overflow sections and the one between the fourth and the fifth overflow sections (Figure 2). The powerhouse is installed with 3 vertical Kaplan turbines. The engine room is the main facility, the structural behavior is monitored in 8 experimental points allocated on the south wall (4) of the engine room, on the turbine shafts (3) and on the landing (1) connecting the turbine shafts (Figure 2 and Figure 3). Detailed information on the experimental points is given in Table 1.

In our research we use Polytec LDV-100, a fully portable device, which enables high vibrational velocity resolution in the frequency range from 0.5 Hz to 22 kHz. The technology is based on the detection of the Doppler shift of the laser light. Since the frequency shift is proportional to the surface velocity, the technology enables non-contact measurement of the vibration velocity (Rothberg *et al.*, 2017). The LDV offers many advantages over accelerometer-based measurements: non-contact data acquisition with high precision, no permanent installation required, high temporal resolution, relative measurement. Since the technology has not been used in dam engineering, the use of a vibrometer was combined with the use of accelerometers and speed transducers. Methodology

In order to establish a solid baseline for long-term structural health monitoring, the dynamic properties of a dam should be quantified shortly after the completion of construction work. The first measurements using the LDV were made in May 2016. On the day of the measurements,

Figure 1.   Brežice dam (source: www.he-ss.si).

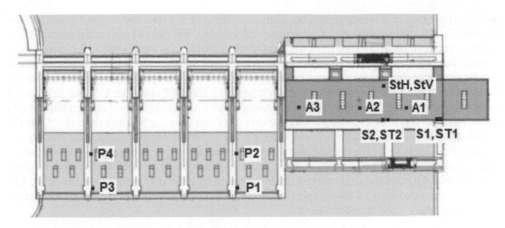

Figure 2.   Layout of the experimental points (black dots) on the Brežice dam.

Figure 3.   (a) The engine room, (b) An example of the measurement site.

Table 1.   Description of, ST1 measuring points.

Experimental point	Location	Measuring distance [m]
P1	Spillway (1st pillar)	16
P2	Spillway (1st pillar)	16
P5	Spillway (5th pillar)	16
P6	Spillway (5th pillar)	16
ST1	Engine room	12
ST2	Engine room	12
S1	Engine room	12.5
S2	Engine room	12.5
A1	Engine room	3
A2	Engine room	3
A3	Engine room	3
StH, StV	Engine room	0 (only speed transducers)

most of the work on the site was focused on the installation of components for the hydraulic gate and electrical equipment in the powerhouse. The following measurements captured the effect of mining of the diversion dyke and the state of the dam before the first impoundment. After the initial acceptance of all three generating units and filling of the reservoir to the nominal height (August 2017), the start-up tests of generating units started. This was a long period when various operational regimes with hydro units were tested: start-stop cycles, steady state operation, power regulation, emergency shut down. During these tests, measurements were taken in all 8 points in the powerhouse (two points measured simultaneously). With the conclusion of the tests and the final acceptance of the generating units in October 2017, the HPP entered into a one-year period of trial operation. In January 2018, an additional test on the electrical grid also included rapid shutdowns of the units, including rapid shutdowns of multiple units simultaneously. After the conclusion of the trial operation period, the HPP will enter the longest period in its lifecycle, i.e. regular operation. Our investigation also aims to capture the response during regular operation. The conclusion of the trial operation also concludes the learning period, which offered insight into the initial state of the structure and the response under different loading scenarios. Time histories are recorded with a 2400-sample frequency. MATLAB 2016a software is used for further data manipulation and transformation.

3 RESULTS AND DISCUSSION

The measurements during startup tests were concentrated in the engine room. Due to the structural design of the dam, we do not expect the influence on operation with turbines to be transferred into the spillway sections. Besides the structural vibration, they vibrations were also captured on the turbines. Figure 4 shows the typical time-history and spectrogram measured on turbine 2. The amplitude of the vibration velocity is up to 1.3 cm/s (Figure 4(a)) and frequency notation of the turbine during regular operation shows the dominant frequencies at 4 Hz, 28 Hz, 43 Hz, and 100 Hz (Figure 4(b)).

Turbines are the source of the excitation of the dam structure. A typical time series of the engine room wall when the structure is at rest is shown in Figure 5(a). The experimental point under consideration is presented in Figure 3(b). Amplitudes are within the limits of 0.4 mm/s and the presence of higher frequency oscillations is also evident. Figure 5(b) shows the influence of the construction work on the response on the engine room wall. We can notice the rise in the amplitudes up to 1 mm/s. Figure 6(a) shows the response of the wall during the launch of turbine 1 and the increase in the surface oscillations is evident, up to 5 mm/s (16 times higher for the case when the structure vibrates naturally). Transformation in the frequency spectrum reveals the frequency notation of the measurement. Raw data were filtered using an ecliptic bandpass filter, while the transformation to the frequency spectrum was done using fast Fourier transformation. Figure 6(b) shows the frequency spectrum. Peaks that mark dominant frequencies are detected at 3 Hz, 13 Hz, 41 Hz, 100 Hz, and 130 Hz.

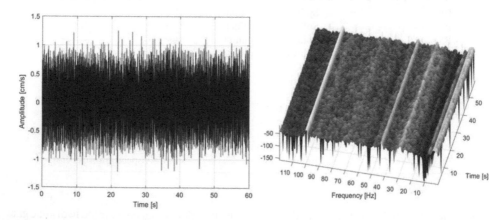

Figure 4.   (a) Time series on the turbine during regular operation, (b) Spectrogram of the measurement.

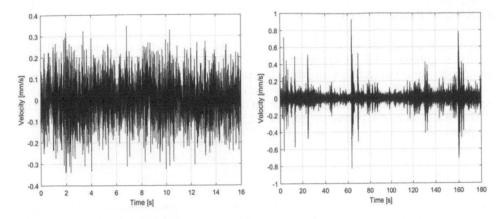

Figure 5. (a) Time series of the engine room wall at rest, (b) Time series of the engine room wall under the influence of construction work.

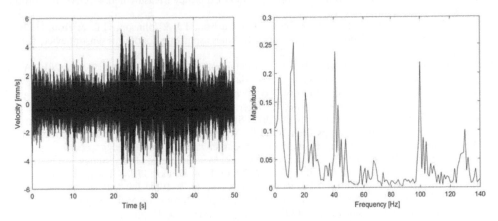

Figure 6. (a) Time series of the engine room during the launch of turbine 1, (b) Frequency spectrum.

4 CONCLUSIONS

In this paper, implementation and initial results of the experimental work at the Brežice dam are presented. The majority of experimental data was obtained using non-contact measurements with the use of laser vibrometry. The measured time histories reveal the influence of the operation on the dam structure. A structure vibrates with much higher amplitudes when subjected to the effect of launching a turbine, while the structure vibrates naturally. Furthermore, the influence is recognized in the frequency domain. If we compare the frequency domain of the structure with the frequencies detected at the turbines, we notice that the operation of the turbines is the source of excitations with frequencies that are close to the frequencies of the structure, especially in the range of low frequencies, between 40–50 Hz, and at around 100 Hz. We have also recognized the magnitude of the effect of construction work on the dam. This investigation is the first step in the long-term structural health monitoring, which includes observation of the aging phenomenon caused by operational dynamic loading. The investigation has captured the reference state of the structure right after the completion of the construction work, before the impoundment of the reservoir, and during the testing of the hydraulic machinery. The investigation furthermore allowed for testing the possibilities of using laser vibrometry and the advantages of the methodology over conventional monitoring equipment used on dams.

ACKNOWLEDGEMENTS

This study is part of the work by the Faculty of Civil and Geodetic Engineering of the University of Ljubljana (UL FGG) under the research programme P2-0180 Water Science and Technology, and Geotechnics, financed by the Slovenian Research Agency (ARRS). The support is gratefully acknowledged. The project was also supported by the hydro power company Hidroelektrarne na spodnji Savi, d.o.o. (HESS). The company allowed the publication of the results. The professional support of HESS is greatly acknowledged by the authors of the paper.

REFERENCES

Castellini, P. & Martarelli, M. & Tomasini, E.P. 2006. Laser Doppler Vibrometry: Development of advanced solutions answering to technology's needs, *Mechanical Systems and Signal Processing*, 20(6), pp. 1265–1285.

Deschênes, C. & Fraser, R. & Fau, J.-P. 2002 New Trends in Turbine Modelling and New Ways of Partnership, International Conference on Hydraulic Efficiency Measurement—IGHEM, Toronto, Ontario, Canada, July, pp. 1–12.

Fu, T. & Deng, Z.D. & Duncan, J.P. & Zhou, D. & Carlson, T.J. & Johnson, G.E. & Hou, H. 2016. Assessing hydraulic conditions through Francis turbines using an autonomous sensor device, *Renewable Energy*. Elsevier Ltd, 99, pp. 1244–1252.

Governmental Decree for on Detailed Plan of National Importance for HPP Brežice area. 2012. Available at: http://www.pisrs.si/Pis.web/pregledPredpisa?id=URED6213.

Goyal, R. & Gandhi, B.K. 2018. Review of hydrodynamics instabilities in Francis turbine during off-design and transient operations, *Renewable Energy*. Elsevier Ltd, 116, pp. 697–709.

Klun, M. & Zupan, D. & Kryžanowski, A. 2017. Structural measurements of dynamic response of hydraulic structures, Proceedings 85th ICOLD Annual Meeting International Symposium, 85th Annual Meeting of International Commission on Large Dams. Prague, Czech.

MATLAB Release 2016b, The MathWorks, Inc., Natick, Massachusetts, United States.

Rothberg, S.J. & Allen, M.S. & Castellini, P. & Di Maio, D. & Dirckx, J.J.J. & Ewins, D.J. & Halkon, B.J. & Muyshondt, P. & Paone, N. & Ryan, T. & Steger, H. & Tomasini, E.P. & Vanlanduit, S. & Vignola, J.F. 2017. An international review of laser Doppler vibrometry: Making light work of vibration measurement, *Optics and Lasers in Engineering*. Elsevier Ltd, 99, pp. 11–22.

Trivedi, C. & Gandhi, B. & Michel, C.J. 2013. Effect of transients on Francis turbine runner life: A review, *Journal of Hydraulic Research*, 51(2), pp. 121–132.

Trivedi, C. & Gandhi, B.K. & Cervantes, M.J. & Dahlhaug, O.G. 2015. Experimental investigations of a model Francis turbine during shutdown at synchronous speed, *Renewable Energy*. Elsevier Ltd, 83, pp. 828–836.

Advances and Trends in Engineering Sciences and Technologies III – Al Ali & Platko (Eds)
© 2019 Taylor & Francis Group, London, ISBN 978-0-367-07509-5

Assessment of sway buckling influence for laced built-up portal frames

M. Kováč & Zs. Vaník
Faculty of Civil Engineering, STU in Bratislava, Bratislava, Slovakia

ABSTRACT: This paper is focused on the sway buckling influence in portal frames with laced built-up members. For frames sensitive to buckling in a sway mode the second order effects with initial sway imperfection should be taken into account in analysis. In portal frames with the laced compression columns under sway imperfection the second order effects cause additional axial forces in chords. A parametric study was performed to assess these additional chord forces. A simplified method, which was proposed and verified in previous papers, was used here to produce results for profiles with various shapes and dimensions.

1 INTRODUCTION

Portal frames which consist of laced built-up members are frequently used in praxis for longer spans. Two-hinged portal frame, the most commonly used statical system, is sensitive to buckling in a sway mode. If the second-order effects and imperfections shall be considered in the global in-frame analysis of such structure, an equivalent imperfection in the form of an initial sway imperfections and individual bow imperfections of columns should be introduced. The second order effects cause additional sway deformation of such frame, which then causes additional axial forces in chords of laced built-up columns.

The imperfections and second order effects are mostly carried out in praxis by equivalent column method. Individual stability checks of members by this method (clause 6.3 of EN 1993-1-1) require using appropriate buckling length according to the global buckling mode of the structure. In the stability check of chords of laced built-up columns, if the length of the chord between nodes of lacings is used as the buckling length in the equivalent column method with the first order internal forces, then these additional axial forces due to the additional sway deformation of frame as the second order effect are neglected. In such case only the local bow imperfection between the lacing nodes for the second order effects is taken into account in the equivalent column method.

2 SIMPLIFIED PROCEDURE

2.1 Conception

In order to take into account the additional chord axial forces due to additional sway deformation of portal frames as the second order effects, one have to use equivalent imperfection in the form of initial sway imperfections and local bow imperfection of columns and carried out the second order analysis. Only after that it is possible to assess values of these additional forces. However, there is a simpler way than creating imperfect model of the structure and performing second order analysis in some of FEA based softwares.

According to clause 6.4 of EN 1993-1-1 it is possible to assess laced built-up columns which is supported by hinges at the ends and have imperfection in the shape of half-wave of sinus function (Figure 2d)). As this shape represents the global buckling mode of the such column the additional deformation obtains same shape and the second order effects can be expressed

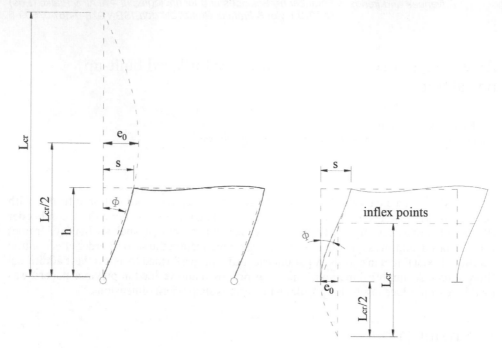

Figure 1.   Imperfections for simplified procedure.

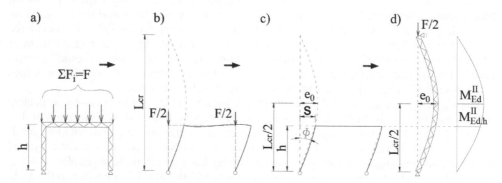

Figure 2.   Conception of simplified procedure.

by simple relations which lead to formulae (6.69) of EN 1993-1-1. Besides, it takes into account also the shear stiffness of the lacings which is equally distributed along the column.

This clause can be utilized for the assessment of the additional chord forces due to the second order effects in laced built-up columns of portal frames. Utilization of clause 6.4 of EN 1993-1-1 for this purpose was presented in Vaník et al. (2015) and Kováč et al. (2016) and it will be referenced as the simplified procedure. In order to be able to use the half-wave sinus imperfection it was necessary to determine the amplitude of such imperfection from the equivalent imperfection.

2.2   *Imperfection for simplified procedure*

The equivalent imperfection in the form of initial global sway imperfection and local bow imperfection of the columns according clause 5.3.2 of EN 1993-1-1 is intended to approximate global elastic in-plane buckling mode of structure. The shape of half-wave of sinus function on the buckling length of the columns in the simplified procedure is supposed to approximate it better than the equivalent imperfection does it. However, it is necessary to transform equivalent imperfection into half-wave sinus imperfection with appropriate amplitude. Transformation for two-hinged portal frame was presented in Vaník et al. (2015), where resulted amplitude was compared with the L/500 amplitude according to clause 6.4 of EN 1993-1-1. One of the possible transformations from global sway imperfection for portal frames into half-wave sinus imperfections is illustrated in Figures 1 and 2c).

2.3 *Equivalent portal frame*

Because of the clause 6.4 of EN 1993-1-1 is only for columns simply supported, it can be used only if the column length is such to be equal to the critical length of the laced built-up column being assessed. Then the two neighboring inflex points in the global buckling mode of portal frame would be hinge supports of the column being assessed by clause 6.4. For the global buckling mode determination, it is reasonable to transform portal frame from laced built-up members into an equivalent portal frame from compact cross-section members. Then it is possible to obtain the one relevant global buckling mode without any incidence of many buckling modes where only local bow buckling modes on the lacing's and chord's domains occur in the buckling analysis.

The equivalent portal frame have to has members with bending rigidities equivalent to bending rigidities of laced built-up members of portal frame. The effective second order moment of area of laced built-up members may be taken according formula (6.72) of EN 1993-1-1. Then the compact cross-sections of members of the equivalent portal frame should have the same effective second order moment of area in order to obtain the same bending rigidities of members.

For this equivalent portal frame defined, the buckling analysis may be performed. According to the global buckling mode, with the sway deformation occurrence, the buckling length of columns may be determined (see Figure 2b)).

2.4 *Equivalent laced built-up column*

Equivalent laced built-up column (Figure 2d)) consists from the same chords and lacings (and has the same spacing between lacings) as the column of the portal frame being assessed (Figure 2a)). Its length is derived from buckling length from previous buckling analysis and loading should be equivalent to loading of portal frame. Finally, such equivalent laced built-up column may be assessed according to clause 6.4 of EN 1993-1-1. But instead of the second order bending moment at the middle-span M_{Ed}^{II} by formula (6.69) the bending moment $M_{Ed,h}^{II}$ at relevant height h according to sinus function (Figure 2d)) will be used. The first order bending moment M_{Ed}^{I} due to primal loading will be applied too, if it exists.

3 PARAMETRIC STUDY

The estimation of the sway buckling influence by the simplified procedure discussed above was used here in the parametric study. The two-hinge portal frame from Figure 2a) was considered here with various ratios of horizontal to vertical member lengths to obtain buckling length factors of columns to be equal to 2, 2.5 and 3 values.

In the assessment of chords by the equivalent column method according EN 1993-1-1 with first order theory normal forces and the length of chord between lacing nodes as the buckling length, the utilization depends on the relative slenderness of chord. In the portal frame with sway imperfection, the additive second order theory normal forces of chords depend on the relative slenderness of the whole laced built-up column according its buckling length from global buckling mode. So, it was expected that the most important parameter for the assessment of the sway buckling influence would be the ratio of relative slenderness of the chord to the relative slenderness of the laced column. In order to obtain great variability in this ratio the cross-sections of chords were chosen to be varied in the parametric study. The whole ranges of IPE and UPE from open cross-sections and CHS, SHS and RHS from hot-finished hollow cross-sections were used in the study. The cross-sections parameters were the variable parameters of the study. The other parameters were chosen as follows: the height of the frame $h = 15$ m; distance between centroids of chords $h_0 = 1$ m; the length of chord between the lacings $a = 2$ m and the shear stiffness of the lacings $S_V = 2.5 \times 10^5$ kN. Steel grade S235 was considered.

For loading at the level of design buckling resistance the utilization by the equivalent column method is 100% and it is same as the utilization by the second order theory with the bow imperfection (considering same amplitude). At the lower level of loading, the utilization by the equivalent column method tends to be higher than the real utilization by the second order theory with same imperfection. To correctly assess the sway buckling influence, it is necessary to set the loading at the level of design buckling resistance. In the study, for each given cross-section shape and size such specific loading was find and for such loading the sway buckling influence was determined.

The utilization by the simplified procedure, with the influence of sway imperfection on the column domain as well as the bow imperfection on the chord domain, is $U_{wi} = 1$ for such loading. The utilization by the equivalent column method, with only the bow imperfection between lacing nodes, is $U_{wo} < 1$. Then, the sway buckling influence can be expressed as difference between the utilizations:

$$dif = \left(U_{wi} - U_{w0} \right) / U_{wi} \tag{1}$$

The relative slenderness of equivalent laced built-up column (Figure 2d)) is:

$$\bar{\lambda}_{eq.l.c} = \sqrt{2A_{ch}f_y / N_{cr}} \tag{2}$$

where A_{ch} is cross-section area of chord and N_{cr} is critical force related to sway buckling mode from global buckling analysis of equivalent portal frame (Figure 2b)). The relative slenderness of chord is:

$$\bar{\lambda}_{ch} = \sqrt{A_{ch}f_y / N_{cr.ch}} \tag{3}$$

where $N_{cr.ch}$ is the smallest critical force of chord related to buckling mode on the chord domain between lacings, concerning flexural, torsional or torsional-flexural buckling mode depending on cross-section type.

The results of parametric study are shown in Figures 3 and 4. The vertical axis denote the sway buckling influence expressed by formulae (1). The horizontal axis denotes the ratio of relative slenderness of the chord to the relative slenderness of the equivalent laced column:

$$r = \bar{\lambda}_{ch} / \bar{\lambda}_{eq.l.c} \tag{4}$$

The lower indexes denote the type of cross-sections and the numerical value denote the buckling length factor, which when applied on the column height the buckling length L_{cr} of laced built-up column (Figure 2b)) should be obtained. Each point in the Figures 3 and 4 represent one value of difference between utilizations of chord with and without accounting the sway imperfection of the frame for one specific cross-section.

From the results in Figures 3 and 4 can be concluded that with the increase of the relative slenderness of the equivalent laced built-up column and (or) with the decrease of the relative slenderness of the chord, the sway buckling influence is increasing, considerably more in the slendernesses ratio by formulae (4) of interval (0, 1.5) and for greater buckling lengths of columns. The maximal sway buckling influence could reach 25% influence. It can be stated that even above the ratio 1.5 the influence couldn't be neglected as it almost reaches 10%.

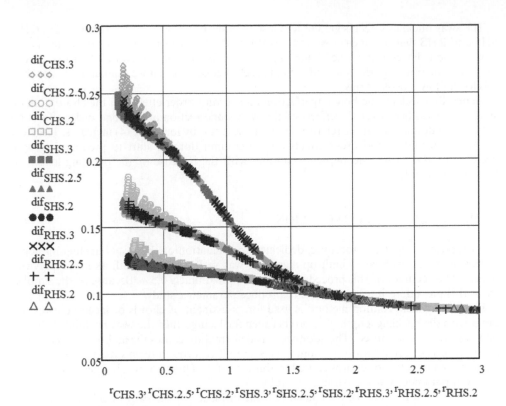

Figure 3. The sway buckling influence for CHS, SHS and RHS chords' cross-sections.

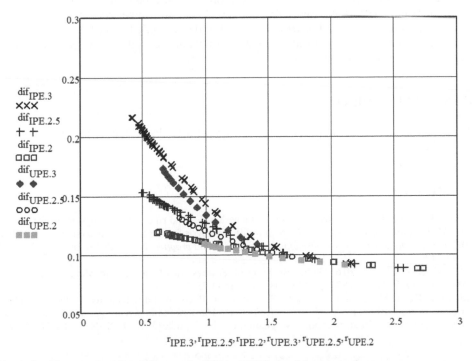

Figure 4. The sway buckling influence for IPE and UPE chords' cross-sections.

The sway buckling effect is greater for closed hollow hot finished type cross-sections CHS, SHS and RHS than for open cross-sections IPE and UPE. This is caused by the difference in cross-section shapes and by the higher imperfection factors for IPE and UPE cross-sections for which the bow imperfection and second order effects on the chord domain is greater than it is for hollow cross-sections.

Generally speaking the bow imperfection and second order effects on the chord domain start to predominate over the effects of the sway imperfection and second order effects on the column domain when the relative slendernesses ratio by formulae (4) increases. The sway imperfection and second order effects on the column domain start to predominate with decrease of the relative slenderness ratio and with the increase of the buckling length of laced built-up column.

4 DISCUSSION AND CONCLUSION

The proposed simplified procedure, dedicated to evaluation of the sway buckling influence of portal frames with laced built-up members, was verified by the authors in the previous papers. The accuracy of this procedure reaches approximately 0.5%. Because of the simplicity of this procedure, it was used here to produce parametric study.

If the equivalent column method is used for assessment of chords of laced built-up columns with the buckling length of chord between the lacings, then the sway buckling influence on portal frames is omitted. The additive sway deformation causes the additive normal forces in the chords of laced built-up column. This simplified procedure was used to evaluate the error in the utilization of chords caused by this omission. This error was in interval (5%, 25%) for given parameters of the parametric study.

From the results it can be concluded that the influence of the sway buckling of the sway type portal frames from laced built-up members is not negligible and should be always checked in praxis. The proposed simplified procedure is simple and provides a sufficient prediction of the sway buckling influence on the chords of laced built-up columns. Whether to choose the commercial FEA based software to create the sway imperfect model of the portal frame and perform the second order analysis or use the simplified procedure to evaluate the additive chords' normal forces from sway deformation and assess the chord by the equivalent column method is upon the designer.

ACKNOWLEDGEMENTS

The authors acknowledge support by the Slovak Scientific Grant Agency under the contract No. 1/0773/18.

REFERENCES

EN 1993-1-1. 2005. Design of steel structures—General rules and rules for buildings. Brussels: CEN.
Kováč M. & Vaník Zs. & Magura M. 2016. Simplified analysis for frames with laced built-up members. In *International journal of interdisciplinarity in theory and practice.* No.11 (2016), pp 222–226. ISSN 2344-2409.
Kováč, M. & Vaník, Zs. 2016. Global Buckling of Frames with Compression Members. In *Applied Mechanics and Materials* 837: pp 103–108. ISSN 1660-9336.
Vaník Zs. & Kováč M. & Magura M. 2015. Second order effects and imperfections of frames with laced compression members. In *MMK 2015. Mezinárodní Masarykova konference pro doktorandy a mladé vědecké pracovníky.* Hradec Králové: Magnanimitas, 2015, pp 2612–2619. ISBN 978-80-87952-12-2.

Deformation characteristics of masonry structure subjected to horizontal loads

M. Kozielová & L. Mynarzová
Faculty of Civil Engineering, VSB—TU Ostrava, Ostrava, Czech Republic

ABSTRACT: This contribution deals with effects of horizontal loads to masonry structures. It is focused on estimation of masonry parameters needed for modelling of horizontally loaded structures and their proper application when choosing the prestressing force and anchors for prestressing device. Test equipment for three-axial state of stress is used for determination of deformation characteristics. Potentiometers record the local consolidation. They read only the values of masonry displacement at the place of local stress and they are not affected by deformation of the whole sample. Measured values serve as input parameters for calculation of masonry modulus of elasticity which could be included in simplified numerical models for modelling of huge parts of masonry required for practical assessments. Furthermore, the models could be used for determination of prestressing forces and losses of prestress in anchors during the restoration of masonry structures.

1 INTRODUCTION

Masonry is a composite material whose components (bricks, mortar) are characterized by various geometric and material parameters. Modulus of elasticity as one of the basic mechanical parameters is important for determination of stiffness of the structure just before cracking.

Possibilities for determination of the elastic modulus are presented in the standard EN 1996-1 (2007) or in EN 1052-1 (2000) which introduce the procedure on the standardization masonry sample. In both standards, the methods are related to the modulus of elasticity perpendicular to the bed joints. For analysis of deformation characteristics of masonry, these procedures are not convenient. Experiments for pressure of various masonry units in longitudinal and transversal direction are mentioned in Schubert & Graubohm (2004). Within these tests, masonry units were connected to various types of mortars and head joints were filled or unfilled with mortar.

Currently a new method for determining of masonry local stresses or deformation parameters parallel to the bed joints is being carried out: "flat jack" is a semidestructive testing of current masonry buildings that need reparation, Łatka (2017). One or two flat steel hydraulic presses are applied in the mortar bed joint. Oil is pumped into the flat jacks and displacements between installed sensors are recorded. According to the used methodology, the interpretation of obtained data allows to calculate the elastic modulus of masonry that could be used at analysis of masonry structure before and after the strengthening of masonry, Terzijski (2012) and Witzany (2008).

2 MATERIAL PROPERTIES AND EXPERIMENTAL TEST OF MASONRY

2.1 *Experimental equipment*

Experiment was carried out on masonry representing the corner of a building so its plan is in a shape of a letter L (Figure 1). Experimental equipment for measuring of masonry three-axial state-of-stress was made at Faculty of civil engineering, VŠB—TU Ostrava. Plan dimensions are 900 × 900 mm, height 1550 mm. It was possible to build the masonry into the equipment only to the height of 900 mm. Load plate for application of vertical load was also a part of the equipment. This plate was made of steel 11 373 with thickness of 12 mm. For better distribution of a load, steel braces were placed on the plate.

2.2 *Used materials and principle of measurement*

Deformation characteristics of masonry were determined on the masonry corner. Height of masonry was 870 mm, width 440 mm (without plaster) and length 900 mm. Masonry was made of solid bricks with dimensions of 290 × 140 × 65 mm. Lime mortar for bed and head joints was mixed with sand in the rate of 1:4. Resulting strength of a mortar was very low to correspond with quality of a mortar in restored structures. Strength of mortars in older buildings was often under 0.4 MPa. Characteristic strength of masonry used for testing in direction perpendicular to bed joints was 1.366 MPa, calculating according to EN 1996-1 (2007). Strength parallel to bed joints was determined as 50% of perpendicular strength (then there was a violation of the masonry).

During construction procedure, prestressing bar with diameter of 26 mm from steel 11 523 was put into bed joints. Steel anchor plate 150 × 150 × 40 mm was fixed on the prestressing bar. For uniform distribution of a load, the plate was placed on the layer of cement mortar, see (Figure 2).

Measuring of deformation parameters was executing parallel to bed joints with the help of prestressing device. In the first phase prestressing forces was applied in steps of 10% from total strength of masonry perpendicular to bed joints. Local consolidation was recorded using

Figure 1.   Model of masonry corner and laying up of masonry.

Figure 2.   Installation of prestressing force (left), positions of potentiometers (right).

8 potentiometers that read the values of displacement in places of local stress and they are not affected by deformation of the sample (sensors are fixed to the frame of testing equipment).

Records of local consolidation of masonry were used for evaluation of results and modeling of masonry corner in software based on FEM.

Vertical load was applied with the help of hydraulic press placed on the load plate. Value of vertical load was 0.1 MPa.

2.3 Determination of elastic modulus of masonry

On the basis of measured deformation it could be seen that deformations occur close to the anchor plate—local deformations. At the same time, during prestressing, bottom edge of the anchor plate was displaced and the anchor plate inclined. The highest recorded displacement at 50% of total masonry strength was measured as 1.305 mm at the place of prestressing bar. Values of maximum deformations for particular prestressing forces are presented in Table 2 and in Figure 3.

For determination of elastic modulus of masonry parallel to bed joints, the standard EN 1052-1 (2000) was used. This standard prescribes to consider at least 3 same steps of loading at least until 50% of expected compressive strength of masonry is reached. At this experiment, loading was distributed into five levels by 10% of the masonry strength.

According to the theory of elasticity, elastic modulus is defined as a ratio of normal stress and particular extension. For determination of elastic modulus in direction parallel to bed joints, stress from prestressing force is acting on the area of the anchor plate. Extension is given as a ratio of masonry deformation and length of masonry in direction of bed joint. Resulting local short term elastic modulus parallel to bed joint is then 0.471 GPa.

Table 1. Stress under the anchor plate and prestressing forces.

Strength of masonry in %	Stress under the anchor plate in MPa	Prestressing force in kN
10%	0.137	3.08
20%	0.273	6.14
30%	0.410	9.22
40%	0.546	12.28
50%	0.683	15.36

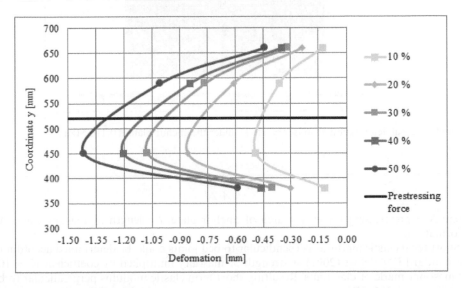

Figure 3. Masonry deformation course.

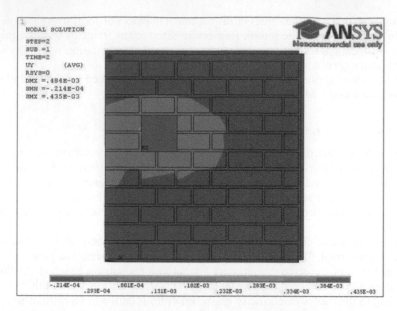

Figure 4. Local deformation of masonry; stress under the anchor plate is 0.137 MPa.

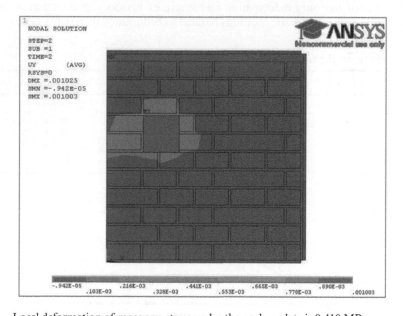

Figure 5. Local deformation of masonry; stress under the anchor plate is 0.410 MPa.

$$E = \frac{N \cdot l}{A \cdot \Delta l} \qquad (1)$$

where N = prestressing force; A = area of anchor plate; l = length of masonry; and Δl = deformation.

Short term elastic modulus perpendicular to bed joints could be determined according to the standard EN 1996-1 (2007) as strength of masonry multiplied by coefficient K_E = 1000 for masonry made of clay units. Resulting short term elastic modulus perpendicular to bed joints is then 1.366 GPa.

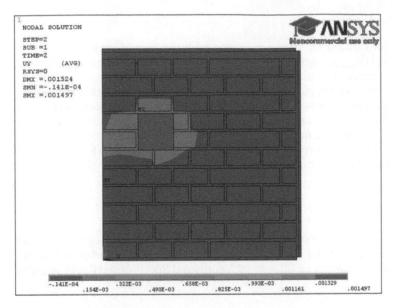

Figure 6.   Local deformation of masonry; stress under the anchor plate is 0.683 MPa.

Table 2.   Comparison of results from experiment and numerical model.

Strength of masonry in %	Stress under the anchor plate in MPa	Deformations obtained from experiment in mm	Deformations calculated from numerical model in mm
10%	0.137	0.450	0.435
20%	0.273	0.790	0.744
30%	0.410	0.990	1.003
40%	0.546	1.110	1.251
50%	0.683	1.305	1.497

3   NUMERICAL MODELING OF MASONRY

Optimal solution for modeling of masonry is a large issue. It is introduced in many texts, for example Lofti & Shing (1994), Lourenco & Rots (1997). Modeling of a masonry structure is performed in the ANSYS application, based on FEM. Masonry structure is considered as one homogenized unit without regard to material parameters of bricks and joints. This type of model is called a macromodel. Firstly, geometry of the structure was created with fine mesh that is suitable for places with local extremes. Boundary conditions important for pregnancy of calculated deformations were defined. The macromodel was solved using elastic modulus obtained from experiments described in Chapter 2.3. Shear modulus was considered as 40% of elastic modulus of masonry. Value of Poisson's ratio is 0.2. Then calculated deformations were compared to the values measured during experiment.

The numerical model was created according to the experiment. The tested masonry corner is modeled by element SOLID45. Anchoring plates for insertion of tensile forces are modeled with the finite element SOLID45. Pre-stress is applied in the model using the effective areas. Effective area for thickness of anchor plate 40 mm is the area of anchor plate 0.0225 m². Behavior of macromodel is close to the concrete as it could be observed in the courses of deformations. Surveyed values were highest near the anchor plate. They were spreading uniformly farther through the structure and were decreasing with higher distance from the anchor plates. Deformations of masonry from numerical models are given in the Table 2.

4 CONCLUSION

Deformations of masonry were measured with the help of prestressing bar placed in bed joint during the execution of the masonry. Within the experiment, prestressing force was applied to induce the masonry deformations. Measured values were used for determination of the elastic modulus parallel to bed joints. The resulting elastic modulus corresponds to c. 35% from elastic modulus perpendicular to bed joints. The determined elastic modulus was used for a numerical model. Comparison of the maximum deformations obtained from the experiment and the numerical model shows good agreement. After exceeding of 10 kN of prestressing force, deformations from the numerical model are higher. Results calculated from numerical model could be considered as on the safe side.

ACKNOWLEDGEMENT

The works were supported from sources for conceptual development of research, development and innovations for 2018 at the VŠB-Technical University of Ostrava which were granted by the Ministry of Education, Youths and Sports of the Czech Republic.

REFERENCES

EN 1052-1. 2000. Methods of Test for Masonry, Part I, Determination of Compressive Strength, CEN Brussels 2000.

EN 1996-1-1. 2007. Eurocode 6: Design of Masonry Structures: Part 1-1, General rules for buildings – Rules for reinforced and unreinforced masonry. CEN Brussels 2007.

Lofti, H.R. & Shing, P.B. 1994. Interface Model Applied to Fracture of Masonry Structures. *Journal of Structural Engineering*. Vol. 120, iss. 1, s. 63–80. DOI: 10.1061/(ASCE)0733-9445(1994)120:1(63).

Lourenco, P.B. & Rots, J.G. 1997. A Multi-surface Interface Model for the Analysis of Masonry Structures. *Journal of Engineering Mechanics*. Vol. 123, iss. 7, s. 660–668. DOI: 10.1061/(ASCE)0733-9399(1997)123:7(660).

Sayari, A. 2012. Mechanical properties of masonry samples for theoretical modeling. *Proceedings of the 15th International Brick and Block Masonry Conference*. June 03rd to 06th, 2012, Florianópolis, Santa Caterina, ISBN 9788563273109. Brazil. Florianópolis: Federal University of Santa Catarina.

Schubert, P. & Graubohm, M. 2004. Druckfestigkeit von Mauerwerk parallel zu den Lagerfugen. Mauerwerk [online], ISSN 1432–3427, 8(5), 198–208. DOI: 10.1002/dama.200490074.

Terzijski, I., Klusáček, L., Bažant, Z. et al. 2012. Determination of Strain Properties of Masonry. The civil engineering journal. 1/2012, ISSN 1210–4027. Faculty of civil engineering CVUT, Prague.

Łatka, D. 2017. Stress state laboratory verification in masonry structures according to the flat jack method. *Czasopismo Techniczne* [online]. DOI: 10.4467/2353737XCT.17.022.6215.

Witzany, J. & Čejka, T. & Zigler, R. 2008. Determination of residual capacity of existing brick structures. The civil engineering journal. 9/2008, ISSN 1210-4027. Faculty of civil engineering CVUT, Prague.

Advances and Trends in Engineering Sciences and Technologies III – Al Ali & Platko (Eds)
© 2019 Taylor & Francis Group, London, ISBN 978-0-367-07509-5

Shear resistance of steel and GFRP reinforced beams

D. Lániová & V. Borzovič
Faculty of Civil Engineering, Slovak University of Technology in Bratislava, Slovakia

ABSTRACT: The objective of the presented research is to compare the performance of the beams reinforced with steel and GFRP (Glass Fiber Reinforced Polymer) reinforcement supplied by local producer. The shear resistance of the beams was verified through a load test. Together 26 beams with the length of 1.5 m were tested. Along with the testing of the beams the strength and deformation characteristics of the applied materials were measured. The concrete strength was determined on three cubic and cylinder specimens, the elastic modulus was examined on three prisms. The steel and GFRP reinforcement bars were subjected to a tensile strength measurement. Based on the results of the experiment, the comparison of different predictions for shear resistance was carried out.

1 INTRODUCTION

Shear is one of the basic loading modes of structural members. In terms of designing steel reinforced structural members, it gains attention by its brittle failure mode in comparison to ductile flexural failure. Although the factors influencing the shear resistance are very well known, the models for prediction of shear resistance are often empiric and set according to processing of multitude of experiments. In this manner the shear resistance without shear reinforcement was determined by the currently valid European standards—Eurocode 2.

Reinforced concrete members are present in many construction areas. Thereby these elements are exposed to many different requirements in addition to the load capacity itself. In particular service life, durability, but also aesthetic requirements. In case of application of reinforced concrete in areas with higher environmental load, such as road bridges, it is necessary to avoid corrosion of the steel reinforcement. This is mainly due to the presence of deicing salts in combination with temperature and humidity varying throughout the year. In this environment degradation of the concrete cover layer and subsequent corrosion of steel reinforcement occur. These problems can be eliminated by using composite materials (FRP—fiber reinforced polymer), which in many cases appear to be a suitable substitute for conventional steel reinforcement. The work focuses on composite reinforcement made of glass fibers, known as GFRP (glass fiber reinforced polymer), which can serve as a substitute for conventional steel reinforcement. In the experimental part of the work a series of test beams reinforced with steel and GFRP (reinforcement) was designed for the purpose of measuring shear resistance without shear reinforcement and comparison of different reinforcement types, steel and GFRP.

2 SHEAR STRESS

An element with the length of Δx is considered on the beam which is subjected to bending moment and shear force. On one side of the element the value of the bending moment is M, on the other side this moment rises by $\Delta M = V.\Delta x$. The ΔM addition increases the pressure force F_c in the compression zone by $\Delta F_c = \Delta M/z$ and the tension force F_t in tension zone by $\Delta F_t = \Delta F_c$, where z is the lever arm of internal forces. Thus a shear stress $\tau_c = \Delta F_c/(\Delta x.b)$

develops in a cross-section. The shear crack is formed in a direction perpendicular to the main tensile stress σ_1 (Figure 1).

The shear resistance of members without shear reinforcement is ensured by: the aggregate interlock in shear crack, compressed concrete in the area above neutral axis, longitudinal reinforcement—dowel action (Figure 2). It is difficult to evaluate the individual contribution of these effects. They depend on the properties of concrete, the applied reinforcement and interaction between these two materials. Considering the steel reinforced members without shear reinforcement, the aggregate interlock in shear crack is decisive, which is influenced by the shear crack width and tensile strength of the concrete. The longitudinal reinforcement prevents the opening of the shear crack. These principles are taken into account in the following empiric formula which determines the design value of shear resistance of the cross section according to EN 1992-1-1 (Bilčík et al., 2008).

$$V_{Rd,c} = [C_{Rd,c}\, k\, (100\, \rho_1 f_{ck})^{1/3} + 0.15\, \sigma_{cp}]\, b_w\, d \tag{1}$$

where:
$C_{Rd,c}$ = constant, $C_{Rd,c} = 0.18/\gamma_C$ [MPa], $\gamma_C = 1.5$;
k = represents the size factor, $k = 1 + \sqrt{200/d} \le 2$, d in mm;

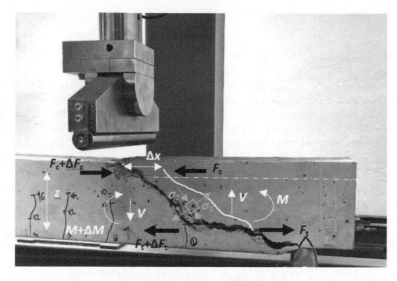

Figure 1.   Shear stress in the cracked reinforced concrete member.

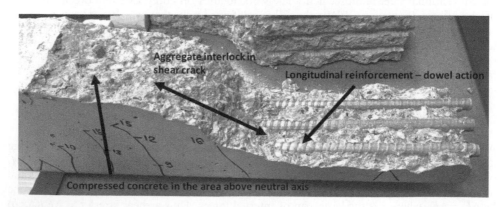

Figure 2.   Contribution of individual effects to shear resistance.

174

ρ_l = longitudinal reinforcement ratio;

f_{ck} = characteristic compressive strength of concrete [MPa];

σ_{cp} = axial stress induced by the design values of the axial force $\sigma_{cp} = N_{Ed}/A_c \le 0.2 f_{cd}$;

b_w = width of the cross section;

d = effective depth.

The new proposed model from Cavagnis provides an additional insight into the shear resistance of the members. Compared to Equation 1, the Equation 2 at its basic form takes into account not only the considered/calculated cross-section but also the shear stiffness i.e. the position of the load on the element or the type of the aggregate used (Cavagnis 2017, Cavagnis & Fernández Ruiz & Muttoni 2018).

$$V_{Rd,c} = 1/\gamma_C \, (100 \, \rho_l f_{ck} \, d_{dg}/a_v)^{1/3} \, b_w \, d \qquad (2)$$

where:

d_{dg} = coefficient to account for concrete strength class and aggregate's properties

= 32 mm for normal weight concrete with strength $f_{ck} \le 60$ MPa and $D_{lower} \ge 16$ mm

= $16 + D_{lower} \le 40$ mm for normal weight concrete with strength $f_{ck} \le 60$ MPa and $D_{lower} < 16$ mm

= $16 + D_{lower} \, (60/f_{ck})^2 \le 40$ mm for normal weight concrete $f_{ck} > 60$ MPa;

a_v = refers to the mechanical shear span

= max $(a_{cs}; 2.5d)$; a_{cs} is the effective shear span with respect to the control section, for reinforced concrete members without normal force, it may be calculated $a_{cs} = M/V$.

The other notations are the same as in Equation 1.

Shear resistance of GFRP reinforced members assumes the same principles of shear resistance initiation. The difference in properties of GFRP and steel reinforcement affects however the resulting shear resistance of the cross section. In general, bigger deformations of reinforcement caused by lower modulus of elasticity result in greater deformations and wider cracks. This effect is accounted for by an equivalent reinforcement ratio of GFRP to steel reinforcement. Adjustment for the EC2 Equation 1 comprises the different type of reinforcement by using modulus of elasticity ratio (Guadagnini, 2016).

$$V_{Rd,c} = [C_{Rd,c}k \, (100 \, \rho_l \, (E_f/E_s) f_{ck})^{1/3} + 0.15\sigma_{cp}] \, b_w \, d \qquad (3)$$

where:

E_f = modulus of elasticity of composite reinforcement;

E_s = modulus of elasticity of steel reinforcement, $E_s = 200$ GPa.

The reinforcement type is considered also through modulus of elasticity ratio in Equation 2.

$$V_{Rd,c} = 1/\gamma_C \, (100 \, \rho_l \, (E_f/E_s) f_{ck} \, d_{dg}/a_v)^{1/3} \, b_w \, d \qquad (4)$$

Formulas in Equation 1 to 4 were used for the theoretical calculation of the shear resistance of tested beams. These relations have been modified to compare calculated and experimentally measured values of shear resistance. Partial safety factors for materials were considered 1.0 and the strength of the materials were considered as a mean value of the measured values. The size factor k was accounted for with two values, firstly the value is limited to 2 and in the other case the value is based on the relation $k = 1 + \sqrt{200/d}$.

3 EXPERIMENTAL PROGRAM

The aim of the experiment was to determine the shear resistance of beams without shear reinforcement in a range of reinforcement ratios, one series of beams being reinforced with a steel reinforcement and a second series of beams with GFRP. In total, 26 beams were prepared, of which 14 pieces with steel and 12 pieces with GFRP reinforcement. There were 2 pieces of beams in each reinforcement ratio in order to obtain 4 measured values of resistance per each reinforcement ratio. The first requirement for the design of dimensions of the beams was their shear failure prior to the rupture in bending. For the considered loading scheme (Figure 3), the beam cross section dimensions were set to width $b = 95$ mm and height $h = 125$ mm. At higher levels of reinforcement ratio, it was necessary to place the reinforcement into two rows to maintain the effective height of the beam therefore the height of these beams increased to 145 mm. The effective height was designed as $d = 100$ mm for all beams. The following reinforcement (reinforcement ratio) was prepared for steel reinforced beams: 3ø 6 (0.89%), 2ø 8 (1.06%), 2ø 8 + 1ø 6 (1.36%), 2ø 10 (1.65%), 2ø 10 + 1ø 6 (1.95%), 2ø 12 (2.38%) and for reinforcement bars placed in two rows 2ø 8 + 2ø 6 (1.65%). For the beams with GFRP, the designed reinforcement (reinforcement ratio) was: 2ø 10 (1.34%), 3ø 10 (2.01%), 3ø 12 (3.00%), 2ø 16 (3.72%) and for reinforcement bars placed in two rows 2ø 16 + 2ø 10 (5.06%), 2ø 16 + 2ø 16 (7.44%).

The strength and deformation test of the used materials were performed to specify the inputs into the calculation formulas. Based on three specimens (cubes, cylinders, prisms), the average cube strength of concrete was $f_{c,cube} = 42.54$ MPa, the average cylinder strength of concrete was $f_{c,cyl} = 31.51$ MPa and the average value of modulus of elasticity of concrete was $E_c = 35.7$ MPa. The maximum size of grain aggregates in concrete was 16 mm. The steel

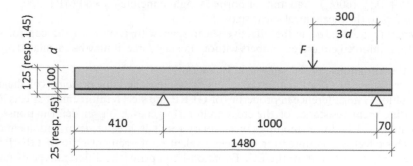

Figure 3. Scheme of 3-point loading test of the beams, dimension in mm.

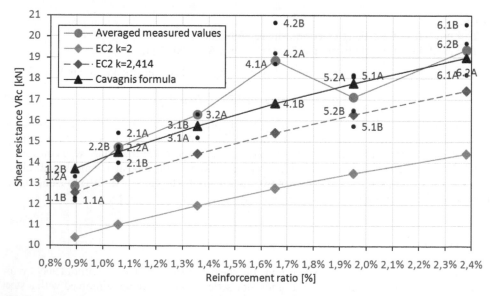

Figure 4. Evaluation of shear resistance of beams with steel reinforcement V_{Rc} [kN] – calculated and measured values.

176

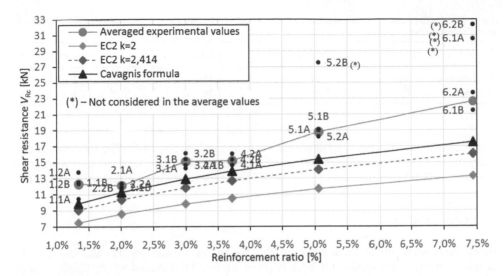

Figure 5. Evaluation of shear resistance of beams with GFRP reinforcement V_{Rc} [kN] – calculated and measured value.

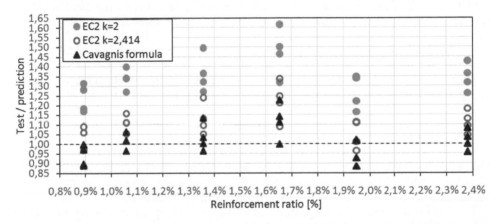

Figure 6. Evaluation of reliability of models for shear resistance of beams with steel reinforcement.

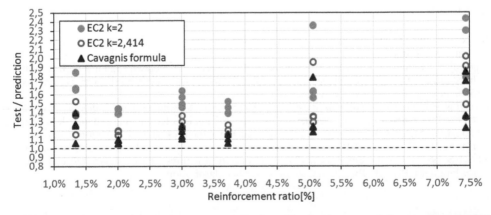

Figure 7. Evaluation of reliability of models for shear resistance of beams with GFRP reinforcement.

Table 1. Average value of ratio measured value to calculated value of shear resistance for individual shear resistance predictions.

Theoretical model	EC2, $k = 2$	EC2, $k = 2.414$	Cavagnis formula
Steel reinforced beams	1.34	1.11	1.02
Beams with GFRP	1.64	1.36	1.25

reinforcement was of class B500B with a yield strength $f_y = 550$ MPa. The modulus of elasticity of steel reinforcement was taken with value $E_s = 200$ GPa. The short-term strength in tension of GFRP reinforcement $f_{fk} = 1200$ MPa and the modulus of elasticity $E_f = 50$ GPa.

4 CONCLUSION

Based on observation of the loading tests of the beams, we can say that the failure of the elements in shear was similar for both types of beams. After exceeding the tensile strength of the concrete, the first bending cracks appeared which gradually developed. The graphs in Figures 4 and 5 show the measured and calculated values of the shear resistance of the beams with steel reinforcement and beams with GFRP, depending on the reinforcement ratio. It is obvious that the shear resistance increases with the rising reinforcement ratio. Considering the steel reinforced beams around reinforcement $\rho_s = 2\%$, the shear strength increases more slowly.

The results were also evaluated in form of the ratio of measured and calculated shear resistance for individual predictions to determine the reliability of the individual models (Figures 6 and 7, Table 1). Based on these values we can say that the formula according to Cavagnis was the closest to the experimentally measured values for both types of the beams.

ACKNOWLEDGMENTS

This work was supported by the Scientific Grant Agency VEGA under the contract No. VEGA 1/0810/16. This work was supported by the University Science Park (USP) of the Slovak University of Technology in Bratislava (ITMS: 26240220084).

The authors appreciate the suggestion and the support of experimental work by company STRABAG. The work was further supported by company Armastek.

REFERENCES

Bilčík, J. & Fillo, Ľ., et al. 2008. *Concrete structures, design according to STN EN 1992-1-1 (in Slovak)*. Bratislava: Vydavateľstvo STU.

Cavagnis, F., Fernández Ruiz, M., Muttoni, A. 2018. A mechanical model for failures in shear of members without transverse reinforcement based on development of a critical shear crack. *Engineering Structures*, 157:pp. 300–315. Elsevier.

Cavagnis, F. 2017. *Shear in reinforced concrete without transverse reinforcement: from refined experimental measurements to mechanical models, Thesis doctoral*. Ecole Polytechnique Fédérale de Lausanne.

Guadagnini, M. 2016. Reinforcing and strengthening of structures with FRP reinforcement. *Training Course*. 25–29. January 2016. University of Sheffield, dept. of Civil and Structural Engineering.

STN EN 1992-1-1 (2004) Eurocode 2: design of concrete structures – Part 1–1: general rules and rules for buildings. Brussels: European Committee for Standardization.

Advances and Trends in Engineering Sciences and Technologies III – Al Ali & Platko (Eds)
© *2019 Taylor & Francis Group, London, ISBN 978-0-367-07509-5*

Evaluation of expressions for calculating the cooling time in carbon steels welding

V. Lazić & D. Arsić
Faculty of Engineering Kragujevac, University of Kragujevac, Serbia

R. Nikolić
Research Center, University of Žilina, Žilina, Slovakia
Faculty of Engineering, University of Kragujevac, Kragujevac, Serbia

Lj. Radović & N. Ilić
The Military Technical Institute, Belgrade, Serbia

B. Hadzima
Research Center, University of Žilina, Žilina, Slovakia

ABSTRACT: The cooling time, in the most important temperature range from 800 to 500°C–$t_{8/5}$, can be calculated according to several different expressions. That temperature range is important since all the transformations in steel have to be done within it. Knowing the cooling time and relevant material properties enables prediction of characteristics of the welded joint and the heat affected zone. In the recent times, there is an increase of application of analytical and empirical expressions for calculating $t_{8/5}$. The accuracy of several expressions was evaluated comparing their results to the experimental ones, obtained by measurements with thermocouples. The objective of this analysis was to show that expensive experiments can be avoided if analytical and/or empirical expressions can provide sufficiently good results for value of the cooling time $t_{8/5}$.

1 INTRODUCTION

Prescribing the optimal welding or hard-facing technology of certain types of steels can be a very complex task. Execution of the adequate welded joint is, in majority of cases, accompanied by voluminous experimental and numerical investigations, Arsić et al. (2015). However, today exist ways to predict the weldability and expected properties of the welded joint by application of the analytical and empirical expressions, Lazić et al. (2010). If satisfying results of are obtained from those calculations, the process can be significantly shortened by eliminating the expensive and tedious experimental research. The objective of this paper is to point to those possibilities, as well as to accuracy of data that could be expected, after the execution of the welded joint. In some previous research, the authors have shown that this method can be reliably used in certain situations, Lazić et al. (2010).

Research presented in this paper was conducted on examples of hard-facing of the carbon and tempering steels, while the formulas of Rikalin (1951), Ito and Bessyo (1972) and formula based on limiting thickness Arsić et al. (2015), Lazić et al. (2010) were analyzed and their results were compared to the experimental ones. The formulas used are as follows:

– The Rikalin formula:

$$T_{(x,y)} = \frac{q}{v \cdot s \cdot \sqrt{4 \cdot \pi \cdot k \cdot c \cdot t}} \cdot e^{\left(-\frac{y^2}{4at} - bt\right)}. \tag{1}$$

– The Ito-Bessyo formula:

$$t_{8/5} = \frac{k \cdot q_l^n}{\beta \cdot (T_s - T_0)^2 \cdot \left[1 + \frac{2}{\pi} \cdot arctg \left(\frac{s - s_0}{\alpha} \right) \right]} \, , \, s. \tag{2}$$

– The limiting thickness formula:

$$s_{\lim} = \sqrt{\frac{q_l \cdot N_3}{2 \cdot \rho \cdot c} \cdot \left(\frac{1}{500 - T_0} + \frac{1}{800 - T_0} \right)}, \tag{3}$$

where: q is the effective power of the arc, q_l is the heat input, a is the thermal diffusivity coefficient, b is the heat transfer coefficient, r is the material density, v is hard-facing velocity, k is the heat conduction coefficient, s is the part thickness, s_0 is the initial thickness, T_s is the temperature on the surface, T_0 is the initial temperature, c is the specific heat, N_3 is the joint type factor, α and β are the empirical coefficients depending on type and shape of the welded joint.

2 PREPARATION (HARD-FACING) OF SAMPLES

Samples for experiments (measurements of temperature) were actually the plates made of carbon steels (one structural and four tempering steels): S355 J2G3, C15, C35, C45 and C45E. Plates were of various thicknesses (s = 7.4, 8, 10, 20 and 30 mm) and dimensions (430 × 200, 410 × 60, 394 × 192, 400 × 200 mm), and the holes were drilled in plates of diameter 1.7 ± 0.05 mm (Figure 1). The thermocouples were placed and fixed in those holes (shielded or manually-held) of diameter 1.5 mm. The glass tubes of diameter 0.4 to 0.6 mm were drawn over the thermocouples prior to their insertion in the holes for the purpose of reducing the measurement errors (Figure 2).

Both thickness of the hard-faced materials and values of the input energy were varied during the process. For that purpose, experiments were performed with various electrode diameters (Ø3.25 – PIVA 430 B (ISO, E1-300), Ø4.0 – PIVA 440 B (ISO, E1-400) and Ø5.0 – PIVA 460 B (ISO, E2-60), FIPROM (2015). The hard-facing rate was determined by measuring the hard-faced caterpillar length and time needed for each pass, i.e. layer, while the interlayer temperature was measured directly by the digital measuring device Tastotherm D1200 or indirectly from the obtained T-t diagram. The hard-facing parameters, given in Table 1 are

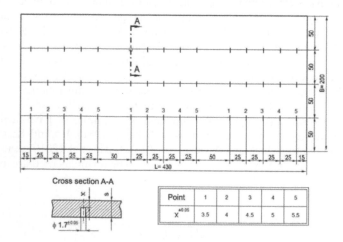

Figure 1.   Appearance of plates for temperature measurements.

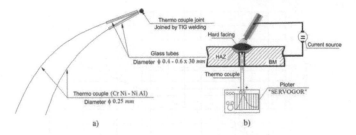

Figure 2. Preparation of manual-held thermocouples (a) and ways of fixing (b).

Table 1. Hard-facing parameters.

Thickness s, (mm)	Electrode diameter d_e, (mm)	Input energy I, (A)	Working voltage U, (V)	Hard-facing rate v_z, (cm/s)	Input heat $q_i = \dfrac{U \cdot I}{v_z} \cdot \eta$, (J/cm)
10	4	140	25.6	0.162–0.136	17650–21101
20 30	5	210	28.5	0.286–0.098	16736–48610

showing the conditions in which the temperature cycles in the heat affected zone (HAZ) were recorded.

3 SELECTION AND PREPARATION (CALIBRATION) OF THERMOCOUPLES

For determination of the cooling time $t_{8/5}$ it is necessary to know the temperature cycles in the hard-faced layer's HAZ, where the temperatures range between the solidus temperature (T_s) and the A_{C1} temperature. Application of the shielded thermocouples enables measuring the temperature in that narrow zone, but taking into account their price and possibility of damages it was decided to use the thermocouples of unknown characteristics (from the class of type "K") of diameters 0.25 mm and 0.40 mm, whose tips are joined by the TIG welding procedure in the argon protected atmosphere. Detailed description of the procedure can be found in references, Lazić (2001). The biggest problem in those thermocouples is the unknown characteristics E-T (E is an electromotor force in mV while T is temperature in °C), which was to be determined. According to the corresponding calibrating schemes, Lazić (2001), the large number of such prepared thermocouples was selected, then their characteristics were determined and compared to a thermocouple of the known characteristics, i.e. thermocouples of type K (Chromel-Alumel, NiCr-NiAl). The thermocouple type K characteristic is:

$$E = 0.041 \cdot T , \qquad (4)$$

while characteristics of the applied, unknown thermocouples was determined based on calibration, both on lower temperatures (T ≤ 100°C), as well as on significantly higher temperatures (T= 300°C, 500°C and 800°C). The linear dependence E-T was established, which, after the statistical processing of obtained results, was:

$$E = 0.031 \cdot T + 0.314 . \qquad (5)$$

In Figure 3 are shown the E-T relationships for both used thermocouples. Checking of the measured values was also performed by the digital thermometer with the two measuring scales DTIM, with thermocouple NiCr-Ni.

181

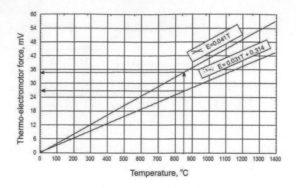

Figure 3. Dependence of the thermo-electromotor force on temperature.

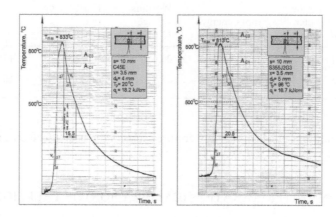

Figure 4. Temperature cycle of the HAZ (a) for steel C45E; (b) for steel S355 J2G3.

Table 2. Comparative values of the cooling time $t_{8/5}$ ($s = 10$ mm).

Electrode diameter d_e (mm)	Hard-facing driving energy q_p (J/cm)	Preheating temperature T_o/T_p (°C)	Cooling time $t_{8/5}$, (s)				Base metal
			$(t_{8/5})^J$	$(t_{8/5})^{Slim}$	$(t_{8/5})^{EXP}$	$(t_{8/5})^R$	
4.00	20082	20	19.6	57.7	20.5	31.0	
4.00	17650	20	16.1	44.6	18.5	25.5	
5.00	29400	138	54.6	245.4	44.5	79.0	C15
5.00	24758	36	28.3	95.3	29.0	44.0	
4.00	18200	20	16.9	47.4	15.5	26.8	
4.00	19413	36	19.6	58.6	20.0	31.5	C45E
5.00	16736	96	19.7	61.1	20.5	34.0	
5.00	40551	80	69.8	326.3	57.5	84.0	
5.00	34087	62	50.3	208.2	47.5	68.0	S355 J2G3
5.00	34588	20	44.2	171.2	40.0	60.0	

$(t_{8/5})^J$ – Ito-Bessyo formula; $(t_{8/5})^{Slim}$ – Limiting thickness formula; $(t_{8/5})^{EXP}$ – experiment; $(t_{8/5})^R$ – Rikalin's formula.

4 EXPERIMENTAL RESULTS OF TEMPERATURE CYCLES RECORDING DURING THE HARD-FACING

Connecting of the manual-held thermocouple with the recorder's poles (SERVOGOR S RE 541) enables continuous registering of the temperature variation with time. By adjusting the corresponding characteristics on the recorder (type of the scale, range of the scale, paper

Table 3. Comparative values of the cooling time $t_{8/5}$ ($s = 20$ mm and 21 mm).

Electrode diameter d_e (mm)	Hard-facing driving energy q_l (J/cm)	Preheating temperature T_o/T_p (°C)	Cooling time $t_{8/5}$, (s)				Base metal
			$(t_{8/5})^J$	$(t_{8/5})^{Slim}$	$(t_{8/5})^{EXP}$	$(t_{8/5})^R$	
4.00	19809	20	7.6	14.0	8.5	10.0	C45
4.00	17975	50	7.3	13.5	7.5	10.1	($s = 20$ mm)
4.00	21101	20	8.4	15.9	9.5	12.1	
5.00	37807	20	19.4	46.4	23.5	33.0	
5.00	48610	20	28.3	76.7	28.0	49.5	
5.00	42533	20	23.1	58.7	18.0	40.0	C45
5.00	35692	20	17.8	41.3	16.0	30.0	($s = 21$ mm)
5.00	35817	100	24.1	65.0	28.0	45.0	

Table 4. Comparative values of the cooling time $t_{8/5}$ ($s = 30$ mm).

Electrode diameter d_e (mm)	Hard-facing driving energy q_l (J/cm)	Preheating temperature T_o/T_p (°C)	Cooling time $t_{8/5}$, (s)				Base metal
			$(t_{8/5})^J$	$(t_{8/5})^{Slim}$	$(t_{8/5})^{EXP}$	$(t_{8/5})^R$	
5.00	28356	71	13.1	16.1	12.5	11.5	
5.00	38667	97	23.0	36.4	19.0	16.2	
5.00	34027	20	14.3	18.4	14.7	11.2	C35
5.00	34255	20	14.4	18.7	17.5	11.5	
5.00	26863	20	10.0	11.5	10.3	9.0	

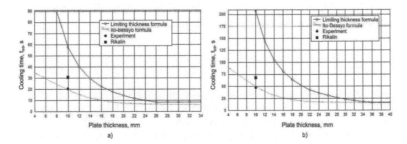

Figure 5. Comparative results of the cooling time (a) C15, $s = 10$ mm, $d_e = 4$ mm, $T_p = 20°C$, $q_l = 20082$ J/cm; (b) S355J2G3, $s = 10$ mm, $d_e = 5$ mm, $T_p = 62°C$, $q_l = 34017$ J/cm.

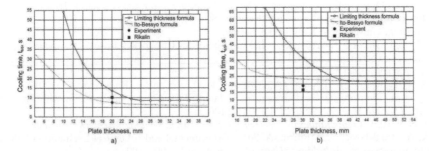

Figure 6. Comparative results of the cooling time (a) C45, $s = 20$ mm, $d_e = 4$ mm, $T_p = 50°C$, $q_l = 17975$ J/cm; (b) S355J2G3, $s = 30$ mm, $d_e = 5$ mm, $T_p = 97°C$, $q_l = 38667$ J/cm.

feeding rate, etc.) the optimal combination of those characteristics was obtained. In Figure 4 are shown the two characteristic temperature cycles obtained from recording by the manual-held thermocouples for two different materials, Lazić et al. (2014), Sedmak et al. (2017).

In Tables 2 to 4 and on diagrams in Figures 5 and 6 are presented the comparative values of the cooling time obtained for different hard-facing conditions.

5 CONCLUSIONS

Weldability of the carbon steels is generally good, however, almost always it must be checked by calculations. In calculating the cooling time $t_{8/5}$ a dilemma arises which formula to choose, while the experimentally obtained values are considered as the most accurate and reliable, Arsić et al. (2017). In other words, until now it was not known which formula gives results that are the closest to the curve of experimental cycle, or the curve of temperature cycle obtained by other methods.

By analyzing results from Tables 2 to 4 one can notice the unacceptable differences between the cooling time calculated from the formula $t_{8/5} = f(s_{lim})$ and experimental results. The best agreement with experimental results was obtained by the Ito-Bessyo formula. This conclusion is valid for hard-facing of the flat plates, while results for hard-facing of other surfaces were not considered.

ACKNOWLEDGEMENTS

This research was financially supported by European regional development fund and Slovak state budget by the project ITMS2014 + 313011D011 "Research Centre of the University of Žilina- 2nd phase" and by the project ITMS2014 + 313011D011T426 "Research and development activities of Žilina University in Žilina for 21st century industry in the field of materials and nanotechnologies" and by the Ministry of Education, Science and Technological Development of Republic of Serbia—grant TR 35024.

REFERENCES

Arsić, D., Lazić, V., Samardžić, I., Nikolić, R., Aleksandrović, S., Djordjević, M., Hadzima, B. 2015. Impact of the hard facing technology and the filler metal on tribological characteristics of the hard faced forging dies. *Technical Gazette*. 22(5): 1353–1358.

Arsić, D., Lazić, V., Nikolić, R., Hadzima, B. 2017. Weldability estimation of steels for hot work by the CCT diagrams, Advances and Trends in Engineering Sciences and Technologies, Editors: Mohamad Al Ali and Peter Platko, Leiden, The Netherlands, CRC Press: 9–14.

FIPROM Catalogue of Filler Materials, 2015. Jesenice: SŽ Fiprom.

Ito, Y., Bessyo, K. 1972. Weld crackability Formula of High Strength Steels. *Journal of Iron and Steel Institute*. 13: 916–930.

Lazić, V. 2001. Optimization of the Hard Facing Procedures from the Aspect of Tribological Characteristics of the Hard Faced Layers and Residual Stresses, Ph. D. Dissertation, Faculty of Mechanical Engineering, University of Kragujevac, Serbia.

Lazić, V., Sedmak, A., Živković, M., Aleksandrović, S., Čukić, R., Jovičić, R., Ivanović, I. 2010. Theoretical-experimental determining of cooling time (t8/5) in hard facing of steels for forging dies. *Thermal Science*. 14(1): 235–246.

Lazić, V.N., Ivanović, I.B., Sedmak, A.S., Rudolf R., Lazić, M.M., Radaković, Z.J. 2014. Numerical Analysis of Temperature Field during Hard facing Process and Comparison with Experimental Results. *Thermal Science*. 18(Suppl. 1): S113-S120.

Rikalin, N.N. 1951. *Computations of the Thermal Processes in Welding*. Moscow: Mashgiz.

Sedmak, A., Tanasković, D., Murariu, A. 2017. Experimental and Analytical Evaluation of Preheating during multi-pass repair welding. *Thermal Science*. 21(2): 1003–1009.

Numerical modeling of load-bearing structures of silos

M. Magura, J. Brodniansky Jr. & J. Brodniansky
Faculty of Civil Engineering, Slovak University of Technology in Bratislava, Bratislava, Slovakia

ABSTRACT: The most often used load bearing structures of silos are welded from plain sheets, or structures from corrugated galvanized sheets connected by bolts. Depending on the size and volume, they may be self-supporting or reinforced with vertical or horizontal ribs. For safe serviceability, and to meet the requirements related to economy, a suitable computer design model with correct loads is important. The correct model of the structure must capture the action of the structure and correctly present the results of the analysis. The ideal case occurs when the computer design can be verified by experimental measurement directly on the realized structure. This paper will summarize the basic types of silo structures and recommendations for their design using FEM computer modeling.

1 INTRODUCTION

Silo is a structure for storing bulk materials. Silos are used in agriculture to store grain (see grain elevators) or fermented feed known as silage. Silos are commonly used for bulk storage of grain, coal, cement, carbon black, woodchips, food products and sawdust. Structures welded from plain sheets (Figure 1), or silos from corrugated metal plates with bolt connections (Figure 2) are most the used options in tssswhe present time. These structures can be constructed either with or without vertical reinforcement ribs.

The computers and software enables us to model entire structures with their load in a number of load situations required by the engineering standards or the demands of the investor. Creating the proper model requires the experience of a designer who can choose the optimal model with consideration for all the structural and design details. It is also necessary to obtain relevant results from the program so that the structure can be economically designed and assessed. Incorrect interpretation of the results may result in over-dimensioning of the structure or, in the opposite case, in insufficient load-bearing capacity. In order to verify the functionality of the program, it is appropriate to verify the theoretical models based on in-situ experimental measurement.

Figure 1. Silo welded from plain sheet.

Figure 2. Silo screwed from corrugated sheets with vertical reinforcement ribs.

The load forces are described in detail in the current standards, but there is also space for research in this area—however, the presented article does not deal with this issue.

2 SILO LOAD BEARING STRUCTURE MODELLING

The following subchapter presents the various ways of modelling corrugated silo structure using the FEM. The presented results are in all cases from the same fragment of the structure with the same load. The results vary only as a result of a change in the coarseness of the finite element network, the type of solution (plastic/elastic), and the evaluated parameter (nodal/element stress). It is important to know when the results are as close as possible to the real behaviour of the structure.

2.1 *Geometry of FEM model of the silo from corrugated plates*

This subsection demonstrates the possibility of modelling a fragment of a silo with a diameter of 32.8 m and with a wall thickness of 8 mm. Structure is supported by line-hinge. The height of the model fragment is 3 m. Along the perimeter, vertical reinforcement ribs are applied at a distance of 1.264 m from each other, 80 pieces together. The horizontal pressure acting perpendicular to the wall is 72.6 N/m². The vertical force induced by friction of the load is 39.8 kN/m². These values are calculated by formulas from EN 1991-4: Eurocode 1 from real structure with height of 30 m. The material of walls is steel class S355. The bolt connection between plates is not modelled. Plates are between each other rigidly connected in place of ribs. Three types of models were assessed: Model A – the easiest way, replacing corrugated sheet metal with a smooth sheet (Figure 4). Model B – the waves are modelled as polygons (Figure 5). Model C – the surface of the sheet metal is curled, forming a sinusoid (Figure 6).

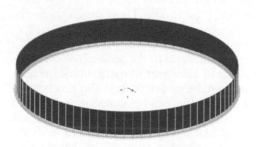

Figure 3.   Model of whole silo fragment.

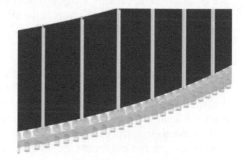

Figure 4.   Model A – flat plates.

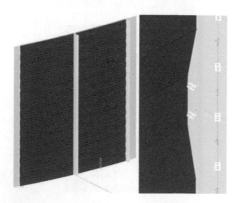

Figure 5.   Model B – trapezoidal plate cross section.

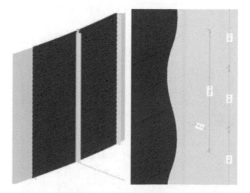

Figure 6.   Model C – sinusoidal plate cross section.

Table 1.   Model A – Comparison of a 3 m long vertical plate panel, elastic material properties.

Interpolated nodal results

Standard mesh generation (500 mm)	Forced mesh generation	Standard mesh generation (20 mm)

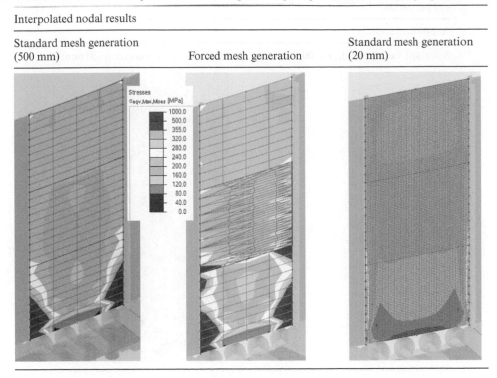

Table 2.   Model A – Comparison of a 3 m long vertical plate panel, elastic material properties.

Element centre results—average on element

Standard mesh generation (500 mm)	Forced mesh generation	Standard mesh generation (20 mm)

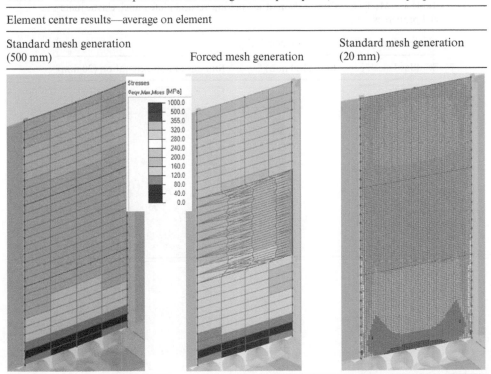

Table 3.   Model B – Comparison of a 3 m long vertical plate panel, elastic material properties.

Interpolated results

Standard mesh generation (100 mm)   Standard mesh generation (20 mm)

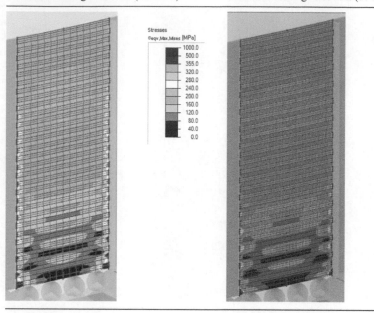

Table 4.   Model C – Comparison of a 3 m long vertical plate panel, elastic material properties.

Interpolated results

Standard mesh generation (100 mm)   Standard mesh generation (20 mm)

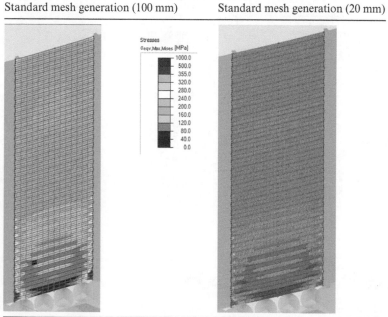

Table 5. Model A – Comparison of elastic and plastic solution.

Elastic—Interpolated nodal results	Elastic element centre results-average on element	Plastic element centre results-average on element

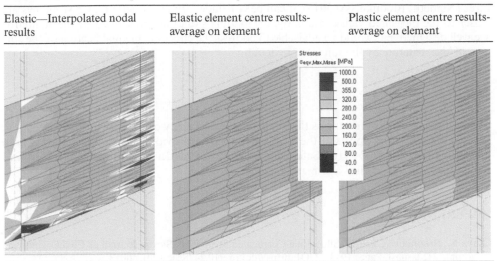

2.2 *FEM mesh and type of solution*

Another parameter influencing the quality of the results obtained from the solution is the choice of the size of the finite element mesh. All programs generate the network automatically, but in order to obtain more accurate and correct results, user intervention is required. Even more accurate results can be obtained by manually changing the shape and size of the elements. Several examples are shown in the following part.

It is obvious from presented results that the most disadvantageous is a triangular network with large elements with sharp side angles. If the engineer evaluated the structure only on the basis of elastic nodal results, he would be forced to increase the thickness of the sheets and the structure would thus be uneconomical. In case of serving for the assessment of an existing structure, it would provide negative results. By smoothening the FEM mesh and using the correct results, the structure meets the requirements of the standard.

When refining the network, node and element results begin to converge.

3 CONCLUSION

When designing new silo structures, as well as when assessing an existing one, it is necessary to consider their specificities. A suitable calculation model should not only capture the global geometry of the structure (diameter, height, sheet thickness), but also the geometry of individual structural elements. For example, it is incorrect to replace the corrugated sheet with a plane. The stress distribution and stiffness differ considerably when comparing corrugated sheet and plane sheet (Table 1 and Table 2). It is also necessary to take into account the means attaching the reinforcement elements.

Another factor influencing the calculation representativeness is the choice of the correct FEM network. Using large elements with bad geometry results in the occurrence of stress peaks which are particularly misleading when evaluating through nodal forces. A square mesh seems to be the most effective way to cover the corrugated sheet metal surface. Large areas with high stress levels are eliminated by smoothening the network. With a fine mesh, the difference between the results on nodes and elements is lost. The most realistic model is the calculation with plastic material. However, this solution demands significantly more calculation time.

Based on long-term measurements provided by Department of Steel and Timber Structures (Nemeth, Brodniansky, 2013), it can be concluded that well-deflated FEM models are closer to the measured values. Similarly, the loading situations prescribed by the standards are well designed, but would be advisable to address the effect of unequal heating of the surface of the structure by the sun in more detail.

ACKNOWLEDGEMENT

The authors would like to express thanks to the Grant agency of the Ministry of Education, Science, Research and Sports of the Slovak Republic for providing a grant from the research program VEGA Nr.1/0747/16 – Safety and reliability of modern load-bearing members and structures from metal, glass and membranes.

REFERENCES

Csaba N., Brodniansky J. 2013. Silo with a Corrugated Sheet Wall, Slovak Journal of Civil Engineering Volume 21: Issue 3, Pages 19–30, DOI: https://doi.org/10.2478/sjce-2013-0013.

EN 1991-3: Eurocode 3. Design of steel structures.Part 1–3: General rules. Supplementary rules for cold-formed members and sheeting.

EN 1991-4: Eurocode 1: Actions on structures – Part 4: Silos and tanks.

EN 1993-1-5: Eurocode 3: Design of steel structures. Part 1–5: Plated structural elements.

Iwicki, P., Rejowski K., Tejchman J. 2015. Stability of cylindrical steel silos composed of corrugated sheets and columns based on FE analyses versus Eurocode 3 approach, Engineering Failure Analysis Volume 57, November 2015, Pages 444–469, https://doi.org/10.1016/j.engfailanal.2015.08.017.

Kowalski, L. 2012. Theoretical and experimental analysis of the cylindrical shell construction. Diagnostics and monitoring of exposed steel structures – PhD. work, Bratislava, Slovakia 2012.

Tang, G., Yin, L., Guo, X. 2015. Finite element analysis and experimental research on mechanical performance of bolt connections of corrugated steel plates, International Journal of Steel Structures, vol. 15, Issue 1, pp. 193–204.

Advances and Trends in Engineering Sciences and Technologies III – Al Ali & Platko (Eds)
© 2019 Taylor & Francis Group, London, ISBN 978-0-367-07509-5

Detection of concrete thermal load time by acoustic NDT methods

M. Matysík, I. Plšková & Z. Chobola
Faculty of Civil Engineering, Brno University of Technology, Brno, Czech Republic

ABSTRACT: When increasing the temperature of the concrete, there are various chemical processes that cause mechanical and chemical changes of material. These changes have a significant impact on the physical and mechanical properties of concrete structures (modulus of elasticity, toughness, surface hardness), which are essential to meet the basic building requirements, especially mechanical strength, stability and fire safety. The paper presents the results from non-destructive testing of concrete specimens degraded by high temperatures. The test specimens were heated to 600°C. This temperature was then maintained for one, two and five hours. After heating, the samples were slowly cooled down to room temperature and then tested by acoustic non-destructive methods (Impact-echo method and one of the non-linear acoustic spectroscopy methods). The objective was to verify whether the parameters obtained by these methods are able to describe material degradation due to high temperature and residence time at this temperature.

1 INTRODUCTION

1.1 *Basic information*

Mechanical parameters of concrete are reduced due to high temperatures. Therefore, it is important to find NDT methods for fast testing of concrete damaged by high temperature. For concrete testing, we chose two acoustic methods: impact echo method and nonlinear acoustic spectroscopy method with one exciting signal. We have experience with the applications of both testing methods in civil engineering.

1.2 *Nonlinear acoustic spectroscopy methods*

Nonlinear acoustic spectroscopy is one of the newer methods of non-destructive acoustic testing, but it brings many new possibilities in the field of diagnostics and defectoscopy (Van Den Abeele 2001). The method of nonlinear acoustic spectroscopy deals with the study of nonlinear effects induced by propagation of elastic waves. It has been found that the non-linear response of the material is very sensitive to the presence of structural disorders and thus makes it possible to detect them. As a tool for diagnostics in the building industry it is accompanied by some specifics related to the great inhomogeneity of the material or the complexity of the shapes of the examined samples. Due to its composition concrete shows a certain degree of non-linearity in the undamaged state (Zardan et al. 2010). However, if defective zones are present in the test specimen, they become another source of non-linearity that ordinarily exceeds the specimen's own non-linearity (Novak et al. 2012).

Elastic waves that propagate through the material under investigation and in the presence of very small structural elements or structural interfaces generate strong nonlinear dynamic phenomena. These nonlinear phenomena are observable already at the beginning of material degradation, before changes to the linear parameters, which use classical methods of acoustic testing (Yim et al. 2012). When studying nonlinear phenomena, we observe some of their parameters, such as resonance frequency shift, highlighting of some harmonic frequencies, amplitude-dependent attenuation, etc. (Lesnicki et al. 2011).

A continuous signal or pulse excitation can be used to generate the elastic stress in the material. When using a continuous signal, the sample is excited by electroacoustic transducers. When using pulse excitation, we can use radio pulses or mechanical pulses.

In addition, we can divide these methods according to the resonance response of the test specimens, to resonance methods (for high resonance specimens) and non-resonance methods (for specimens with suppressed resonance expressions).

Nonlinear acoustic spectroscopy methods can be further divided by the number of excitation signals. If one excitation signal is used, we monitor the formation of other harmonic components and their change due to the presence of defects (Antonaci et al. 2010). Amplitudes of these harmonic frequencies generally decrease, but due to a small asymmetry of nonlinearity, some components may be suppressed or highlighted. For measurements with two or more excitation signals, a higher number of harmonic components result from nonlinearities (defects). The sum and differential components of both harmonic signals are generated. Mostly we monitor the amplitude of this difference component (Kober & Převorovský 2014).

1.3 *Impact-echo method*

The Impact-echo method is a classical non-destructive method. During the Impact-echo method, a short mechanical pulse is excited in the measured object. It can be excited by, for example, a piezoelectric exciter or a hammer stroke (Garbacz et al. 2017). The elastic wave spreads from the point of the strike to all sides. Here we can distinguish three basic types of waves, namely longitudinal, which are associated with pressure and tensile characteristics, transverse, which are associated with shear characteristics, and surface waves (Pazdera et al. 2015). As the mechanical wave passes through the solid mass, it encounters inhomogeneities, cracks and defects. At these interfaces, a partial absorption and reflection of the mechanical wave occurs. This will change the wave energy, which we can measure with a properly positioned sensor (Timcakova-Samarkova et al. 2016). The received signals are studied in the frequency domain. In this way, cavities, cracks, nests and other defects can be detected (Krzemień & Hager 2015).

2 EXPERIMENTAL SECTION

2.1 *Testing specimens*

Concrete beams were prepared according to the following mix design (composition for 1 m³): 339 kg of Portland cement CEM I 42.5 R Mokra, 161 kg of water, 2.9 kg of super-plasticizer Sika Viscocrete 2030, 849 kg of quartz sand 0/4 mm Zabcice and 981 kg gravel aggregate 8/16 Olbramovice. The dimensions of the beams were 10 × 10 × 40 cm. The specimens were 28 days soaked in water and then dried. First in the laboratory temperature and then 48 hours in the electric dryer at temperature 110°C.

The concrete beams were heated in programmable laboratory furnace Rhode KE 130B. The samples were heated to 600°C. The heating rate was 5 °C/min. The target temperature was kept for one, two and five hours. We had six pieces of samples from each temperature, 24 concrete beams together (6 of them without temperature loading).

2.2 *Measuring equipment for nonlinear acoustic spectroscopy method*

Due to acoustic attenuation of concrete samples, their shape and size, we have chosen non-resonance method with single continuous harmonic ultrasonic signal for measurement.

The transmitting part of apparatus consists of four functional blocks: a controlled-output-level harmonic signal generator, a low-distortion 100 W power amplifier, an output low-pass filter to suppress higher harmonic components and a sufficiently powerful piezo-ceramic exciter.

Receiving part consists of piezoceramic sensor, low noise preamplifier, amplifier with band—pass filters and oscilloscope HandyScope3 TPHS3-25.

The measuring equipment must meet: high dynamic range, very good resolution in the frequency domain, optimized sensor and transmitter location and high linearity of all instruments (generator, amplifier, sensor and transmitter). The entire measurement is computer controlled (Štoudek et al. 2016).

2.3 Measuring equipment for Impact-echo method

The mechanical impulse is caused by the impact of a steel hammer. The steel hammer is launched from a predetermined height. Measurement is still under the same conditions. The shock results in a low frequency stress wave which propagate through structure of material and is reflected by internal defects and external surfaces. The response in the form of surface waves is captured by the piezoelectric MIDI sensor. The sensor is attached to the sample with beeswax. The response from sensors is captured by a 16-bit TiePie engineering Handy scope HS3 two-channel oscilloscope. The recorded signal is then digitized, stored and processed using the control computer and the software package (Matysik et al. 2015).

3 EXPERIMENTAL RESULTS

3.1 Nonlinear acoustic spectroscopy method results

Figures 1–4 shows the typical frequency spectra of concrete beams obtained by non-resonance method with single continuous harmonic ultrasonic exciting signal. Figure 1 shows the frequency spectrum of the beam which has not been exposed to high temperature. In Figure 2 you can see the frequency spectrum of the beam which has been exposed to high temperature (600°C) for one hour, Figure 3 two hours, Figure 4 five hours.

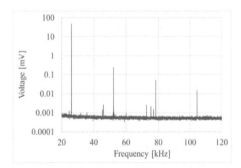

Figure 1. Frequency spectrum of the beam which has not been exposed to high temperature.

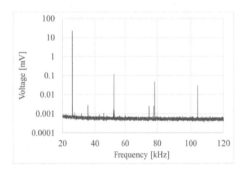

Figure 2. Frequency spectrum of the beam which has been exposed to 600°C for one hour.

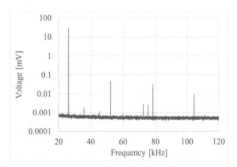

Figure 3. Frequency spectrum of the beam which has been exposed to 600°C for two hours.

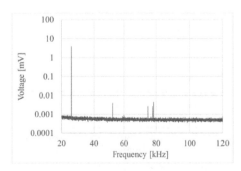

Figure 4. Frequency spectrum of the beam which has been exposed to 600°C for five hours.

We have focused our attention on the amplitude characteristics of the higher harmonic components (2f, 3f). The most appropriate parameter that describes the increasing nonlinearity and therefore characterizes the degree of damage is the amplitude ratio of the spectral components of the third and second harmonics 3f/2f (Hettler et al. 2012). The aggregate (for all beams) average results of parameter 3f/2f are listed in Figure 5.

3.2 *Impact-echo method results*

Figures 6–9 shows demonstration of concrete beams frequency spectra obtained by Impact-echo method. Figure 6 shows the frequency spectrum of the beam which has not been exposed to high temperature. In Figure 7 you can see the frequency spectrum of the beam which has been exposed to high temperature (600°C) for one hour, Figure 8 two hours, Figure 9 five hours. The data from all Impact-echo measurements are summarized in Figure 10. There is a clear shift in the dominant frequency with the increasing heat load time.

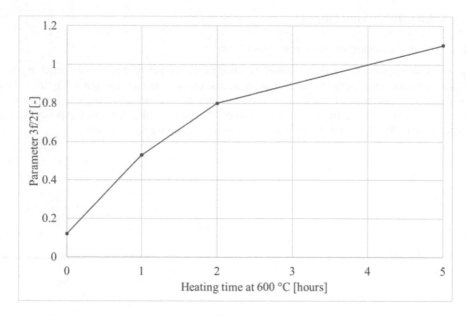

Figure 5.   The average results of parameter 3f/2f for all beams.

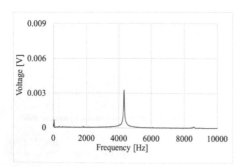

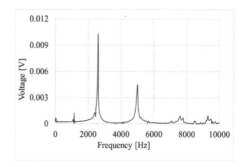

Figure 6.   Impact-echo freq. spectrum of the beam which has not been exposed to high temperature.

Figure 7.   Impact-echo freq. spectrum of the beam which has been exposed to 600°C for one hour.

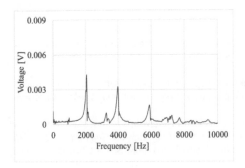

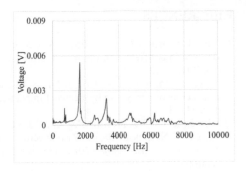

Figure 8. Impact-echo freq. spectrum of the beam which has been exposed to 600°C for two hours.

Figure 9. Impact-echo freq. spectrum of the beam which has been exposed to 600°C for five hours.

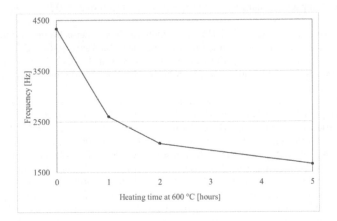

Figure 10. The average position of dominant frequency positon depending on temperature.

4 CONCLUSION

The aim of the paper was to document the sensitivity of two acoustic non-destructive methods for monitoring the condition of concrete elements damaged by high temperature. For nonlinear acoustic spectroscopy method described in the text, a suitable method for characterizing the damage was the analysis of higher harmonic frequencies. We especially concentrated on the ratio of third and second harmonic components (3f/2f). Thermal damage in concrete specimens caused an increase in even and suppression of odd amplitudes of harmonic frequencies. This is the reason for the increase of the 3f/2f parameter. The results obtained by this method very well agree with the results obtained by the Impact-echo method. For Impact-echo, increasing duration of thermal stress caused a reduction of the dominant frequency. The increasing damage of the samples was observable also visually. During longer periods of thermal stress, macroscopic cracks appeared on the surface of the samples.

ACKNOWLEDGMENT

This paper has been worked out under the project No. LO1408 "AdMaS UP—Advanced Materials, Structures and Technologies", supported by Ministry of Education, Youth and Sports under the "National Sustainability Programme I" and the project GAČR No.16–02261S supported by Czech Science Foundation.

REFERENCES

Antonaci P., Bruno, C.L.E., Gliozzi, A.S. & Scalerandi, M. 2010. Monitoring evolution of compressive damage in concrete with linear and nonlinear ultrasonic methods. *Cement and Concrete Research.* 40 (7): 1106–1113.

Garbacz, A., Piotrowski, T., Courard, L. & Kwaśniewski, L. 2017. On the evaluation of interface quality in concrete repair system by means of impact-echo signal analysis. *Construction and Building Materials.* 134: 311–323.

Hettler, J., Kober J. & Převorovský, Z. 2012. Monitoring of the Damage Evolution in Concrete Slabs by Means of Nonlinear Elastic Wave Spectroscopy. *Proc. of the 42nd International Conf. on NDT 'NDE for Safety – Defektoskopie 2012', October 29 – November 1, 2012.* 301–308, Seč u Chrudimi, Czech republic.

Kober, J. & Převorovský, Z. 2014. Theoretical investigation of nonlinear ultrasonic wave modulation spectroscopy at crack interface. *NDT & E International.* 61: 10–15.

Krzemień, K. & Hager, I. 2015 Post-fire assessment of mechanical properties of concrete with the use of the impact-echo method. *Construction and Building Materials.* 96: 155–163.

Lesnicki, K.J., Kim, J.-Y., Kurtis, K. & Jacobs, L.J. 2011. Characterization of ASR damage in concrete using nonlinear impact resonance acoustic spectroscopy technique. *NDT & E International.* 44 (8): 721–727.

Matysik, M., Plskova, I. & Chobola, Z. 2015. Assessment of the impact-echo method for monitoring the long-standing frost resistance of ceramic tiles. *Materiali in Tehnologije.* 49 (4): 639–643.

Novak, A., Bentahar, M., Tournat, V., El Guerjouma R. & Simon, L. 2012. Nonlinear acoustic characterization of micro-damaged materials through higher harmonic resonance analysis. *NDT & E International.* 45(1): 1–8.

Pazdera, L., Topolář, L., Smutný, J., Timčaková, K. 2015. Nondestructive Testing of Advanced Concrete Structure during Lifetime. *Advances in Materials Science and Engineering.* 2015 (7): 1–5.

Timcakova-Samarkova, K., Matysík, M. & Chobola, Z. 2016. Possibilities of NUS and impact-echo methods for monitoring steel corrosion in concrete. *Materiali in Tehnologije.* 50 (4): 565–570.

Van Den Abeele, K., Sutin, A., Carmeliet, J. & Johnson, P.A. 2001. Micro-damage diagnostics using nonlinear elastic wave spectroscopy (NEWS). *NDT & E International.* 34 (4): 239–248.

Yim, H.J., Kim, J.H., Park S.-J. & Kwak, H.-G. 2012. Characterization of thermally damaged concrete using a nonlinear ultrasonic method. *Cement and Concrete Research.* 42 (11): 1438–1446.

Zardan,J.-P., Payan,C., Garnier,V. & Salin, J. 2010. Effect of the presence and size of a localized nonlinear source in concrete. *The Journal of the Acoustical Society of America.* 128 (1): 38–42.

Štoudek, R., Trčka, T., Matysík, M., Vymazal, T. & Plšková, I. 2016. Acoustic and Electromagnetic Emission of Lightweight Concrete with Polypropylene Fibers. *Materiali in tehnologije,* 50(4): 547–552.

Advances and Trends in Engineering Sciences and Technologies III – Al Ali & Platko (Eds)
© 2019 Taylor & Francis Group, London, ISBN 978-0-367-07509-5

The analysis and effects of flow acoustic in a commercial automotive exhaust system

B. Mohamad
Faculty of Mechanical Engineering and Informatics, University of Miskolc, Miskolc, Hungary

S. Amroune
Université Mohamed Boudiaf, M'sila, Algérie

ABSTRACT: One of the most valuable criteria for vehicle quality assessment is based on acoustic emission levels. Unstable exhaust gas at high temperature flowing from internal combustion engine manifold may cause noise and disturbance conflicting with the high standard of acoustic comfort requested by this kind of vehicle. This research was involved in carrying out flow simulation model and analysis on an exhaust system design, and based on virtual data output from CFD tools and AVL-Boost software aiming at identifying the natural frequencies and mode shapes of its structural components, to ensure that external excitation sources do not lead to resonances. The objective of this study was to analyze the dynamic behaviour and to identify the overall characteristics of the exhaust gas, so that components sensitive to induced turbulent vortices may be identified and assessed relative to acceptance criteria for acoustic levels requested by technical specification.

1 INTRODUCTION

The exhaust system collects the exhaust gases from the cylinders, removes harmful substances, reduces the level of noise and discharges the purified exhaust gases at a suitable point of the vehicle away from its occupants. The exhaust system can consist of one or two channels depending on the engine. The flow resistance must be selected so that the exhaust backpressure affects engine performance as little as possible. Barhm et al. (2017) used 1D Boost software to describe the effect of using different blend fuels on engine performance and exhaust properties. The result show variation of outlet temperature and flow distribution by using different volume percentage of alcohol-gasoline blends. Barhm et al. (2018(studied the effect of Ethanol-Gasoline blend fuel on engine power output and emissions, the literature's results show great improvement in combustion process and exhaust gas characteristics. Barhm et al. (2017) presented in their technical paper a review of a muffler used in the industry, and this review depicts flow and temperature distribution along the muffler ducts. The techniques for different methods used in the design, calculation and construction of muffler both experimentally, practically and transmission loss characteristics were described. 1D calculations are much faster, and still give a good overview of the system under investigation. By being fast, they are also more suitable to run in optimization loops, which is a desirable feature in the area of exhaust system design, since mufflers are becoming more and more complex (Tonković et al. 2012). There are two basic different principles here: reflection of the sound waves and dissipation of the acoustic energy in the muffler. The mufflers based on the first principle are said to be reactive and are used to mitigate sound consisting of discrete tones, especially in low frequency region. The mufflers based on the second principle are called resistive, and are most suited dealing with high frequency, broadband noise. These two principles are generally combined in a single muffler (Macchiavello et al. 2016). This acoustic level resembles the crack of an explosion and must be reduced by approx. Catalytic converters also have a sound-absorbing effect. The exhaust system is itself a system subject to vibration, it produces noise through natural frequencies and vibration, which are transmitted to the car body (Walker Autolexikon 2016).

2 MATHEMATICAL MODEL

2.1 *Governing equations*

The sound usually generated because of the coupling between the turbulent average flow field and the acoustic field is said to be self-excited. Under the condition of multidimensional compressible steady flow, the Mass and momentum conservation equation are as follows (Jianmin et al. 2014).

$$\frac{\partial}{\partial X_j}\left(\rho u_j\right) = 0 \tag{1}$$

$$\frac{\partial}{\partial X_j}\left(\rho u_j u_i - \tau_{ij}\right) = -\frac{\partial P}{\partial X_{ij}} + Si \tag{2}$$

where S_i is source term, it represents perforate part resistance. τ_{ij} is stress tensor, for Newtonian flow there are:

$$\tau_{ij} = 2\mu\left(S_{ij} - \frac{1}{3}\frac{\partial u_k}{\partial X_k}S_{ij}\right) - \overline{\rho u_i u_j} \tag{3}$$

where μ is molecular dynamic viscosity coefficient, S_{ij} is Kroneker number, $\overline{\rho u_i u_j}$ is Reynolds stress tensor, and fluid deformation rate tensor is given by the following formula:

$$S_{ij} = \frac{1}{2}\left(\frac{\partial u_i}{\partial X_j} + \frac{\partial u_j}{\partial X_i}\right) \tag{4}$$

2.2 *Turbulent model*

When the gas flows between the chambers, through the perforated ducts, part of the acoustic energy convert into turbulent vortices that form at perforations in duct. Using a standard k-ε model to calculate the Reynolds stress to solve the flow control equations above (Elnady et al. 2010 and Alfredson et al. 2006), there are:

$$\overline{\rho u_i u_j} = -2\mu_t S_{ij} + \frac{2}{3}\left(\mu_t\frac{\partial u_k}{\partial x_k} + \rho k\right)S_{ij} \tag{5}$$

where μ_t is the turbulent viscosity, it is given by the following formula:

$$\mu_t = \frac{C_{ij}\rho K^2}{\varepsilon} \tag{6}$$

k and ε are turbulent kinetic energy and turbulent energy dissipation rate respectively. Their transport equations are:

$$\frac{\partial}{\partial X_j}\left(\rho\mu_j k - \frac{\mu_{eff}}{\sigma_k}\frac{\partial k}{\partial x_j}\right) = \mu_\tau s_{ij}\frac{\partial u_i}{\partial x_j} - \rho\varepsilon - \frac{2}{3}\left(\mu_\tau\frac{\partial u_i}{\partial x_i} + \rho k\right)\frac{\partial u_i}{\partial x_j}$$

$$\frac{\partial}{\partial x_j}\left(\rho\mu j\varepsilon - \frac{\mu_{eff}}{\sigma_\varepsilon}\frac{\partial\varepsilon}{\partial x_j}\right) = C_{\varepsilon 1}\frac{\varepsilon}{k}\left\{\mu_\tau s_{ij}\frac{\partial u_i}{\partial x_j} - \frac{2}{3}\left(\mu_\tau\frac{\partial u_i}{\partial x_i} + \rho k\right)\frac{\partial u_i}{\partial x_i}\right\} \qquad (7)$$

$$- C_{\varepsilon 2}\rho\frac{\varepsilon^2}{k} + C_{\varepsilon 4}\rho\varepsilon\frac{\partial u_i}{\partial x_i}$$

In the formula above $\mu_{eff} = \mu + \mu_t$; the empirical coefficients $C_\mu, \sigma_k, \sigma_\varepsilon, C_{\varepsilon 1} C_{\varepsilon 2} C_4$ of determined according to (Table 1).

2.3 Geometrical model

The entire model of the system was in (iges) files, representing exhaust system. The muffler has both resistive and reflective parts. The reflective part consists of one perforate chamber, and one channel separated by partitioning wall. The process of creation of the CFD model, starting from the (iges) files of the exhaust system, was entirely developed taking advantage of 3D Ansys—Fluent.

2.4 Mesh generation

Grid generation-the sub-division of the domain into a number of smaller subdomains: a grid (or mesh) of cells (or control volumes or elements).

Selection of the physical and chemical phenomena that need to be modelled and the definition of fluid properties. For this research work. As shown in Figure 2, the muffler, porous and ducts the mesh was generated with 9.45 million elements.

Table 1. Coefficient of experience (Jianmin et al. 2014).

Cμ	σk	σε	Cε1	Cε2	Cε4	Cμ
0.09	1.0	1.22	1.44	1.92	−0.33	0.09

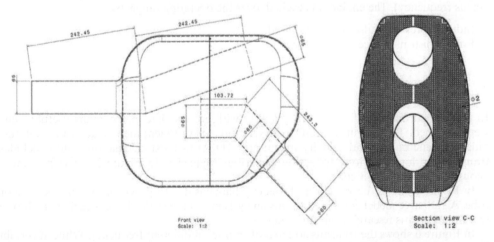

Figure 1. (a) Three-dimensional model of exhaust duct and muffler (left), (b) Porous plate part in middle of chamber (right).

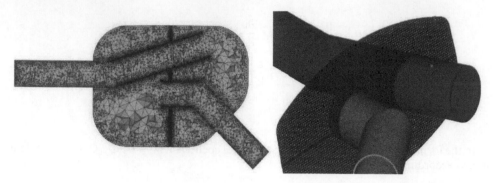

Figure 2. (a) Tetrahedral mesh generated for the entire model (left), (b) Porous plate mesh (right).

Table 2. Main data.

Boundary conditions	
Inlet temperature	480 K
Inlet velocity	220 m/s
Inlet pressure	211000 kPa
Mass flowrate	0.05 kg.s^{-1}

3 BOUNDARY CONDITIONS

3.1 *Main data*

The boundary conditions considered for the flow acoustic analysis are representative of the exhaust system advancing at constant engine speed. At the inlet into the geometry the velocity (u) and the turbulence quantities k and ϵ were assigned. The pressure p = 0 bar was assigned at the outlet as relative (Ansys solver 2010). Then same parameter was adding to 1D AVL-Boost software.

The flow acoustic analysis was performed as a frequency response analysis (amplitude versus frequency). The engine was excited, over the operating range, by:

- Induced pressure pulses from manifold,
- Engine pistons induced heat energy.

4 RESULTS

In the current analysis of automotive muffler with porous plate inside, many useful results were obtained. The output from 3D analysis using Ansys-Fluent include acoustic level, turbulent kinetic energy and velocity magnitude. 1D AVL-Boost calculation results includes transmission loss and noise reduction at various frequency. In Figure 3 shows maximum acoustic energy at outlet zone.

In Figure 5 shows the variation of velocity happens due to provision of porous plate or tube. A radiative model combined to this analysis may address this issue, which is the future work plan of this research from this analysis.

In Figure 6 shows the transmission loss of muffler at different frequency. While at certain frequency the existence of the porous plate started to influence the transmission loss of the muffler that frequencies related to the design type.

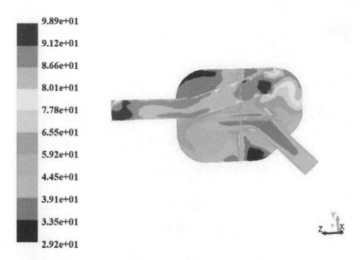

Figure 3.   Component of acoustic energy level (dB) versus frequency for the model.

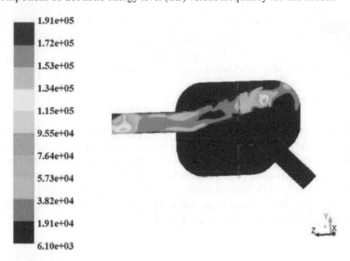

Figure 4.   Contours of turbulent kinetic energy (m²/s²).

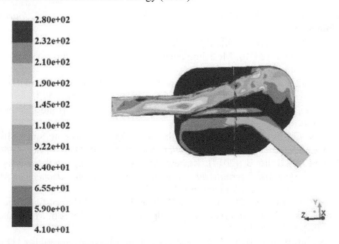

Figure 5.   Velocity variation of muffler (m/s).

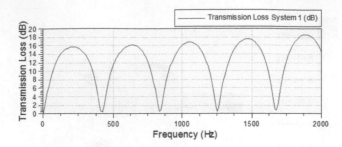

Figure 6. Transmission loss versus frequency for the model.

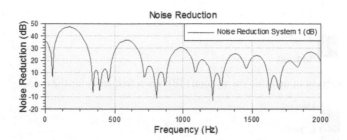

Figure 7. Noise reduction, pulse—frequency response function.

5 CONCLUSIONS

A 3D CFD simulation was done to a part of the exhaust system which is the muffler for describing acoustic energy distribution and turbulent kinetic energy condition. The effect of variation of velocity as properties of exhaust system was show in the results. Now it's necessary to define all the micro-aerodynamic geometries of every single component in detail. Afterwards, a further 1D optimization phase will define the final layout of the exhaust system performance. Further research in this area can also be extended to test the engine exhaust muffler with different design and absorptive materials.

REFERENCES

Alfredson J., Davies P. (2006). The radiation of sound from an engine exhaust. *Journal of Sound and Vibration*. Virginia, U.S.A.

ANSYS (2010). Solver Theory Guide, ANSYS FLUENT Release 14.0 Manual.

Elnady T., Abom M., Allam S (2010). Modeling perforates in mufflers using two-ports, ASME Journal of *Vibration and Acoustics*; 132, 1–11.

Jianmin X., Shuiting Z. (2014). Analysis of flow field for automotive exhaust system based on computational fluid dynamics. *The Open Mechanical Engineering*: 8, 587–593.

Macchiavello S., Tonelli A. (2016). Pressure vessel vibration and noise finite element analysis. *CAE Conference*. Parma: Italy.

Mohamad B., Szepesi G., Bolló B. (2017). Combustion Optimization in Spark Ignition Engines. *MultiScience—XXXI. microCAD Scientific Conference*. University of Miskolc: Hungary.

Mohamad B., Szepesi G., Bolló B. (2017). Review Article: Modelling and Analysis of A Gasoline Engine Exhaust Gas Systems. *International Scientific Conference on Advances in Mechanical Engineering*. University of Debrecen: Hungary. pp. 345–357.

Mohamad B., Szepesi G., Bolló B. (2018). Review Article: Effect of Ethanol-Gasoline Fuel Blends on the Exhaust Emissions and Characteristics of SI Engines. Lecture Notes in Mechanical Engineering. 29–41. 10.1007/978-3-319-75677-6_3.

Tonković D., Putz N., Juretić F. (2012). 3D exhaust system (muffler) design tool for 1D CFD simulation purposes. *International design conference*. Dubrovnik: Croatia.

Walker Autolexikon Co. (2016). Modern design and effectiveness of vehicle exhaust system. *Annual report*.

Advances and Trends in Engineering Sciences and Technologies III – Al Ali & Platko (Eds)
© *2019 Taylor & Francis Group, London, ISBN 978-0-367-07509-5*

Application of the firefly algorithm for the optimization of cranes

Sz. Nagy & K. Jármai
University of Miskolc, Miskolc, Hungary

ABSTRACT: Evolutionary algorithms are well used for solving nonlinear engineering optimization problems. The firefly algorithm can be used to efficiently compare to another technics. We have made comparisons to different metaheuristic technics on standard test functions and we use FA for solving real life problems like optimization of overhead traveling cranes. We have considered welded box main girders. The objective function is cost the crane girder including the material, the welding preparation, the real welding and the additional welding costs. The design constraints are static stress, local buckling, fatigue and deflection. The unknowns are the four sizes of the box beam. Parametric inspections have been made changing the span length, load size, number of load cycles and steel grade.

1 INTRODUCTION

Metaheuristic and nature inspired evolutionary algorithms are efficiently used for solving non-linear engineering problems, such as multi dimensional optimization problems. In the last few years, many algorithms were developed, for example random search technics RS, differential evolution DE (Storn et al. 1997), firefly algorithm FA (Yang, 2010), the hybrid combination hFADE (Zhang et al. 2016) and the multilevel combination mRSFA (Kota et al. 2018).

The most evolutionary algorithm is developed for solving continuous, unconstrained optimization problems. In the real word most engineering problems are constrained.

$$\min f(\bar{x}) \quad \bar{x} = \left\{ x_1, x_2, \ldots, x_n \right\} \in \mathbb{R}^n$$
$$g_i(\bar{x}) \leq 0 \quad i = 1, \ldots, q \tag{1}$$
$$h_j(\bar{x}) = 0 \quad j = q+1, \ldots m$$

where $\bar{x} = \left\{ x_1, x_2, \ldots, x_n \right\}$ is vector of solution, design variables. There are q inequality and $m - q$ equality constraints, $f(\bar{x})$ is the objective function.

There are several approaches proposed to handle constrained problems (Yeniay, 2005). One of the groups of these approaches is the penalty function. In this paper the death penalty function is used.

$$min\, f(\bar{x}) + \sum_{i=1}^{q} P_i(\bar{x}) \quad P_i(\bar{x}) = \begin{cases} 0 & g_i(\bar{x}) \leq 0 \\ \infty & else \end{cases} \tag{2}$$

2 FIREFLY ALGORITHM

Firefly algorithm is inspired by the flashing behaviour of tropical fireflies. This algorithm uses the following three idealized rules:

− all fireflies are unisex. One firefly will be attracted to other firefly.
− the attractiveness is proportional to the brightness and distance between two fireflies
− the brightness of firefly depend from quality of objective function's fitness

Pseudo code of basic steps is summarized in (Figure 1).
Light intensity defined following equation

$$I = I_0 e^{-\gamma r^2} \tag{3}$$

where γ is a light absorption coefficient, r is distance and $I_0 \propto f(\bar{x})$ is light intensity at $r = 0$. It could be faster than calculate exponential function if we use any simpler $I \propto r, f(\bar{x})$ function. Attractiveness between two fireflies is

$$\beta(r_{ij}) = \beta_0 e^{-\gamma r_{ij}^2} \tag{4}$$

where r_{ij} is Cartesian distance between two fireflies i and j and β_0 is attractiveness at $r_{ij} = 0$.
The movement of a firefly to brighter is

$$x_i = x_i + \beta(r_{ij}) + \alpha\left(rand - \frac{1}{2}\right) \tag{5}$$

> *Generate initial population*
> *Calculate light intensity I_i at x_i.*
> *Define light absorption coefficient γ*
> **while** *exit criteria is not true*
>   **for** *i=1 → all fireflies*
>   **for** *j=1 → all fireflies*
>   **if** *$I_j > I_i$*
>   *Move firefly i towards j*
>   **end if**
>   *Attractiveness varies with distance r via $e^{-\gamma r}$*
>   *Evaluate new solutions and update light intensity*
>   **end for** *j*
>   **end for** *i*
>   *Rank the fireflies and find the current best*
> **end while**
> *Postprocess results and visualization*

Figure 1.   Data and cross.

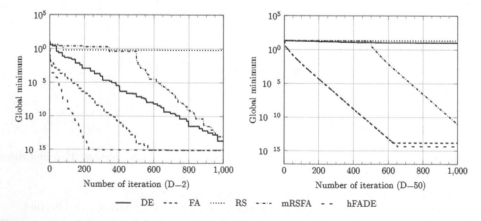

DE --- FA ······· RS ---- mRSFA -- hFADE

Figure 2.   Result of simulation with Ackley's function.

where α is randomization parameter, that generates some random noise in the optimum search process.

The second term of (Equation 5) is the attraction, which can be imagined as primitive communication between two fireflies. This property makes the firefly algorithm to an effective metaherustic search algorithm. Performance simulation with standard functions (Szilárd et al. 2018) and other metaherustic algorithms are shown in (Figures 2–5).

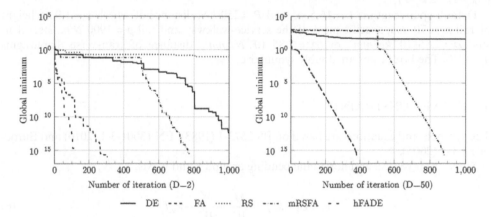

Figure 3.   Result of simulation with Griewank function.

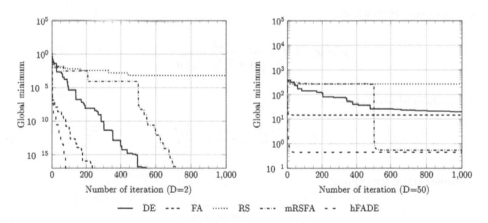

Figure 4.   Result of simulation with Levy function.

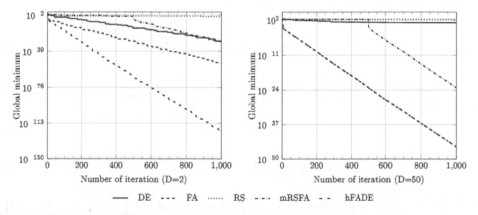

Figure 5.   Result of simulation with Sphere function.

3 THE TREATED CRANE

In this paper main girder of overhead travelling crane is designed as a double box beam. The rail placed over the inner web of the box beam.

Dynamic factor $\Psi_d = 1.3$ of workshop crane is selected from BS 2573-1 (1983). The coefficient of the spectrum is according to EN 13001-3-1 (2010) $s_3 = 2$. The safety factor for fatigue is $k_x = k_y = k_\tau = 1$.

Default span length is $L = 16,5\ m$, load $P = 250\ kN$, distance of wheels $k = 1,9\ m$, height of rail $h_s = 70\ mm$, specific mass of the service-walkway and rail $p = 1900\ N/m$, steel density $\rho = 7,85 \cdot 10^{-6}\ kg/mm^3$ or $\rho_0 = 7,85 \cdot 10^{-5}\ N/mm^3$, distance of transverse diaphragms $a = L/10$. The box beams are doubly symmetric.

4 ACTIVE CONSTRAINTS

The symbols and equations are based on BS 2573-1 (1983), EN 13001-3-1 (2010) and Eurocode 3-1-9 (2005).

Stress from vertical and horizontal bending around x and y axis:

$$g_1 = \sigma_z - k_x f_y = \frac{M_x}{W_x} + \frac{M_y}{W_y} - k_x f_y \tag{6}$$

Compression from wheel:

$$g_2 = \sigma_y - k_y f_y = \frac{2F}{\left[50 + 2\left(h_s + t_f\right)\right]t_w} - k_y f_y \tag{7}$$

Torsional and shear stress:

$$g_3 = \tau - \frac{k_\tau f_y}{\sqrt{3}} = \tau_V + \tau_t - \frac{k_\tau f_y}{\sqrt{3}} \tag{8}$$

Complex static stress limit:

$$g_4 = \sqrt{\sigma_z^2 + \sigma_y^2 - \sigma_z \sigma_y + 3\tau^2} - f_y \tag{9}$$

Local buckling limits for web plate and flange:

$$g_5 = \frac{2h}{t_w} - 0.67 \cdot 28.42\varepsilon\sqrt{k_{\sigma x}}; \varepsilon = \sqrt{235 / f_y} \tag{10}$$

$$g_6 = \frac{2h}{t_w} - 31\varepsilon\sqrt{k_\tau}; \ k_\tau = 5.34 + \frac{4}{\alpha^2}; \ \alpha = \frac{L}{10h} \tag{11}$$

$$g_7 = \frac{2h}{t_w} - 60.67 \tag{12}$$

$$g_8 = \frac{b}{t_f} - 0.67 \cdot 28.42\varepsilon\sqrt{k_{\sigma y}} \tag{13}$$

$$g_9 = \frac{b}{t_f} - 31\varepsilon\sqrt{k_{\tau b}}; \ k_{\tau b} = 5.34 + \frac{4}{\alpha_b^2}; \ \alpha_b = \frac{L}{10b} \tag{14}$$

206

Fatigue constraints for the weld under the rail:

$$g_{10} = \left(\frac{\sigma_z}{\Delta\sigma_{Rd}}\right)^3 + \left(\frac{\sigma_y}{\Delta\sigma_{Rd}}\right)^3 + \left(\frac{\tau}{\Delta\tau_{Rd}}\right)^5 - 1 \tag{15}$$

5 THE COST FUNCTION

The cost function is consisting of more components.

$$f(\bar{x}) = K_m + \sum_i K_{wi} + K_p \tag{16}$$

where K_m is material cost, K_p is post welding cost and K_{wi} is cost of i-th welding according to (Farkas & Jármai 2015)

$$K_{wi} = k_w \left(\Theta_i \sqrt{\kappa_i \rho V_i} + 1.3 C_i a_{wi}^n L_{wi}\right) \tag{17}$$

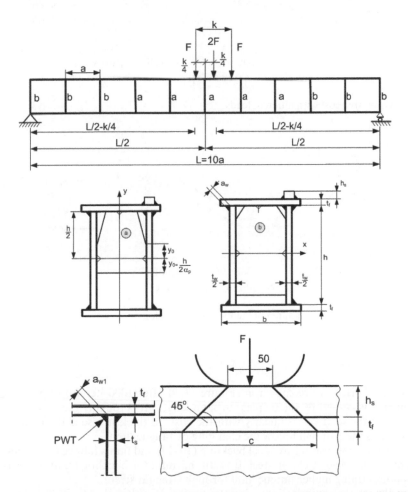

Figure 6.   Data and cross-section of the crane beams. Diaphragms are used in the middle of beams for high bending stresses. Diaphragms are used near the beam ends. The PWT is used for the welds joining the diaphragms. Load distribution in the beam web from the crane wheel.

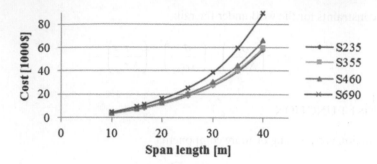

Figure 7.   Optimized cost with different length.

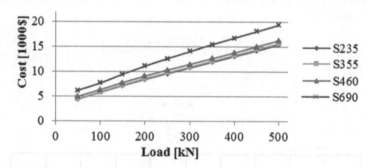

Figure 8.   Optimized cost with different loads.

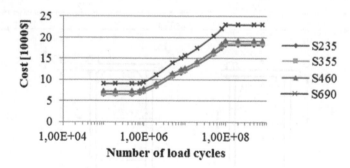

Figure 9.   Optimized cost with different number of load cycles.

6   CONCLUSION

Results of optimisation are shown in (Figures 7–9) after 1000 iteration steps. The largest change in cost is caused by the increase in span length. It is roughly growing exponentially. The change in the load varies linearly with the cost. In the last case if the number of load cycles increases the cost will follow the changing of fatigue limit.

The governing constraints are local buckling (10–14) and limits fatigue (15). These are not dependent on the yield stress of steel. It is still harmful for buckling limits. Eurocode 3-1-9 (2010) is not defining higher fatigue limit for high strength steel.

The results of the optimization using the firefly algorithm along with the previous reasons are shown that is no advantage in using more expensive high strength steel in these applications.

REFERENCES

BS 2573-1:1983. Rules for the design of cranes. Part 1: Specification for classification, stress calculations and design for structures.

EN 13001-3-1:2010. Cranes – General design – Part 3-1: Limit states and proof competence of steel structure.

Eurocode 3-1-9:2005. Design of steel structures. Fatigue strength of steel structures.

Farkas, J. & Jármai, K. 2015. *Fémszerkezetek innovative tervezése.* Miskolc: Gazdász-ElasztikKiadóés Nyomda. (Innovative design of metal structures). (In Hungarian), 625p.

Kota, L. & Jármai, K. 2018. Application of Multilevel Optimization Algorithms *Advances in Structural and Multidisciplinary Optimization* 12:710–715.

Storn, R. & Price, K. 1997. Differential Evolution – A Simple and Efficient Heuristic for global Optimization over Continuous Spaces Journal of Global Optimization 11: 341–359.

Szilárd, N. & Jármai, K. 2018. Alap, hybrid éstöbbszintűevolúciósalgoritmusok (Basic, hybrid and multilevel evolutionary algorithms) *GÉP* (In Hungarian) 2: 44–51.

Yang, X.S. 2010. Firefly algorithms for multimodal optimization *Lecture Notes in Computer Sciences* 5792: 169–178.

Yeniay, Ö. 2005. Penalty function methods for constrained optimization with genetic algorithms *Mathematical and Computational Applications* 10: 45–56.

Zhang, L., Liu, L., Xin-She, Y. & Dai, Y. 2016. Novel Hybrid Firefly Algorithm for Global Optimization PLoS ONE 11: 1–17.

Advances and Trends in Engineering Sciences and Technologies III – Al Ali & Platko (Eds)
© 2019 Taylor & Francis Group, London, ISBN 978-0-367-07509-5

Numerical analysis of soil-slab interaction—influence of the input parameters

Z. Neuwirthova & R. Cajka
VŠB-Technical University of Ostrava, Ostrava, Czech Republic

ABSTRACT: The problematic of soil-structure interaction has been under research for many years, but the optimal procedure of numerical modeling of such task is still not known and the results are insufficient. For this reason, parametric study was created to clarify influence of the input parameters of boundary conditions as well as dimensions of the selected model to the final deformation of the slab on the elastic sub-soil. For this purpose, a computational model of the slab on the flexible subsoil was created in Ansys 18.0. Soil model is based on the elastic half-space theory assuming that large enough cube can substitute the half-space. Soil was represented as a finite element cube with different dimensions. Finally, three different types of boundary condition were applied and compared. Influence of the boundary conditions is observed.

1 INTRODUCTION

Interaction between structure and subsoil has been the subject of many researches in the recent decades. Experimental tests go hand-in-hand with computer simulations, which is a consequence of the rapidly evolving area of computer technology. Scientists have been trying to simulate the behavior of the soil using computers for many years now (Labudkova & Cajka 2016, Sadecka 2000, Shams et al. 2018) and the development of interaction models is still the subject of research as a sufficient precision of the modeling has not been reached (Cajka & Labudkova 2016). For this reason, experimental measurements are carried out not only in the Czech Republic (Cajka et al. 2011), but also around the world (Alani & Aboutalebi 2012, Wijffels et al. 2017) to compare the model results and refine calculation methods.

Therefore, we are trying to find the optimal procedure and parameters of the calculation in order to simulate the behavior of the subsoil as reliably as possible. The main problem is that the behavior of the model is affected by many parameters, including the choice of the computing models, size of the finite element mesh, type of contact but also dimensions of the modeled region or boundary conditions (Cajka & Labudkova 2014). Therefore, it is necessary to examine these parts and calculation parameters individually to know how they influence the results before solving large-scale tasks. This step is crucial for better understanding of the investigated phenomena. Influence of the boundary conditions and influence of the size of the modeling region to the total deformation is the subject of this article.

2 METHODS

The computational 3D model was created using the finite element method in the computing environment Ansys 18.0. The numerical model (Figure 1) consists of two parts—a slab and a soil.

The behavior of the soil is based on the homogenous elastic half-space theory. The elastic half-space is replaced by the cube with finite dimensions. The dimensions vary to observe their dependence to the results of the calculations. Concrete slab has dimensions $2 \times 2 \times 0.15$ m and lies on the soil. A load of 150 kN was applied to the slab through the distribution area of 0.2×0.2 m centered on the slab. Since it is a contact task, a contact element was used to connect both parts. The contact is face to face type.

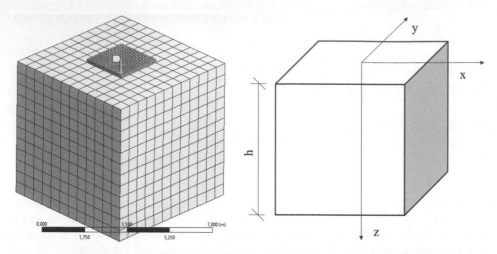

Figure 1. 3D model created in Ansys (left), graphical indication of coordination system (right).

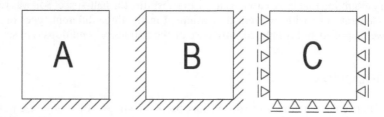

Figure 2. The boundary conditions used in the numerical models: Variant A (u = v = w = 0); Variant B (u = v = w = 0); (u = 0; v = 0 or w = 0).

The linear stress-strain diagram of the concrete slab was selected. The modulus of elasticity of concrete E is 31 GPa and Poisson coefficient is $v = 0.2$. The subsoil consists of loess loam of class F4. Volumetric weight of soil is $\gamma = 20.4$ kNm^{-3}, Poisson coefficient is $v = 0.25$, Young's modulus is $E = 957.6$ MPa. The friction was not taken into account. Self-weight of both parts was neglected.

Both parts of model are made from 3D cubic elements which are 8 nodes elements where every element has 6 degrees of freedom—the displacements and rotations are allowed in all three axes. Finite element mesh was chosen as tetrahedral or hexahedral and its size was 0.1×0.1 m in the slab and 0.1 to 1.0 m with gradually sizing with increasing depth of the soil. Different types of boundary conditions were used.

3 SCOPE OF SOLUTION

Two aspects were taken into account to investigate the influence of the input parameters. First aspect is type of the boundary condition.

Three types of boundary conditions were applied to the model (Figure 2). First boundary condition (A) is represented by the fixed support applied to the bottom surface of the soil while all other surfaces were left free. Second type (B) was created by applying fixed support to the all surfaces of the soil except the top one, which was left free. In the third condition (C) were selected the same surfaces as in the B variant, but in this case they were pinned against the movement only in the perpendicular direction to the plane of the surface.

Second aspect that was studied is the dependence to the dimensions of the soil; 4 different dimensions were chosen ($h = 3 \times 3 \times 3$; $4 \times 4 \times 4$; $6 \times 6 \times 6$ and $8 \times 8 \times 8$ m).

4 RESULTS

Overall 12 models were created while using 4 different dimensions of the subsoil for three different boundary conditions. Results were subtracted in different depths in the center of the soil (beneath the force load). From the results the maximum deformations were selected to create tables. From the values, the range of variation was evaluated.

Results are evaluated for each dimension separately. For the cube size of $3 \times 3 \times 3$ m the maximum detected deformation from all boundary conditions can be found in the Table 1. The progress of deformation with rising depth of the subsoil is denoted in the Figure 3. Same tables and graphs were created for the task with the soil dimensions of $4 \times 4 \times 4$ m (Table 2, Figure 4), $6 \times 6 \times 6$ m (Table 3, Figure 5) and $8 \times 8 \times 8$ m (Table 4, Figure 6).

Table 1. Deformation of soil the dimensions of the soil are $3 \times 3 \times 3$ m.

	Variant A	Variant B	Variant C	Range of variation
Max deformation ($z = 0.00$ m) [mm]	−17.27	−14.63	−17.4321	2.802

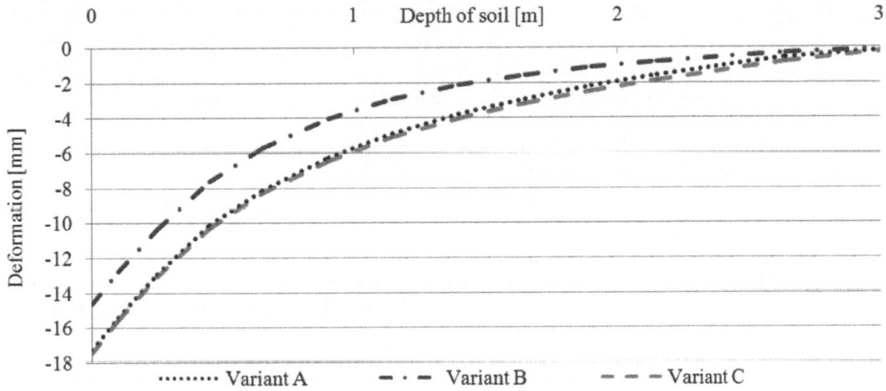

Figure 3. Deformation of soil by depth, when the dimensions of the soil are $3 \times 3 \times 3$ m. The diagram shows three rows—each for one boundary condition as shown in Figure 1.

Table 2. Deformation of soil the dimensions of the soil are $4 \times 4 \times 4$ m.

	Variant A	Variant B	Variant C	Range of variation
Max deformation ($z = 0.00$ m) [mm]	−17.661	−15.707	−17.78	2.073

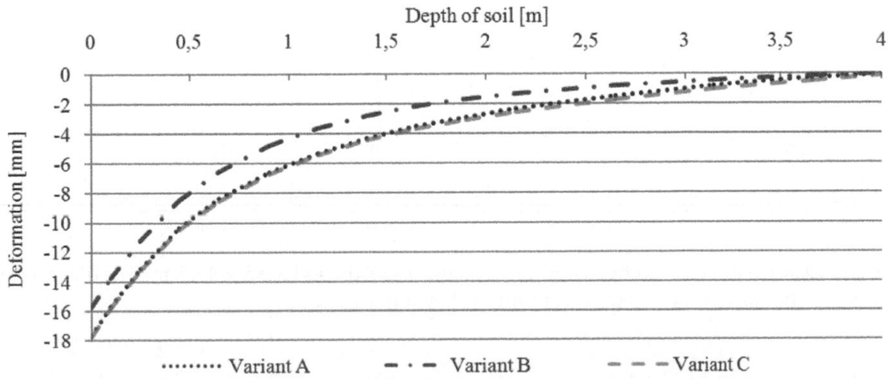

Figure 4. Deformation of soil by depth, when the dimensions of the soil are $4 \times 4 \times 4$ m. The diagram shows three rows—each for one boundary condition as shown in Figure 1.

Table 3. Deformation of soil when the dimensions of the soil are $6 \times 6 \times 6$ m.

	Variant A	Variant B	Variant C	Range of variation
Max deformation ($z = 0.00$ m) [mm]	−18.445	−17.155	−18.522	1.367

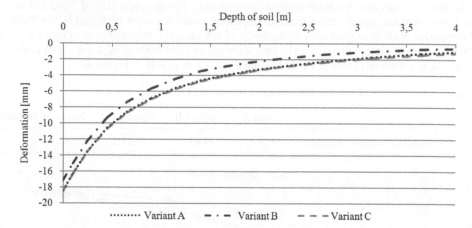

Figure 5. Deformation of soil by depth, the dimensions of the soil are $6 \times 6 \times 6$ m. The diagram shows three rows—each for one boundary condition as shown in Figure 1.

Table 4. Deformation of soil the dimensions of the soil are $8 \times 8 \times 8$ m.

	Variant A	Variant B	Variant C	Range of variation
Max deformation ($z = 0.00$ m) [mm]	−18.662	−17.693	−18.721	1.028

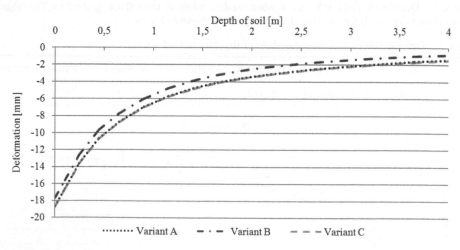

Figure 6. Deformation of soil by depth, the dimensions of the soil are $8 \times 8 \times 8$ m. The diagram shows three rows—each for one boundary condition as shown in Figure 1.

After comparison of the tables we can notice, that the range of variation decreases with increasing dimensions of the cube. It means that the influence resulted from the choice of the boundary conditions is greatest in the case of smallest dimensions of the soil.

Table 5. Surface deformations ($z = 0$ m) for different boundary conditions as well as different soil dimensions.

Dimensions of subsoil [m]	Variant A [mm]	Variant B [mm]	Variant C [mm]	Range of variation [mm]	Percentage difference [%]
$3 \times 3 \times 3$	−17.270	−14.630	−17.432	2.802	19.15
$4 \times 4 \times 4$	−17.661	−15.707	−17.780	2.073	13.19
$6 \times 6 \times 6$	−18.445	−17.155	−18.522	1.367	7.97
$8 \times 8 \times 8$	−18.662	−17.693	−18.721	1.028	5.81

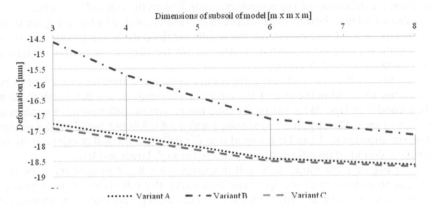

Figure 7. Graph of development of the surface deformations ($z = 0$ m) due to the change of the boundary conditions and the size of the subsoil.

5 DISCUSSION

Two dependences were investigated. Based on results published in the previous chapter we can conclude that both parameters have influence to the final amount of the calculated deformation. It is obvious, that the influence of both parameters is the most prominent on the surface and becomes less apparent once the depth of soil gets bigger.

For this reason, another comparison was created to better observe the results and to find another dependency. Table 5 summarizes all the maximal deformations from the Tables 1–4 were created. Also the maximal deformations were recorded into Figure 7. Again the range of variation is mentioned and a percentage difference of the maximum deformation is stated for better comparison of the mentioned phenomenon. The minimum deformation was assumed as base – 100% for percentage difference.

It is obvious that the influence of the boundary conditions is greater in the case of smallest dimensions of the soil area of $3 \times 3 \times 3$ m as the percentage difference between the values as 19.15%. The difference becomes less significant with increasing size of the cube and goes down to the 5.81% for the soil with dimensions $8 \times 8 \times 8$ m. That means that the smallest model has an insufficient dimension.

Another phenomenon which can be noticed is that with the increasing size of the model area of soil the deformation raises as is graphically inscribed in the Figure 7. This phenomenon is same for every type of boundary condition but the value of increase is slightly different. The increase emerging from the boundary condition A and C is very similar and has the same trend. Differed from them the boundary condition type B is steeper. To be assumed this tendency will be preserved even in bigger models.

6 CONCLUSION

Totally 12 models were created to investigate the dependence of the type of boundary conditions and the size of the subsoil that is needed to replace the area of the elastic half-space.

From the published results we can conclude that the influence of the boundary conditions is the greatest on the surface of the soil model and becomes less substantial with the increasing depth of the model. Also the influence of the dimensions has the greatest impact in the surface of the soil and becomes less substantial with the increasing depth. That is caused by the specification of the boundary conditions itself, because the deformation value at the bottom edge of the cube was predefined as zero.

It can also be stated that the influence of the boundary conditions is the greatest when selecting small area while replacing elastic half-space and becomes less substantial with the increasing dimensions of the area. This phenomenon is also caused by the values resulting directly from the definition of the boundary conditions on the sides of the cube.

It was also found that the differences between the results obtained while using the boundary conditions type A and C are minor, but the results obtained while using boundary condition B show significant difference compared. Besides this the slope of the curve is significantly steeper than in the variant A and C.

When examining the size of the area, the results' behavior is same for every type of boundary conditions. Decreasing the size of the modeled area leads to less deformation no matter boundary condition type. With the larger cube the maximum deformation increases as well as the affected area. The largest area examined was $8 \times 8 \times 8$ m but based on the tendency of the approximation curve it can be assumed that the deformation will continue to grow. This phenomenon should be investigated in the future through larger models.

From the Figure 7 the model with soil dimensions $8 \times 8 \times 8$ m appears to be the most accurate due to inclination of the approximation curve. But this claim has no support. Since there is so many results based only on the numerical model parameters, the comparison with experiments is crucial for finding the best model which leads to the most realistic results.

ACKNOWLEDGEMENTS

This paper was supported by the Student Grant Competition held at Faculty of Civil Engineering, Technical University of Ostrava within the project No. SP2018/76 "Numerical modeling of fiber-composite slab in soil interaction with HPC" and by the Moravian-Silesian Region under the program "Support for Science and Research in the Moravia-Silesia Region 2017" (RRC/10/2017).

REFERENCES

Alani, A.M. & Aboutalebi. M. 2012. Analysis of the subgrade stiffness effect on the behaviour of ground-supported concrete slabs. *Structural Concrete:* 102–108.

Cajka, R., Krivy, V. & Sekanina, D. 2011. Design and Development of a Testing Device for Experimental Measurements of Foundation Slabs on the Subsoil. *Transactions of the VSB—Technical University of Ostrava, Construction Series:* 1–5.

Cajka, R. & Labudkova J. 2014. Dependence of deformation of a plate on the subsoil in relation to the parameters of the 3D model. *International Journal of Mechanic:* 208–215.

Cajka, R. & Labudkova J. 2016. 3D numerical model in nonlinear analysis of the interaction between subsoil and sfcr slab. *International Journal of GEOMATE:* 120–127.

Labudkova, J. & Cajka R. 2016. Experimental measurements of subsoil-structure interaction and 3D numerical models. *Perspectives in Science:* 240–246.

Sadecka, L. 2000. Finite/infinite element analysis of thick plate on a layered foundation. *Computers and Structures* 76(5): 603–610.

Shams, M., Shahin, M. & Ismail, M. 2018. Simulating the behaviour of reactive soils and slab foundations using hydro-mechanical finite element modelling incorporating soil suction and moisture changes. *Computers and Geotechnics* 98: 17–34.

Wijffels, M.J.H., Wolfs, R.J.M., Suiker, A.S.J. & Salet, T.A.M. 2017. Magnetic orientation of steel fibres in self-compacting concrete beams: Effect on failure behaviour. *Cement and Concrete Composites* 80: 342–355.

Advances and Trends in Engineering Sciences and Technologies III – Al Ali & Platko (Eds)
© *2019 Taylor & Francis Group, London, ISBN 978-0-367-07509-5*

Verification of the ability of selected acoustic methods to detect the amount of steel fibers in concrete

I. Plšková, L. Topolář, T. Komárková, M. Matysík & Z. Chobola
Faculty of Civil Engineering, Brno University of Technology, Brno, Czech Republic

P. Stoniš
Faculty of Military Technology, University of Defence, Brno, Czech Republic

ABSTRACT: The use of different fiber types (polypropylene, steel, glass, natural fiber) to cement based materials for better performance is becoming more and more frequent. The addition of fibers increases the mecsshanical characteristics of these composites. This paper examines the cement based composites reinforced by steel fibers. We focused on the relationship between the amount of steel fibers (40, 80 and 120 kg on 1 m³ of concrete) and changes of parameters of acoustic waves which pass through the material. For non-destructive evaluation the two pulse-echo methods were used. First of these pulse-echo methods used as impactor the steel hammer and the second one used electronic ultrasound transmitter. The aim was to verify the ability of these methods to determine the amount of steel fibers in concrete.

1 INTRODUCTION

1.1 *Fiber-reinforced concrete*

Fiber concrete is a composite material formed by joining a concrete matrix and a short reinforcement dispersed in the matrix. The fibers occupy only a small part of the volume of the resulting composite. Plain concrete is characterized by a relatively high compressive strength, brittleness, low tensile bending strength and low tensile strength (Malhotra & Carino 2004). Dispersed reinforcement captures mainly tensile stress and prevents the formation of microcracks from shrinkage and the development of tension cracks in the structure (Štoudek et al. 2016). In this paper, we examine the reinforcement impact on parameters of acoustic waves which pass through the material. During production of concrete with steel fibers sometimes the fibers cannot evenly mix. It is therefore important to choose and test NDT methods which are able to determine the amount of fibers. We chose two pulse-echo methods, which we have been engaged in for a long time (Matysík et al. 2015, Timčaková et al. 2016).

1.2 *Pulse-echo methods*

In principle, pulse-echo methods belong historically among the oldest methods of NDT. "Spectral analysis" by hearing was used by potters, bellfounders and many other professions. But it is a fact that such testing is subjective and less sensitive to small defects (Bajer & Hájek 2011). That is why, with the development of technology and computers, it was replaced by the discrete Fourier spectral analysis (Fast Fourier Transformation—FFT), which has eliminated subjectivity and increased the frequency resolution and thus theoretical sensitivity. The practical sensitivity is then further increased by creating a more stable test method (electronic exciter, impact shock stability, stability of the test body location), and scans for as many reference signals as possible from undamaged specimens (Sansalone & Street 1997, Garbacz et al. 2017). It is then important to create the correct test criteria for allowing deviations from the calculated frequency spectrum (Hsiaoa et al. 2008.). For this paper two pulse-echo methods were used, first used as impactor the steel hammer and the second one used electronic ultra-sound transmitter.

2 EXPERIMENTAL SECTION

2.1 *Production and treatment of test specimens*

For the purpose of the experiment, concrete test specimens were made that differed from each other by the amount of steel fibers. We chose concentrations 40 kg, 80 kg and 120 kg of steel fibers on 1 m³ of fresh concrete. The composition of the individual mixtures and properties of the used steel fibers are given in Table 1 and Table 2. Test specimens were core samples (cylinders) with a diameter of 10 cm and a height of 10 cm.

2.2 *Measuring equipment for pulse-echo methods*

Pulse-echo methods are based on a pulse response analysis that excites harmonic waves in the studied sample at its own frequency and at higher harmonic frequencies (Azari et al. 2014.). The resulting response is scanned as surface waves by acoustic piezoelectric sensors. The captured signal is then digitized and stored by the data acquisition system. The signal is further processed and evaluated by control computer. Time course of response signals is converted to the frequency spectrum by the Fast Fourier Transformation (Po-Liang et al. 2008). In the pulse-echo method, a pulse source applied to the surface of the sample may be a mechanical shock, or a generated signal transmitted by the electronic exciter (Zoidis et al. 2013.). If a mechanical pulse is used for testing, we are talking about the Impact-echo method. The first NDT method (Impact-echo) in this paper used as impactor the steel hammer and the second one used as exciter the ultrasound piezoelectric transmitter.

When measured by Impact-echo, the test bodies were placed on rubber pads. The boost pulse was realized by a metal hammer strike. The resulting surface wave response was captured by a piezoelectric sensor of the MIDI type. The sensor was attached to the test specimen using beeswax. The two-channel TiePie engineering Handy scope HS3 oscilloscope was used for processing.

For method with piezoelectric exciter, the pulse generator Agilent 33220 was set to the following output values: period 5 s, amplitude 10 Vpp, pulse width 500 μs. These pulse signals were captured after passing through the sample by the piezoelectric sensor DAKEL IDK 09 and then amplified with a preamplifier AS3 K 433 with a gain of 35 dB and subsequently recorded by the DAKEL XEDO device where they were again amplified by internal circuits (if necessary).

Table 1. Composition of the mixtures.

SFRC mixture	Cement 42.5 R [kg/m³]	Water [kg/m³]	Water ratio [–]	Aggregates			Plasticizer [kg/m³]	Steel fibers [kg/m³]
				0–4 [kg/m³]	4–8 [kg/m³]	8–16 [kg/m³]		
T 1	390	160	0.41	900	250	670	4	40
T 2	390	160	0.41	900	250	670	4	80
T 3	390	160	0.41	900	250	670	4	120

Table 2. Properties of the steel fibers.

Fibers	Length [mm]	Diameter [mm]	Tensile strength [GPa]	Young's modulus [GPa]
DRAMIX RC-65/35-BN	35	0.55	1.345	210

3 EXPERIMENTAL RESULTS

3.1 *Impact-echo experimental results*

Figures 1–3 present the results of the Impact-echo acoustic measurement. Figure 1 shows the frequency spectrum of the T1 set with a 40 kg/m³ steel fiber concentration. The measurements were made in the longitudinal direction, which means that the sensor was placed at the end of the concrete beam using a beeswax and steel hammer blow was performed on the opposite side. Figure 2 presents the results of T2 set samples with a 80 kg/m³ steel fiber concentration. Figure 3 is the frequency spectrum for a T3 set with the highest steel fiber

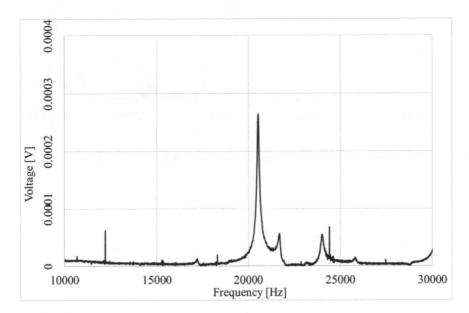

Figure 1.   Frequency spectrum of the T1 set with a 40 kg/m³ steel fiber concentration.

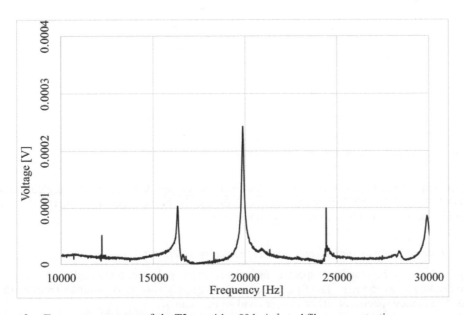

Figure 2.   Frequency spectrum of the T2 set with a 80 kg/m³ steel fiber concentration.

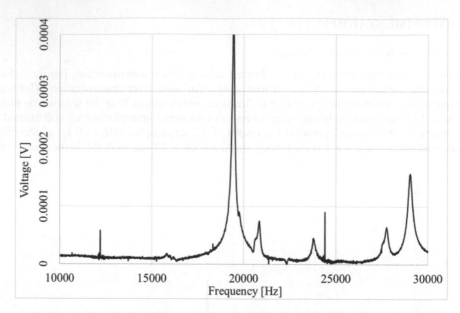

Figure 3.   Frequency spectrum of the T3 set with a 120 kg/m³ steel fiber concentration.

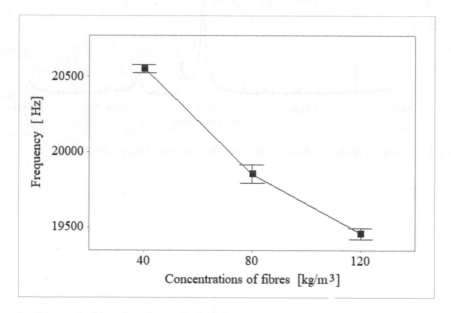

Figure 4.   Summarized impulse-echo method results.

concentration (120 kg/m³). As can be seen from the graphs, with the increasing concentration of steel fibers, the dominant frequencies of the test bodies decrease. The results of the Impulse-echo method are summarized in Figure 4.

3.2   *Pulse-echo (with piezoelectric exciter) experimental results*

Figure 5 shows the frequency spectra obtained from the pulse-echo method (with piezoelectric exciter) measurement. The figure shows the frequency spectra of samples from all three sets. Frequency spectra of other samples had similar course.

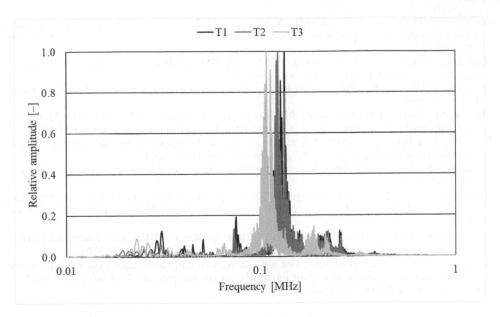

Figure 5. Frequency spectra obtained from the pulse-echo method with piezoelectric exciter.

4 CONCLUSION

The aim of the paper was to document the sensitivity of two Pulse-echo methods to the amount of steel fibers in concrete. During the testing of steel fiber reinforced concrete test specimens by Pulse-echo methods we reached the following findings. When observing the obtained signal, the highest dominant frequencies were recorded for all test specimens with a steel fiber concentration of 40 kg/m³. With increasing concentrations of steel fibers, shifts of dominant frequencies to lower values occurred. This phenomenon is probably due to the increased number of transitions between steel fibers and cement matrix. The results of both pulse-echo methods are similar. The piezoelectric exciter method is more difficult for practical realization, since the driver must be well attached (acoustically well attached) to the surface by the coupling agent. This of course also applies to sensors. If this requirement is met, the measurements are well reproducible and both methods are applicable.

ACKNOWLEDGMENT

This paper has been worked out under the project No. LO1408 "AdMaS UP—Advanced Materials, Structures and Technologies", supported by Ministry of Education, Youth and Sports under the "National Sustainability Programme I" and the project GAČR No.16–02261S supported by Czech Science Foundation.

REFERENCES

Azari, H.; Nazarian, S. & Yuan, D. 2014. Assessing sensitivity of impact echo and ultrasonic surface waves methods for nondestructive evaluation of concrete structures. *Construction and Building Materials.* 71: 384–391.

Bajer, J. & Hájek K. 2011. Differential Spectral Analysis of the Impact-Echo Signal. *Proc. of the 41st International Conf. on NDT 'NDE for Safety—Defektoskopie 2011', 9–11 November, 2011.* Ostrava, Czech republic.

Garbacz, A.; Piotrowski, T.; Courard, L. & Kwaśniewski, L. 2017. On the evaluation of interface quality in concrete repair system by means of impact-echo signal analysis. *Construction and Building Materials*. 134: 311–323.

Hsiaoa, C.; Cheng, C.C.; Liou, T. & Juang, Y. 2008. Detecting flaws in concrete blocks using the impact-echo method. *NDT & E International*. 41(2): 98–107.

Malhotra, V. & Carino N. 2004. *Handbook on nondestructive testing of concrete*. 2nd ed. Boca Raton, Fla.: CRC Press.

Matysík, M.; Plšková, I. & Chobola, Z. 2015. Assessment of the Impact-echo Method for Monitoring the Long-standing Frost Resistance of Ceramic Tiles. *Materiali in tehnologije*. 49(4): 639–643.

Po-Liang, Y. & Pei-Ling, L. 2008. Application of the wavelet transform and the enhanced Fourier spectrum in the impact echo test. *NDT & E International*. 41(5): 382–394.

Sansalone, M.J. & Street W.B. 1997. *Impact-Echo: Nondestructive Evaluation of Concrete and Masonry*. Ithaca NY: Bulbrier Press.

Štoudek, R.; Trčka, T.; Matysík, M.; Vymazal, T. & Plšková, I. 2016. Acoustic and Electromagnetic Emission of Lightweight Concrete with Polypropylene Fibers. *Materiali in tehnologije*, 50(4): 547–552.

Timčaková, K.; Matysík, M. & Chobola, Z. 2016. The Possibilities of NUS and Impact-echo Methods for Steel Corrosion Monitoring in Concrete. *Materiali in tehnologije*. 50(4): 565–570.

Zoidis N.; Tatsis, E.; Vlachopoulos, C.; Gotzamanis, A.; Clausen, J.S.; Aggelis, D.G. & Matikas, T.E. 2013. Inspection, evaluation and repair monitoring of cracked concrete floor using NDT methods. *Construction and Building Materials*. 48: 1302–1308.

Advances and Trends in Engineering Sciences and Technologies III – Al Ali & Platko (Eds)
© 2019 Taylor & Francis Group, London, ISBN 978-0-367-07509-5

Monitoring the shrinkage of concrete slabs reinforced by steel and FRP reinforcement

S. Priganc, D. Kušnírová & Š. Kušnír
Faculty of Civil Engineering, Institute of Structural Engineering, Technical University of Košice, Košice, Slovakia

ABSTRACT: The article is aimed at monitoring the shrinkage of concrete slabs under laboratory conditions. Different types of reinforcement (material and amount of bars) were applied. GFRP reinforcement was used because of its material characteristics (strength and modulus of elasticity), which are significantly different from that of the conventional steel reinforcement. Process of research preparing and principle of measurement are described. Result of shrinkage influence on the strain and deflection of concrete slabs reinforced by steel and GFRP reinforcement are included in this paper.

1   INTRODUCTION

In building practice, new materials are constantly introduced. They should have better and more beneficial properties than the traditionally used and time-tested materials. To prove that their properties and interaction with other materials are really better, their experimental verification in each action sphere and drawing the right conclusions are necessary. Such a material is the fiber-reinforced polymer (FRP). We focused on the glass fiber-reinforced polymers (GFRPs) and specifically on their action in concrete caused by concrete shrinkage. The modulus of elasticity of GFRP is significantly lower than modulus of conventional steel reinforcement. Therefore the influence of concrete shrinkage to mainly strains and deflections in constructions reinforced by GFRP should be different. We decided to find out it.

2   RESEARCH PREPARING

The aim of the experiment is to find out the impact of shrinkage on the strains and deflection of five concrete elements reinforced by the GFRP reinforcement (two different types—Armastec (3 ps.) and ComBAR (2 ps.)). For comparison, three elements reinforced by the conventional steel reinforcement (B500) as well as a specimen of unreinforced concrete were observed. These types of reinforcement differ not only in strength, but also in modulus of elasticity (for steel 200 GPa, for Armastec 50 GPa, for ComBAR 60GPa), what is in our tests very important (Schöck ComBAR 2013). As the test specimens the slabs of dimensions 600×1800 mm and 120 mm thickness were selected. The concrete slabs were reinforced not only by the different type of reinforcement but also by the different reinforcement ratio.

Reinforcement was placed to only one surface (simulation of simply supported plate) with a cover of reinforcement 10 mm. In order to obtain the various reinforcement ratios, we used different amount and different diameter of the bars, namely 4 ϕ 8 mm, 7 ϕ 8 mm and 7 ϕ 10 mm. As a secondary reinforcement, the bars of 6 mm diameter (steel or GFRP) at a distance of 200 mm were used. Exact slabs reinforcing plans are in Figure 1.

A specimen storage chamber and stands for slabs hanging were prepared prior to concreting. The chamber should prevent from sudden change of moisture and temperature at work in laboratory (e.g. during door opening).

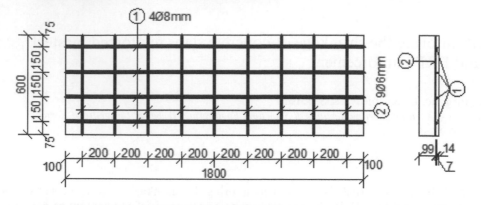

Steel, ComBAR and ARMASTEC reinforcement

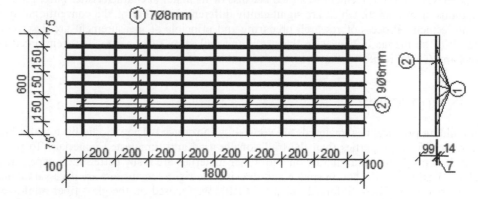

Steel, ComBAR and ARMASTEC reinforcement

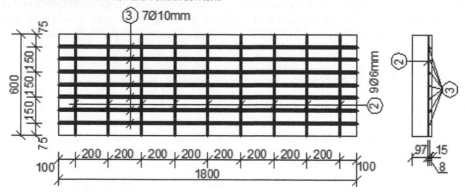

Steel and ARMASTEC reinforcement

Figure 1. Slabs reinforcing plans.

The slabs were made from C 20/25 concrete (from concrete plant) to large steel moulds divided by wooden boards on smaller parts (Figure 2 in left).

For the need for later manipulation and storing of the plates, it was necessary to insert the elements for hanging the slabs and also for placing the measuring frame to the prepared forms. On Figure 2 (right) the final reinforcement arrangement can be seen.

After concreting, samples were cured by spraying for 7 days and covered with foil. On seventh day, the samples were picked up from forms, placed on the stands in the prepared chamber and marked and also monitored parameters were measured for first time (Figure 3).

Figure 2. Forms prepared for reinforcement placing (left) and final reinforcement arrangement for slab reinforced by 7 φ 10 mm ARMASTEC reinforcement (right).

Figure 3. Samples placed on stands in prepared chamber.

3 PRINCIPLE OF MEASUREMENT

The slabs are stored on stands in the vertical position (Figure 3). In such a position, the self-weight of the slab acts in the center line. The high cross-section stands against the accrued bending moment and also due to the position of the hanging eyes, small stresses (in the order of 0.1 MPa) are formed in the sample. Therefore, when evaluating, it is possible to neglect the impact of self-weight, only the net effect of shrinkage will be evident.

Measuring of strains was carried out by the steel attachment frame equipped with mechanical strain gauges for direct measurement of the deflection and by glued targets which

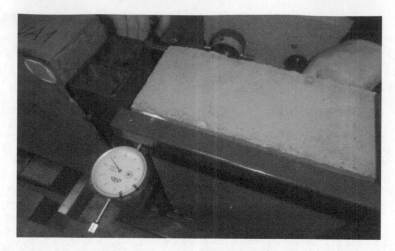

Figure 4.   Real measuring by the steel frame.

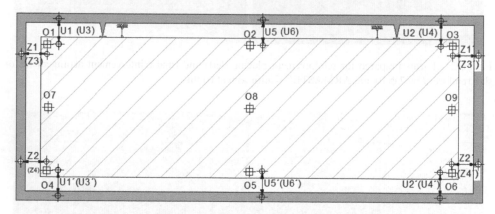

Figure 5.   The principle of strain measuring by the steel attachment frame (O_i – mechanical strain gauges, Z_i – measuring base for monitoring by the attachment deformer).

represented one end of the measuring bases monitored by the attachment deformer. The second end of the measuring base was on the two border surfaces of the concrete slabs (Figure 4, Figure 5) (Fecko 1986, Hájek et al. 1983).

4   RESULTS OF MEASUREMENT

Measurements of strains and deflections caused by shrinkage have a long-standing character. The article only shows the partial results of the current measurement that will continue for a longer time.

From the measured values of strains, the time developments of shrinkage at both plate surfaces (on the reinforced side and on the non-reinforced side) were arranged. As an example in Figure 6 the time development of shrinkage at two surfaces of the slab VO1, reinforced by 4 bars of diameter 8 mm is shown. At the beginning of measurement, perhaps to tenth day, concrete samples have been swelling and then have been shrunk. That is visible in the first part of graphs on Figures 6, 7 and 8. This fact was caused by relative humidity, which in first days increased, respectively was stabile (on the value of approximately100%) and then began decrease (to the value of 61%). In the long term it can be seen that the surface without reinforcement shrinks more. The reinforcement prevents free concrete shrinking. If the element is

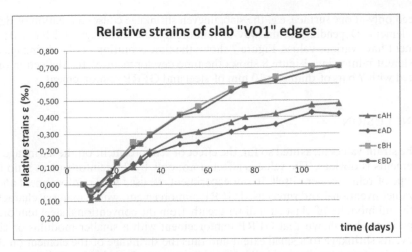

Figure 6.   Time development of shrinkage at two surfaces (A and B) of the slab VO1.

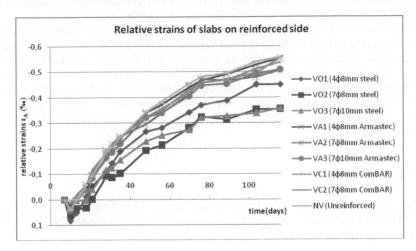

Figure 7.   Time evolution of the shrinkage of slabs with different reinforcing.

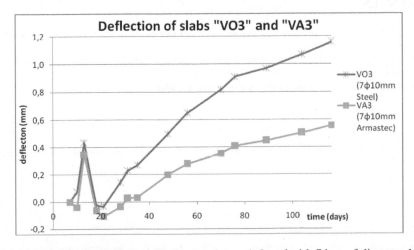

Figure 8.   Time development of the deflection on plates reinforced with 7 bars of diameter 10 mm of steel and GFRP (Armastec) reinforcement.

reinforced only at one surface, e.g. the slab, uneven shrinkage causes a creation of deflection on the element. Depending on the way of reinforcing, the shrinkage and the deflection of the element have various values. Figure 7 shows the time evolution of the shrinkage of slabs with different reinforcing. Figure 8 shows the time development of the deflection on plates reinforced with 7 bars of diameter 10 mm of steel and GFRP reinforcement.

5 CONCLUSION

From the strain values measured so far, the effect of reinforcement on the shrinkage of concrete elements is obvious. The value is affected not only by the reinforcement ratio but also by the type of reinforcement itself. In our experiment, we used steel reinforcement and in our country unconventional and rarely used GFRP reinforcement, whose material characteristics (strength and modulus of elasticity) differ greatly from the conventional steel reinforcement. The experiment has shown that GFRP reinforcement with a smaller modulus of elasticity only prevents shrinkage to a small extent and thus the shrinkage of the element on the reinforced side is greater (similar to shrinkage on the non-reinforced side of slab). Because the shrinkage values on both surfaces are similar, the element deflection due to concrete shrinkage is smaller.

No deflection caused by shrinkage is taken into account in construction practice. Even though a low value can affect the final bend value and cause problems in designing elements for serviceability limit states. Moreover, as a result of shrinkage, additional tension stresses arise in the reinforcement of reinforced elements. These stresses should be considered in the reinforcement design (for ultimate limit states). A comparison of stresses in the reinforcement caused by shrinkage will be the subject of our further research. Its aim is to find the advantages or disadvantages of using GFRP reinforcement in building practice in terms of the impact of shrinkage.

ACKNOWLEDGEMENT

The article was supported by the scientific grant agency MŠVVaŠ SR and SAV by project VEGA 1/0661/16 Behaviour of load bearing elements from ordinary and light concrete affected by temperature.

REFERENCES

Fecko, L. 1986. Sledovanie pretvorení železobetónových dosiek od zmrašťovania (in Slovak). *Stavebnícky časopis 34(8)*: 615–630.
Hájek, J. et al. 1983. Dlhodobé pretvorenia železobetónových dosiek pri rôznych hladinách zaťaženia (in Slovak). *Stavebnícky časopis 31(č. 6–7)*:517–531.
Schöck ComBAR. 2013. Technical Information. Baden-Baden.

Advances and Trends in Engineering Sciences and Technologies III – Al Ali & Platko (Eds)
© *2019 Taylor & Francis Group, London, ISBN 978-0-367-07509-5*

Influence of fissures in timber on the charring rate

J. Sandanus & Z. Kamenická
Department of Metal and Timber Structures, Faculty of Civil Engineering, Slovak University of Technology in Bratislava, Slovakia

P. Rantuch, J. Martinka & K. Balog
Institute of Integrated Safety, Faculty of Materials Science and Technology in Trnava, Slovak University of Technology in Bratislava, Slovakia

ABSTRACT: The charring depth, rate and residual cross-section have a significant influence on the mechanical resistance of timber members in fire. In case of various dimensions of fissures, it is possible that as the cross-section reduces to the residual one, the mechanical resistance will be reduced as well. In the absence of rules that allows incorporating the influence of fissures on the charring rate, this paper is focused on the numerical analysis and experimental measurements of selected dimensions and number of fissures. In terms of results, the smaller fissures can be neglected, whilst the bigger fissures should be taken into account by means of increased value of the charring rate, or by advanced calculations.

1 INTRODUCTION

The charring depth and rate are factors (Werther, 2016) relevant in determination of the residual cross-section, i.e. the cross-section capable of carrying the load after a period of fire exposure. The position of the char-line is considered as the position of the 300-degree isotherm according to Eurocode 5, Part 2 (STN EN 1995-1-2, 2008). Constant value of the notional design charring rate of solid timber is $\beta_n = 0.8$ mm/min and of glued laminated timber is $\beta_n = 0.7$ mm/min according to simplified methods in the standard (STN EN 1995-1-2, 2008). This difference is caused by the greater number of fissures in solid timber. Fissures and gaps with a width greater than 4 mm (Fornather, 2001) (or 2 mm according to (Fornather, 2004)) should be taken into account, but no proposal to quantify this effect is given (Frangi et al., 2010).

2 EXPERIMENTAL ANALYSIS

2.1 *Generally*

The experiment focused on the effect of fissures on the charring rate of timber structural members. It was conducted by Faculty of Civil Engineering in cooperation with Faculty of Materials Science and Technology in Trnava.

Two different measuring approaches were used for the experimental analysis. The first is measuring of the charring depth from a split cross-section as in (Xu et al., 2015). The second is measuring of temperatures by means of thermocouples like as in (Tsantaridis & Östman, 1998). There were two types of samples. 30 samples had dimensions $60 \times 100 \times 100$ mm and 50 samples $140 \times 100 \times 100$ mm (Figure 1). Samples were made of timber with strength class C24. Samples were divided into six groups. The first group of samples had the fissure 4×16 mm, the second group was without fissures, the third group had a fissure 4×32 mm, the fourth group 6×24 mm, the fifth group 6×60 mm (this fissure was not in samples $60 \times 100 \times 100$ due to dimensions and this group was also without fissures) and the sixth group had three fissures 4×32 mm. The average density of samples $60 \times 100 \times 100$ was

445,05 kg/m³ and of samples 140 × 100 × 100 was 398,76 kg/m³. The average moisture content of samples 60 × 100 × 100 was 7,4% and of samples 140 × 100 × 100 was 7,61%. The average temperature of air during tests was 27°C and relative ambient humidity was 30%.

Both types of samples were tested in a cone calorimeter (Figure 2). The heater and samples were oriented in a horizontal direction. Thermal radiation from the cone heater and action of spark igniter were applied on samples according to the standard (ISO 5660-1, 2002).

2.2 Samples 60 × 100 × 100

Samples 60 × 100 × 100 were exposed to a constant heat flux 25 kW/m². These samples were split in a half (Figure 3) and the charring depth was determined visually (Figure 4 left), thus allowing the calculation of the charring rate (Figure 4 right). The highest values of the charring depth and rate are for samples with three fissures 4 × 32 and the lowest values for samples without fissures.

2.3 Samples 140 × 100 × 100

Samples 140 × 100 × 100 were exposed to a constant heat flux 50 kW/m² and were used with thermocouples to measure temperatures in specific points (different distances from the

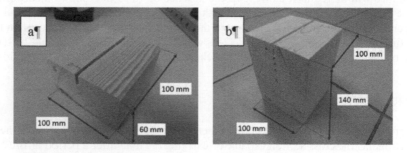

Figure 1.   Samples 60 × 100 × 100 (a) and samples 140 × 100 × 100 (b).

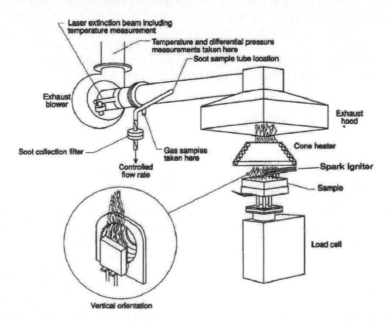

Figure 2.   Parts of the cone calorimeter (Fires in Mass Transit Vehicles Guide for the Evaluation of Toxic, 1991).

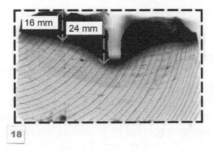

Figure 3. The charring depth on the edge of fissure and in distance from the fissure.

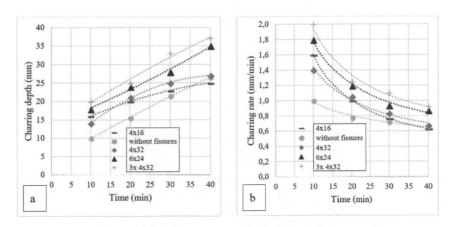

Figure 4. The charring depth (a) and the charring rate (b) of samples $60 \times 100 \times 100$.

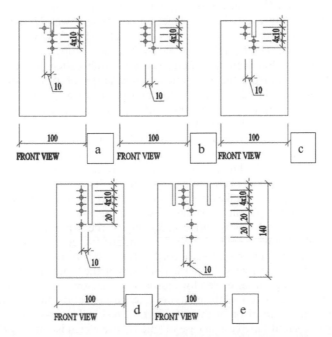

Figure 5. Scheme for the location of thermocouples in samples $140 \times 100 \times 100$ – samples with fissure 4×16 (a), 4×32 (b), 6×24 (c) and 6×60 (d) and with three fissures 4×32 (e).

exposed surface) (Figure 5). These thermocouples were used to identify the time of reaching the temperature 300°C in different depths and to compare the numerical model with the experiment.

A time to ignition of samples $140 \times 100 \times 100$ and $60 \times 100 \times 100$ was monitored. This time was bigger for samples $60 \times 100 \times 100$, because the time to ignition increases with lower heat flux (Rantuch et al., 2016).

According to (Tsantaridis & Östman, 1998), there is a ratio between charring depths in the cone calorimeter with the heat flux 50 kW/m^2 and the furnace with the standard fire exposure during 30–40 min, which can be expressed by means of this equation:

$$\frac{d_{char,cone}}{d_{char,furn}} = 1,997e^{-0,019t} \tag{1}$$

where $d_{char,cone}$ = charring depth from the cone calorimeter (mm); $d_{char,furn}$ = charring depth from the furnace (mm); and t = time (min).

3 NUMERICAL MODELS

There were created several numerical models on a base of samples from the experimental analysis. For the purpose of this study, the software Ansys Workbench was used. The used elements types were PLANE77, CONTA172, TARGE169 and SURF151. The element PLANE77 is an eight-node element with one degree of freedom (temperature at each node). The element CONTA172 represents a deformable surface in contact with a 2-D "target" surface (TARGE169) with the possibility of sliding. This element is located on the surface of 2-D solid elements (PLANE77). The target segment element TARGE169 is used to represent 2-D "target" surfaces for the associated contact elements (CONTA172). The target surface is discretized by a set of target segment elements. The two- to four-node element SURF151 may be used for various types of load (including thermal load) and surface effect applications (Thermal Analysis Guide, 2009). The element size was 1 mm for the timber part and 0.5 mm for the air part. Emissivity was 0.8 (according to STN EN 1995-1-2, 2008). Film coefficient (convection coefficient) was 25 W/(m^2 K). The time step settings of the transient thermal analysis were defined as follows: 60 initial substeps, 30 minimal substeps and 6000 maximal substeps. Thermal properties of timber according to (STN EN 1995-1-2, 2008) were modified to gain minimal differences of temperatures in time compared to results from the experiment. Thermal properties of air were according to (Property tables and charts, [without date]) and values of a thermal conductivity were multiplied by a factor from (Erchinger, 2009). More detailed information is presented in a thesis of the author (Kamenická, 2018).

Numerical models of local fissures were simplified to "smeared" models with thermal properties, which were averaged according to an area of timber and area of air. These models were created with different thickness of the layer with averaged properties and they are represented by the Figure 6, where t is the thickness and h is the depth of the fissure. "Smeared" models are more appropriate to take into account the effect of fissures on the charring rate in consideration of the whole perimeter of the cross-section.

At first, the thermal load was specified according to the experiment with constant heat flux. Then, other models were created with the standard fire curve ISO 834 as the thermal load (thermal properties were modified accordingly) to satisfy the standard requirements. These models with the curve ISO 834 and the thickness of the "smeared" layer which is equal to a half of the fissure depth were used for mutual comparison of the charring rates for different fissures. The comparison of the charring rate with effect of different fissures is stated in the Table 1 and the comparison of this effect on the residual cross-section is in the Table 2.

The charring rate should increase in case of one fissure which is not very deep (< 35 mm), as is derived from this equation:

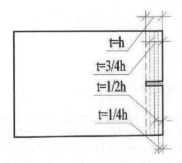

Figure 6. "Smeared" layers with different thickness.

Table 1. Values of the charring rate from models with fissures compared to the model without fissures (β/β (model without fissures)).

Time (min)	Without fissures	4×16	4×32	6×24	6×60	$3 \times 4 \times 32$
5	100%	148%	157%	200%	215%	273%
10	100%	132%	159%	181%	228%	235%
15	100%	118%	149%	149%	248%	184%
20	100%	111%	133%	135%	209%	160%
25	100%	108%	125%	128%	186%	148%
30	100%	106%	120%	123%	171%	139%
35	100%	105%	116%	120%	161%	134%
40	100%	104%	113%	117%	153%	129%
45	100%	103%	111%	115%	147%	125%

Table 2. Values of the residual cross-section from models with fissures compared to the model without fissures (A_{res}/A_{res} (model without fissures)).

Time (min)	Without fissures	4×16	4×32	6×24	6×60	$3 \times 4 \times 32$
5	100%	99%	99%	98%	97%	96%
10	100%	98%	96%	95%	92%	92%
15	100%	98%	95%	95%	86%	92%
20	100%	98%	96%	95%	85%	92%
25	100%	99%	96%	95%	85%	92%
30	100%	99%	96%	95%	85%	92%
35	100%	99%	96%	95%	85%	92%
40	100%	99%	96%	95%	85%	92%
45	100%	99%	96%	95%	84%	92%

$$\beta = \begin{cases} \beta . 2,0 \text{ if } t \leq 20 \text{ min} \\ \beta . 1,3 \text{ if } t > 20 \text{ min} \end{cases} \quad (2)$$

where β' = increased charring rate (mm/min); β = charring rate of timber without fissures (mm/min); and t = time (s).

The charring rate should increase in case the one fissure which is deeper (≥ 35 mm) or in case there are more fissures, as can be seen in this equation:

$$\beta = \begin{cases} \beta.2{,}8 \ if \ t \le 20 \ min \\ \beta.1{,}9 \ if \ t > 20 \ min \end{cases} \tag{3}$$

4 DISCUSSION AND CONCLUSIONS

The lowest charring rate applies for timber samples without fissures and the highest charring rate is for samples with fissure 6×60 and with three fissures 4×32. Results from the experiment corresponded with the results from numerical models. "Smeared" models are more appropriate for the consideration of the effect of fissures on the charring rate and the residual cross-section compared to models with local fissures. Differences in the charring rate between structural members are bigger in the initial minutes compared to later minutes. Fissures can increase the charring rate by 3 to 173% and decrease the residual cross-section by 1 to 16% depending on time, dimensions and number of fissures. The charring rate should increase with one fissure which is not very deep (< 35 mm), as defined by the equation (2). The charring rate should increase with one fissure which is deeper (≥ 35 mm) or more fissures, as stated by the equation (3). In terms of results for the residual cross-section, the smaller fissures can be neglected, whilst the bigger ones should be taken into account by means of increased value of the charring rate, or by advanced calculations.

REFERENCES

Erchinger, C.D., 2009. Zum Verhalten von mehrschnittigen Stahl-Holz-Stabdübelverbindungenim Brandfall, Zürich, PhD. Thesis, Institutf ür Baustatik und Konstruktion Eidgenössische Technische Hochschule Zürich. http://e-collection.library.ethz.ch/eserv/eth:2352/eth-2352-01.pdf.

Fires in Mass Transit Vehicles Guide for the Evaluation of Toxic, 1991. Washington: National Academies Press. ISBN 9780309078405.

Fornather, J. et al., 2001. Versuchsbericht – Kleinbrand versuchsreihe 2 Teil 1 (KBV 2/1) – Versuchemit Rissen. Universität für Bodenkultur, InstitutfürkonstruktivenIngenieurbau.Vienna.

Fornather, J. et al., 2004. Brennbarkeit und Brandverhalten von Holz, Holzwerkstoffen und Holzkonstruktionen: Zusammenfassung und Erkenntnisse für die Bemessungs praxis; ein Forschungs projekt des Fachverbandes der Holzindustrie Österreichs. 2. Aufl. Wien: proHolz Austria. ISBN 3902320036.

Frangi, A. et al., 2010. Fire safety in timber buildings: Technical guideline for Europe. http://eurocodes. jrc.ec.europa.eu/doc/Fire_Timber_Ch_5–7.pdf.

Kamenická, Z., 2018. Selected Problems in Determining the Fire Resistance of Timber Structural Members. Bratislava. PhD. Thesis. Slovak University of Technology in Bratislava. Supervisor J. Sandanus.

Property tables and charts (SI units), [without date]. https://www.researchgate.net/file.Post-FileLoader. html?id=54c8f917cf57d749248b4689&assetKey=AS%3A273740741447683%4014422-76287686.

Rantuch, P. et al., 2015. Vplyv hustoty tepelného toku na termický rozklad OSB. ACTA FACULTATIS XYLOLOGIAE ZVOLEN. 57(2), 125–134. DOI: 10.17423/afx.2015.57.2.13. https://www.research-gate. net/publication/283677358_The_influence_of_heat_flux_density_on_the_thermal_decomposition_of_ OSB.

STN EN 1995-1-2, 2008. Eurocode 5: Design of timber structures – Part 1–2: General – Structural fire design.

Thermal Analysis Guide. Canonsburg: ANSYS, 2009. https://www.ansys.com/products/struc-tures/ thermal-analysis.

Tsantaridis, L.D. & Östman, B.A.L., 1998. Charring of protected wood studs. Fire and Materials: Fire Mater. https://onlinelibrary.wiley.com/doi/full/10.1002/%28SICI%291099-1018%28199803/04% 2922%3A2%3C55%3A%3AAID-FAM635%3E3.0.CO%3B2-T.

Werther, N., 2016. External and internal factors influencing the charring of timber – an experimental study with respect to natural fires and moisture conditions: Structures in fire. Lancaster: DEStech Publications. http://www.destechpub.com/product/structures-fire-2016/.

Xu, Q., L. et al., 2015. Combustion and charring properties of five common constructional wood species from cone calorimeter tests. Construction Building Materials, 96: 416–427. DOI: 10.1016/j. conbuildmat.2015.08.062.

Advances and Trends in Engineering Sciences and Technologies III – Al Ali & Platko (Eds)
© 2019 Taylor & Francis Group, London, ISBN 978-0-367-07509-5

Study on ligaments' areas differences between different sizes of modCT concrete specimens

J. Sobek & J. Klon
Faculty of Civil Engineering, Institute of Structural Mechanics, Brno University of Technology, Brno, Czech Republic

T. Trčka
Department of Physics, Faculty of Electrical Engineering and Communication, Brno University of Technology, Brno, Czech Republic

H. Cifuentes & J.D. Ríos
School of Engineering, Department of Continuous Medium Mechanics and Theory of Structures, Universidad de Sevilla, Sevilla, Spain

ABSTRACT: The topic of the paper is aimed at the initial study of different types of surface roughness—depending on the three different sizes used during destructive loading experiments. The specimens subjected to modified compact tension tests (modCT) are used for tensile mode loading. Same material is used—concrete, so the quasi-brittle fracture is represented. Change of roughness (through fracture surface area) is investigated by 2D laser profilometer scanning.

1 INTRODUCTION

This study is aimed at the problematic of different surface shapes (surface roughness) of the specimen's ligament after its testing during real experiments conducted on concrete specimens. At the end of 2016, with a cooperation of Universidad de Sevilla, two types of test specimens were cast, designated to comprehensive study about the fracture process zone (FPZ) in quasi-brittle materials. The three-point bending (3PB) (RILEM Technical Committee 50 FMC 1985) and modified compact tension (modCT) test specimens (Veselý & Sobek 2013). Almost one year later (in November 2017), a wide set of these specimens was tested in laboratory of the Department of Continuous Medium Mechanics and Theory of Structures in Spain. Next step of the comprehensive analysis became the evolution of the loading-deflection diagrams of the all variants (differing in size) and their impact on the obtaining of basic fracture parameters like fracture energy G_f. This parameter can be obtained from the area under the loading curve and at this moment, the area of the test specimen's ligament (where the work of fracture W_f is consumed) A_{lig} is crucial. Usually in the case of quasi-brittle materials, the area of ligament differs in the way of intended (plane) and damaged (three dimensional) breach. The question of how size of the test specimen and different ways of the loading can affect the test specimen's ligament area is the object of this paper, which focuses on the second group of the test specimens tested—modCT. For this analysis, the 2D optical profilometry (scanning of the ligament area) is used (Ficker & Martišek 2012, Erdem & Blankson 2013, Micro-Epsilon 2018b).

2 THEORETICAL BACKGROUND

2.1 *Modified compact tension test specimen (modCT)*

Test configuration subjected to determination of fracture parameters for quasi-brittle and alkali-activated composites was developed at Brno University of Technology. It is called modified compact tension test specimen (modCT) (Veselý & Sobek 2013, Klon & Sobek & Keršner 2017) and differs in the position (eccentricity) of the inputted loading force. This modification causes significant changes in stress distribution (by change of constraint of stress distribution near the crack tip) through whole specimen body.

The scheme of modCT test configuration is illustrated in Figure 1 accompanied with the photographs of the test procedure in Figure 2 and Figure 3. As is shown, this configuration consists from main specimen and two steel platens, which are glued to the specimen in bottom and top part. Steel grips serve to input the loading force from testing equipment (machine) through the steel platens to body of the test specimen. Eccentricity can be set in three different positions marked by letters A, B or C as is evident from Figure 1. Position A indicates loading in the close vicinity of the initial crack (pure tension mode with the crack opening in the direction of specimen body). Other one, marked as B position, is given by the shift of loading force close to the crack-tip. Last position (C) is significant for position of loading force behind the initial crack-tip.

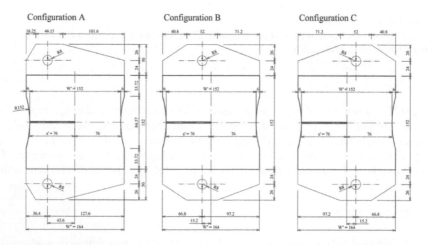

Figure 1. Scheme of *modCT* test configuration, XL size of test specimen is displayed with the dimension in [mm], (Klon & Sobek & Keršner 2017).

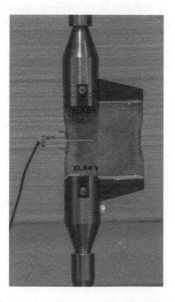

Figure 2. Test specimen of XXL size placed in the test machine.

Figure 3. Test specimen of XL size placed in the test machine, (Klon & Sobek & Keršner 2017).

There was planned wide range of test specimen sizes—to investigate phenomenon of so called size effect (Bažant & Planas 1998, RILEM Committee FMT 89 1990). From the small ones with the width of $W = 19$ mm to the largest with $W = 304$ mm (in total 5 different sizes: S, M, L, XL, XXL, all specimens with the same thickness of 20 mm) whose can be found in Figure 4. It can be also seen that all specimens have "dog bone" shape which serves as stress concentrator (of circular shape) on both sides.

In experimental campaign in total of 47 test specimens were tested but for analysis by the 2D optical profilometry (scanning of the ligament area) only sizes XXL, XL and L were used in amount of 4 from each size (in total of 12 specimens), made with an initial chevron notch.

2.2 Experimental testing procedure

Used testing machine for all experiments was a servo-hydraulic closed-loop machine with a maximum load of 25 kN (see Figure 3). The linear variable displacement transformer (LVDT) transducer with a range of 10 mm and a clip-gauge transducer of 4 mm were used.

Itself experimental test of modCT specimens was managed by displacement control of testing machine.

2.3 Laser 2D profilometry

To determine the surface roughness and topography of body's ligament the optical profilometry is widely used in a lot of industrial applications. It is non-destructive non-contact way how to analyze changes in a surface plane.

Laser 2D profilometer operates generally with a defined wavelength semiconductor laser and uses principle of the laser triangulation for two-dimensional profile detection on different target surfaces. By using special lenses, a laser beam is enlarged to form a static laser line and is projected onto the target surface. The diffusely reflected light from the laser line is replicated on a sensitive sensor matrix by a high quality optical system and evaluated in two dimensions. In addition to distance information (usually marked as z-axis), the exact position of each point on the laser line (usually marked as x-axis) is also acquired by an integrated controller. Obtained values are finally output in a two-dimensional coordinate system that is fixed with respect to the current sensor.

A commercial laser profilometer (scanCONTROL 2750-100 made by Micro-Epsilon Company) was used to precise scanning of modCT test specimen's ligaments after loading experiments done. It contains a semiconductor laser which has a wavelength of 660 nm (visible red line). Manufacturer declares minimum resolution in the order of 25 μm and scanning speeds up to 4000 single profiles per second (with the 640 points in total per single profile). The 3D surface topography is obtained by the movement of the samples or laser scanner. In set-up, which was used for modCT test specimens' ligaments, the laser scanner (2D profilometer) is mounted on a tripod and a motion of current sample is realized by a precise linear sliding of platform with stepper motor. Designed system is capable of the smooth and adjustable motion (controlled by a developed microprocessor driver that is linked to the master control

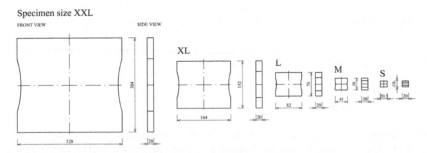

Figure 4.   Scheme of different sizes of modCT test specimens, from the largest to the smallest, with the dimension in [mm], (Klon & Sobek & Keršner 2017).

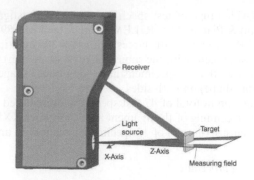

Figure 5.   Scheme of scanCONTROL 2750-100, (Micro-Epsilon 2018a).

Figure 6.   Experimental set-up with L size specimen.

PC) with the step of 2.5 μm and the maximum weight of samples up to 10 kg. The distance of 50 μm was selected between the scanned lines/profiles (distance on the y-axis) and the spacing between adjacent points was also 50 μm within one line/profile (x-axis). Typical photos of the experimental set-up can be seen in Figure 5 and Figure 6.

The measuring process is controlled by a utility developed in MATLAB (MATLAB 2018) code. It allows movement of mentioned sliding platform and measurement of current profile of specimen's ligament by the 2D profilometer. The integrated FireWire interface enables complete control of the used laser scanner via computer, as well as high-speed data acquisition. Utility provides also the post-processing package for data obtained, including their storage in the resulting data matrix. Thereafter the final 3D sample topography can be displayed, then some additional operations can be performed (cropping of the monitored area, rendering of 2D and 3D surfaces, saving results in different data formats, etc.).

The next module allows to calculate some basic surface roughness parameters, that can be evaluated from individual 2D profiles or the whole 3D surfaces. Another separate module is focused on determination of the fracture surface area of the ligament's surfaces scanned. Its solution is done by a simple algorithm which allows approximate calculation of surface area between the four adjacent points in the data (z-axis) matrix by dividing the resulting pattern into two triangles, then the area is calculated (on the basis of the Heron's formula or selected vector's products). The summation of all calculated triangles areas in intended region forms the fracture surface (real) area marked as A_{lig}^{*}.

3 DISCUSSION OF RESULTS

All twelve ligament's surfaces were scanned and typical illustration of the reconstructed surface can be seen in Figure 7 and Figure 8. Original shape of the initial chevron notch was removed from the calculation to make a reconstruction more comparative for each of test specimens. Table 1 shows results of the areas for all variants of test specimens used. Denotation of test

Figure 7.   Typical illustration of the fracture surface area for test specimen denoted as LA1SB.

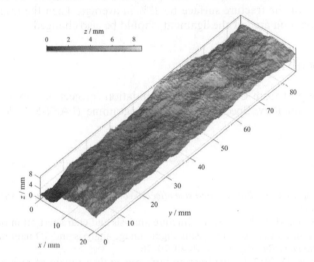

Figure 8.   Typical illustration of the fracture surface area for test specimen denoted as XLA2SC.

Table 1.   Ratio between "basic" and determined area of the specimen's ligament for all analysed modCT test specimens of three different sizes in [mm²].

Specimen	Basic area A_{lig}	Calculated area $A_{\text{lig}}{}^{*}$	Ratio ($A_{\text{lig}}{}^{*}/A_{\text{lig}}$)
LA1SA	720.16	881.84	1.225
LA1SB	647.97	786.76	1.214
LA1SC	593.23	712.04	1.200
LA2 LC	579.44	722.64	1.247
XLA1SA4	1112.28	1382.93	1.243
XLA1SA	1813.54	2252.13	1.242
XLB2SB	1763.06	2168.94	1.230
XLA2SC	1788.80	2266.12	1.267
XXLA1A2	2286.56	2813.00	1.230
XXLA1SA	3838.71	4662.05	1.214
XXLB2SB	3641.24	4464.46	1.226
XXLB2SC	3786.91	4718.63	1.246

specimens can be seen in the first column—sorted by size firstly, then by used eccentricity. Second and third columns contain values of basic and calculated (determined) areas. The last column shows ratio between these two areas, basically, the difference between each other. For example, the first line belongs to ligament of test specimen denoted as LA1SA. Size of this specimen is L, A1 means label of mixture of the concrete used, S is for short initial notch (proportionally the same one for all variants used) and the last letter, at the end of denotation, is the variant of eccentricity, which can be taken from Figure 1. It is apparent that average difference between assumed area of ligament and "real" area of fracture surface is around 23%. It means that real areas of fracture surface, used in calculations of fracture parameters, like fracture energy G_f, should be taken into account with 20% – 25% increase. However, if we compare the three chosen sizes with each other, there are no significant differences. Same can be noted in the case of different concrete mixtures (their composition is not part of this paper).

4 CONCLUSIONS

To conclude this study, no significant differences can be found between the three sizes used (L, XL and XXL) in the case of fracture area surface. However, some suggestions should be mentioned—with the usage of modCT test specimens, one should notice the increase in the value of real area of the fracture surface by 23% in average. Then the fracture parameters, which are dependent on area of the ligament, should be also changed.

ACKNOWLEDGEMENT

The financial support from Czech Science Foundation (project No. 18-12289Y) and Brno University of Technology, Specific Research programme (FAST-S-18-5614) is gratefully acknowledged.

REFERENCES

Bažant, Z.P. & Planas, J. 1998. *Fracture and size effect in concrete and other quasi-brittle materials*. Boca Raton: CRC Press.

Erdem, S. & Blankson, M.A. 2013. Fractal-fracture analysis and characterization of impact-fractured surfaces in different types of concrete using digital image analysis and 3D nanomap laser profilometery. *Construction and Building Materials* 40: 70–76.

Ficker, T. & Martišek, D. 2012. Digital fracture surfaces and their roughness analysis: Applications to cement-based materials. *Cement and Concrete Research* 42: 827–833.

Klon, J. & Sobek, J. & Keršner, Z. 2017. Modelling of modified compact tension test of fine-grained cement based concrete specimens using FEM software. Key Engineering Materials 754: 329–332.

MATLAB 2018. MATLAB and Statistics Toolbox Release. The MathWorks, Inc., Natick, Massachusetts, United States.

Micro-Epsilon 2018a. Instruction Manual scanCONTROL 27 × 0 [online]. Available from: https://www.micro-epsilon.com/download/manuals/man--scanCONTROL-2700--en.pdf.

Micro-Epsilon 2018b. 2D/3D laser scanner (laser profile sensors) [online]. Available from: https://www.micro-epsilon.com/download/products/cat--scanCONTROL--en-us.pdf.

RILEM Technical Committee 50 FMC 1985. Determination of the fracture energy of mortar and concrete by means of three-point bend test on notched beams. *Materials & Structures* 18: 285–290.

RILEM Committee FMT 89 1990. Size effect method for determining fracture energy and process zone size of concrete. *Materials and Structures* 23: 461–465.

Veselý, V. & Sobek, J. 2013. Numerical study of failure of cementitious composite specimens in modified compact tension fracture test. *Transactions of the VŠB – Technical University of Ostrava: Civil Engineering Series* 13(2). DOI: 10.2478/tvsb-2013-0025.

Advances and Trends in Engineering Sciences and Technologies III – Al Ali & Platko (Eds)
© 2019 Taylor & Francis Group, London, ISBN 978-0-367-07509-5

Determination of basic dynamic characteristics of a tower structure from ambient vibrations

M. Sokol, M. Venglár, K. Lamperová & Z. Šišmišová
Department of Structural Mechanics, Faculty of Civil Engineering, Slovak University of Technology, Bratislava, Slovakia

R. Ároch
Department of Steel and Timber Structures, Faculty of Civil Engineering, Slovak University of Technology, Bratislava, Slovakia

ABSTRACT: The paper describes measurement of dynamic effects on an industrial steel tower structure and determination of dynamic characteristics from ambient vibrations. A numerical model of construction was created. The calculated and experimentally acquired natural frequencies and mode shapes were compared to verify the behaviour of the load-bearing system and the numerical model.

1 INTRODUCTION

Ambient vibrations and determination of dynamic characteristics is a topic followed by many researchers, see e.g. (Wenzel & Pilcher, 2005), (Birtharia & Jain, 2015), (Ivanovič & Todorovská & Trifunac, 2000). Such an approach was also used for an industrial tower-like structure.

The steel tower is in the industrial area of Považie – in Western Slovakia. After years of operation vibrations of the structure have emerged that affect the production. The purpose of the research was to measure the dynamic effects on the industrial tower.

2 DESCRIPTION OF THE STRUCTURE

The supporting structure of the industrial tower is a 3D steel truss frame. The ground plan dimensions are 13.5 m × 13.5 m according to the project documentation and the height of the structure is approximately 58 m (Figure 1).

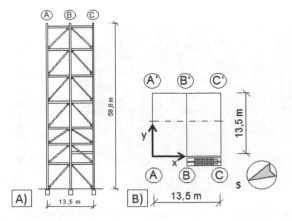

Figure 1. Structure of the industrial tower, A – side view, B – ground plan.

3 CALCULATED MODE SHAPES AND NATURAL FREQUENCIES

A numerical model of construction was created according to documentation. Natural frequencies and mode shapes were calculated by a modal analysis (Table 1).

The second column shows calculated values by means of modal analysis. The third column shows the values obtained from experimental measurements. These variables were then verified experimentally verified.

4 DYNAMIC MEASUREMENTS

4.1 Measuring equipment

For the experimental measurements a National Instruments (NI) measuring system, consisting of the NI CompactRIO with 4 NI 9234 input modules (analogue-to-digital converter), was used (Figure 2). An original LabVIEW program according to (National Instruments 2013) guide was written. The number of modules allowed to measure 16 different locations on the structure. Accelerometers PCB Piezotronics 393B31 were used, with an acceleration range of ± 4.9 m/s^2. The measurement also needed more than 1 km of cable and temperature sensor of the structure.

4.2 Location of sensors

Altogether, 13 accelerometers were placed on the structure of the tower to the locations needed to determine the global modes-shapes.

The accelerometers were placed horizontally in two perpendicular axes; one control sensor was in the vertical direction (acc. Q). Horizontal sensors were placed on both corners of the structure (acc. G, I, O, P, R, U, V, Z). Further accelerometers (acc. J, N, S and T) were placed perpendicular to them. In this way, four top floors were instrumented. A schematic view of

Table 1. Natural frequencies and mode shapes – comparison.

Mode (i)	f_i – calculated [Hz]	f_i – measured [Hz]	Description
1	0.89	0.88	1st mode in Y axis direction
2	0.93	0.94	1st mode in X axis direction
3	1.80	1.74 to 1.89	1st torsion mode
4	3.16	2.75 to 3.25	2nd torsion mode

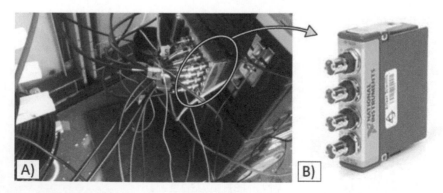

Figure 2. Measuring equipment, A – NI CompactRIO, B – NI 9234 module.

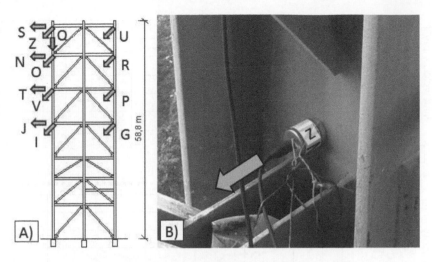

Figure 3.   A – locations of accelerometers, B – located accelerometer by magnet.

the accelerometer deployment is shown in Figure 3A. The accelerometers were attached to the steel structure of the industrial tower by the magnets (Figure 3B).

Wind effects were recorded using the ALMEMO 2590 A with a digital sensor for measuring air temperature, atmospheric pressure and wind velocities. Wind speed measurements were recorded on the upper tower platform, approximately 58 m above ground level, with a sampling frequency ~1 Hz.

5   EXPERIMENTAL MEASUREMENTS

Measurements took place during a windy day where the wind blew predominantly from the southeast direction – accelerations and wind speeds were recorded. In the following figure (Figure 4), the wind velocity, recorded during one measurement with 120 seconds duration, is shown.

Figure 4A shows wind velocity for 120 seconds of acceleration measurement. From the time record a section was selected (Figure 4B – section), where the wind velocity began to rise slightly and then stabilized at 6.5 m/s (Polčák & Šťastný, 2010). At the same time, an acceleration record was made (Figure 5A).

Interesting was the section between 30th and 37th second, Figure 5. It clearly shows the line of the period (highlighted line), from which the frequency of the 1st mode shape can be determined using the basic relation ($f = 1/T$). The period T_1 is approximately 1.14 s. The value of the 1st natural frequency is then $f_1 = 0.88$ Hz.

In Figure 6 is shown an enlarged scale of the acceleration amplitude spectrum of accelerometers Z, O and V. From this Figure it is evident that the deflections of the structure increase with the height at the same frequency and in the same phase (Figure 7), thus confirming that this is the 1st mode shape.

Figure 6 and 7 prove that it is a vibration in the 1st mode shape, which was caused by the wind load. In addition to the above measurements, six further measurements were performed and the 1st mode shape of structure was verified more than once. Table 2 shows the frequency values. Time of the records was always 120 seconds long and Table 2 shows the beginning of the measurement record. Natural frequencies were found for each record. It is clear from the above values that the presented measurement (highlighted in Table 2) corresponds to the other measurement data obtained and therefore it is not necessary to present them in more detail.

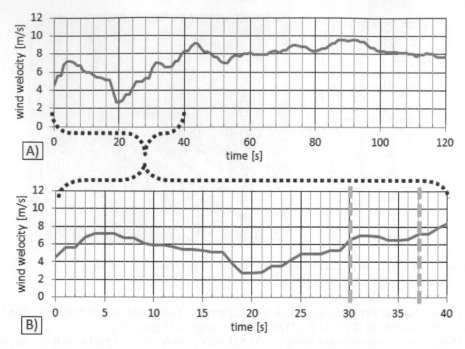

Figure 4. Record of wind velocity, A – total record of 120 s, B – section 0 s to 40 s.

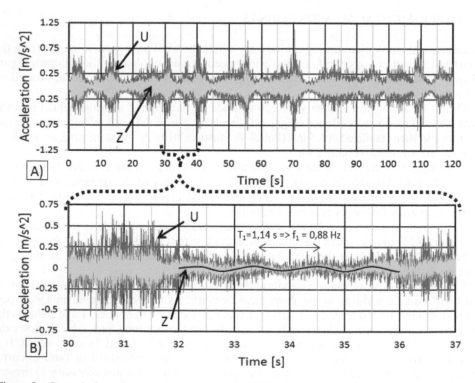

Figure 5. Record of acceleration, A – total record of 120 s, B – section 30 s to 37 s.

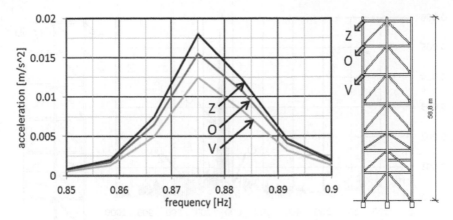

Figure 6.   Amplitude spectrum of accelerometers Z, O, V – 1st mode shape.

Table 2.   Comparison 1st natural frequency from multiple measured records.

Beginning of record	f_i [Hz]
13:01:53	0.87
13:03:54	0.86
13:13:22	0.88
13:15:23	0.88
13:17:31	0.87
13:21:12	0.88
13:23:13	0.88

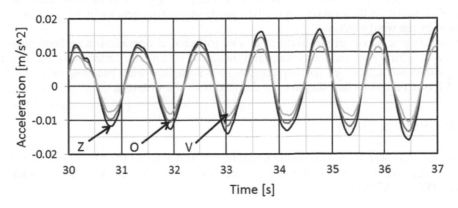

Figure 7.   Filtered record of acceleration (low pass 0.88) – accelerometers Z, O, V.

6   DYNAMIC EFFECTS OF THE TECHNOLOGICAL EQUIPMENT

In Figure 5. A we can see also accelerations that are for an order of magnitude higher than the acceleration caused by the wind load. Based on the prepared spectrum (Figure 8) the actual effects on the records are also from the operating equipment. However, these frequencies are much higher than the actual frequencies of the measured structure. In this case, it was approximately 640 Hz.

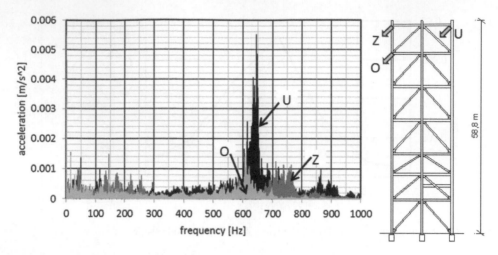

Figure 8. Amplitude spectrum - accelerometers U, Z, O.

For this reason, acceleration values that are caused by the wind loads and which have a significant effect on the global dynamic response of the structure may be taken into consideration for possible further calculations.

7 CONCLUSIONS

A good match of measured frequencies with numerical values has been achieved.

From the comparison of synchronized acceleration records and wind velocity records measured directly on the structure, it can be seen that the overall vibration of the structure relatively correlates with the dynamic wind velocity change. It can be confirmed that the global vibration of the structure in the first mode shapes (frequencies up to 2 Hz) is mainly caused by the wind effects.

Considerable accelerations, likely to be due to technical operation, have been measured on the structure. At these frequencies and measured maximum accelerations, e.g. Figure 5 ($a_{max} = 0.75$ m/s²), however, the displacements are negligibly small, reaching only one thousandth of a millimeter and they are not dangerous for the structure.

ACKNOWLEDGMENT

This paper has been supported by the Scientific Grant Agency of the Ministry of Education, science, research and sport of the Slovak Republic and the Slovak Academy of Sciences – grant VEGA 1/0773/18.

REFERENCES

Birtharia, A. & Jain S.K. 2015. Applications of Ambient Vibration Testing: An Overview. *IRJET International Research Journal of Engineering and Technology*. Vol. 2, No. 5: pp. 845–852.

Ivanovič, S.S.; Todorovská, M.I. & Trifunac, M.D. 2000. Ambient Vibration Test of Structures – A Review. *ISET Journal of Earthquake Technology*. Vol. 37, No. 4: pp. 165–197.

National Instruments (Austin, Texas) 2013. *NI LabVIEW for CompactRIO Developer's Guide.*

Polčák, N. & Šťastný, P. 2010. *Impact of the Relief on the Wind Conditions of the Slovak Republic*. (In Slovak) Bratislava: Slovak Hydro-meteorological Institute.

Sokol, M. & Tvrdá, K. 2011. *Dynamic of Structures*. (In Slovak) Bratislava: Publishing house of STU.

Wenzel, H. & Pilcher, D. 2005. *Ambient Vibration Monitoring*. Chichester: John Wiley & Sons Ltd.

Advances and Trends in Engineering Sciences and Technologies III – Al Ali & Platko (Eds)
© 2019 Taylor & Francis Group, London, ISBN 978-0-367-07509-5

Influence of joint stiffness on design of steel members

L. Šabatka, D. Kolaja & M. Vild
IDEA StatiCa, Brno, Czech Republic

F. Wald & M. Kuříková
Czech Technical University in Prague, Prague, Czech Republic

J. Kabeláč
Hypatia Solution, Brno, Czech Republic

ABSTRACT: The paper explains the types of joints classified according to stiffness in a steel structure. Applying the joint stiffness to the structural model brings a number of problems and inaccuracies. The article shows possibilities for solving entire member, including the joints. A brief study describes the influence of the joint stiffness on the redistribution of internal forces in the beam.

1 INTRODUCTION

Most steel structures are analyzed in a bar model. For designing common bar elements, this idealization is quite sufficient. Traditionally, it is assumed that the connections are perfectly rigid or ideally pinned. This allows for a simple structural model with linear behavior of joints, but the reality is somewhere between these boundary assumptions. The design of a joint is essential for the behavior of the entire member. The team of authors focuses on development of methods and models for more accurate analysis of the whole "joint-to-beam-to-joint" subsystem and its influence on the beam behavior.

Component based finite element method (CBFEM) (Wald et al., 2016) is used for the simulation of such subsystem. The steel plates are modelled as shell elements and bolts and welds are modelled as springs with nonlinear behavior according to Component method in EN 1993-1-8 (CEN, 2005b). This way, the main problems of Component method are avoided; the point of rotation is calculated automatically, the shape and number of profiles is arbitrary and loading can be general. So far, CBFEM was successfully used in the design of connections and the development is underway to use this method for the design of members. The main advantage is in bringing the model closer to reality, especially in boundary conditions, the application of point loads or determination of the behavior of non-uniform beam, e.g. with haunches, stiffeners and openings – see Figure 2. The model may be used for determination of critical buckling force or critical bending moment.

2 JOINT ROTATIONAL STIFFNESS

Joints are classified according to EN 1993-1-8 – 5.2.2 (CEN, 2006) by stiffness as rigid, semi-rigid and pinned, see Figure 1. Structures with rigid joints can be considered as continuous, pinned joints transfer only moments that can be neglected and freely rotate thanks to their rotational capacity, but semi-rigid joints do not fall into any of these two categories, they transmit bending moments partially. Connections are classified according to the initial stiffness, which is considered linear up to 2/3 of the joint load-bearing capacity, $M_{j,Rd}$. For semi-rigid joints, secant stiffness is important for designing the ultimate bearing capacity. After exceeding 2/3

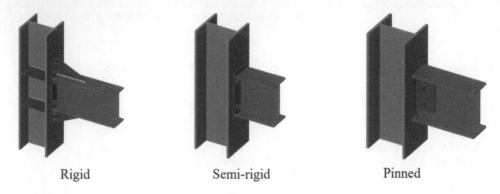

| Rigid | Semi-rigid | Pinned |

Figure 1.   Examples of joints with different classification according to stiffness.

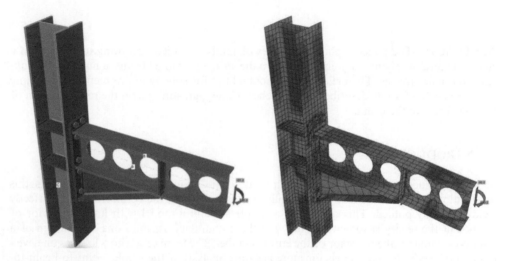

Figure 2.   CBFEM method -design of non-uniform beams.

of $M_{j,Rd}$, the stiffness of the semi-rigid joint is assumed to be $S_{j,ini}/\eta$, according to Article 5.1.2, where the stiffness coefficient in accordance with Table 5.2 is assumed to be between 2 and 3.5 in dependence on the type of connection. By using Article 5.1.2, an iterative process can be avoided. When using the secant stiffness for a specific load, the history of the loading of the joint is considered. If the load has been previously higher, the joint may exhibit plastic deformations and the beam bends more and transmits a higher bending moment in the span.

Initial stiffness can be calculated by the component method in Chapter 6.3. The component method allows to calculate the stiffness of each component, and the position of the axis of rotation is estimated, usually at the compression flange. The component method considers the combination of loads, such as bending moment, shear and normal force, the calculation of each load-bearing capacity with interactions, or interactions when compiling the components. The method was validated by experiments (Baniotopoulos et al., 2006), but the differences in stiffness evaluation are in tens of percent. Differences are due to simplifications in determining component deformations and their assembly. Another factors are measurement errors. Particularly with rigid joints, it is difficult to measure deformations with sufficient precision and to derive the deformation of individual members, strong-floor, load-bearing frames, etc.

In addition to the stiffness of the joint, sufficient rotational capacity must be guaranteed. The code provides simplified models for rotation capacity estimation in Chapter 6.4. Generally, fragile components, such as welding or concrete cone breakout in anchoring, should not be considered for rotational capacity. Higher deformations can occur safely in the tensioned parts in an end plate, column panel in shear or in a connected member.

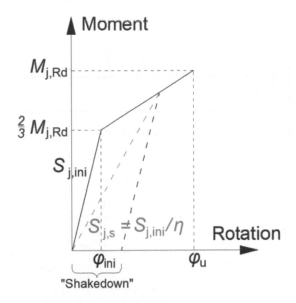

Figure 3. *Moment*-rotation behaviour of a semi-rigid joint.

A monotonically loaded semi-rigid joint is exhibiting initial stiffness up to 2/3 $M_{j,Rd}$. Afterwards, the plastic deformations develop and the stiffness is decreased. After unloading the joint which underwent the plastic deformation, the joint does not return to its original state but is rotated and the so called shakedown occurred. The history of loading is therefore important and if anytime in the life-span of the joint plastic deformation occurred, the joint is initially rotated. As a simplification, for the use in models using linear elastic calculation, the reduced stiffness is used. EN 1993-1-8 suggests use of $S_{j,ini}/\eta$ for semi-rigid joints which are loaded by a higher bending moment than 2/3 $M_{j,Rd}$ – see Figure 3.

Perhaps the biggest problem is implementation of correct stiffness into 3D bar structural model. In the complex model, there are usually tens or even hundreds of load combinations. The stiffness of semi-rigid joints of beam-to-column joints are not so complicated. The stiffness $S_{j,ini}$ should be used if the bending moment did not exceed 2/3 $M_{j,Rd}$ in any load combination. If this is not the case, the stiffness $S_{j,ini}/\eta$ should be set. The most problematic are column bases where the stiffness varies for dominant tensile or compressive force or bending moment.

3   CASE STUDY

The effect of joint bending stiffness can be illustrated on the example in Figure 4. The beam with length L_b, moment of inertia of the cross-section I_b, and material with elastic modulus E is loaded with uniform load q. The beam is attached to the columns by connections with linear stiffness $S_{j,ini}$. For simplicity, it is assumed that the columns are infinitely rigid.

Depending on the stiffness of the joint, the moment diagram corresponds to values according to the rigid type of support up to the pinned type of support.

The stiffness of the joint can be classified according to EN 1993-1-8 by engineering estimation or by reference to the stiffness of the connected beam. The dimensionless stiffness k_b is defined as:

$$k_b = \frac{S_{j,ini} \cdot L_b}{I_b \cdot E}$$

(1)

The influence of the dimensionless stiffness defined in this way is shown in Figure 5. It is clear from the graph that the influence of stiffness of the joint is essential. Depending on

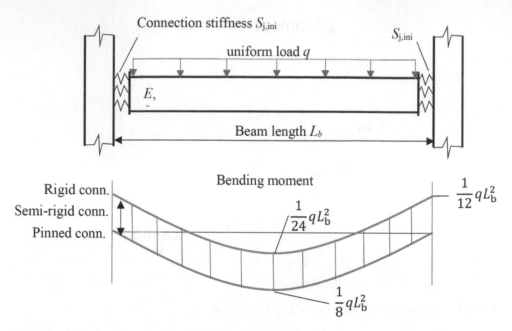

Figure 4. Influence of connection stiffness on the bending moment diagram along a beam.

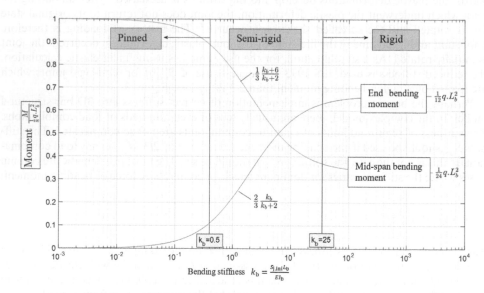

Figure 5. Influence of the relative bending stiffness of the joint in logarithmic scale on the relative mid-span and end bending moment.

the stiffness of the joint, the bending moment in the beam differs by tens of percent. Three areas of the joint stiffness classification are clearly visible from the graph. EN 1993-1-8 classifies joints with stiffness $k_b < 0.5$ as pinned and joints with stiffness $k_b > 25$ as rigid. This corresponds well to the analytical solution in Figure 5. For semi-rigid joints, determining the stiffness for the moment derivation along the beam is important for its design. Conversely, for rigid and pinned beams, the exact determination of stiffness is not significant. With increasing or decreasing stiffness, the moment does not change substantially, as shown in Figure 5. The exact determination of stiffness is problematic for rigid joints. It is a conservative estimate of the stiffness values of the individual components, which consists of an estimated arm of the internal forces. Even the orderly error in the determination of the stiffness of the joint does not have an effect on the moment and deformation of pinned and rigid joints. The beam is designed for the bending moment. When designing for ULS a 5% error is admitted and a 20% error is proposed when designing for SLS.

In Figure 6, stress analysis is shown on the example of beam IPE 330. The beams are 5 m long and are loaded by uniform load 30 kN/m. The pinned connection is always a conservative estimate for the beam design. The pinned and rigid joints are represented by a fin plate connection on the web and a welded connection, respectively.

In Figure 7, the results of linear buckling analysis are shown. The boundary conditions have profound effect on the buckling of steel members, in this case lateral-torsional buckling. Linear buckling analysis provides a factor of set loads to reach the critical load—a point of bifurcation due to stability of a perfect structure. A real structure contains imperfections— e.g. bow imperfection, eccentricity in supports, thickness variation. Therefore, the buckling resistance of a real structure is lower than the critical load. EN 1993-1-1 – Cl. 5.2.1 (CEN, 2005a) states that for critical load factor, $\alpha_{cr} \leq 15$, buckling cannot be neglected and thorough analysis must be performed. EN 1993-1-1 – Cl. 6.3.4 shows a guide for general method for lateral and lateral torsional buckling of structural components. The method uses the combination of geometrically linear and materially nonlinear analysis to obtain $\alpha_{ult,k}$ and linear buckling analysis to obtain $\alpha_{cr,op}$. The global non dimensional slenderness, $\overline{\lambda}_{op}$, for the structural component is determined from the following equation:

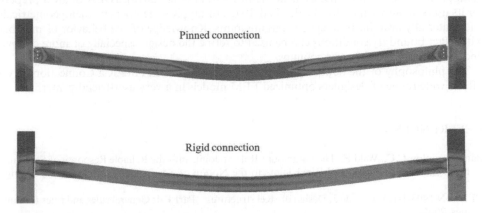

Figure 6. Output of von Mises stress on a beam supported by pinned and rigid connections in the IDEA StatiCa Member application.

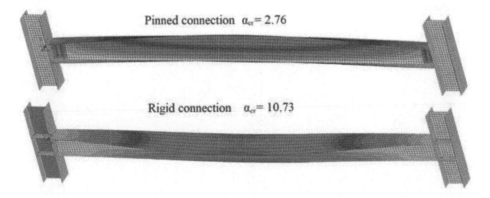

Figure 7. Output of normalized deformation from linear buckling analysis on a beam supported by pinned (fin plate) and rigid (welded) connections in the IDEA StatiCa Member application.

$$\overline{\lambda}_{op} = \sqrt{\frac{\alpha_{ult,k}}{\alpha_{cr,op}}} \qquad (2)$$

The general method can serve as an approximation of the buckling resistance of generally loaded complicated members. Both geometrically linear and materially nonlinear analysis and linear buckling analysis are very fast methods. However, the final step of the design should be geometrically and materially nonlinear analysis with imperfections (GMNIA). This analysis is time consuming. A guide for setting imperfections is provided by EN 1993-1-1.

4 SUMMARY

Changing the joint stiffness causes the bending moments to be redistributed in the attached member. Even a large change in joint stiffness at a rigid or pinned joint causes only a slight redistribution. Changing stiffness is quite important in the range of semi-rigid joints.

The vast majority of members in steel structures can be easily and reliably designed using standard procedures. The design of standard steel members on their own is thoroughly researched and understood but the boundary conditions are often only guessed. The successful IDEA StatiCa Connection software, which is becoming the world standard for the design of joints (used by 1 500 offices), has put into practice a good way of designing joints using the component method and finite element method (Wald et al., 2016). IDEA StatiCa prepares for its users an application for the detailed design of atypical members by their geometrically and materially non-linear analysis including joints. Knowledge of the behavior of members with exact boundary conditions can be used to refine the design especially of more complex and non-standard beams and columns.

The philosophy of the new program is the same as for IDEA StatiCa Connection. It will offer a wide range of designers optimized FEM models in a very user-friendly interface.

REFERENCES

Baniotopoulos, C.C., Wald F., The Paramount Role of Joints into the Reliable Response of Structures, From the Classic Pinned and Rigid Joints to the Notion of Semi-rigidity, NATO series, Springer, 2000.
CEN, EN 1993-1-1: Eurocode 3: Design of steel structures—Part 1–1: General rules and rules for buildings, 2005.
CEN, EN 1993-1-8: Eurocode 3: Design of steel structures—Part 1–8: Design of joints, 2005.
Wald, F., Šabatka, L., Bajer, M., Barnat, J., Gödrich, L., Holomek, J., Jehlička, P., Kabeláč, J. et al., Benchmark cases for advanced design of structural steel connections, Prague: Českátechnika—nakladatelství ČVUT, 2016.

Advances and Trends in Engineering Sciences and Technologies III – Al Ali & Platko (Eds)
© 2019 Taylor & Francis Group, London, ISBN 978-0-367-07509-5

Non-linear analysis of deck slabs subjected to a concentrated load

A. Vidaković, J. Halvonik, R. Vida & T. Augustín
Faculty of Civil Engineering, Slovak University of Technology, Bratislava, Slovakia

ABSTRACT: The shear resistance of bridge deck slabs without shear reinforcement is currently a decisive structural property for design of the slab thickness. The dominant shear action is represented here by the concentrated load due to the wheel contact pressure. The missing model for the shear distribution in case of a concentrated load became the major motivation for an experimental campaign and afterwards for a non-linear analysis of this phenomenon. Together 12 slabs were tested in Laboratory with concentrated load and the obtained results were used for calibration of the non-linear model. The non-linear Finite Element (FE) analysis was carried out using the ATENA software. The paper will present the obtained results and recommendations concerning non-linear modeling of bridge deck slabs subjected to a concentrated load.

1 INTRODUCTION

The shear resistance of RC slabs without shear reinforcement is a discussed topic because of their brittle mode of failure. Bridge deck slabs subjected to a concentrated load have the ability to distribute the load in transverse direction, which contributes to their shear resistance.

One-way shear failures are generally associated with distributed or line loads and linear supports, while two-way shear failure is associated with the introduction of punctual supports such as columns or concentrated loads due to wheel pressure (Vaz Rodrigues et al. 2008).

The aim of this study is to investigate the shear force distribution of cantilever slabs subjected to a concentrated load. In order to reflect the influence of shear crack, a non-linear finite element analysis using three-dimensional elements that take into account the out-of-plane shear response was carried out. Furthermore, the influence of factors in one-way and two-way shear behavior, such as the layout of reinforcement and geometry of the deck slabs were investigated.

2 EXPERIMENTAL PROGRAM

To study the shear resistance of concrete haunched slabs representing the overhang slabs of concrete road bridges, an experimental campaign (Vida & Halvonik 2018) has been carried out.

The experimental campaign consisted of two parts. First part represented experimental investigation of slab strips with T shaped cross section and width 0.5 m (type ST). These specimens have been loaded along the whole width to ensure uniform shear flow along the whole width. Second part was focused on the slabs with the same cross section, but the width of the slabs was 2.5 m and they were loaded with concentrated load (type SL).

Both specimens consisted of two symmetrical overhang slabs with central beam resting on a steel base, see Figure 1. Test load was generated by a hydraulic jack supported by a strong steel frame anchored to the strong floor. The T shaped specimens allowed to test each specimen two times while the second side (not tested) has been supported by a couple of steel beams and a synchronized hydraulic jack. After the test on the first side (A), the setup was rearranged and the second side (B) was tested.

Figure 1.   Experimental setup of SL specimen (left) and ST specimen (right).

Table 1.   Reinforcement material properties of investigated specimens.

Specimen		d^* [mm]	Reinforcement layout	ρ_l [%]	f_{ym} [MPa]	E_s [GPa]
SL 0.1			Ø14 s125	0.73	608	
SL 1.1	ST 1.1	168	Ø14/16 s125	1.69	576	200
SL 2.1	ST 2.1		Ø14/12 s125	1.27	580	

*Effective depth at the face of the support.

In case of SL type specimen, the load was applied through a pair of steel plates with dimensions 250 × 250 mm and thickness 40 mm positioned on top of each other. The position of the loaded area was chosen so that the front edge was at the distance of 335 mm (twice the effective depth of the slab) from the face of the intermediate support (the central beam under the slab). This distance is assumed as the most effective position of load for shear and at the same time, the width of the slab carrying the load is the smallest.

The setup of the ST specimen was almost identical as the SL specimen, with the exception of the loading plate. In this case, the width of the plate was the same as the width of the sample, the second dimension was 250 mm and the distance from the support's face remained 335 mm. This setup ensured constant flow of shear forces over the whole specimen's width and also the same position of load transmission.

The specimens were cast from concrete with measured cylindrical strength (f_{cm}) between circa 30–36 MPa and modulus of elasticity (E_{cm}) between 32–35 GPa. The maximum aggregate size of 16 mm was used in all specimens. The layout and amount of reinforcement was variable, values for each specimen can be seen in Table 1. Specimens with the same numerical label were cast at the same time from the same concrete mix.

3 NUMERICAL ANALYSIS

The numerical analysis presented in this paper was performed by ATENA (Cervenka Consulting), non-linear FE software, designed specifically for non-linear analysis of reinforced concrete structures. The geometry of tested specimens was created in ATENA 3D, while analysis execution, post-processing and data extraction was performed in ATENA Studio software.

Concrete tension is based on fracture model, which employs the Rankine failure criterion, exponential softening and it can be used as a rotated or fixed crack model. The compression hardening and softening behavior is computed using the plasticity model based on Menétrey-William

(1995) failure surface. The model can be used to simulate concrete cracking, crack closure and crushing under high confinement (Cervenka & Papanikolau 2008).

3.1 Material input parameters

The concrete material was defined by concrete cylinder strength based on values obtained from tests performed on material samples taken from laboratory. The fracture energy G_f, was derived using the relation given in fib Model Code 1990 (1).

$$G_f = G_{f0} \left(\frac{f_{cm}}{f_{cm0}} \right)^{0.7} \quad if \ f_{cm} \leq 80 MPa \tag{1}$$

where f_{cm} = the concrete mean compressive strength; f_{cm0} = a reference value for the concrete compressive strength, set to 10 MPa; and G_{f0} = the fracture coefficient as a function of the maximum aggregate size, as shown in Table 2.

All other concrete parameters were automatically calculated by the program and adjusted according to the results obtained from laboratory (Table 3).

The reinforcement was assigned according to a bilinear elastic-perfectly plastic stress-strain diagram with perfect bond to the concrete. This was based on the assumption that every reinforcement bar was continuous over the area of the failure zone, thus eliminating possible effect of bond slip between reinforcement and concrete. Reinforcement arrangement was set to maintain the reinforcement ratio for all types of specimen as it is indicated in Table 1.

3.2 Element properties

The concrete part of the specimen was modeled using hexahedral (brick) 8-node elements with *CC3DNonLinCementitious2* material. The mesh size of 30 mm was chosen through a convergence study for the following parametric study. Seven elements were used over the slab thickness for both SL and ST specimen. Tetrahedral 4-node elements were used for steel plates with *CC3DElastIsotropic* material. All elements were linear.

Table 2. Fracture coefficients based on aggregate size.

Maximum aggregate size d_{max} [mm]	Fracture coefficient G_{f0} [Nmm/mm²]
8	0.025
16	0.030
32	0.058

Table 3. Material properties for the concrete.

Property	Specimen		
	SL 0.1	SL 1.1, ST 1.1	SL 2.1, ST 2.1
Compressive strength [MPa]	36.05	34.80	30.47
Poisson's ratio [–]	0.2	0.2	0.2
Specific weight [kN/m³]	23.0	23.0	23.0
Tensile strength [MPa]	2.919	2.851	2.609
Modulus of elasticity [GPa]	34.79	34.33	32.60
Specific fracture energy [Nm/m²]	73	71	65
Crack shear stiffness factor [–]	20	20	20
Fixed crack coefficient [–]	1.0	1.0	1.0

3.3 Boundary conditions and loading

A correct modeling of supports is important to reproduce the actual structural behavior of slabs (Shu et al. 2017). The bottom surface of the central beam of the tested sample was supported by linear elastic compression-only springs, representing the stiffness of the steel base

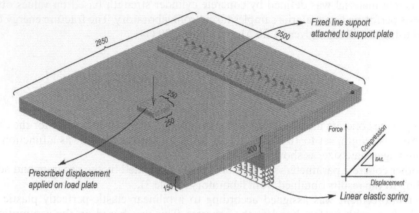

Figure 2. Geometry of the numerical model of SL specimen.

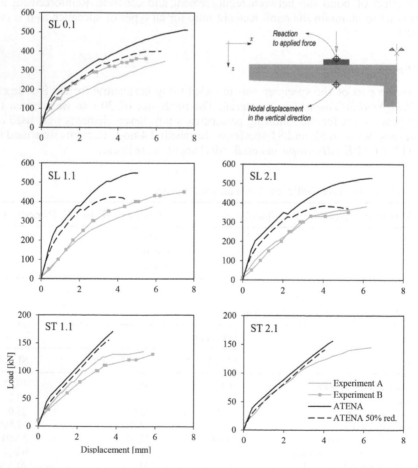

Figure 3. Load-displacement relations obtained from numerical analysis with comparison to experiments and the location of the monitoring points on the slab specimen.

in compression and having zero stiffness in tension to allow uplifting where it occurs. The boundary and loading conditions were identical for both SL and ST models.

A controlled prescribed vertical deformation step by 0.2 mm was used to load the steel plate, and the full Newton-Raphson iterative method based on force and energy convergence criteria with a tolerance of 0.01 was used to solve the structural response. This solver uses a fixed displacement for each step and obtains the load-displacement response through an iterative procedure and using the tangential stiffness from previous iterations.

Two monitoring points were selected to monitor force and displacement under the applied load. The geometry of the numerical model is shown in Figure 2.

3.4 *Results*

The measured shear capacity and load-displacement relations obtained from the non-linear FE analysis and the comparison to experiments are shown in Figure 3 and Table 4. It can be observed that the stiffness of SL slabs was slightly overestimated by model, while the stiffness of ST slabs was reflected in the model more accurately. It is known that both the tensile and compressive strength of concrete increases up until a certain point in time as the concrete hardens and ages. While the development of the compressive strength ends at certain time and stays essentially the same, the tensile strength will begin to decrease due to micro cracks originating from the effects of shrinkage. Thus, as a result of this, the tensile strength of the concrete material model was reduced by 50%, representing an approximate tensile strength of 1.459, 1.425 and 1.305 MPa, respectively. As can be seen from the load-displacement relations (dashed line in Fig. 3), the model with the 50% tensile strength reduction provides results more accurately.

Figure 4 and Figure 5 show the crack patterns that were predicted by FE model with good accuracy.

Table 4. Measured shear capacity of SL and ST specimen.

Specimen	Experiment V_{EXP} [kN]		ATENA V_{FEM} [kN]	ATENA 50% red. $V_{FEM0.5}$ [kN]	V_{EXP}/V_{FEM}	$V_{EXP}/V_{FEM0.5}$
	A	B				
SL 0.1	347	360	511	400	0.70	0.90
SL 1.1	373	450	550	424	0.82	1.06
SL 2.1	380	350	528	384	0.72	0.99
ST 1.1	135	130	171	156	0.79	0.87
ST 2.1	145	–	156	141	0.93	1.03
Average					0.79	0.97
Coefficient of variation					0.11	0.09

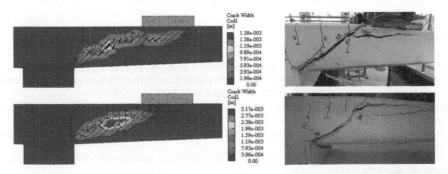

Figure 4. The crack pattern of ST 1.1 (top) and ST 2.1 (bottom) specimen obtained from numerical model and the comparison to experiment (0.1 mm crack filter in ATENA crack output).

Figure 5. The crack pattern of SL 0.1 specimen obtained from numerical model and the comparison to experiment (0.1 mm crack filter in ATENA crack output).

4 CONCLUSIONS

An experimental campaign was carried out and afterwards the numerical models with non-linear FE analysis based on the smeared crack approach were developed with the aim to predict the response of the reinforced bridge deck slabs without shear reinforcement. Some conclusions regarding structural behavior of these specimens can be drawn.

– The reduction of the default value of the concrete tensile strength given by ATENA significantly improved conformity with the experimental behavior of the tested SL specimens, while it had no large significance on the ST specimens.
– It can be observed in Figure 3 that the SL specimens behave more ductile and therefore some flexural yielding might occur before the shear failure is reached.
– The numerical results confirmed that the shear strength increases with increasing reinforcement ratio. However, this fact has not been quite confirmed by the experimental results.
– Based on the crack patterns obtained from the numerical and experimental results, critical shear crack in the ST specimens was developed between the load and the support (one-way shear failure), while in SL specimens, both tangential and radial cracks have occurred, following with a certain mode of failure that is somewhat between one-way and punching (two-way) shear failure.

The following analysis was carried out in order to better understand the behavior of the tested specimens and to evaluate some modifications to the deck slab design. The model calibration was based on monitoring of the shear capacity and load-displacement relations obtained from the tests. FEM models calibrated in this way can be used for further parametric studies of this phenomenon.

ACKNOWLEDGEMENT

This work was supported by the Scientific Grant Agency of the Ministry of Education, science, research and sport of the Slovak Republic and the Slovak Academy of Sciences No 1/0810/16 and by the Slovak Research and Development Agency under the contract No. APVV-17-0204.

REFERENCES

CEB-FIP. *fib* Model Code for Concrete Structures 1990. Lausanne, Switzerland (1993).
Cervenka, J. & Papanikolaou V.K. (2008). Three dimensional combined fracture-plastic material model for concrete. *International Journal of Plasticity*. pp. 2192–2220.
Cervenka, V. et al. (2018) ATENA Program Documentation Part 1 – Theory. *Cervenka Consulting s.r.o.* Prague, 26. January.
Shu, J. et al. (2017). Internal force distribution in RC slabs subjected to punching shear, *Engineering Structures*. 153 (2017), pp. 766–781.
Vaz Rodrigues, R. et al. (2008). Shear strength of R/C bridge cantilever slabs, *Engineering Structures*. 30 (2008), pp. 3024–3033.
Vida, R. & Halvonik, J. (2018). Tests of shear capacity of deck slabs under concentrated load, *Proc. of the 12th fib International PhD Symposium in Civil Engineering*. Prague, 29.–31. August 2018.

Advances and Trends in Engineering Sciences and Technologies III – Al Ali & Platko (Eds)
© *2019 Taylor & Francis Group, London, ISBN 978-0-367-07509-5*

An interface damage model in applications with cyclic loading

R. Vodička, K. Krajníková & I. Katreničová
Faculty of Civil Engineering, Technical University of Košice, Košice, Slovakia

ABSTRACT: A layered structure exposed to cyclic loading is numerically analysed within a quasi-static interface damage model leading to propagation of interface cracks. The crack growth is controlled by a damage parameter whose evolution guarantees stress-separation laws as in known cohesive zone models, including additionally a kind of hysteresis pertinent to the evolution. The cyclic load makes the process to have a fatigue-like character, where the crack appears for a smaller magnitude of the cyclic load than for pure uploading. Presented numerical results demonstrate the properties of the model and reveal its characteristic response to cyclic loading.

1 INTRODUCTION

Many civil engineering problems require analysis of rupturing interfaces. The present paper provides a model which simulates damage evolution and crack propagation in such structures under a cyclic load with a constant amplitude which may cause the damage and crack growth for the amplitudes substantially smaller than under pure uploading conditions. The simulation of such fatigue processes requires special stress-separation laws. One of the possibilities based on the implementation of cohesive zone model (CZM) was introduced by the cyclic variant of CZMs in (Roe & Siegmund 2003, Roth et al. 2014).

The present approach is physically based on state evolution expressed in terms of energy functionals which include stored elastic energy, energy of the external load and also dissipation due to damage. The evolution of damage is closely related to the dissipation potential which includes a kind of viscosity associated to damage, see also (Roubíček et al. 2013), in order to provide a model in which the unloading and reloading paths are uncoincident. Besides, it was required to decrease transferable force with increasing damage, and to maintain a lower bound of stress for damage evolution. Also, the implementation of CZMs, in the context of the energy evolution implemented e.g. in (Vodička 2016), may lead to various endurance limits.

The solution evolution is approximated by a semi-implicit time stepping algorithm (a staggered scheme) providing a variational structure to the solved problem. In time steps, it recursively solves two minimisations with respect to separated deformation and damage variables, described in the previous author's works (Vodička et al. 2014, Vodička 2016, Vodička & Mantič 2017). The numerical procedures also comprehend the Symmetric Galerkin BEM (SGBEM) (Vodička et al. 2007) to provide the complete displacement and stress data for the solids and sequential-quadratic programming algorithms to calculate the solutions for energy minimisations in the variationally based approach (Dostál 2009).

The presented numerical results demonstrate the properties of the proposed model in two simple configurations with two kinds of cracks: an opening crack and a shearing crack, and reveal a characteristic response of the model to cyclic loading.

2 DESCRIPTION OF THE MODEL

Let us consider the energies which are taken into account in the model for two domains Ω^A and Ω^B with boundaries Γ^A and Γ^B, respectively. The stored energy functional is defined as

$$E(t;u,\varsigma) = \sum\nolimits_{\eta=A,B} \int_{\Gamma^\eta} \frac{1}{2} u^\eta \cdot p^\eta d\Gamma + \int_{\Gamma_c} \frac{1}{2} \left[k_n \phi(\varsigma) [\![u]\!]_n^2 + k_t \phi(\varsigma) [\![u]\!]_t^2 + k_g \left([\![u]\!]_n^-\right)^2 \right] d\Gamma \qquad (1)$$

for the admissible (time t dependent) displacement $u_D^\eta = g^\eta(t)$ on a part of boundary Γ_D^η with $\eta = A$, or B, and where p^η denotes pertinent surface tractions. The first integral, representing the elastic strain energy in the subdomains Ω^η, is expressed in its boundary form. The last integral is the interface term which defines dependence of the interface stiffnesses k_n and k_t, on the actual state of damage by the function $\phi(\varsigma)$, such that $\varsigma = 1$ pertains to the initial undamaged state and $\varsigma = 0$ means total damage (a crack). The terms $[\![\cdot]\!]_n$ and $[\![\cdot]\!]_t$, respectively, are the normal and tangential separation of opposite interface points, i.e. $[\![u]\!] = u^A - u^B$. The last $[\![\cdot]\!]_n^-$ term, denoting the negative part of the relative normal displacement, reflects normal-compliance contact condition being understood as a reasonable approximation of Signorini contact conditions.

The potential energy of external forces, acting along a part of the boundary Γ_N^η, is given as

$$F(t,u) = -\sum\nolimits_{\eta=A,B} \int_{\Gamma_N^\eta} f^\eta(t) \cdot u^\eta d\Gamma, \qquad (2)$$

where f^η are the given tractions on Γ_N^η. Finally, the dissipated energy is introduced by a pseudopotential which reflects the rate-dependence and unidirectionality of the debonding process

$$R(\dot\varsigma) \int_{\Gamma_c} -G_c \dot\varsigma + \frac{1}{2} \alpha \dot\varsigma^2 \, d\Gamma, \quad \text{for } \dot\varsigma \le 0, \qquad (3)$$

with G_c being related to the instant of damage initiation and with α as a visco-damage parameter. The relations which govern the evolution of damage can be written in form of nonlinear variational inclusions with initial conditions

$$\partial_u E(t;u,\varsigma) + \partial_u F(t,u) \ni 0, \quad u|_{t=0} = u_0,$$
$$\partial_{\dot\varsigma} R(\dot\varsigma) + \partial_\varsigma E(t;u,\varsigma) \ni 0, \quad \varsigma|_{t=0} = \varsigma_0, \qquad (4)$$

denoting ∂ the (partial) subdifferential of a non-smooth convex function which can be replaced by the Gateaux differential for a sufficiently smooth functional, as e.g. F. The initial condition for damage is usually $\varsigma_0 = 1$, i.e. the undamaged state. The energy release rate is obtained from the second inclusion of Eq. 4 as it expresses a change of energy E due to a change of damage ς.

3 EXAMPLES

The numerical procedures proposed for the solution of the problem Equation 4 consider the time discretization and the spatial discretization separately. The time discretization scheme is defined by a semi-implicit algorithm with a fixed time step size τ and introducing the time instants t^k. In order to obtain such an algorithm from Equation 4, the damage rate is approximated by the finite difference $\dot\varsigma = (\varsigma^k - \varsigma^{k-1})/\tau$, where ς^k denotes the solution at the instant t^k. For details, see e.g. (Vodička 2016). The role of the SGBEM in the present computational procedure is to provide a complete boundary-value solution from the given boundary data for each domain in order to calculate the elastic strain energy in these domains. Thus, the SGBEM code calculates also unknown tractions along the interface, assuming the displacements gaps to be known from the used minimization procedure, in the same way as proposed and tested in (Vodička et al. 2007, Vodička et al. 2014). The model formulation is implemented in a MATLAB computer code.

3.1 *A beam test*

We use the model to find solutions for various amplitudes of the same cyclic load applied to the beam shown in

Figure 1(left), see also (Vodička 2016). The material properties of both layers in our example are: $E = 70$ GPa, $v = 0.35$. The beam is loaded by a time dependent displacement or force load with the amplitude $g_m = vt_0$ or $f_m = pt_0$, respectively. It is also shown in Figure 1(right), where $v = 1$ mms^{-1}, $p = 2800$ MPas^{-1}, t_0 varies in value.

The characteristics of the interface needed for the functionals in Equations 1 and 3 are $G_c = 1.01$ Jm^{-2}, $\alpha = 100$ Jsm^{-2}, $k_n = 2k_t = 2$TPam^{-1} and $\Phi(\zeta) = 100\zeta/(101-\zeta)$, cf. (Vodička & Mantič 2017).

The discretization is uniform such that the BE mesh size (element length) is $h = 2$ mm and the pertinent time step is $\tau = 5$ ms. First, we had solved the problem with this mesh until we reached a crack of the length $a = 22$ mm. Table 1 summarises number of time steps Δn needed to propagate the crack from length 10 mm to a, where the crack is considered whenever ζ falls below 0.1. With given τ, the number of cycles ΔN which leads to supposed crack length is $\Delta N = \Delta n . \tau/(2t_0)$. By the linearised regression, we found that $\Delta a/\Delta N = 852.2\, g_m^{2.872}\, (R^2 = 0.9985)$ for the displacement load and $\Delta a/\Delta N = 2.0 \cdot 10^{-8} f_m^{3.416}\, (R^2 = 0.9965)$ for the force load, which is a good approximation of the Paris power law of fatigue crack growth, (París & Erdogan 1963).

The relations that can be checked at the interface where a crack propagates include traction-separation relation and evolution of damage. Both of them are shown in

Figure 2 for displacement loading and in Figure 3 for force loading. The interface point is that of $x_1 = 78$ mm corresponding to the crack of $a = 22$ mm. It should be noted that particular shape of the curves is influenced by the neighbourhood of the selected point, the decrease of the stress appears to be faster as the crack tip approaches that point.

Table 1. Number of time steps needed for propagating the crack by a specified length, the beam test.

	Displacement loading				Force loading			
t_0 [s]	0.025	0.05	0.1	0.2	0.025	0.05	0.1	0.2
Δn	3013	734	186	63	1646	340	81	10

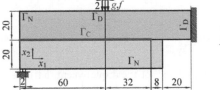

Figure 1. Geometry of the beam test together with prescribed cyclic loading.

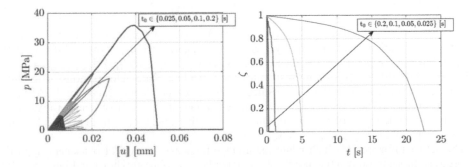

Figure 2. The stress-separation relation (left) and evolution of ζ (right) for the point at $x_1 = 78$ mm in the case of displacement loading, influence of the amplitude $g_m = vt_0$.

261

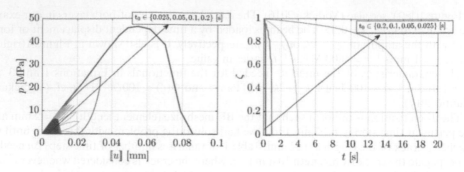

Figure 3. The stress-separation relation (left) and evolution of ζ (right) for the point at $x_1 = 78$ mm in the case of force loading, influence of the amplitude $f_m = pt_0$.

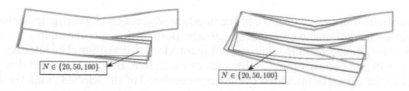

Figure 4. The deformation of the beam (magnified 200 times) at three instants of both applied loads: displacement (left) and force (right), $t_0 = 0.05$ s.

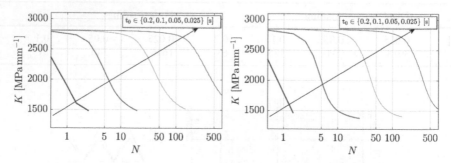

Figure 5. The stiffness K of the beam considered for the displacement (left) the force (right) load as $f = Ku_f$ and $p_g = Kg$, respectively, where u_f, p_g are averaged calculated.

The evolution of deformation is compared for both load cases. Here, we used the results of one particular amplitude given by $t_0 = 0.05$ s. The left graph of Figure 4 shows the maximal displacements for the displacement load at three instants of loading after N cycles, where pertinent Ns are displayed in the picture. The displacement type of loading causes that the deformation of the upper layer is more or less the same for any instant, while the bottom layer deforms differently in the selected instants due to the growing crack. The right graph provides the corresponding results for the force load. The deformations of the upper beam are now increasing as the stiffness decreases due to increasing crack length. At the last used instant the stiffness would be approximately a half of the original one as the crack has extended to the left part of the beam. Quantities at the same locus, influence of the amplitudes $g_m = vt_0$ and $f_m = pt_0$, respectively.

The actual stiffnesses K which relate the average force and average displacement applied or enforced at the same locus in the middle of the upper layer top face are shown in Figure 5 for both type of loads. As we expected and as we have already mentioned above, the stiffness decreases faster for force loading which is in close connection to the faster crack propagation in this case.

3.2 A shear test

Another test includes a shear type loading and crack propagation as shown in Figure 6. The material properties of both layers in this example are the same as in the previous one. The upper layer is loaded by a time dependent displacement or force load with the amplitude $g_m = vt_0$ or $f_m = pt_0$, respectively, where $v = 0.818$ mms^{-1}, $p = 100$ MPas^{-1}, and t_0 varies.

The discretization is uniform: the BE mesh size is $h = 2.5$ mm, the time step is $\tau = 5$ ms. The solution of the problem had been evolved until a crack in the whole interface was reached.

Table 2 summarises number of time steps Δn needed to propagate the crack (i.e. ζ smaller than 0.1) from length 15 mm to 35 mm so that $\Delta a = 20$ mm. The number of cycles $\Delta N = \Delta n \cdot \tau / (2t_0)$ which includes the propagation of the crack by Δa obeys the relation $\Delta a/\Delta N = 3.69 \cdot 10^4 g_m^{3.2932}$ ($R^2 = 0.9974$) for the displacement load, and $\Delta a/\Delta N = 9.11 \cdot 10^{-6} f_m^{3.8898}$ ($R^2 = 0.9902$) for the force load, which is again a satisfactory approximation of the París power law of fatigue crack growth.

The relations of traction-separation law and evolution of damage demonstrating rupturing of the interface are shown in Figure 7 and Figure 8 for displacement and force loading, respectively. The selected interface point is $x_1 = 170$ mm corresponding to the crack of $a = 30$ mm. Both load cases provide similar results, nevertheless the force load naturally terminates earlier.

The evolution of deformation is compared for both load cases in Figure 9 where the maximal displacements for the displacement and force load at selected instants of loading after N cycles are shown. Here, we used the results of the amplitude determined by $t_0 = 0.2$ s.

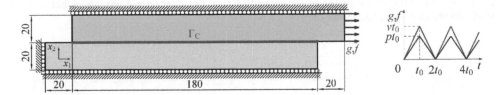

Figure 6.   Geometry of the shear test together with prescribed cyclic loading.

Table 2.   Number of time steps needed for propagating the crack by a specified length, the shear test.

	Displacement loading					Force loading				
t_0 [s]	0.2	0.25	0.3	0.4	0.6	0.2	0.25	0.3	0.4	0.6
Δn	2874	1602	994	504	232	1704	993	615	332	67

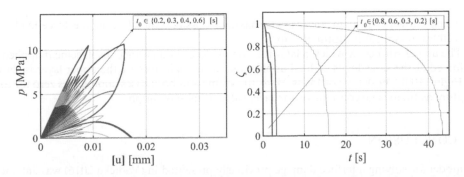

Figure 7.   The stress-separation relation (left) and evolution of ζ (right) for the point at $x_1 = 170$ mm in the case of displacement loading, influence of the amplitude $g_m = vt_0$.

263

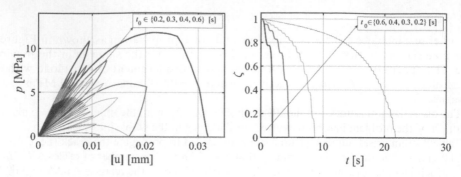

Figure 8. The stress-separation relation (left) and evolution of ζ (right) for the point at $x_1 = 170$ mm in the case of force loading, influence of the amplitude $f_m = p t_0$.

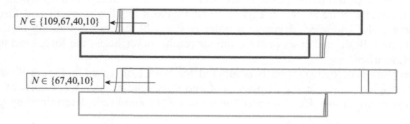

Figure 9. The deformation of the layers (magnified 1000 times) at several instants of both applied loads: displacement (top) and force (bottom), $t_0 = 0.2$ s.

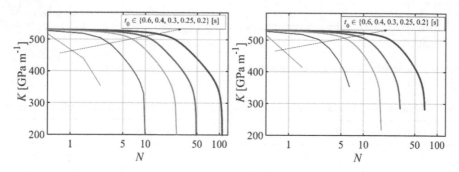

Figure 10. The stiffness K (as in Figure 5) of the layered structure under shear loading.

The last shown instant in any case is maximal possible, therefore in the force-load case $N = 109$ is missing as it ruptured before.

The actual stiffness K which relates the average force and average displacement applied or enforced at the same locus of load application in the right face of the upper layer is shown in Figure 10 for both type of loads. Here, the stiffness in both cases approaches zero for total rupture as there is no coupling of the layers after total separation.

4 CONCLUSION

A model for solving interface damage previously presented in (Vodička 2016) was enhanced here to include some hysteresis in damage to observe its behaviour under cyclic loading. The presented results confirm that the model is capable to evolve the damage in the sense of fatigue life. Of course, there are many parameters in the model that should be appropriately set to follow satisfactorily some experimental results. Nevertheless, it might be expected that the proposed approach turns to be useful in practical engineering calculations.

ACKNOWLEDGEMENT

Authors acknowledge support from the Slovak Ministry of Education by the grants VEGA 1/0078/16 and VEGA 1/0477/15.

REFERENCES

Dostál, Z. 2009. *Optimal Quadratic Programming Algorithms,* volume 23 of Springer Optimization and Its Applications. Springer, Berlin.

París, P. & Erdogan, F. 1963. A critical analysis of crack propagation laws. 85(4):528–533.

Roe, K.L. & Siegmund, T. 2003. An irreversible cohesive zone model for interface fatigue crack growth simulation. *Eng. Frac. Mech.*, 70:209–232.

Roth, S., Hütter, G. & Kuna, M. 2014. Simulation of fatigue crack growth with a cyclic cohesive zone model. *Int. J. Frac.*, 188:23–45.

Roubíček, T., Souček, O. & Vodička, R. 2013. A model of rupturing lithospheric faults with reoccurring earthquakes. *SIAM J. Appl. Math.*, 73(4):1460–1488.

Vodička, R., Mantič, V. & París, F. 2007. Symmetric variational formulation of BIE for domain decomposition problems in elasticity—an SGBEM approach for nonconforming discretizations of curved interfaces. *CMES—Comp. Model. Eng.,* 17(3):173–203.

Vodička, R. Mantič, V. & Roubíček, T. 2014. Energetic versus maximally-dissipative local solutions of a quasi-static rate-independent mixed-mode delamination model. *Meccanica*, 49(12):2933–2963.

Vodička, R. 2016. A quasi-static interface damage model with cohesive cracks: SQP-SGBEM implementation. *Eng. Anal. Bound. Elem.,* 62:123–140.

Vodička, R. & Mantič, V. 2017. An energy based formulation of a quasi-static interface damage model with a multilinear cohesive law. *Discrete and Cont. Dynam. Syst. – S,* 10(6):1539–1561.

Advances and Trends in Engineering Sciences and Technologies III – Al Ali & Platko (Eds)
© 2019 Taylor & Francis Group, London, ISBN 978-0-367-07509-5

Strength parameters of polyester reinforced PVC coated fabric after short term creep loading in biaxial mode

K. Zerdzicki
Faculty of Civil and Environmental Engineering, Gdansk University of Technology, Gdansk, Poland

Y.M. Jacomini
Institut National des Sciences Appliquées—Centre Val de Loire, Bourges, France

ABSTRACT: This study addresses the analysis of tensile strength parameters of the technical fabric VALMEX, which is composed of two orthogonal polyester thread families (named the warp and fill) and both sides PVC coated. The material was firstly subjected to 48-hour biaxial creep loading with the equal stress level in both orthogonal directions of the fabric. The stress levels were established as follows: 4.6 kN/m, 10.4 kN/m, 16.4 kN/m, 22.4 kN/m, 28.4 kN/m, 34.4 kN/m. The samples after creep loading were left unloaded in laboratory conditions for the subsequent 6 months and then subjected to biaxial tension till rupture. For all tests, the basic tensile properties of the fabric have been identified for the warp and weft directions separately. The evolution of parameters demonstrates how the level of biaxial material prestressing affects tensile strength properties of the VALMEX fabric.

1 INTRODUCTION

1.1 *Architectural fabrics*

Architectural fabrics (also known as technical fabrics or coated textile membranes) are one of the most impressive and challenging building materials. The high strength, low dead weight, simple production (prefabrication), low construction costs (usually without form-works or decks) are the crucial advantages of the membrane structures (Kazakiewicz M.I., Miełaszwili J.K. 1988). Two structural models are possible to implement architectural fabrics (Stanuszek M. 2002): the air-supported structures (Figure 1) and the tensile cable-membrane structures (Figure 1).

Figure 1. Examples of architectural fabrics applications: air supported structure – Air dome, China (left), tensile cable-membrane structure – Forest Opera, Poland (right).

They are often assembled in the form of wall coverings or roof structures over large size public gathering places, e.g. sport stadiums, entertainment halls, open-air theatres, shopping malls, communication terminals, pavilions and in smaller cubature buildings, like car parks, temporary shelters or decorative objects. They give designers almost endless possibilities to form canopies of different shapes.

The so called "hanging roofs" are characterized by geometric non-linearity, which is manifested by change of the covering original configuration under an applied load. Shape reconfiguration is accompanied by the redistribution of acting forces and occurrence of wrinkles. To overcome this problem the form finding analysis must be performed and the initial tensioned configuration of the canopy established (Bridgens and Birchall 2012). In order to assure proper conditions of the tensioned canopy in a long time domain, it is necessary to regard the viscoelastic behavior of the technical fabrics and the global structural dynamics, due to its sensitivity to wind load.

1.2 Composite nature of architectural fabrics

The structure of standard architectural fabric is presented in Figure 2. The reinforcing fiber yarns are responsible for material tensile strength, while exterior coating protects yarns from damage, stabilizes weave structure, as well as provides water resistance and shear stiffness. The reinforcement base can be manufactured in two variants, resulting in woven or knitted fabrics. Typical weave patterns include plain, 2/2 twill, leno, mock leno, panama, 4H satin and 8H satin, while knitting patterns contain mainly the warp-knit fill inserted method and raschel knit (Air, Tent & Tensile Structures, Fabric Specifier's Guide, Fabric Architecture 2013).

The geometric non-linearity of the twisted fiber structure, complex interactions of orthogonal yarns under the bi-axial in plane stress state (friction, crimp interchange, locking effect) and influence of coating lead to the time-dependent, hysteretic and anisotropic material behavior of technical fabrics (Allaoui et al. 2012; Boisse et al. 2001).

During prefabrication the threads in one or two major directions (called commonly the warp and fill direction) are prestressed, while covering (e.g. PVC layer) is placed on the reinforcement base. Next, during the membrane structure assembly, the architectural fabrics must be pre-tensioned to a given level to guarantee stress equilibrium in the material prior to loading. The level of prestress seems to have impact on the material behavior. The problem becomes more complex, when the biaxial character of stress distribution in the fabric is taken into account. Some studies were realized to evaluate the behavior of textile composites under biaxial cyclic loading (Ambroziak and Kłosowski 2014b; Ambroziak 2015b; Ambroziak 2015a), in high and low temperatures (Ambroziak and Kłosowski 2014a) and after natural and laboratory ageing (Zerdzicki, Klosowski, and Woznica 2018). This study aims to analyze the influence of the biaxial prestressing of the ready-to-use material on the basic tensile properties of the polyester reinforced PVC coated VALMEX fabric.

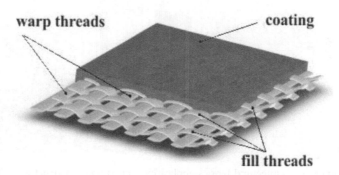

Figure 2. Composite structure of plain weave architectural fabric.

2 EXPERIMENTS

2.1 Samples and laboratory equipment

The technical fabric **VALMEX FR 1000** was examined in this research. It is a polyester reinforced PVC-coated fabric with the P 2/2 weave pattern, which means that two threads always go together composing orthogonal reinforcing mat with the major anisotropic directions approximately parallel with the warp and fill of the material. The warp direction threads are straight, while the fill direction threads are interspersed through them (also going in pairs). The producer's specification of the **VALMEX FR 1000** is summarized in Table 1.

All laboratory tests were realized on the Zwick biaxial testing machine with data registration obtained by the video extensometer following four markers on the samples surface (Figure 3). The samples had a shape of cross, with directions of orthogonal arms coincided with the warp and fill material directions. The sample arm width was 10 cm and the total length of the cross arm was 40 cm. The markers were set in the middle of the cross (gauge length about 40 mm in both directions), in the anticipated biaxial stress location.

2.2 Testing protocol

The working stress usually ranges between 0.1 and 0.2 of the uniaxial strength in particular fabrics direction, depending on the fabrics' surface curvature and weave pattern (Heidrun Bogner-Baltz 2009). Taking 5 as the safety factor, the prestress would be five times lower than the working stress. Thus, taking the data from Table 1, it was calculated that the prestress should be 120 kN/m /5/5 = 4.8 kN/m and 110 kN/m /5/5 = 4.4 kN/m, for the warp and fill directions, respectively. The identical for both directions prestress value of 4.6 kN/m was finally taken for further analysis as the first prestress level. The following subsequent prestress levels were: 10.4 kN/m, 16.4 kN/m, 22.4 kN/m, 28.4 kN/m and 34.4 kN/m. It is seen, that most of the loading levels are below the working stress level, only cases of 28.4 and 33.4 kN/m

Table 1. Basic information of the VALMEX FR 1000 technical fabric (Mehler Texnologies 2010).

Trade name	VALMEX FR 1000 MEHATOP F – type III
Total weight	1050 g/m^2
Tensile strength (warp/fill)	120/110 kN/m*
Base fabric material	PES
Base fabric yarn count	1670 dtex
Base fabric weave type	P 2/2 (Panama 2/2)
Type of coating	PVC
Type of finish layer	PVDF-lacquer on both sides

*Thickness material is usually not established for technical fabrics (Bridgens and Birchall 2012; Ambroziak 2015b), therefore all mechanical characteristics in this paper are defined for the unitary thickness e.g. kN/m^2 * m = kN/m.

Figure 3. Biaxial testing machine with cross-shaped sample during test.

exceed the working stress level and are beyond elastic region of stress-strain relation for both directions. The prestress was performed in the form of creep loading of 48 hours duration with the constant force control, executed on the Zwick biaxial testing machine. After the creep, the samples were unloaded and left in the laboratory, at room temperature for the subsequent six months. Then, the samples were tensioned to rupture with the constant force increase of 20 N/s in both orthogonal directions using the same Zwick biaxial testing machine and the video-extensometer. For each prestress level at least three samples were tested and the presented results were calculated always as the mean value obtained from all the tests performed.

3 ANALYSIS OF THE TENSILE STRENGTH PROPERTIES

The demonstrative stress-strain curves for the uniaxial and biaxial tensile tests (force increase of 20 N/s till rupture) are presented in Figure 4a. Comparing the obtained outcomes, it is clearly seen, that the biaxial loading resulted in lowering the uniaxial ultimate tensile strength (*UTS*) for both material directions significantly. For the warp direction case it dropped from 110 kN/m (uniaxial) to about 76 kN/m (biaxial), and for the fill one from 90 kN/m (uniaxial) to 76 kN/m (biaxial). The obtained uniaxial results are lower than the ones stated by the producer (see Table 1) that can be a result of different experimental protocol.

Another remark concerns the curves' trajectories. For the warp direction we can distinguish three, and for the fill direction four particular ranges, where the stress-strain relation is almost linear. It has been confirmed before that for the same material type and producer, for the warp direction the tangent of the line found in the first strain range 0–0.01 is related to the Young's modulus of the material (Klosowski, Zerdzicki, and Woznica 2017). For the fill direction, the line found in the strain range 0–0.03 is related to the stiffness of the PVC covering, while in the strain range 0.03–0.06 the Young's modulus for the fill direction can be evaluated. However, for accurate identification of the elasticity moduli, the cyclic load-unload tests are necessary.

The illustrative stress-strain curves for biaxial tensile tests till rupture, for different prestress levels, for the warp and fill directions of the VALMEX fabric, are presented in Figure 4b. It can be observed, that increase of the prestress level resulted in stiffening of the material in both directions, while the curve shapes remain similar to each other. Only the highest prestress level of about 33.4 kN/m resulted in different stress-strain curve above the stress of 30 kN/m, which is fortunately far above the working stress 22–24 kN/m, which is an acceptable limit in real membrane structures. Figure 5a shows that the *UTS* for the warp and fill directions are similar (about 76 kN/m), regardless of the prestress level.

It could be concluded, that biaxial loading used in this research uniformed the *UTS* values for both directions to the comparable level. On the other hand, the elongation at break (ε_{ult}) found for biaxial tests decrease for both material directions with the prestress level increasing (Figure 5b). The evolution of ε_{ult} and *UTS* values over prestress level can be described by linear functions of satisfactory high determination coefficients between the

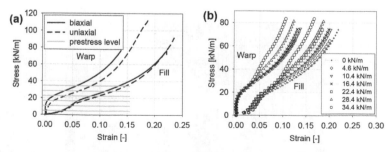

Figure 4.   The uniaxial and biaxial results of tension till rupture of the VALMEX fabric. Gray lines indicate the prestress levels tested in the study (a); Comparison of the tension till rupture stress-strain curves for samples with different prestress levels (b).

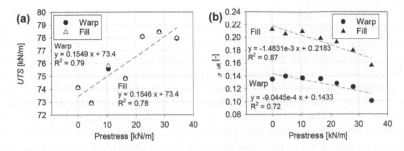

Figure 5. Ultimate tensile strength UTS (a) and elongation at break εult (b) for different prestress levels, for the warp and fill directions of the VALMEX fabric.

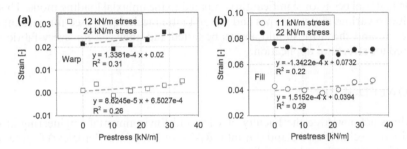

Figure 6. Strains at the working stress and for half of the working stress level for the warp (a) and fill (b) directions of the VALMEX fabric.

fitted lines and the obtained results, independently for the warp and fill directions of the VALMEX fabric (Figure 5a, b). The established equations can be used for calculating the ε_{ult} and *UTS* for different prestress levels limited by the 33.4 kN/m value, for the warp and fill directions separately. The ratio between elongation at break for the fill direction ($\varepsilon_{ult\text{-}FILL}$) and the warp direction ($\varepsilon_{ult\text{-}WARP}$) is constant throughout different prestress levels and approximately equals $\varepsilon_{ult\text{-}FILL}/\varepsilon_{ult\text{-}WARP} = 1.5$.

In order to calculate the elasticity moduli for the technical fabrics the cyclic load-unload tests are necessary, but not provided in the current research. Therefore, to highlight the influence of biaxial prestress on tensile properties of the VALMEX fabric, an alternative approach was introduced. The strain values corresponding to the working stress (24 kN/m and 22 kN/m for the warp and fill directions, respectively) and to the half of the working stress (12 and 11 kN/m for the warp and fill directions, respectively) were found for different prestress levels. The obtained results are plotted in Figure 6a and Figure 6b, for the warp and fill directions, respectively. The values of the achieved strains for different prestress levels always oscillate in the strain range of 1% not composing any particular, clear tendency (for instance low determination coefficient of the fitted straight lines). This observation stays true for the warp and fill directions of the fabric and for working stress level and its half. It could be an evidence that the prestress level does not affect the elongation level when considering the stress range below working stress. However, for confirmation of this conclusion, the analysis of the cyclic load-unload tests is suggested and it will be performed at the next research stage concerning the same fabric.

4 CONCLUSIONS

In the presented study of the tensile strength properties of the VALMEX fabric, the biaxial prestressing of the levels 4.6 kN/m, 10.4 kN/m, 16.4 kN/m, 22.4 kN/m, 28.4 kN/m, 34.4 kN/m. (equal in both orthogonal material directions) was realized as the creep loading of 48 hours duration. Firstly, it was observed that the biaxial results, when compared with the uniaxial outcomes, led to reduction and uniforming of the ultimate tensile strength (*UTS*) for the warp and fill directions of the fabric. Next, taking into account the prestress influence, it was witnessed that the greater the prestress level was, the lower the elongation at break (ε_{ult}) and greater the *UTS* were. The material stiffness raised up with the prestress increment. Additionally, the variations of ε_{ult} and *UTS* with the prestress levels can be easily described by linear functions providing a methodology to interpolate these values for another prestress levels. However, it seems that the biaxial prestress proposed in this research did not affect the material parameters below the working stress level, equals 24 kN/m and 22 kN/m for the warp and fill directions, respectively. Nonetheless, the cyclic load-unload tests are suggested to confirm the observed behavior of the VALMEX fabric.

The overall conclusion is that when designing the structure made of technical fabrics, the biaxial tests results should always be taken into account, as the biaxial loading mode influences the material behavior significantly compared to the uniaxial loading mode. This impact differs due to various material types and woven/knitted techniques used for particular fabric manufacture and therefore should always be tested and analyzed for every fabric material considered for structural application.

ACKNOWLEDGEMENTS

The authors acknowledge the Faculty of Civil and Environmental Engineering at Gdansk University of Technology, Poland (Grant GRAM for Krzysztof Żerdzicki) for the financial support of the research and its publication.

REFERENCES

Air, Tent & Tensile Structures, Fabric Specifier's Guide, Fabric Architecture. 2013.
Allaoui, S., G. Hivet, A. Wendling, P. Ouagne & D. Soulat. 2012. "Influence of the Dry Woven Fabrics Meso-Structure on Fabric/Fabric Contact Behavior." *Journal of Composite Materials* 46 (6). SAGE Publications Ltd STM: 627–39. doi:10.1177/0021998311424627.
Ambroziak, A. 2015a. "Mechanical Properties of Polyester Coated Fabric Subjected to Biaxial Loading." *Journal of Materials in Civil Engineering* 27 (11): 1–8. doi:10.1061/(ASCE)MT.1943-5533.0001265.
Ambroziak, A. 2015b. "Mechanical Properties of Precontraint 1202S Coated Fabric under Biaxial Tensile Test with Different Load Ratios." *Construction and Building Materials* 80. Elsevier Ltd: 210–24. doi:10.1016/j.conbuildmat.2015.01.074.
Ambroziak, A & Kłosowski, P. 2014a. "Influence of thermal effects on mechanical properties of PVDF coated fabric." *Journal of Reinforced Plastics and Composites* 33 (7): 663–73.
Ambroziak, A. & Kłosowski, P. 2014b. "Mechanical Properties of Polyvinyl Chloride-Coated Fabric under Cyclic Tests." *Journal of Reinforced Plastics and Composites* 33 (3): 225–34. doi:10.1177/0731684413502858.
Boisse, P., K. Buet, A. Gasser & J. Launay. 2001. "Meso/Macro-Mechanical Behaviour of Textile Reinforcements for Thin Composites." *Composites Science and Technology* 61 (3): 395–401. doi:10.1016/S0266-3538(00)00096-8.
Bridgens, B. & Birchall M. 2012. "Form and Function: The Significance of Material Properties in the Design of Tensile Fabric Structures." *Engineering Structures* 44. Elsevier Ltd: 1–12. doi:10.1016/j.engstruct.2012.05.044.
Bogner-Baltz, H. 2009. "Report on Biaxial Tests. Mehler VALMEX FR 1000 Mehatop F Type III."
Kazakiewicz M.I., Miełaszwili J.K. & Sułabierdze O.G. 1988. *Aerodynamika Dachów Wiszących*. Arkady.
Klosowski, P., Zerdzicki, K. & Woznica, K. 2017. "Identification of Bodner-Partom Model Parameters for Technical Fabrics." *Computers and Structures* 187. doi:10.1016/j.compstruc.2017.03.022.
Mehler Texnologies. 2010. "Technical Datasheet VALMEX FR 1000."
Stanuszek M. 2002. "Computer Modelling of Cable Reinforced Membranes." *Comput. Assist. Mech. Eng. Sci.* 9: 223–37.
Zerdzicki, K., Klosowski, P. & Woznica K. 2018. "Influence of Service Ageing on Polyester-Reinforced Polyvinyl Chloride-Coated Fabrics Reported through Mathematical Material Models." *Textile Research Journal*. doi:10.1177/0040517518773374.

Part B
Buildings and structures
Construction technology and management
Environmental engineering
Heating, ventilation and air condition
Materials and technologies
Water supply and drainage

Advances and Trends in Engineering Sciences and Technologies III – Al Ali & Platko (Eds)
© 2019 Taylor & Francis Group, London, ISBN 978-0-367-07509-5

The use of high voltage discharge in treatment of water saturated with air bubbles

M.Ju. Andrianova
Peter the Great Saint-Petersburg Polytechnic University, St. Petersburg, Russia

D.A. Korotkov & S.V. Korotkov
Ioffe Physical-Technical Institute of the Russian Academy of Sciences, St. Petersburg, Russia

G.L. Spichkin
Megaimpulse Ltd., Saint-Petersburg, Russia

ABSTRACT: Plasma generated by high voltage discharge can be applied in drinking water or wastewater treatment. The plasma treatment of deionized and tap water was performed in an experimental unit. Electric discharge was formed in air bubble flows spreading between the electrodes. Electric potential had amplitude 33 kV, frequency 500 Hz, impulse front 8 mksec, current in electric breakdown had amplitude up to 400 A. Specific Electric Conductivity (SEC), concentrations of hydrogen ions (from pH), concentrations of total iron and sum of nitrates and nitrites in water increased with time of treatment due to formation of nitric and nitrous acids and sedimenting iron-containing particles (the tap water after 3 minutes of treatment had 8 mgN/L, 5 mgFe/L, pH = 3.6). Concentrations of other ions in the treated tap water remained approximately the same as before treatment. Optical density of water in the range from 200 to 800 nm decreased or increased with the time of treatment depending on the wavelength. SEC and fluorimetry were suggested for express-monitoring of plasma treatment efficiency.

1 INTRODUCTION

High-voltage electric discharges generating plasma in water and air result in several effects in the media: electrons flow, electromagnetic radiation in a wide wavelength range, strong electromagnetic field, local heating and consequently shockwaves and acoustic waves. These factors influence the chemical structure of the media due to the dissociation of molecules and formation of active radicals, inducing various chemical reactions. As a result, new substances are produced (e.g. ozone, nitric acid) or existing substances undergo degradation (e.g. residual hormones, antibiotics and other active pharmaceuticals, which can affect the human health or aquatic life). Also the mentioned physical impacts provide disinfection of the media. These effects make plasma methods perspective for application in drinking water and wastewater treatment among other advanced oxidation processes (Stefan 2018).

Various types of discharge (corona, arch, spark, barrier, etc.) and reactor types are applied in plasma treatment methods (Stefan 2018). Spark discharge formed in the air bubbles moving through water is a very efficient variant, because the initiation of discharge in air requires much lower breakdown fields than in liquid, and the media are mixed by shockwaves enriching water with products of electrochemical reactions (Stefan 2018). Such type of reactor was studied in the presented work.

The aim of the work was to study changing of chemical and optical parameters of the water after treatment with high voltage discharge generating plasma in air bubbles in water. The following tasks were set: to study changing of concentrations of main ions, organic matter and iron in water during the process of plasma treatment; to study changes of water optical density (OD) and fluorescence intensity (I), and suggest parameters for express-monitoring of water treatment efficiency in the experimental unit.

2 MATERIALS AND METHODS

2.1 *Experimental device and water treatment*

Generator of high-power current pulses was constructed in Ioffe institute (Korotkov et al. 2011). It had the following parameters: electric potential between the electrodes had amplitude up to 33 kV, frequency 500 Hz, impulse front up to 8 mksec, current in electric breakdown had amplitude up to 400 A.

The reactor for water treatment with submerged electrodes had volume 1 L (the scheme is shown on Figure 1). High-voltage electrode (HV) was configured as a disk. Ground electrodes were configured as stainless steel parallel tubes. Air was pumped through these tubes, organizing parallel bubbles flows (air flow rate 3 L/min). When electric potential was applied to electrodes, powerful spark discharge was formed in these air flows. Visually, sparks were characterized by bright flashes with approximately same intensity, showing even distribution of energy among discharge channels. Media (air and water) were intensively mixed by shock-waves. Time of water treatment varied from 0 to 7 minutes.

2.2 *Chemical analysis of water samples*

Deionized water and tap water were used as model solutions. Deionized water was obtained using distillation, ion exchange and carbon filters with final control of water purity by its specific electric conductivity (SEC).

Tap water was taken in St. Petersburg from water supply system in the building of Ioffe institute. Before collection, water was let to run for 5 minutes washing out residual impurities from pipe walls. Water samples were collected from the reactor after treatment, transported to the laboratory, stored at room temperature and analyzed in 1 or 2 days after the experiment.

SEC of water was measured immediately after the experiment by conductometer HI 8733 (HANNA Instruments, Austria). Measurement of pH was done by pH-meter I-500 (Akvilon, Russia) with glass and silver-chloride electrodes. Concentrations of total organic carbon (TOC), total inorganic carbon (IC) and total nitrogen (TN) were measured by analyzers TOC-Lcpn and TNM-L (Shimadzu, Japan).

Concentrations of anions (nitrates, nitrites, chlorides, sulfates) and cations (ammonium, sodium, potassium, magnesium, calcium) were measured by capillary electrophoresis analyzer Capel 103R (Lumex, Russia). Concentration of total iron was measured by colorimetric method based on reaction of Fe(III) with sulfosalicic acid.

Optical density (OD) of water in ultraviolet and visible region (UV-Vis) was measured by spectrophotometer Specord 40 (AnalytikJena, Austria) in quartz cuvette with optical path length of 1 or 2 cm in wavelength (λ) range from 200 to 800 nm.

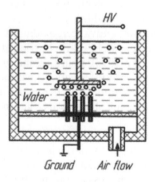

Figure 1.   Scheme of the reactor for water treatment.

Fluorescence intensity (I) spectra of undiluted water samples were obtained by analyzer "RF 5301 PC" (Shimadzu, Japan) at two excitation wavelengths 230 nm and 270 nm near to maximums of excitation spectra for humic substances and proteins. Fluorescence was registered at emission wavelengths from 220 to 650 nm. Data were not corrected for inner filter effects.

3 RESULTS AND DISCUSSION

3.1 Study of deionized water treated with electric discharge

For deionized water, maximal period of treatment was 5–7 minutes, at which SEC stopped increasing and reached 280 mkSm/cm (see Figure 2). After water treatment, TOC and IC remained at the same level (near zero for TOC, near 2–3 mg/L for IC) in the frames of the method error, while pH decreased and concentrations of TN, nitrate-ion and total iron increased. Results for these parameters are shown on Figure 3 (in molar concentrations).

It is clear from the data that nitric acid has been formed during treatment, because the molar concentrations of TN and nitrates are very close, showing that no other forms of nitrogen were present in the water. Concentrations of hydrogen ions and nitrates have also changed in the similar way and were almost equal in many measured points. Lower concentrations of H^+ after 5 and 7 minutes of treatment can be explained by partial neutralization with iron hydroxide. In general, SEC and concentrations of TN, nitrates, hydrogen ions and total iron increased with time of water treatment. Dependences of these parameters from time can be described by polynomial trend lines with almost linear parts at the first 3 minutes. Capillary electrophoresis revealed only nitrate-ions in water samples (without nitrites and other anions), no ammonium-ions, and no other cations.

During the treatment transparent water acquired opalescence and rusty-brown color. It was supposed that iron-containing particles were formed during treatment because of electrochemical corrosion of electrodes. The particles settled to the bottoms of the bottles with water samples during the night.

Optical density spectra of one water sample (treatment time 7 min) were obtained at various times of storage (in a cylinder) after thorough mixing. The results are shown at Figure 4.

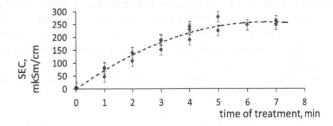

Figure 2. Changing of SEC during deionized water treatment (with standard error bars).

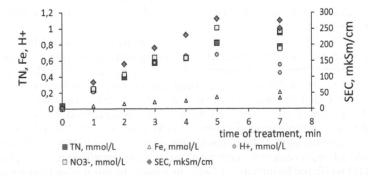

Figure 3. Changing of chemical parameters during deionized water treatment.

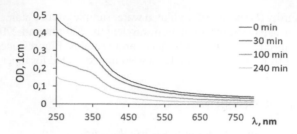

Figure 4. Spectra of optical density for deionized water sample (treated for 7 minutes) at various times of settling after mixing of the sample.

It can be seen from the spectra that OD of water decreased with time due to particles settling. After 30 minutes of settling OD at 250–800 nm decreased by 17–23%, while concentration of total iron in water (above the bottom) decreased about 60% (from 7.5–8.1 mgFe/L to 3.0–3.4 mgFe/L).

3.2 Study of tap water

Before the plasma treatment tap water had the following parameters: pH 6.57–6.70, SEC 113–116 mkSm/cm, concentrations of substances: TN 0.5–0.6 mg/L, TOC 3.5–3.6 mg/L, IC 7.6–7.7 mg/L, total iron 0.2–0.3 mg/L. The tap water samples were treated in experimental unit for maximal time of 3 minutes. During this period, SEC have reached the values of 250–260 mkSm/cm, which are close to the last values in experiment with deionized water, and pH decreased to 3.56. Results of the tap water analysis are shown on Figures 5–6.

Changing with time of treatment can be described by strait line for SEC, for other chemicals – by polynomial lines. Unlike the results in deionized water, two types of N-containing anions were registered – nitrates and nitrites. Molar concentrations of H^+ were less than sum of nitrogen contained in nitrates and nitrites. It means, that nitric and nitrous acids formed in the reactor were partly neutralized by tap water. Nitrite anions formed up to 77% of anionic TN at short period of water treatment (0.5 minutes); in the samples treated for 3 minutes nitrites formed 6–14% of anionic TN. Presence of ammonium ions that could be formed due to plasma treatment (Judée et al. 2018) were not registered by capillary electrophoresis. However, before the treatment tap water contained small amounts of ions of ammonium (0.1 mgNH$_4^+$/L) and nitrates (1.3 mgNO$_3^-$/L), sum of TN in these substances was less than 0.6 mgN/L.

Concentrations of other ions did not change much after treatment. The values before and after treatment were the following: chlorides 9–10 mg/L, sulfates 25–28 mg/L, potassium 1.1–1.5 mg/L, sodium 6–7 mg/L. These results were in the ranges of measurement error (±15%). For bivalent cations decrease by 36–37% was observed after 3 minutes of the plasma treatment: for magnesium from 25 to 16 mg/L, for calcium from 65 to 41 mg/L. However, these variations are close to the ranges of measurement errors and these results cannot be unequivocally explained by elimination of temporary hardness ions after local heating during treatment (Davis 2010).

Spectra of OD for tap water at various times of treatment are shown on Figure 6a. The main difference between spectra before and after treatment is seen at 200–400 nm where absorbance of nitrates and nitrites appear. According to Thomas & Burgess (2007) absorbance maxima are observed for nitrates at 205 nm (bigger peak) and 302 nm, for nitrites at 213 (bigger peak) and 354 nm. On Figure 6a they cannot be distinguished clearly because they are summarized with absorbance of the other substances, mainly organic matter, about half of which remains after river water treatment at waterworks (Andrianova et al. 2014). It can be seen from Figure 6a that the wavelength 254 nm, commonly used for water quality monitoring (Thomas & Burgess 2007), is inconvenient for control of water treatment efficiency with plasma. At this wavelength OD is affected by increase of light absorbance by nitrates and nitrites and decrease of light absorbance due to destruction of chromophores in organic matter. From this point

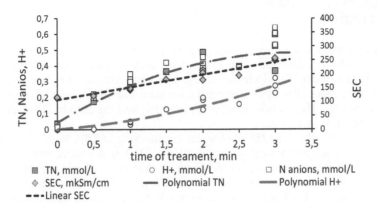

Figure 5. Changing of SEC, concentrations of anionic nitrogen (nitrates and nitrites), TN, and H+ during tap water treatment.

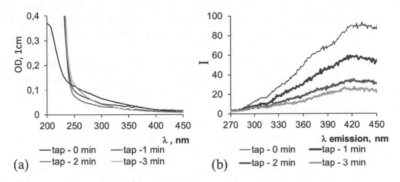

Figure 6. Spectra of optical density (OD) (a) and spectra of fluorescence intensity (I) at excitation wavelength 230 nm (b) for tap water after different times of plasma treatment.

of view longer wavelengths used to determine such parameter as color of water (380–410 nm) (Rice et al. 2012) could be more useful. However in this part of spectrum light scattering on the particles significantly increases OD. Colloid particles are contained in tap water of Saint Petersburg; their number concentrations reach 10^8 cm^{-3}. Colloids form up to 30% of OD at 254 nm and up to 80% of OD at 413 nm in tap water (Andrianova 2008). Also iron-containing suspended matter is formed during plasma treatment because of steel electrodes corrosion. Concentration of iron in tap water rose to 5 mgFe/L after 3 minutes of plasma treatment. The iron-containing particles remained suspended in water for more than 30 minutes after treatment, thus increasing OD of water.

In such a case another method to control treatment efficiency should be suggested. Unlike OD of water in UV-Vis region (which is affected by at least three factors in our case), fluorescence of water in UV-Vis is formed only by aquatic organic matter without direct influence of inorganic substances (Coble et al. 2014). Thus degradation of fluorophoric groups in organic matter can indicate efficiency of treatment. For rapid analysis two excitation wavelengths were chosen (230 and 270 nm) that had shown informativity in study of surface water in Saint Petersburg (Il'ina et al. 2017).

Fluorescence spectra of the tap water samples at various times of treatment at excitation wavelength 230 nm are shown on Figure 6b. Analogous results were received at excitation wavelength 270 nm, values of I were about 1.8 fold lower than I at excitation wavelength 230 nm (data not shown). It is clearly seen from the spectra that intensity of humic-like peak of fluorescence (with maximum at about 430 nm) definitely decreases with time of treatment (3 times from initial value after 3 minutes of treatment for both excitation wavelengths). These data demonstrate applicability of fluorimetry for water control after plasma treatment.

4 CONCLUSIONS

Samples of deionized water and tap water were treated with plasma generated by high voltage discharge. The treatment was done in a reactor with volume about 1 L where electrodes were submerged in water. The electric discharge was formed in the air bubble flows spreading between the electrodes. The electric potential between the electrodes had amplitude up to 33 kV, frequency 500 Hz, impulse front up to 8 mksec, and the current in the electric breakdown had amplitude up to 400 A.

Specific electric conductivity (SEC) of water increased up to 250–280 mkSm/cm due to the formation of nitric and nitrous acids from N_2, O_2 and H_2O. In deionized water these values of SEC were reached after 5–7 minutes of treatment, in tap water – after 3 minutes of treatment.

In deionized water, the concentration of total nitrogen (TN) increased up to 13 mg/L, and the pH decreased to 3.17. The concentration of total iron (formed due to electrodes corrosion) increased up to 11.5 mg/L after treatment.

In the tap water, SEC changed from 112 mkSm/cm (before treatment) to 257 mkSm/cm (after 3 minutes of treatment). TN increased from 0.5 to 7.5 mg/L (after the treatment TN was present in form of nitrates and nitrites), pH decreased from 6.65 to 3.56. Concentrations of total organic carbon, ions of chlorides, sulfates, sodium, potassium, magnesium and calcium changed in or near the frames of measurement error. Concentration of total iron increased up to 5 mg/L.

From the obtained data it was concluded that SEC is a suitable parameter for express-monitoring of plasma treatment efficiency. Spectra of optical density (OD) showed that the commonly used wavelengths for river water quality control (254 nm, 380 nm, 410 nm) are unsuitable for monitoring because OD was influenced by several factors which increased it (light absorption of nitrates and nitrites, light scattering from iron-containing particles) or decreased it (destruction of chromophores of organic matter). Measurement of fluorescence intensity (I) had no such drawbacks for this signal was formed only by organic matter. In tap water values of I at the excitation wavelengths of 230 or 270 nm and emission wavelength 430 nm decreased 3 times after 3 minutes of treatment. These data proved informativity of fluorimetry in express-control of plasma treatment efficiency.

REFERENCES

Andrianova, M. Ju. 2008. *Control of bioorganic impurities in surface water source and drinking water supply system basing on spectrofluorimetry*. PhD thesis. SPb. (in Russian).

Andrianova, M.J., Molodkina, L.M., Chusov, A.N. 2014. Changing of contaminants content and disperse state during treatment and transportation of drinking water. *Applied Mechanics and Materials*. 587–589: 573–577.

Coble, P.G., Lead, J., Baker, A., Reynolds, D.M., Spencer, R.G.M. 2014. *Aquatic Organic Matter Fluorescence*. Cambridge University Press.

Davis, M.L. 2010. *Water and wastewater engineering*. McGraw-Hill.

Il'ina, Kh.V., Gavrilova, N.M., Bondarenko, E.A., Andrianova, M.Ju., Chusov, A.N. 2017. Express-techniques in study of polluted suburban streams. *Magazine of Civil Engineering*. 8:241–254.

Korotkov S.V., Aristov Yu. V., Kozlov, A.K., Korotkov, D.A., Rol'nik, I.A. 2011. A generator of electrical discharges in water. *Instruments and Experimental Techniques*. 54(2): 190–193.

Judée, F., Simon, S., Bailly, C., Dufour, T. 2018. Plasma-activation of tap water using DBD for agronomy applications: Identification and quantification of long lifetime chemical species and production/consumption mechanisms. *Water Research*, IWA Publishing. 133:47–59.

Rice, E.W., Baird, R.B., Eaton, A.D., Clesceri, L.S. (eds.) 2012. *Standard Methods for the Examination of Water and Wastewater*, 22th edition. American Public Health Association, American Water Works Association, Water Environment Federation NW Washington.

Stefan, M. (ed.) 2018. *Advanced Oxidation Processes for Water Treatment. Fundamentals and Applications*. IWA Publishing.

Thomas, O. & Burgess, C. (eds.) 2007. *UV-Visible Spectrophotometry of Water and Wastewater*. Elsevier.

Advances and Trends in Engineering Sciences and Technologies III – Al Ali & Platko (Eds)
© 2019 Taylor & Francis Group, London, ISBN 978-0-367-07509-5

The protection of environment during cleaning ETICS with biocides

N. Antošová, B. Belániová, B. Chamulová, K. Janušová & J. Takács
Faculty of Civil Engineering in Bratislava, STU, Bratislava, Slovakia

ABSTRACT: Cyclic cleaning is a conservative technology for liquidation of algae, mosses and fungi on ETICS surface. The technology includes biocide decontamination processes, mechanical cleaning, rinsing of surface by a pressure cleaner and a preventive coating with a biocidal preparation. In this maintaining process, ambient environments are loaded with running water with detergents which are usually based on heavy metals. The article presents the results of research which is based on the reduction of the impact of soil and water contamination by this technology.

1 INTRODUCTION

The basic concept of thermal insulation system and complex building reconstruction is the minimizing of energy usage. The thermal insulation is connected with the development of technologies, the way of realization, the development of construction products and also the development of maintenance and treatment of the ETICS. Recently we can often see short-comings on the surface of ETICS, which are created by a biological film composed of green and black stains. These microorganisms: fungi, algae and moss can change the properties of contaminated surfaces of the thermal insulation system. They affect mainly the stability of coating, its permeability, colour, hygroscopicity, hydrophobicity or density. These microorganisms are able to create crusts and slimes. Due to chemical reactions these substances are able to penetrate under the coating, which lowers the lifespan and quality of the composite construction. In order to ensure functionality of the contact thermal insulation system, an active maintenance is needed throughout the whole lifespan of ETICS.

2 METHODOLOGY

Methodology of work consisted from research focused on technologies designed to eliminate microorganisms from surface thermal insulation system, from research on environmental effect of mostly used technologies we draw ideas for further measures. The result of the work is a plan of operative measure for active maintenance of ETICS with secure protection of the environment.

2.1 *Technologies for maintenance thermal insulation system*

Long term solution on how to eliminate green stains from the facade of the thermal insulation system is not yet known. Building material as an inorganic compound is not able to develop a defense mechanism against the colonization. We can protect inorganic building materials (ETICS also belongs to this category) against the negative activities of microorganisms by using biocides. Most effective are the chemical methods, which use specific toxic properties and biocides effects on microorganisms, which are to be eliminated or inhibited (Wasserbauer, R. 2000). It is a direct chemical intervention, its purpose is to eliminate the entire microorganism colony by combining mechanical and chemical reactions, what is the desired way of cleaning (Antošová, N. 2015). Protective methods and technologies eliminating biodegradation of building materials are difficult to execute, financially demanding and ecologically problematic (Ledererová, J. et al., 2009). In order to eliminate microorganisms living on facades according to (Minarovičová, K. 2016), many different chemicals are being used, some of them containing toxic compounds of heavy metals that have basis on active silver.

When using biocides for protecting ETICS we have two types of technology:

- free (unencapsulated) – form of biocides in paint with low absorbability and high water repellency. From the market analysis of the products available and recommended applications of the products with active chemical substance implies, that application of biocides, is used in combination with pressured water rinse. If followed by rainfall there is a risk of runoff resulting in contamination of surface waters and canalisation by heavy metals. By using natural mineral materials, we can minimize the negative effect at cleaning proposal ETICS on the environment (Morela, J.C., et al., 2001). Several other studies are also investigating with different types of bio-purification. According to (Valentinia, F., et al., 2010) a research in Italy was studying cleaning of microorganisms from building materials by a product based on glucose oxidase (GOx).
- encapsulated – form of application, microcapsules of biocides in paint and in final surface layer of ETICS, when the components of biocides inserted as micro capsules (size of 10–20 ηm) directly into the plaster or into the renewing paint and the release of effective substances is controlled by diffusion of water vapor, surface moisture of plaster or by activity of the microorganisms. Biocides runoff causes less environmental damage (Breuer, K. et al. 2012).

In terms of environmental protection an encapsulated form of ETICS protection is preferred. If this form of protection is used it is important to make this decision during the ETICS planning phase. Statistics imply that the most used form to eliminate microorganisms on ETICS is by using chemicals available on the market. Almost all manufacturers require method, which consists of: preparation of underlying structure for ETICS, application of effective chemical substance, inspection and provision of protection of ETICS after application of effective chemical substance. Every manufacturer is obliged to, according to legislature, provide a technical sheet of the product and security data card from which we are able to find out: toxicological information, ecological information, possible dangers, how to use, store, transport and dispose of. Manufacturer is also obliged to define a way of applicating the product. In none of the manufacturer's technological regulations or technological sheets talking about eliminating microorganisms is information regarding disposal of wastewater, that is created during cleaning.

2.2 Security of environmental protection in Slovakia according to legislature in context of ETICS maintenance

Evaluation of environmental effect of construction industry is a topic of interest for many. It is starting to use methods, that evaluate according to sustainability as the criteria throughout the whole lifespan of the product (Švajlenka, J., et al. 2018). During an elimination of microorganisms with chemical products a risk of contaminating soil and water occurs. Biocidal products used to maintenance ETICS are a potential risk for humans, animals, and the whole fauna and flora. Due to this reason a strong set of regulations and procedures was created in the EU region. These regulations ensure controlled usage of these chemical substances and guarantee a high level of protection for the environment (Commission Delegation (EU) No. 1062/2014). The intention directive is to achieve a condition, that holds companies responsible for the consequences of their behaviour, which means they are also responsible for the damage they have incurred. In Slovakia for the any building action you have to follow law No. 39/2013 statute book, that talks about integrated preventions and inspections on environmental damage. Based on this law a principle in environmental politics applies. In case of contamination of water, a person or organisation that caused the contamination are hold responsible and will cover the damages. It is also important to follow a law regarding water (No. 409/2014).

The amendment is based in particular on the need to incorporate the European Commission's comments on the transposition of Directive 2006/11 /EC of the EU Parliament and of the Council of 12th December 2006 on the protection of groundwater against pollution and deterioration quality. The principle of "the polluter pays" is in environmental politics of advanced countries considered as a principle of general responsibility of polluter to cover the damages he caused. That's why in case of planned or operative maintenance of ETICS it is expected from the owners to provide protection for the environment (mostly soil, water, flora) from chemically contaminated water created during the cleaning.

Table 1. Experiment researching amount of released biocide substances has been done on both identical objects with facade with contact thermal insulation system, where for the first object a technology with encapsulated biocides was used and for the second an unencapsulated biocides were added to a paint and then used on object. Total [mg/m²] and scaled [%] value of released effective substances (run-off) compared to amount of effective substance used (1600 mg/m²), (Breuer, K. et al. 2012).

Active substance	Encapsulated		Unencapsulated (abandoning)	
	mg/m²	%	mg/m²	%
Terbutryn (CAS-Nr. 886-50-0)	24	1,5	59	3,7
IPBC (CAS-Nr. 55406-53-6)	75	4,7	135	8,4
OIT (CAS-Nr. 26530-20-1)	78	4,9	189	11,8
DCOIT (CAS-Nr. 64359-81-5)	10	0,6	13	0,8
Diuron (CAS-Nr. 330-54-1)	29	1,8	187	11,7

Table 2. The investment costs for the planned (permanent) solution of drainage systems. Created by program CENKROS, (authors).

Itemization	M.J.	Material price EUR per m²	Price of work EUR per m²	Total price EUR per m²
Excavation work in the rock 3 to 100 m³	m³	–	51,11	51,11
Insulation against earth moisture	m²	1,64	1,53	3,18
Filling with gravel	m³	1,50	0,62	2,11
Installation of a drainage plastic drain trap	m	48,10	2,14	50,24
Imbed the concrete curb into of plain concrete	pcs	5,66	1,76	7,42
			Total costs in all:	114,06
Bin for rainwater	pcs	570,00	24,46	594,46
			Total costs in all:	708,52

Table 3. The investment costs of operational solution of catching system. Created by program CENKROS, (authors).

Itemization	M.J.	Material price EUR per m²	Price of work EUR per m²	Total price EUR per m²
Wooden prisms of different sizes	pcs	0,45	–	0,45
Installation of a drainage plastic drain trap	m	30,03	2,14	32,17
Film of PVC-P, width of 1,3 m	m²	2,45	–	2,45
Attaching the insulation to the vertical surface of the base by anchorage strips	m	–	5,87	5,87
			Total costs in all:	40,94
Collecting container for wastewater, 100 l	pcs	51,00	–	51,00
Renting the pump with a collection tank		40,00 €/day		40,00
			Total costs in all:	131,94

3   REDUCING ENVIRONMENTAL DAMAGE DURING ETICS CLEANING

Basic rules for cleaning surface of ETICS contaminated by microorganisms implies, that the most optimal time interval for application of the product is when the algae and microorganisms go through vegetation period, spring and fall. The disposal of microorganisms can be realized within of the building options and the site construction from scaffolding, or from the assembly platform. Construction site must include source water designed to rinse surface of the facade and also electrical energy source. Construction site also must include collecting mechanisms for runoff water with applied active substance according to the legislation. Collected wastewater must be drained into collecting tanks, which need to be always accessible on the construction site. Contaminated water must be disposed of according to legislation in force. It is unacceptable to drain the contaminated water with active chemical substance into a canalisation system without the permission of the canalisation network administrator. Maintenance cycle and protection of ETICS that is relative to biocorrosion resistance is recommended in 3 to 5 year intervals, depending on various factors (Antošová, N. 2015). Expected lifespan of ETICS according to (ETAG 2004) is 25–30 years. It follows, that disposal of wastewater from cleaning of the surface needs to be done at least 4 to 5 times during the entire lifespan of ETICS. This information is important for complex project solution for renewal external cladding, which have to include maintenance suggestions and ways to implement them.

3.1   *Plan of measures to reduce environmental damages*

According to (Minarovičová, K., & Dlhý, D., 2018) there are more ways to dispose of contaminated water drained from the facade: Gutter leaching system and line drainage channels (temporary structure) (Figure 1), Operational solution of drainage systems (Figure 2, 3).

A suitable way is to make line drainage channels around the building, into which this water can be collected and safely take off. In the case of a permanent solution, filtration can be used for cleaning contaminated water, which can then be connected to the canalization system or drained into the surrounding area. The drainage system has to in every case be designed with regard to the amount of water required for washing the facade, for regular cleaning of the biocide product solution on the façade or the amount of precipitation water in the area.

In this case, the authors have focused on the permanent solution that is included in realization of the thermal insulation system to the renewal of around the house, or it uses the additionally created gutter system at the plinth profile of the thermal insulation system. The installation of hooks carrying gutters sloping into the collecting containers. Even in this case, however, it is necessary to consider the concept of additional suspension of the collecting channel and safe draining of water already during the project of renewal of the building or realization of ETICS.

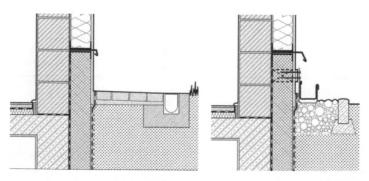

Figure 1.   Gutter leaching system and line drainage channels (temporary structure) – water drainage system from the ETICS surface maintenance (Minarovičová, K., & Dlhý, D., 2017).

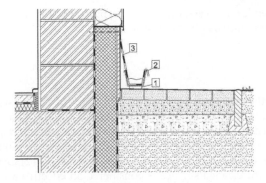

Figure 2.   Proposal of temporary solution for draining water from cleaning (detail in a plinth) (authors).
1. wooden (concrete) pads, 2. collecting gutter in inclination, 3. membrane attached to the object.

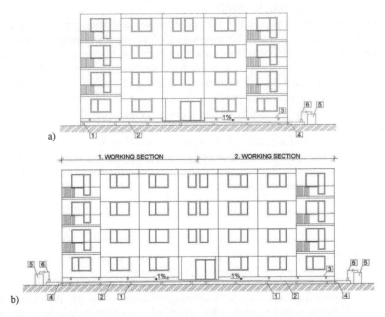

Figure 3.   Proposal of temporary solution for draining water from cleaning (view of the object), a) design for smaller objects, b) design for larger objects (authors): 1. wooden (concrete) pads, 2. collecting gutter in inclination, 3. membrane attached to the object, 4. collecting container for contaminated waste water, 5. collecting tank for pumping water, 6. Pump.

3.2   Proposal of operational measures for reducing environmental damages during ETICS cleaning

There are multiple factors that affect the water draining plan such as: purpose of the object, construction of the object, slope of the land, availability of disposal containers, foundation of the object, availability of using filtration tanks. All these factors need to noted in project. In case of not being able to provide plans for collecting wastewater, it is a possible to use the solutions mentioned. It is a collecting wastewater method has been designed for there is an operative and also temporary solution, that can be used on whole section of ETICS with bio-corrosion. In this case the gutter is folded up on the place, portable and then laid on wooden or concrete pads of different sizes that were produce decline gutter and drain wastewater to collection tank. On ETICS surface near the plinth block part we attach waterproof mem-brane that is cca 1 m wide (can be adjusted to a situation). Second end of the membrane is placed into the gutter. This way a safer transfer of contaminated water from ETICS surface into the gutter. Collection of wastewater can be provided by gravity or by pumping with small submersible pump into a collection tank.

4 CONCLUSION

In fact, there is no other than regular biocide protection of façade surfaces, but cannot be considered as the ultimate solution for façade protection. This popular trend of using biocides in building industry, without verifying and looking into its effects, can cause unpredictable damage to the environment. Most biocides are released when rainfall occurs a few days after their application (called "first wash") (Breuer, K. et al., 2012). System designed to collect wastewater from the cleaned surface is considered an effective way to reduce unwanted effects of biocidal substances on the environment throughout the whole lifespan of ETICS. There is a great potential for using this system, due to its easy application and multi usability, which reduces the capital costs of the project. Unpredictable costs affect the costs of the building's maintenance (Ďubek, S., 2015).

The comparison of investment costs therefore shows that after multiple uses, the cost of the operational solution is lower. The high costs of planned drainage systems are mainly influenced by the cost of the collection tank that needs to be connected to the system and placed underground. A case study has been done within the experiment, on a four floor object in the town of Tlmače (Levice region), the compared total of water caught from cleaning with water used for cleaning the proposed system and has shown positive results. Ninety percent of the water used for cleaning of the ETICS surface after the application of biocides was collected. There are a few of objects where the planned maintenance was forgotten or the location, design, financial status of the construction does not allow the application of a permanent solution for catching water with an effective chemical substance. That's why we consider this solution as an effective alternative for reducing the unwanted effects that biocidal substances have on the environment.

REFERENCES

Antošová, N. 2015. The methodology for the selection of technology for eliminating microorganisms on the EITCS. *Czech Journal of Civil Engineering 1*: 6–14, Brno.
Breuer, K. et. al., 2012. Wirksamkeit und Dauerhaftigkeit von Bioziden in Bautenbeschichtungen. *Ernst & Sohn Verlag für Architektur und technische Wissenschaften GmbH & Co. KG*. Berlin, Bauphysik 34.
Ďubek, S., 2015. Vplyv nepredvídateľných nákladov pri návrhu objektov zariadenia staveniska. (The impact of Unpredictable Costs at Design Buildings of Construction Site Equipment). *Journal Almanach znalca 2*: 24–26. Bratislava.
ETAG 004, 2004 Smernica pre Európske technické osvedčenie kontaktných zatepľovacích systémov. (Guideline for European Technical Approval of External Thermal Insulation Composite System with Rendering).
Krueger, N., Schwerd, R., Hofbauer, W., 2016. Einsatz biozidfreier Komponenten in Wärmedämmverbundsystemen, Fraunhofer-Institut für Bauphysik, Dessau-Roßlau.
Ledererová, J. et al., 2009. Biokorózní vlivy na stavební díla, In Czech. (Biocorrosion effects on construction works). *Silicate union*.
Minarovičová, K. 2016. Údržba ETICS šetrná k životnému prostrediu. (Maintenance of ETICS environmentally friendly.) In Slovakia. *Journal Správa Budov 03*, Bratislava: Jaga.
Minarovičová, K., Dlhý, D., 2017. Environmentally safe system for treatment of bio corrosion of ETICS. MATEC Web of Conferences 146, 03005. Building Defects 2017.
Morela, J.C., Mesbaha, A., Oggerob, M., Walkerc, P., 2001. Building houses with local materials: means to drastically reduce the environmental impact of construction. *Building and Environment 36*:1119–26.
Švajlenka, J., & Konzlovská, M., 2018. Houses Based on Wood as an Ecological and Sustainable Housing Alternative – Case Study, *Sustainability in Civil Engineering, Volume 10, Issue 5*.
Valentinia, F., et al., 2010. Diamantia, A., Palleschi, G., New bio-cleaning strategies on porous building materials affected by biodeterioration event. *Applied Surface Science 256*, 6550–6563.
Wasserbauer, R., 2000. Biologické znehodnocení staveb. (Biological degradation of buildings) In Czech. Publisher: ARCH a.s.

Advances and Trends in Engineering Sciences and Technologies III – Al Ali & Platko (Eds)
© *2019 Taylor & Francis Group, London, ISBN 978-0-367-07509-5*

The possibility of heating water by photovoltaic panels

R. Baláž
Faculty of Civil Engineering, Institute of Architectural Engineering, Technical University of Košice, Košice, Slovakia

ABSTRACT: The submitted article offers one of the possible options of using photovoltaic panels for domestic hot water preparation with an option to extra heating a heating system, and followed by recalculation of amount of the energy produced in regards to the total return of the assembled system. An electric heating boiler combined with a solid propellant based heating option was chosen for the experiment. The most frequently used combined electric storage tank was chosen as a classic option to heat domestic hot water.

1 IMPLEMENTATION PHOTOVOLTAIC

Solar energy systems include photovoltaic (PV) materials and devices that convert sunlight into electric energy; PV cells are commonly called solar cells (http://www.energy.gov/energybasics/rticles/photovoltaics).

A large study in California completed by Lawrence Berkeley National Laboratory compared sales of homes with and without PV to determine what premium, if any, existed on homes sold with PV systems. Their results indicated that depending on whether the home was a new or existing construction, the price per watt premium varied between $2.30 – $2.60 /watt, and $3.90 and $6.40 /watt, respectively for homes with PV systems, as compared to comparable homes without PV.

Sample Comparison of Energy Savings Value to better understand some factors that affect energy savings, and therefore value to the consumer, an example follows comparing identical 5 kW PV systems—one in Colorado and one in Louisiana. This example illustrates the difference in energy savings as a function of production potential, utility electricity rates, and typical electricity consumption patterns. The example uses average conditions and is only intended as an illustration, not a real-world scenario. In Colorado, the typical household uses around 711 kWh of electricity per month, or approximately 8.532 kWh/year, according to 2011 US Energy.

Information Administration (EIA) data.16 Using the typical PV system size of 5 kW, a PV system in Denver will produce approximately 7.594 kWh in the first year, 17 offsetting approximately 89% of the household usage. Using average electricity rates in Denver of 11.1 cents/kWh, 18 the price typically paid by the homeowner is around $947 /year, and the value of electricity produced by the PV system is approximately $843 in the first year (11.1 cents/kWh × 7.594 kWh), effectively reducing the amount paid by the homeowner from $947 to $104 in the first year, an 89% reduction (Klise, G.T. et al. 2013).

A photovoltaic system by itself and its behavior should be introduced at the beginning. The article will follow the characteristics of the system's equipment. Subsequently, we will evaluate amount of produced electric energy used for the production of domestic hot water and amount of produced electric energy used for additional heating-up the heating system, also with calculation of payback from designed solution for heating.

The photovoltaic is a technical section, which concerns with a process of a direct transformation of light into the electric energy. The term is derived from the word photo (light) and volt (electric current unit). The transformation process runs in a photovoltaic cell. The photovoltaics were discovered by Alexander Edmond Becquerel in 1839. The photovoltaic cells for energy production in cosmic programs were used for the first time in 1958. The photovoltaic cells were first used for energy production in cosmic programs in 1958. Since then, it has become an inseparable part of the cosmic program. The article will not analyze the transformation process more deeply.

The basis of the photovoltaic (PV) systems are the photovoltaic (PV) cells merged into the photovoltaic (PV) panels. The most popular photovoltaic panels are made from silicium. The different ways of silicium treatment can create monocrystalline, polycrystalline and amorphous (non crystalline) photovoltaic cells. The monocrystalline cell is a black octagon and the polycrystalline cell is a blue square.

The monocrystalline cells are more efficient than polycrystalline, however, the area usage in the polycrystalline cells is not so perfect due to its shape, so both types have similar performance in conclusion.

The polycrystalline cells effectivity is 12–14%. The monocrystalline cells effectivity is 12–16%. Price and durability are the same. The photovoltaic panels are capable to produce an electric energy without a direct sunlight based on diffuse equipment, which is dominant in the Slovakia.

Types of used photovoltaic (PV) systems:

It is necessary to describe a division of photovoltaic systems and their functioning before we describe the actually used and tested system.

Photovoltaic systems:

Stand-alone systems: They are used when the costs of the connection and its operation are higher than a single photovoltaic system. An exemplary case is when the distance from the connection point is longer than 500–1000 m.

Stand-alone systems are divided:

Direct connection: It is a simple reconnection of a photovoltaic panel and a device. The device works only when there is a sufficient intensity of sunlight. The usage of this type is mostly for charging small devices, pumping water for irrigation, powering ventilation fans.

Hybrid systems:

They are used when a yearlong operation is necessary, and sometimes a device with a high input power is used. It is possible to create a lot less energy from the photovoltaic panels in the winter than in the summer. It is necessary to design the systems for winter operation, which has an effect on a higher production power, and essentially higher costs at the beginning. The preferable choice is to extend the system with an additional source of electricity, which covers the need of electric energy during months with an insufficient sunlight and during operation of devices with a higher input power. The additional source for example can be a wind generator, a generator, a cogeneration machine, and so on.

Storage system:

It is used when there is a need for electricity, though there is no sunlight. The stand-alone systems have special storage batteries made for slow charging and discharging due to reasons mentioned above. Optimal charging and discharging of batteries is ensured by a regulator. It is possible to connect devices powered by a direct current (system current is usually 12 or 24 V alternately 48 V) and common network devices (230 V /~ 50 Hz) connected by an inverter.

Systems connected to the network:

They perform automatically thanks to a microprocessor, which runs an inverter and transforms direct current from the panels into an alternating current powering devices. The connection to the network is subjected to an authorization approval of energy distributions. It is necessary to comply with given technical parameters.

2 DESIGNED SYSTEM

The designed system is with a direct connection. This system was chosen due to its quick payback, what was the initiative though of a system design.

- Used system elements: Photovoltaic (PV) panels Kyoto 275 kwp – 16 pieces
- MPPT regulators DC/AC designed for PV system – 2 pieces
- Combined boiler Mora K 120 L 2 × 1000 W or 2 × 1800 W
- Storage tank Regulus 200 L
- Electric heating spirals *2 × 2000 W*

Existing heating system combining electric boiler and heating by a solid propellant, it is necessary to mention, that the photovoltaic system was installed into the existing fully functional system of domestic water heating. It was ensured by an electric energy from the network in a summer and the heat exchanger connected to the heating system in a winter.

The combined boiler was used MORA K 120 L by reason of having dry spirals installed in the flange with changeable input power 2 × 1000 W or 2 × 1800 W /LD TOPTEL/. The combination of the spirals settings 2 × 1000 W was used during connection to the network 230 V/~ 50 Hz, combination 2 × 1800 W was used during connection of the MPPT regulators DC/AC determined for medium solar PV system in a number of 2 pieces. They are connected to the photovoltaic panels in number of 8 pieces with a performance 275 Wp by piece on every regulator, what gives together approximately 4.4 kWp.

3 MEASUREMENT OF THE CONSUMPTION OF AN ELECTRIC ENERGY TAKEN FROM THE NETWORK

I did not measured the real consumption of an electric energy from the network, I keep up to the information from manufacturer, who officially indicate 7.64 kWh with input power 2 × 1000 W and temperature up to 65°C from initial entry temperature of water 15°C (Table 1). This information is sufficient for a functionality demonstration and a payback of the whole system. It is possible to measure the real consumption of an electric energy with given type of the boiler and boundary conditions in a case of deeper examination.

It is required to mention the spirals connection to the boiler 2 × 1000 W was changed to 2 × 1800 W to use the most input power from the photovoltaic panels in the shortest time. We watched it as an advantage during winter, when there is a less sunlight.

3.1 *Display of the system connection*

The domestic hot water heating system is solved by 16 pieces of the photovoltaic panels FV Kyoto 275 kwp, which are connected by 8 pieces in a series with 2 pieces of regulators MPPT OPLOCKY. The panels are directly connected into to the regulators, where a switch is designed by a capillary thermometer in a case of overheating the combined boiler. An excessive energy after overheating of the combined boiler goes straight to the heating system

Table 1.   Technical parameters of energy consumption.

Type	K 120 L
Volume	120 l
Pressure	0.6 MPa
Weight/filled with water	62/66 kg
Anticorrosive protection	Enamel/Mg anode
Input power	2 × 1000 W
Voltage	230 V~
Protection class	I.
Protection level	IP25
Time to heat water to 65°C	3.82 h
Energy consumption to temperature 65°C	7.64 kWh
Quality of water with 40°C	228 l
Heat loss	1.77 kWh/24h

Figure 1.   Photovoltaic panels.

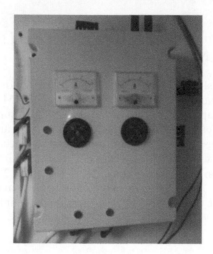

Figure 2.   Regulator MPPT OPLOCKY.

Table 2.   Input costs.

Part	Price
Photovoltaic panels (16 pieces)	2 240 €
Connection accessories & cables	387 €
Regulator MPPT OPLOCKY (2 pieces)	300 €
Cartridge heater (2 pieces)	105 €
Assembly material on the roof	274 €
Total price	3285 €

via resisting heating spirals assembled to the storage tank connected to the existing heating system. The excessive energy after overheating the combined boiler will be discussed some other time. The combined boiler is directly connected to the **MPPT OPLOCKY** regulators with classic 16 A plug box, although, there is a change against the classical electric energy network. The dry spirals with a performance 2×1000 W were changed for the heating device for contact heating **LD-TOPTEL**, cartridge heater with caliber 12.5 mm, L = 400 mm, and outlet length Lv = 250 mm, U = 230 V and performance 1800 W for one piece.

Figure 3.   Designed system.

Table 3.   Consumption of the combined boiler.

Consumption	kWh
From manufacturer/day	7.64
Real/day*	11.46
Real/year	4182

*Real consumption of an electric energy reported by a manufacturer multiplied by ratio 1.5, because of need of hot water was mostly more than 1 heat up of the combined boiler in a day.

Table 4.   Calculation of a total electricity price*.

	Yearly	Monthly
Invoicing items	€	€
Payment for electricity	186.68	15.56
Fixed monthly payment—keeper	7.80	0.65
Fixed payment for distribution (breaker, emergency capacity)	50.83	4.24
Variable payment for distribution	105.80	8.82
Loss	25.05	2.09
System services	28.82	2.40
System operation	109.57	9.13
Nuclear fund	13.42	1.12
DPH	105.60	8.80
Total	633.59	52.80

*For East Slovakia, product DD2 (amount of energy import 1T/VT – 4 182 kW/h).

3.2   *Measurement methodology*

The basis of my research is the question: "Will the given photovoltaic system fulfill needs of a domestic hot water with correct operation and without any feeling of missing comfort due to lack of hot water, or any necessary physical switching to a classical electric network?" The measurement was designed for reading the values based on a boiler's water temperature, which was set on a maximum operational temperature. The reading was always at 5 pm. The measurement of the really produced electric energy was made similarly, however, we will not represent it in this article (http://www.energybulletin.net/node/17219).

3.3 *Payback of domestic water heating*

It is necessary to bring up, that particular photovoltaic system would also run with a half number of panels and one regulator **MPPT OPLOCKY** for a half price of photovoltaic panels with accessories. However, my though was to design a heating system, which will still work during winter months—December, January. It could not be done with a system of 8 pieces of panels and only one regulator **MPPT OPLOCKY** (the measurement implemented in 2016 and 2017).

The payback of the whole system after total calculation of all costs is 5.18 year. However, with half number of the panels and only one regulator is payback 2.59 year.

4 CONCLUSION

The photovoltaic system for hot water heating was designed with the thought to speed up money payback from the initial capital. The time of *5.18* years might look long, yet, we did not focus on an excessive energy after overheating the combined boiler. This energy could be used to heat water in a heating system, though this topic will be discussed in some other article.

REFERENCES

Hoen, B. et al. 2011. *An Analysis of the Effects of Residential Photovoltaic Energy Systems on Home Sales Prices in California*. Lawrence Berkeley National Laboratory.
Klise, G.T. et al. 2013. *Valuation of Solar Photovoltaic Systems Using a Discounted Cash Flow Approach*. Appraisal Journal.
http:/www.energy.gov/energybasics/articles/photovoltaics.
http://www.energybulletin.net/node/17219, Energy Payback of Roof Mounted Photovoltaic Cells by Colin Bankier and Steve Gale.
http://www.ld-toptel.sk/index.php?page=uvod.

Advances and Trends in Engineering Sciences and Technologies III – Al Ali & Platko (Eds)
© *2019 Taylor & Francis Group, London, ISBN 978-0-367-07509-5*

Green roofs and their measurement

R. Baláž & S. Tóth
Faculty of Civil Engineering, Institute of Architectural Engineering, Technical University of Košice, Košice, Slovakia

ABSTRACT: The submitted article is devoted to the research of the green roofs located at Faculty of civil engineering of Technical University of Kosice. The research was implemented in the climate chamber during quasi stationary state of the internal environment with variable meteorological conditions. The sample of the green single skin roof without irrigation and the sample of green single skin roof with a water storage gutter in its composition are in the climate chamber module. Physical-technical measuring described in the article is implemented on the mentioned samples.

1 IMPLEMENTATION GREEN ROOFS IN TEST CHAMBERS

The article presents the physical and technical measurement of two different green roofs in the module of the climate chamber during quasi stationary conditions of the internal environment with dynamic external conditions. The S1 green roof is classical single skin roof with above rafter insulation and vegetative layer without irrigation. The S2 sample is designed as single skin roof with above rafter insulation, containing auto irrigation vegetative layer ensured by a plastic gutter installed in soil. The though of this solution came by after measuring the S1 sample of the green roof, where the vegetative grid was drying out during a summer season. Modernization of the S2 sample's green roof in the climate chamber is represented in the article, following its measurement of the physical and technical parameters.

The roof design accrues from the consistent analysis. The roof model with dimensions 1 700 mm × 2 200 mm was missing the middle support, although, two pieces of identical roof panels had to be spread over the module to create two different parts. The structure of both

Figure 1.   Situation of test chambers measure.

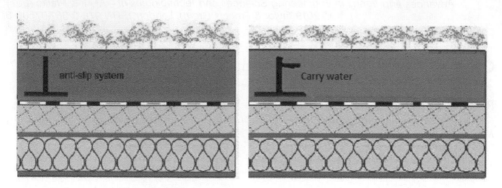

Figure 2.   Previous roof layers composition (left), current roof layers composition (right).

roof elements is different. The roof construction was designed as a single skin 45° pitched roof oriented toward South. A purpose of this design was to achieve the most adverse conditions for an appropriate roof functioning. The bearing roof structure was designed as a rafter system with shuttering, insulation between rafters, and additional insulation above rafters from PIR panels. Waterproofing layer was made of modified bitumen thick 4.5 mm topped with sprinkle (Katunský et al. 2014). Vegetative layer of the roof structure consisted of reversed plastic T pieces used to ensure vulcanic soil in a place, thereafter, the vegetative grid was planted. The same layer concept was used in the both green roof samples. The precise measurement during a summer season brought us to the conclusion, there is a properly designed roof structure from thermal aspect. However, the dry out of the vegetative grid was caused by high temperatures of the external environment during a summer, what led to absolute dry up of plants. Modernization of the S2 sample's roof construction had to achieve the auto irrigation of the roof, which was ensured by installing a plastic gutter filled with a geotextile to carry water in the sample. The detailed analysis of the used elements is possible to see in the Figure 2.

2   DISTRIBUTION AND TYPE OF SENSOR IN THE TEST SAMPLE ROOF COVER

The construction in the roof research module was divided into construction S1 and construction S2. S1 contains a vapour barrier JUTAFOL N in contrast to S2 which is illustrated in Figure 3 Sensors in the roof construction sample were positioned in individual layers of the roof element, mirroring each other at 300 mm from the centre line defined by a load bearing beam. The first temperature sensor was mounted closest to the exterior underneath the bitumen layer of S1 A3.7. as with S2 sensor A3.8. the second layer of the roof cover consists of thermal insulation (polyurethane) with a sensor A3.9 and A3.10 placed in the centre line of the insulation i.e. 40 mm. The third layer of the roof cover consists of a connection layer which placed on OSB particle board.

They house sensors A3.11 and A3.12. Sensors A3.7, A3.8, A3.9, A3.10, A3.11 and A3.12 are temperature sensors based on the resistance without a tip with capabilities from –50 to +125°C. Internal insulation consists of mineral wool *125 mm* thick where a sensor I1/6 was placed at its centre which mirrors v I1/2. These sensors measure air temperature, relative humidity, dew point and absolute humidity. Finally sensors are D1/13 and D1/14. They were positioned on the internal drywall on the interior surface. This sensor measures the surface temperature at two points. The structure S2 S1 (green roof construction) was mounted sensor surface temperature directly below the vegetation mat D1/41, 42 (Baláž 2016).

Measured parameters are outside and inside air temperature and relative humidity, temperature and relative humidity inside the structure, surface temperature and heat flux. For those purposes is used temperature sensors based on resistance without nib, type NTC and

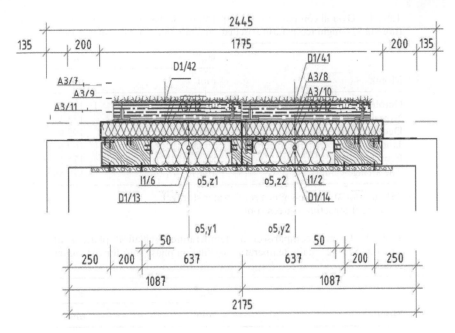

Figure 3. Deployment of sensors.

NiCr. Weather station measures wind direction, average and maximum wind speed, atmospheric pressure, air temperature and relative humidity, rainfall and we also monitor global radiation by Pyranometer (Flimel 2013, Chuchma & Kalousek 2014).

3   MEASUREMENT RESULTS OF THE CLIMATIC CHAMBER MODULE

All measured data of a test roof module except for data obtained by the meteorological station were recorded in one minute intervals over duration of four winter months at the Civil Engineering Faculty of The Technical University of Kosice.

The Meteorological station fixed to the given module of the climatic chamber supplied real time data regarding the outside air temperature, wind speed and the duration and intensity of global solar radiation. Values of outside air temperature used in this work were obtained from meteorological station sensors. Licensed software was used to process data. In order to improve visibility and reduce overlapping of individual curves in charts a maximum of three sensor readings were used in graph.

3.1   *Phase shift of temperature oscillation in the roof structure S1 and S2 in the quasi-steady state of the internal environment and the changing external environment*

Comparison of the phase of the temperature oscillation was assessed in the months October 2017, November 2017, December 2017, January 2018 and February 2018. For the calculation itself is to be noted that as the inner surface temperature was used value from the sensor D/13 to construct sensor S1 and S2 D/14 for design. (Sensor measuring the surface temperature of the plasterboard). In individual months were selected values of minimum and maximum outdoor temperature with their associated values of surface temperature in the roof structure and roof structure S1 and S2 they were subsequently compared.

A full assessment of the phase shift of the temperature oscillation in the construction of the roof shell S1 and S2 shown in Table 1 and Table 2.

Comparison of the phase of the temperature oscillation was assessed in the months October 2017 November 2017, December 2017, January 2018, February 2018. For the calculation

Table 1. Overall comparison of temperature oscillation phase shift structure S1a S2 – night temperature oscillation phase shift for the winter season.

Month	ψ_{S1}*	ψ_{S2}*
	min	min
October 2017	70.0	180.1
November 2017	90.3	173.1
December 2017	90.0	195.0
January 2018	133.0	160.6
February 2018	80.3	175.5
The average value	105.75	176.86

*S1-the roof structure—green roof (water storage).
S2-the roof structure—green roof.

Table 2. Overall comparison of temperature oscillation phase shift structure S1a S2 – daily temperature oscillation phase shift for the winter season.

Month	ψ_{S1}*	ψ_{S2}*
	min	min
October 2017	100.1	156.2
November 2017	115.9	135.5
December 2017	95.0	150.6
January 2018	85.0	145.5
February 2018	90.7	46.2
The average value	97.30	126.81

*S1-the roof structure—green roof (water storage).
S2-the roof structure—green roof.

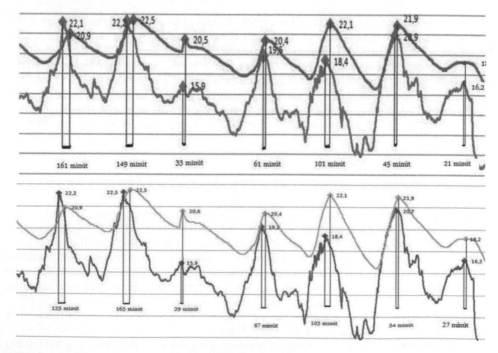

Figure 4. Day phase shift for winter period November 2017 (day).

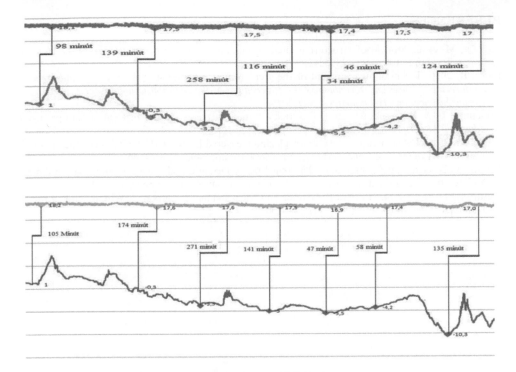

Figure 5. Day phase shift for winter period February 2018 (day).

itself is to be noted that as the inner surface temperature was used value from the sensor D/13 to construct sensor S1 and S2 D/14 for design. (Sensor measuring the surface temperature of the plasterboard.) In individual months were selected values of minimum and maximum outdoor temperature with their associated values of surface temperature in the roof structure and roof structure S1 and S2 they were subsequently compared.

A full assessment of the phase shift of the temperature oscillation in the construction of the roof shell S1 and S2 (green roof) shown in Table 1 and Table 2 (Katunský et al. 2013, Bagoňa et al. 2014).

Surface temperature on Drywall roof S2 green roof, surface temperature on Drywall roof S1.

4 CONCLUSION

From the confrontation of tested measurement in climate chamber module in quasi-stationary state of indoor environment and in changing state of external environment of structure S1 and structure S2 it is possible to state, that our green roof layer in roof structure S2 has an impact on the roof structure from the phase shift temperature oscillation as follows:

- Roof structure S2 due to the use of green roof layer shows, that it has 71.11 minutes better daily temperature oscillation phase shift in winter time (2017–2018) than roof structures S1. These measurements were realised in quasi-stationary indoor environment in climate chamber.
- Roof structure S2 due to the use of green roof layer shows, that it has 29.51 minutes better daily temperature oscillation phase shift in winter time (2017–2018) than roof structures S1. These measurements were realised in quasi-stationary indoor environment in climate chamber.

REFERENCES

Bagoňa, M. et al. 2014. Roof structure evaluation in climatic chamber module. *Advanced Materials Research* volume 1041: pages 239–242.

Baláž, R. 2016. Technical measurement of the green roof in a winter season, *Advances and Trends in Engineering Sciences and Technologies II – Proceedings of the 2nd International Conference on Engineering Sciences and Technologies, ESaT 2016.*

Chuchma, L. & Kalousek, M. 2014. Electricity storage in passive house in Central Europe region. *Advanced Materials Research* volume 899: pages 213–217.

Flimel M. 2013. Differences Ug – values of glazing measured in situ with the influence factors of the internal environment. *Advanced Materials research* volume 649: pages 61–64.

Katunský, D. et al. 2013. Analysis of Thermal Energy Demand and Saving in Industrial Buildings: A Case Study in Slovakia. *Building and Environment* volume 67, no. 9: pages 138–146.

Katunský, D. et al. 2014. Numerical Analysis and Measurement Results of a Window Sill. *Advanced Materials Research* volume 899: pages 147–150.

Cement screeds—selected methods of humidity measurement

M. Bederka, P. Makýš, M. Ďubek & M. Petro
Slovak University of Technology in Bratislava, Slovakia

ABSTRACT: Wet processes are an integral part of the construction on the majority of today's buildings, whether they are part of load-bearing or non-load-bearing structures. Moisture can be the cause of failures. Deformities in the form of failures are the most likely to be manifested as corrugation of wooden floors, formation of bubbles on linoleum, colour stains on floor surfaces, or non-adhesion of surface treatment to the substrates as well as adhesion of surface treatment to substrates, etc. It is necessary to deal with the possibility of the occurrence of these failures, for example, when the floor constructions are made of cement screed and other floor layers. The article is focused on the moisture measurements in the cement screed by several devices directly on the building site—the IN-SITU method. These measurements determine the shortest required time to dry the floor layer at a certain thickness. This time is eventually representing.

1 INTRODUCTION

1.1 *Moisture*

The completed screed, like other structures, must be inspected. It is recommended to check the mechanical properties of the screed, but it is advisable to add a visual inspection of the surface and the accessible edges, checking the thickness of the layer by making several probes, and checking the screed moisture before storing the following layers. The standard procedure, gravimetric measurement, is defined according to ČSN EN ISO 12570. This method is based directly on the definition of the moisture content of the material, which is the ratio of the weight of moisture contained in the material and dried material. It is necessary to alert that the sample drying temperature is normally 105°C, but it is only 40°C for plaster-based materials (e.g. anhydrite). In the flooring work, the CM method was approved. This method uses a capsule with calcium carbide that is broken in a sealed container containing a sample of the test material. Its reaction with water gives rise to acetylene, whose pressure in the test container varies. This method is fast and provides fairly accurate results. In addition to these two methods, methods based on the measurement of electrical quantities (conductivity, capacity, etc.) are used. These methods were primarily developed to measure the wood moisture. However, when we are measuring the moisture of silicate materials, we deal with the problem of the transition relation of the measured quantity to moisture. It is influenced by the properties of the structure of the monitored material, for example by the amount of cement, the type and size of the aggregates, etc. (Bentz 2005, Breugel 1995; Jennings 2000, Černý 2009). At higher temperatures, a considerable amount of chemically bound moisture is released. (Tuma 2009, Dohnálek 2008). The technological break and build of moisture in the floor structure may result as the wear layer failures of the floor. Moisture in new buildings occurs in all wet processes (Špak 2016). Failures are most often manifested by the deformation of the surface layers (STN 73 1316) (buckle, warp—especially in the case of wooden floors (Švajlenka 2018) or air bubbles under the wear layer of the floor—in laminate floors) or its colour changes. The apartment owner complains about the aesthetic side of the floor to the manager (Hanko 2018; Kantova 2014), but the problem is in its construction. Therefore, it is important to correctly determine the length of technological breaks after wet processes, but also to keep these time slots in practice. The time of technological breaks for optimizing the moisture value also depends on the concrete additives and their density because the fibres or wires added to the concrete influence the need for care and their concentration is not easy to determine (Gregorová 2016; Gregorová 2017).

There are several methods for measuring moisture in concrete (Vinkler 2016), of which four there are experimentally tested: the gravimetric method, the relative moisture measurement of the concrete, the CM method, and the method by means of a surface hygrometer. Some of these methods are destructive, requiring time-consuming calibration, long measurement, and considerable electricity or radioactive material use. Not all methods are portable and economically suited for smaller companies (Ralph 2004; Wirtanen 2001). However, all methods are based on the base mass method (weighing of wet and dry materials). Destructive and nondestructive methods were selected. The choice of methods was not accidental. The gravimetric method makes it possible to achieve the most accurate results and to create the basic values with which the results of the other methods will be compared. The CM method is a method that is easy to use in practice, with little damage to sampling. Relative moisture measurement by means of hygrometer probe is a method used mainly in the US and England, and it is also a simple method where it is necessary to make negligible boreholes into the structure. The last method chosen is the surface hygrometer measurement. This measurement method is the easiest to realize, low cost, but represents the highest measurement uncertainty.

2 METHODS OF MOISTURE MEASUREMENT

2.1 *Moisture measurement with a surface hygrometer*

The capacitor formed by the spring contact is placed on the building sample. An alternating electric field is applied and depending on the shape, penetrates the sample to a greater or lesser depth. The standard shape of the sensor is the hemispherical elastic contacts. The penetration depth of the electric field of the capacitor into the measured material is between 2 and 5 cm. This depends on the shape and structure of the layer. The water contained in the building material has a significant effect on the electric field. Field changes are therefore evaluated as the water content. In case of inhomogeneity (tubes, mortar in joints, mixed plasters), they cause fluctuations and "mixed values". It is understandable that the measuring instruments do not give the water content directly, but the output is the electrical voltage. This usually changes to a dimensionless indication, "unit" or "number", or by LED's diode indicating that the conditions are "dry", "damp" or "wet". Top devices allow for individual scale adjustment by user. Although it is possible to determine the water content of U from the displayed values, it is necessary to perform a laboratory calibration on the material with its own specific weight by mass method (Kysel 2008).

The method is based on the measurement of electrical conductivity or resistance. It is a frequently used method, because it allows evaluating the IN-SITU measurement locally and is non-destructive when the measured sample is not damaged. The result of the measurement is moisture only on the surface of the structure to a depth of approximately 20 to 30 mm, depending on the material to be measured and used instrument. When the allowable moisture content of the screed is measured by the surface hygrometer, it is necessary to check it by gravimetric method as the measurement at only such depth is insufficient, since only the surface moisture of the screed is measured. However, in the entire cross section, moisture may increase with increasing depth. For example, if convenient moisture content of the screed is measured by a surface hygrometer and the sample (construction) is subsequently covered by a steam impermeable layer (floor layers), the moisture is equalized throughout the sample cross section. From the lower part with the higher water content, the moisture spreads upwards to the lower moisture region and consequently the moisture equalized in the cross section of concrete (Polder 2000, Strangfeld 2016, Strangfeld 2018). It is very likely that this balanced moisture will reach much higher values than those permitted by STN 74 4505. The measurement was also performed according to the manufacturer's instructions and available literature with similar issues (Greisinger, Proceq 2015, Šmotlák 2014, Brown 2005).

2.2 Method of relative moisture measurement

This method of moisture measurement in concrete is used mainly in the US and therefore we have to follow ASTM F2170-11, which prescribes the measurement procedure. ASTM F710 specifies the values of relative moisture that need to be achieved. It is a destructive method, but damage to the structure is minimal as the holes with cross section of less than 7 mm are drilled into it. The advantage is the possibility of retrieving results over a longer measuring period of several weeks or months. The disadvantage is the influence of the results by ambient conditions. The openings for probe positioning can be made in two ways, by drilling or by installation on embedding the sample. The second method was chosen for the experiment, as during the first one, there is a warming of the hole on the bore probe that could affect the results. The depth of probe storage is set at 40% of the depth of tested sample. In this case, the screed is 70 mm thick, so the depth of measurement is 30 mm. Goran Hadenblad introduced the concept of "equivalent depth". According to the author's theory, there is an "equivalent depth" measured from the surface of the structure, in which the moisture is exactly equal to the moisture content that the structure would have in its entire thickness after laying the floor layer (Šmotlák 2014).

2.3 Gravimetric method

This moisture measurement test is prescribed in STN EN ISO 12570, STN 73 1316. It is a destructive test that is carried out in the laboratory. In this test, the moisture content of the sample is given in% by weight of the dried sample. Testing is carried out by comparing the weight of the sample before drying and after drying, the mass difference is the amount of physically bound water (moisture) present in the sample before drying. An advantage is the accuracy of the results. The disadvantage is the difficulty and duration of the test, the necessity of cutting the samples, the necessary laboratory equipment and the impossibility of repeating the measurement from the sample. There may be a change in the results if the samples were made by drilling (sample heating) or by improper sample transport, when the sample's moisture could escape (STN 73 1316).

The test samples should have a volume of at least 0,001 m^3 using the biggest aggregate grain with size up to 22 mm or 0,003 m^3 with aggregate grain size above 22 mm. The wet test sample is weighed. It is then dried to a constant weight at 110 ± 5°C and weighed again. Concrete moisture in % by weight is given by reference to the standard (STN 73 1316).

2.4 Measurement method with CM instrument

One of the extended methods is the carbide method. It is measured by a CM instrument, by simply measuring a sample taken directly from building construction to ensure high accuracy of the measured results. Sample removal is done by hand cutting the part of the concrete, in which the moisture is to be measured. The sample can also be picked up by drilling, but it may distort the measured results by heating the sample during drilling and reducing the humidity in it. Samples are crushed in a steel bowl, weigh the exact amount of sample. The weighed amount is placed in the pressure vessel together with the calcium carbide ampoule, and the precision-weight steel balls are also attached. The pressure vessel closes and, for a few minutes, the calcium carbide ampoule is shattered with steel balls. Pressure in the container starts to rise due to the water reaction in the test sample and calcium carbide. This pressure then rises slowly and it is necessary to repeatedly shake the container to completely stir and release the moisture from the sample. The resulting pressure was read on the manometer, which also includes the scales according to the quantity of the inserted sample. From these scales, the final moisture content was counted off by this test as a percentage of CM. These percentages of CM are not identical to percentages by weight (in the gravimetric test), so it is necessary to convert them to percentages by weight according to Table 1 (Janser 2015).

3 COMPARISONS OF MEASURED VALUES

3.1 *The results*

Measurement by the methods described above was made on the construction of a family house. In total, 3 measurements were made at 28 days, 42 days and the last again for another 14 days. Individual measurements took place over one day in different rooms. In a particular room, two measuring locations were set during the concreting, in which the rods were placed for relative moisture test by means of a hygrometer probe. The moisture and air temperature was also monitored for 56 days in the house. The average moisture in room A was 66.3% and the temperature reached an average of 20.5°C, room B had a higher average moisture level of 68.8%, with the ambient temperature being identical. In Room C there were worse conditions for drying the cement screed and therefore the results show a slower leakage of moisture from the screed, the relative moisture humidity of the air was 70.6% on average and the temperature was 19.0°C. Closer to the rods, the moisture content was measured three times by surface hygrometer. The measurement with surface hygrometer probe was measured as indicated by the standard (ASTM F2170-11). In close proximity to the built-in rods, specimens of cement screed were taken along the entire thickness of the screed. These samples were used to determine the moisture content of the cement screed by gravimetric method and CM instrument. After removal, the samples were placed in steam impermeable bags. Immediately, the test was performed using the CM instrument according to the instructions. The rest of the samples was taken to the laboratory premises and a gravimetric test was completed in accordance with STN 73 1316. The average of these values was determined at measured values by surface hygrometer for comparison with the other methods of measurement.

The measurement results are presented in Tables 1, 2, 3, which show the distribution of the measurement period, the individual rooms and subsequently the 2 measuring location. In the tables, the values are already calculated as a percentage by weight so they can be compared with one of the most accurate of these methods—a gravimetric test. The surface hygrometer listed only verbal phrase according the instructions, as moisture cannot be accurately determined. It only describes whether the concrete is dry (0–5), damp (6–9) or wet (9 or more).

3.2 *Summary*

In the above-mentioned Tables 1, 2, 3, it is possible to monitor the moisture course over time. Based on these results it can be stated that according to the standard STN 74 4505, almost all types of the wear layers besides the layer most sensitive to residual moisture in the screed, can be placed in the A and B rooms on the screed for 56 days. Thus, the wear layers cannot be made of wood and laminate flooring in the room A and B (see Table 4). Room C reached a worse level of moisture after 56 days than the previous two rooms, so PVC, linoleum, rubber and cork flooring cannot be made here (see Table 4). A graph showing the moisture course over time measured by the gravimetric test (see Figure 1) was also produced.

Table 1. Comparison of the measured values of moisture in the cement screed by different methods of testing – 28th daym.

| Measurement method | Cement screed moisture (% by weight) at 28 days | | | | | |
| | Room A | | Room B | | Room C | |
	Probe 1	Probe 2	Probe 1	Probe 2	Probe 1	Probe 2
Gravimetric test	5,84	5,22	5,98	6,02	6,23	6,26
Surface hygrometer	wet	wet	wet	wet	wet	wet
CM instrument	5,70	5,25	5,90	6,00	6,20	6,20
Hydrometer probe	5,50	5,10	5,80	5,80	6,00	6,10

Source: Author.

Table 2. Comparison of the measured values of moisture in the cement screed with different methods of testing – 42nd daym.

Measurement method	Cement screed moisture (% by weight) at 42 days					
	Room A		Room B		Room C	
	Probe 1	Probe 2	Probe 1	Probe 2	Probe 1	Probe 2
Gravimetric test	4,22	4,18	4,32	4,35	4,65	4,57
Surface hygrometer	damp	dry	damp	damp	damp	damp
CM instrument	4,20	4,15	4,25	4,25	4,50	4,45
Hydrometer probe	4,05	3,95	4,20	4,25	4,35	4,40

Source: Author.

Table 3. Comparison of the measured values of moisture in the cement screed with different methods of testing – 56th day.

Measurement method	Cement screed moisture (% by weight) at 56 days					
	Room A		Room B		Room C	
	Probe 1	Probe 2	Probe1	Probe 2	Probe 1	Probe 2
Gravimetric test	3,42	3,33	3,47	3,49	3,63	3,59
Surface hygrometer	dry	dry	dry	dry	damp	damp
CM instrument	3,40	3,30	3,45	3,45	3,60	3,55
Hygrometer probe	3,35	3,25	3,35	3,40	3,50	3,45

Source: Author.

Table 4. Maximum permissible moisture content of cement screed in percentage by weight at the time of application of the wear layer [STN 74 4505].

Wear layer	Cement screed, concrete
Stone and ceramic tiles	5,0%
Cement-based casting floor	5,0%
Synthetic casting floor	4,0%
Steam permeable fabric	5,0%
PVC, linoleum, rubber, cork	3,5%
Wooden floors, parquet, laminate flooring	2,5%

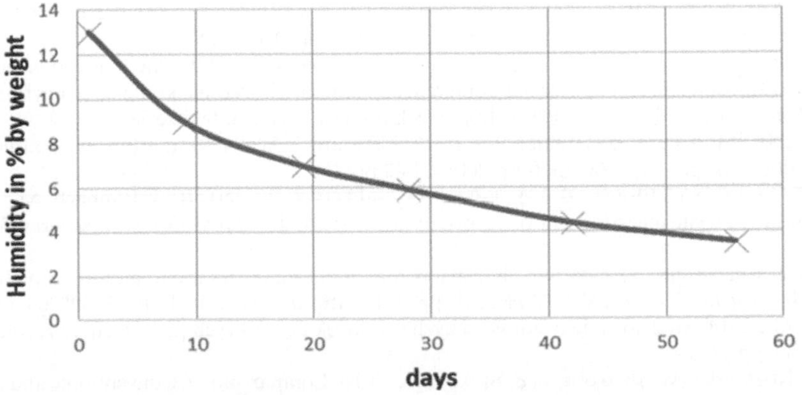

Figure 1. The course of moisture measured by the gravimetric test in time.

4 CONCLUSIONS

Together 6 experimental probes were performed at different locations of the cement screed. The moisture of the cement screed has been determined by four different methods, which are currently used these days. The results obtained were compared. Measurements were performed in 28, 42, and 56 days. The drying process of the screed, as well as the indoor moisture was relatively identical for all rooms. The analysis of the results shows that similar results have been achieved by the different ways of determining the moisture of the screed and that the methods can be used to determine the moisture in the base and to determine the length of the technological break. A surface hygrometer is suitable for rapid measurement. Other methods are suitable for exact determination of humidity. In terms of price, instrumentation and complexity of testing, we recommend the CM method.

REFERENCES

Bentz D.P. 2005. *CEMHYD3D: A three dimensional Cement Hydratation and Microstructure Development Modelling Package, Version 3.0. Building and Fire Research Laboratory*. National Institute of Standards and Technology. Gaithersburg. Maryland.

Breugel K. van, 1995. Numerical simulation of hydratation and microstructural development in hardening cement-based materis (I) Theory. In *Cement and Concrete Research* vol. 25 no. 2. Pergamon. 319–331.

Brown, R. & Kanare, H. 2005. Desiccant dehumidification puts slab drying on the fast track. In *Concrete Construction*. Pg. 86, ProQuest Central.

Černý, R. 2009. Time-domain reflectometry method and its application for measuring moisture content in porous materials: a review, In *Measurement* 42 (3) 329–336, https://doi.org/10.1016/j.measurement.2008.08.011

Dohnálek, J. & Tůma, P. 2008. Nové znění normy ČSN 74 4505 Podlahy – Společná ustanovení. In: *Konference Podlahy 2008*, Praha 25.–26. září 2008, pp. 141–147, vydal Betonconsult s.r.o.

Gregorová V. Ledererová M. & Štefunková Z. 2016. *Investigation of Influence of Recycled Plastics from Cable, Ethylene Vinyl Acetate and Polystyrene Waste on Lightweight Concrete Properties*. In Procedia Engineering: 18th IC Rehabilitation and Reconstruction of Buildings, CRRB 2016. Brno, Czech Republic, 24–25 November 2016. Vol. 195, (2017), online, pp. 127–133. ISSN 1877–7058.

Gregorová, V. & Štefunková, Z. 2016. *Properties of Lightweight Concrete Made from Waste Polystyrene and Polypropylene*. In Construmat 2016 – Conference on Structural Materials: [elektronický zdroj] sborník příspěvků z 22. mezinárodní konference o stavebních materiálech. Stará Živohošť, ČR, 1. – 3. 6. 2016. 1. vyd. Praha: České vysoké učení technické v Praze. pp. 69–76. ISBN 978-80-01-05958-6.

GREISINGER electronic GmbH, 2014. *Operating manual for moisture indicator for wood and buildings GMI 15*. H42.1.01.6C-05.

Hanko, M. & Ďubek, S. Údržba stavebných konštrukcií budov v systéme facility managementu. In *Buildustry*, 20. pp. 20–22.

JANSER, s.r.o., 2015. *Návod na použití k CM-přístroji*.

Jennings, H.M. & Tennis, P.D., 2000 A model for the microstrucutre of calcium silicate hydrate in cement paste. In *Cement and Concrete Research* vol. 30, Pergamon 2000, pp. 855–863.

Kantová, R. & Motyčka, V. 2014. Construction Site Noise and its Influence on Protected Area of the Existing Buildings, příspěvek na konferenci Elektronický sborník konference enviBuild. ISSN 1022-6680, ISBN 978-80-214-5003-5, Trans Tech Publications Ltd, Switzerland.

Kiseľ, D. 2008. Meranie vlhkosti materiálov. In: AT&P journal. Available at: https://www.atpjournal.sk/buxus/docs/casopisy/atp_2008/pdf/atp-2008-09-40.pdf.

Polder, R. Andrade, C. Elsener, B. Vennesland, Ø. Gulikers, J. Weidert, R. & Raupach. M. 2000. Test methods for on site measurement of resistivity of concrete. In *Materials and Structures*, 33(10): pp. 603–611

Ralph, D. & Burkinshaw, F. 2004. Which instruments should surveyors use to monitor moisture condition?. In *Structural Survey*, Vol. 22 Issue: 1, pp. 7–19, https://doi.org/10.1108/02630800410530891

Šmotlák, M., 2014. Modelovanie technologických prestávok po mokrých procesoch. Dizertačná práca, Svf STU, pp. 60–64.

Špak, M. Kozlovská, M. Struková, Z. & Bašková, R. 2016. Comparison of Conventional and Advanced Concrete Technologies in Terms of Construction Efficiency. In *Advances in Materials Science and Engineering*. Vol. 2016 (2016), pp. 3729–3729. ISSN 1687-8442.

STN 73 1316: 1989. *Stanovenie vlhkosti, nasiakavosti a vzlínavosti betónu.*

STN 74 4505:1987. *Podlahy. Spoločné ustanovenia.*

Strangfeld, C. Johann, S. Müller, M. & Bartholmai. M. 2016. Moisture measurements by means of rfid sensor systems in screed and concrete. In *8th European Workshop on Structural Health Monitoring.* Bilbao. Spain. Unpublished.

Strangfeld, C. Prinz, C. Hase, F. Kruschwitz, S. 2018. Humidity data of embedded sensors, sample weights, and measured pore volume distribution for eight screed types, In *Constr. Build. Mater.* 1–3, https://doi.org/10.4121/uuid:d2ba436f-78c0-4105-8a1f-5422fcb37851.

Švajlenka, J. & Kozlovská, M. 2018, Houses Based on Wood as an Ecological and Sustainable Housing Alternative – Case Study. In *Sustainability.* https://doi.org/10.3390/su10051502.

Tůma, P. 2009. Podlahové potery a najčastejšie príčiny ich porúch, In: *ASB*, available at: https://www.asb.sk/stavebnictvo/stavebne-materialy/podlaha/podlahove-potery-anajcastejsie-priciny-ich-poruch.

Vinkler, M. & Vítek, J.L. 2016. Drying concrete: experimental and numerical modeling. In *J Mater Civil Eng* 28 (9) 1–8, https://doi.org/10.1061/(ASCE)MT.1943-5533.0001577.

Wirtanen L. 2001. *The moisture content and emissions of ground slabs and floors between storeys subjected to a moisture load (in Finnish).* Licentiate's thesis. Helsinki University of Technology, Espoo, Finland. 161 pages.

Standards of the universal design to stabilize the housing quality

A. Bílková, P. Kocurová & R. Zdařilová
VSB-Technical University of Ostrava, Ostrava, Czech Republic

ABSTRACT: Current international research shows that most of the population wants to live in private house known community. In order to ensure this requirement, the accessibility of living space itself must be, both in its internal disposition and in its outer space, fully and correctly designed. The article focuses on the design of residential buildings using the principles of the universal design. For the apartment itself, it is primarily about the adaptability of the space in order to minimize or simplify the necessary adaptations resulting from the changing needs of its users throughout their lifetime. A few principles of universal design were monitored, which were implemented on the internal environment of the apartment, along with barrier-free requirements. Defining specific requirements can make a significant contribution to enhancing the adaptability and the quality of housing, especially in the area of flexibility and universality of the dwelling and the subsequent laying down of basic technical standards.

1 INTRODUCTION

Recent international researches have shown that the majority of population wishes to continue living in their existing houses or communities where they have established close relationships with their families and neighbors. Other researches show that, at the present time, 40 per cent of the European population are people with disabilities, seniors or persons with temporary reduced mobility, such as those with a child in pram, pregnant women, injured (or convalescing) people, passengers with heavy luggage, etc, as described by Navrátilová (2011).

The issues concerning the improvement of living environment adaptability, apartment modification and worthy and adequate living should be addressed if we are to meet the needs of all users. The accessibility of living environment is a basic precondition for the creation of an environment of equal value, usable by any user (resident) regardless of their gender, age or health condition.

The creation of an adaptable, flexible and universal residential environment providing for simple and minimum alterations is a basis and solution. It also accepts the universal design principles and concepts, in connection with housing policy priorities and other documents of the respective states.

Many centers and organisations have been concerned with these housing requirements since the 1990's, applying their findings in the above mentioned universal design (also, design for all; in the Czech legislation and documents, the term of life-long housing can frequently be found, too), as described by Zdařilová (2011). The basic idea of the universal design is to enable adequate and comfortable life to all people for the duration of their lifetimes, regardless of their physical or mental conditions. At the same time it appears that the products and environment designed on the basis of the universal design principles are equally comfortable even for those facing no handicap at the first sight. Therefore, working with the universal design principles should become a natural part of many professions across all fields.

2 OVERVIEW OF CONTEMPORARY DESIGN DEVELOPMENT

Universal design roots can be found as early as in the first half of the 20th century, in the period of functionalism. In Scandinavian countries, a so called 'ergonomic design' appeared in the 1950's, the best known representative of which being Finnish architect and designer Alvar Aalto. During the 1960's, the first civic movements campaigning for the disabled rights appeared.

2.1 Universal design—design for all

What is the meaning of universal design? The universal design, or design for all (Fig. 1), is the approach to the products, buildings and environment designing that has developed from barrier-free concepts. Focused on all people whatever their age, health condition, physical abilities, nationality, culture, religion or social background, it aims to meet the needs of peoples' diversity and social equality and to create equal opportunities for people in all spheres of life. In many areas, we are often surrounded by universal design without even being aware of it.

Various centers and organisations all over the world have been concerned with the design for all since the 1990's. These include, for example, the Center for Universal Design at the North Caroline State University in Raleigh (since 1989), European Institute for Design and Disability (EIDD) or Design for All Europe, European Design for All eAccessibility Network (EDeAN), and some others.

Generally, we can say that the changes in society and development in science and technologies were the key factors for the universal design philosophy emergence. The factors can be subdivided into the following five groups, as described by Navrátilová (2011):

a. Demographic development of society (rise in population or ageing of society) – With the population increase, the number of disabled or otherwise handicapped persons increases. (In 2000, there was 10.3 million of population of which 1/10 disabled and 14% of age over 65.)
b. Social factors – Changes in people's lifestyles in the context of globalisation seem to be the most important factor. Everything is faster, distances shorter, and elderly generation wants to stay active in old age.
c. Technological development – First and foremost it is the development in medical and rehabilitation technologies. Much more people are surviving formerly fatal diseases or severe injuries, however, often with permanent consequences.

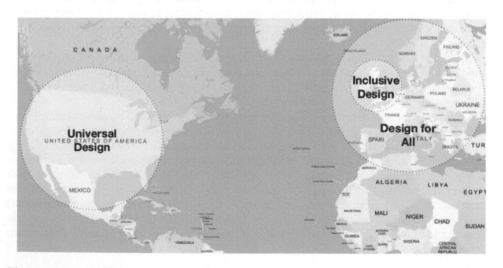

Figure 1.   Geographical representation of the term use.

d. Legislation (civic movements, laws) – In Czechoslovakia, Decree No. 53/1985 Coll., on general technical requirements for the use of buildings by people with impaired mobility, was the first comprehensive legal act dealing with barrier-free access. Nowadays, Decree No. 398/2009 Coll., on general technical requirements for the barrier-free use of buildings (as amended) is in effect, together with its relevant methodologies, such as: The Methodology of Accessible Housing Environment—LIFE-LONG HOUSING, developed within the frame work of research project WD-05-07-3, "Regional Disparities in the Availability and Affordability of Housing, their Social and Economic Effects and Proposals for the Regional Disparity Reduction Measures" of 2011.

2.2 *Universal design ideas and principles*

Enabling fully enjoyed and comfortable life to all people, inclusive those with handicaps of mental, sensory or physical character, whether permanent or temporary, is the basic idea of the universal design. The key intention of universal (inclusive) design is to design such products and environments that will be functioning properly for all the people, as described by Váňová (2009).

The 7 Principles of Universal Design (hereinafter the "UD") were defined in 1997 in the Center for Universal Design. They are based not only on the needs of the persons with physical disability or sensory impairments, but, most importantly, on the demand for comfortable life of all people's groups. Sometimes it may be difficult to define the handicap since anybody, e.g. a tall or short person, or that having no knowledge of any foreign language may feel as handicapped, as described by Navrátilová (2011).

1. Equitable use: The UD is useful—It provides the same means of use for all users, avoids segregation or stigmatization of any users, provides for privacy, security and safety equally to all users, and is appealing to all users.
2. Flexibility in use: The UD should accommodate a wide range of individual preferences and abilities, provide choice in methods of use, suit the right—as well as left-handed, and facilitate the user's accuracy and precision.
3. Simple and intuitive use: The UD is easy to understand, regardless of the user's experience, knowledge, language skills, or current concentration level; it is intuitive.
4. Perceptible information: The UD communicates effectively important information to the user, regardless of ambient conditions or the user's sensory abilities, using various modes for the presentation of essential information (pictorial, verbal, tactile).
5. Tolerance for error: The UD minimizes hazards and adverse consequences of accidental or unintended actions.
6. Low Physical Effort: The UD provides for an efficient and comfortable use with a minimum of fatigue, minimizes repetitive actions and sustained physical effort, and allows the user to maintain a neutral, natural body position.
7. Size and space for approach and use: Appropriate size and space is provided for approach, reach, manipulation, and use regardless of user's body size, posture or mobility, together with a clear line of sight and reach to all important components for any seated or standing user, and adequate space for the use of assistive devices or personal assistance.

3 TARGET GROUPS FOR THE APPLICATION OF UNIVERSAL DESIGN

The philosophy and practical solutions of universal design are oriented to all people regardless of their age, size, physical or mental abilities, or cultural and religious background. For all that, there are groups of people for which the solutions based on the universal design principles do not mean just some sort of luxury or extra comfort but essentially determines how full and independent life is such person able to lead.

This refers especially to the disabled people or people with reduced mobility or orientation, or people with temporary handicap, such as persons with a baby pram or passengers with heavy luggage. These groups of people usually have bigger spatial demands but, at the same time, reduced reach.

The handicapped persons can be grouped depending on the type of needs for the barrier-free use of environment. This mainly refers to people with intellectual or mental impairments and disabled people or persons with partially reduced mobility, or combination. Disabilities include mobility, visual and hearing impairments. The mobility impairments broke down to severe impairments (wheel-chaired persons) and reduced mobility (such as persons walking on crutches), visual impairments (blind and partially-sighted persons), and hearing impairments (deaf and hard-of-hearing persons), as described by Zdařilová (2007).

4 UNIVERSAL DESIGN APPLICATION FOR THE IMPROVEMENT OF THE LIVING ENVIRONMENT ADAPTABILITY

Universal design provides simple solutions based on anthropometry, with respect to human physiology and psychology. The products and environment designed on the basis of universal design principles have proved to be appealing and comfortable for all people including those facing no handicap at first sight.

In apartments, the goal of space adaptability is to minimize adaptations for the changing needs of users during their lifetime. On the basis of these assumptions, the universal design concept and principles can be developed, followed by the specification of basic technical standards, thus contributing significantly to the increase in the environment adaptability and housing quality stabilization, especially as for the apartment flexibility and universality, as described by Šestáková (2012).

The research was focused on the residential buildings typology and design with the application of universal design principles to full extent, with a minimum of errors usually associated with such designing. The above described principles were examined (see 2.2).

Table 1. Application of universal design principles on the currently available housing.

Application of universal design principles on the currently available housing	
Universal design principles	Description for the application to currently available housing.
1. Equitable use—accessibility	The apartment/house accessibility from its inside area, and especially from outside (from the entire public space, pavement, corridor, lift, etc.). The continuity and complexity of performed adaptations are emphasized.
2. Flexibility in use	The apartment arrangement should facilitate a wider range of individual preferences and abilities and provide choice in the methods of use and functions of single spaces.
3. Simple and intuitive use	In relation with principle (2.). Single spaces should be so arranged that any user's problems with orientation that might complicate his/her everyday activities are avoided.
4. Perceptible information	This principle is especially important in mass types of residential buildings with a big share of common areas and more floor levels.
5. Tolerance for Error	The apartment should be furnished and furnishing elements so arranged that they are usable without the need for high concentration, thus avoiding irritating or hazardous situations and increasing the living environment safety in general.
6. Low physical effort	Any excessive physical effort means complications for a user with physical disabilities (whether temporary or permanent). Care should be taken that the apartment interior includes minimum pieces of equipment requiring sustained physical effort in handling, or such effort is eliminated with the help of advanced technologies.
7. Size and Space for Approach and Use	Consists particularly in ensuring approach, reach, manipulation, and use regardless of user's posture, mobility, view, or size.

Table 2. General determining requirements for the universal living standards in apartment interior environments.

General determining requirements for the universal living standards in apartment interior environments	
General requirements	Description for the application to apartment interior environment.
1. Universality	The need to live in an unrestrictive and adaptable environment. The user does not experience any change in housing quality with change in his/her marital status, sickness or old age. The same apartment is usable with minimum or even no adaptations, regardless of variations in life situations.
2. Types of buildings	New building, redevelopment, detached one-family house, etc. The type of building to high extent affects possible apartment variability and changes in interior.
3. Exterior-interior connection	Design of elements and arrangement of interior for persons with specific needs is important for meeting the needs of these users and other target groups.
4. Apartment layout	Relationships and connections between single rooms, clearly distinguished private and non-private spaces and fluent transition between these spaces.
5. Floor area	Defining requirements and demands for single apartment rooms and specification of needed circulation and manipulation areas.
6. Furniture and furnishings	Specification of any necessary apartment furnishings, accessibility of furnishing elements and the elements handling in relation to the group of users. Accent on as much use variability as possible.
7. Surface and furniture materials	Accent on safety, simple, healthy, and easy use and maintenance.

Generally, attention was paid to spatial needs and the arrangement of areas such as entrance, entrance hall, kitchen, toilet, bathroom, types of buildings (a new building, redevelopment, detached one-family house, apartment house, or apartment); flexibility and universality; exterior-interior connection; apartment layout; furniture and furnishings; and materials and surfaces.

The goal of the research was to define basic technical criteria for the creation of conditions capable of ensuring the needs for worthy and adequate housing usable by any user, regardless of his/her age, gender or disability. In the building assessment, accent was put on the barrier-free approach in connection with meeting the universal design principles. Attention was further paid to the specification of parameters as necessary for the improvement of the environment adaptability, flexibility and universality.

4.1 *Identification of issues and application of the universal design principles on currently available housing*

With the changes in society, the perception of the disabled people's needs has enhanced in our country (as well as all over the world). After examining the needs of all people's groups, basic universal design parameters and rules have been specified. The universal design principles need to be implemented in their complexity.

Based on the examinations and comparisons, within the framework of the research, solutions for the interior environment of adaptable apartment have been defined for every universal design principle. Some of the principles mingle or are impossible to be classified unambiguously. For this reason, the below tabled descriptions are to be taken as general only.

The application of basic universal design principles to housing (to apartment interiors in this case) has revealed a few general fields to which consideration needs to be given in order that universal design standards for the stabilization of housing quality may be improved.

311

5 CONCLUSION

As mentioned above, the creation of flexible and universal living environment is the basis and solution for the enhancement of living environment adaptability. Such environment makes simple and minimum adaptations possible, thus responding to people's variable and changing needs), as described by Zdařilová (2011).

The apartment accessibility within its interior layout and as well as from its exterior is a basic condition. A few generally defined fields in the universal design standards for the stabilization of housing quality can have specific parameters further defined. These particularly include the floor area of the individual/respective spaces (based on users' needs and limitations), type of the building (a new building, redevelopment, detached building, apartment building, etc.), selection and modification of the existing apartments, universality (a user experiences no change in housing quality with the change in his/her marital status, sickness or old age), the connection between the interior and the exterior, the apartment layout (relationships and connections between the individual rooms, clearly differentiated private and non-private spaces and fluent transition between these spaces), furniture and furnishing, and surface materials.

Detailed specification and definition of the above described requirements can highly facilitate in the further development of new adaptable housing or the adaptation of that existing, making it possible to practise reasonable planning and application of flexibility and universality, that is, housing for all people regardless of their life cycle phases or situations, and thus enhance living comfort not only in the newly developed residential buildings.

ACKNOWLEDGMENT

The work was supported by the Student Grant Competition VŠB-TUO. Project registration number is SP2018/129.

REFERENCES

Navrátilová, B. 2011. Univerzální Design. Prague: Czech Technical University. Project FRVŠ 2011. [online]. 6. 5. 2009. [Cited 15 January, 2018]. Available from: <http://univerzalnidesign.sweb.cz/>.

Navrátilová, B. 2011. Univerzální Design a přístupnost staveb veřejné hromadné dopravy osobám s tělesným nebo smyslovým handicapem. Prague: Czech Technical University. Project FRVŠ 2011.

Šestáková, I. 2011. Bydlení pro lidi (nejen) se zdravotním postižením. Prague: Ministerstvo práce a sociálních věcí České republiky.

Váňová, L. 2009. Universal Design in Germany. Czech design [online magazine]. 10. 11. 2009. [Cited 1 July, 2017]. Available from: <http://www.czechdesign.cz/temata-a-rubriky/univerzalni-design-v-nemecku>.

Zdařilová, R. 2007. Bezbariérové užívání staveb základní principy přístupnosti – TP 1.4 technické pomůcky k činnosti autorizovaných osob. Prague: Informační centrum ČKAIT.

Zdařilová, R. 2011. Metodika přístupného prostředí bytového fondu – Celoživotní bydlení. Prague: Informační centrum ČKAIT.

Advances and Trends in Engineering Sciences and Technologies III – Al Ali & Platko (Eds)
© *2019 Taylor & Francis Group, London, ISBN 978-0-367-07509-5*

Examination of electricity production loss of a solar panel in case of different types and concentration of dust

I. Bodnár, P. Iski, D. Koós & Á. Skribanek
University of Miskolc, Miskolc-Egyetemváros, Hungary

ABSTRACT: In this study, energy generation of a solar panel in case of different surface contaminants is experimentally examined. The energy production of a solar cell is significantly influenced by the contamination of the panel's surface. Not only objects in the environment, but pollutants on the panel's surface also can create shadow effect. The energy production and the efficiency of a solar panel is experimentally investigated in case of different types of contamination, such as leaves, dust and simulated bird excrements. Measurements are done in case of different polluted areas and amount of pollution. A connection between the power generation and the contamination is observed. Following the observation, it is possible to determine the loss of energy production, caused by the surface contaminants, while the optimal cleaning period can be determined.

1   INTRODUCTION

In today's fast-paced world, people's energy consumption has become enormous. Over the last century, the energy demand of an ordinary person has grown about fivefold, which is attributed to the spread of machinery and electronics. In order to avoid the excessive consumption of our energy, the increase of efficiency became important in both energy production and consumption. The operating efficiency of a solar panel is influenced by the installation environment and the weather conditions. Among these factors, the intensity of illumination, the temperature of the solar panel's surface and its ambient temperature together with the surface pollution of the solar panel and its shadow effect are the most significant ones. In this research, We present the effect of a solar panel's surface temperature on its electrical parameters. The main goal of the research is to establish a correlation between the surface contamination of the solar panel and its electrical parameters.

2   ENVIRONMENTAL CIRCUMSTANCES AND SURFACE POLLUTION

Apart from the maximum efficiency of the technology and the use of materials, the external factors that may occur when using the finished solar cells can also reduce the efficiency of the modules. Polluted cells, temperature and load levels can be classified into these environmental conditions. Solar modules are virtually maintenance-free, but we do not neglect the efficiency of the surface contamination during operation. The annual electricity loss can be significant, its value can reach up to 17%. The characterization of the process is not simple, because many factors influence it. In the followings, we will outline the most important parameters that are related to the surface contamination of the solar panel and the degree of efficiency deterioration [Abderrezek et al. (2017), Malik et al. (2003), Bhattacharya et al. (2015), Gürtürk et al. (2018)].

2.1   *Pollutants and resources*

The most commonly occurring impurities in nature are the followings:

– bird excrements,
– dust, pollen, sand- and soil particles,
– leaves on the surface of the solar panel.

Surface contamination by human activity:

– during the operation of industrial plants the deposition of airborne pollutants,
– carbon black, fly ash from population heating,
– dust from agricultural or other human activities,
– dirt from road traffic (for example, rubber, carbon black).

2.1.1 *2.2 Forms of deposits on the surface of the solar cell*

Local impurities (such as bird excrements) may not have a negligible influence on the solar energy performance. For crystalline solar panels, the presence of this type of contamination is even more critical, since the efficiency of a unit, consisting of cells in a row is significantly reduced by the spot-like contamination. Powder or other contaminants that cover the surface of the entire solar panel also have a detrimental effect on the proper operation. These deposits can accumulate on the others due to inadequate cleaning, over the years becoming thicker and more durable "blanket" layers on the surface of the solar panel. In the corners, accumulation of pollutants can be extremely important because the cleaning effect of rain on these surfaces does not prevail. Thick deposits can cause the panel's overheating, because they act as heat insulators. If a cell is damaged, the cells that are in series, can also be cut off from power generation, so the efficiency and lifetime of the solar panel are reduced. Experiences had shown that the particle size of the pollutant is important for the change of efficiency. The connection can be seen in Figure 1. The solar panel is not cleaned by rain. A higher rain can wash the surface sufficiently, but a mild shower can have additional negative effects. A small amount of precipitation on the deposited powder layer can form a sludge layer that can further damage the efficiency of the solar panel [Abderrezek et al. (2017), Bhattacharya et al. (2015)].

2.1.2 *Correlation between the energy production and contamination*

The value of contamination is described by the climatic conditions and the impacts of polluting sources. These factors can be extremely varied, depending on the type of climatic conditions and impurities. Over the last decades, global warming has caused climatic belts to shift and more frequent extreme weather conditions (extreme rainfall, wind storms, high temperature fluctuations, etc.). As a result, pollutants also appeared in climatic zones where they had fewer characteristics (eg desert sand in Europe) [Adinoyi et al. (2013), Malik et al. (2003)].

Categories of Climatic Characteristics (K):

K1. Moderate climatic, climatic climate:
– Annual precipitation is 800 mm or above.
– The distribution of precipitation over the year is relatively even.
– (for example, a significant part of Germany's territory).

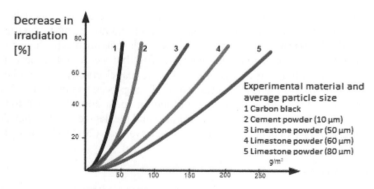

Figure 1.   Decrease in irradiation different pollutants.

K2. Dry summer continental climate:
- Annual precipitation is 450–550 mm.
- The summer semester is dry.
- (This is the southern part of Hungary and Romania).
  K3. Extremely dry, semi-desert, desert climate:
- The annual precipitation is below 300 mm.
- (This includes Iraq, Arizona, Saharan Africa).
  K4. Seaside climate:
- Annual precipitation is 800 to 900 mm or more.
- Strong salt vapor, polluting, corrosive.
- (such as Malta).

Dust intensity categories (S):
S1. Solar Powered System with Moderate Pollution:
- the number of polluting sources and their impacts are small, and significant amounts of rainfall in the summer season are used to scrub most of the impurities.
  S2. Medium Impact Areas:
- the solar plant is located away from sources of pollution (industrial, agricultural, busy roads).
  S3. Intensive contamination areas:
- close to the solar system, there are several pollutants that can cause significant contamination. (Significant industrial and agricultural activity, road traffic, bird trapping zone). The cleaning effect of rain has only a slight effect.
  S4. Extremely high contamination areas:
- powerful sources of pollution from many sources. These may be, for example, semi-arid, desert areas with frequent sand storms. (Large amounts of sand deposited on solar cells, with significant power losses).

In Table 1. The estimated annual energy losses as a percentage can be seen while taking two factors into account.

2.1.3 *Reduction in solar power and energy production*

The power and energy production of the solar cell can be described by one function. The effects of surface impurities appear indirectly in the function. Firstly, we thought that the proportion of the solar panel's useful surface as a result of the surface contamination will have a similar proportion of energy loss. However, this is not entirely true. The power of the solar cell is the multiplication of the current and the output voltage. The amperage of the solar panel is linearly influenced by the illumination. Due to surface contamination, the useful surface of the solar cell decreases linearly, however, each impurity has partial light transmittance, so the decrease is not always linear. The output voltage of the solar panel depends on the temperature. If impurities are seized on the surface of the solar cell, the heat radiation decreases, resulting in a further heating of the solar cell surface. The voltage of the solar panel terminal is in the ratio of temperature, so lower voltage can be measured on the warmer solar panel. While the current increases as a result of temperature increase, however, this increase is negligible because the constant of the current is one order of magnitude smaller

Table 1. The conditions of climatic conditions and pollution intensity.

Categories of pollution intensity	Categories of climatic conditions			
	K1	K2	K3	K4
S1	2.5% – 4.0%	4.0% – 6.5%	4.0% – 6.5%	2.5% – 4.0%
S2	2.5% – 4.0%	7.5% – 10.0%	7.5% – 10.0%	4.0% – 6.5%
S3	4.0% – 6.5%	8.5% – 12.0%	8.5% – 12.0%	7.5% – 10.0%
S4	7.5% – 10.0%	9.0% – 14.0%	11.0% – 17.0%	7.5% – 10.0%

Table 2. Electrical parameters of the solar panel.

Parameter	Symbol	Value	Measurements
Year of manufacture	–	2008	–
Intensity of illumination	I_{ill}	861	W/m^2
Peak Power	P_{max}	85	W
Max. power current	I_M	4.88	A
Max. power voltage	U_M	17.45	V
Short circuit current	I_{SC}	5.40	A
Open circuit voltage	U_{OC}	21.20	V
Nominal fill factor	φ	0.74	–
Serial resistance	R_s	0.0035	Ω
Parallel resistance	R_P	10,000	Ω
Number of serial connected cells	N_S	18	piece
Number of parallel connected cells	N_P	2	piece
Temperature co-efficient for P_{max}	K_{PM}	–0.391	W/°C
Temperature co-efficient for I_{sc}	K_{ISC}	0.001674	A/°C
Temperature co-efficient for U_{oc}	K_{UOC}	–0.073776	V/°C
Percentage Temperature co-efficient for P_{max}	μ_{Pm}	–0.460	%/°C
Percentage Temperature co-efficient for I_{sc}	μ_{Isc}	0.031	%/°C
Percentage Temperature co-efficient for U_{oc}	μ_{Uoc}	–0.348	%/°C
Efficiency (maximal power)	η	12.75	%
Nominal operating temperature	T_N	25	°C

Figure 2. The experimental composition.

than the voltage constant. So, the effect of surface pollution is further reduced [Malik et al. (2003), Bhattacharya et al. (2015)].

3 THE EXPERIMENTAL COMPOSITION

The solar panel was placed on a table with the same size during the measurement [Siddiqui et al. (2016)]. The temperature of the solar panel's surface was measured by a four channel YC-747D digital thermometer. The four sensors were placed on four different parts of the solar panel. The previous measurements proved that the back surface of solar panels heat up likewise the absorber surface. The voltage and current of the solar panel was measured at the same time by a Protek DM-301 and a METEX M-365OD digital multimeters. The temperature of the lit solar panel was around 76.6°C. The ambient temperature was 27.7°C.

Figure 2. Illustrates the measurement composition. The left figure shows a solar cell contaminated with 50 g/m² of sand. In the upper figure there is cement with a concentration of 70 g/m² and in the lower figure a solar cell contaminated with 100 g/m² of wood ash. During the measurements, it has been found that low density ash may have a significantly larger surface area than the sand of higher mass density at the same concentration [Ndiaye et al. (2018), Gürtürk et al. (2018)].

4 MEASURMENTS RESULTS

During our laboratory measurements, three different powder-based contaminants were examined. These were: cement, sand and wood. Our basic hypothesis was that due to larger particle size of the smaller dust particles or the lower density powder, the covered surface of the solar cell is larger in proportion and therefore results in higher power losses. Figure 3 shows the effect of the decrease in performance as a function of the dust concentration on the surface of the solar cell. It can be observed that coverage with sand and tree ash has a near linear effect on the solar power drop, while in case of cement, there is nonlinearity. On the basis of the measurement results it can be said the case of surface dust concentration of 50 g/m², the cement is reduced by 20.91%, sand by 12.95% and ashes by 35.81%.In the case of surface dust concentration of 100 g/m², the power of the solar cell decreases by almost two thirds (63.89%). For Cement contaminated solar cells, this value is 40.89% and sand is only 22.41% of the power loss. The hypothesis has been proved by our measurements. Similar results can be observed in the works of Rao et al. (2013) and Ndiaye et al. (2018).

The fourth figure illustrates a decrease in irradiation. Reduction in solar power is due to two factors: surface roughness and temperature increase. The surface shield reduces the amount of irradiation, reducing the current delivered by the solar panel, while, on the other hand, the solar panel can drop heat and therefore it heats up, causing the voltage to drop. Reduction in irradiation can be determined from the current data. In the present results, the trend shown in Figure 3 is also observed. Cement and wood-ash have a higher irradiation effect than sand. The most important reason for the loss of performance is therefore the magnitude of the blindness and the temperature increase, which is less than one order of magnitude, so negligible.

The fifth figure shows the increase in the surface temperature of the solar cell as a function of the dust concentration. The temperature increase can be approached with the polynomial in

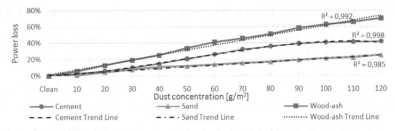

Figure 3. Power reduction with Sand, Cement and Wood-ash contaminated solar cell.

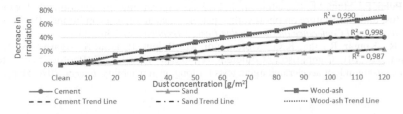

Figure 4. Decrease of irradiation at various pollutants and concentrations.

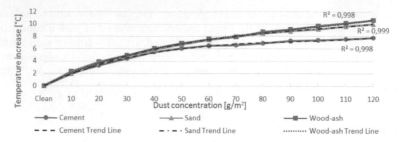

Figure 5. Rate of increase in temperature depending on the concentration of dust.

case of the three port types. It can be observed, that the greatest temperature increase is caused by the tree ash. At a dust concentration of 100 g/m², the solar cell temperature increased by 9.67°C. This is followed by sand at 9.17°C. The lowest temperature increase at cement is 7.36°C. Our hypothesis has been confirmed. Because sand has a larger particle size than cement, and wood has a lower density, it forms a thicker thermal insulation layer than the solar cell surface, which is the reason, why the solar cell temperature increased more [Ndiaye et al. (2018)].

5 CONCLUSIONS

Summarized, it can be said that the solar panel's power dropped and temperature increased significantly in all three types of dust examined. The drastic differences between the individual types of ports can be traced back to the difference in particle size and density. The smaller particle size of sand and the lower density of the tree have a higher surface area and thus have a stronger effect on the solar cell's performance and losses at the same concentration. As a result, the energy produced by the solar cell is reduced and the payback time is extended. Laboratory tests have also shown that the temperature of the dusty solar cell increases, which results in a long-term loss of life durability, so removal of impurities is a factor in the solar cell's operation.

REFERENCES

Abderrezek, M., Fathi, M. 2017. Experimental study of the dust effect on photovoltaic panels' energy yield. *Solar Energy* 142: 308–320.

Adinoyi, M.J., Said, S.A.M. 2013. Effect of dust accumulation on the power outputs of solar photovoltaic modules. *Renewable Energy* 60: 633–636.

Bhattacharya, T., Chakraborty, A.K., Pal, K. 2015. Influence of Environmental Dust on the Operating Characteristics of the Solar PV Module in Tripura, India. *International Journal of Engineering Research.* 4(3): 141–144.

Gürtürk, M., Benli, H., Ertürk, N.K. 2018. Effects of different parameters on energy – Exergy and power conversion efficiency of PV modules. *Renewable and Sustainable Energy Reviews.* 92(9): 426–439.

Malik, A.Q., Damit, S.J.B.H. 2003. Outdoor testing of single crystal silicon solar cells. *Renewable Energy* 28:1433–1445.

Ndiaye, A., Kébe, C.M.F., Bilal, B.O., Charki, A., Sambou, V., Ndiaye, P.A. 2018. Study of the Correlation Between the Dust Density Accumulated on Photovoltaic Module's Surface and Their Performance Characteristics Degradation. *Innovation and Interdisciplinary Solutions for Underserved Areas.* pp. 31–42.

Rao, A., Pillai, R., Mani, M., Ramamurthy, P. 2013. An experimental investigation into the interplay of wind, dust and temperature on photovoltaic performance in tropical conditions', *Proceedings of the 12th International Conference on Sustainable Energy Technologies.* 2303–2310.

Siddiqui, R., Kumar, R., Jha, K.G., Morampudi, M., Rajput, P., Lata, S., Agariya, S., Nanda, G., Raghava, S.S. 2016. Comparison of different technologies for solar PV (Photovoltaic) outdoor performance using indoor accelerated aging tests for long term reliability. *Energy.* 107(15): pp. 550–561.

Advances and Trends in Engineering Sciences and Technologies III – Al Ali & Platko (Eds)
© 2019 Taylor & Francis Group, London, ISBN 978-0-367-07509-5

Thermal effects around a heated circular cylinder in horizontal, parallel and contra flow

B. Bolló, P. Bencs & Sz. Szabó
Department of Fluid and Heat Engineering, University of Miskolc, Miskolc, Hungary

ABSTRACT: The laminar vortex shedding of airflow behind an electrically heated circular cylinder was experimentally and numerically investigated at low Reynolds numbers (Re < 200). The Z-type Schlieren technique was applied to measure the temperature distribution. The experiments were carried out at free and forced convection in a 0.5×0.5 m cross-section wind tunnel. For numerical simulations a heated horizontal circular cylinder is exposed to approaching flow stream, in horizontal (cross-flow), parallel (parallel flow) and opposing (contra flow) directions to the buoyant force. The results show significant changes of the wake flow structures behind the heated cylinder due to the thermal effects at different approaching flow arrangements.

1 INTRODUCTION

There are several applications of the combined forced and free convection heat transfer from the bluff bodies in nature and industry such as electrical transmission lines, cartridge heaters, pipes of heat exchangers, factory chimneys etc. The structure of the flow around prismatic bodies has been already examined for a long time. Also the z-type Schlieren measurement technique has also been investigated by many researchers (Settles (2001), Baranyi et al. (2009)). One example for the use of the Schlieren technique is the visualization of shockwaves in a supersonic tunnel (Bencs et al. (2009)). The system is basically adapted for 2D measurements, as there are many problems with 3D measurements (Settles (2001)). The method has also been used for general visualization of heat transport processes (Settles (2001)). Our experimental tests were conducted at low Reynolds numbers (Re < 200), thus the flow field was approximately two dimensional (same flow in every normal plan of the heated circular cylinder). In this work the adaptability of a Schlieren system is investigated to the described setup, which should be ultimately applicable in a wind tunnel in order to investigate free and forced convection flows.

The heat transfer process from a heated cylinder to the surrounding fluid can be either free, forced or mixed convection, depending on the geometry of the flow and can be characterized by critical value of the Richardson number (Ri = Gr/Re^2). Here Gr is the Grashof number defined as Gr = $g\,\beta\,(T_w - T_\infty)\,D^3/\nu^2$, where g is the acceleration due to the gravity, β the temperature coefficient for volume expansion and ν the kinematic viscosity. For Ri >> 1 free convection effects dominate while for Ri << 1 it is forced convection problem. The criterion for mixed convection to occur in the cross flow situation is Ri ≈ 1 (Janna (1986)). Additional complications arise depending upon the direction of the forced flow with reference to the direction of gravity. Badr (1984) numerically investigated the thermal effects on the wake behind a heated cylinder in the mixed convection regime when the forced flow is directed either vertically upward or vertically downward. He found that the flow structures behind the heated cylinder change at different approach arrangements.

In the present study, the thermal effects on the wake flow behind a horizontal heated circular cylinder exposed to a horizontal-flow are investigated experimentally and numerically. We compare temperature contours in the wake of the cylinder for free and forced convection and investigate the average Nusselt number by varying the Richardson number. The temperature fields around heated cylinder are compared at different approach flow arrangements.

2 EXPERIMENTAL SETUP

During the experiments, an electrically heated circular cylinder was placed perpendicular to the flow in a 0.5×0.5 meter cross-section wind tunnel. The diameter of the heated circular cylinder was 10 mm. The temperature of the cylinder was 300°C and the range of velocity was from 0–0.3 m/s. The temperature of the forced flow was 20°C.

The basic optical Schlieren system uses light from a single collimated source shining on or from behind a target object. Variations in the refractive index, being caused by density gradients in the fluid, distort the collimated light beam. The light source is placed in the focal point of the first mirror (see Figure 1). The knife-edge (color filter) is placed in the focal point of the second mirror, so that it blocks about half of the light. In flows of uniform density this will make the photograph half as bright. However, in flows with density variations the beam is deviated and focused in an area covered by the knife-edge and thus is blocked. The result is a set of lighter and darker patches corresponding to positive and negative fluid density gradients in the direction normal to the knife-edge. When a knife edge is used, the system is generally referred to as a Schlieren system, which measures the first derivate of density in the direction of the knife-edge. Temperature is determined from the density (refractive index) (Settles (2001)). The principle of the Schlieren technique is shown in Figure 1.

2.1 Calibration curve

In a Schlieren system, the blockage of light by a knife-edge (color filter) placed at the exit focal plane is due to the deviation of light by an inhomogeneous medium. This blockage can be similarly obtained in a homogeneous medium by a knife-edge which is allowed to be translated laterally by a quantity Δx. By considering this, we can establish a relationship between the light levels at the observation plane (the exit focal plane) to the corresponding transverse knife-edge position. This transverse position may cover the conditions from the minimum 'Hue' degree to the maximum. 'Hue' is one of the main properties of a color, defined technically (in the CIECAM02 model) (Fairchild (2012)). Relationship between Δx and light color was determined by 3-axis stage (see in Figure 1).

2.2 Z-type Schlieren system

General recommendation for the applications of a z-type Schlieren system is include in Table 1 (Settles (2001)).

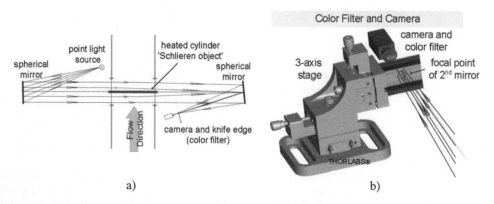

a)   b)

Figure 1.   Principle of Schlieren measurement technique (color filter).

Table 1. Properties of the applied components in the Z-type Schlieren system.

Property	Value
Mirror thickness	25% of the diameter
Optical quality of the mirrors	$\lambda/8$
Offset angle	$\theta = 3.81°$
Power of lens	$f/10$
Distance between the mirrors	$L = 4500$ mm

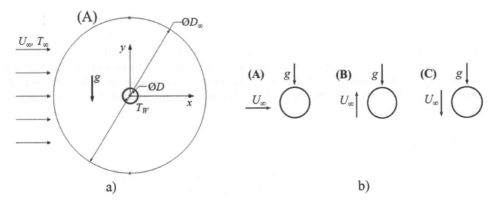

a)   b)

Figure 2. (a) Computational domain; (b) direction of flow and gravity.

3 NUMERICAL SOLUTION

The commercial software Ansys Fluent is used for the numerical simulations, based on the finite volume method (FVM). The two-dimensional (2D), unsteady, laminar, segregated solver is used to solve the incompressible flow for the collocated grid arrangement. The second order upwind scheme was used to discretize the convective terms in the momentum equations. The semi-implicit method for the pressure linked equations (SIMPLE) scheme is applied for solving the pressure-velocity coupling.

The computational domain is characterized by two concentric circles: the inner represents the cylinder surface with diameter D, the outer the far field with diameter D_∞ (Figure 2a). The origin of the coordinate system is in the center of the cylinder. Three different cases are investigated: the positive x-axis is in the downstream direction and gravity force perpendicular to it (A); flow stream for opposing (contra flow) (B) and parallel (parallel flow) (C) directions to the gravity (Figure 2b).

The accuracy of the computed results depend on the computational mesh, the time step, the size and shape of the computational domain. For uniform flow past an unheated stationary circular cylinder the effect of domain size, mesh and time step was investigated to determine a combination at which the solution is roughly parameter independent (Bolló (2010)). The computational method has been validated by comparing dimensionless coefficients such as means of integral quantities like lift, drag and the base-pressure coefficients and the Strouhal number. The numerical results were in good agreement with data of literatures (Bolló et al. (2010)). Based on these studies the computational domain size is chosen $D_\infty/D = 160$ and here a 360×292 (azimuthal $\times$ radial) grid is used. In the physical domain logarithmically spaced radial cells are used, providing a fine grid scale near the cylinder wall and a coarse grid in the far field. A dimensionless time step of $\Delta t = 0.001$ is used.

At the inlet uniform velocity distribution (U_∞) and constant temperature (T_∞) are prescribed. The absolute temperature of the cylinder surface is assumed to be constant, T_w.

For flows over a heated cylinder the fluid properties such as viscosity, density and thermal conductivity vary with the temperature. The dependence of the viscosity on temperature is given by Sutherland's formula (White (1999)) and further fluid properties are obtained from (VDI-Wärmeatlas (2002)).

The heat transfer between the cylinder and the surrounding fluid is determined using the dimensionless, or local Nusselt number,

$$Nu = h \, D/k,\tag{1}$$

where k is the thermal conductivity of the fluid, and h is the local convective heat transfer coefficient defined as

$$h = q/(T_w - T_\infty),\tag{2}$$

where q is the convective heat transfer rate per unit area from the cylinder wall to the fluid. In the present work fluid properties are not constant for heated cylinders so the thermal conductivity of the fluid k also depends on the temperature, which influences the Nusselt number value. The physical properties of the working fluids are evaluated at the free stream temperature T_∞.

4 RESULTS

The results will first focus on the temperature distribution on the wake behind a heated cylinder exposed to horizontal cross-flow, where the direction of approach flow is perpendicular to the direction of the thermally induced buoyancy force. The temperature of the cylinder surface is assumed 300°C, respectively. The range of air flow velocity in the wind tunnel has been set to 0–0.3 m/s. For different velocities the temperature distributions are shown by the Schlieren images and numerical simulations in Figure 3. In the figure the 'dark parts' are actually the warm fluid shedding and around the heated cylinder a boundary layer appears, which becomes thinner with increasing inlet velocity. As seen in the figure, the three cases are differentiated: when free convection dominates ($U_\infty = 0$) in Figure 3a. In the next case if a weak, externally forced flow is present ($U_\infty = 0.15$ m/s), it plays an important role in moving fluid near a heated surface, although free convection might still be dominant (Figure 3b). When the externally forced flow is quite large (here $U_\infty = 0.3$ m/s), forced convection effects dominant. In this case the alternate shedding of 'heat packages' associated with the 'Kármán' vortices can be seen clearly (Figure 3c).

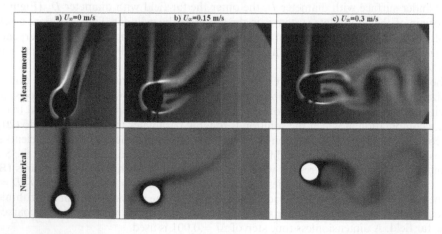

Figure 3. The temperature distributions by measurement and numerical simulations for (a) $U_\infty = 0$ m/s, (b) $U_\infty = 0.15$ m/s and (c) $U_\infty = 0.3$ m/s.

Now it will be necessary to fine tune the Schlieren system, in order to get more contrasted and fine structured images, by adjusting the light intensity and position, the color filter.

In the present study, the average Nusselt numbers (Nu) of the heated cylinder were calculated at $T_w = 300°C$ and for different inlet velocity (Reynolds number change). The average Nusselt number decrease with the increasing Ri number, as can be seen in Figure 4.

Figure 5 shows the temperature distribution around the heated cylinder at the different approach flow direction: (A) cross-flow, (B) parallel-flow and (C) contra flow for free and forced convection. As seen in the figure, vortex structure changes with the angle between the approach flow direction and the thermally induced buoyancy force. For cross-flow (A) the vortex shedding begin to appear at $U_\infty = 0.15$ m/s, causing 'heat packages' spread behind the cylinder. With increasing forced flow the 'Kármán' vortex street formation appears as in the wake of the unheated cylinder.

For parallel flow (B) periodic vortex shedding behind the heated cylinder can be not observed. When the cylinder temperature or the inlet velocity is decreased (eg. $U_\infty = 0.05$ m/s), then periodic flow can be observed behind the heated cylinder.

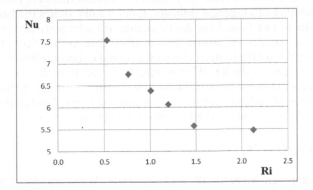

Figure 4. The average Nusselt number versus Richardson number at $T_w = 300°C$.

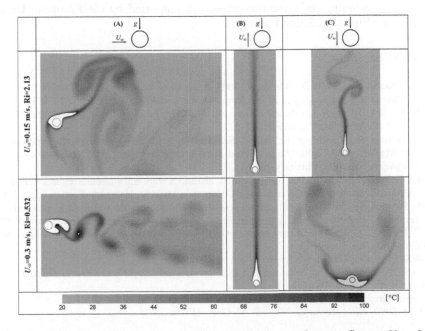

Figure 5. Temperature distribution for cross-flow, parallel flow and contra flow at $U_\infty = 0.15$ and 0.3 m/s.

323

For contra flow (C) the vortex shedding cannot appear at low inlet velocity, the heat streams almost evenly upwards. The forced flow is opposite to the thermally induced buoyancy force, therefore at increasing inlet velocity ($U_\infty = 0.3$ m/s) the forced flow tries to return the 'heat packages' so it spreads around the cylinder.

5 CONCLUSION

The thermal effects on the wake flow behind a heated circular cylinder operating and exposed to a cross-flow were investigated experimentally and numerically. The measurements results presented in this work confirm that the z-type Schlieren system is in principle suitable to visualize and quantitative analyze the temperature field in a wind tunnel. However, considerable improvement (such as precision color filter) is still required in the existing system to make more precise and really accurate measurements. In order to analyze the images in a further step, the recording quality must be increased to get more meaningful images. These temperature field results will be good to validate our results of numerical simulations. The first Schlieren results and numerical simulations are similar; therefore our Schlieren system is suitable for validation of numerical simulation, but it still requires further development.

The heat transfer from a heated horizontal circular cylinder is studied numerically for the three cases of cross, parallel and contra flow regimes. Due to the effect of buoyancy force, the wake instability behind the heated cylinder was found to become much more complicated than behind an unheated cylinder. With increasing forced flow the numerical simulations revealed significant modifications of the wake flow structures behind the heated cylinder due to the thermal effects at different approach flow arrangements. Further studies will include the investigation of flow at lower cylinder surface temperature.

ACKNOWLEDGEMENTS

The research was supported by the EFOP-3.6.1–16–00011 "Younger and Renewing University—Innovative Knowledge City—institutional development of the University of Miskolc aiming at intelligent specialisation" project implemented in the framework of the Széchenyi 2020 program. The realization of these two projects is supported by the European Union, co-financed by the European Social Fund.

REFERENCES

Baranyi, L., Szabó, Sz., Bolló, B., Bordás, R. 2009. Analysis of Low Reynolds Number Flow around a Heated Circular Cylinder. *Journal of Mechanical Science and Technology* 23: 1829–1834.

Bard, H.M. 1984. Laminar combined convection a horizontal cylinder – parallel and contra flow regimes. *Int. J. Heat Mass Transfer* 27: 15–27.

Bencs, P., Bordás, R., Zähringer, K., Szabó, Sz., Thévenin, D. 2009. Towards the Application of a Schlieren Measurement Technique in a Wind-Tunnel. Proc. *MicroCAD International Computer Science Conference*, Miskolc, Hungary, pp. 13–20.

Bolló, B. 2010. Grid independence study for flow around a stationary circular cylinder. Proc. *MicroCAD 2010, International Computer Science Conference*, Miskolc, Hungary, Section F, pp. 1–6.

Bolló, B., Baranyi, L., 2010. Computation of low-Reynolds number flow around a stationary circular cylinder. Proc. *7th International Conference on Mechanical Engineering*, Budapest, pp. 891–896.

Fairchild, M. 2012. Color Appearance Models: CIECAM02 and Beyond. *Tutorial slides for IS&T/SID 12th Color Imaging Conference*.

Janna, W.S. 1986. *Engineering heat transfer.* 2nd ed., PWS Publishers, U.S.A.

Settles, G.S. 2001. *Schlieren and Shadowgraph Techniques: Visualizing Phenomena in Transparent Media.* Springer-Verlag, Berlin, Heidelberg.

VDI-Wärmeatlas 2002. 9th ed., Springer-Verlag.

White, F.M. 1999. *Fluid Mechanics.* 4th ed., McGraw-Hill.

Advances and Trends in Engineering Sciences and Technologies III – Al Ali & Platko (Eds)
© 2019 Taylor & Francis Group, London, ISBN 978-0-367-07509-5

Fluorimetric characteristics of river waters compared with pollutant concentrations

E.A. Bondarenko, E.D. Makshanova & M.Ju. Andrianova
Peter the Great Saint-Petersburg Polytechnic University, St. Petersburg, Russia

N.V. Zueva & E.S. Urusova
Russian State Hydrometeorological University, St. Petersburg, Russia

ABSTRACT: The fluorimetric and photometric properties were studied in water samples collected from the city river Okhta and the countryside river Oredezh. The Okhta water had higher concentrations of organic matter (esteemed as COD) and consequently higher values of optical parameters connected with it: optical density D at 254 nm (D_{254}) and intensity of fluorescence I (humic-like and protein-like). Correlation coefficient (r) between D_{254} and COD is 0.66, which is more than 3 times higher than between COD and D at wavelengths used for determination of color (380 or 410 nm). In the samples from river Okhta, correlation coefficients were between 0.69 and 0.90 for concentration of total nitrogen (TN) and I (humic-like or protein-like). Concentrations of chlorides and inorganic nitrogen gave higher values of r with protein-like fluorescence (0.58–0.71), concentrations of ammonium-ion and total phosphorus—with humic-like fluorescence (0.54–0.83). BOD_5 showed negative correlation with I. These data show that I at studied wavelengths can be used for monitoring of the water quality in the urban river.

1 INTRODUCTION

The surface water pollution due to wastewater effluents is still a problem in many cities. In reality it is impossible to eliminate the risk of water pollution. Discharge of contaminants can happen as a result of an accident in the wastewater treatment plant, direct connection of domestic sewerage (e.g. from new buildings) to the drainage system opening to a river, etc.

In Saint Petersburg, river water sampling and analysis is done according to the RD 5224.309-2011 (2011) once a month. Such periodicity is insufficient to reveal and manage water contamination. Optical express-methods of analysis can be used in online devices or mobile laboratories to make management decisions rapidly. The measurement of the optical parameters such as fluorescence intensity (I) and optical density (D) of water can be applied.

D of water in ultraviolet and visible range (UV-Vis) is influenced by the concentrations of organic and inorganic substances. In water quality control, several wavelengths are applied. The color is determined using measurements at selected wavelengths ranging from 360 to 700 nm (depending on the method and correction on water turbidity) (Rice et al. 2012). D at 254 nm (D_{254}) is used to esteem concentration of dissolved organic matter (Matilainen et al. 2011, Hansen et al. 2016). However, in some cases values of D or even entire spectra do not provide enough information for the distinguishing between water from polluted and not polluted parts of the river (Thomas & Burgess 2007).

The fluorescence spectra are more informative. The organic matter produces two main types of peaks in the fluorescence spectra of water. Protein-like peaks are registered at excitation wavelength of 210–300 nm and emission wavelengths of 300–310 nm (tyrosine-like peak) and 330–350 nm (tryptophan-like peak). Humic-like peak has the maximal emission at 410–470 nm. In unpolluted river waters, humic-like peak is prevailing over protein-like peak. However, humic-like peak is also present in spectra of oils, landfill leachates and domestic wastewaters along with high protein-like peaks. As a result pollution can increase I for both types of peaks (Coble et al. 2014).

At certain excitation (ex) and emission (em) wavelengths ($I_{ex,em}$), I shows correlation with several chemical parameters in natural and polluted waters. Positive correlations (with Pearson correlation coefficient r = 0.68–0.97) were found between biochemical oxygen demand (BOD) and I of tryptophan-like peak, BOD and I of humic-like peak, total organic carbon (TOC) and I of humic-like peak or tryptophan-like peak. Similarly the strong positive correlations were found between fluorescence parameters and concentrations of not fluorescing compounds typical for the ewage pollution (phosphates, nitrates, total nitrogen, and ammonium) (Coble et al. 2014). However, in some studies correlation between TOC, COD or BOD and peaks intensities was weak or not observed (Comber et al. 1996, Coble et al. 2014). Obviously, properties of real water and informativity of exact wavelengths should be examined and taken into account before applying express-techniques for pollution monitoring.

The aim of this work was to study informativity of D and I at several wavelengths in monitoring of river waters. Two different rivers were taken for this study—the urban river Okhta and the rural river Oredezh from the North-West of Russia.

2 MATERIALS AND METHODS

2.1 *Water sources and sampling*

The Okhta is the most polluted river in Saint Petersburg (Serebritckii 2016). It has total length 90 km. Water samples were collected from the last 17 km of the river where it flows through Saint Petersburg. Sampling points are marked from O1 (before the inflow of the Okhta to the Neva) to O14 (at the Chelyabinsky bridge at the city boundary). Several samples were collected from the two tributaries of Okhta—rivers Okkervil (Ok1, Ok3, Ok5) and Lubja (L1, L3, L5), the least number is the closest to the river`s mouth. Inflow of the Okkervil is located between sampling points O2 and O3, inflow of the Lubja—between O11 and O12.

The rural river Oredezh flows in the Leningrad region and has the length of 192 km. Five samples were collected from the part of the river with 20 km length (near it source) from village Bolshoye Zarechje to settlement Vyra.

2.2 *Water analysis*

Samples of each series were collected in July 2017 at one day. The hydrochemical parameters were measured in Russian State Hydrometeorological University. The following parameters were determined: pH, specific electric conductivity (SEC), total hardness (TH), BOD_5, COD, concentrations of dissolved O_2, chlorides (Cl^-), sulfates (SO_4^{2-}), ammonium (NH_4^+), nitrites (NO_2^-), nitrates (NO_3^-), total nitrogen (TN), phosphates (PO_4^{3-}), total phosphorus (TP).

Optical analysis of the water was done in Peter the Great Saint-Petersburg Polytechnic University. Optical parameters were determined in 10-fold diluted samples in order to reduce the influence of light absorption on the fluorescence signal.

Optical density (D) was determined by spectrophotometer "SF-56" (OKB Spectr, Russia) in 1 cm cuvette. Fluorescence data were obtained by analyzer "RF 5301 PC" (Shimadzu, Japan) at two excitation wavelengths 230 nm and 270 nm near to maximums of excitation spectra for humic substances and proteins. Fluorescence spectra were registered at the emission wavelengths from 220 to 650 nm. For further comparison, the following emission wavelengths were chosen: 300, 320, 350 nm (for protein-like fluorescence) and 420 nm (for humic-like fluorescence).

3 RESULTS AND DISCUSSION

3.1 *General characteristic of waters by chemical parameters*

The results of chemical analysis are presented on Figures 1–2. Several parameters, such as various forms of nitrogen (nitrates, nitrites, and ammonium) and phosphorus (phosphates and

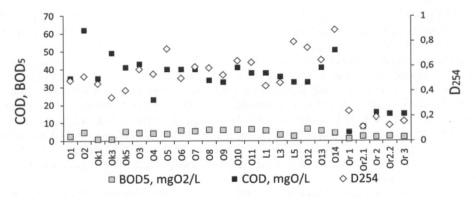

Figure 1.   Values of D$_{254}$ and parameters of organic matter in the water samples.

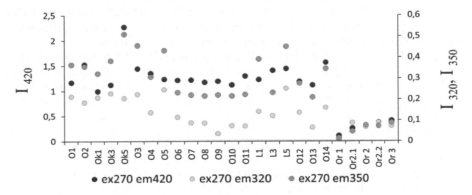

Figure 2.   Values of I at the excitation wavelength 270 nm in the water samples.

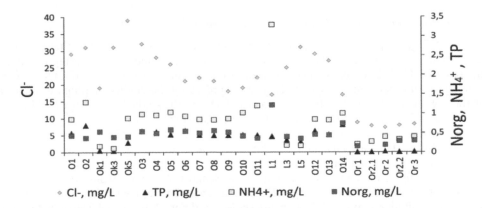

Figure 3.   Parameters of pollutants in the water samples.

TP), chlorides, sulfates, dissolved solids (esteemed by SEC) and pH were within the allowed limits and lower than maximal allowable concentrations (MAC) of SanPiN 2.5.980-00 (2000) and GN 2.1.5.1315-03 (2003). The only exception was a sample from the Lubja river, where concentration of ammonium ion reached 3.3 mg/L (equal to 1.7 MAC).

Values of COD overcame the MAC in the Okhta and its tributaries (23–62 mgO/L) and in one sample from the Oredezh (16 mgO/L). This can be explained not only by pollution but also by the high natural background. Rivers in the North-West of Russia can have COD as much as 142 mgO/L (Vodogretckii 1972). Cases of definite pollution were registered in the last 7 km of the Okhta, where concentrations of dissolved O$_2$ were lower than allowed limits

(2.2–3.1 mgO$_2$/L) and in the sample near the city boundary where it fell to 0.5 mgO$_2$/L. BOD$_5$ in all samples of the Okhta and several samples of its tributaries (2.6–6.7 mgO$_2$/L) were also higher than the MAC showing results of river pollution with wastewaters.

General difference between the rivers Okhta and Oredezh was observed in the concentrations of organic matter and dissolved mineral substances. Both rivers originate in the forest and swampy areas which supply them with natural organic matter and increase the values of COD. However, COD in the samples from one river is significantly lower than in the samples from another (5–16 mgO/L in the Oredezh, 23–62 mgO/L in the Okhta and its tributaries). This can be explained by addition of organic matter with wastewaters to the Okhta, which led to the high values of BOD$_5$ (2.6–6.7 mgO$_2$/L) and low values of dissolved O$_2$ due to its consumption during microbial decomposition of easily degradable organic matter from domestic wastewaters.

However, another mechanism of water quality formation should be considered as well. Samples from the Oredezh have higher content of dissolved salts (especially the cations of hardness), than the waters from the Okhta and its tributaries (SEC and hardness are 455–603 mkSm/cm and 5.7–6.6 meq/L in the Oredezh, 232–267 mkSm/cm and 1.3–1.6 meq/L in the Okhta). Due to this aquatic organic matter in river Oredezh is involved in process of natural coagulation and as a result part of organic substances is removed from waters with sediments.

In spite of lower values of total mineralization (measured as SEC), the Okhta is characterized by higher concentrations of chlorides than the Oredezh (17–39 mg/L in the Okhta, the Lubja and the Okkervil, 7.3–8.0 mg/L in the Oredezh). Also the Okhta possesses higher levels of phosphates and TP (0.44–0.52 mgTP/L), which are almost absent in the Oredezh (0.01–0.02 mgTP/L). Situation with nitrogen compounds is more complicated because the Oredezh has high natural concentrations of nitrates: 4.5–6 mg/L as reported in (Vodogretckii 1972), up to 18 mg/L measured in our work. Due to this TN in its waters is higher than in the Okhta (1.8–2.2 mgTN/L). However, the values for ammonium-ion which is a marker of fresh wastewater pollution are higher in the Okhta (0.8–1.3 mgNH$_4^+$/L), than in the Oredezh (0.2–0.4 mgNH$_4^+$/L). In general, according to the given data, waters of the Oredezh can be considered less polluted by wastewaters than waters of the Okhta.

3.2 Optical characteristics of water in connection with chemical data

Values of D$_{254}$ are presented on the Figure 1. It can be seen from the picture that values of D$_{254}$ are in general higher in the Okhta than in the Oredezh, and their changing resembles changing of COD and BOD$_5$. The correlation coefficient between D$_{254}$ and COD is 0.66 (among all samples from all rivers). This result proves the possibility of rough estimation of the organic matter content from the values of D$_{254}$ in the studied water objects. The correlation coefficients were also calculated between COD and D at other wavelengths used for determination of water color (380 and 410 nm) (Rice et al. 2012). The values were 0.23 for COD and D$_{380}$, 0.05 for COD and D$_{410}$, showing poor informativity of these wavelengths in estimation of organic matter content.

The changing of I at two studied excitation wavelengths (230 and 270 nm) was similar. The signal intensity at excitation wavelength 230 nm was about 2 times higher than at 270 nm. Values of I at excitation wavelength 270 nm are shown on Figure 2. In general, values of D$_{254}$ and I changed in the similar way, representing the organic matter that was able to absorb light in UV, and the part of it which was able to emit fluorescence. It is difficult to attach optical data directly to concentrations of exact pollutants, but Figures 1–3 show that the increase of some parameters characteristic for pollution with wastewaters (concentrations of chlorides, organic nitrogen, ammonium ion, TP, and BOD$_5$) coincide with the increase of I (e.g. compare data for points Ok1 and Ok1, L1–L5, Or1 and Or2, O14 and O13, and gradual change from O11 to O2).

In order to find the connection between optical and chemical parameters correlation coefficients (r) were calculated for the samples from the Okhta (Table 1). As it can be seen from the data, water pollutants showed several different types of relationship with optical parameters.

TN showed moderate or strong correlation (r = 0.69–0.90) with all fluorimetric parameters and no correlation with D$_{254}$. These results correspond to the data received earlier for the Okhta

Table 1. Correlation coefficients between parameters in the Okhta.

	TP	TN	NH_4^+	COD	BOD_5	Cl^-	Norg	Ninorg
D_{254}	0.25	0.14	0.02	0.20	0.13	−0.29	0.55	−0.16
$I_{230,300}$	0.45	0.90	0.26	0.26	−0.79	0.67	0.20	0.71
$I_{230,320}$	0.52	0.89	0.24	0.27	−0.83	0.63	0.23	0.69
$I_{230,350}$	0.75	0.87	0.41	0.51	−0.72	0.44	0.34	0.61
$I_{230,420}$	0.80	0.81	0.54	0.52	−0.47	0.22	0.38	0.54
$I_{270,320}$	0.46	0.89	0.26	0.28	−0.78	0.62	0.23	0.69
$I_{270,350}$	0.46	0.91	0.30	0.29	−0.76	0.58	0.28	0.68
$I_{270,420}$	0.83	0.69	0.62	0.58	−0.34	0.18	0.33	0.45

(Andrianova et al. 2014). According to the data of present study, inorganic forms of nitrogen (ions of ammonium, nitrates and nitrites) formed 65–83% of TN and this also explained the noticeable correlation coefficients between inorganic nitrogen and I ($r = 0.45$–0.71).

NH_4^+ and COD showed moderate correlation ($r = 0.52$–0.62) with I of humic-like fluorescence and weaker correlation ($r = 0.24$–0.51) with I of protein-like fluorescence. TP showed close type of relationship with the higher values of r. It should be noticed here that the pollution with domestic wastewaters increases both humic-like and protein-like fluorescence; correlation between the concentrations of pollutants and humic-like peak was also registered before in the Okhta and in other rivers (Andrianova et al. 2014, Coble et al. 2014).

Chlorides are known to be one of the stable markers of anthropogenic pollution (De Sousa et al. 2014). MAC for chlorides in wastewaters allowed for discharge to rivers (350 mg/L) is 1–2 orders of magnitude higher than natural concentrations in the studied rivers. It means that, even after discharge of treated wastewaters, increase of chlorides in river waters can be expected. Municipal wastewaters contain 30–100 mgCl⁻/L (Metcalf & Eddy 2003). Reported variation of natural values at the control site of the Okhta near the city boundary was 4–16 mgCl⁻/L (Vodogretckii 1972). Values registered in the present study are from 17 to 32 mgCl⁻/L. These values show that the increase of chlorides concentration in the Oktha has an anthropogenic cause. Correlation of chlorides concentrations was positive with I of protein-like fluorescence ($r = 0.58$–0.67); weak negative correlation was found with D_{254}.

Close values of r were received in our previous study for 30 samples from the larger part of Okhta in 2013–2014 ($r = 0.50$–0.77 for Cl⁻ and I, $r = −0.51$ for Cl⁻ and D_{254}). Negative correlation was also found between changes of SEC and D_{254} ($r = −0.48$) in that study (Bondarenko & Andrianova, unpublished). The close value of r was received in the present study ($r = −0.58$ for SEC and D_{254}). So, these results show that the river pollution marked by increase of chlorides can lead to the decrease of D_{254} and increase of I.

The negative correlation between protein-like fluorescence and BOD_5 was rather unexpected. However, such situation was observed in some other studies (Comber et al. 1996). Maybe it is so because a significant part of BOD_5 (50% or more according to (Metcalf & Eddy 2003)) in wastewaters is present as suspended matter, while optical parameters (D_{254} and I) are formed mainly by dissolved substances. Generally, these data show that fluorimetry at the studied wavelengths can be used for monitoring of water quality in the urban river. Usage of protein-like fluorescence is more reasonable for this purpose.

4 CONCLUSIONS

Water samples from the urban river Okhta and two its tributaries (in Saint Petersburg) and the rural river Oredezh (in Leningrad region) were investigated. Along with chemical parameters of the waters optical density D and fluorescence intensity I at several wavelengths were studied in order to find informative optical parameters for water quality monitoring.

Waters of the Oredezh were less polluted, demonstrating lower concentrations of BOD_5, TP and phosphates, ammonium-ion, and the higher concentrations of dissolved O_2. COD was also lower in the Oredezh in spite of the fact that both the Okhta and the Oredezh have enough source of organic matter from the forests and swamps in their catchment areas. This was explained by more intensive natural coagulation of the organic matter in the Oredezh due to the higher content of salts (especially multivalent cations) in water.

Changing of D_{254} and fluorescence intensity I in the water samples resembled variation of COD. Correlation coefficient r between D_{254} and COD was 0.66, between COD and D_{380} or D_{410} (used for determination of the color in waters): 0.23 and 0.05, correspondingly, showing poor informativity of the color in estimation of the organic matter concentration in these waters.

For the samples from river Okhta correlation coefficients were calculated between optical and chemical parameters. D_{254} had noticeable positive correlation only with concentration of organic nitrogen (r = 0.55). $I_{230,\ 300-350}$ and $I_{270,\ 300-350}$ (protein-like fluorescence) had moderate or strong correlation (r = 0.52–0.90) with concentrations of total phosphorus, chlorides, inorganic nitrogen, total nitrogen. $I_{230,\ 420}$ and $I_{270,\ 420}$ (humic-like fluorescence) had moderate or strong correlation (r = 0.52–0.81) with concentrations of NH_4^+, total phosphorus, total nitrogen, COD. In general these data show that fluorimetry at the studied wavelengths can be used for water quality monitoring in the urban river. Using of protein-like fluorescence is more reasonable for it is only present in high intensities in the wastewaters, while humic-like fluorescence can be present both in the wastewaters and the natural unpolluted waters.

ACKNOWLEDGEMENT

Hydrochemical analysis of water samples was fulfilled in the frames of work funded by RFBR according to the research project № 16–35–00382 mol-a.

REFERENCES

Andrianova, M.J., Bondarenko, E.A., Krotova, E.O., Chusov, A.N. 2014. Comparison of chemical and optical parameters in monitoring of urban river Okhta. *EESMS 2014 - 2014 IEEE Workshop on Environmental. Energy and Structural Monitoring Systems. Proceedings 6*. Pp. 198–202.

Bondarenko, E.A. & Andrianova M.Ju, unpublished data.

Coble, P.G., Lead, J., Baker, A., Reynolds, D.M., Spencer, R.G.M. 2014. *Aquatic Organic Matter Fluorescence*. Cambridge University Press.

Comber, S.D.W., Gardner, M.J., Gunn, A.M. 1996. Measurement of Absorbance and Fluorescence as Potential Alternatives to BOD. *Environmental Technology* 17: 771–776.

De Sousa, D.N.R., Mozeto, A.A., Carneiro, R.L., Fadini, P.S. 2014. Electrical Conductivity and Emerging Contaminant as Markers of Surface Freshwater Contamination by Wastewater. *Science of the Total Environment* 484: 19–25.

GN 2.1.5.1315–03. 2003. *Maximum Allowable Concentrations (MAC) of Chemical Substances in Water Bodies for Potable, Nonpotable and Recreational Water Use*. (in Russian).

Hansen, A.M. et al. 2016. Optical properties of dissolved organic matter (DOM): Effects of biological and photolytic degradation. *Limnol. Oceanogr.* 61(3): 1015–1032.

Matilainen, A., Gjessing, E.T., Lahtinen, T., Hed, L., Bhatnagar, A., Sillanpää, M. 2011. An overview of the methods used in the characterization of natural organic matter (NOM) in relation to drinking water treatment. *Chemosphere* 83: 1431–1442.

Metcalf & Eddy. 2003. *Wastewater Engineering: Treatment and Reuse*. McGraw-Hill.

RD 52.24.309–2011. 2011. *Guidance document. Organization and conduct of regime observations of the state and pollution of surface waters*. (in Russian).

Rice, E.W., Baird, R.B., Eaton, A.D., Clesceri, L.S. 2012. *Standard Methods for the Examination of Water and Wastewater*, 22th edition. American Public Health Association. American Water Works Association. Water Environment Federation NW Washington.

SanPiN 2.1.5.980–00. 2000. *Hygienic requirements for surface waters protection. Sanitary norms and rules* (in Rusian).

Serebritckii, I.A. (Eds.) 2016. *Report on ecological situation in St.Petersburg in year 2015*. SPb: OOO Sezamprint. (in Rusian).

Thomas, O. & Burgess C. (Eds.) 2007. *UV-Visible Spectrophotometry of Water and Wastewater*. Elsevier.

Vodogretckii, V.E. (Eds.) 1972. *Resources of surface waters in the USSR. Vol.2. Kareila and North-West*. Leningrad: Gidrometeoizdat.

Advances and Trends in Engineering Sciences and Technologies III – Al Ali & Platko (Eds)
© *2019 Taylor & Francis Group, London, ISBN 978-0-367-07509-5*

Implementation of virtual reality in BIM education

R. Bouska & R. Schneiderova Heralova
Faculty of Civil Engineering at Czech Technical University in Prague, Prague, Czech Republic

ABSTRACT: This paper deals with the topic of the implementation of VR (Virtual Reality) into the teaching syllabus of building information modelling. Virtual reality is a very popular and effective visualization technology. It offers new means of interaction with virtual models in virtual environments. In Civil engineering, virtual reality is mostly used for all sorts of project visualizations. It can be utilized for architectural visualizations, which helps to present the architects' ideas to the client. Structural engineering data can be visualized to help optimize the structural design of the building. Another great example is model error checking and collision detection. This paper aims to present one of the possible ways of applying virtual reality in university tuition, namely in the teaching of building information modelling. It shows how to navigate models and extract information. It is evident that students will be using these technologies in their future careers as civil engineers and project managers.

1   INTRODUCTION

The topic of building information modelling is very broad. It is one of the most anticipated global shifts in how civil engineering projects are delivered. This paper focuses on the implementation of virtual reality into the syllabus of building information modelling at university level. The advancement in interactive and immersive technologies can have a noticeable impact on the various styles of teaching and learning. Virtual reality is a technology that is attractive to the student community (Abdul-Hadi G. Abulrub 2011).

In recent years, virtual reality has become a very popular technology, which embodies the newest research achievements in the fields of computer technology, computer graphics, sensor technology, ergonomics, and the human-machine interaction theory. More and more people devote into their time and energy to this field and commit themselves to its research, development and application. Virtual reality as a technology has become one of the means for humans to explore the world. What does virtual reality bring us? Firstly, it has changed our ideas. Nowadays, it is the person and not the computer who has become the main body of information technology. Secondly, it has improved the human-machine interactive manner. Currently, the natural interactive means of perception, i.e. by hand and sound, is fully replacing the passive mode that existed in the past. Thirdly, it has changed the way of life and forms of entertainment. In a word, the virtual reality technology is an important technology to consider, as it will make a great impact on our lives and work (Xinxing 2012).

BIM applications have also grown tremendously, from a tool used to design in three dimensions and with the use of components, to a tool used for model analysis, clash detection, product selection, and the whole project conceptualization. Although many benefits can be gained by the implementation of BIM (Matějka & Tomek 2017) – such as increasing constructability; reducing conflict; simplifying the process of requesting information owing to the high quality visualization approach (Cernohorsky & Matejka 2017, Hromada 2017); reducing the time for cost estimation (Strnad 2017, Vitásek & Matějka 2017) and increasing smooth coordination and information among parties—the pace of integrating BIM into VR is still rather slow (Wang et al. 2018).

Building modelling improves 2D drafting by allowing designers to view the building and its contents from all angles, and revealing problems at earlier stages to allow for correction without costly change orders. A truly parametric design saves time by creating and editing multiple design portions simultaneously. Sections, elevations and three-dimensional views can be created instantly, reducing the need for check plots. Changes to any one of these elements affect all of the others, including materials, costs and construction schedules. The two-dimensional printed documentation becomes the quick and accurate by-product of parametric design.

Main aim of this paper is to describe the process of designing a VR aided learning environment as a support for educating civil engineering students on building information modelling. The whole process includes preparation of the model, transformation of the model into the VR software and execution of some basic tasks by students. The paper also describes the process of the designing of a VR laboratory and all the aspects of the design that need to be taken into consideration.

2 OVERVIEW OF RELEVANT EDUCATIONAL TOOLS

There are many aspects that need to be considered, when creating the appropriate process of how to educate students through virtual reality. One of the first steps is choosing the right hardware and software tools for VR.

The hardware for VR can be divided into 2 main components. One is a compatible PC and other a VR head-mounted display (VR headset). There are also other VR technologies like a VR room (Bouška & Heralová 2017) etc., but this paper will focus only on VR headsets compatible with the PC platform.

2.1 Overview of relevant software tools

Firstly, proper software tools capable of working with information models had to be selected. The most common way of utilizing 3D models from BIM is to transfer 3D objects into a game engine, like Unity or Unreal engine 4. (Wang et al. 2018) Since this is a rather complicated process, which is not suitable for typically non-IT oriented civil engineers, it is better to focus only on user friendlier solutions. Secondly, the other main requirement is that this software must work with Autodesk Revit, which is the main tool used for teaching BIM at the Czech Technical University (CTU). And thirdly, an important parameter is the price of the product. Based on these criteria the selection of possible tools was reduced to two. The first product that was taken into consideration is directly from Autodesk. It was the Revit Live cloud service that turns Revit models into an immersive experience, helping architects to understand, explore, and share their designs. (Revit Live 2018) The other software that met the initial criteria was IrisVR Prospect. Prospect reads 3D files and automatically turns them into an immersive, navigable virtual reality environment. It can manage VR files within the Prospect Library plus schedule and host virtual meetings. (IrisVR Prospect 2018) Both these products are very capable. Their biggest advantage is that the process of generating the virtual environment is done automatically through add-ons installed directly into Revit. A detailed comparison of these two products is presented below in Table 1.

2.2 Overview of relevant hardware tools

Based on supported hardware, two VR platforms were taken into consideration. HTC Vive features a precise 360-degree controller and headset tracking, realistic graphics, directional audio and HD haptic feedback mean realistic movement and actions in the virtual world. (VIVE™ 2018) Oculus Rift (Oculus Rift 2018) offers a similar experience. The biggest difference between these two tools is the availability of room scale feature out of the box in the case of HTC Vive. This gives students more freedom of movement and is more suitable for BIM applications. If students want this feature on Oculus Rift, they need to buy an additional

Table 1. Comparison of VR software.

Main features	Revit Live	Iris VR prospect
Supported HW	HTC Vive, Oculus Rift	HTC Vive, Oculus Rift
Supported SW	Autodesk Revit	Autodesk Revit, Sketchup, Fbx, Rhino
Cloud model conversion	Yes	No
Mark up tool	Yes	Yes
Multiuser meetings	No	Yes
Model scaling	Yes	Yes
BIM data information	Yes	No
Dynamic sun positioning	Yes	Yes
Dynamic interior lightning	No	Yes
Capturing screen shots	Yes	Yes
Layers manager	No	Yes
Measuring tools	Yes	Yes
3D sectioning	No	Yes
Price	2755 $/year	2700 $/year
Educational license	No	Yes

Table 2. Comparison of VR hardware.

Main features	HTC Vive	Oculus Rift
Display	OLED, 2160 × 1200, 90 Hz	OLED, 2160 × 1200, 90 Hz
Field of view	110°	110°
Maximum tracking area	5 × 5 m	2,5 × 2,5 m
Built-in audio/mic	Yes	Yes
Controller	Vive controller, PC compatible gamepad	Oculus Touch, Xbox controller
Compatible graphic card	NVIDIA GeForce GTX 970/AMD Radeon RX 480 or greater	NVIDIA GeForce GTX 960/AMD Radeon RX 470 or greater
Compatible processor	Intel Core i5-4590 equivalent or greater	Intel Core i3-6100/AMD FX4350 or greater
System memory	4GB+ of RAM	8GB+ RAM
Connections	HDMI 1.3, 1x USB 2.0 port	HDMI 1.3, 2x USB 3.0 ports
Price	499 $ (*HTC*)	399 $ (*Oculus*)

sensor that however in comparison with the former still offers a smaller space where they can move around and yet still be trackable by the VR sensors. It should also be noted that HTC is about to launch HTC Vive Pro (VIVE Pro 2018), which will be introducing a wireless headset and improved sound and display technology. Likewise, HTC Vive Pro will also include wireless technology that shall significantly improve user experience. Since it is not available yet, this option could not have been taken into account during the design of the VR lab. A detailed comparison of these two products is presented below in Table 2.

In order to build a VR PC, minimal specifications for the selected VR platform were used. The next step in the implementation of VR into the teaching syllabus at CTU was to design a suitable environment for students to try virtual reality. Both the size of the group and type of lecture assignment are directly related to the number of VR headsets and supported features of the chosen VR platform. All this will be discussed in the following part of the paper.

3 DEVELOPMENT OF VIRTUAL REALITY LAB AND ASSIGNMENT

The comparison of available VR software solutions was complicated. Both products offer similar functionality and performance. As is shown in Table 1. above, the main advantage of Live is its more advanced BIM information viewer, which however, is also planned to be improved in the future versions of Prospect. On the other hand, Prospect offers offline model conversion, multiuser VR meetings, dynamic interior lighting, layer management and 3D sectioning. (IrisVR Prospect 2018) All these functions are so far missing in Live. Last but not least, Prospect offers an educational license, which is very important for educational institutions.

Therefore, once Prospect had been selected as the most suitable for our educational purposes, the next step was selecting a VR headset. As Table 2. above shows, again both platforms offer a very similar experience. Oculus rift seems cheaper, however, the price does not include a 3rd base station for VR room experience. On the other hand, HTC Vive offers more freedom of movement, a bigger maximum tracking area right out of the box.

3.1 Design of virtual reality lab

After selecting a combination of Prospect + HTC Vive as the basis for creating the VR laboratory, the next phase was to make a spatial plan of the VR laboratory, as show on Figure 1, below.

The following basic requirements for the space were set:

- Minimum size of the space was set to 6,5 × 7 m.
- Place for other students to wait for their turn
- Data projector with screen to show spectators progress of the VR assignment
- Dedicated place for VR PC and storage for VR gear.

3.2 Preparation of student assignment

The final part of the preparation for including VR into BIM related courses was to prepare an individual assignment for each student. As all the students are asked to create their own information models, the VR assignment can be more general and the scope can be modified by the lecturer. Figure 2 shows the basic algorithm for individual student assignments.

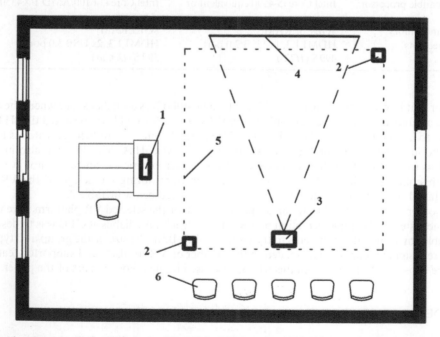

Figure 1. Scheme of the VR Lab: 1 – VR PC; 2 – VR base stations; 3 – data projector; 4 – screen; 5 – border of tracking area; 6 – chairs for waiting students.

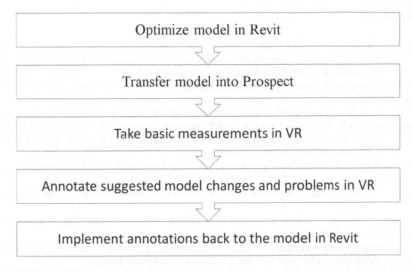

Figure 2.   Scheme of basic student assignment workflow.

These student assignments require the students to first clean and optimize the model, that includes purging the model from unnecessary data.

The following step is simply using Prospect plugin to transfer the model into the Prospect app. The subsequent parts are already done in virtual reality. Students are asked to do some specific measurements, like the distance between the floor and ceiling. They are also requested to redline their models and make some comments and annotations. This includes visually checking if the model contains unwanted geometry (for example some backup copies of objects).

The final part of the assignment is to implement all the suggestions back into the Revit model and possibly repeat the algorithm.

The possibility of doing other tasks is strongly limited by the current sophistication of the virtual reality software, which is at present more focused on delivering presentations to clients, than actual modelling tasks as such.

4   CONCLUSIONS

Out of the two possible tested virtual reality solutions, IrisVR Prospect, allows the highest versatility with regard to student assignments and likewise offers an educational license at a significantly reduced price. At an almost identical price, HTC Vive offers more freedom of movement and interaction with the BIM model, which is better suited for the teaching environment. The use of VR with BIM models is still in its beginnings, there is a rapid development of new tools that can improve the learning and understanding of the construction process. Assignments for the students are limited by the current state of the technology. They should enable the initial introduction into the virtual reality and its applications in civil engineering. The follow-up research will focus on improving the learning tools for the BIM/VR environment and bring more functionality to the existing tools. Mainly not only manipulating 3D objects, but also manipulating and showing information data attached to those 3D objects.

ACKNOWLEDGEMENTS

This work was supported by the Grant Agency of the Czech Technical University in Prague, grant No. SGS17/121/OHK1/2T/11.

REFERENCES

Abdul-Hadi G. Abulrub 2011. *IEEE Global Engineering Education Conference (EDUCON), 2011: 4–6 April 2011, Princess Sumaya University for Technology in Amman, Jordan.* Piscataway, NJ, IEEE.

Bouška, R. & Heralová, R.S. 2017. Opportunities for use of Advanced Visualization Techniques for Project Coordination. *Procedia Engineering.* [Online] 196, 1051–1056. Available from doi:10.1016/j.proeng.2017.08.061.

Cernohorsky, Z. & Matejka, P. 2017. Initial investment to 3D printing technologies in a construction company. *Business & IT.* [Online] VII (1), 14–19. Available from doi:10.14311/bit.2017.01.03.

Hromada, E. 2017. Analysis of the real estate market in the Czech Republic. *Business & IT.* [Online] VII (1), 32–37. Available from doi:10.14311/bit.2017.01.05.

Matějka, P. & Tomek, A. 2017. Ontology of BIM in a Construction Project Life Cycle. *Procedia Engineering.* [Online] 196, 1080–1087. Available from doi:10.1016/j.proeng.2017.08.065.

Oculus Rift | Oculus. [Online]. Available from: https://www.oculus.com/rift/ [Accessed 23rd March 2018].

Pricing Plans for Prospect and Scope | IrisVR. [Online]. Available from: https://irisvr.com/pricing [Accessed 23rd March 2018].

Revit Live | Immersive Architectural Visualization | Autodesk. [Online]. Available from: https://www.autodesk.com/products/revit-live/overview [Accessed 25th March 2018].

Strnad, M. 2017. Building information modeling in budgeting. *Business & IT.* [Online] VII (2), 10–17. Available from doi:10.14311/bit.2017.02.02.

Vitásek, S. & Matějka, P. 2017. Utilization of BIM for automation of quantity takeoffs and cost estimation in transport infrastructure construction projects in the Czech Republic. *IOP Conference Series: Materials Science and Engineering.* [Online] 236, 12110. Available from doi:10.1088/1757-899X/236/1/012110.

VIVE Pro | The professional-grade VR headset. [Online]. Available from: https://www.vive.com/us /product/vive-pro/ [Accessed 23rd March 2018].

VIVE™ | VIVE Virtual Reality System. [Online]. Available from: https://www.vive.com/us/product/vive-virtual-reality-system/ [Accessed 23rd March 2018].

Wang, C., Li, H. & Kho, S.Y. 2018. VR-embedded BIM immersive system for QS engineering education. *Computer Applications in Engineering Education.* [Online] 112, 197. Available from doi:10.1002/cae.21915.

Welcome To IrisVR Prospect (Desktop App). [Online]. Available from: https://help.irisvr.com/hc/en-us/articles/216406967-Welcome-To-IrisVR-Prospect-Desktop-App- [Accessed 25th March 2018].

Xinxing, T. 2012. *Virtual Reality—Human Computer Interaction.* InTech.

Advances and Trends in Engineering Sciences and Technologies III – Al Ali & Platko (Eds)
© *2019 Taylor & Francis Group, London, ISBN 978-0-367-07509-5*

Impact of flow rate on the water film thickness of water wall

K. Cakyova, F. Vranay & M. Kusnir
Faculty of Civil Engineering, Institute of Architectural Engineering, Technical University of Kosice, Kosice, Slovakia

ABSTRACT: The presented paper investigates the impact of the water film thickness on the ability of evaporation from the falling water film of the water wall. The physics of these processes is quite complex, involving conjugate heat transfer among solid-liquid film-gaseous phase together with the change of phase associated with evaporation. These processes depend not only on air and water temperatures, but also on the thickness of the water film and air velocity. For this purpose, a water wall prototype was designed and constructed, where water runs down the glass pane and forms the water film. The experiment was carried out in three measurement cycles where the water flow rate was set in a range of 300 l/h to 500 l/h. There is an assumption that with different flow rate the water film thickness and evaporation ability of the water wall is different. The experiment is carried out under laboratory conditions, temperatures of water, air and relative humidity are similar.

1 INTRODUCTION

Humidity and air temperature are the main parameters of the six primary physical factors that must be achieved for optimal thermal comfort in the environment, and have attracted considerable attention in the field of Heating, Ventilating, Air Conditioning, and Refrigeration (ASHRAE, 2010).

Humidity of indoor air is an important factor influencing energy consumption of buildings, durability of building components, and the perceived air quality. In summer, due to relatively high temperatures, water vapour content in the outside air is significant; the higher the air temperature, the more it is able to absorb water. The air supplied to the interior is then, after cooling to the internal temperature, almost saturated with water vapour, its relative humidity is too high and people can find the air unpleasantly humid. In winter, due to low temperatures, water vapour content in the outside air is low, because it condenses or freezes and falls to the ground. The air supplied to the interior, after heating to the internal temperature, is too dry, and oftentimes the relative humidity falls below 20%. A low level of indoor humidity leads to dry air and is associated with SBS (Sick Building Syndrome) symptoms such as dryness, irritation or itching of the skin and eyes, dry throat and nose, and irritation in the upper airways (Jokl, 2011). In March 2016, a series of measurements focusing on the quality of the environment in the university classroom in Kosice, Slovakia, were carried out. The measurement results showed low relative humidity; the minimum air humidity in the classroom ranged from 27.1% to 38.1% with a mean value of 32.57% (Burdova et al., 2016). Similar research was carried out in primary school classes in October 2016. Measurements demonstrated that the RH values were in the range of 30% to 70% in accordance with official regulations of the Ministry of Health of the Slovak Republic. 210/2016 Coll. However, 60% of pedagogical staff and 40% of students sometimes felt the air dry, which causes symptoms such as dry or sore throat and itching, burning or irritation to the eyes (Vilcekova et al., 2017). Other surveys show that in office areas with and without air humidification the employees suffer from the various symptoms caused by dry air. If air humidification is used, the discomfort is reduced by nearly half (Rief et al., 2014).

Indoor humidity depends on several factors, such as moisture sources, airflows, moisture exchange with materials and the state of water vapour in the exterior. One of the sources can be also water elements. Design of water elements in the interior can vary, either in shape or material. A water wall, where water flows on the glass surface, is a universal solution in the case of already existing buildings, because it can be added in the interior without requiring changes. The vertical position of the water wall does not take much space in the interior and brings out the natural beauty of falling water. The small footprint gives great possibilities for the construction of water walls in any space; most often they are used in communal areas of hotels, but also in the areas of shopping centres and for the positive effects of water in the working environment too.

The application of falling film evaporation in air conditioning can bring environmentally friendly products and reduce energy consumption. A number of researchers have already studied film evaporation under various boundary conditions. Falling water film evaporation devices have been used in many industrial processes, for example, to cool water, moisten air or as a refrigerant in a refrigeration or air conditioning system (Sosnowski et al., 2013, Sun et al., 2018). Problems related to the heat and mass transfer processes have received considerable attention. The presented document focuses on the water wall where the water circulates without modification to its temperature but with modification to the flow rate. The water film may change and improve the relative humidity and air temperature via the evaporation process. These processes are suitable in a low humidity environment, such as in the case of the Slovak republic in winter season. The water wall acted like a humidifier. During the experiment, the parameters (temperature of water, air, and relative humidity) using different water flow rates were monitored. It is assumed that the thickness of the water film will change with different water flow rates. On the basis of this, the hypothesis was determined that a different water flow rate changes the evaporation ability of the water wall.

2 BOUNDARY CONDITIONS OF EXPERIMENT

2.1 Experimental prototype

For the purpose of verifying the hypothesis, the prototype has been designed. It defines the substantial parts of the water wall. One possible way to create the water film is by overflow over the edge. The construction through which water overflows is called the spillway, the largest part of the spillway is the spillway edge.

The upper collection tank (Figure 1) consists of the water reservoir of rectangular shape made of a polypropylene bonded by melt welding in order to achieve waterproof. Into the

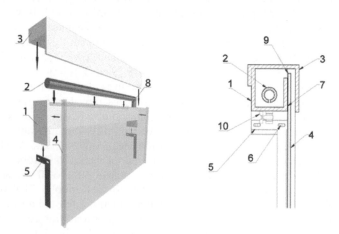

Figure 1. Arrangement scheme of main components of the upper part of prototype and section: collection tank (1), perforated pipe (2), a cover (3), glass pane (4), supporting metal structure (5), screws (6), silicone adhesive (7), water supply (8), spillway edge (9), ball valve (10).

water reservoir is inserted a water supply pipe, which is perforated. These holes are oriented downwardly, this allows a uniform flow of water along the entire length of the reservoir to form the same water film across the entire width of the water wall. The spillway edge with a rectangular cross-section is a perfect overflow (the overflow is not affected by the down water level). The effective water film area is 1 m².

The spillway edge is formed by the glass pane which is connected to the collection tank with a silicone adhesive. The spillway edge must be in a horizontal position. The upper collection tank is anchored to the supporting metal structure, the anchor position can be changed in the horizontal direction. In order to achieve a uniform appearance of the water wall and at the same time to prevent undesired evaporation of water, the tank is covered by a cover. The technical solution is characterized by a certain degree of variability: the possible tilting of the glass pane by a different anchorage position; there is the possibility of replacing the inner pipe with perforation for others, but the perforation condition must be respected.

2.2 Physical assumption and hypothesis

Adjustment of air temperature and humidity is a basic priority and requirement to achieve optimum environment. For the modification of thermal-humidity microclimate, the water wall can be used in two different ways. In the first case, the water wall is used for cooling and dehumidification of air in the space. In this case the water film is cooling and its temperature is below the dew point, so there is condensation. Formed condensate along with a film of water is diverted to the collection tank. In the second case the temperature of the water film is the same as the air temperature in the room so evaporation occurs, which increases the air humidity (Künzel, 2010). The presented experimental prototype of the water wall is focusing on the second case, and so humidification of air by the evaporation process. The physics of these processes is quite complex, involving conjugate heat transfer among solid-liquid film-gaseous phase together with the change of phase associated with evaporation. These processes depend not only on air and water temperatures, but also on the thickness of the water film and air velocity (Min, 2015). There is a premise that the change of flow rate will change the thickness of the water film. However, it is necessary to verify whether the impact of the water flow on the change of the water film thickness is significant in the scale of the water wall.

2.3 Methodology of measurement

Experimental verification of the hypothesis was carried out under laboratory conditions in a climatic chamber where stable conditions can be maintained, the temperature range is from −20°C to + 125°C and the climatic range is from 20% to 95%. Its internal dimensions are 3.95 × 1.60 × 2.85 m and the volume of the air in the chamber is 18.012 m³. The walls of the chamber are made of stainless steel, so there is no mutual exchange of mass between the air and the chamber. During the experiment, the climate chamber was inactive, the experiment focused on the interaction of the water wall with the volume of air, with the possibility of excluding the absorption of surrounding components.

To measure the amount of evaporated water, a simple solution was proposed with the measuring cylinder (Figure 3). Once the experiment is started, the amount of water in the system will be recorded after flooding parts of the water wall. Subsequently, during the experiment, the water from the lower collection tank will flow into the measuring cylinder. From this the water is pressed towards the upper part of the water wall by means of a pump. The globe valve is used to regulate the water speed and the flow meter is used to monitor the exact amount. During the running of the water wall, it has been found that the optimal rate of water flow is from 300 l/h to 500 l/h, at lower speed, the water film was not formed in the whole width and at a higher speed, water bounced off the water film.

In the chamber was installed a set of measuring sensors (Figure 4). The temperature and relative humidity sensors are connected with the AHLBORN control unit. The AMR Win Control software was used to gain and collect measured data. This has ensured continuous recording of the measured values in a time step of 5 min. Two sensors measured the water

Figure 2.   Final prototype of testing water wall.

Figure 3.   Water wall measurement system.

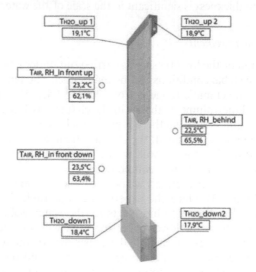

Figure 4.   Set of sensors.

temperature at the bottom tank and two measured the water temperature in the upper tank. From these sensors, the average water temperature for the top and bottom part was determined. Two sensors were then installed to measure the temperature and the relative humidity of the air in front of the water film and behind the glass.

3   RESULTS AND DISCUSSION

The series of measurements was performed with approximately the same initial boundary conditions (air temperature 22°C and RH 70%, water temperature 17.5°C). At the beginning of the measurement, it was not possible to ensure the same initial conditions as well, the difference in the initial conditions was minimal. The walls of the chamber are made from stainless steel, thanks to which the moisture absorption to the walls was prevented during the experiment.

The three measuring cycles were carried out at a flow rate of 300 l/h, 400 l/h and 500 l/h. The experiment was 6.5 hours in duration with two repetitions for each flow. Graphs (Figure 5–7) show that the water temperature and the air temperature equalize after 2 hours for each flow. At the end of the experiment, the average temperature, RH and specific air humidity at flow rates 300 l/h were 23.19°C, 91.83% and 16.17 g/kg; at flow rate 400 l/h were 23.02°C, 92.57% and 16.47 g/kg; at flow rate 500 l/h were 23.41°C, 92.77%, 16.91 g/kg. Versus the initial state the specific humidity increased by 4.5 g/kg at a flow rate of 300 l/h, 5.39 g/kg at a flow rate of 400 l/h and 4.9 g/kg at a flow rate of 500 l/h. These different values are caused

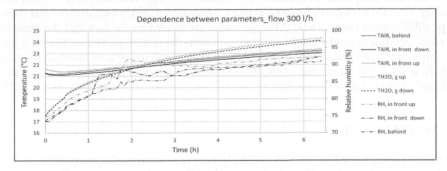

Figure 5.   Evaporation process from the surface of the water film at a flow rate of 300 l/h.

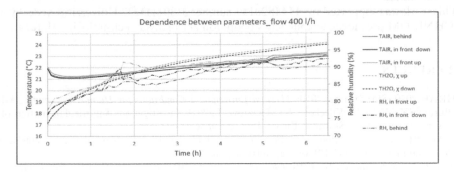

Figure 6.   Evaporation process from the surface of the water film at a flow rate of 400 l/h.

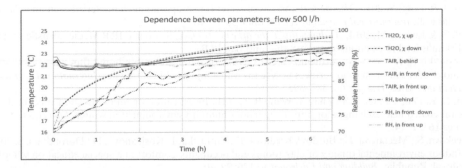

Figure 7.   Evaporation process from the surface of the water film at a flow rate of 500 l/h.

341

by the fact that it was not possible ensure the same initial conditions. However, the data show that the resulting temperature and relative humidity of air are almost the same at all three test flow rates.

4 CONCLUSION

Nowadays, when energy consumption in architecture is growing, it is important to use every element that is in the building. The water wall is a decorative element, but in addition to the decorative character it has the potential for changing the indoor climate parameters, but it is necessary first to experimentally test it. The article has presented testing of the water wall prototype at three different flow rates. Testing is based on the assumption that different flow rates strongly affect the water film thickness and therefore the amount of water evaporation. It has been demonstrated that evaporation ability of the water wall at different flow rates changes only minimally which is negligible in this scale. At the end of the measured cycles, the water temperature reached approximately 24°C, air temperature 23°C and relative humidity 92%. Comparison of graphs showed minor deviations, which are low and therefore it is possible neglect them. These different values are caused by the fact that the same initial boundary conditions could not be secured, however, the results show that the final parameters (air temperature, relative humidity, water temperature) are almost identical for all three tested flow rates. The presented contribution is part of the doctoral study and another part will focus on the condensation ability of the water wall under different boundary conditions.

ACKNOWLEDGEMENTS

This work was supported by: VEGA 1/0202/15 Bezpečné a udržateľné hospodárenie s vodou v budovách tretieho milénia/Sustainable and Safe Water Management in Buildings of the 3rd. Millennium.
    This work was supported by: ITMS2014+ number 313011D232 Univerzitný vedecký park TECHNICOM pre inovačné aplikácie s podporou znalostných tech-nológií—II. fáza.

REFERENCES

ASHRAE 2010. Standard 55–2010: *Thermal Environment Conditions for Human Occupancy.* Atlant-aBurdova Kridlova, E.; Vilcekova, S.; Meciarova, L. 2016. Investigation of particulate matters of the university classroom in Slovakia. Energy Procedia 96: 620–627.
Jokl, M. 2011. *Teorie vnitřního prostředi budov.* <Available online: http://www.ib.cvut.cz/sites/default/files/Studijni_materialy/TVPB/Teorie_vnitrniho_prostredi.pdf>
Künzel, H.M. & Sedlbauer, K. 2010. *Chilled water wall.* IBP – FRAUNHOFER. <Available online: https://www.ibp.fraunhofer.de/content/dam/ibp/en/documents/Informationaterial/Departments/Hygrothermics/Produktblaetter/IBP_255_PB_Klimabrunnen_neu_en_rz_web.pdf>
Min, J. & Tang, Y. 2015. Theoretical analysis of water film evaporation characteristics on an adiabatic solid wall. *International journal of refrigeration* 53: 55–61.
Rief, S. & Jurecic, M. 2014. *Air humidity in the office workplace.* Stuttgart: IRB Mediendienstleistungen Fraunhofer-Informationszentrum Raum und Bau IRB.
Sosnowski, P.; Petronio, A.; Armenio, V. 2013. Numerical model for thin liquid film with evaporation and condensation on solid surfaces in systems with conjugated heat transfer. *International Journal of Heat and Mass Transfer* 66: 382–395.
Sun, X.Y., Dai, Y.J.; Ge, T.S., Zhao Y. Wang R.Z. 2018. Investigation on humidification effect of desiccant coated heat exchanger for improving indoor humidity environment in winter. *Energy and Buildings* 165: 1–14.
Vilcekova, S.; Meciarova, L.; Burdova Kridlova, E.; Katunska, J.; Kosicanova, D.; Doroudiani, S. 2017. Indoor environmental quality of classrooms and occupants' comfort in a special education school in Slovak Republic. *Building and Environment* 120: 29–40.

Advances and Trends in Engineering Sciences and Technologies III – Al Ali & Platko (Eds)
© 2019 Taylor & Francis Group, London, ISBN 978-0-367-07509-5

Using wooden materials for heavy metals removal from waters

S. Demcak, M. Balintova & Z. Kovacova
Faculty of Civil Engineering, Košice, Slovakia

M. Demcakova
Slovak Academy of Science, Košice, Slovakia

ABSTRACT: The aim of this article is a study of the heavy metals removal by wooden material. The adsorption experiments were carried out using wood sawdust and bark from poplar, cherry, and spruce trees for the removal of Cu(II) and Zn(II) from the model solutions with initial concentration of 10 mg.L^{-1}. The FT-IR spectra of wooden materials confirmed the presence of functional groups that have potential for heavy metal binding from the model solutions. The poplar wooden sawdust had the best efficiency of approximately 80.0% for Cu(II) and Zn(II) removal from the model solutions. The changes of the pH values after the adsorption experiments were also measured.

1 INTRODUCTION

The intensification of industry in last decades had a negative impact on all parts of the environment, especially on aquatic environment. The discharging of industrial effluents is a major source of contamination of water resources by heavy metals (Igwe, 2007). Heavy metal ions must be removed from contaminated waters due to their mobility in natural aquatic ecosystems and due to their toxicity, because they are stable and persistent environmental contaminants that cannot be biologically degraded and destroyed. These metal ions can be harmful or toxic to living organisms in water and can also cause a serious health problem for human (Tangahu et al., 2011; Demcak et al., 2017b). There are several methods available to achieve the reduction of heavy metals concentration in wastewater that can increase the quality of the environment in the affected localities and thus prevent adverse effects on living organisms (Babel and Kurniawan, 2003). The most commonly used methods for the heavy metals removal from aqueous solutions are chemical precipitation, membrane filtration, electro extraction, ion exchange and sorption (Fu and Wang, 2011; Zinicovscaia and Cepoi, 2016).

The heavy metal removal by the chemical precipitation may be costly in many cases. The precipitation requires relatively large amounts of space for the clarifier, produces a wet, bulky sludge and it generally requires final purification of the residual levels of heavy metals (Hsu et al., 2008). Other available processes such as ion exchange, reverse osmosis, adsorption on activated carbon, and solvent extraction are relatively expensive, involving either elaborate and costly equipment or high costs of operation and energy requirements. The ultimate disposal of the contaminants may also be a problem with some of these techniques (Blázquez et al. 2011). To achieve wide-spread removal of heavy metals from water sources, a more efficient and low-cost process is needed. Recently, biosorption has attracted growing interest. Using inexpensive sorbents, biosorption can achieve high purity in treated wastewater (Kratochvil and Volesky, 1998; Demcak et al., 2017b).

The natural sorbent materials are available in large quantities, easily regenerable, and cheap. The sorbents bind molecules by physical attractive forces, ion exchange, and chemical binding (Hashem, 2007). The application of natural sorbents such as zeolite, clay, sawdust, bark, leaves, rice husk, and peat in environmental treatment of wastewaters contaminated by heavy metals has become a significant research area in the last decade (Demirbas, 2008). Physic-chemical processes based on adsorption of natural organic materials are cheap and effective techniques for metals removal from wastewater (Barakat, 2011). The natural organic sorbent materials originated from agricultural wastes or by-products from timber industries can be used for effective removal and recovery of heavy metal ions from wastewater streams (Bailey et al., 1999; Crini, 2006; Keränen et al., 2016). A major benefit of the bio-sorption technology is the effectiveness in reducing the concentration of heavy metal ions to very low levels and the use of inexpensive natural organic sorbent materials (Demcak et al., 2017a).

The heavy metals removal from aqueous solutions using wooden materials such as sawdust and bark is a relatively new process which has been proven very promising in the removal of contaminants from aqueous effluents (Blázquez et al. 2011, Balintova et al. 2016). The major benefits of wooden materials as adsorbents over conventional treatment methods include: low-cost, high efficiency, minimization of chemical and/or biological sludge, regeneration of bio-sorbent, no additional nutrient requirement and possibility of metal recovery (Ahalya et al., 2003). Local availability of various kinds of wooden sawdust and bark give prerequisites for their use as a promising low-cost sorbent material for the removal of heavy metals from contaminated aquatic environment.

In this study the removal of Cu(II) and Zn(II) by using poplar, cherry, and spruce wooden sawdust and barks was investigated. For characterization of functional groups, which can be responsible for metal binding on the wood sawdust and bark of poplar, the infrared spectrometry was used.

2 MATERIAL AND METHODS

The sawdust and bark of poplar, cherry, and spruce trees were dried and sieved. For the adsorption experiments were used the fractions with a particle size under 2.0 mm (sawdust) and max. 8.0 mm (bark). The FT-IR measurements of the wooden materials were performed on a Bruker Alpha Platinum-ATR spectrometer (BRUKER OPTICS, Ettingen, Germany). A total of 24 scans were performed on each sample in the range of 4,000 to 400 cm^{-1}.

The model solutions with initial concentrations of Cu(II) and Zn(II) 10 mg.L^{-1} were prepared by dissolving of $CuSO_4.5H_2O$ and $ZnSO_4.7H_2O$, respectively in deionised water. Concentrations of the Cu(II) and Zn(II) were determined using the colorimetric method with a Colorimeter DR890, (HACH LANGE, Germany) and the appropriate reagents. The input pH value of model solutions was measured by pH meter inoLab pH 730 (WTW, Germany).

The batch adsorption experiments were carried out on static conditions, where 1 g of each dry adsorbent material was mixed with 100 mL of each model solution. The sorbent-sorbate interaction time was 24 hour. After the end of experiments, wooden sawdust and bark were removed by filtration through a laboratory filter paper. The residual concentrations of appropriate ions in filtrate were determined by colorimetric method and pH changes were also measured. The efficiency of ion removal η (in%) was calculated using the following equation (Equation 1):

$$\eta = \left(\left(c_0 - c_e \right) / c_0 \right) \cdot 100\%, \tag{1}$$

where c_0 is the initial concentration of appropriate ions [mg.L^{-1}] and c_e equilibrium concentration of ions [mg.L^{-1}]. All adsorption experiments were carried out in triplicate under the batch conditions and results are given as arithmetic mean values.

3 RESULTS AND DISCUSSION

3.1 Infrared spectra of wooden sawdust and bark

The utilization of IR spectroscopy is related to the basic characterization of functional groups of sorbents that are involved to the mechanism of heavy metal binding. The metal ions adsorption capacity is strongly influenced by the surface structures of methyl, carbonyl, carboxyl, and hydroxyl functional groups which are present in natural wooden organic materials (Ricordel et al., 2001). Functional groups of poplar, cherry, and spruce wood sawdust and bark were determined using FTIR spectroscopy and their IR spectra are shown in Figures 1–3.

Poplar wooden sawdust was detailed studied and characterized by Demcak et al. (2017b). As can be seen from Figure 1, the IR spectrum of the poplar bark (the deadwood part of the wood pole) has a similar pattern to sawdust. A strong representation of hydroxyl functional groups is visible in both structures (3,650–3,000 cm^{-1}). The characteristic IR absorption bands of bark and sawdust indicating that the functional groups present of poplar on the surface (bark) and in the core (sawdust) of poplar are similar. The IR spectra of bark showed a strong band at 3,287 cm^{-1} indicating the presence of hydroxyl groups. The peaks at 2,918 and 2,890 cm^{-1} are due to the C–H stretching frequency and the peak at 1,622 cm^{-1} is due to C = O stretching mode of the primary and secondary amides (NH$_2$–C = O) (Reddy et al., 2010). The peak at 1,512 cm^{-1} is indicative of the N–H stretching of the primary and secondary amides, and the presence of amide or sulphamide band, respectively. The band at 1,317 cm^{-1} indicates a presence of carboxylic acids (Reddy et al., 2012). Weak band at 1,512 cm^{-1} is attributed to aromatic CC and two sharp peaks in area from 1,750 to 1,600 cm^{-1} which are characteristic of carbonyl group stretching were also observed. The strong C–O band at wavenumber 1,156 cm^{-1} was also observed in poplar bark (Reddy et al., 2010).

The FT-IR spectra shown on Figure 2 revealed very similar spectra for cherry sawdust and bark that confirmed that both have similar functional groups like poplar sawdust and bark (Figure 1). On the other hand, two changes in the bark FT-IR spectrum were also observed. In the cherry bark spectrum was found presence of the more intensive peaks of symmetric and asymmetric aliphatic C–H bond at wavenumber 3,000–2,850 cm^{-1} (Blázquez et al., 2011). The second change in spectrum was observed on the peaks intensity at wavenumber area 1,750–1,250 cm^{-1}. These changes of peaks intensity could be caused by different ratio of hemicellulose, cellulose and lignin in the cherry bark.

The spectra of spruce sawdust and bark are shown on Figure 3. In bark FT-IR spectrum was revealed significant changes. The strong deformation bands at wavenumbers 2,931;

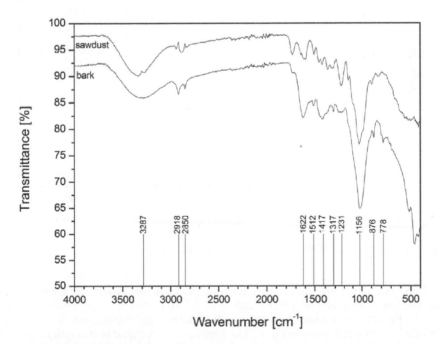

Figure 1.   Infrared spectra of poplar wooden sawdust and bark.

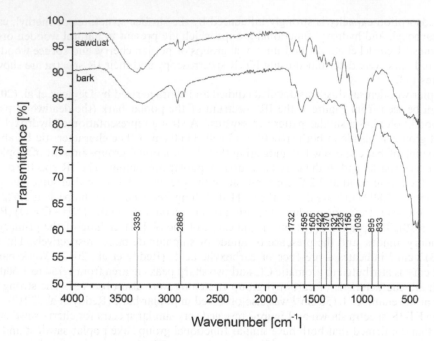

Figure 2.   Infrared spectra of cherry wooden sawdust and bark.

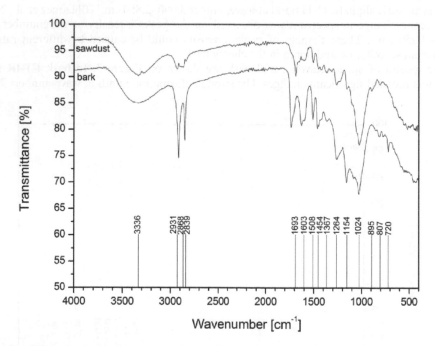

Figure 3.   Infrared spectra of spruce wooden sawdust and bark.

2,868; 2,839 and cm^{-1} were due to the stretching mode of symmetric and asymmetric C–H bonds. The intensive peaks at wavenumber area 1,700–1,200 cm^{-1} may be due to the presence of aromatic rings or rings with CC bonds. Additionally, the deformation at wavenumber 895 cm^{-1} can be also assigned to C–Hn aliphatic or aromatic bonds (Bilba et al., 2007; Blázquez et al., 2011).

Table 1. Results of sorption experiments with selected wooden sawdust and bark.

Sorbents		Cu(II) Initial pH = 4.6			Zn(II) Initial pH = 4.7		
		c_e [mg.L^{-1}]	η [%]	pH	c_e [mg.L^{-1}]	η [%]	pH
Poplar	sawdust	1.58	84.2	5.4	1.92	80.8	4.8
	bark	2.97	70.3	5.2	2.05	79.5	4.6
Cherry	sawdust	3.68	63.2	4.6	2.36	76.4	4.9
	bark	4.34	56.6	5.2	3.28	67.2	4.8
Spruce	sawdust	4.38	56.2	4.7	1.37	86.3	4.5
	bark	3.44	65.6	4.7	2.09	79.1	4.6

3.2 Adsorption study of wooden sawdust and bark

The results of the static sorption experiments by the selected wooden sawdust and bark are shown in the Table 1. In the case adsorption by wooden sawdust, the spruce sawdust had the best efficiency (86.3%) of Zn(II) removal from model solution. On the other hand, the poplar sawdust showed a very good efficiency (over 80%) for the removal of both studied ions from model solutions.

The adsorption properties of the selected wooden sawdust were compared with results obtained at adsorption experiments by selected wooden barks. It was found, that the poplar, cherry, and spruce barks had the higher efficiency for Zn (II) removal in comparison to adsorption of Cu (II) from model solutions. The poplar bark had the best efficiency on Cu(II) and Zn(II) removal from model solutions η = 70.3% and η = 79.5%.

Changes of pH values were also measured in the solutions after adsorption experiments by poplar, cherry, and spruce sawdust and bark. The monitoring of pH changes is an important parameter for the characterization of adsorption processes and mechanism of ion-exchange (Bulut and Zeki, 2007). After the adsorption experiments by wooden sawdust and barks the slightly pH changes was observed. It could be caused with the ion exchange mechanism that was occurred during the adsorption process (Demcak et al., 2017a; Demcak et al. 2017b).

4 CONCLUSIONS

Bio-sorption is a relatively new process that has proven very promising in the removal of contaminants from wastewaters. Wooden materials such as sawdust and barks are inexpensive and for removal of metal ions they can be used in their natural form or after modification by some physical or chemical processes.

The FT-IR analysis of the studied wooden sawdust and barks confirmed the presence of methyl, carboxyl, carbonyl, hydroxyl, and amino functional groups that are responsible for binding of heavy metal ions.

The best efficiency (over 80%) for Cu(II) and Zn(II) removal from model solutions was reached by the poplar wooden sawdust. In all cases the wooden sawdust had better absorption properties in comparison with wooden barks. It could be caused by to their higher specific surface area and smaller fraction. For using of bark as adsorption materials, it would be appropriate to mill it to a smaller fraction and thereby to increase the surface area of the absorbent. The problem of using natural bark as an adsorbent was the leaching of organic-dyes into the solution.

ACKNOWLEDGEMENTS

This work has been supported by the Slovak Grant Agency for Science (Grant No. 1/0563/15).

REFERENCES

Ahalya, N., Ramachandra, T.V. & Kanamadi, R.D. 2003. Biosorption of heavy metals. Research Journal Of Chemistry And Environment 7(4) 71–79.

Babel, S. & Kurniawan, T.A. 2003. Low-cost adsorbents for heavy metals uptake from contaminated water: a review. Journal of hazardous materials 97(1): 219–243.

Bailey, S.E., Olin, T.J., Bricka, R.M. & Adrian, D.D. 1999. A review of potentially low-cost sorbents for heavy metals. Water research 33(11): 2469–2479.

Balintova, M., Demcak, S. & Pagacova, B. 2016. A study of sorption heavy metals by natural organic sorbents. Environments 2(11): 189–194.

Barakat, M.A. 2011. New trends in removing heavy metals from industrial wastewater. Arabian Journal of Chemistry 4(4): 361–377.

Bilba, K., Arsene, M.A. & Ouensanga, A. 2007. Study of banana and coconut fibers: botanical composition, thermal degradation and textural observations. Bioresource technology 98(1): 58–68.

Blázquez, G., Martín-Lara, M.A., Dionisio-Ruiz, E., Tenorio, G. & Calero, M. 2011. Evaluation and comparison of the biosorption process of copper ions onto olive stone and pine bark. Journal of Industrial and Engineering Chemistry 17(5–6): 824–833.

Bulut, Y. & Zeki, T.E.Z. 2007. Removal of heavy metals from aqueous solution by sawdust adsorption. Journal of Environmental Sciences 19(2): 160–166.

Crini, G. 2006. Non-conventional low-cost adsorbents for dye removal: a review. Bioresource technology 97(9): 1061–1085.

Demcak, S., Balintova, M., & Demcakova, M. 2017b. Study of heavy metals removal from model solutions by wooden materials. In IOP Conference Series: Earth and Environmental Science, 92(1), p. 012008. IOP Publishing.

Demcak, S., Balintova, M., Hurakova, M., Frontasyeva, M.V., Zinicovscaia, I. & Yushin, N. 2017a. Utilization of poplar wood sawdust for heavy metals removal from model solutions. Nova BiotechnologicaetChimica 16(1): 26–31.

Demirbas, A. 2008. Heavy metal adsorption onto agro-based waste materials: a review. Journal of hazardous materials 157(2): 220–229.

Fu, F. & Wang, Q. 2011. Removal of heavy metal ions from wastewaters: a review. Journal of environmental management 92(3): 407–418.

Hashem, M.A. 2007. Adsorption of lead ions from aqueous solution by okra wastes. International Journal of Physical Sciences 2(7): 178–184.

Hsu, T.C., Yu, C.C. & Yeh, C.M. 2008. Adsorption of Cu2+ from water using raw and modified coal fly ashes. Fuel 87(7): 1355–1359.

Igwe, J.C. 2007. A Review of Potentially Low Cost Sorbents for Heavy Metal Removal and Recovery. Terrestrial and Aquatic Environmental Toxicology 1(2): 60–69.

Keränen, A., Leiviskä, T., Zinicovscaia, I., Frontasyeva, M.V., Hormi, O. & Tanskanen, J. 2016. Quaternized pine sawdust in the treatment of mining wastewater. Environmental technology 37(11): 1390–1397.

Kratochvil, D. & Volesky, B. 1998. Biosorption of Cu from ferruginous wastewater by algal biomass. Water Research 32(9): 2760–2768.

Reddy, D.H.K., Seshaiah, K. & Lee, S.M. 2012. Removal of Cd (II) and Cu (II) from aqueous solution by agro biomass: equilibrium, kinetic and thermodynamic studies. Environmental Engineering Research 17(3): 125–132.

Reddy, D.H.K., Seshaiah, K., Reddy, A.V.R., Rao, M.M. & Wang, M.C. 2010. Biosorption of Pb 2+ from aqueous solutions by Moringaoleifera bark: equilibrium and kinetic studies. Journal of Hazardous Materials 174(1): 831–838.

Ricordel, S., Taha, S., Cisse, I. & Dorange, G. 2001. Heavy metals removal by adsorption onto peanut husks carbon: characterization, kinetic study and modelling. Separation and purification Technology 24(3): 389–401.

Tangahu, B.V., Sheikh Abdullah, S.R., Basri, H., Idris, M., Anuar, N. & Mukhlisin, M. 2011. A Review on heavy metals (As, Pb, and Hg) uptake by plants through phytoremediation. International Journal of Chemical Engineering 2011(1): 1–31.

Zinicovscaia, I. & Cepoi, L. 2016. Cyanobacteria for bioremediation of wastewaters. Switzerland: Springer. ISBN: 978-3-319-26749-4.

Advances and Trends in Engineering Sciences and Technologies III – Al Ali & Platko (Eds)
© 2019 Taylor & Francis Group, London, ISBN 978-0-367-07509-5

Usage analysis of the information systems for valuation of the construction output

H. Ellingerová, E. Jankovichová, S. Ďubek & J. Piatka
Slovak University of Technology in Bratislava, Slovakia

ABSTRACT: The issue of calculating the prices of construction works, budgeting bidding and contract prices for construction production, are activities considered time and professionally extremely demanding. The aim of the calculations used in construction is determining the amount of the price or the cost of partial or final production. From existing documents and respecting the specifics of construction production. These specifications must be respected and considered in the valuation process. The processor of the budget or of the calculations must have at its disposal an enough range of relevant data and information and support system with the appropriate databases. The processor of the budget often is not able to identify the material or design of the construction. Therefore, as an example in the paper, the trellis is solved on the facade of a building. Individual calculation must not violate the basic calculation principle that only economically justifiable costs can be deducted in the price but only once.

1 INTRODUCTION

Prices are generated in a comprehensible manner by gradually calculating all economically justified costs for individual calculation elements in construction. These form the price structure by the calculation formula. These costs must be based on the volume of construction works, design and construction technology solutions. Free use of the calculation formula causes, that the unit pricing of the companies, have a different content in each item of direct and indirect cost (Hanák 2018).

For example, the cost of procurement of materials can be included in someone in the item cost of materials and with someone in another item of the calculation formula. They are included in other direct costs. (Ostrowski, 2013) Overhead costs may or may not include the cost of construction site or part of it (operating and social part). Also, other titles of subordinate budget costs, etc. This can affect the correct selection of item, but also the overall structure of the budget. When using the guide prices, we need to know their content and structure. The calculation must include all current input components of the price. The basic calculation principle must apply. The price may calculate only economically justified costs and reasonable profit, but only once (Towey 2013).

2 BIDDING METHODOLOGY

Still debated issue is the process of procurement of construction contracts. The Public Procurement Office addressed a professional public in drafting new criteria for the selection of construction contracts when preparing a new Public Procurement Act. In most contests, the lowest supply will win, which is certainly not sufficient for a construction contract as the only evaluation criterion. Consequently, there are situations when bidders take a bold management decision to award the contract only at the level of direct costs without overhead and zero profits. (Kuda et al. 2016) At present, architects are increasingly proposing innovative (atypical) design elements in their design solutions. Designers, budget processors, get into the problem of compiling the budget statement and control budget by which items to appreciate the innovative design.

The annual changes in the legislative business environment and the persisting problems in the construction sector also have an impact on the hitherto prevailing practices in the valuation of construction production. In particular, the level of detailed individual calculation used primarily by contractors or subcontractors for the process of contractual price reconciliation.

When calculating the prices of construction works, it is important to precisely define the performance to be valued. (Nguyen Tien et al. 2017) In this area, we can expect new trends coming to us from abroad, which construction companies gradually have to adapt in Slovakia. For example, in the UK, new measurement rules are used to process the budget statement.

In Germany and Austria, there are already known and regularly updated standards and norms of standard performance, or the price book published by the companies at the level of direct cost rates. None of them are prescribed by law, but strictly respected by professional construction publics.

In the process of free pricing is not a methodology for compiling price and cost calculations, and thus the budget statement, prescribed. Only in public procurement is it necessary to apply from 1st January 2006 for all public construction works contracts the provisions of the methodological guideline MCRD (Ministry of Construction and Regional Development) SR, No. 1/2004 on the use of the Statistical Classification of Construction Work. At present, we are still using our calculation system in construction to compile the budget statement in relation to the Building Structures and Works Class (BSWK = TSKP) and the Building Works Class (BWC = TSP). They give a budget statement and budget a basic unified framework.

3 CASE STUDY

3.1 *Input data*

The architect has designed a construction that is not available in our existing price lists of construction works. The budget processor was thus faced with the difficult task of creating a unit price for such an innovative construction. The problem arose as soon as the budget statement was drawn up for the control budget. At the stage when only project documentation with a simple description of this construction was available only in detail for the construction procedure.

In this situation the budget processor has two options:

a. create an item at its discretion and budgeting experience, or
b. to address companies that already have experience with its production and installation.

Budget processors, when creating such an item, often use existing price lists and choose a technologically similar item. Add the classifier code used (TSKP, TSP), for example, by adding an alphanumeric character (number, letter) to highlight the fact that the item requires an individual calculation.

Analysis of items mostly removed, and the unit price of an item manually override. Such an orientated price is often unrealistic and sometimes the unit of measurement is chosen incorrectly, without the ability to report.

For the follow-up analysis of the case study, we state that this is specifically the delivery and installation of an innovative shielding construction called "Trellis made of WPC wooden prisms (Woodplastic). Trellis is a building supplement for a garden or a balcony of a residential building, creating a decorative effect. Trellis can be grouped by other structural elements. They have a different size, a design with an arched and straight segment. Figures 1 and 2 shows an example of a construction visualization and a part of a floor plan showing trellis location on the site.

Figure 1. View of the construction, showing the location of trellis from WPC, XY's internal source.

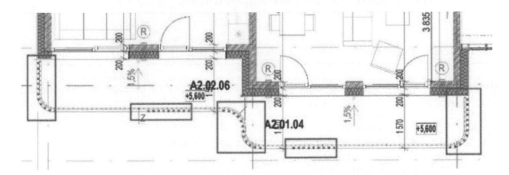

Figure 2. Floor plan of the construction, showing the location of trellis from WPC, XY's internal source.

3.2 *Solving budget statement*

The figure shows that it is a structure that consists of an arched and straight segment. Budget processor, which creates a budget statement and control budget for the building structure, the trellis appreciated the "traditional" assembly items and separate specifications of the material. Both items are expressed in units of measure in meters. The installation item is retrieved from the existing pricing database with the modified TSKP sorting code but retained by the original item description. The specification has an altered material code as well as a description that has been downloaded from the architect (Table 1).

The budget processor instinctively used the budget item relating to the installation of wooden battens but have not been verified in practice the unit price of an existing database is current for such a structure. The general contractor, who was selected by public procurement as the winning tenderer, found from the production dossier only when calculating his own contract costs (in production calculations) that the present structure was entered incorrectly and inadequately specified in the contract. He addressed the specialized company that implements such structures and asked for a quotation. The bid prepared by a specialized company is in Table 2.

To evaluate this innovative technology, the company divided the price tender into three aggregated items that have a specific description, an internal code and a unit of measurement (Table 2). Used aggregate items and their unit prices are appreciated for delivery plus installation of the structure and are separately separated for arched and straight trellis. This technological factor was neglected by the budget processor that compiled the budget statement and the control budget.

Table 1. Part of the control budget, with a budget statement of summaries compiled as a basis for preparing the bidder's bid, in the procurement process.

No.	Item code	Description	Units	Quantity	Unit price (€/unit)	Total price without VAT (€)
		763 – Construction – Wood buildings				107,939.580
10	763792101,1	Installation of other parts of sheet, cloth, with a cross—section up to 25 cm^2	m	9970.080	2.500	24,925.200
11	6051713000,1	WPC trellis 50 × 50 mm	m	10,468.584	7.500	78,514.380
12	998763201	Transfer materials for wooden buildings up to 12 m high	%	1000.000	4.500	4,500.000

Table 2. Part of tendering budget—bid from specialized company.

No.	Item code	Description	Units	Quantity	Unit price (€/unit)	Total price without VAT (€)	Comment
						188,364.96	
218	7666-trellis_a	Delivery and installation of WPC trellis – type a – arc segment, radius 500 mm	pc	226.000	320.760	72,491.760	The connecting material is measured in galvanized steel
219	7666-trellis_b	Delivery and installation of WPC trellis – type b – straight segment, radius 0.9 m	pc	166.000	339.660	56,383.560	
220	7666-trellis_c	Delivery and installation of WPC trellis – type c – straight segment, radius 1.25 m	pc	129.000	461.160	59,489.640	

4 RESULTS

The material basis for cost creation is the consumption of the production factors that must be available in the required quantity and structure for the implementation of planned outputs. When analyzing the structure of costs entering the unit price of the item "Installation of other parts of the sheet, the material with a cross-sectional area of up to 25 cm^2", the unit of measure standard (length of the trellis), selected by the budget processor in the budget for the control budget, was found to enter prices of materials and labor costs of manufacturing workers. An item created by a budget builder from a technologically similar construction work (Table 3).

The calculated unit cost of assembly adjusted list price technology of similar construction (in the view budget processor) is 2.50 €/m (the length of the slats).

The individual unit price calculation assembly trellisWoodplastic(type c – straight segment) is based on input from a specialist company's supply of these types of constructions. Analyzing the cost structure entering to the unit price of the item "Woodplastic trellis installation – type c – straight segment, radius 1,25 m", the unit – piece was found to enter the price of the item. Item created according to inputs of the specialized company and the price base of the contractor (Table 4). Thus, the individually calculated unit cost of installation is 257.31 €/piece (straight segment, type c). After conversion to the unit of measurement of 1 meter, the total length of the slats 10 × 2.7 m = 27 m (which are located in the flat type trellis c), is 9.53 €/m. The calculated unit price (€ 9.53/m) represents approximately 3.8 times the unit cost of the budget processor calculated using a comparable item from existing databases

Table 3. Analysis of item created by budget processor.

Materials	Units	Quantity (Nsm)	Unit price (incl. transport) (€/unit)	Total price (€)
Dowel 8 mm	pc	4.000	0.02	0.10
Bolt 8 mm	tpc	0.00400	98.21	0.39
Total				0.49
Labor cost (profession)				1.28
Additional wages + fees:				0.04 + 0.46
Total labor cost:				1.78
Indirect costs (production s and administrative overheads)				0.16
Total cost:				2.43
Profit				0.06
Unit price [€/m]				2.50

Table 4. An item analysis created by a specialized company's inputs.

Materials	Units	Quantity (Nsm)	Unit price (incl. transport) (€/unit)	Total price (€)
Chemical capsule (concrete anchor)	pac	0.10000	415.47	41.55
Screw combiM 8 * 60 mmM 8*60 mm	pc	30.0	0.07	2.16
Support bracket (system mounting bracket)	pc	10.0	8.23	82.26
Anchor bolt galvanized	pac	0.10000	210.00	21.00
Welding electrode D 4 mm × long. 350 mm, unalloyed	tpc	0.00100	244.29	0.24
Cartridge VS trel. R 9 mm, red	tpc	0.00099	96.43	0.10
Total:				147.31
Labor cost (profession)				38.45
Additional wages + fees:				5.77 + 5.56
Total labor cost:				59.78
Machinery:				35.87
Total cost:				242.96
Profit				14.35
Unit price [€/pc]				257.31

(€ 2.50/m). By incorrect calculation, the contractor's costs for the construction of this construction increased by € 80,425.38 without VAT, which is 74.5% more than the budget processor in the control budget.

The cause was a lack of detailed information on the new construction at the time of processing specification measurement as tender documents for the selection of the contractor. Before the production or realization of a product or performance begins, the whole production process must be prepared in terms of design, technology and organizational. In the construction industry, detailed technical and technological data are always required.

From the quality of preparation depends on the level of planned future cost of performance and the search for solutions for the individual calculation of the bid prices by the contractor.

5 DISCUSSION

Each calculation formula represents a specific algorithm for the practical application of a cost model calculation. When calculating the cost breakdown, costs are allocated according to their relationship to the production process. Calculation allows you to track costs by the purpose of their spending and location. This method is already recognized as a traditional cost and profit calculation that provides information where costs arise and who is responsible for costs. There are several calculation formulas in the history of free pricing of construction works:

Union Calculation Formula—Union Calculation Formulas have been introduced for individual unions. This calculation formula was introduced by the FMF Decree No. VII/224224000/85 valid from January 1st, 1986.

Type Calculation Formula—was declared by FMF Decree no. 21/1990 Coll. of January 30th1990 with effect from 1st February 1990. The Decree was abolished by the new Act on Accounting in 1992.

Uniform Calculation Formula—used in the CENKROS plus and KALKULUS program, which calculates the unit prices of construction and assembly work in the traditional database of items according to the Classification of Structures and Works (TSKP), but also in the new database according to the Classification of Construction Works (TSP). (Systematic s.r.o. 2018)

The algorithm for calculating direct cost items is the same as in the previous formulas, except for wage costs and other direct costs. (Cenekon a.s. 2018)

New Calculation Formula—it is part of the new ODIS price system, which calculates the unit prices of construction and assembly works in the traditional database of items according to the Classification of Construction Structures and Works from 2014. (Odis s.r.o. 2018)

Costs are classified for their purpose of calculating, planning, recording and managing, as well as evaluating the level of individual cost items, uncovering reserves and reducing them.

With the new trends in direct and indirect cost calculations in the unit price of a construction work for example, the new calculation formula is based on the simplified structure of the calculation formula used in the UK. Their calculation formula consists only of items that we refer to as direct costs (labor, materials and goods, machinery and equipment) in our practice. Work" includes basic salary rates, rewards, and other personal expenses.

Both the Unified and the New Calculation Formulas are aimed at accurately calculating direct costs. Indirect costs (overhead) and profit need to be negotiated individually for each order, so adjust to real competitive conditions in a given segment of the region's market.

6 CONCLUSION

A comprehensive system of cost and prices modeling the construction output throughout the procurement process means that all partners have to make the price formation. Mutual satisfaction will be expressed by the investor receiving the expected benefit of the invested funds, the contractor will achieve the expected profit on the realized contract and the designer will have paid his performance at the agreed rate. This "Satisfaction" process has several weaknesses and, therefore, the motives for further solutions. In the paper the authors point to this issue in a case study, which clearly indicates the need to define the details of the procurement object. The responsibility of the contracting authority, particularly in the public sector, is the accuracy and completeness of the quantity take off processing. Thereby preventing an increase in actual costs compared to the calculated cost in the control budget. In the preparation of tender documents by the contracting authorities, the importance of quantity takes off the report is largely underestimated. This must be processed as a specification of the subject of the procurement when selecting the contractor in the procurement process. Information system for the valuation of construction production database using list prices. These are inefficiently complemented by new constructions, which are being used more and more frequently when designing new buildings at present. The databases still contain outdated items, which can cause errors for budget processors and complicate the whole construction production. Innovative designs from architects will increasingly put pressure on individual calculations of new items. This will lead to the necessary change of the information systems for the valuation of the construction production.

REFERENCES

Cenekon a.s. 2018. *Rules for the Use of Indicative Orientationprices Valuation and Calculation Tools and a list of Investment Titles and Costs not Covered, resp. Partially Contained in the SON.* Cenekon, a.s.. Bratislava, ISBN 978-80-971324-3-9.

Hanák, T. 2018. Electronic Reverse Auctions in Public Sector Construction Procurement: Case Study of Czech Buyers and Suppliers. In *TEM JOURNAL—Technology, Education, Management, Informatics*, vol. 7, iss. 1, pp. 41–52. ISSN 2217-8309.

Kuda, F., Wernerová, E., & Endel, S. 2016. *Information Transfer Between Project Stages in the Life Cycle of a Building.* In Vytapeni, Vetrani, Instalace, 25(3), 156–159.

Nguyen Tien, M., Jankovichová, E., Ďubek, S. & Ďubek, M. 2017. The Implementation of Wind Energy Source in the Design of Construction Equipment. In *SGEM 2017. 17th International Multidisciplinary Scientific Geoconference. Vol. 17. Nano, Bio and Green—Technologies for a Sustainable Future.* Conference proceedings. Sofia, Bulgaria: STEF92 Technology, pp. 701–708. ISSN 1314-2704. DOI: 10.5593/SGEM2017H/63/S26.088.

Odis s.r.o. 2018. *Pricing Reports for Construction.* Žilina: Odis s.r.o.

Ostrowski, S.D.C. 2013, *Estimating and Cost Planning Using the New Rules of Measurement.* Wiley-Blackwell.

Systematic s.r.o. 2018. [cit. 20.07.2018] available at: http://www.systematic.sk/web/index.php.

Towey, D. 2013. *Cost management of construction projects.* Wiley-Blackwell.

Advances and Trends in Engineering Sciences and Technologies III – Al Ali & Platko (Eds)
© 2019 Taylor & Francis Group, London, ISBN 978-0-367-07509-5

Using computer simulations in building information modeling

M. Faltejsek & B. Chudikova
Faculty of Civil Engineering, VŠB-Technical University of Ostrava, Ostrava, Czech Republic

ABSTRACT: Building Information Modeling (BIM) is a concept that has gained its position worldwide not only in the field of construction. Its width and absorption of all processes throughout the life cycle of the building supports multidisciplinary cooperation across the building sector. It is a concept that is constantly evolving and shifts its use. One of the neglected elements that surely belong to this concept are the simulations in the 3D BIM model, e.g. wind flow or sunshine/shading simulations. These simulations can help us in different situations throughout the life cycle of the building. They will find their application in the planning, construction, facility management or building's demolition. Simulations have a very wide range and their results help predict future problems or crisis situations, which may occur. They help us decide on the materials, disposition, construction solutions or location of the building. This article describes several such of these simulations and their use in the BIM concept.

1 INTRODUCTION

BIM—Building Information Modeling is a current trend that shifts the construction industry into the sphere of digitization, cooperation and innovative solutions. The term 'construction 4.0' was coined from the industry 4.0 concept, which is regarded as the 4. Industrial revolution. For many years the quiet waters of the construction industry have been stirred by the coming of the construction 4.0. concept. The BIM brings a new look at the construction industry through the entire life cycle of the construction, from its planning stage through the design, implementation and operation to its demolition/reconstruction. The BIM concept is based on a 3D model and data that corresponds with the actual state of the building. Information is important for all phases of the life cycle of buildings and can now be of greater utility value than the 3D model itself. In this regard, the BIM can be interpreted as the Building Information Management. (Kuda, F., Beránková, E., Soukup, P. 2012).

However, the 3D model is still a very useful output and an added value of the BIM concept which cannot be ignored. That includes the availability of a very precise visualization, easy editing, changes to the model transferred directly into the drawing documentation, bill of quantities, etc., but also as a basis for simulation. If we have a model of a construction—its true digital image—we can subject it to many simulations.

With the development of new technologies, software packages and increasing options, the simulations are increasingly becoming the forefront of the information modeling. Their utility value increases with this progress due to the increased accuracy of the results, the possibilities of creating complicated simulations and the linking of the results to the BIM data model. Individual stages of the life cycle of constructions can draw on the results of simulations for their decision-making, which will be supported by relevant results and can prevent the occurrence of crisis situations, increase of the cost of repairs or changes, and save time with correctly set methodologies and procedures.

Simulations can be of many types and can be divided into several categories. Simulation and analysis can be for example applied to building processes, creation of time scale or financial simulations, etc. However, this article will deal with simulating substance movements. These are simulations based on structures, materials, layouts or surroundings of constructions and the effects of temperature, wind, or acoustics imposed on them. The results of the simulation of the propagation of these substances are subject to the knowledge of pertaining conditions, such as wind speed and its direction or temperature at a given moment and at a given point. There are models that can be simplified while you can achieve still achieving relevant results. For example, we can use average or extreme values from the long-term statistics.

Non-standard simulations, which are currently not used widely but can have a high added value include for instance, for instance in simulating the wind flow around the building or simulating the shading created by surrounding built-up area. Such simulations concentrate on the placement of a construction into a built-up area and therefore do not focus solely on the construction itself. Location can have a significant impact on the future operation and functioning of a construction.

2 BUILDING INFORMATION MODELING/MANAGEMENT

Information modeling, alternatively Building Information Management is a modern concept of construction industry across all disciplines. It presupposes their mutual cooperation and maximization of the value of the information in order to reduce costs improve quality and save time. The Information Management means that the main component of our intent is not the 3D visualization model, but the information contained therein and the follow-up work with these data. Effective use of information across all disciplines, mutual exchange, up-to-datedness and availability for all is a key element in increasing the efficiency of the entire process.

The BIM is a modern, efficient and intelligent process of creating projects. The approach facilitates the creation, recording and exchange of data throughout the building lifecycle, from idea/design, to complex project preparation and operation, to its demolition. Its main benefits include more efficient, faster, more economical and more transparent creation and management of any constructions. (Černý, M. a kol. 2013) BIM can be naturally applied to entire areas, complexes or urban development.

2.1 Creating a model for simulation

The model can be created in many ways and with many different software packages. If we are creating a model for complex use throughout the life cycle of a building, a standard BIM model is used. However, if we are creating a model for simulation only, we can make the model much simpler and do not have to have some parts in detail. It always depends on what simulation we want to use the model for and what results we want to get. Whether we are after results of the exterior, interior or just part of the building.

We also differentiate whether we are creating a model for existing constructions (for example, by scanning point cloud, by re-measuring, etc.) or whether we are modeling an entirely new building. (Loyd's Register. 2016) If we are creating a model of an unrealized construction, the model creation is usually in the hands of an architect and his ideas. He or she creates a model in a clean environment of any CAD/BIM software. For the subsequent simulations, we can use such completed and comprehensive BIM model.

The second option is to model existing constructions. We must convert such available construction to the most realistic digital form. The easiest and technically easy-to-use way is to physically go through the object, measure it, take pictures and then create a 3D model from these sources. Another option is to get quality drawing documentation, if it is available and matches the as-built measurements and parameters, and convert it to digital form.

The most accurate alternative to creating a model is to use for instance laser scanning using a point cloud or drones. (Azhar S. 2011) This method is suitable for scanning the entire site or part of an urban development. The desired output is to get a very accurate picture of a large space. The obtained outputs can be used for simulations and they do not have to be difficult to modify into a BIM model. For simulation of e.g. wind flow through a built-up area, we do not have to distinguish individual structural components or particulars of an interior, because all we need is a construction as a solid mass and its surroundings. This also reduces the size of the resulting file.

2.2 Level of development for simulation

The BIM is based on a model that should be identical to the representation of the object in digital form and the information contained in the model itself and the individual elements. Currently we recognize two basic detail levels on which the final Level of Model Definition (LoMD) is based. They are Level of Development (LoD) and Level of Information (LoI) with values between 100 and 500. The information has its own structure and is very important for the longest life cycle of the construction—the operation. [6] However, they can also provide a number of important indications in the decision-making process. It is very important to correctly determine the degree of detail and adhere to the established value. (Černý, M. a kol. 2013).

Only an identical digital model can effectively serve its purpose. It is therefore very important at the start of the model creation to determine in what form is the model going to be made—in terms of both its details and the elaboration. (Charles-Edouard, T., Castaing, Ch., Diab, Y. and Morand, D. 2017).

We can determine the level of elaboration for the whole model or its parts and define those that can be critical and important to us. Too low a detail level may not have a sufficiently predicative value, and it may not be good enough to make decisions and to work with the model. Conversely a too high level of detail might not be necessary and turn out to be too costly, while complicating corrections and updating of the model. Such a model is also data intensive. (Townsend, Anthony M. 2014) For many simulations, however, we need only a simplified model, that is, with a low level of LoMD.

2.3 Simulation software and formats

There are quite a number of formats available. The best-known exchangeable format for use with BIM is the IFC format. (Eastman, Ch., Teicholz, P. and Liston, K. 2011) This format is suitable for exchanging a BIM model between software that directly supports the BIM format and is listed at BuildingSMART, which supports this format globally. However, software for demanding and detailed simulations does not usually support this format. Therefore, you need to export the BIM model to a supported format, such as *.stl or *.obj.

Formats of these types are supported by simulation software because they take the model as a composition of bulk bodies to which different properties, boundary conditions and other attributes important for simulations can be assigned.

Detailed simulations can be performed in software such as COMSOL Multiphysic or ANSYS (fluent). Open-source software for very demanding simulations is for example OpenFOAM.

3 EXAMPLES OF SIMULATIONS APPLICABLE TO THE BIM MODEL

There are many types of simulation software packages already in use for instance for technical facilities of buildings (HVAC), the correct setting of indoor environment in buildings, establishing the category of buildings energy performance certificates etc. However, there are simulations that can help us determine the ideal design solution, material choice or building location in the surrounding built-up area, but are still not made use of. These mainly involve the simulation of exterior types, under which the model is subjected to the BIM simulation as a whole, including the surrounding built-up area and outdoor weather conditions. These are not always easy to determine but one can always turn to the extreme or average values of long-term statistics. (Porter, S., Tele Tan, Xiangyu W. a Vishnu P. 2018).

3.1 Wind flow

The applicability of wind flow through built-up areas simulation is wide. The results indicate the spread of pollution, smoke or pollen through the built-up area. However, simulations can be carried out on much larger scale, for example of the entire region. By evaluating this simulation, we can see if the wind carries odour or pollution from a nearby source directly "to our front door". A supercomputer infrastructure is used in cases of large and complicated areas simulated with very accurate results based on complex and detailed computations. The results can help us identify risk areas emanating from very strong sources of pollution and/or odour.

If we have a BIM model available for solitary, smaller type constructions we can use the simulation of the wind flow, for example, to design a small local wind turbine or to solve the correct placement of air conditioning.

3.2 Sunshine/shading

By placing the construction in a future built-up area we can easily determine whether the construction will or will not be overshadowed by surrounding buildings and to what extent. For example this information is very important to us when looking for a suitable location for

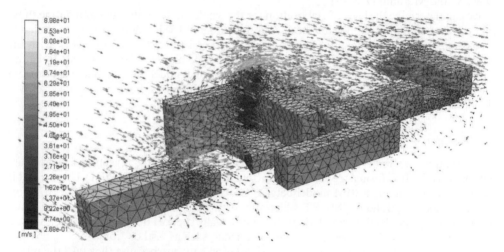

Figure 1.   Wind flow through VŠB-TUO, FAST and risk areas (ANSYS fluent, author).

Figure 2.   Sunshine/Shading model VŠB-TUO, FAST, northern view (ArchiCAD, author).

a solar system. On the basis of the data obtained, we can also decide on the layout, correct arrangements of rooms as well as the size, shape and the cardinal direction windows should face.

According to ČSN 73 4301 Residential buildings have to be designed so that all living rooms are sunny. Calculation of sunlight and insolation is more complex, but in short we can say that according to the standard 1/3 of the floor areas of all living rooms must be reached by sunshine. At least 1/2 of the living quarter floor areas must be covered by sunshine in cases of single residential houses, semi-detached houses and end town-houses. (ČSN 73 4301. 2004).

4 SIMULATION USABILITY

There are many ways to exploit the results obtained by simulations and depend on the type of simulation, the level of complexity and the complexity of the calculation, as well as the individual situation and simulation. Many simulations are commonly used in practice and are supported by legislative requirements. However, there are still many simulations, the value of which is underestimated and misjudged in terms of time or financial constrains. However, even these simulations have their place in the BIM concept. (Vyskočil, V. K., Kuda, F. a kol. 2011).

If we want to know the behaviour of constructions throughout their entire lifecycle, to predict problems that may arise, to know the possible risks or all the costs associated with the construction, operation and demolition of constructions, it is inevitable that many of these simulations become an automated BIM concept process.

The concept itself, which creates a 3D BIM model, prepares very suitable conditions for creating simulations, and unlike standard original 2D drawings the model can be immediately subjected to simulations. Just transfer the model through the exchangeable formats to specialized software.

With technology development, many BIM design software packages will contain these simulations automatically and their results will be more accurate, offer shorter processing times and much easier accessibility. All this leads to a more efficient use of the BIM-related information and the further development of this global concept of building digitization and more effective work with construction information.

5 CONCLUSION

The BIM abbreviation does not just stand for modeling and information. This concept expands every year and its substance, based on the life cycle of constructions, absorbs all the processes that construction's life cycle offers. This also includes simulations and the results obtained from them: about the future behaviour or state of the construction that we get to know before constructions are implemented. The processes that BIM absorbs through the entire life cycle of constructions are also very skillfully and effectively interconnected, creating synergy, thereby increasing the efficiency of individual parts. Basic building information has its own utility value. However, if we know other connectedness the benefits can exponentially increase and conversely the costs, labour or time needed decrease noticeably.

Simulation of future states and constructions behaviour can support information management and create quality decision-making conditions that are presently based on prescribed constants, experiences, or estimates. But these do not have to correspond with real situations. We will never get the perfect real state by simulation when we make allowances for weather conditions because it is impossible to determine their exact value. However, we can count on long-term statistical data, averages or extremes to obtain the necessary oversized results.

The BIM is the concept applied throughout the life cycle of the construction and projection of individual phases into the project phase to guarantee the creation of an ideal model of future construction. If we include the placement of the future construction—the location, the surrounding built-up area, the weather conditions and other related information, we can predict possible problems and modify the ideal design based on the acquired values. This may be, for example, the design of a solar system, the layout of a structure and the cardinal direction windows should face, or the location of a small wind farm.

Decision-making should always take place on the basis of relevant information. That is why we are currently moving towards the trend of the Building Information Management in place of the Building Information Modeling, where we prefer descriptive information over the graphic one. However, all alternatives should be made use of, thus maximum benefit gained. It is clear that complex and demanding simulations can increase the cost of the project itself, but with the rapid development of technologies and options the availability of simulations is becoming increasingly easier. It's all about knowing their added value, purpose and skills needed to work with them effectively.

ACKNOWLEDGEMENT

This work was supported from the funds of the Students Grant Competition of the VSB-Technical University of Ostrava. Project registration number is SP SP2018/113.

REFERENCES

Azhar S. 2011. *Building Information Modeling (BIM): Trends, benefits, risks, and challenges for the AEC industry. Leadership and Management in Engineering, 11 (3): 241–252.*

Černý, M. a kol. 2013. *BIM příručka, 1. vydání, Praha,, ISBN 978-80-260-5297-5, s. 80.*

ČSN 73 4301. 2004. *Obytné budovy. Praha: Český normalizační institute.*

Eastman, Ch., Teicholz, P. and Liston, K. 2011. *BIM Handbook: A Guide to Building Information Modeling for Owners, Managers, Designers, Engineers and Contractors. 978-0-470-54137-1.*

Kuda, F., Beránková, E., Soukup, P. 2012. *Facility management v kostce pro profesionály i laiky, nakladatelství FORM Solution, první vydání, ISBN 978-80905257-0-2.*

Loyd's Register. 2016. *Building information modeling (BIM). Utilities and Building Assurance Schemes [online]. [cit. 2018-07-15]. Available: http://www.lr.org/en/utilities-building-assurance-schemes/building-information-modeling*

Porter, S., Tele Tan, Xiangyu W. a Vishnu P. 2018. *LODOS—Going from BIM to CFD via CAD and model abstraction. Automation in Construction, 94, ISSN 09265805.*

Tolmer, Charles-Edouard, Christophe CASTAING, Youssef DIAB a Denis MORAND. *Adapting LOD definition to meet BIM uses requirements and data modeling for linear infrastructures projects: using system and requirement engineering. Visualization in Engineering [online]. 2017, 5(1). DOI: 10.1186/s40327-017-0059-9. ISSN 2213-7459. Available at: http://viejournal.springeropen.com/articles/10.1186/s40327-017-0059-9.*

Townsend, Anthony M. 2014. *Smart Cities: Big Data, Civic Hackers, and the Quest for a New Utopia. New York: W.W. Norton & Company.*

Vyskočil, V.K., Kuda, F. a kol. 2011. *Management podpůrných procesů Facility management, 2. vydání, Příbram, ISBN 978-80-7431-046-1, s. 492.*

Advances and Trends in Engineering Sciences and Technologies III – Al Ali & Platko (Eds)
© *2019 Taylor & Francis Group, London, ISBN 978-0-367-07509-5*

Management of raw-energy recovery of waste in Slovakia

V. Ferencz
Ministry of Economy of the Slovak Republic, Bratislava, Slovakia

M. Majerník, N. Daneshjo & G. Sančiová
University of Economics in Bratislava, Košice, Slovakia

ABSTRACT: Current global production waste, about 1.3 billion tons/year, is a global environmental problem and a key element of critical infrastructure of individual regions and countries. International and European development strategies are therefore aimed at progressive methods of recycling and recovery of waste, materials, raw materials, energy. Within the paper, an innovative logistics and legislative model of raw material and energy recovery of waste is based on evaluation of the management of waste as a valuable source of raw materials in Slovakia and abroad, in the EU, with an emphasis on progressive trends, designed for the purposes of preventive replacement of exhaustible raw material resources. As a part of decomposition of the proposed model are dimensioned individual material flows through strategic quantified goals, supporting legislation of identification of the flows and other supporting measures of implementation of the model. Authors of papers to proceedings have to type these in a form suitable for direct reproduction by the publisher. In order to ensure uniform style throughout the volume, all the papers have to be prepared strictly according to the instructions set below. The publisher will reduce the camera-ready copy to 75% and print it in black only. For the convenience of the authors template files for MS Word 6.0 (and higher) are provided.

1 INTRODUCTION

Waste as a risk factor for the environment and health is nowadays regulated by strict legislation worldwide. It is the largest environmental polluter and a key element of critical infrastructure in the EU. Waste is a highly potential source of materials, raw materials and energy electrical or thermal, especially for its availability, permanence and regularity, recoverability and high raw material energy content. Based on this, and from the comprehensive assessment of waste and municipal waste management in Slovakia and the European Union, with an emphasis on progressive trends in this area, the contribution presents an innovative model of raw material—energy recovery of waste in Slovakia. The proposed model reflects the objectives of the European Development Strategies, considering the specific aspects of the Slovak Republic, for which the long-term "environmental debt" and low "environmental awareness" belongs.

2 CURRENT APPROACHES TO WASTE RECOVERY

Throughout the world, it is now possible to find many huge landfills as mankind's today produce about 1.3 billion tons of waste per year. The largest landfill in Asia is located in South Korea. The Sudokwon dump, which was established in 1992, will bring about 18 to 20 thousand tons of waste each day, representing about 6.3 million tons per year. The US is the world's biggest waste producer, producing as much as 620 thousand tons of waste a day, or 713 kg per capita and year. Europe has no such large landfills. The biggest landfill in the EU is the Malagrotta landfill in Rome. It produces 4 to 5 thousand tons of waste per day. European nations produce about 503 kg of waste per person. A long-term negative trend in the waste management is waste disposal through landfilling. Since 2010 the quantity of the waste going to landfills has increased to as many as 5 million tons in 2013, which means that waste disposal on landfills in proportion to total waste management has exceeded the level of 50%.

3 EU WASTE MANAGEMENT LEGISLATION

The basic legal norms in the field of waste management in the legislation of the European Union include Directive of the European Parliament and of the Council no. 2008/98/EC on waste. The most important change it brings is a change in attitude towards waste. Waste is no longer a matter that the holder wants or must dispose of, waste is, in particular, a valuable resource that saves primary resources and can also be a significant source of energy. This new attitude, which should actually lead to a recycling society and a circulating economy, is embedded in the waste management hierarchy introduced by the new Framework Directive (Vandák, 2010).

Achieving sustainable social and economic development requires a sound and economical and environmentally effective waste management. We can also declare this based on historical developments in the EU Member States, including Slovakia (Pabian, 2014; Millward-Hopkins et al., 2018; Kolcun, 2009). Expansion of landfilling of all types of waste is unacceptable from the point of view of the creation and protection of the environment and also from the point of view of non-renewable resources is currently unacceptable. The Waste Management Program of the Slovak Republic (WSH SR) is the most important strategic document in waste management for the years 2016 to 2020. It is prepared in line with indicators for sustainable socioeconomic growth. Its content complies with the requirements set out in the legislative regulations of the Slovak Republic and the European Union (EU), particularly in Act no. 79/2015 Coll. Waste Act and European Parliament and Council Directive no. 2008/98/EC (POHSR, 2015).

4 WASTE MANAGEMENT—STRATEGY FOR SR

The logistics of waste management is very closely related to the management of waste collection. Waste management is a sector which directly concerns all stages of production and consumption cycle from raw material extraction through production, transportation and consumption of products to their disposal. Waste Management is a global complex of factors, primarily reflecting the level of utilization of raw material inputs and environmental care. Some of the questions that were previously considered as a local matter are now matters of international and global nature (ISO 9000:2005). To meet the requirements of waste management abroad are processed integrated waste management systems, called IWM (Integrated Waste Management) and ISWM (Integrated Sustainable Waste Management) (Čermák, 2007).

The IWM includes a waste management hierarchy by considering direct impacts (transport, collection, treatment and disposal of waste) and indirect impacts (the use of waste materials and energy outside of the waste management system). It is a framework that can be built on optimizing existing systems as well as designing optimization of new waste management systems (Seadon, 2006). In accordance with generally accepted definition, recycling means the return of solid, liquid and gaseous waste substances into circulation and the re-use of waste energy and heat. As a characteristic sign of recycling, the aspect of dual mitigation of environmental burden is emphasized, i.e.:

- On the side of production system inputs (by using waste, the consumption of natural resources of primary raw materials and energy is reduced).
- On the side of production system outputs (reduction of the quantity of harmful substances released to the environment).

The European Union generates some 2.5 billion tons of waste every year. When comparing the municipal waste generation expressed on a per capita basis, most of the waste was produced in Denmark (759 kg), Cyprus (626 kg) and Germany (618 kg) in 2014. The lowest figures were reported in Romania (254 kg), Poland (272 kg), Latvia (281 kg), the Czech

Republic (310 kg) and the Slovak Republic (321 kg). In the EU, average municipal waste generation per capita has gradually declined from 497 kg in 2011 to 475 kg in 2014. In the period under review, municipal waste increased in Slovakia from 275 kg in 2004 to 335 kg in 2010 and subsequently dropped to 304 kg in 2013 while increasing again to 321 kg in 2014.

The reported quantity of municipal waste generation per capita in the Slovak Republic stands at 67% of the EU average and is the fifth lowest from among the 27 EU Member States under review. Waste disposal on landfills has been a negative trend in the long term. Since 2010 the quantity of waste discarded in landfills has increased to as much as 5 million tons in 2013, which means that waste disposal on landfills in proportion to total waste management has exceeded the level of 50%. Every year, 55,000 tons of waste on average are disposed of by incineration without energy utilization. The share of energy recovery of waste in total waste management is 3%, accounting for some 300 thousand tons of waste a year.

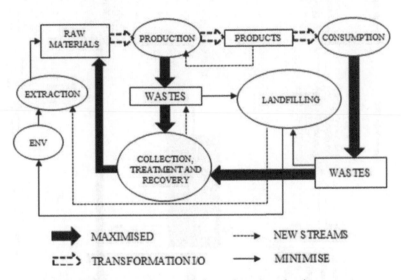

Figure 1.   Raw-material and waste streams—recycling-oriented production.

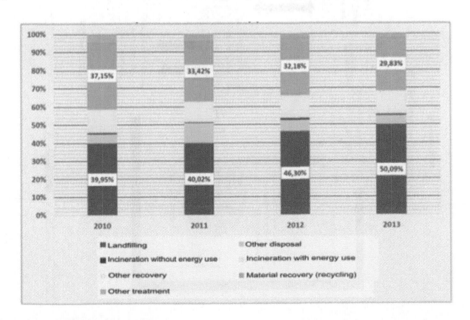

Figure 2.   Material recovery of waste.

Material recovery of waste showed a year-on-year increase, but is not reaching the expected figures and, in comparison with the previous years, the quantity of recycled waste has dropped by as much as 0.5 million tons. In 2013, material recovery accounted for only 30% of waste. The share of other recovery of waste in total waste management represents 10%. Other waste disposal methods have a five percent share in total waste management and other waste management activities accounted for 2% of total waste management in 2013, Figure 2.

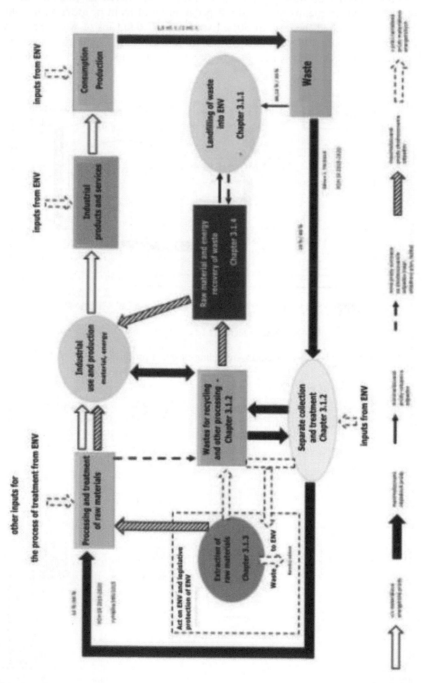

Figure 3.   Logistical-legislative model for material and energy recovery of waste in Slovakia until 2020.

5 MODEL FOR MATERIAL AND ENERGY RECOVERY OF WASTE

Figure 3 shows a logistical-legislative model for material and energy recovery of waste in Slovakia until 2020. The model is based on progressive global and European trends in particular as regards environmental needs in the management of waste with a particular focus on municipal waste; these needs are currently considered inevitable on a nation-wide basis.

The systemic logistical-legislative model for material and energy recovery of waste involves the logistics of waste management from separate collection, additional separation or mixing with coal, wood chips or co-incineration as part of incineration processes for energy purposes, including the relevant legislative support for the quantification of streams and identification of environmental-economic effectiveness of waste recovery. The most recently presented waste quantities from 2014 were compared with the projection and the target quantities to be achieved in 2020. (Ferencz et al., 2017).

To comply with the 50% municipal waste recycling target, it is necessary to substantially increase the level of separate collection of recyclable components of municipal waste, in particular paper and paperboard, glass, plastic, metals and biodegradable municipal waste. Because the separated components of municipal waste are not 100% recyclable due to the quality of raw materials for the recycling process, the targets for the separate collection rate for municipal waste must be higher than the recycling target itself.

6 PROPOSAL OF SUPPORT MEASURES FOR IMPLEMENTING THE MODULE

Based on advanced European and global trends and, for the most part, Slovakia's environmental needs, we deem it necessary—in particular as regards innovation and management in the treatment of waste and predominantly municipal waste—to implement key changes as follows:

- More thorough monitoring (and, where necessary, imposing sanctions) of compliance with statutory requirements for municipal waste separation and ensure that this system is introduced by every municipality and organization.
- Developing additional support methodologies and legislative conditions for intensifying the energy recovery of waste not processed as part of material recovery. By laying the groundwork for better energy recovery and disposal of waste, a synergy-oriented requirement concerning better protection of the environment will also be supported through reducing the share of the use of primary energy sources by substituting them with renewable sources.
- The prices for the goods, products or services represent a very important instrument affecting every area of the economy. In this case this involves the amount of the fee for discarding waste into landfills in the Slovak Republic. At present, this fee is relatively low in the Slovak Republic and does not sufficiently incentivize entities that discard waste into landfills to seek other waste treatment options, in particular its recovery.

7 DISCUSSION

The proposals made within this paper in the form of possibilities to improve effective recovery of municipal waste in Slovakia were compared against the solutions of other authors as well as with the applicable advanced legislation with a view to justifying their feasibility and reasonability. One of the proposals is aimed at amending the legislation governing the preparation of WMP for 2016–2020. Over the past period, there was no legal framework for the strategic planning of waste management; this brought a stagnation rather than improvement of the situation in particular as regards the percentage increase of the share and other areas related to waste management, and these expectations were not met also due to the quality of innovation and project management. The key objective of waste management of the Slovak Republic until 2020 is the minimization of negative effects of waste generation and management on human health and the environment. In order to achieve the set targets, it will be necessary to more vigorously promote and comply with the binding waste hierarchy in order to increase the waste recycling and recovery rate primarily in the area of municipal, construction and demolition waste in accordance with the requirements of the Waste Framework Directive. In waste management it is necessary to continue to apply the principles of proximity, self-sufficiency and, for selected waste streams under our model, also the extended producer responsibility for new waste streams, in addition to the generally established "polluter pays" principle. Building and managing the waste treatment infrastructure requires the application of the Best Available Techniques (BAT) or the Best Environmental Practices (BEP) requirement. For the period between 2016 and 2020, a strategic objective of waste management in the Slovak Republic will be to essentially divert the waste streams from disposal by landfilling towards strengthening the material and energy recovery in particular as regards municipal waste.

8 CONCLUSION

At present, the importance of raw material—energy recovery of waste, which has priority in the waste management hierarchy, is increasing in European practice. Reuse and maximized material waste recovery should be followed by energy recovery as a substitute for exhaustible raw material resources and the final phase of waste management. Minimized waste from these technologies can then be disposed of by storage in the sense of an innovative model of raw-energy recovery of waste. The long-term goal of the strategy should be the fulfilment of the principle of the circulation economy at the landfill 0% of waste.

The current, globalized challenge is to promote more efficient and effective energy use of waste to the level of material use (waste to energy processes). Material and raw-energy use of waste is a unique strategy for efficient waste management and the fulfilment of demands for maximum independence from non-renewable energy sources.

ACKNOWLEDGMENTS

The paper was supported by KEGA 026EU-4/2018 and VEGA 1/0251/17.

REFERENCES

Čermák, O. 2007. Odpadové hospodárstvo. STU v Bratislave, 2007, s. 106, ISBN 978-227-2662-7. http://www.svf.stuba.sk/docs/dokumenty/skripta/cermak_odp_hospodarstvo.pdf Access on: 10.1.2018.
Eurostat: Generation of waste. http://appsso.eurostat.ec.europa.eu/nui/show.do Access on: 10.1.2018.
Ferencz, V., Majerník, M., Chodasová, Z., Adamišin, P., Danishjoo, E. et al. 2017. Sustainable socio-economic development. A&A Digitalprint GmbH Karlstrasse 31, Düsseldorf, Germany.
ISO 9000:2005 Quality management systems—Fundamentals and vocabulary.
Kolcun M., Szkutnik J., Future Ecological Problems in European Energy Sector, in: Rynek Energii, vol. 1/2009.
Millward-Hopkins, J., Busch, J., Purnell, P., et al. 2018. Fully integrated modelling for sustainability assessment of resource recovery from waste. Science of the Total Environment. Vol. 612, p. 613–624.
Pabian A., Pabian B. 2014. Sustainable Management of an Enterprise—Functional Approach. Polish Journal of Management Studies. Vol. 10, no. 1.
Program odpadového hospodárstva SR na roky 2016–2020schválený dňa 14.10.2015 vládou Slovenskej republiky číslo uznesenia 562/2015. http://www.minzp.sk/files/sekcia-enviromentalneho-hodnoteni-ariadenia/odpady-a-obaly/registre-a-zoznamy/poh-sr-2016-2020_vestnik.pdf Access on: 10.1.2018.
Seadon, J.K.: Integrated waste management—Looking beyond the solid waste horizon. In Waste management. Elsevier Vol. 26, Issue 12, p. 1327–1336. ISSN 0956-053X.
Vandák, R. 2010. Vplyv rámcovej smernice 2008/98/ES na nakladanie s odpadmi v SR. In Odpady: odborný časopis pre podnikateľov, organizácie, obce, štátnu správu a občanov. Bratislava: EPOS, 2010, roč. 10, č. 12, 48 s. ISSN 1335-7808.

Advances and Trends in Engineering Sciences and Technologies III – Al Ali & Platko (Eds)
© 2019 Taylor & Francis Group, London, ISBN 978-0-367-07509-5

Optimization method of elevator selection for the realization of construction processes

J. Gašparík, S. Szalayová & B. Alamro
Slovak University of Technology in Bratislava, Slovakia

M. Gašparík
POMAKS Bratislava, Slovakia

ABSTRACT: The obtaining of fuels, their economization, and their effective utilization belong among the most complex problems of the present and the future. Energy saving is one of the most important environmental factors in a developed and implemented Environment Management System (EMS) according to ISO 14001. Our contribution tries to contribute to a suitable selection of machines for building processes to lead to a lowering of their energy consumption. This paper describes the structure of an optimizing machine selection method for construction processes and provides a model example with results concerning the selection of construction elevators for the vertical transport of building materials during finishing processes on site. It also suggests effective uses of this method in practice. This paper presents an optimizing elevator selection method concerning the minimization of energy consumption along with its application to 5 different construction elevators with software support.

1   INTRODUCTION

The optimal selection of a machine or machine group for building processes is a very important task for a building planner during the process of planning a building. During this process the building planner must analyze several factors that can influence the final effective decision concerning this problem.

During the process of planning a building, the planner must analyze the suitable selection of construction machines and groups to ensure an effectively designed mechanized building process.

There are several criteria for the selection of construction machines. Our contribution analyzes the following aspects: the ability of the machines to execute the designed building process (quality aspect) and minimize the energy consumption (its cost and environmental aspects).

Several theories and methods were implemented in our research work. The most important ones were the theory of systems (Štach, J.,1983), the principles of the optimizing method (Niederliňski, A. 1983), the scientific analysis method, and a final synthesis. These methods were implemented into the optimal selection of construction elevators from the point of view of minimizing energy consumption with software support.

Applications of these methods and software will increase the effectiveness of the selection of construction elevators from the point of view of the key criterion of optimization: electrical energy consumption, thus speeding up the whole process and avoiding exhausting calculations and experiments. Several authors have analyzed similar problems concerning the economic aspects (Chodasová, Z, Tekulová, Z, Kralik, M., 2016) and environmental aspects (Tekulová, Z., Chodasová, Z, Králik, M., 2017), Strukova, Z. and Kozlovska, M., 2013) and Myungdo Lee, T., Kim, H.K., Jung, U.K., Lee, H. and Cho, K.I.K., 2014).

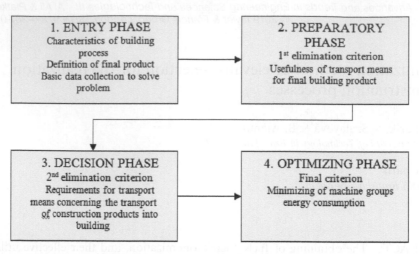

1. ENTRY PHASE	2. PREPARATORY PHASE
Characteristics of building process	1st elimination criterion
Definition of final product	Usefulness of transport means
Basic data collection to solve problem	for final building product

3. DECISION PHASE	4. OPTIMIZING PHASE
2nd elimination criterion	Final criterion
Requirements for transport	Minimizing of machine groups
means concerning the transport	energy consumption
of construction products into	
building	

Figure 1. The OES method phases and criteria.

2 OPTIMIZATION ELEVATOR SELECTION (OES) METHOD

By suggesting the Optimization Elevator Selection (OES) method, we have developed the present state of knowledge of the purpose of elevators for transporting building materials and of elements of finished processes and also information which has been obtained by studying the theory of systems and the optimization theory of the process.

The "OES Method" consists of four phases (Figure 1): entry, preparations, decision and optimization. An analysis of all these phases except for the entry phase is examined (Gašparík, J. (2013):

– the input universe of the system, i.e., the set of the transport means submitted for analysis in the given phase,
– the criterion, according to which the input universe of the system of the given phase is analyzed,
– the procedural steps necessary to realize the appreciation of the input universe of the system according to the criterion of the given phase,
– the output universe of the system, i.e., the set of the transport means fulfilling the criterion of the given phase.

A detailed description of this method is in (Gasparik, 2013).

3 APPLICATION OF THE OES METHOD

The OES method was applied to the selection of the transport means for the vertical transport of building materials during the finishing processes in terms of the minimum consumption of electrical energy. The objective is to select a means of transport for the vertical transport of building materials at a construction facility in term of minimizing energy consumption.
Input data:

– 9-story building structure (see Figure 2),
– operational project of the building structure.

The input universe of the system (possible means of transport) analyzed in the building are the following 5 elevators: WBT 5-600, VS 5, VS 6.01, NOV 500 and NOV 1000 A.
Model example:

- $M_i = 10\,000$ kg for $i = 1, \ldots, 9$, (typical floor)
- $h_1 = 3$ m, $h_2 = 6$ m,
- $h_9 = 27$ m (Figure 2),
- $\bar{x} = 70\%$,
- E_c [Wh/tm] (Figure 5).

An assessment of the elevators in terms of the height of the transport (Table 1):

$$H_r \leq H_e \tag{1}$$

where H_r = required transport elevation of an elevator in m; and H_e = actual transport height achieved by elevator in m.

An assessment of elevators in terms of their load capacity (Table 2):

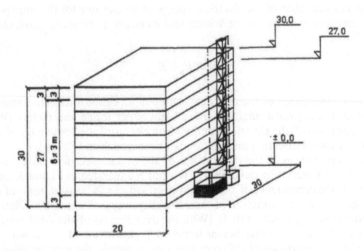

Figure 2.   Building structure analyzed for optimal selection of transport means.

Table 1.   Assessment of elevators in terms of height of transport.

N.	Type	H_r(m)	H_e(m)	Assessment
01.	WBT 5-600	27	33	+
02.	VS 5	27	30	+
03.	VS 6.01	27	48	+
04.	NOV 500	27	60	+
05.	NOV 1000 A	27	100	+

Note: Assessment (+ satisfy, – fail to satisfy).

Table 2.   Assessment of elevators in terms of load capacity.

N.	Type	M_p (kg)	LC_e (kg)	Assessment
01.	WBT 5-600	200	600	+
02.	VS 5	200	600	+
03.	VS 6.01	200	600	+
04.	NOV 500	200	500	+
05.	NOV 1000 A	200	1000	+

Note: Assessment (+ satisfy, – fail to satisfy).
Note: All the elevators satisfy the requirements defined in the 2nd criterion.

$$M_p \leq LC_e \tag{2}$$

where M_p = mass of the heaviest piece from all types of transported materials in kg; and LC_e = load capacity of elevator in kg.

The consumption of the electrical energy of an elevator for a tonometer of transported material in Wh/tm could be calculated as follows:

$$E = \sqrt{3}UI\,(3600\,vm)^{-1} \tag{3}$$

where U = nominal line voltage of a three-phase distribution system in volts (380 V); I = actual intensity of electric current withdrawn by electromotor of elevator in ampers; v = transport velocity of elevator in meters per second; and m = load mass transported by elevator in tonnes.

The overall consumption of the electrical energy of an elevator for the transport of material to a 1 m elevation (tonometer) in Wh/tm and its return to its initial position can be calculated as follows:

$$E_c = E_h + E_{pd} \tag{4}$$

where E_h = the consumption of the electrical energy of an elevator for a tonometer of transported material in an upward direction in watt hours per tonne and meters (Wh/tm); and E_{pd} = the consumption of the electrical energy of an elevator for a tonometer of transported material in a downward direction and not loaded by materials in Wh/tm.

A graphic interpretation of the dependence of the overall electrical energy consumption on the transport of material by elevators E_c [Wh/tm] is illustrated in Figure 3, and a graphic interpretation of the dependence of the overall electrical energy consumption of elevators E_c [Wh/tm] on the percentage extraction of their load capacity is illustrated in Figure 4.

For an approximate calculation of E_v [Wh], the overall mass of the ABP materials, which were transported by elevator to particular floors M_i [kg], where i = 1, ..., n, was determined.

Based on the characteristics of different types of materials, the average value of the percentage of the extraction of the load capacity of elevator x was subsequently estimated, based on this value, the E_c [Wh/tm] is defined as seen in Figure 4.

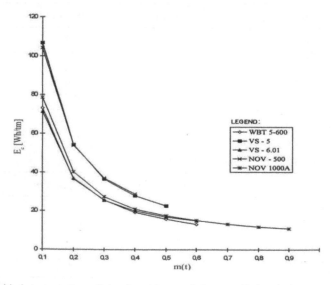

Figure 3. Graphic interpretation of the dependence of the overall electrical energy consumption on the transport material by the elevators for tonnes and meter Ec [Wh/tm].

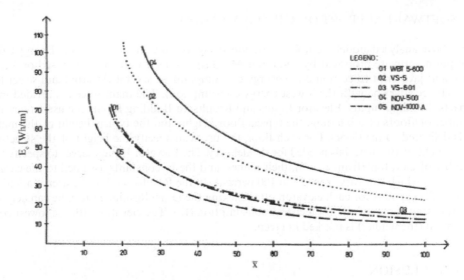

Figure 4. Graphic interpretation of the dependence of the overall electrical energy consumption of elevators Ec [Wh/tm] on the percentage extraction of their load capacity.

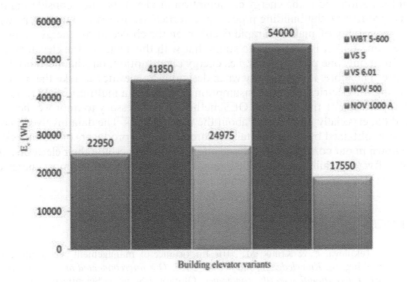

Figure 5. A comparison of the elevator variants in terms of the overall electrical energy consumption during the process of the vertical transportation of building materials to the building structure analyzed.

The resulting electrical energy consumption of elevators E_v [Wh] is calculated as follows:

$$E_v^k = E_c^k \sum_{i=1}^{n} M_i h_i 10^{-3} \qquad (5)$$

where E_c^k = overall electrical energy consumption of the "k" type of elevator for transport of 1 tonne of material to a 1 m height [Wh/tm], where k = 1, ..., p; M_i = mass of the material, which is transported to "i" floor [kg], where i = 1, ..., n; and h_i = height of the "i" floor [m], where i = 1, ..., n. The solution of our model example is shown in Figure 5.

4 SOFTWARE SUPPORT OF THE OES METHOD

The above analysed model example can be effectively solved using the Elevators Energy Consumption software proposed by co-author M. Gasparik as a web solution based on html, php, and javascript to compute the energy consumption of several elevators and select the best choice (the one with the lowest energy consumption). The main screen is divided into 3 parts: Building Inputs, Elevator Inputs and Results. In Building Inputs the user enters the number of floors of a building, the typical floor height, and the typical height of the materials delivered to the floors. For each floor, the user can specify the height of the structure and weight of the materials needed for the delivery. The Elevator Inputs serves to specify the number of elevators (from 1 to 5), their names, and the E_c constants assigned to the capacity values, which can vary from 25 to 100 percent. In the Results section the user is asked to choose the capacity for each elevator; after clicking the Draw Results button, he can acquire the resulting values of the total energy consumption (E_v). The elevator with the lowest consumption (the winner) is marked in green.

5 CONCLUSION

The most-important factor in the OES method is its ability to eliminate energy variants of the elevators during the process of planning a building. With the use of the software it can provide information about the energy consumption of elevators when considering their use in the final product of the building process (the vertical transport of building materials) as well as the possibility of making a rapid decision on the choice of an energy optimal elevator. At the same time it is necessary to stress that with this method, the elevators are being evaluated from only one point of view, i.e., energy consumption, but that need not be crucial in every case. Therefore, it is necessary when designing an elevator to take the point of view of minimizing the electrical energy consumption as a part of a multi-criterial proposal. For a practical application of the proposed OES method, it is necessary to improve the quality of the input data, especially information about the requirements. The data involved in our contribution were obtained by experimental measurements. The producers of elevators can offer the data shown in our contribution as basic information concerning their elevators, and with the use of software, a building planner can quickly and easily select the most energy optimal variant of the elevators available for construction processes.

REFERENCES

Chodasová, Z., Tekulová, Z. & Kralik, M. 2016. Importance of management accounting knowledge in decision making. In: *Knowledge for market use 2016: Our interconnected and divided world [electronic source]: International scientific conference.* Olomouc: Societas Scientiarum Olomucensis II, pp. 162–167.

Gašparík, J. 2013. *Automated system of optimal machine group selection implemented into soil processes.* Brno: TRIBUM EU.

Myungdo Lee, T., Kim, H.K., Jung, U.K., Lee, H. & Cho, K.I.K. 2014. Green construction hoist with customized energy regeneration system. *Journal Automation in Construction.* Vol. 45, Elsevier, pp. 66–71.

Niederliński, A. 1983. *Numerical systems of control technologic processes.* Prague: SNTL, 1983.

Štach, J. 1983. *Teória systémov.* Prague: SNTL.

Strukova, Z. & Kozlovska, M. 2013. Environmental impact reducing through less pollution from construction equipments. *13th SGEM GeoConference on Ecology, Economics, Education and Legislation, SGEM Conference proceedings.* Vol. 1. pp. 401–408.

Tekulová, Z., Chodasová, Z. & Králik, M. 2017. *Environmental policy as a competitive advantage in the global environment.* Institute for Computer Sciences, Social Informatics and Telecommunications Engineering. The LNICST series. Smart Technology Trends in Industrial and Business Management [electronic source]. London: Springer.

Advances and Trends in Engineering Sciences and Technologies III – Al Ali & Platko (Eds)
© 2019 Taylor & Francis Group, London, ISBN 978-0-367-07509-5

Full-scale tests for the simulation of fire hazards in the building with an atrium

M.V. Gravit & O.V. Nedryshkin
Peter the Great St. Petersburg Polytechnic University, Saint-Petersburg, Russia

ABSTRACT: Modeling of dangerous fire factors is an important element in the system of modern fire safety assessment of buildings and structures. The article presents a comparative analysis of design models for the development of fire in a room. A comparative analysis of the characteristics of the software systems Sigma PB, Fenix, PyroSim, Fogard, Toxi Risk is presented. The verification of the fire model in the FDS program, performed based on full-scale tests conducted by Professor Chow W.L. was analyzed. The analysis of empirical and calculated data on modeling of fire in the atrium is made, the conclusion is made about the accuracy of simulation in the FDS program and the coincidence of the experimental data with the calculated ones.

1 INTRODUCTION

Determination of the calculated values of fire risk is implemented by various software systems. One of the key parameters in the calculation of fire risk is to determine the time of blocking the escape routes by dangerous fire factors. Time is determined by means of mathematical modeling, which are based on integrated, zone or field propagation dangerous fire factors models (McGrattan, 2005). Modeling and forecasting of fires in premises and buildings is of great interest. In this paper, the authors analyze the results of full-scale fire tests in the atrium, conducted under the guidance of Professor Chow, and a mathematical model of the fire in the atrium, made based on one of the most common programs (Fryanova, 2017).

2 TOOLS FOR CALCULATING FIRE RISK IN THE ROOM

It is now widely distributed as a tool for modeling dangerous fire factors—program Fire Dynamic Simulation (hereafter FDS). First the estimated model FDS (McGrattan, 2005) was demonstrated in 2000. FDS numerically solves the Navier-Stokes equations for low-velocity temperature-dependent flows, with attention to smoke propagation and heat transfer in fire (Gravit, 2016). The model is a system of partial differential equations, including the equations of mass, moment and energy conservation. It is solved on a three-dimensional regular grid and describes the spatiotemporal distribution of the temperature and velocities of the gas medium in the room, the concentrations of the components of the gas medium (oxygen, combustion products, etc.), pressures and densities (Gravit, 2018).

The development of calculation models has fundamental importance for fire risk assessment and there are a few number of solutions for determining the calculated values of fire risk and modeling dangerous fire factors in the form of various software. Table 1 presents a summary of the calculation models implemented in the software.

Table 1. Comparison of calculation models implemented in software systems.

Characteristic	Software					
	Fogard-PM	SITIS Flammer	Fenix + Fenix +2	PyroSim	Sigma PB	Toxi + FireRisk
Mathematical model						
Integral	+	+	–	–	–	+
Zone	+	+	–	–	–	–
Field	+	+	+	+	+	–
The complexity of the subject						
One-storey object	+	+	+	+	+	+
A multi-storey facility	+	+	+	+	+	–
Unique (non-standard constructions)	+	+	+	+	+	–
The method of constructing building model						
Import AutoCAD files	–	–	–	+	–	–
Import Revit files	–	–	–	–	+	–
Import of finished drawings in DXF and DWG formats	–	–	+	+	–	–
Own GUI Builder, 2D/3D model of the object	2D/3D	2D/3D	2D/3D	2D/3D	2D/3D	2D
Possibility to use the substrate	+	+	+	+	+	–
Developer of the field model and visualizer design core						
Russia	–	–	–	–	+	+
Foreign	+ (FDS)	+ (FDS)	+(FDS)	+ (FDS)	–	–

3 SIMULATION OF THE FIRE FIELD MODEL (FDS)

To study the FDS field model, the initial conditions of the full-scale test of V.K. Chow (Chow 2009) were set in the program PyroSim (FDS graphical interface). The full-scale test was carried out on the designed installation, which had the following geometric parameters: length 35 m, width 9 m, height 28 m the Design consisted of a metal frame and translucent panels as enclosing walls. Thermocouples located in the left part of the unit at the level of 0 m with a step of 2 m to a height of 22 m. Temperature was measured (Krivtcov, 2016) by 12 thermocouples is indicated in Figure 2 by red dots. The design has 6 ventilation openings: 4 horizontal at 0 m and 2 verticals at a height of 28 m (ceiling ventilation holes). Ventilation openings have the shape of a square. Ceiling ventilation openings are located at 13.5 m from the upper left point of the structure, the area of ceiling openings of 9 sq. m (each). Ventilation openings at the level of 0 m have an area of 3.61 sq. m (each) and located at 4.7 and 20.7 m from the leftmost point design with front and rear sides. In the full-scale test, the fire load was methanol, which is in 6 steel pallets (pallet area S = 0.51 sq. m, volume V = 16 l, burnout rate is set based on reference materials PyroSim) located on the right side of the structure, as shown in Figure 1.

The geometric parameters of the construction were reproduced in the program PyroSim. The modeling area in PyroSim was divided into 144,768 calculated cells in the form of a cube, the size of the edge of one cell was 0.5 m (Figure 1). The decision on the size of the cell was based on the previous experience of modeling in FDS, the accuracy of modeling at a given cell size and the approximate time of modeling FDS – 48 h (computer CPU—Intel Core i5-7400, 3 GHz).

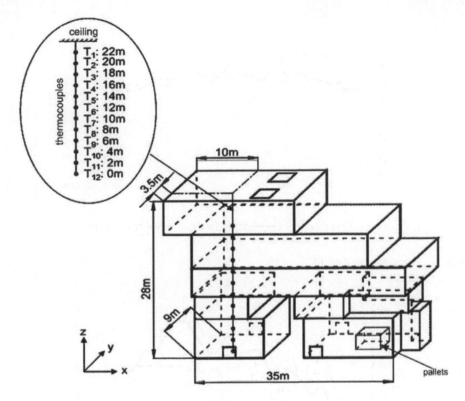

Figure 1. The scheme of construction of the full-scale experiment.

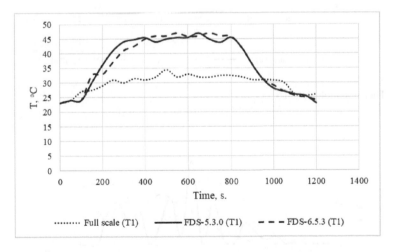

Figure 2. Thermocouple data T1.

4 RESULTS

In Figures 2–5 shows some the graphs with the data of thermocouples (control points) in full-scale testing and modeling in FDS 5.3.0 and FDS 6.5.3.

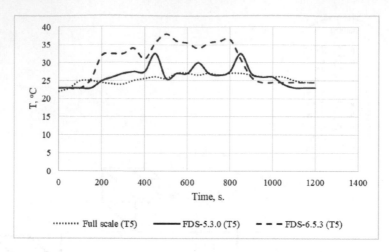

Figure 3.   Thermocouple data T5.

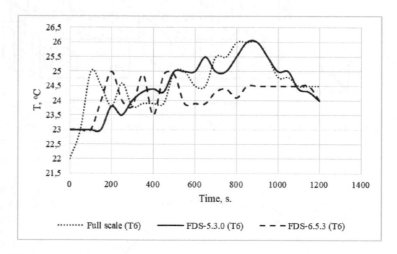

Figure 4.   Thermocouple data T6.

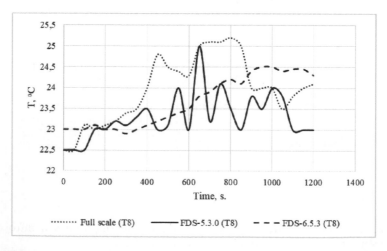

Figure 5.   Thermocouple data T8.

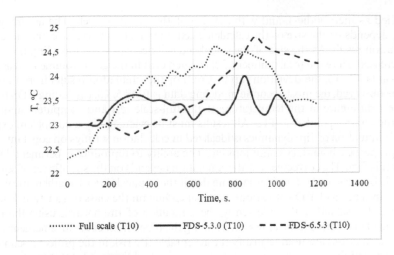

Figure 6.   Thermocouple data T10.

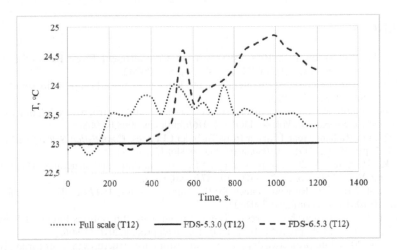

Figure 7.   Thermocouple data T12.

5   DISCUSSION AND CONCLUSION

The deviation (difference) of the temperature of the full-scale test and the simulation data (FDS) for the thermocouple T1 (h = 22 m) does not exceed 15°C. Such a deviation re-main to the level h = 14 m and for thermocouples T2–T5. The temperature difference for thermo-couples T6–T12 is 5°C.

The reference points (thermocouples) T6–T12 show the most accurate simulation data, the results of calculating the FDS for T6 are comparable with the full-scale test data.

The results of the fire hazard (Babrauskas, Williamson 1978. Mudan 1995. Burns 1999. Cox 1977. Vedat 1995) modeling experiment in the PyroSim FDS graphical interface and the com-parative analysis of the full-scale fire experiment in the atrium confirm the relevance of math-ematical model of fire development. The parameters of the temperature curves deviate from the values of the actual experiment. The results of modeling the heat transfer process under given conditions have a certain degree of correspondence depending on the location of the control points. It can be assumed that the deviations of the simulation results from the full-scale test data for the T1–T5 reference points are due to the lack of heat exchange conditions for the walls with internal structural elements (in the PyroSim program, the structural elements of the model are specified as inert bodies), and with the size of the calculated cell. The accuracy of FDS modeling depends on the size of the calculated cell, the smaller the cell size, the more accurate the simulation result. As the size of the calculated cell decreases, the modeling time is significantly increased, since the total number of calculated cells in the given volume increases. At the control points T1–T5, the deviation (up to 15°C) of the simulation results from the full-scale test data is observed, the most accurate modeling data is FDS 5.3.0. (including FDS 6.5.3.) and the smallest difference in the temperature of the full-scale test and computer simulation are noted at the control points T6–T12. All results must be evaluated by a competent expert. The numerical simulation of fire dynamics is idealized in calculation settings on constant values for material properties, boundary conditions etc. The ability to reproduce experimental results be related to this fact. The correctness of the simulation depends on the specific set of source data. Probably, it is necessary to consider in more detail the applicability of mathematical model of fire in the simulation of FDS at various sites, depending on the class of functional fire hazard (Rybakov, 2017, Saknite, 2016). Currently, the simulation of fire hazards using the FDS program is considered correct and allows determining the time for blocking an emergency exit. The results of calculations are used to justify the acceptable fire risk in the project documentation, during construction, reconstruction and declaration of buildings and structures.

REFERENCES

Babrauskas, V., Williamson R.B. 1978. Post-flashover Compartment Fires: Basis of a Theoretical Model, *Fire and Materials*. Vol. 2. DOI: 10.1002/fam.810020202.

Burns, S.P. 1999. Turbulence radiation interaction modeling in hydrocarbon pool fire simulations, *Sandia Report*. DOI: 10.2172/752015.

Chow, W. 2009. Numerical studies on atrium smoke movement and control with validation by field tests, *Building and Environmental*. Vol. 44. DOI:10.1016/j.buildenv.2008.08.008.

Cox, G. 1977. On Radiant Heat Transfer from Turbulent Flames, *Combustion Science and Technology*. Vol. 17:1–2, pp. 75–78. DOI: 10.1080/00102209708946815.

Fryanova, K., Perminov, V. 2017. Impact of forest fires on buildings and structures, *Magazine of Civil Engineering*. Vol. 75 (7). pp. 15–22. DOI: 10.18720/MCE.75.2

Gravit, M. et al. 2016. Fire Resistant Panels for the Tunnel Linings, *MATEC Web of Conferences*. Vol 73. DOI: 10.1051/matecconf/20167304007.

Gravit, M. et al. 2016. Research Features of Tunnel Linings with Innovations Fireproof Panels, *Procedia Engineering*. Vol. 165. pp. 1651–1657. DOI: 10.1016/j.proeng.2016.11.906.

Gravit, M. et al. 2018. Fire protective dry plaster composition for structures in hydrocarbon fire, *Magazine of Civil Engineering*. Vol. 79 (3). pp. 86–94. DOI: 10.18720/MCE.79.9.

Gravit, M. et al. 2018. Transformable fire barriers in buildings and structures, *Magazine of Civil Engineering*. Vol. 77 (1). pp. 38–46. DOI: 10.18720/MCE.77.4.

Krivtcov, A. et al. 2016. Calculation of Limits of Fire Resistance for Structures with Fire Retardant Coating, *MATEC Web of Conferences*. Vol. 53. DOI: 10.1051/matecconf/20165301032.

McGrattan, K. 2005. *Stand. Technol. Spec. Publ.* Vol. 1019. DOI: /10.6028/nist.ir.6469.

Mudan, K.S., Croce P.A.1995. SFPE Handbook of Fire Protection Engineering, *Elsevier Sequoia*. DOI: 10.1016/0379-7112(86)90055-x.

Rybakov, V. et. al. 2017. Stress-strain state of composite reinforced concrete slab elements under fire activity, *Magazine of Civil Engineering*. Vol. 74 (6). pp. 161–174. DOI: 10.18720/MCE.74.13.

Saknite, T. et al. 2016. Fire design of arch-type timber roof. Magazine of Civil Engineering. Vol. 64 (4). pp. 26–39. DOI: 10.5862/MCE.64.3.

Vedat, S. 1995. Turbulent Forced Diffusion Flames, *Combustion Fundamentals of Fire*. pp. 170–178. DOI: 10.1016/0010-2180(95)00026-3.

Advances and Trends in Engineering Sciences and Technologies III – Al Ali & Platko (Eds)
© 2019 Taylor & Francis Group, London, ISBN 978-0-367-07509-5

Indicators of profitability in the building enterprise

Z. Chodasová
Institute of Management of the Slovak University of Technology, Bratislava, Slovakia

M. Králik
Faculty of Mechanical Engineering, Slovak University of Technology, Bratislava, Slovakia

ABSTRACT: Development process influences the overall advancement of the property market, which is reflected also in the proper operation of the financial sector. Uncontrolled development of the property market may have a negative impact on a country´s finance sector, but also the economics in general. It is, therefore, necessary that the management of the development process includes such methods as finance controlling, and others that enable substantial elimination of risks of the given process. The present paper is therefore aimed to justify the employment of the Indicators of profitability in the building enterprise.

1 INTRODUCTION

Maintaining and increasing competitiveness is becoming the prime role of the management of all businesses because competitiveness at all levels is a prerequisite for prosperity, but sometimes even simple survival. Competitiveness, which in the current market environment is the basic attribute of the success of every business, raises the constant pressure to improve and make radical changes in all the contexts where the enterprise is located. One of the most important factors of competitiveness, which mainly applies to globalization is, therefore, an innovation, which is generally considered to be an accelerator of the economy.

Competitiveness belongs into the constantly changing market environment of the present time. It is one of the leading requirements in the economic market, therefore, it is considered to be one of the basic characteristics of an enterprise. The key feature of the market economy is the customer's freedom to decide not only what product they buy, but from who. This leads to competition among potential suppliers to the customer. The company should also strive to create competitive advantages that enhance the competitiveness of the offered products. The competitiveness of the products is mainly due to their quality, and the position of the company on the given market is determined by the customer's interest in the offered products (Tekulová et al., 2016).

In order to be able to respond to these changes in the business environment and to be competitive, it is above all necessary to moderately manage the business in the companies, i.e. introduce new approaches, tools and techniques, set ambitious goals, put emphasis on sustainable performance, and follow a comprehensive set of performance criteria. It is also essential to create a dynamic and flexible organizational management structure, continually improve key competencies and to look for opportunities to increase the competitiveness of the business.

2 THE OBSERVED COMPETITIVENESS FACTORS IN INDUSTRIAL ENTERPRISES

Business competitiveness factors can be divided into external and internal. External factors include those that an enterprise can not influence, or its ability to influence these factors is either limited or indirect. This includes the negotiating power of suppliers and buyers, competitive struggle, product market, environmental corruption, job interest in the company, and support from local and state authorities. Internal factors include those that an enterprise can influence and which are within the enterprise. These include science and technology development factors, marketing and distribution factors as well as factors of production and its management, labor resources, and financial and budgetary aspects of the business (Bontis, et al., 2002).

Not all competitiveness factors are objective. They are not often measurable but they are subjectively perceived by confronting customers with their requirements, values, or only by their moods. In a broad sense, competitiveness is a superposition for the following factors:

- A product and portfolio of offers that should be as close as possible to the needs and expectations of the customer.
- Business behavior towards customers, especially in the level and quality of communication skills.
- Reflection of business timing (a speed of response to customer requirements, the speed of processes and changes).
- The aggregate work of an organization, which is partly determined by the relatively fixed features of the business entity, but partly also by the image.
- Additional competitive criteria such as capital strength and financial capacity also engage in competitive business conditions, application of quality management systems, the productivity of work, etc. are also emerging in relation to large customers.
- In some cases, competitiveness is also matched by the membership of a region equipped with the appropriate infrastructure, legal environment and other macroeconomic criteria of the region.
- The latest development of globalization creates a new factor of competitiveness—the ability to engage in networks, the ability to create strategic alliances.

The uniqueness of one of these factors creates a competitive advantage which is appreciated by customers and motivates them to establish a business relationship. Other factors of competitiveness include employee motivation and education, performance, logistics, quality, innovation, cost reduction, advertising, information technology, etc. (Kucharčikova, et.al., 2015).

3 INDICATORS OF PROFITABILITY

Management is often in a situation where it has to decide whether to produce or not produce, which particular product to prefer at a given time, respond flexibly to market requirements, and so on. For such a decision, managers need an extensive but flexible information system with the ability to select high-quality information (Kucharčíková, 2014). The above-mentioned management methods allow us to have a better quality:

- evaluate how the company's planned goals are being met,
- identify risks, draw attention to imminent and real deviations from desirable developments,
- analyze and evaluate the effects of business activities and decisions,
- plan and program the development of the company in both aggregate and analytical indicators,
- inspire business leadership to uncover new business activities that have an economic effect.

Such methods have an important role to play in selecting high-quality information, thus helping to make the company even more competitive. Today's modern management relies on different classifications of thought directions. It should be noted, however, that there is no recipe for unified, universally valid management. There are more types of management, represented by several schools and directions, but in any case, management follows the intention of employees interested in this type. And, of course, they also have the qualities needed, because they have to react quickly and promptly to the day-to-day changing situation on the labor market, capital, and goods. Rapid, accurate and correct decisions lead to company's prosperity (Tokarčíková, 2011).

The Balanced Scorecard—the philosophy of the method is to consolidate strategic and thus corporate thinking and behavior at every level of enterprise management. Interactive and interdisciplinary cooperation between different areas and functions in strategy design and implementation makes it possible to balance strategic objectives within and between the perspectives. The main idea was to use the Balanced Scorecard as a template for proposing objectives and indicators in each of the following perspectives:

- The financial perspective includes the financial goals of the company and enables managers to monitor financial success and value for shareholders. It is looking for the answer to the question "How should an enterprise looks like from an investor perspective?"
- Customer's perspectives relate to goals such as customer satisfaction, market share, and product and service attributes. It is looking for the answer to the question "Which customer needs have to be satisfied?"
- The perspective of internal processes includes internal goals and the results of key processes necessary to achieve the goals of customers, shareholders and other stakeholders. It is looking for the answer to the question "In what processes and which products do we have to perform special performances?"
- The Perspective of learning and growth relates to the intangible driving force of future success, such as human capital, corporate capital and information capital, including skills, education, enterprise culture, management, systems, and databases.

After the Balanced Scorecard has been introduced, the double feedback by which the management monitors whether the activities selected actually lead to the fulfillment of the indicators and whether the indicators set help to achieve the strategic objectives. Otherwise, business leaders must proceed with a reasonable change concerning the indicators or activities. By introducing the Balanced Scorecard method to measure and track indicators, the company gets much more information than before it was introduced. The information thus obtained enables the company to more thoroughly assess the success of the implementation of the chosen strategy. Business reporting becomes more complex, so reporting both financial and non-financial data makes it possible to provide the market and investors with more relevant information, making the enterprise more transparent. When implementing the Balanced Scorecard method, the benchmarking method (copying and retrieving experience from companies that work with this method with certain (Pilkova, Schomburg, 2010).

4   INVESTMENT EVALUATION OPTIONS

The biggest challenge facing Slovakia and the EU in the field of innovation today is the inability to make full use of and share the R & D results and then translate them into economic and social values. The gap between the results of research and their application in practice is still large. The following major factors will have a determining influence on the development of the economy and innovation in the medium term:

- Continuing globalization, which will manifest in Slovakia in forms of foreign investments, increase in the share of export and import, and using the potential of growing markets in Asia.
- The emergence of a knowledge-based economy characterized in particular by growth in R & D, knowledge intensive services, high technology, and innovation. The increasing impact of the knowledge economy in the growth of Slovakia is strongly supported by the innovation policy of the Slovak Republic, the support from the structural funds and changes in the attitudes of entrepreneurial subjects, especially in the managerial activities.
- The development of industrial production and its structural changes, moving from cost production to a production supported by knowledge and innovation.
- The growing importance of environmental policy and sustainable development. Different developments are expected in this area.

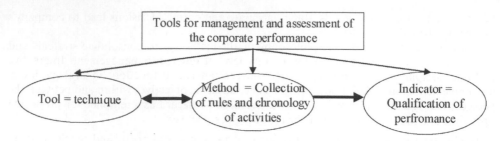

Figure 1.   Tools for management and assessment of the corporate performance. Source: Own design.

The Innovation Strategy of the Slovak Republic presents an ambitious plan to create a national innovation system that will lead to the sustainability of the knowledge-based development of the Slovak Republic. The implementation of such designed innovative model should among other things, increase the competitiveness of Slovak enterprises. For reasons of long-term effect and frozen capital formation, the investment management, decision-making and implementation are counted among the most important tasks in the company (Jacková, 2016). Investments are linked to the risk of economic outcomes and liquidity. The role of controlling is to ensure systematic planning, control and management of investments and their profitability. The main mission is to ensure liquidity and to determine the short-term and long-term need for capital. Equally important is ensuring the effectiveness of the resources available for this purpose. Investment decision making is typical because it is a long-term decision, which also takes into account the time factor, the risk of market changes during the preparation, implementation, and life of the project. Investments affect economic performance very well for several years (profit, profitability, liquidity), and all this has an impact on market value of the firm. Other important indicators are ability to compete, maintain quality criteria, corporate performance and so on. It is necessary to know, manage and evaluate the corporate performance. Enterprise tools, methods, and indicators can be used to manage and evaluate corporate performance (Ďurišová, 2011). Tools represent techniques, a way of examining a corporate's performance as an economic reality Figure 1. Tools for management and assessment of the corporate performance.

For these reasons, it is important to analyze all factors of time, risk, market change, competition demand, quality preferences in evaluating investment projects, which will in the long term ensure the success of the company and its market position with the right decisions. Due to the demanding character, complexity, and continuousness of the investment process in an enterprise, it is necessary to use the complex of methods applicable to the evaluation of investments in terms of their overall effectiveness (Tokarčíková, 2011).

5   APPROACHES TO EVALUATION OF INVESTMENTS

Investment decision making is among the most important decisions the entrepreneur has to take. The importance of deciding how much, to what and when to invest, is highlighted in particular by the fact that entrepreneurs have relatively low private capital and mostly limited access to foreign capital. Investments increase the share of fixed costs in total costs. Lack of capital limits the scale of investments. Every investment is a financial problem both at the moment of its acquisition and also when it is rejected. The aim of controlling investment management is to ensure that strategic goals are met by strategic investment plans, allowing to use the chances and avert the risk of bankruptcy. Depending on the purpose, initial or expansion investments are linked to the establishment of an enterprise or to expansion—by increasing the production capacity of facilities, renewal—replacement investments ensure the maintenance of the company's economic performance, rationalization—allow for improved efficiency and performance.

The aim should be to reduce material costs, energy costs and labor savings, which should be reflected in overall environmental and social reductions—for environmental protection, small businesses are mostly driven by the demands of the external environment (Bažik, Gašparik, 2014). The main trends in investment are based on portfolio analysis, i.e. the competitive force matrix—the attractivity of the market, or the technology portfolio in which technology and resource intensity stand against each other. When analyzing opportunities and threats, we look for ideas to avert threats to core business and development opportunities.

In the case of investment decisions of any kind, the entrepreneur must take into account the following factors, namely:

- the factor of time—the expected benefits do not arise today, but only in the future. Benefits are in most cases divided over several years, and the cost is usually created by the one sum, already at present,
- the risk factor—mainly linked to the estimate of price developments and the resulting estimate of future earnings. When deciding on investments (realization of an investment action, a project), a number of implementation options are a prerequisite.

The particular alternatives differ in various technical and technological parameters. Only the analysis of technical performance is not enough, because for assessment of financial variables are important financial quantities such as capital expenditures, cash flow, return on investment—profitability market (Virlanuta et al., 2012).

At the investment conversion procedures, it is about supportive decision-making methods to determine the continuity of individual investment projects. In the case of a greater number of alternatives, individual investment objects can be selected taking into account the return criterion. Investment recalculations also serve to assemble investment programs. Investment recalculations thus contribute to reducing uncertainty in investment decisions. Individual methods and procedures have their advantages and disadvantages, which must be taken into account when applying them (Kucharčikova et al., 2015).

6 CONCLUSION

The survey results confirmed that more than 80% of enterprises use several methods to recalculate their investment plans. Small and Medium Sized Enterprises use static investment evaluation methods and procedures for evaluation of the investments for simpler investment objectives with short-term validity, due to their simplicity, speed, and availability of information. Dynamic methods and procedures are used by larger corporations that prefer the net present value of the internal yield percentage. Among the methods of investment recalculation, I recognize the methods that are used for the assessment of individual investment objects and the methods used to compile investment programs.

The current market situation in Slovakia leads many manufacturing companies to focus their attention primarily on those components of the production system that are easily measurable and changeable. A prerequisite for success and long-term viability and a firm market position is the response to the company environment conditions that affect the production program, pressure on new production technologies, and quality improvement. The assumption of success from the strategic point of view is the management and decision making on investments. Enterprises invest in new technologies, reduce energy consumption, force suppliers to lower sales prices, recruit new employees, and so on. When the key machine fails, the company immediately invests into repairing it or makes a decision to buy a new device. For almost 70% of domestic manufacturing companies, investment in technology is a priority. Investing in technology is followed by investment in buildings and infrastructure. The investment decision-making process is demanding and long-term and bad decisions result in a long-term impact on the company's profit, rentability, and liquidity. The paper discusses the importance of controlling investment management and offers, at the same time, familiar methods of evaluating investment projects with a delimitation of their use, pros, and cons.

In order to be able to respond to these changes in the business environment and to be competitive, it is above all necessary to moderately manage the business, i.e. introduce new approaches, tools and techniques, set ambitious goals, put emphasis on sustainable performance, follow a comprehensive set of performance criteria, and strive for strategic innovation. It is also essential to create a dynamic and flexible organizational management structure, continually improve key competencies and to look for opportunities to increase the competitiveness of the business, in particular through its profitability.

ACKNOWLEDGEMENTS

This article was created as part of application of project VEGA No 1/0652/16 Impact of spatial location and sectorial focus on the performance of businesses and their competitiveness in the global market and project KEGA No. 035STU-4/2017.

REFERENCES

Bažik, P., Gašparik, J., 2014. Efficient method of an optimum construction company supplier selection supported by software. In *ISARC 2014 – Automation, construction and environment: 31st International Symposium. Sydney, Australia, 9.–11.7.2014*. Sydney: University of Technology, 2014, s. 918–924. ISBN 978-0-646-54610-0.

Bontis, N., Fitz-enz, J., 2002. Intellectual capital ROI: a causal map of human capital antecedents and consequents. *Journal of Intellectual Capital*, Vol. 3 Issue: 3, 2002, p. 223–247.

Ďurišová, M., 2011. Application of cost models in transportation companies. *Periodica Polytechnica: social and management sciences.* ISSN 1416-3837, Vol. 19, iss. 1, 2011, pp. 19–24.

EC 2013, Decision no 1386/2013/EU of the European Parliament and of the Council of 20 November 2013 on a General Union Environment Action Programme.

EC 2014, Communication from the Commission to the European Parliament, the Council, the European Economic and Social Committee and the Committee of the Regions "Towards a circular economy: A zero waste programme for Europe", COM/2014/0398 final.

Jacková, A., 2016. Production Management and Engineering Sciences, 2016 Taylor and Francis Group London, p. 113–118, ISBN: 978-1-138-02856-2/.

Kucharčíková, A., 2014. Investment in the Human Capital as the Source of conomic Growth. *Periodica Polytechnica, Social and Management Sciences*. Volume 22, Issue 1, 2014, Pages 29–35. ISSN 14163837.

Kucharčikova, A., Konušikova, L., Tokarčíková, E., 2015. The Quantification of Human Capital Value in Digital Marketing Companies. Marketing Identity: Digital, Pt I, Conference: International *Scientific Conference on Marketing Identity—Digital Life*, Smolenice, Slovakia, nov. 10–11, 2015, pp. 151–163, ISBN:978-80-8105-779-3, ISSN: 1339-5726.

Lapidus, B. 2017. FP&A Survey: How Relevant is Your Cost of Capital? *Association for financial professionals*. (online). (Pub. 2017-08-12) (Cit. 2018-06-10). Available at: https://www.afponline.org/ trends-topics/topics/articles/Details/fp-a-survey-how-relevant-is-your-cost-of-capital.

Pilkova, A., Schomburg, J., 2010. *Loss given default scenario calculations as one of the contributors to future mortgage financing within the banking regulatory framework* In: Terra Spectra STU, Planning Studies č.1 /2010s. 37–46, ISSN 1338-0370.

Scarborough, H., Swan, J., Preston, J. 1999. Knowledge Management. A literature review, 1999 London, Institute of personnel and Development.

Tekulová, Z., Králik, M., Codasová, Z., 2016. *Analysis of produktivity in enterprises automotive production*, Production Management and Engineering Sciences, 2016 Taylor and Francis Group, London, p. 545–549, ISBN: 978-1-138-02856-2/.

Tokarčíková E., 2011. *Influence of social networking for enterprise´s activities,* Periodica Polytechnica, Social and Management Sciences, Budapest, 19/1. 37–41 doi: 10.3311/pp.so.2011 – 1.05 – ISSN 1416-3837.

Virlanuta, O. F., Muntean, M. C., Nistor, C. 2012. Strategic alternatives for increasing the competitiveness of Romanian tourism. In: *18th IBIMA: From Regional Development to World Economies*; Istanbul; Turkey; 9–10 May 2012, 1303–1307, Code 103297.

Advances and Trends in Engineering Sciences and Technologies III – Al Ali & Platko (Eds)
© *2019 Taylor & Francis Group, London, ISBN 978-0-367-07509-5*

Modern management methods in the corporate process

Z. Chodasová
Institute of Management of the Slovak University of Technology, Bratislava, Slovakia

A. Kucharčíková & M. Ďurišová
Faculty of Management Science and Informatics, University of Zilina, Zilina, Slovakia

ABSTRACT: Value management involves thinking which aims at identification and creative solution of problems. This process is organised with the purpose to identify and eliminate unnecessary costs from a project while achieving optimum balance between function and costs. This facilitated creation of a programme and methodology in value management, which improves efficiency and profitability of examined projects. Since value management focuses on an examined project as a whole, decision making process is improved, routine approaches are questioned, realistic budgets for required functions are made, which has a positive effect on project designs while ensuring function. The paper analyzes the value management, it is a special type of management, clearly focused on creating and increasing value for the customer while increasing the value of the company.

1 INTRODUCTION

Value management is a process of managerial activities through approaches and methods used at different levels of value-driven enterprise management. It contributes to meeting the ultimate goal of the business, which is to maximize value for the owners, respecting the constraints given by the interested business groups. Profiling of the value management is based on the value analysis. By its development and interconnection of principles and methods of value analysis with managerial functions and activities was created the subject, methods and specific activities of value management. Their basis is a cost-effective (value) approach initially market-oriented (primarily customer). The essence of the function-cost principle is to maximize the relationship between meeting the needs and the resource consumption. Satisfaction of needs is expressed in functions and consumption of resources in costs. Value management is primarily focused on motivating people, acquiring skills and promoting synergies and innovation to maximize the organization's overall performance. As experts say the current approach to business management focuses on increasing the performance of business processes.

The globalization processes in Europe have created a new business environment for business entities entering the Slovak Republic into the European Union, reflecting a number of positive and negative business changes. Due to the extent of the changes, a new competitive environment has also emerged. In order to be able to respond to these changes in the business environment and to be competitive, it is above all necessary to moderately manage their business, i.e. introduce new approaches—value management, tools and techniques, set ambitious goals, put emphasis on sustainable performance, follow a comprehensive set of performance criteria, and strive for strategic innovation (Plowman, 2001). The paper analyzes the value of value management for an enterprise, it is to maintain and intensify the competitiveness of the company by building a long-term sustainable relationship with the stakeholder groups. A value management enterprise will gain a competitive advantage by focusing on value creation.

2 VALUE MANAGEMENT

Value management can be considered as a framework for deploying methods and tools for improving performance. It deals with management through value, while functions and value being taken into account in management and decision making. A value management enterprise will gain a competitive advantage by focusing on value creation. Elements of the Value Management System are contained in Figure 1.

At the top level in the enterprise, value management is based on a value culture of the business, and takes into account value for stakeholder groups and customers. Value culture expresses a relationship, awareness and sufficient knowledge of the meaning of value for the enterprise and its stakeholders, and of the factors that can influence that value. This includes adequate knowledge of available methods and tools and awareness of managerial and environmental conditions. An enterprise where management has decided to use value management is based on a compiled value management program. The Value Management Program is a planned and organized set of activities that enable to develop, implement, and maintain a value management policy in a sustainable manner. It is deployed as an organizational framework, such as specific programs, such as projects and individual project studies (Tokarčíková, 2011).

The Value Manager is responsible for planning, organizing, supervising and implementing a value management program. Value managers are specialist managers because they are professional managers in the field of value creation. A substantial part of their work is teamwork, therefore it is necessary for them:

- focus on functional principle,
- communication and support of mutual communication,
- encouraging people to work together to achieve the goal,
- creating an atmosphere stimulating creativity and innovation,
- recognition of team members' results,
- convergence of members through interaction,
- providing support for collective decision-making, etc.

The value is the quantification of utility or rate of satisfaction of the needs in relation to the spent resources through the function of the given object (Vlček, 2008) indicates:

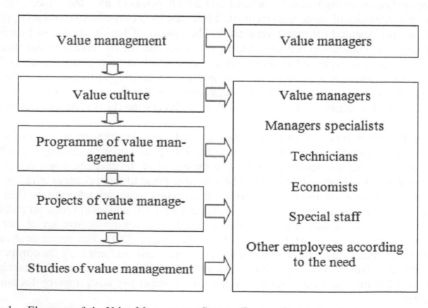

Figure 1.   Elements of the Value Management System (Source: Own processing).

- Functionality or benefit does not exist in itself. It only exists if the need for this functionality to meet the needs exists.
- A benefit can be measured only by a need since the benefit is a measure of satisfaction.
- Use or functionality consists of quality and efficiency. Effectiveness is the importance of the benefit of the object being investigated, able to meet the customer's need. Quality is the con version of performance features.

The concept of value management, where business aims to increase the value of an enterprise, is based on the idea that the value of an enterprise depends on its ability to generate future revenues (Brigham et al., 2014).

3 FUNCTIONAL ANALYSIS—THE SPECIFIC METHODS OF VALUE MANAGEMENT

Functional analysis is a process describing functions and their interrelations, which systematically characterizes, classifies and evaluates. The functional analysis combines functions and relationships. Based on it, an abstract model of a new product is created, which is a technical-physical system. It contains a basic function and secondary functions. Its creation can influence both secondary and basic function by determining shape and material (Marinič, 2014). Using a functional analysis, it first determines what the product should do (what need to satisfy) and then add the parts (components) that complement the product's functions. Using functional analysis provides space for new product creators not to be affected by known products and components, but looking for answers to questions about what the new product is doing, what to do, how it can do. Functional analysis has its rationale, especially when generating ideas. The more abstract is the function defined, the more opportunities for diversified thinking and the creation of unconventional solutions are opened.

Value analysis is an organized and creative approach utilizing a functional and economical design process for increasing the value of an existing product or service. The subject of value analysis is usually an existing product or service. Fundamentals of value analysis were created by Lawrence D. Miles in 1947, who characterized it as an organized approach to locating, finding and removing unfounded costs. Value analysis is a systematic and creative exploration of all items of product or service cost, with aim of reducing or eliminating those that do not deliver acceptable value from the customer's point of view, while maintaining the quality and performance requirements (Pollak, 2005).

The value analysis distinguishes:

- Functional costs, which are all expected or actual expenditures necessary to bring the function into the object of value analysis, into the product.
- Target costs are the costs of the end product under the specified conditions of its implementation.
- Life cycle costs are the costs of acquiring and possessing the product over a defined life cycle. They may include the costs of development, acquisition, user training, operation, support, decommissioning and disposal of the product (Strack, R., 2001).

Managing and evaluating value creation in an enterprise declares heavy and complex business activity as it ensures the coordination of individual efforts. It includes a wide range of planning, organizational and control activities along with staffing and staff development and leadership. For these activities, it is necessary to take into account the different viewpoints of the owners, managers, customers and other interested groups of the company.

4 VALUE MANAGEMENT AND MEASURING THE CORPORATE PERFORMANCE

Performance measurement is a demanding process that must be done by every company. Based on performance measurement, an enterprise may detect mistaken processes or practices that hamper the performance of individual processes as well as the enterprise as a whole. Perform-ance gaps may reveal a variety of weaknesses for which increased costs are incurred, resulting in increased spending of resources that could be used elsewhere in the business process. Based on performance measurement, the efficiency of the production process is measured, as well as the quality of the workforce involved in the production process. Performance measurement is performed by qualified employees from the company's management. The measured perform-ance of individual business processes as well as the enterprise as a whole is compared to estab-lished standards. In the case of deviations, it is necessary to identify the causes and propose improvements to this situation or to propose measures that permanently remove the errors. Prior to measuring performance and before commencing operations, an enterprise must set the objectives it wants to achieve. They must also establish standards to compare the results achieved. Standard is the level of activity that serves as a model for enterprise performance evaluation. Essentially, standards are measures that determine whether corporate perform-ance is average or not (Gašparik et al., 2015).

From the point of view of the quality management system, performance measurement is performed on the business side when the difference between the target and the quality pro-vided is identified. This means that the business compares what it has foreseen in advance to give to its customers what really provided. If the quality of the target is higher than that provided, there is dissatisfaction on the part of the customer, which is very negative for the company. Such a fact can result in a loss of customers and, in turn, a decline in sales. The second case is positive, in which case the quality is higher than the target. Customers are satis-fied, become loyal customers, repeat their purchases and recommend to their acquaintances. From an economic point of view, this fact has a positive impact not only on the growing sales but also on the good name of the business (Enzinger, 2004).

Performance measurement is an activity carried out by a particular entity or group of entities, a set of activities that are demanding on the quality of the operation and coordi-nation. These functions help you better navigate throughout the process. Measurement of corporate performance is realized through classical and modern methods. Classical methods based primarily on maximizing profits as the primary business objective use a large number of frequently incompatible indicators of ex-post and ex-ante analysis to express the goal. The shortcomings of classical methods stem from their exclusive orientation to accounting data and do not take into account the impact of the market on the value of capital, and in particular, do not take into account the risk, the impact of inflation, the time value of money, the profit is affected by valuation techniques used, reserve and provisioning, depre-ciation policy, cost and income differentials (Rajnoha R. et al., 2013). Modern performance evaluation methods are based on the assumption that an enterprise is powerful if it is able to achieve predefined strategic goals (Tokarčiková, 2011).

A modern performance indicator should meet the following requirements:

- enable the use of as much as possible information provided by accountancy;
- calculate the risk and take into account the extent of the tied capital,
- enable a clear and transparent identification of the link to all levels of corporate governance,
- enable performance and valuation of the business.

The corporate performance management system needs to be developed through planning, decision-making, organizational and control activities. For these activities, it is necessary to take into account the different viewpoints of the owners, managers, customers and other enti-ties of the business environment. Depending on performance factors, an enterprise is created by a group of indicators with the highest reporting ability depending on business conditions (Pavelkova et al., 2005). The intent is to create business performance measurement variants as shown in Table 1.

The essence of decision-making is the choice of the most appropriate option to achieve the stated goal in planning the performance of an enterprise, taking into account the various constraining conditions resulting in particular from the economic environment. The follow-ing sequence shall be followed in deciding:

Table 1. Variants of measuring the eneterprise performance (Source: Own processing).

1. VARIANT	GROUP OF INDICATORS	EVA – Economic Value Added
		MVA – Market Value Added
		SVA – Share Value Added
		CFROI – Cash Flow Return On Investment
	SUITABILITY FOR THE ENTERPRISES	the enterprises whose shares are publicly traded on the market
	FOCUS	on performance related to the market
		intended to express the ability to bring value to shareholders
	RECOMMENDTIONS	processing of time and professionally demanding indicators
2. VARIANT	GROUP OF INDICATORS	EVA – Economic Value Added
		CFROI – Cash Flow Return On Investment
		RONA – Profitability Of Net Assets
		GROGA – Cash Return On Gross Assets
	SUITABILITY FOR THE ENTERPRISES	enterprises with a high volume of assets that yield a certain return over time periods
	FOCUS	on the assessment of the company's internal activities
	RECOMMENDTIONS	the processing of indicators requires reasonable time and expertise
3. VARIANT	GROUP OF INDICATORS	EVA – Economic Value Added
		SVA – Share Value Added
		GROGA – Cash Return On Gross Assets
		CFROI – Cash Flow Return On Investment
	SUITABILITY FOR THE ENTERPRISES	the enterprises whose shares are not publicly traded on the market
	FOCUS	to express the ability to bring value to the investor (the owner of capital) and to manage the business
	RECOMMENDTIONS	a comprehensive view of business performance in terms of

- defining the problem that impedes the performance of the business,
- information analysis for decision making,
- determination of solution variants, the criteria for choosing the optimal option,
- assessment of the variants,
- selecting the optimal variant, implementation of the selected variant,
- verifying the selected variant in terms of the set target.

Planning of value creation is based on the stated goal of achieving performance for publishers. According to (Tokarčíková, 2011), "manager's decisions need to be made effectively and wisely under varying circumstances with different level of knowledge about alternatives and consequences". The essence of decision-making is the choice of the most appropriate option to achieve the stated goal in planning value creation, taking into account the various restrictive conditions resulting, in particular, from the economic environment.

5 CONCLUSION

Current value management concepts are based on the assumption that the capitalization of shareholders' equity is only possible through increasing of the value for other stakeholders. The value for owners can only be increased if the company has competitive advantages compared to competitors. The common goal of all concerned groups is the long-term existence of the business. The value creation process for interested groups can be defined by following steps:

- Identification of significant stakeholder groups in the company,
- An expression of value for identified stakeholders,

– Measurement of value and its expression through financial and non-financial indicators,
– Identifying factors influencing the value creation.

Value management is a special type of management, clearly focused on creating and increasing value for the customer while increasing the value of the company. The value of value management for an enterprise is to maintain and intensify the competitiveness of the company by building a long-term sustainable relationship with the stakeholder groups. A value management enterprise will gain a competitive advantage by focusing on value creation.

Decision-making in an enterprise is subject to uncertainty and risk. Value creation is the ability of an enterprise to value invested capital, the ability to increase market value for owners while respecting the interests of other stakeholders in the business. It is a complex economic category that determines the economic added value and corporate performance and so on.

For the long-term development of the company, it is necessary to meet the expectations of the owners who want their capital to be valorized sufficiently, the expectations of the employees who are interested in the wage level, the employment perspective, the possibility of professional growth, and the assumption of the business partners about the stable development of the company.

ACKNOWLEDGEMENTS

This article was created as part of the application of project VEGA No 1/0652/16 Impact of spatial location and sectorial focus on the performance of businesses and their competitiveness in the global market and project APVV-16-0297 Updating of anthropometric database of Slovak population.

REFERENCES

Brigham F.E., Ehrhardt C.M., 2014. *Financial management: theory & practice*. 14th ed. Mason, Ohio: SouthWestern, 2014. 1163s. ISBN 978-1-111-97220-2.

Ďurišová M., 2011. Application of cost models in transportation companies. *Periodica Polytechnica: social and management sciences*. ISSN 1416-3837, Vol. 19, iss. 1, 2011, pp. 19–24.

Enzinger, M., 2004. *Verknüpfung der Balanced Scorecard mit dem Value Based Management*, Controller. News, 2004, Nr. 5, s. 162–164. ISSN 1214-5149.

Freeman, R.E. 1984. Strategic Management: *A stakeholder approach*. Boston: Pitman. 1984.

Gašparik, J., Szalayová S., 2015. Optimal method of building elevator selection from the point of their energy consumption minimizing. In *ISARC 2015 – Connected to the Future [elektronický zdroj]: proceedings of the 32nd International Symposium on Automation and Robotics in Construction and Mining. Oulu, Finland, 15. – 18. 6. 2015*. Oulu: University of Oulu, 2015, online, [7] s. ISSN 0356-9403. ISBN 978-951-758-597-2. (A – Scopus Elsevier).

Marinič, P., 2014. Hodnotový manažment ve finančním řízení. *Hodnota verzus financie*. Praha: Wolters Kluwer, a. s., 2014. 260 s. ISBN 978-80-7478-405-7.

Miles, L.D., 1961, *Technique of Value Analysis and Engineering.*, New York, 1961 McGraw-Hill.

Pavelkova, D., Knapkova, A. 2005. *Výkonnost podniku z pohledu finančního manažera*. Praha: LINDE nakladatelství.

Plowman, B. 2001. Activity Based Management: *Improving Processes and Profitability*. Hampshire: Gower Publishing Limited, 2001. 228 s. ISBN 978-0566081453.

Pollak, H., 2005. Jak odstranit neopodstatněné náklady. *Hodnotová analýza v praxi*. Praha. Grada Publishing, a.s.

Strack, R., Villis, M. 2001. RAVE™- Die nächste Generation im Sharehol der Value Management. *Zeitschrift für Betriebswirtschaft*, 2001, Heft 1, s. 67–84. ISSN 0044-2372.

Tokarčíková E. 2011. *Influence of social networking for enterprise's activities*, Periodica Polytechnica, Social and Management Sciences, Budapest, 19/1 (2011) 37–41. doi: 10.3311/pp.so.2011-1.05-ISSN 1416-3837.

Vlček, R. 2008. *Management hodnotových inovací*. Praha, 2008. Management Press.

Advances and Trends in Engineering Sciences and Technologies III – Al Ali & Platko (Eds)
© 2019 Taylor & Francis Group, London, ISBN 978-0-367-07509-5

Application of the virtual reality and building information modeling to solar systems

B. Chudikova, M. Faltejsek & I. Skotnicova
Faculty of Civil Engineering, VSB-Technical University of Ostrava, Ostrava, Czech Republic

ABSTRACT: Civil engineering exponentially heads towards renewable energy sources in all types of buildings. The use of solar energy is one of the alternatives that can be applied in a good ratio of demands for space, price and resultant benefits. Building information modelling is a modern and highly effective way of solution of buildings with regard to all aspects of the life cycle. Virtual reality allows people to immerse themselves in artificial worlds, to create visual, auditory and tactile experiences that affect humans through computers. Augmented reality is a combination of real world with additional 2D or 3D digital information. Blending of digital elements into the real space can be a key to improve the design practice, subsequent maintenance or more effective reconstruction. The article describes benefits that arise from the use of modern technologies when applied to the management, maintenance and implementation of solar systems.

1 INTRODUCTION

After 20 years of using information technology, the technologies has undergone major changes. Smartphone and tablet sales are already breaking the sales of computers and laptops. The new era constantly give us new opportunities in other dimensions and easier access to information. Today's young generation is trying to combine all of its actions with the use of applications. As people are increasingly clinging to the graphical user interface, it is certainly a good decision to include the physical world into computer experience. And since the world is not flat or composed of written documents, it is also necessary to develop the existing user interface.

Virtual reality and augmented reality are often used as tools in several disciplines. With the arrival of new industrial revolution and the effort to digitize construction globally and expand the application of building information modelling in this sector, there are many possibilities of how to apply and exploit new technologies. Augmented reality has a great potential to be on the first positions during the revolution of the current user interface, as its quality may be excellent and it may provide a direct link between physical reality and virtual information about this reality. Therefore, our whole world can become a user interface.

Solar thermal systems fall under the use of renewable energy sources which are the subject of discussions, especially in connection with the implementation of the obligations stated in Directive 2009/28 /EC. It is primarily about the fulfilment of the binding goal of producing 20% of energy consumption through renewable sources by 2020 (Donova, Kucerikova, Zdrazilova, 2014).

Building information modelling (BIM) has been growing in connection with global digitization of all industries, including the construction industry. This is a process throughout the building life cycle which links construction participants in order to collaborate on one information model. A result of this collaboration is the creation of an ideal building design and comprehensive drawing of all information that maximizes utility. A 3D model containing detailed information at the level of individual elements is a basic element of the **BIM** method. The process of ideal building design using the **BIM** method also includes simulations and analyses that use a building model, surrounding area and the environment into which we place the building. That way we can eliminate bad practices and anticipate problems that might arise during the life of the building (Eastman, Teicholz, Sacks, Liston, 2011).

2 VIRTUAL AND AUGMENTED REALITY

Nowadays, virtual reality is a very popular topic and it enables us to dive into computer-generated worlds, to create visual, auditory and tactile experiences that give us subjective impressions and can be perceived by special helmets or glasses. Although virtual reality can be used extensively in construction, it may not be strictly the most appropriate tool for all construction sub-sector. Attention should be paid to the user interface which allows a combination of occasional use and provides information in small, easy-to-understand parts. Virtual reality has reached a point where virtual images are very accurate and often indistinguishable from the real world. Basically, anything, without any limitation, is possible after entering virtual reality.

We use the term augmented reality to create automatic links between physical real world and electronic information that contains digital objects created by a computer. It could be said that this is an improved reality where a user can see the real world, but with the addition of the computer-generated images, objects or information. It can be seen through special glasses, a tablet or smartphone. The added information that appears in the augmented reality will help the user to perform their activity or to better imagine the space itself with new elements.

Augmented reality always meets three conditions:

- It combines real and virtual environments,
- It is interactive in real time,
- It is recorded even in 3D.

Augmented reality has a feedback between the user and the visualization system. The user controls the view by its movement, while the system monitors the movement of the view and adapts the virtual content placed in the real world (Aukstakalnis, 2016).

All technological devices are becoming more and more complex which greatly affects their planning, implementation and operation. Architectural structures, infrastructure and technologies using renewable energy sources are designed using various software. During the implementation, a lot of changes are usually made and when software is used for designing, it is not a problem to record these changes into the model so that it can be easily and clearly documented in the as-built building implementation. Current practice, however, usually does not record these changes back into the drawing documentation well, and later there are problems, for example, during reconstructions or failures, either in structures or in technologies, or in facility management in general. Another option is to place the proposed building into a planned area using augmented reality and then to evaluate the impact on the surrounding area.

The greatest benefit of augmented reality is the spatial visualization itself which is created by the interconnection of the real environment and embedded digital elements or information. We can instantly recognize and identify problems directly in the given area and solve them on the spot. The collaboration between multiple allows professionals during the construction process through viewing equipment, such as a tablet, an immediate solution of the given situation between all participants with spatial visualization and added information.

3 SOLAR SYSTEMS

Solar systems are characterized by the conversion of solar radiation into heat, which we can use further. The heat obtained by absorption must be led to a point of its use. The heat transport is usually solved by a piping system in which heat transmission fluid circulates. General requirements for these piping systems result in an effort to make the system as short as possible and well insulated against heat losses. An integral part of the active system is also an accumulation tank, the volume of which needs to be adapted to the number of people in the building and the anticipated use of water. The tank also has to be insulated against heat losses. Water heating is always indirect, i.e. via a heat exchanger in the tank. The reason is a type of heat transfer fluid that contains poisonous antifreeze fluid and other ingredients against corrosion of pipes.

The connection of solar collectors to a distribution system in the building can be used for the preparation of hot water or for heating. The choice of an ideal type of solar collector also depends on the type of connection. Solar thermal collectors are divided into flat plate collectors, evacuated tube collectors and concentration collectors. The use of concentration collectors is mainly for industrial production, for civil buildings we can choose between flat plate and evacuated tube collectors.

4 USE OF BIM IN THE DESIGN OF SOLAR SYSTEMS—THE IMPORTANCE OF 3D MODEL

The BIM method is based on information. Data contained in the model reflect the future state of the building, and users, using the model at different stages of the building's life cycle, may effectively use the information. Complete information and an overview of the entire system can be generated from the model as a list of elements, including prices, materials and other specifications that we can define. Based on this information, we know actual costs of acquiring a solar system, and if we link the BIM model to the CAFM system, we can use it, including enclosed documentation and information on revisions, inspections and potential repairs. The model carries information about the system manufacturer, the warranty, or the service life of individual elements (Sanchez, Hampson, Vaux, 2016).

4.1 Model planning and use

Another benefit is a complete design of the as-built state of the system. At the planning stage, before construction, we know a complete list of elements (pipe fittings), lengths and other specifications for the proposed piping. The distribution, location or e.g. pipe fastening, is designed with regard to the structural design of the building. By creating a complete building model, we know all the problem areas, structurally load-bearing structures or other installations in the building, so we can modify the given design into an ideal design.

Figure 1. Augmented reality (author).

Figure 2.   Space notes on the maintenance of an expansion tank (author).

Figure 3.   Information about the solar collector in augmented reality (author).

It is very easy to put possible changes in the implementation of the building into the model and simultaneously, the documentation of the as-built implementation of the building is created, and after the building is done, the documentation will tell us about its real state. The model should serve throughout the life of the building and it should be handed over to the operator of the building after the implementation.

5   SIMULATION OF SOLAR SCOPE AND INFLUENCES OF THE SURROUNDING BUILDINGS

As standard, a study of the solar scope (shadowing) evaluates in which daytime an apartment or an outdoor space for recreation has the solar scope and whether the quantity is sufficient. In this case, the solar scope study allows you to assess an impact of adjacent buildings on the maximum possible solar exposure of the solar collectors to determine their most suitable position on the roof of the building and thus the highest efficiency of the system (Skotnicova, Galda Tymova, Lausova, 2014). For example, the ArchiCAD software can be used to study the solar scope, in this software it is possible to model a real situation and create a simulation of the day course of sun exposure. From the day course of shadows of the surrounding buildings, for example, we can re-evaluate the original design of the solar collectors, which was based on the requirement for the shortest piping. Using the simulation, we can find that by placing collectors directly above the technical room, we can lose from few minutes to several hours of solar exposure every day, and hence we can lose considerable profits that the system would otherwise have produced (Act No. 406/2000). Once the model is created, it would also be possible to integrate virtual or augmented reality. In the augmented reality, we can go through the real environment, add planned solar systems with collectors and, for example, evaluate urban suitability or inappropriateness of the proposed solution. Virtual reality will allow you, for example, to walk on the planned roof and to check the connection of the entire system, or to determine the correct distance between solar collectors.

6 APPLICATION OF VIRTUAL AND AUGMENTED REALITY TO SOLAR SYSTEMS

Solar systems fall into a category of technical equipment of buildings which is already an integral part of our buildings and of our lives. Technical equipment of buildings includes water distribution systems, sewerage systems, gas-water pipelines, heating systems, air-conditioning equipment, wiring and therefore facilities for the use of renewable energy sources. All these technologies have their own specifics and require expertise while designing and handling them. Virtual and augmented reality can make it easier to check the right design and execution of the technical equipment of the building, allowing you to view pipelines and technologies that lead under the surface of plasters and to prevent possible problems.

6.1 Use in designing

Virtual reality can facilitate both designing, and implementation of technology systems. Orientation in the design intent can be difficult with large technological systems. Imagine a large and complicated engine room that is now being complicatedly drawn into a diagram, which even the people who created it do not understand and a 3D model of this engine room that we can go through, check for the correctness of all connections, and carefully evaluate the accuracy of all designed parts. This 3D model, of course, serves to check the as-built implementation, and has other benefits when it is connected to the CAFM system.

Augmented reality can serve as an analysis of contradictions, for example, in large technology operations. In the case of a tour through a completed engine room, the tablet shows the planned design intent and it is possible to see different connections or missing components. This can prevent big problems that might arise when the operations start. For example, a missing or insufficiently dimensioned safety valve can cause considerable financial damage.

6.2 Use in system maintenance

Understanding of how technologies work or how to dismantle, repair or replace them is quite challenging in some professions. Maintenance technicians often spend a lot of time studying manuals of individual components. Augmented reality can provide technicians with instructions that appear directly at their workspace. Instructions created directly by the manufacturer of the given device or component in augmented reality may be more effective than some training. If a technician needed to deal with something unusual, he could address the operator. Augmented reality can provide shared visual space for remote collaboration on specific physical actions. Using this approach, a remote expert can solve the given problem by exploring and communicating through space notes that are instantly visible in augmented reality in real time. Augmented reality therefore combines the benefits of live video conferencing and a remote scene into real-life environment, creating an ideal way of collaboration.

6.3 *Element information*

Using augmented reality, it would be possible to find information about parts of the solar system. This information can be useful especially when there is a need to replace certain element. A solar collector may be damaged several years after installation and it needs to be replaced. In this case, by using augmented reality, it would be enough to focus on the collector, and the information about collector dimensions, total area, aperture area, efficiency, losses, type of connection or absorber material would show up. According to this information, it is easy to select a similar collector to replace the damaged one. If we did not have this information and chose a collector with a lower efficiency, it might be that this installation would bring the whole system out of balance and it would not work properly.

7 CONCLUSION

The use of sunshine is one of the most popular possibilities of how to use renewable resources, it is relatively easy available financially and relatively easy to solve technically. Their use and application can be found across all types of buildings, from family or apartment buildings to community facilities. The design of such system basically depends on an optimal setting for the sun motion in the sky. General rules for the slope of panels, direct adjustment to the south and the requirement for the shortest piping system located best above the technical room, however, may not always be the most efficient solution. Proper analysis of the situation and evaluation of various design alternatives in the planning phase allows us to choose an ideal and effective solution in terms of technical and economic issues as well as cost-effective, with maximum system performance. Building information modelling (BIM) and the involvement of virtual and augmented reality is an option how to create, evaluate and analyse these simulations on a realistic model. It allows us to plan and model the actual future state of the structure and installation with information on dimensions, prices and other specifications and to set an ideal concept in the future location of the building, taking into account all factors, current and future. Last but not least, due to easy control, it eliminates errors and reduces costs during the entire lifecycle of the building.

ACKNOWLEDGEMENTS

This article was created with the support of the Student Grant Competition of VŠB-Technical University Ostrava, registration number of the project SGS SP2018/66.

REFERENCES

Act No. 406/2000 Coll. *on Energy Management, as amended.*
Aukstakalnis, S. 2016. *Practical Augmented Reality: A Guide to the Technologies, Applications, and Human Factors for AR and VR (Usability). Indiana: Addison-Wesley. ISBN 978-0134094236.*
Donova, D., Veronika K. and Zdrazilova N. 2014. *Thermal analysis of performance of the building envelope. In: SGEM: Albena, Bulgaria, pp. 401–408. ISBN 978-619-7105-21-6. ISSN 1314-2704, DOI: 10.5593/sgem2014B62.*
Eastman, Ch., Teicholz, P. and Liston, K. 2011. *BIM Handbook: A Guide to Building Information Modeling for Owners, Managers, Designers, Engineers and Contractors. 978-0-470-54137-1.*
Sanchez A. X. and Hampson K. D. and Vaux S. 2016. *Delivering Value with BIM, A whole-oflife approach. London a New York: Routledge. ISBN: 978-1-138-11899-7.*
Skotnicova, I., Galda Z., Tymová P. a Lausová L. 2014. *Evaluation of indoor environment and energy efficiency of a passive house. International Multidisciplinary Scientific GeoConference-SGEM. 543–548. DOI: 10.5593/SGEM2014/B62/S27.070. ISSN 1314-2704.*

Advances and Trends in Engineering Sciences and Technologies III – Al Ali & Platko (Eds)
© 2019 Taylor & Francis Group, London, ISBN 978-0-367-07509-5

Characteristic determination of solar cell by simulation and laboratory measurement

P. Iski, I. Bodnár, D. Koós, Á. Skribanek & Cs.B. Boldizsár
University of Miskolc, Miskolc-Egyetemváros, Hungary

ABSTRACT: The efficient operation of a solar cell is influenced by several factors. Some of these factors are the intensity of illumination and the spectral composition of it, as well as the ambient temperature together with the temperature and contamination of the solar cell and the atmosphere. This study presents the voltage, amperage, and power change of a commercially available solar cell, caused by the temperature transient, by the help of simulations and laboratory measurements. Temperature transient investigations allow us to know more about cooled and non-cooled solar panel behavior, in case of constant intensity of illumination. During the measurements, we have concluded that increasing the temperature decreases the maximum power of the solar cell. Compared to the simulation results we experienced good tendential.

1   INTRODUCTION

The use of solar panels is rising nowadays, but it should be mentioned that the operation of these devices is highly influenced by several weather conditions. Solar cells convert the energy, coming from the sun light into electrical power. The temperature of a solar cell can reach over 80°C during normal operation, which can lead to the decrease of the cell's efficiency. Considering this fact, the main goal of this research work is to determine the connection between the temperature change and electrical parameters of solar cells, basing on laboratory measurements and numerical simulations. Measurements and numerical simulations in case of a constant cell temperature are presented in numerous literature. In contrast, this study demonstrates the measurement and simulation results of the behavior of a solar cell in case of temperature transient.

2   EQUIVALENT ELECTRICAL MODEL OF SOLAR CELL

For the simulation of the operation of a solar cell, the first step is the determination of the solar cell's electrical model. Numerous equivalent electrical models are mentioned by the relevant literature. The ideal electrical model of solar cell consists of a current generator and a parallelly connected diode. The important feature of this model is that the current generator simulates properly the DC current from the p-n junction and the diode substitutes the resistance which is parallel to the load. The extended model completes the ideal model with a series resistance, which represent the electrical loss in case of the real solar cell constructions. Mathematical model of the extended equivalent circuit contains four parameters. If we continue to improve the model, the internal resistance of the solar cell should be simulated as a resistance parallel to the shunt diode. During numerous implementation of this model, the value of the internal resistance is infinite, ergo the resistance is replaced with an electrical discontinuity. This model is considered to be the most accurate model which simulates the operation of a solar cell. With the help of simplifications, other described equivalent circuits can be derived from this model [Bodnár (2018)].

U_{oc} open circuit voltage and I_{sc} short circuit current can be measured on a solar cell. If any load is connected to the solar cell, the measureable I current and U voltage will always be lower than in case of no load. The I current is the difference between the dark current (I_D) and the photo current (I_{ph}). The dark current exponentially depends on the temperature and linearly on theIs saturation current because of the semiconductor characteristic of the solar cell. This is described by equation (1) [Bodnár (2018), Ishaque et al. (2011)]:

$$I = I_{ph} - I_D - I_P, \tag{1}$$

$$I_D = I_s \left[exp\left(\frac{e \cdot U_D}{n \cdot k \cdot T} \right) - 1 \right], \tag{2}$$

$$I_P = \frac{U_D}{R_p} = \frac{U + I \cdot R_S}{R_p}, \tag{3}$$

$$U = U_D - U_s, \tag{4}$$

where I = current; I_{ph} = photo current; I_D = dark current; I_p = current of parallel resistance; I_s = saturation current; U = voltage; U_D = diode voltage; U_S = voltage of serial resistance; R_p = parallel resistance; R_S = serial resistance; e = electron charger; n = diode ideality factor; k = Boltzmann constant; T = Temperature [Bodnár (2018), Ishaque et al. (2011)].

The short circuit current (5) and the open circuit voltage (6) can be expressed by the substitution of $U = 0$ and $I = 0$. Open circuit voltage logarithmically depends on current values and linearly depends on U_T temperature dependent thermic voltage. According to these [Bodnár (2018), Ishaque et al. (2011), Munoz-Garcia et al. (2012)]:

$$I_{SC} = I_{ph}, \tag{5}$$

$$U_{OC} = \frac{k \cdot T}{e} \cdot ln\left(\frac{I_{ph}}{I_s} + 1 \right) = U_T \cdot ln\left(\frac{I_{ph}}{I_s} + 1 \right), \tag{6}$$

where I_{SC} = short circuit current; U_{OC} = open circuit voltage; U_T = thermic voltage.

The effective electrical power (P) of a solar panel can be determined by the multiplication of I amperage and U voltage, measured on the R resistance [Bodnár (2018), Ishaque et al. (2011), Munoz-Garcia et al (2012)]:

$$P = I \cdot U = I_{SC} \cdot U - I_s \cdot U \cdot exp\left(\frac{U}{U_T} - 1 \right), \tag{7}$$

where P = power [Bodnár (2018), Ishaque et al. (2011), Munoz-Garcia et al. (2012)].

The T_S operating temperature of the solar panel can be expressed by equation (8) [Bodnár (2018), Ishaque et al. (2011), Munoz-Garcia et al. (2012)]:

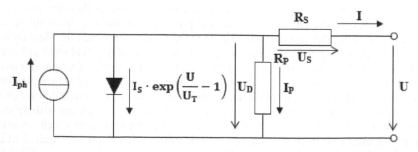

Figure 1.   The real circuit model of the solar panel.

$$T_S = (T_N - T_A) \cdot \frac{E_{ill}}{E_{STC}} + T_A, \qquad (8)$$

where E_{STC} = the intensity of standard illumination (1,000 W/m²); E_{ill} = the intensity of standard illumination (1,000 W/m²); T_S operating temperature of the solar cell; T_N = the nominal temperature of the solar cell; T_A = the ambient temperature [Bodnár (2018), Ishaque et al. (2011), Munoz-Garcia et al. (2012)].

Taking these into account, the photocurrent can be determined as a function of temperature [Bodnár (2018), Singh et al. (2012)]:

$$I_{ph} = I_{SCN} \cdot \left[1 + \mu_{ISC} \cdot (T_S - T_A) \right], \qquad (9)$$

where: I_{SCN} = nominal sort-circuit current; μ_{Ir} = the percentage coefficient of the short-circuit current [Bodnár (2018), Singh et al. (2012)].

If the intensity also changes, the photocurrent value can be written as follows [Ishaque et al. (2011), Munoz-Garcia et al. (2012), Radziemska (2003)]:

$$I_{ph} = \frac{E_{ill}}{E_{STC}} \cdot I_{SCN} \cdot \left[1 + \mu_{Ir} \cdot (T_S - T_A) \right] \qquad (10)$$

The saturation current value, as a function of temperature can be calculated basing on the two diode models [Ishaque et al. (2011), Munoz-Garcia et al. (2012)]:

$$I_s = \frac{I_{ph}}{\left[exp\left(\left(\frac{e \cdot U_{OC}}{n \cdot k \cdot T \cdot N_s} \right) \cdot \left(1 + \mu_{UOC} \cdot (T_S - T_A) \right) \right) \right] - 1}, \qquad (11)$$

where μ_{OC} = the percentage coefficient of the open-circuit voltage.
The temperature dependence of idle voltage [Munoz-Garcia et al. (2012)]:

$$U_{OCT} = U_{OCN} \cdot \left[1 + \mu_{UOC} \cdot (T_S - T_A) \right], \qquad (12)$$

where U_{OCN} = nominal open-circuit voltage [Munoz-Garcia et al. (2012)].

3 THE EXPERIMENTAL COMPOSITION

Our experimental investigation can be separated into two parts. On one hand during temperature transient, the change of the solar cell's electrical parameters is measured, on the other hand in case of constant temperature and light intensity the electrical characteristic

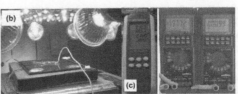

Figure 2. Measurement composition (a), investigated solar cell (b), temperature-, voltage- and current level measurement (c).

of the solar cell is determined. The temperature of the solar cell is influenced by a cooling-system, which system is based on Peltier-modules. The temperature measurement of the cell is granted by a four-channel digital thermometer, type of: Voltcraft PL-125-T4. Current and voltage level of electricity is measured by two digital multimeters, type of: METEIX MX 59H. The applied solar simulator is our own development, which is suitable for standard C-class illumination according to the ASTM E975 standard. Figure 2 represents the measurement system.

4 BOUNDARY CONDITIONS OF THE SIMULATIONS

The numerical simulations were made by using the equation system of the two-diode model, mentioned in the previous chapters. During the simulation, we started from the simplified circuit model of the solar panel. Table 1. contains the constants used during the simulation, while the electrical parameters of the solar panel can be seen in Table 2.

During the simulation, we have had the following considerations and neglects:

- we reduced the solar module to one cell,
- we neglected the serial and parallel resistance,
- we took the integrated mean of the intensity of the illumination,
- we calculated with the help of the open circuit voltage, short circuit current and temperature constant, which were given by the manufacturer,
- we considered the difference between the spectral composition (spectral energy density) of the halogen and the sunlight as a constant [Bodnár (2018)].

Table 1.   Constant parameters [Bodnár (2018)].

Parameter	Symbol	Value	Measurements
The solar irradiation intensity at standard test conditions	E_{STC}	1,000	W/m^2
Intensity of illumination	E_{ill}	1,000	W/m^2
Diode reverse bias saturation current (according to the two-diode model)	I_s	$1 \cdot 10^{-11}$	A/cm^2
Electron charger	e	$1.60 \cdot 10^{-19}$	C
Boltzmann constant	k	$1.60 \cdot 10^{-19}$	J/K
Diode ideality factor	n	2	–
Constant of the light spectral composition	C	0.9267	–

Table 2.   Electrical parameters of the solar cell.

Parameter	Symbol	Value	Measurements
Year of manufacture	–	2018	–
Peak Power	P_{max}	0.68	W
Max. Power Current	I_M	0.094	A
Max. Power Voltage	U_M	7.2	V
Short Circuit Current	I_{SC}	0.115	A
Open Circuit Voltage	U_{OC}	8.4	V
Nominal Fill Factor	φ	0.7	–
Percentage Temperature co-efficient for P_{max}	μ_{Pm}	−0.43	%/°C
Percentage Temperature co-efficient for I_{SC}	μ_{Isc}	0.047	%/°C
Percentage Temperature co-efficient for U_{OC}	μ_{Uoc}	−0.32	%/°C
Nominal Operating Temperature	T_N	80	°C

5 COMPARASION OF MEASUREMENT AND SIMULATION RESULTS

During our measurements, the behavior of the solar cell's electrical parameters is examined in case of temperature transient, therefore the mathematical model of the solar cell should be used to describe similar phenomenon's. During the temperature transient measurements, the open-circuit voltage and the short-circuit current of the unloaded solar cell are registered in case of permanent light intensity and transient temperature. Based on the previous described mathematical equations the measured electrical parameters as a function of temperature change can be calculated. With the help of the numerical simulations, beyond the open-circuit voltage and short-circuit current, theoretical- and real electrical power of solar cell can be specified. Simulation model is implemented into MATLAB environment.

Temperature data from the transient measurements are used for the transient simulations. Short-circuit current, open-circuit voltage and theoretical power are represented as a function of time and temperature, in case of temperature transient of cooled and non-cooled cell. The graphs show the measured and simulated data at the same time, under standard measurement conditions.

Figure 3 describe the short-circuit current as a function of temperature. In case of low temperature, the difference between the measured and calculated result is more significant, but in case of higher temperature, great agreement between the data can be seen. The slope of the measured values curve is greater than the curve from the simulation. The reason of this is the thermal inertia of the solar cells, and the whole cell warmed at lower temperatures.

Figure 4 shows the open-circuit voltage change as a function of temperature. We can observe that the slope of the measured value's curve is lighter than the curve from the simulation.

Theoretical power of the solar cell is represented in Figure 5. As in our basic assumption, the temperature increase leads to the cell's power decrease. Required tendency can be observed in the results. The ripple of the measurement curve is caused by the measurement inaccuracies.

Agreement between the measured and simulated values are summarized in Table 3. Within the whole examined temperature range, the developed simulation model in case of short-circuit current overestimates-, while in case of open-circuit voltage and theoretical power underestimates the real values. Most accurate results in case of short-circuit current can be calculated. Similar result is given by Radziemska (2003).

Furthermore, two cases were examined: cooling-system at 50% power and cooling-system at 100% power. In non-cooled case the cells temperature reaches 70°C, in case of 100% cooling power, the cell's temperature decreased by 18°C at steady state. The applied cooling-system is our own development, which contains four Peltier modules and two fans.

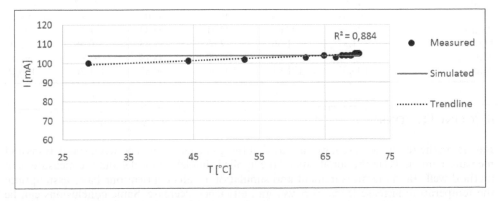

Figure 3.   Short-circuit current as a function of temperature.

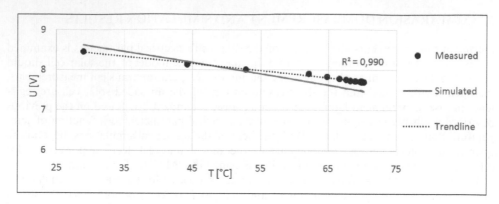

Figure 4. Open-circuit voltage as a function of temperature.

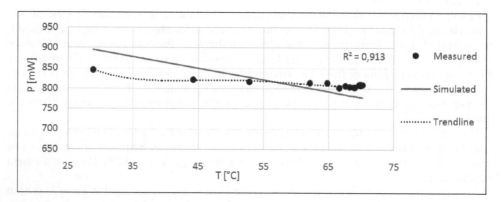

Figure 5. Power as a function of temperature.

Table 3. Agreement between the measured and simulated results.

	Non-cooled			Cooled 100%		
	I ratio	U ratio	P ratio	I ratio	U ratio	P ratio
Statistic parameter	%	%	%	%	%	%
Modus	99.08	–	–	–	99.13	–
Median	99.08	97.10	96.27	99.08	99.13	98.22
Avarage	100.3	97.67	97.71	99.84	99.27	99.12
Deviation	1.37	1.36	2.69	1.64	0.48	2.09
Min	99.08	96.89	96.00	98.14	98.99	97.15
Max	103.81	101.97	105.86	100.87	101.26	106.19

6 CONCLUSIONS

Basing on the results achieved, we can say that the good agreements between simulations and measurements indicate the suitability of the applied simulation model and the measurement method well. Both the measurement and simulation results confirm our basic assumption; the temperature increase leads to power and efficiency decrease. Same conclusions can be seen in the work of Ishaque et al. (2011) and Singh et al. (2012).

ACKNOWLEDGMENT

Supported by ÚNKP-18-2-I.-ME/27. New National Excellence Program of the Ministry of Human Capacities.

REFERENCES

Bodnár, I. 2018. Electric parameters determination of solar panel by numeric simulations and laboratory measurements during temperature transient. *Acta Polytechnica Hungarica.* 15(4): 59–82.

Dubey, S., Sarvaiya, J.N., Seshadri, B. 2013. Temperature Dependent Photovoltaic (PV) Efficiency and Its Effect on PV Production in the World A Review. *Energy Procedia* 33: 311–321.

Gürtürk, M., Benli, H., Ertürk, N.K. 2018. Effects of different parameters on energy – Exergy and power conversion efficiency of PV modules. *Renewable and Sustainable Energy Reviews.* 92(9): 426–439.

Ishaque, K., Salam, Z., Taheri, H., Syafaruddin. 2011. Modelling and simulation of photovoltaic (PV) system during partial shading based on a two-diode model. *Simulation Modelling Practice and Theory.* 19(7): 1613–1626.

Munoz-Garcia, M.A., Marin, O., Alonso-García, M.C., Chenlo, F. 2012. Characterization of thin film PV modules under standard test conditions: Results of indoor and outdoor measurements and the effects of sunlight exposure. *Solar Energy.* 86(10): 3049–3056.

Radziemska, E. 2003. The effect of temperature on the power drop in crystalline silicon solar cells. *Renewable Energy.* 28(1): 1–12.

Singh, P., Ravindra, N.M. 2012. Temperature dependence of solar cell performance – an analysis. *Solar Energy Materials and Solar Cells.* 101: 36–45.

Load-bearing structure of Tokaj observation tower

J. Kanócz
Faculty of Art, Technical University of Košice, Košice, Slovakia

P. Platko
Faculty of Civil Engineering, Institute of Structural Engineering, Technical University of Košice, Košice, Slovakia

ABSTRACT: Tokaj observation (or viewing) tower is a relatively new landmark of the Tokaj wine region erected in the vineyards between the villages Malá Tŕňa and Černochov. The construction of this modern and attractive tower is 12 m high and its unusual shape resembles a wooden wine barrel. The tower offers beautiful views on the Tokaj vineyards and their surroundings. Load carrying structure of the Tokaj observation tower is created by an interesting and effective combination of several materials. This combination includes i) glued laminated timber members, ii) steel members with flexural rigidity, iii) steel members without flexural rigidity and iv) ThermoWood members.

1 INTRODUCTION

The design of wooden towers (as well as steel towers or towers made of combined materials), such as observation towers, was viewed as a predominantly engineering task in the past and the main emphasis was given to their functionality and simplicity. Their construction usually made use of a simple load-bearing structure and a square or triangular ground plan.

Nowadays the demand for an architectural rendition of these buildings is becoming increasingly common in order to create works with valuable aesthetic parameters and at the same time to reflect the current level of technical and constructional requirements and material options. Therefore, when designing modern observation towers, it is necessary to take into account the following design principles (Kanócz, Mihaľák, Platko, 2015):

– Architectural—designing an object with a certain personal presentation by the author, while respecting the general rules of architectural design,
– Structural—designing a construction which is an inherent part of the architectural expression of the object, one that is simple but at the same time reflects the most current design knowledge and trends, meets all current normative requirements and is safe throughout its entire lifetime,
– Material—using the appropriate materials, or structural elements made of these materials (for example wood-based materials such as glued laminated timber, densified wood, laminated veneer lumber, etc.).

The paper presents an example of observation tower with combined steel-timber load-bearing structure, design of which made use of the above mentioned principles.

2 GENERAL INFORMATION

The Tokaj observation tower was designed as a part of the "Tokaj je len jeden" (Tokaj is the only one) cycling route project in the Tokaj vine growing area. It is a part of a rest stop of a cycling path that passes between the vineyards in the hillside above the village of Malá Tŕňa.

The observation platform provides views of the surrounding region. Visitors are also provided with binoculars and information tables with panorama descriptions. The tower is lit at night.

The basic architectural concept is based on the idea that during a cycling trip, "one encounters the need to get on top of something", from where the wider area is seen. In the wine region, of course, a barrel seems natural, which also gives rise to the architectural concept (Kanócz, Mihaľák, Platko, 2015). The lookout itself is situated on the altitude of +12.0 m.

General information:

– authors of architectural solutions: Ing. arch. Michal Mihaľák, doc. Ing. Ján Kanócz, CSc.,
– chief designer: doc. Ing. Ján Kanócz, CSc.,
– designer of architectural parts: Ing. arch. Michal Mihaľák,
– designers of load-bearing structures: doc. Ing. Ján Kanócz, CSc., Ing. Peter Platko, PhD.,
– main contractor: Ing. Jaroslav Pčola, REINTER,
– developer: "Tokajská vínna cesta" (Tokaj wine route),
– total investment costs (excluding VAT): 0.108 mil. €,
– construction period: 05/2014 – 05/2015.

3   DESCRIPTION OF THE LOAD-BEARING STRUCTURE

The ground plan of the observation tower is triangular (with the sides slightly rounded) with dimensions of approximately 8.9 m × 8.6 m and its total height is approximately 13.1 m. The load-bearing structure is designed as a combined steel-timber structure (Figures 1 and 2).

The load-bearing structure of the Tokaj observation tower is founded on relatively massive concrete gravity based foundations (Figure 3) and it is formed of the following subparts (Kanócz, Mihaľák, Platko, 2011):

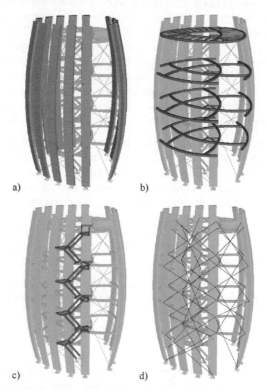

a)   b)

c)   d)

Figure 1.   Subparts of the 3D computational model of the main load-bearing structure.

Figure 2. Top view of the 3D computational model of the main load-bearing structure.

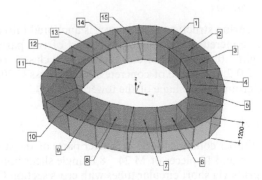

Figure 3. 3D computational model of the tower foundations.

– perimeter timber columns made of glued laminated timer (Figure 1a),
– horizontal steel grids and terrace (Figure 1b),
– inner steel staircase (Figure 1c),
– tension-only bracing system (Figure 1d).

It follows that the load-bearing structure of the tower makes use of not only a material combination, but also a combination of structural elements with flexural rigidity and structural elements without flexural rigidity. The steel part of the load-bearing structure (except for the attachment of tension-only bracing system) is designed as whole welded. The execution class of steel structures is EXC2. The surface of all steel members is protected by galvanizing (the welded joints executed on the site were treated with zinc paint for repairs).

3.1 Perimeter timber columns

Fifteen columns made of glued laminated timber of strength class GL 24h are situated around the tower's perimeter (five on each side of the triangle). The columns measure 700/180 mm in cross-section and their length is approximately 13.1 m, they are slightly curved along the height (the internal radius of the curvature is approximately 22.8 m). The columns also create a quasi-sheathing of the steel parts of the load-bearing structure.

3.2 Horizontal steel grids

The perimeter timber columns are connected with steel grids located at the levels of +3.0 m, +6.0 m, +9.0 m, and a terrace grid at +12.0 m. Individual parts of the grids are designed from square tubes with cross section SHS 180/180 mm, with wall thickness 5 or 6 mm. Structural steel of grade S 235 is used.

3.3 Inner steel staircase

Between the two horizontal grids is placed a three-flight steel staircase with two side string-ers and two landings. Rectangular tubes with cross section RHS 150/100 × 5 mm, made of structural steel S 235 are used.

3.4 Tension-only bracing

In order to ensure the spatial stability of the tower a system of diagonally intersecting ten-sion-only rods is used. The bracing system consisting of 48 members and it is located in the space between the staircase and the perimeter timber columns. All the members are made of solid bars with a diameter of 22 mm or 26 mm from structural steel of grade S 235 (some parts are made of steel of grade 8.8). Each of the members is equipped with rectification and they are connected to the base structure using pin joints.

3.5 Gravity based foundations

The steel-timber load-bearing structure of the tower is anchored into a massive foundation structure made of plain concrete. The foundations consist of two parts—a strip foundation of width 1.6 m and height 1.0 m is made of concrete strength class C 16/20, and a slab foun-dation with thickness of 0.2 m is made of concrete strength class C 30/37. The shape of the foundations copies the ground plan shape of the tower.

3.6 Characteristic details

The connection of the timber columns to the perimeter parts of the steel grids is carried out using a 20 mm thick plate and 12 pieces of M 24 – 8.8 single shear bolts. The steel plates are connected to the steel grids via short circular tubes with cross section CHS 89/10 mm.

The connection of the timber columns to the steel bases, which are embedded into the foundation structure, is solved analogously. But in this case the steel plate is used as the cen-tral member of a double shear connection.

3.7 Other structures

For the flooring of the observation platform at the level of +12.0 m and for the staircase treads spruce ThermoWood boards are used. The thickness of the boards is 21 mm and their strength class C 24. Railings of the stairs and terraces are made from steel square tubes and welded wire mesh panels.

4 3D COMPUTATIONAL MODEL

In order to determine the internal forces and deformation response of the steel-timber load-bearing structure of the observation tower, a 3D beam computational model was created (Figures 1 and 2). The concrete foundation structure was modeled separately (Figure 3).

Basic characteristics of the 3D computational model:

– the model consists of 636 nodes (contacts) and 804 elements (rounded shape of the struc-tural members is replaced by adequate facet curves),
– joints of the steel members (except for the attachment of the tension-only bracing) are modeled as rigid,
– the connections of the timber columns to the to the perimeter parts of the steel grids are modeled using short rods simulating fixed/joint bond,
– the load-bearing structure is supported with the hinged supports,
– fifteen load states were created,
– 196 combinations for the ultimate limit state (ULS) and 98 combinations for the service-ability limit state (SLS) were generated,

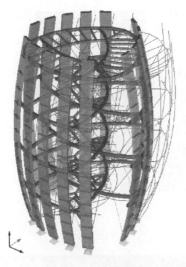

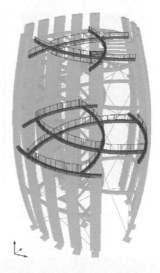

Figure 4. Maximum horizontal displacement in the second limit state (approximately 37 mm).

Figure 5. Normal forces at the horizontal steel grids at the levels of +6.0 m and +12.0.

Figure 6. Views on the entire tower from different sides.

- global imperfections in the form of an initial global inclination of columns were also taken into account,
- geometrically nonlinear and physically linear model was solved using the incremental Newton-Raphson procedure (large deformation analysis was performed).

Some illustrative results of the numerical modeling are presented in Figures 4 and 5.

5 PHOTODOCUMENTATION

The following pictures show various views on the combined steel-timber load-bearing structure of the Tokaj observation tower as whole and some of its construction details.

Figure 7.   Various views on the tower and its parts.

Figure 8.   Column anchoring and the connection of the tension-only member with the rectification.

Figure 9.   Top view on segment of the tension-only bracing system.

6   CONCLUSION

The above listed example demonstrates that the involvement of an architect in the process of designing of observation towers with timber, steel or material combined load-bearing structure is a trend that brings a significant improvement to the resulting work. It gives rise to objects which are not only functional but also visually attractive and aesthetic with the potential to become an interesting landmark and an attraction for visitors of the region.

The fact that the Tokaj observation tower meets at least part of the attributes described above is also confirmed by its place in the competition "Stavba roka 2015" (Construction of the Year 2015), where it received i) the Prize of the Public, ii) the EUROSTAV Publishing Prize, and iii) nomination for the Main Prize (www.stavbaroka.eu). Additionally, it also earned the second place in the Czech and Slovak galvanizing award 2018 competition (www.konstrukce.cz).

REFERENCES

http://www.stavbaroka.eu/portfolio/2015/.
http://www.konstrukce.cz/clanek/komenskeho-most-v-jaromeri-vitez-czech-and-slovak-galvanizing-award-2018/.
Kanócz, J., Mihaľák, M., Platko, P. 2015. "Návrh rozhľadne pri Malej Tŕni". In Dřevostavby 2015. ISBN 978-80-86837-72-7. Volyně: VOŠ a SPŠ, s. 165–168, Czech Republic (in Slovak).
Kanócz, J., Mihaľák, M.- Platko, P. 2011. "Vyhliadková veža Tokaj, Projekt stavby". Kanócz CONSULTING, s.r.o. (in Slovak).

Advances and Trends in Engineering Sciences and Technologies III – Al Ali & Platko (Eds)
© *2019 Taylor & Francis Group, London, ISBN 978-0-367-07509-5*

The benefits of vegetated roofs in reducing the excess heat in three urban areas with different climate conditions

S. Konasova
Faculty of Civil Engineering, CTU in Prague, Czech Republic

ABSTRACT: Green roof infrastructure is gaining popularity all over the world considering its many benefits and advantages. Green roofs can play an important role in cooling buildings and cities. They may assist in the adaptation to climate changes and ameliorate the urban heat island effect. In order to quantify these benefits, it is essential to conduct experimental and modeling studies. The objective of this paper is to evaluate effects of green roofs on reduction of the excess heat in three urban areas with different climate conditions; Prague, Rio de Janeiro, and Sydney, based on the evidence of occurrence of the heat islands. The effect is examined at a precinct scale, 600 m × 600 m. Modeling in ENVI-met testes three scenarios; conventional roof, extensive green roof, and intensive green roof. The results of this research show that the implementation of vegetated roofs is an effective strategy in tackling excess heat by reducing the air temperature by almost one degree Celsius.

1 INTRODUCTION

Growing cities are developing both vertically and horizontally, resulting in releasing more anthropogenic heat and higher absorption of solar radiation due to the implementation of impervious materials, such as concrete and asphalt. High-rise buildings and narrow streets trap warmth and reduce air flow. Therefore, cities are hotter than their surrounding rural areas. This phenomenon is called the "Urban Heat Island" (UHI) effect, "island" of higher temperatures in the area (Li et al., 2014).

One of the opportunities to mitigate the UHI effect comes through the integration of vegetation into the building envelope, for example in form of green roofs. The use of green roofs can provide multiple benefits such as shade, cooling effect through evapotranspiration, and serve as an insulating layer reducing energy consumption and improving thermal comfort. All these factors of a green roof can reduce the incoming solar energy up to 90% (Getter et al, 2006).

2 LITERATURE REVIEW

2.1 *Urban heat island effect*

The urban microclimate is influenced by urban form and their surfaces. Cities are characterized by impervious surfaces with a high concentration of anthropogenic activities leading to significant increases in the air temperatures and the surface temperatures, which are higher than the temperatures of countryside. Such effect is known as the urban heat island (Oke, 1987), which its magnitude depends primarily on the size of the city and local climatic conditions.

According to Rizwan et al. (2008), urban heat islands are generated by factors that can be categorized as controllable and uncontrollable factors. The controllable factors include anthropogenic heat, air pollutants, sky view factor, green areas, and construction materials. Uncontrollable factors include cloud cover, wind speed, seasons, daylight conditions, and other conditions.

A range of factors influence how urban surfaces interact with the atmosphere including: moisture available for evapotranspiration, anthropogenic heating, and albedo (Dousset & Gourmelon 2003). Many studies established the correlation between an increase in green areas and a reduction in local temperature (Takebayashi & Moriyama, 2007), recommending the augmentation of urban green areas as a possible mitigation strategy for the urban heat islands. Due densely urbanized areas, there are few spaces that can be converted into green areas and therefore one of the possibilities is to turn conventional roofs into green ones.

Green roofs have the same energy providers as conventional roofs, but they have additional energy consumers of shading, photosynthesis and evapotranspiration that set it apart as a living system (MoL, 2008). By shading, the plants and growing media of green roof block sunlight from reaching the underlying roof membrane and reduce the surface temperature below the plants. Following this, these cooler surfaces reduce the sensible heat re-emitted into the atmosphere (EPA, 2008).

Gaffin (2005) reported that green roofs could cool as effectively as the brightest white roofs. Albedo of green roof is between 0.25 and 0.30, but taking into consideration evapotranspiration, the value of "equivalent albedo" is generally around 0.70 to 0.85.

The mitigation potential of green roofs in Singapore was evaluated by study of Wong et al. (2003). The study claimed that the cooling effect of the green roofs is restricted by distance from the roof. They found that the reduction of the maximum surface temperature measured under vegetation was about 30°C. The maximum temperature difference of the ambient air was 4.2°C around 18:00.

3 COMPARATIVE ANALYSIS

The potential of green roofs to mitigate the urban heat island effect is determined based upon simulations conducted by ENVI-met simulation software. The investigation of this paper is focused on simulation of the potential of green roofs in specific simplified urban conditions, one square section of Holesovice district in Prague (temperate climate), one square section of Copacabana district in Rio de Janeiro (tropical climate), and one square section of Parramatta district in Sydney (subtropical climate), to measure their effectiveness in real conditions. These urban areas were selected based on the evidence of occurrence of the UHI effect.

3.1 *Computer simulation*

ENVI-met is selected due to the capabilities of this software in modeling greenery and urban fabrics. ENVI-met is a three-dimensional microclimate numerical model that can simulate the surface-plant-air interactions, solar path, buildings and vegetation within urban environment and is based on the fundamental laws of fluid dynamics and thermodynamics. It shows a good resolution that satisfactorily models small-scale interactions between buildings, surfaces, and plants.

The first step in order to use ENVI-met is to create the area input file. This file combines the height and location of buildings, location of plants, distribution of surface materials and soil types, location of gas sources, position of receptors, database links and geographic position of the location on the Earth. The next step is the establishment of the configuration file which defines the settings for the simulation to run (midsummer of northern hemisphere in case of Holesovice and midsummer of southern hemisphere in case of Rio de Janeiro and Parramatta). These settings are the area input file, the name of the output file, the day the simulation runs, the meteorological settings and the plant database. The inserted data is specified in Table 1.

The model requires a user-specified area input file that defines the 3D geometry of the model environment such buildings heights, type of vegetation, soil and surface types (Table 2). The exanimated areas of case studies are 600 m × 600 m based on limitation of free version of ENVI-met.

In the end, the simulation runs and gives temperature results for every period of the time selected by the configuration file. Leonardo is used to visualize and analyze the received

Table 1. Meteorological inputs.

Scenarios	Czech Republic Prague Holešovice	Brazil Rio de Janeiro Copacabana	Australia Sydney Parramatta
Start simulation at day	20.06.2016	21.12.2016	21.12.2016
Start simulation at time	9:00	9:00	9:00
Total simulation time in hours	24	24	24
Save model state each min	60	60	60
Wind speech in 10 m above ground (m/s)	4	6	4
Wind direction (0:N, 90:E, 180:S, 270:W)	292	90	130
Roughness	0.1	0.1	0.1
Initial temperature atmosphere (°C)	16	28	26
Specific humidity in 2500 m (g water/kg air)	9	21	13
Relative humidity in 2 m (%)	79	90	61

Source: Author.

Table 2. Selected input parameters of buildings and vegetation.

Category	Czech Republic Prague Holešovice	Brazil Rio de Janeiro Copacabana	Australia Sydney Parramatta
Building inputs			
Buildings average height	15	36	12 – 26 – 38
Material of conventional roof	Concrete slab	Concrete slab	Concrete slab
Type of vegetation used on GR			
Extensive GR	Grass 50 mm	Grass 50 mm	Grass 50 mm
Intensive GR	Bushes 500 mm	Bushes 500 mm	Bushes 500 mm

Source: Author.

results. This software is the interactive visualization and analysis tool for ENVI-met and BOT world. It can turn simulation results from charts into simple line charts to complex 3D animations. It includes a special interface to ENVI-met data files which allows a simple navigation through the data. After selecting the variables needed for the simulation, the map can be extracted either in 2D cut or 3D cut.

4 RESULTS

After running the simulation for selected case study areas in three scenarios: conventional roof (CRS), extensive green roof (EGRS), and intensive green roof (IGRS), Leonardo tool is used to visualize the results. The simulation was programmed to make measurements every hour but in finale figures, the results are presented at 16h (afternoon) and 24h (night) when the urban heat island effect has a strong impact on quality of sleep and causes the necessity of utilization of air conditioners, thus causes also an increase of consumption of electricity. Moreover, it can be seen how warmth is trapped inside buildings. The measurements of the air temperature were taken in height of 1.5 m above the ground because it is nearly an average human height.

The following Figure 1 shows the comparison among three scenarios in Holesovice district, Prague, Czech Republic in degree Celsius on 20th of June, 2016. Based on the figures, implementing greenery on the existing roofs in this area has a significant effect on reducing the air temperature during the observed period.

Figure 2 demonstrates the comparison among three scenarios by a temperature receptor P at 1.5 m above ground level in Holesovice, Prague, Czech Republic during 20.6.2016.

Figure 3 demonstrates the comparison among three scenarios in Copacabana district, Rio de Janeiro, Brazil, in degree Celsius on 21st of December, 2016.

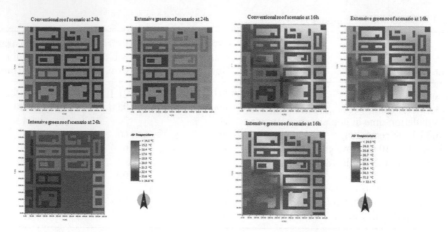

Figure 1.   Air temperatures, 20.6.2016 in Holesovice, Prague, Czech Republic. Source: Author.

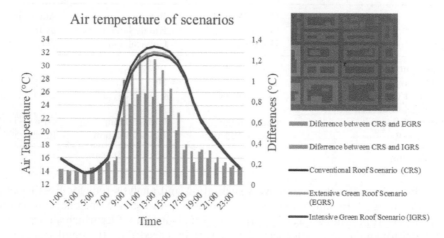

Figure 2.   The receptor (P) results of scenarios, Holesovice, Prague, Czech Republic. Source: Author.

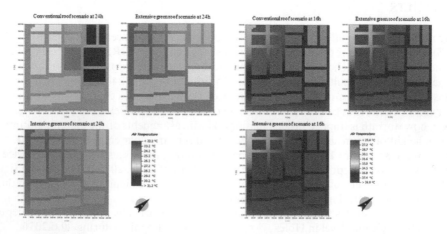

Figure 3.   Air temperatures, 21.12.2016 in Copacabana, Rio de Janeiro, Brazil. Source: Author.

Figure 4 shows the comparison among three scenarios by a temperature receptor P at 1.5 m above ground level in Copacabana, Rio de Janeiro, Brazil.

The next Figure 5 shows the comparison among conventional roof scenario, intensive green roof and extensive green roof scenarios in Parramatta district of 600 m × 600 m, Sydney, Australia in degree Celsius on 21st of December, 2016 to investigate the role of greenery in ameliorating the urban heat islands.

Figure 6 demonstrates the comparison among three scenarios by a temperature receptor P at 1.5 m above ground level in Parramatta, Sydney, Australia.

The air temperatures could seem to be high but it is important take into consideration that ENVI-met calculate the air temperature at 1.5 m above asphalt surface in build-up areas and is directly illuminated by the sun in contrast to case of measurement of the air temperature by weather station when the thermometer must be placed in the shade.

The results of study areas suggest if traditional materials of roofing system are covered or replaced by green roofs, it helps in mitigating the urban heat island effect. Moreover, overheating problems related to high solar absorption surfaces can be further solved if the green roofs are installed instead of conventional ones to increase the sunlight reflectance and humidify the air. Therefore, the installation of vegetation on existing roofs, which does not retain so much heat as in case of conventional roofs, can ameliorate the urban heat islands in all three selected urban areas.

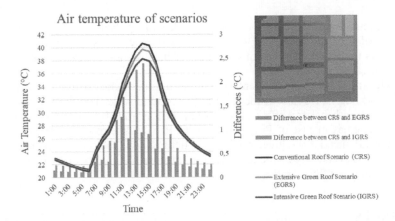

Figure 4.   The receptor (P) results of scenarios, Copacabana, Rio de Janeiro, Brazil. Source: Author.

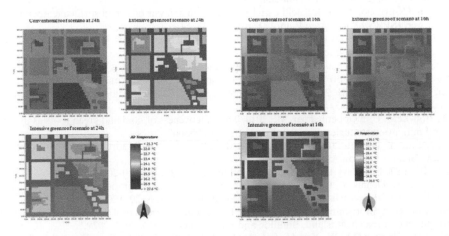

Figure 5.   Air temperatures, 21.12.2016 in Parramatta, Sydney, Australia. Source: Author.

417

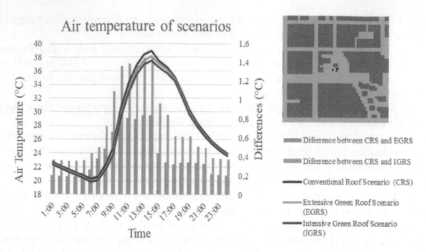

Figure 6. The receptor (P) results of scenarios, Parramatta, Sydney, Australia. Source: Author.

5 CONCLUSION

This paper focused on the analysis of the possibility of application of precinct scale green roofs as one of the most promising measures for reducing excess heat in urban areas. This validation was undertaken by the consideration of heat islands in three specific urban areas with different climate conditions.

The results show that the implementation of vegetation on the existing roofs is an effective strategy in tackling excess heat at the local neighborhood scale. The data demonstrates that green roofs have a beneficial effect on the urban heat islands by lowering the air temperature within the cities, in case of Prague by 0.49°C, Rio de Janeiro by 0.92°C, and Sydney by 0.72°C in the summer. Hence, green roof infrastructure should be considered an effective tool by local authorities and city councils in order to improve the quality of living in urban areas and creating energy efficient cities.

REFERENCES

Bruse, M. & Team. 2010. Envi-met manual.
Dousset B, Gourmelon F. 2003. Satellite multi-sensor data analysis of urban surface temperatures and landcover. ISPRS J Photogram Remote Sensing 58:43–54.
EPA, Environment Protection Agency 2008. Reducing Urban Heat Islands: Compendium of Strategies: Green Roofs, EPA, [Online], Available.
Gaffin, S.R. 2005. Energy balance modelling applied to a comparison of white and green roof cooling efficiency, Greening Rooftops for Sustainable Communities, Washington, DC.
Getter, K.L. & Rowe, D.B. 2006. The role extensive green roofs in sustainable, Development. Hort-Science 41(5), 1276–1285.
Li, D., Bou-Zeid, E. & Oppenheimer, M., 2014. The effectiveness of cool and green roofs as urban heat island mitigation strategies. Environ. Res. Lett. 9 (5), 055002.
MOL, Mayor of London 2008. Living Roofs and Walls. Technical Report: Supporting London Plan Policy, Greater London Authority.
Oke, T., R. 1987. Boundary Layer Climates, London: Routledge, ISBN 0-203-71545-4.
Rizwan, A.; M., Dennis, Y., C., L. & Lio, C. 2008. A review on the generation, determination and mitigation of Urban Heat Island, Journal of Environmental Sciences, 20, 120–128.
Takebayashi, H. & Moriyama, M., 2007. Surface heat budget on green roof and high reflection roof for mitigation of urban heat island, Building and Environment 42 (8), 2971e2979.
Wong, N.H.; Cheong; D.K.W., Yan, H.; Soh, J.; Ong, C.L. & Sia, A. 2003. The effects of rooftop garden on energy consumption of a commercial building in Singapore, Energy and Buildings.

Advances and Trends in Engineering Sciences and Technologies III – Al Ali & Platko (Eds)
© *2019 Taylor & Francis Group, London, ISBN 978-0-367-07509-5*

Designing procedure of LED-halogen hybrid solar simulator for small size solar cell testing

D. Koós, P. Iski, Á. Skribanek & I. Bodnár
University of Miskolc, Miskolc-Egyetemváros, Hungary

ABSTRACT: During the experimental investigation of solar cells, the quality of the illumination is essential. In case of solar simulators, the main goal is the accurate reproduction of the sunlight. The designing procedure and the construction of a halogen-LED hybrid solar simulator is described in this article. One of the main goals are the compliance with the relevant standard (ASTM E972). Small size solar cells can be investigated with the help of the designed solar simulator.

1 INTRODUCTION

The first-generation solar simulators are based on halogen, xenon or simple tungsten light sources and on their combinations, as we can see in the review of Wang et al. (2014). Due to the development of the semiconductor technology, new types of solar simulators can be created, which are based on high power LED units. These devices are energy-efficient and well controllable, the solar spectrum can be approached well by the combination of different color LED units. Kohraku et al. (2006) demonstrate well that one of the main disadvantages of LED solar simulators are the followings: 1,000 W/m^2 light intensity is required for standard measurements and it is difficult to assure that by using LED units only. This problem is eliminated by combining LED and conventional illumination (e.g. halogen). Such constructions are shown by Bliss et al. (2009) in case of AM 1.5 and Kim et al. (2014) in case of AM 0.

During our previous researches, we used halogen solar simulator for solar panel measurements. That device consisted of eight halogen reflectors, 300 W electrical power each and that could provide 1,000 W/m^2 light intensity for a normal sized solar panel. The measured light inhomogeneity of the solar simulator is around 50% and spectral match was not examined. With that illumination, the solar panel measurements produced good agreement with the characteristics from the manufacturer, only the decreased short-circuit current was an exception. Previous solar simulator is mentioned at Koós et al. (2016) and Bodnár (2018).

In case of the current research, our goal was the accurate laboratory measurement of small size solar cells, and for this, a better solar simulator was needed, which can provide 1,000 W/m^2 light intensity with negligible inhomogeneity and a good light spectral match with the sun light. This article describes the designing procedure and construction of a hybrid LED-halogen solar simulator, which is suitable for small size solar cells testing according to the standards.

2 DESIGNING

American Standard for Materials (ASTM) E972 describes the standard specifications of solar simulation for photovoltaic testing, requirements shown by Table 1. Our goal was to design a C-class solar simulator according to the mentioned standard. Adequate illumination of small

Table 1. Solar simulator standard classes and requirements.

Characteristic	Class A (%)	Class B (%)	Class C (%)
Spectral match (Equation 1)	75–125	60–140	40–200
Spatial non-uniformity (Equation 2)	≤2	≤5	≤10
Temporal non-uniformity	≤2	≤5	≤10

size (max. 150 mm × 150 mm) solar cells can be solved by the planed device. The construction is based on the combination of high power LED units with different colors and halogen lamps. During the designing procedure, the optimal supply currents of LED color groups, the optimal position and the quantity of halogen lamps are defined, with the respect of standard spectral distribution and homogeneous light intensity distribution.

2.1 Spectral match

The absolute spectrum distribution of each light sources are determined indirectly. In case of each LED colors, the radiation wavelength range and the wavelength of the radiation peak are known. According to measurements, the average light intensity of LED units is determined in case of various supply currents. With the help of these data, spectral distribution of LEDs is well approached by Gaussian curves. The spectral distribution of halogen lamps are estimated according to the blackbody spectrum in case of 3,200 K color temperature. Superposition of the spectrum curves of each light source gives the spectral distribution of the solar simulator for the given parameters. Spectral match in wavelengths intervals can be calculated by Equation 1.

$$SE(\lambda_a - \lambda_f) = \frac{\int_{\lambda_a}^{\lambda_f} E_{NSz}(\lambda)d\lambda}{\int_{\lambda_a}^{\lambda_f} E_{AM1.5}(\lambda)d\lambda} \tag{1}$$

where $SE(\lambda_a - \lambda_f)$ = spectral match in the actual wavelength range; λ_a = lower limit of wavelength range; λ_f = upper limit of wavelength range; E_{NSz} = intensity of the solar simulator spectrum; $E_{AM1.5}$ = intensity of the AM1.5 spectrum [Kativar et al. (2017)].

During spectral optimization we need to determine the supply currents of the LED color groups (within the allowed limits), which leads to standard C-class spectral match in wavelength intervals. In case of 625 combined current values, the described method is calculated by a MATLAB code and in each case, the spectral match in the wavelength ranges are determined. Figure 1 shows the spectral distribution of sunlight (global AM 1.5) and spectral distribution of each light sources and solar simulator in case of the best spectral match.

2.2 Light intensity distribution

Designing method of the correct light intensity distribution is based on preliminary measurements. In case of LED units, the measurement results of the spectrum analysis are used. Light intensity distribution in case of 625 combined current values are determined by a MATLAB code. For the halogen lamps, the variables are the position and the number of the bulbs. The spots of lamps are fixed (corners and sides of the LED matrix), but the angle with the horizontal, the number of the lamps (4 or 8), the distance from the test plate and from the LED matrix are variables. This time 27 combinations are tested. For each combination, the program calculates the superposition of the light intensity distributions and determines the non-uniformity of the actual light intensity distribution based on Equation 2.

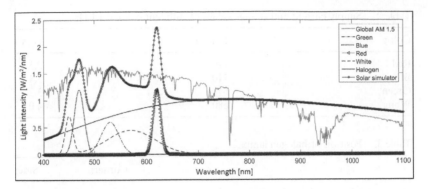

Figure 1. Optimal spectral distribution.

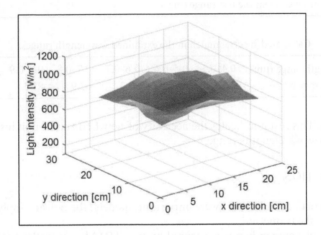

Figure 2. Most homogeneous light intensity distribution in the test plane.

$$T_{EGy} = \frac{E_{max} - E_{min}}{E_{max} + E_{min}} \qquad (2)$$

where T_{Egy} = non-uniformity of light intensity distribution; E_{max} = maximum light intensity E_{min} = minimum light intensity.

The goal is to find the optimal combination, which ensures the smallest light intensity non-uniformity. Most homogeneous light intensity distribution in the test plane is shown in Figure 2.

2.3 Results of designing

The planed solar simulator construction is basing on the combination of 36 high power LEDs (5 W electric power per piece), 8 halogen lamps (50 W electric power per piece) and the subu-nits, which are required for the operation (power supply, control unit, heat sink, fan and frame). Table 2 shows the applicable supply currents in case of each LED color groups. The position of the halogen lamps is described in Table 3. Based on the calculations, the average light intensity value of our solar simulator is 910 W/m² and the spatial non-uniformity is 12%. Table 4 shows the spectral match in case of the investigated wavelength ranges.

Based on the results of designing, we can see that the intensity and homogeneity of illumina-tion is slightly worse than criteria of the standard. Due to the additional optimization during the construction this problem can be solved. For spectral match, the spectral distribution of light sources meets the C-class criteria in each and every wavelength intervals, except for

Table 2. Supply current and voltage values of different colored LEDs.

Color	I (mA)	U (V)
Red	710	2.5
Blue	710	3.4
Green	670	3.2
White	680	4.0

Table 3. Position of halogen lamps.

	Corner:	Side:
Horizontal distance from the LED matrix (mm):	30	34
Angle with the horizontal (°):	30	45
Vertical distance from the test range (mm):	75	60

Table 4. Calculated spectral match in the examined wavelength ranges.

Wavelength range (μm)	0.4–0.5	0.5–0.6	0.6–0.7	0.7–0.8	0.8–0.9	0.9–1.1
Spectral match (%)	12.8	67.0	82.3	67.9	81.5	115.3

400 nm–500 nm. This result indicates the necessity of new LED color unit installation, which is dominant in low wavelengths range.

3 CONSTRUCTION

In case of construction, three tasks were important, namely: the power supply, the cooling of LED matrix and the frame.

Power supply of halogen lamps is ensured by two TRIAK controller type of: Mentavill-Akcent. With the help of this solution, bulbs on the sides and in the corners can be controlled separately. In case of LED power supply the goal is the required light intensity-, ergo the supply current modification. Because of the created LED color strings, each string is supplied and controlled by their own LED driver (type of: Mean Well PCD-60-700B) and TRIAK controller (type of: MentavillAkcent).

Cooling of the high-power LEDs is essential. In our case the chosen heat sink is Stonecold RAD-A6023/190 with 190 mm × 190 mm × 50 mm dimensions, which is of sufficient size to carry the LED matrix. The cooling efficiency of the heat sink is increased by a cooling fan, type of: Artic Cooling F14. With this solution, the LED's temperature remains within the acceptable working temperature range.

In case of the frame the stabile fixing and the good positioning possibility of the bulbs is required. Figure 3 and Figure 4 shows the final construction, we can observe the whole solar simulator and the LED matrix.

4 VALIDATION RESULTS

During the validation, the implemented solar simulator is examined according to the standard for spectral matching and light intensity homogeneity.

For the light intensity homogeneity validation, light intensity measurements are needed at the surface of the test plane, with the help of a solar power meter, type of: PCE-SPM 1. For the measurement, the test plane is separated to 10 mm × 10 mm pixels and the light intensity of each pixel is determined. Figure 5 shows the measured light intensity distribution on the test plane.

Based on the light intensity measurements, we can say the followings: the minimal light intensity value is $I_{min} = 868$ W/m², and the maximal value is $I_{max} = 1,060$ W/m². The average

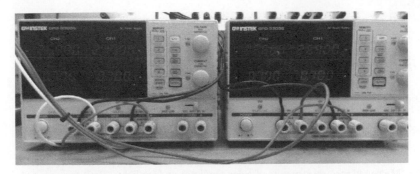

Figure 3.   The power supplies.

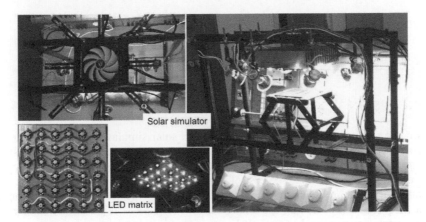

Figure 4.   Implemented solar simulator construction.

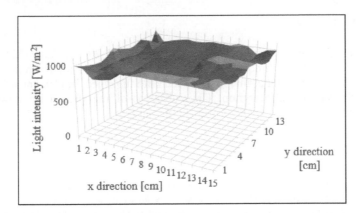

Figure 5.   Measured light intensity distribution on the test plane.

of the measured data is 951 W/m² and the modus is 950 W/m². According to the standard, the calculated light intensity inhomogeneity (Equation 2) is 9.96%, which is below the 10% C-class limit value.

The spectral structure of the illumination is characterized with the help of a spectrometer, type of: Ocean Optics USB 4000. Due to the Figure 6 the solar simulator- and the sun light spectral structure can be compared. Based on the standard (Equation 1) the calculated spectral match in case of the relevant wavelength intervals can be seen in Table 5.

It can be observed that the measured spectral match values are always within the range of 40% to 200%, so the required C-class standard spectral structure can be granted.

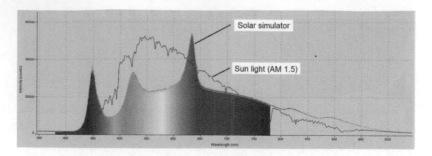

Figure 6.   Measured spectral structure of solar simulator and sunlight.

Table 5.   Real spectral match in the relevant wavelength intervals.

Wavelength range (μm)	0.4–0.5	0.5–0.6	0.6–0.7	0.7–0.8	0.8–0.9	0.9–1.1
Spectral match (%)	120,9	59,0	89,3	116,8	160,6	182,7

5   CONCLUSION

Based on the results we can say that the developed solar simulator meets the requirements of the standard C-class in terms of light intensity homogeneity and light spectrum structure. Good agreement between the validation results and the designing results prove the correctness of the designing method. The cost of the solar simulator is around 350 EUR which is significantly lower than the similar, commercially available devices. As a conclusion we can say that the project is successful, the solar simulator for standard laboratory testing of solar cells is properly designed and constructed.

ACKNOWLEDGMENT

The described article was carried out as part of the EFOP-3.6.1-16-2016-00011 "Younger and Renewing University – Innovative Knowledge City – institutional development of the University of Miskolc aiming at intelligent specialisation project implemented in the framework of the Szechenyi 2020 program. The realization of this project is supported by the European Union, co-financed by the European Social Fund."

REFERENCES

Bliss, M., Betts, R., T., Gottschlag, R.: An LED-based photovoltaic measurement system with variable spectrum and flash speed, *Solar Energy Materials & Solar Cells 93*: pp. 825–830.
Bodnár, I. 2018. *Electric parameters determination of solar panel by numeric simulations and laboratory measurements during temperature transient*. Acta Polytechnica Hungarica. 15(4): pp. 59–82.
Kativar, M., Balkom, M., Rindt, C.C.M., Keizer, C., Zondag, H.A. 2017. Numerical model for the thermal yield estimation of unglazed photovoltaic-thermal collectors using indoor solar simulator testing. *Solar Energy.* 155(10):903–919.
Kim, A, K., Dostart, N., Huynh, J., Krein, T., P. 2014. Low-Cost Solar Simulator Design for Multi-Junction Solar Cells in Space Applications, *Power and Energy Conference at Illinois.* p. 6.
Kohraku, S., Kurakowa, K. 2006. A Fundamental Experiment for Discrete-Wavelenght LED Solar Simulator, *Solar Energy Materials & Solar Cells 90*: pp. 3364–3370.
Koós, D., Szaszák, N., Bodnár, I., Boldizsár, Cs. 2016. Temperature dependence of solar cells' efficiency. *Acta TechnicaCorviniensis—Bulletin of Engineering 9(2):* pp. 107–110.
Wang, W. 2014. Simulate a 'Sun' for Solar Research: A Literature Review of Solar Simulator Technology, *Royal Institute of Technology, Department of Energy Technology*, Swede, Stockholm. p. 7.

Specifications of M&E requirements in the BIM model in the context of use for facility management

M. Kosina & D. Macek
Faculty of Civil Engineering, Czech Technical University in Prague, Prague, Czech Republic

ABSTRACT: The paper addresses the specific requirements of the BIM model that is implemented in the CAFM system, and envisions subsequent implementation for facility management. For facility management, it is important that the BIM model contains comprehensive information and data on both the construction and the technological building systems (M&E). The specifics of the M&E parts in the BIM model to be used in the operational phase of the construction project need to be determined in advance in order to input relevant data into the model during the preparatory and construction phase. This prevents inaccuracies between the project documentation and the actual status or loss of data during the classical management of the construction process and subsequent commissioning. In this paper, the structure of the data and information about the technological systems of the building to be included in the BIM model implemented in facility management are determined.

1 INTRODUCTION

Building information modelling – BIM, is a process where all the data about a building are gathered and managed during its whole life cycle. The result is a BIM model that represents information database including data from the preparation phase, construction phase, operational phase, reconstructions or demolition. It is important for all the participants of a construction project to cooperate on creating and using the BIM model. They can add the data into the BIM model, but the greatest advantage is that the data about every part of the project can be shared between all the participants, which makes it easier to coordinate the whole construction process (Černý et al. 2019).

BIM model represents the building or object including its physical and functional properties. It is possible to track actual position in the schedule of the construction or maintenance and control investment and operating costs. BIM model also allows to make simulations and analysis—operation and optimization of M&E systems and consequently the energy intensity of the building, or dynamic and static behavior of the object (Černý et al. 2013).

For the correct functionality of the BIM model in facility management, it is necessary to specify the data to be included in the model. For this reason, it is advisable for the investor and the facility manager to work closely with the building and technology contractors and to precisely define the level of detail and scope of the data entered into the BIM model (Teicholz 2013, Eastman 2011).

After the decision to implement BIM into facility management, it is necessary to define (Eastman 2011, Reddy 2012):

- Targets that BIM can bring.
  - Coordination with enterprise or organization goals.
- Data scope and use of the BIM model during all phases of the project.
  - List of BIM applications and software used.

- Format and standardize BIM applications and shared data.
- The role of individual project participants and the way data is shared between them.

The benefits of using BIM in facility management throughout the life cycle of a building are as follows (Teicholz 2013):

- Unified Data Base.
- Effective creation and evaluation of analyzes such as energy or cost consumption.
- Inventory of property and equipment, including placement information in an object.
- Accurate surface management.
- User manual of the construction in digital form.
- View real-time data.
- Improving work efficiency due to data centralization.
- Reducing the number of failures by enabling accurate status monitoring and maintenance planning.

2 MATERIALS AND METHODS

One of the problems of using the BIM model in managing a building is that each participant in the construction process uses different software for his own purposes. Each software is working with different data. In order to determine the BIM platform, it is necessary to define how the data will be managed and used in the operational phase. The aim is to create a functioning "ecosystem" that will assist all stakeholders and be usable throughout the life cycle of the building. Any changes made by the individual participants in the model must be shown in the final model to other participants in the construction process. These changes must be made in a coordinated manner. The resulting data from the construction process has to be transferred to the facility management information system. The options for importing data into a CAFM or CMMS system are (Teicholz 2013):

- Manually importing data.
- Using the COBie system.
- Linking the BIM model directly with the CAFM or CMMS system.

The key data for facility management, which the BIM model can contain, is information about areas, equipment and equipment, M&E systems, individual rooms, surfaces, etc. (Teicholz 2013, Kuda & Beránková 2012). These data can be exported from the BIM model as a COBie file and subsequently imported into CAFM or CMMS system.

Exporting data from the database of the CAFM system is also important for facility management. Different level of detail of the data is needed while planning revisions of the system, optimizing energy consumption or tendering a supplier of certain elements of the system under maintenance. The CAFM system must allow the user to generate the outputs

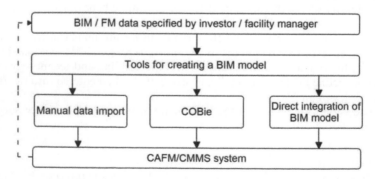

Figure 1.   Data transfer process.

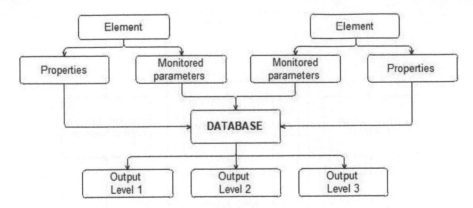

Figure 2. Data output scheme.

from the database in required detail level. Therefore, the structure of the data must be precisely defined as shown on Figure 2.

3 RESULTS AND DISCUSSION

If we consider the data of M&E systems, it is important to realize that not only the information about every separate element is needed. Also their interdependence needs to be mapped at very high detail level. The technological systems are closely linked to each other, and one without the other cannot work at all or very limitedly. To get a general overview of the building systems functions and status, facility managers have to be able to get the information about all the elements linked to the monitored system and their interconnection. The interconnections between M&E systems are shown on Figure 3.

With the knowledge of the information database structure, it is possible to convert available information in advance into the correct format. The output is a structured list of required information on each element. The risk of not complying with this step is both delays in later editing the current data structure, but also the malfunctioning of the system caused by the import of data in the wrong format. The process scheme of implementation of the BIM model into CAFM system is summed up in Figure 4.

After the requirements on the system, data and information structure, specifics of M&E systems and the process of implementation of the BIM model into CAFM system is set, it is advisable to map the risks that can occur during the whole process. For the purpose of this paper, 6 risks were determined and evaluated.

1. Choosing of wrong structure of data and information
If the participants of the project do not participate from the very beginning of the project with the facility manager and the requirements are not set, it is possible that the information and data about elements are input in a wrong format. That can result in the impossibility of implementation of the BIM model from the preparation and construction phase into the operational phase of the project.

2. Loss of data during continuous updates of the BIM model during construction process
While updating the BIM model during the construction process, the loss of data inputted by other participants can occur. Especially while more participants of the project are allowed to input data into the model.

3. Incorrect assignment of interconnections between individual M&E systems
The M&E systems are very complicated and complex. Changes in the systems and its functionality are being made during the construction process. Incorrect or missing

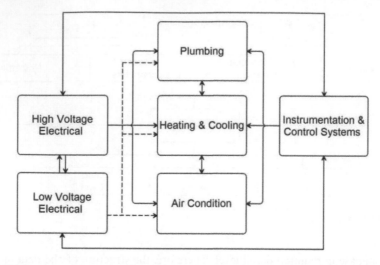

Figure 3.   M&E systems interconnections scheme.

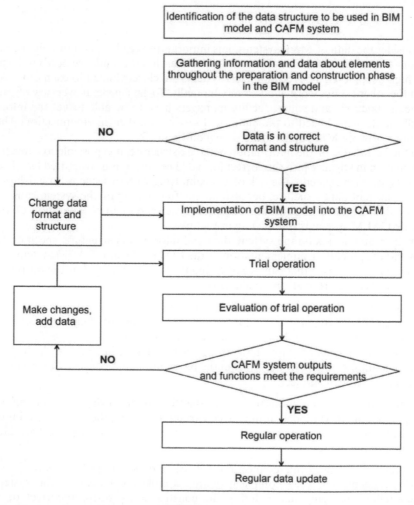

Figure 4.   Process scheme of implementation BIM model into CAFM system.

interconnections between the systems can result in malfunctioning of the BIM model and later CAFM system.

4. Attempt to deliberately disrupt the system
For potential attack on the building, it is easier for the attacker to disrupt the centralized system, especially while all the security and access systems are managed in it.

5. Impossibility of implementing the BIM model into CAFM system
The software for BIM and CAFM are being updated regularly to reflect new technologies and system requirements. Construction process usually takes years from the beginning of the preparation phase until the operational phase. During these years, BIM and CAFM software can be updated in a way that it loses its connectivity with each other.

6. Non-maintenance of actual database during operational phase
For the correct function of the CAFM system throughout the whole operational phase it is necessary to regularly update the data and information after revisions, repairs or replacement of the elements.

After determining the risks, the risks were evaluated by eight facility mangers from the practice. These facility managers work at administrative buildings and manufacturing plants in the Czech Republic. The method of probability and impact was used, where risks were evaluated by Formula 1.

$$Risk = Impact \times Probability \qquad (1)$$

Each risk is assigned with its probability of occurrence (value $1-5$, 1 – highly unlikely to occur; 5 – highly likely to occur) and its impact (value $1-5$; 1 – very low impact; 5 – very high impact) and then evaluated by the Impact vs Probability Matrix in Table 1. From the values specified by facility managers, the mode was selected. The risks are evaluated in Table 2.

Table 1. Impact vs probability matrix.

Probability					
5	5	10	15	20	25
4	4	8	12	16	20
3	3	6	9	12	15
2	2	4	6	8	10
1	1	2	3	4	5
Impact	1	2	3	4	5
1–7	Low risk				
8–14	Medium risk				
14–25	High risk				

Table 2. Risks evaluation.

Number	Risk description	Probability	Impact	Evaluation	Risk
1.	Choosing of wrong structure of data and information	2	5	10	Medium risk
2.	Loss of data during continuous updates of the BIM model during construction process	3	4	12	Medium risk
3.	Incorrect assignment of interconnections between individual M&E systems	3	5	15	High risk
4.	Attempt to deliberately disrupt the system	1	3	3	Low risk
5.	Impossibility of implementing the BIM model into CAFM system	1	5	5	Low risk
6.	Non-maintenance of actual database during operational phase	2	4	8	Medium risk

Identification of input data and in particular their collection is necessary to achieve the required functionality of the system. The output of this step is a detailed list of all input data and the information where these specific data are available. The risk of failure to observe this step is the malfunction of the system due to lack of imported data and delays by later replenishment of the data. Identification of the information database structure is needed for facilitating the import of data into the CAFM system.

While updating the BIM model during preparation and construction phase of the project, the loss of previously updated data can occur. To avoid this, access into the BIM model and the allowance to make changes must be defined for each participant. All the versions of the BIM model should be kept for the case of loss of data during the process.

As shown by the risk evaluation, the highest risk is incorrect assignment of interconnections between individual M&E systems. Therefore the interconnections should be discussed throughout the whole construction process between all the suppliers of individual systems, contractors and facility managers. The information and data must be updated after every change whilst keeping the previous versions for possible reviews.

4 CONCLUSIONS

Investor and the facility manager need to work closely with the building and technology contractor. They need to define the level of detail and scope of the data entered into the BIM model and later into CAFM system. Structure of the data must be also precisely defined, so it is possible to obtain not only the information about the selected element, but also its interconnections with other elements from different systems. In order to get a fully functional system for facility management using BIM, the following steps need to be followed:

- Information about each element of the system must be gathered throughout the whole construction process.
- Level of detail of the outputs from the CAFM system must be variable.
- Interconnections between individual M&E systems must be assigned precisely for correct function of BIM model and CAFM system.
- Updating the data throughout the life cycle of the building and maintaining the current database is essential to obtain correct outputs.

ACKNOWLEDGEMENTS

This work was supported by the Grant Agency of the Czech Technical University in Prague, grant No. SGS18/018/OHK1/1T/11.

REFERENCES

Černý, M. et al. 2013. BIM Příručka. Praha: Odborná rada pro BIM o.s. ISBN 978-80-260-5296-8.
Eastman, C. et al. 2011. BIM Handbook A Guide to Building Information Modeling for Owners, Managers, Designers, Engineers, and Contractors. 2. Hoboken, New Jersey: John Wiley & Sons, Inc. ISBN 978-0-470-54137-1.
Kuda, F. & Beránková, E. 2012. Facility management v technické správě a údržbě budov. Příbram: Professional Publishing. ISBN 978-80-7431-114-7.
Reddy, K.P. 2012. BIM for Building Owners and Developers: Making a Business Case for Using BIM on Projects. Hoboken, New Jersey: John Wiley & Sons, Inc. ISBN 978-0-470-90598-2.
Teicholz, P. ed. 2013. BIM for facility managers. Hoboken, New Jersey: John Wiley & Sons, Inc. ISBN 978-1-118-38281-1.

Advances and Trends in Engineering Sciences and Technologies III – Al Ali & Platko (Eds)
© 2019 Taylor & Francis Group, London, ISBN 978-0-367-07509-5

Safe operation of small municipal water supply systems

S. Krocova
Faculty of Safety Engineering, VSB-TU Ostrava, Ostrava, Czech Republic

ABSTRACT: In the Czech Republic a significant proportion of population lives in small municipalities. The technical equipment of these small municipalities does not differ fundamentally from the average amenities of the cities and large housing stock. One of the basic conditions is enough drinking water, including sanitary households needs. The supply of drinking water in small municipalities is provided by local waterworks and individual wells. Because of aquatic ecosystems' vulnerability, water systems, natural influences and extraordinary anthropogenic events, the article describes these types of threats. It points out what risks can arise and how they can be eliminated at an acceptable and economically manageable level.

1 INTRODUCTION

There are 6245 municipalities registered in the Czech Republic in total. From this number, 4867 of them has less than 1000 inhabitants. This statement shows that a significant part of the population lives in small municipalities (Vaclavik et al., 2013). However the households, technical and sanitary facilities and the relevant small community infrastructure need an equal reliable drinking water source like any other major city or industrial region. The big part of small municipality is connected to a water supply system of group or regional water supply. A large number of small municipalities have not been able to connect to the water supply systems for various technical and operational reasons yet (Hluštík et al., 2016). Subsequently their local water supply systems depend on their own, usually underground drinking water sources. Under the new climate conditions the local underground drinking water sources can be seriously threatened by two effects: hydraulic capacity of the groundwater source and raw material quality.

Both of these factors may cumulate under certain conditions and increase the risk of decommissioning the drinking water source and local water supply. A similar problem can be encountered in private wells in the small municipalities, especially if they draw drinking water from shallow soil waterbeds in the river basin. These groundwater layers are extremely prone to water deficiency during various weather conditions changing and also rainfall changes in every season It is also worthwhile to mention that even small communities have to deal with wastewater treatment. Similar problems arise as in municipalities with more inhabitants and larger sewer networks (Mazák et al., 2017). These problems are dealt with in the following paper.

2 THE LOCAL WATER SUPPLY IMPORTANCE TO THE SMALL VILLAGE

Local water supply is the irreplaceable technical infrastructure of the most municipalities. It is a prerequisite for obtaining water from the natural environment by means of water-technical constructions and the conversion of raw water into drinking water (Dvorský et al., 2016). Its basic structure is illustrated in Figure 1.

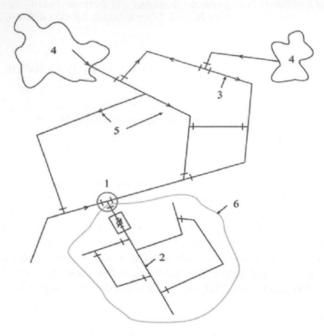

Figure 1. The local water basic scheme with the main internal water connection of a significant drinking water customer. Agenda: 1 – internal water mains safety connection, 2 – internal water supply, 3 – local waterworks, 4 – water source (main/backup), 5 – water flow direction, 6 – operating area.

As shown in the diagram the local municipality waterworks are characterized in particular by following features:

– main or backup water source (wells, galleries),
– water network (circuit, branch or combined),
– control valves (spools, shut-off valves, conical closures),
– one or a set of internal water mains (municipal technical infrastructure),
– fire water supply points (hydrants, spout racks).

The main task of the local water supply system for each municipality is the reliable supply of fresh and health-conscious drinking water to all customers. In many cases especially in the natural source absence of surface water in the village, local waterworks also serve as a significant multipurpose fire water source. Whether the local water supply system is also supposed to fulfill the multipurpose source function of fire water for built-up area always depends on the hydraulic parameters of given water supply networks as defined by the technical standard (ČSN).

3 OPERATIONAL AND SAFETY RISKS OF A SMALL LOCAL WATER SUPPLY SYSTEM

The vast majority of small local water supply systems depend on the local underground water source. Whether the drinking water source is operationally safe in various alternative situations depends primarily on the conditions listed below:

– water catchment (shallow, deep, artesian),
– collection type (technical equipment of raw water supply points),
– raw water quality (physical and chemical properties).

Some underground water to be used for waterworks is not necessary to be technologically adapted into drinking water. Water quality often meets the drinking water parameters set by legislation (Vyhláška 252/2004). Where the physical and chemical properties of raw water do not meet the criteria for drinking water, water must be treated. In both cases a number of local drinking water sources for small municipalities will be at risk in new climates. Threat and security risk will result for at least the following reasons:

Raw water source with secondary treatment for drinking water

– there will be an overall change in the raw water quality above the potential of existing water treatment plant,
– in the season course the water resource yield will be reduced in the short term and at the same time will increase the raw water load,
– the rapid change threat in the raw quality in the source during so called flash rainfall or floods events will increase.

Drinking water source for direct consumption, local waterworks, private well

– volumetric reduction flow of the well to the total drinking water lost,
– drinking water quality change of the drinking water,
– threats to the consumer's health and lives of this water type.

Raw water contamination in a source for natural or anthropogenic reasons

– total physical and chemical composition change of raw water for treatment or drinking water for direct consumption from the soil environment of water source,
– water contamination by anthropogenic origin from agricultural land from old ecological burdens.

Basic threats to the local water supply systems

– low to extremely low hydraulic efficiency of the municipal water supply network or scattered buildings,
– inappropriately dimensioned line of water supply networks due to water flow rates in series, reduced water freshness and health content,
– reducing the hydraulic fire-water sampling points parameters of given multipurpose source or their complete decommissioning.

The above-mentioned cases of threats to small drinking water resources in small local municipal water supply systems and individual water abstraction by natural persons will certainly accumulate in new climate conditions. Given that in practice there is no realistic threat to a number of these threats, the state administration and small municipal communities have to eliminate at least some hazards (Adamec at al., 2016). At the same time it is necessary to define in details the ways and means, public administration prevention and infrastructure managers will be coordinated (Pokorný et al.,). In addition to technical procedures, threats and risk management are also accompanied by ethical issues (Kavan, 2015). The basic options for eliminating these threats are outlined in the following chapter.

4   AND ELIMINATION MEANS OF NATURAL AND ANTHROPOGENIC HAZARDS IN THE WATER SYSTEMS

The primary prerequisite for each local water system in a small village is to keep drinking water sources in operation. These are in particular groundwater resources located in the built-up area in the municipality or in intensively farmed agricultural land in the immediate vicinity of the municipality. In these cases the risk of extraordinary occurrence events will be almost certainly increased throughout the 21st century by natural influences and by anthropogenic events (Bernatik et al., 2013). Natural impacts like hydrological drought or floods will undermine the emergence of extraordinary anthropogenic events, see Figure 2, the main

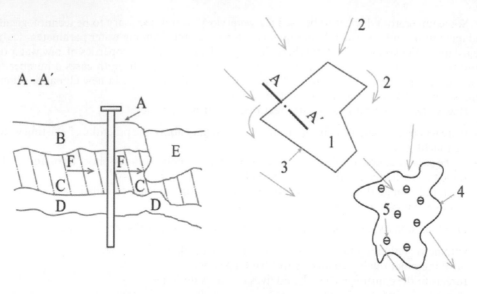

Figure 2. Waterproof water protection against contamination by hazardous substances. Aagenda: A – Milan wall, B – permeable subsoil, C – flooded soil layer, D – impervious soil layer, E – harmful substances landfill, F – water flow direction, 1 – contaminated area, 2 – groundwater flow direction, 3 – Milan wall, 4 – the springfield circuit, 5 – wells.

risk of ensuring availability of the municipal housing stock and their both public and private infrastructure.

Assuming that the region's public administration can reduce or solve the problem, see Figure 2, to eliminate the risk of drinking water source contamination it is assumed that the following formulas will be minimally effective:

– provision of emergency water supplies throughout the small municipality network,
– emergency water supply securing on the hydraulically efficient water supply network lines,
– maintaining the most important buildings in the community dependent entirely on the direct supply of drinking water from the water supply network,
– ensuring the functions of all part of the water supply network fire-fighting points.

However the overall concept of ensuring small municipal water supply safe operation should be based on a comprehensive mind map. The idea of a mind map is that there is no given format. Its purpose and goal is to give the investigator an initial answer to the following questions before starting his own work on solving the risk question on the water system, its part or operating set:

– what problems have been encountered in water facilities in standard and emergency situations,
– what is important for water system operator and what needs to be answered and solver should know how to solve the task,
– what are hypotheses for each problem or query,
– what tools are needed to answer individual questions and their outline,
– which input data are needed to solve the problem,
– what expertise the investigator or team of investigators must have,
– clarification of the decision-making issue and its succession that will need to be done in the process of solving the task,
– perform and input analysis of alternative barriers that have appeared or may appear and complicate the solution (Kročová, 2014).

One of the basic benefits of this type of map is the fact that it provides an effective basis for communicating with people involved in the project, helping to coordinate the assumptions that have been set. As small municipalities have a very different character within the

Czech Republic it is advisable to conduct an intensive national discussion before starting the overall conceptual solution of the small municipalities problems in the water management sphere. The subject could be focused for example on these mentioned topics.

Long term sustainability and reliability of the waterworks of small municipal water systems is also an interesting topic to talk about. The next step which requires an extraordinary attention is small municipal waterworks that is the right choice of used pipe material. At this step it is always necessary to consider at least the following circumstance for example the geological composition of the soil environment, sloping of the area and its susceptibility to soil instability, hydrodynamic pressure ratios in the pipe system, the hydraulic efficiency of the water supply network in the long run. These mentioned factors will consequently affect not only the reliability of drinking water supplies to the customers but also the economic costs associated with the operation of a small local water supply for the entire lifetime of the waterworks what is about ninety till one hundred and ten years.

The long term sustainability and reliability of drinking water supply in each drinking water distribution network is strongly dependent on the correct choice of the pipe material in a relation to the geological soil environment composition. This issue is often underestimated by designers and their investors and it is not taken into account with sufficient respect. The environmental threat underestimation to the durability of the waterworks and its reliability subsequently causes at least the some negative phenomena like a significant increase in the number of accidents in watercourses, unreliability of drinking water supply to consumers, unreliability of fire water supply from a given multipurpose source of fire water, operation costs increasing of a small local water supply system, a significant reduction in the overall life of waterworks by up to fifty percent. For these and other reasons it is necessary to know not only the general requirements for the water work network construction but also the geological composition of the soil environment and its individual parts. This information can be obtained from relevant state institutions or geological surveys in cases where this information is not available in the state information register of concerned area.

When considering the construction concept of a small waterworks, the territory slopes or its parts must not be omitted. In addition to soil aggression which has a potential to reduce the water lines reliability to transfer drinking water from its source to customer, the second factor are the slopes are themselves. At different slopes it is necessary to respect the principle designing different pipe line material and technical measures to reduce the risk. One of the irreplaceable measures should be in the threatened sectors of pipe series, the installation of horizontal pressure compensators, increasing number of the water supply network control valves, monitoring implementation of on-line and off-line elements for monitoring the hydraulic subject efficiency. The exemplary basic measurements not only increase the reliability of drinking water supplies to different types of consumers but also have an economic effect in running a small municipal waterworks in crisis situation caused. For example the reduction of drinking water volumes to be realized from a local or central drinking water source. Drinking water supply reliability does not affect only natural conditions of the small community but also the anthropogenic factors of the water supply network itself. One of the main factors is the hydrodynamic pressures on the water supply network. No pressure levels can be set on the water supply network but only a pressure level which complies with the legislation (Kavan, 2015). The observance of set parameters protects not only the own water supply network from overloading but also individual consumers and their sanitary facilities in the housing stock or the public and private infrastructure technology of the built-up area of the municipality.

The hydraulic water network efficiency was often underestimated in the second half of the 20th century. There was no exception that in some water mains water losses in the system up to forty-nine percent. At the present the situation in hydraulic efficiency vast majority segment of water mains for public use has considerably improved. Nevertheless there is an example from 2016 where water losses represented ninety millions cubic meters, which is fifteen percent of the water to be used. In the water systems most of these water losses represent hidden water leaks. Climate change will already significantly affect the reliability of drinking water supply from its surface or underground sources in the coming years. In order to increase the supply drinking water reliability the following technical and operational measures will be necessary in the shortest possible time. By on-line or off-line monitoring you have an overview of each water supply network and its actual hydraulic efficiency, their parts, pressure zones or separate sectors. Under unsatisfactory conditions there are some following measures implemented. To have a correlator available for detecting hidden failures, to carry out interceptions of network fittings, to monitor and evaluate night flows on the water supply network and to create balance bands. Have an own or an external group of specialist to address the improving hydraulic performance issue of the water network including the ability to progressively plan resources for further long term economic water system sustainability.

The all above mentioned measures are almost always to be realized of course also maintain reliability of drinking water supply to the citizens, the infrastructure of built-up areas and at the same time leading to building fire protection of this type buildings, this type of multipurpose fire water sources (Václavík, 2013). Implementation significantly reduces the failure risk to manage alternative crisis situations and creates a space for further water management development.

5 CONCLUSION

This article dealing with the safe operation of small municipal water supply systems. It is necessary to highlight the small self-government of municipalities and the state of regions to emerging threats. These are not just theoretical threats that may or may not arise. United Nation's forecast and climatologist conclusions point out that climate change will certainly happen. So far there are only doubts about its intensity and its negative impact on ecosystems and individual world continents' water systems. It follows from the above that there is not much time to hesitate and it is necessary to start preparation for the potential consequences of the threat based on outlined risks.

ACKNOWLEDGMENTS

This work was supported by the research project VI20152019049 "RESILIENCE 2015: Dynamic Resilience Evaluation of Interrelated Critical Infrastructure Subsystems", supported by the Ministry of the Interior of the Czech Republic in the years 2015–2019 and project "Development project 2017", supported by the Ministry of the Interior of the Czech Republic in the years 2016–2018.

REFERENCES

Adamec V. et al. 2016. How to assess territory vulnerability. In The Science for popular Protection. p. 35–40.
Bernatik A. et al. 2013. Territorial risk analysis and mapping. In Chemical Engineering Transactions. p. 79–84.
ČSN 730873 Požární bezpečnost staveb- Zásobování požární vodou.
Dvorský T. et al. 2016. Waste water treatment in North Moravia and Silesia from the past to the present.
Hluštík P. et al. 2016. Use of the program SWMM to simulate rainfall runoff from urbanized areas.
Kavan S. 2015. Ethical Aspects of the Work of Rescuers During Extraordinary Events. In the Social Sciences. p. 684–690.
Kročová Š. 2014. The water protection trends in the industrial landscape. In Inžyniera Mineralna. p. 171–174.
Mazák J. et al. 2017. The proposal of recommendations for the operation of vacuum sewerage. In Earth and Environmental Science.
Pokorný J. et al. 2016. Comparison of theoretical method of the glass flow in corridors with experimental measurement in real scale. In Acta Montanistica Slovaca. p. 146–153.
Vaclavik V. 2013. The methods and equipment currently used for sewerage system cleaning.
Vyhláška 252/2004 Sb. Vyhláška, kterou se stanovují hygienické požadavky na pitnou a teplou vodu, a četnost kontroly pitné vody.

Advances and Trends in Engineering Sciences and Technologies III – Al Ali & Platko (Eds)
© *2019 Taylor & Francis Group, London, ISBN 978-0-367-07509-5*

The specifics and criteria of investment appraisal for the investment in human capital within the field of construction technology and management

A. Kucharčíková, M. Mičiak & M. Ďurišová
Faculty of Management Science and Informatics, University of Zilina, Zilina, Slovakia

Z. Chodasová
Institute of Management of the Slovak University of Technology, Bratislava, Slovakia

ABSTRACT: The business environment is nowadays characterized by the continuously rising amount of knowledge, technology advances and skills needed for the application of new technologies. This affects many industries, including the field of construction technology and management. The sum of knowledge, skills and abilities that enterprises have at their disposal, embodied in their employees, represent an asset noted as human capital. This asset has its specifics in comparison with other kinds of assets, such as the tangible fixed assets. However, the investment activity is equally necessary for this intangible asset to be maintained and developed. The aim of this article is to analyse the criteria used for investment appraisal within the investment in other kinds of assets, and to assess their applicability for the investment in human capital. The article presents the possible modifications of investment appraisal criteria applicable for investment in human capital within the field of construction technology.

1 INTRODUCTION

The construction industry is at present a developing sector of the Slovak economy. It is confirmed by the findings in the quarter analysis of this industry based on 103 interviews with the key representatives of the selected construction businesses. The analysis states that construction production in Slovakia rose in the first quarter of 2018 by 7.7 when compared with the previous year. The number of employees in construction businesses rose as well. According to predictions, the construction industry will rise in 2018 by 6. The revenues of businesses here will rise by 5.7 (CEEC Research, 2018). Such period of growth is suitable for using the funds on investment that will bring benefits in the future, helping to overcome weaker periods. Here belongs also the investment in human capital

The methodology applied in the research presented in this article is based on its main aim and the methods needed for the accomplishment of this aim. The aim of this article is to critically examine, assess and analyse the criteria used for the investment appraisal within the investment in other types of assets, and to consider and evaluate their applicability for the investment activity oriented on the human capital (HC). The fulfilment of this aim requires a thorough examination, comparison and analysis of the approaches of researchers, academics and practitioners that can be found in domestic and foreign sources, and the summarization of the substantial features of the appraisal process connected to the selection and assessment of possible investment projects in general. Subsequently, it is necessary to apply the method of deduction to derive the criteria applicable in the investment appraisal within the specific conditions of investment in HC. Using the inventiveness and abstraction, the substantial elements of the appraisal process will be adjusted so they reflect the nature and essence of the HC. All of this will be done while respecting the environment of the construction technology businesses.

2 INVESTMENT IN HUMAN CAPITAL

The investment in human capital of the business enables it to broaden the HC available, increase its value and quality. Without it, it is not possible to achieve high efficiency in the transformation process. The need to invest in HC is created by the technology advances as well as the market.

The core of increasing the value of HC is to spend financial and other resources at present with the aim to achieve financial and other benefits in the future. All costs related to the extending of the scope, increase of efficiency and prolongation of operating of this capital are investments in it. The investment in HC can be done in various forms. The business can focus on:

- Improving employees' health—organizing the reconditioning stays, medical check-ups, visits of the swimming pool and so on.
- Improving the working conditions and ergonomics—providing new, more effective protective equipment to prevent work injuries and to prevent damaging the health of employees. This can be influenced by anthropometric dimensions of people (Hitka et al., 2017).
- Increasing the working abilities, skills, knowledge, changing the attitudes—done via the system of corporate education and development.

In relation to the analysis of HC investment, the economics theory the most often deals with the assessment of investment in education, therefore this is being dealt with in this paper, too. As it was described by Russ (2015), in spite that there is no consensus within the definition of HC or the methods of its valuation, various authors agree that HC is characterized by gathering of investments in education, health and on-the-job training that increase the productivity of an individual in the business and in the labour market. The opinion is in line with Becker's (2009) who wrote that the expenditures on training, education, healthcare and so on are the investments in assets. However, they produces human, not financial or physical capital, because it is not possible to separate the person from the knowledge, skills, health and values the way it is possible to move financial or physical assets.

Krausert (2018) extends the topic by defining the strategic human capital (SHC) and focusing on businesses under the influence of investors on the capital market. He argues that the revenues in the short term create the pressure from the capital market which can impact the decisions to invest in SHC. The investors set quarter objectives of revenues as a stimulus for managers to maximize the revenues in the nearest period. Krausert distinguishes three categories of investment in SHC. The first is the change of structures, processes and routines. The second category is represented by the recruiting and development of qualified, able and skilled workers. The third category are the human resources management practices related to the retaining of employees. It was revealed that the effects of SHC on the performance were dampened by the turnover of employees.

Several research works were revealing the connections between the investment in HC and business results. The research done by García-Zambrano et al. (2018) worked with the sample of 28 Spanish businesses included in the IBEX-35. A positive relationship was revealed between the investment in education and the ratio of the market and the book value of the business. This ratio indicates the importance of intangible assets. The importance of education for success and enterprise growth is also stressed by Kmecová (2017).

Ciriaci (2017) was examining the impact of expenditures used for the on-the-job training oriented on development and deployment of innovations and the HC on the revenues from innovative products (new products in the market). It was tested whether the revenues from this investment differ among the small, medium and large businesses. The data used were from the Community Innovation Survey from 23 European countries. It was confirmed that the investment in training and in the research and development workers had a positive impact on the business's innovativeness. However, the revenues were statistically significantly higher in large businesses. The study confirms the correlation between deployment of new products, expenditures on the training and the HC available.

Onkelinx et al. (2016) focused on small and medium businesses in Belgium exporting to foreign markets (sample of 1,922 businesses). Both variables representing the HC (remuneration and costs on the training) were significantly positively connected with productivity (value added per employee). The businesses that invest in their employees via remuneration and training have a higher probability of achieving higher levels of labour productivity.

3 THE FACTORS INFLUENCING THE GENERATION OF APPRAISAL CRITERIA WITHIN THE FIELD OF HC

The issues of HC are critical in the knowledge economy. They are broad by their very definition. The elements of HC are present in several aspects of the businesses' operation. This complexity influences also the generation of appraisal criteria of investment in HC. It is necessary to consider:

- the structure of HC elements significant for the specific business or industry,
- the influence of the HC elements on the business performance—assessment via key performance indicators (KPIs) (Tokarčíková et al., 2014),
- the influence of HC on the competitive position of the business in the market (Virlanuta et al., 2012),
- the connection of the investment with the HR strategy and with strategies of other functional parts of the business, which are projected in CSR (Tokarčíková et al., 2016, Potkany et al., 2018),
- external trends related to the HC components (knowledge and skills needed for the application of new technologies) and to new forms of their development (new methods of education) (Kozubíková, 2015, 2016),
- the compliance of investment in HC with the expectations of the employees themselves pertaining to their professional development—in the context of career and succession planning (Hitka et al., 2015b, Bartáková et al., 2017).

4 THE APPRAISAL CRITERIA FOR THE INVESTMENT IN FIXED ASSETS OF BUSINESSES

For the appraisal of the investment in fixed assets, economic and qualitative criteria are being used. The economic criteria are the result of a certain method, giving recommendations for the investment's realisation. The procedure of appraisal is based on the determination of investment's costs, costs of financing, prediction of future cash flows, and on the calculation of efficiency. Here are listed the results of researches dealing with the investment appraisal.

Scholleova et al. (2010) used the data from 252 businesses from the Czech Republic. The investment projects are characterized via: cash flows, actual lifespan, and the risk influencing the investment's realisation. Authors divide the appraisal criteria into static and dynamic ones. The static criteria consider cash flows. They are represented by the total revenue from the investment, average annual return, average payback period. The dynamic criteria consider all fundamental factors. They include the Net Present Value (NPV), Internal Rate of Return (IRR), profitability index, benefits-costs ratio, discounted payback period. The selection of appraisal criteria reflects also the preference of the decision-maker, the difficulty of application, and the seriousness of the decision. More than 75 of the respondents used static criteria.

The group of economic investment appraisal criteria considering the time value of money is connected to the problematic determination of the discounting rate. Businesses use the weighted average costs of capital (WACC) connected to the capital asset pricing model. The concept is used to take into account the expectations regarding the revenues perceived by the owners and creditors. To consider the situation on the share market and the situation in the industry, it is necessary to collect and update a great volume of information or to use paid databases. A survey regarding the use of WACC was performed by the Association of financial professionals (2017) with the sample of 606 financial professionals. It was revealed that 30 of businesses do not use this indicator. Publicly traded businesses use this method more often, because it is easier to calculate the components needed. Samset & Christensen (2017) emphasize the distinguishing between ex ante and ex post investment appraisal. According to them, the initial investment appraisal should use the same set of criteria that will be used ex post. Ex ante appraisal provides the information on the possibilities in the phase when the chance to affect the course is the highest. Ex post appraisal provides a learning step for the improvement of decisions in the future. The benefit of ex ante appraisal is the identification of the best solution and avoiding of those inefficient ones. Frank et al. (2013) offer a framework for decision-making on investment when selecting production equipment, technology or product portfolio. The framework considers strategy, quality, and economic aspects. These criteria are assessed using SWOT, QFD (quality function deployment), NPV and payback. Even though the economic criterion is being preferred, it is needed to consider other criteria as well. In the adjusted SWOT matrix, the aim is to identify which weaknesses, opportunities and threats are not solved by the internal strengths. The best investment alternative is the one seizing the opportunities the most and contributing to the reduction of weaknesses. This investment has attributes complicating the assessment of costs and benefits. Investment alternatives were compared using analytical hierarchic process (AHP) and sensitivity analyses. The inputs were represented by the judgements of decision makers on alternatives, appraisal criteria, relationships between them, and relationships between the alternatives. The sensitivity analysis measures the impact on the project's results when the input values change. Benefits of investment in IT are mostly intangible and they manifest themselves during a long time (Buhociu et al., 2009). The listed specifics apply also for the field of investment in HC. Traditional criteria assume that the investment, cash flows, costs of capital and the time horizon are known, and that the effects can be transformed into money units. Since the decision-making on strategic investment in IT, for example distributed information systems (Hrkút et al., 2017), requires the understanding of the business processes, various groups of the affected stakeholders should be involved in the process.

Even though the human capital is denoted as an intangible asset in the literature, due to its specifics, it does not belong to fixed intangible assets from the accounting perspective.

5 CONCLUSIONS—THE APPRAISAL OF INVESTMENT IN HC

When doing the appraisal of investment in HC, it is needed to consider the specifics of its forms and of the HC itself. One such specific is the fact that HC stays in the possession of the employees, and they use it when performing job tasks. The business only hires HC and provides the remuneration in return. The intangible essence influences the determination of all categories of the appraisal. The fundamental categories are the expected costs of investment in HC, expected financial and other benefits from the realisation, and the risk affecting the fulfilment of expectations (Figure 1). This part draws the conclusions for the appraisal of investment in HC with the focus on businesses operating in the construction industry.

Within the category of costs of the investment in HC, the basic part is represented by the financial costs related to the participation in training activities. Other costs are connected to the fact that when engaged in training, the employees do not provide their usual performance. The participation in training courses is just one of the possible forms of HC investment. Another form is the training directly at the place of work, e. g., in the form of mentoring. In this case, the costs also include the time of the person providing this form of HC development of other employees. Then, in the case of self-education, the costs can be viewed in the form of equipment (literature, subscription to online education portals, …), but again also in the form of time that needs to be earmarked for the employee. Another group of the related costs is connecting the involvement in such education with a proper remuneration.

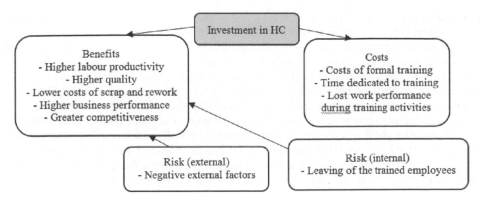

Figure 1.　Fundamental components of the appraisal of investment in human capital.
Source: Own design.

When focusing on the benefits from the realisation of investment in HC, it is needed to look for the connection to the KPIs that are used by the business in the assessment of its operating. It is also needed to study the impacts on strategic goals. Revealing the impacts can be done via benchmarking if the business has several organizational units and the given investment is at first done in one of them. The benefits from the investment in HC, that can be quantified, can reflect in the increase of labour productivity (the business can produce more during the same time) or in the increase of production's quality (which lowers the costs of scrap and rework, or products of a higher quality can be offered for a higher price, alternatively the higher quality can increase customers' satisfaction).

The relevance of the benefits identified in relation to HC investment for the businesses in the construction industry is supported by the findings from the quarter analysis (CEEC Research, 2018). The contracts of construction businesses are in 68 of cases from private investors. Private customers pay heed to a high quality of the results. This forms the prerequisite for choosing the same business in the future and for giving positive recommendations. High quality of works can be performed only by qualified workers, trained for the work with modern technologies and materials. The investment in HC can help businesses to strengthen their market position. It will provide them with stable or growing revenues. The quality is also connected to the selection of a particular construction business by customers. The key criterion is still the price offered (8.8 points out of 10). However, the decision strongly reflects also the own experience with the business (7.8 points) and the previous works done for other customers (7.6 points). Moreover, lower price can also be an effect of HC investment. If the trained workers do the procedures right the first time, there is no waste of material. Finally, there is also the connection to the BIM (Building Information Modelling) processes that are currently being used only by 13 of Slovak construction businesses. Within the context of HC investment, there is a need to acquire new skills for working with this new IT. This will then bring other benefits in the form of simplifying the budget processing and speeding up the changes in the project documentation.

The risk affecting the success of HC investment is specific too. This risk is represented by the leaving of an employee who was trained in the business. Therefore, it is needed to examine the motives making employees stay working for the given business, making them satisfied with their jobs. These motives are connected to the aspirations, work tasks, working team, atmosphere and relationships at work, fringe benefits and so on. The basis for the appraisal of investment in HC is to perceive qualitative aspects of employees in the business as a valuable asset. Only if the managers are able to see people and their skills, knowledge and desirable forms of behaviour as assets, they can pay enough attention to their development and they take it as seriously as the investment in other forms of assets, assessing the efficiency of using the limited funds on these activities. The topic was also studied by Damodaran (2009) according to whom the accounting principles include the rule to separate the capital costs from the operating ones. The accountants follow this rule in case of manufacturing businesses, putting the investment in factories and equipment among capital costs and work and material among operating costs. However, they ignore the principle when it comes to businesses with intangible assets. The result is that businesses with intangible assets report low capital expenditures in relation to their size and growing potential. Finding out how these enterprises manage the retention of their best workers is the key component of their accurate valuation. The opinion is supported by Krausert (2018) stating that the orientation on short-term results refers to investment decisions when managers overvalue the immediate results in comparison with the delayed ones. Expected financial benefits are being discounted beyond the rate that would reflect the opportunity costs and the risks. As a result, the resources are being disproportionally allocated into investments with sooner effects. Therefore, it is being proposed in the literature to capitalize certain expenditures related to HR activities.

ACKNOWLEDGMENTS

This article was created as part of application of project VEGA No 1/0652/16 Impact of spatial location and sectorial focus on the performance of businesses and their competitiveness in the global market and project APVV-16-0297 Updating of anthropometric database of Slovak population.

REFERENCES

Bartáková, G.P., Gubiniová, K., Brtková, J., Hitka, M. Actual trends in the recruitment process at small and medium-sized enterprises with the use of social networking. *Economic Annals*-XXI, Vol. 164, Issue 3–4, pp. 80–84, DOI: 10.21003/ea.V164-18.

Becker, G.S. 2009. *Human Capital: A Theoretical and Empirical Analysis, with Special Reference to Education.* Chicago: University of Chicago Press.

Buhociu, F.M., Moga, L.M., Ionita, I., Virlanuta, F.O., Zugravu, G.A. 2009. Qualitative and quantitative analysis for the evaluation of the informatics systems projected by value-based concepts. *In: IBIMA 2009*, Vol. 1–3, 2009, Cairo, Egypt, 4.1.-6.1.2009, Code 10441, 653–657.

CEEC Research. 2018. Kvartálnaanalýzaslovenskéhostavebníctva Q2/2018. (online). (Pub. 2018-05-29). (Cit. 2018-06-10). Available at: http://www.ceec.eu/research/.

Ciriaci, D. 2017. Intangible resources: the relevance of training for European firms' innovative performance. *Econ. Polit.* 34:31–54.

Damodaran, A. 2009. Valuing Companies with intangible assets. New York: Stern School of Business. Available at: http://people.stern.nyu.edu/adamodar/pdfiles/papers/intangibles.pdf.Cit 2018-06-10.

Frank, A.G., Souza de Souza, D.V., Ribeiro, J.L.D., Echeveste, M.E. 2013. A framework for decision-making in investment alternatives selection. *Int. J. of Production Research* 51(19): 5866–5883.

García-Zambrano, L., Rodríguez-Castellanos, A., García-Merino, J.D. 2018. Impact of investments in training and advertising on the market value relevance of a company's intangibles: The effect of the economic crisis in Spain. *European Research on Management and Business Economics* 24: 27–32.

Hitka, M., Blašková, S., Sedmák, R., Lorincová, S. 2017. Changes of Selected Anthropometric Dimensions of the Adult Slovak Population in the Context of Production Management. In: *Global Scientific Conference on Management and Economics in Manufacturing*, Zvolen, Slovakia, 5–6 Oct. 2017, pp. 174–181, ISBN:978-80-228-2993-9.

Hitka, M., Závadská, Z., Jelačić, D., Balážová, Z. 2015b.Qualitative indicators of company employee satisfaction and their development in a particular period of time. *DrvnaIndustrija*, Vol. 66, Issue 3, 2015, pp. 235–239, DOI: 10.5552/drind.2015.1420.

Hrkút, P., Janech, J., Kršák, E., Meško, M. 2017. A new architectural design pattern of distributed information systems with asynchronous data actualization. In: *AISC* 2017, Vol. 511, 80–90, ISSN: 2194-5357.

Kmecová, I. 2017. Business as orientation value for developing the spirit of enterprise. In: *IBIMA 2017*, 3–4 May, Vienna, Code 129797.

Kozubíková, Z. 2016. Financial literacy in selected groups of the university students. In: *International Scientific Conference on Knowledge for Market Use*, Olomouc, Czech Rep., 222–230, ISBN: 978-80-87533-14-7.

Krausert A. 2018. The HRM-capital market link: Effects of securities analysts on strategic human capital. *Human Resource Management* 57: 97–110.

Lapidus, B. 2017. FP&A Survey: How Relevant is Your Cost of Capital? *Association for financial professionals*.(online).(Pub.2017-08-12)(Cit. 2018-06-10). Available at: https://www.afponline.org/trends-topics/topics/articles/Details/fp-a-survey-how-relevant-is-your-cost-of-capital.

Onkelinx, J., Manolova, T.S., Edelman, L.F. 2016. The human factor: Investments in employee human capital, productivity, and SME internationalization. *J. of International Management* 22: 351–364.

Potkany, M., Gejdoš, M., Debnár, M. 2018. Sustainable innovation approach for wood quality evaluation in green business. In: *Sustainalibility*, 10: 2984.

Russ, M. 2015. *Quantitat. Multidisc. Approaches in HC and Asset Management*. Hershey: IGI Global.

Samset, K. & Christensen, T. 2017. Ex Ante Project Evaluation and the Complexity of Early Decision-Making. *Public Organiz. Rev.* 17:1–17.

Scholleova, H., Svecova, L., Fotr, J. 2010. Criteria for the evaluation and selection of capital projects. *Intellectual Economics* 1(7), 48–54.

Tokarčíková, E., Falát, L., Malichová, E. 2016. Exploitation of corporate social responsibility reports in manager's decision making in automotive company. *20th International Scientific on Conference Transport Means 2016*; Juodkrante; Lithuania; 5–7 Oct. 2016, pp. 259–263 Code 135091.

Tokarčíková, E., Poniščiaková, O., Litvaj, I. 2014. Key performance indicators and their exploitation in decision-making process. In: *8th International Conference on Transport Means, Transport Means2014*; Kaunas; Lithuania; 23–24 Oct. 2014; pp. 372–375, Code 117725.

Virlanuta, O.F., Muntean, M.C., Nistor, C. 2012. Strategic alternatives for increasing the competitiveness of Romanian tourism. In: *18th IBIMA: From Regional Development to World Economies*; Istanbul; Turkey; 9–10 May 2012, 1303–1307, Code 103297.

Advances and Trends in Engineering Sciences and Technologies III – Al Ali & Platko (Eds)
© 2019 Taylor & Francis Group, London, ISBN 978-0-367-07509-5

Analysis of the cost structure of aluminum and glass facades

A. Leśniak & M. Górka
Faculty of Civil Engineering, Cracow University of Technology, Kraków, Poland

ABSTRACT: The architecture of the 21st century requires the designer to create a modern, light and aesthetic shape. Such effect can be obtained by creating an external facade in the form of a light curtain wall as an aluminum and glass facade. Aluminum and glass facades have to meet not only technical requirements such as safety, thermal insulation, sound insulation, but also should be economically viable for the potential recipient. The paper presents three variants of aluminum and glass facades: column—transom, structural and semi—structural. For the application of each, advantages and disadvantages are presented. Additionally, the article presents calculation and cost structure of the aluminum and glass facade. Every option has been assessed for economic viability and the possibility of using it in modern constructions.

1 INTRODUCTION

The newly designed and built constructions have to be attractive with their form and architecture. The modern method used in construction has a great potential to improve the efficiency of production, quality, customer satisfaction, environmental impact (Kozlovska & Mackova & Spisakova. 2016). The body of a contemporary building should be light and modern, as well adapted to the surrounding area. It is achieved by a light curtain wall in the form of an aluminum and glass facade. It meets the requirements of basic technical parameters such as construction safety, thermal insulation, acoustic insulation, etc. In addition, large panes of glass, filling the space between the columns of the curtain wall structure, bring high transparency and reflexivity. Elevations in the form of aluminum and glass facade give the building dynamism and originality, hence are often chosen by investors as a form of external wall cladding (Leśniak & Górka. 2018). High technical parameters and functionality of aluminum and glass facades allow creating curvilinear and sloped shapes. Every aberration from the standard and repeatability, makes the aluminum and glass facade object unique not only for the investor, but also for the user. The curtain wall may take the form of baffles or facade claddings. The investor has to decide which of the systems available on the market will be appropriate. One of the most common criteria include time (Głuszak & Leśniak. 2015), (Plebankiewicz & Juszczyk & Malara. 2015), (Gajzler & Zima. 2017), (Anysz & Zawistowski. 2018) and cost (Juszczyk. 2016), (Juszczyk & Leśniak & Zima. 2018) of the construction. Currently, the impact of technology on environment is also a widely considered criterion (Spisakova & Hyben & Heredos. 2012), (Leśniak & Zima. 2018). The right selection of the system allows to reduce the risk of future investment and benefit from its exploitation (Plebankiewicz & Zima & Wieczorek. 2016), (Skiba & Mrówczyńska & Bazan-Krzywoszańska. 2017), (Spisakova & Kozlovska & Mesaros. 2014).

In their work, the authors present aluminum and glass facade as systemic column—transom, structural and semistructural solutions with consideration of the appropriate glass dedicated for this type of external wall cladding. They show the advantages and disadvantages of each system. Moreover, the authors carried out a cost analysis of the profitability of individual solutions.

2 LIGHT CURTAIN WALL

2.1 *Definition and classification*

The definition of a light curtain wall was first used by W. Dudley Hunt Jr. at The Contemporary Curtain Wall, Its Design, Fabrication and Erection (Hun., 1958). The authors states that light curtain walls are the structures composed of construction and insulation materials selected in a rational and appropriately connected manner; their mass is generally in the range of 10 kg/m^2; they do not carry loads operating in their plane (except for their own weight). PN-EN 13830:2005 standard defines the wall as: The curtain wall is the outer casing of the building with a frame structure usually made of metal, wood or PVC-U, consisting usually of vertical and horizontal structural elements, joined together and attached to the supporting structure of the building.

The curtain wall may take the form of baffles or facade claddings, depending on the material used for the external wall installation (Figure 1).

2.2 *Aluminum and glass facades*

The article selected and analyzed curtain walls made of aluminum profiles and filled with glass. Depending on the method of fixing and connecting the curtain wall with the supporting structure of the building, we can classify them as: suspended type, otherwise called curtain or filler type. The curtain walls are attached to the face of the ceiling, on the outside of the building whereas the filling curtain walls are claddings that fill the space between the ceilings.

Depending on the type of the system, aluminum and glass facades, can be divided into column – transom, structural and semistructural (Figure 2). The column – transom facade can be completed as a curtain wall of the suspended and filling type. The skeletal frame of the structure consists of pillars and bolts, whereas the inner part is filled with glass. The outer part consists of pressure and masking strips. The aesthetics is disturbed by the dividing lines

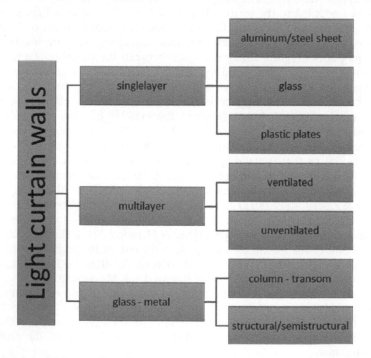

Figure 1.   Division of light curtain walls.
Source: (Urbańska – Galewska & Kowalski. 2016).

  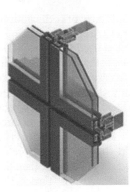

Figure. 2   Aluminum glass facades a) column – transom, b) semistructural c) structural.
Source: Aluprof system catalog.

visible from the outside. High technical parameters, thermal insulation, high impact strength and, above all, functionality are the big advantage of such system. This type of facade allows you shape simple surfaces as well as polyline or inclined surfaces.

The semistructural facade is based on the solution of a column—transom system. It uses the same vertical and horizontal profiles as in column—transom facades. By means of a special, mechanical fastening of the glass to the aluminum structure, a smooth form of the facade can be obtained on the outside. The dividing lines are filled with a suitable silicone with increased insulation parameters. The wide range of colors available for filling silicones gives a new dimension to the visual and aesthetic effects of such a solution.

The structural facade is characterized by the possibility of obtaining an external glass facade without any visible division into horizontal and vertical lines. This effect can be achieved by special connection of the glass pane to the aluminum structure. The assembly takes place by gluing the pane to the aluminum frame using structural silicone. In favour of this form of facade argues the high aesthetics and transparency of the facade. An additional advantage of the structural facade is fast assembly at the construction site. The whole structural bonding process is performed in production plants and ready-made elements are delivered to the construction site. Another aspect that speaks for a facade structural system is the guarantee of the quality of the combination of glass and aluminum. The structural bonding stage is held under strict supervision and quality control, which guarantee trouble free operation of the curtain wall. Unfortunately, not every type and thickness of the glass packet is dedicated to the structural facade, therefore it largely limits the functionality of using this type of solution.

2.3   Glass unit

The aluminum and glass facades and its parameters are influenced by the glass used. The glass pane determines the degree of transparency as well as the visual and aesthetic effect of the facade. In addition, a high degree of glass utilization in external walls allows to keep the user's contact with the external environment and allows daylight to be illuminated, which results in energy saving for the facility. Glass that is used in construction should meet the requirements of thermal protection, safety of use, protection against burglary, protection against penetration of the projectile, protection against noise of fire protection.

In external facades—aluminum and glass panels, usually glass unit is used. It is the arrangement of individual glass panes that are separated from each other by a spacer frame. The space between the glass panes is filled with air or other noble gas. The whole is hermetically closed and sealed with a mass called butyl. We distinguish several basic insulating glass units: heat-insulating, sound-absorbing, solar, fireproof (Pollak. 2011).

3 THE COST ANALYSIS OF AUMINUM—GLASS FACADES

In their work, the authors present the costs of the material of the aluminum and glass facade construction depending on the system (column—transom, semistructural, structural) and the glass used. The valuation was made on the basis of price lists obtained from producers. The first stage showed the cost of materials for individual systems. Costs of aluminum profiles, accessories used as appropriate connectors, gaskets, assembly and masking strips, additionally the cost of insulating materials such as external foil and appropriate glue, individual flashings and structural facades were valued for the cost of structural silicones. The costs of materials are summarized in Table 1.

The average prices per 1 m^2 of the external curtain wall were accepted for the valuation. At the moment of prefabrication and assembly of construction elements, a natural material waste is created, therefore the average material wear factor should be taken into account in the valuation. The coefficients for individual materials have been estimated based on our own research and observations. As a result of optimizing the cutting of aluminum profiles, its consumption increases by approx. 10%, for additional insulation, the additional consumption is about 15%, and for flashings about 10%.

Analysis of the component materials of the skeleton structure of the curtain wall showed that the simplest column – transom system is economically viable.

The second part of the analysis is the calculation of the glass unit, which is the filling of the facade construction. Depending on the parameters we want to achieve for the building,

Table 1. Material cost statement.

Material	Unit price/1 m^2 [EUR]	Wear factor	Total cost [EUR]
Column – transom system			
Aluminum profiles	67.35	1.10	74.09
System accessories	11.09	1.00	11.09
Insulation (foil + glue)	5.88	1.15	6.76
Flashing	14.12	1.10	15.53
Total			107.47
Semistructural system			
Aluminum profiles	67.35	1.10	74.09
System accessories	10.59	1.00	10.59
Insulation (foil + glue)	5.88	1.15	6.76
Flashing	14.11	1,10	15.53
Insulating silicone	3.53	1.15	4.06
Total			111.03
Structural system			
Aluminum profiles	69.25	1.10	76.17
System accessories	9.41	1.00	9.41
Insulation (foil + glue)	5.88	1.15	6.76
Flashing	14.12	1.10	15.53
Insulating silicone	28.24	1.15	32.47
Total			140.34

Table 2. Summary of the glass unit price.

Variant	Unit price/1 m^2 [EUR]
Variant I	80.71
Variant II	114.35

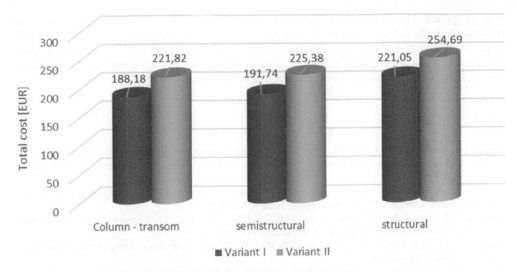

Figure 3. The total cost of aluminum and glass facade materials.

Table 3. The total cost of aluminum and glass facade materials.

Variant/System	Column – transom	Semistructural	Structural
Variant I	188.18	191.74	221.05
Variant II	221.82	225.38	254.69

we need to choose the right glass with the appropriate parameters. Table 2 compares two types of double glazing and price formation for such a package. The two-chamber system comprised of tempered glass with float glass and safety glass was compared. Safety glass consisting of two sheets of glass and double foil. 16 mm spacer frame filled with argon. The individual systems differ in the thickness of the entire glass package. For the first variant, the package consists of 6 mm toughened glass, 6 mm float glass and 4 mm safety glass. The second variant, however, consists of 8 mm tempered glass with 6 mm float glass and 5 mm thick safety glass.

The obtained results are presented graphically in Figure 3. In Table 3 the authors compiled the total cost of making the aluminum and glass facade. It shows that the price of a light outer wall casing, primarily depends on the package of the selected glass. When the investor requires the facade to have sharpened technical parameters, the glass filling affects the cost of making the entire glass facade.

The Figure 3 shows that the column – transom system, in economic terms, is the best solution. The most expensive is structural system. Whereas the cost of the entire facade is determined by the cost of glass filling and its parameters.

4 SUMMARY

Light curtain walls are the trend of the 21st century. The concrete construction seems to be slowly moved away from. The main material used to create exterior walls is glass. It gives the buildings a lightness of form, transparency and unique visual effect. In combination with aluminum as the supporting frame of the curtain wall, the facade becomes the showcase of every object, not only in terms of aesthetics, but also while maintaining all technical and usage parameters. Furthermore, the appropriate combination of the structure with the curtain wall guarantees the complete tightness.

In the work, the authors carried out a cost analysis of the materials used to create aluminum and glass facades. Three systems: column – transom, semistructural, and structural have been proposed. The results show that the simplest solution, the column-transom system, is the best from the economic point of view. What is more, it was pointed out that in the valuation of aluminum and glass facades, the cost of the whole set of materials usually generates the price of glass. However, the price of glass depends mainly on the number of combinations, as well as technical and visual parameters of the glass package.

REFERENCES

Anysz, H., & Zawistowski, J. 2018. *Cost minimization of locating construction machinery park with the use of simulation and optimization algorithms*. In MATEC Web of Conferences, EDP Sciences, Vol. 196, 04088.

Gajzler M. & Zima K. 2017. *Evaluation of Planned Construction Projects Using Fuzzy Logic*. International Journal of Civil Engineering. Issue 4: 641–652.

Głuszak M & Leśniak A. 2015. *Construction Delays in Clients Opinion – Multivariate Statistical Analysis*, Creative Construction Conference. Procedia Engineering, 123: 182–189.

Hunt W.D. Jr. 1958. *The Contemporary Curtain Wall, Its Design, Fabrication, and Erection.* New York. F.W. Dodge Corporation.

Juszczyk M. & Leśniak A. & Zima K. 2018. ANN Based Approach for Estimation of Construction Costs of Sports Fields. Complexity, 7952434.

Juszczyk M. 2016, *Application of PCA-based data compression in the ANN-supported conceptual cost estimation of residential buildings.* AIP Conference Proceedings Vol. 1738, No. 1, p. 200007.

Kozlovska M. & Mackova D & Spisakova M. 2016. *Knowledge database of modern methods of construction.* Advances and Trends in Engineering Sciences and Technologies II – Proceedings of the 2nd International Conference on Engineering Science and Technologies, CRC Press, ESaT 489–494.

Leśniak A. & Zima K. 2018. *Cost Calculation of Construction Projects Including Sustainability Factors Using the Case Based Reasoning (CBR) Method.* Sustainability, 10(5), 1608E.

Leśniak, A., & Górka, M. 2018. *Evaluation of selected lightweight curtain wall solutions using multi criteria analysis.* AIP Conference Proceedings, AIP Publishing.1978, No. 1, p. 240003.

Plebankiewicz E. & Juszczyk M. & Malara J. 2015. *Estimation of task completion times with the use of the PERT method on the example of a real construction project.* Archives of Civil Engineering. Issue 3: 51–62.

Plebankiewicz E. & Zima K. & Wieczorek D. 2016. *Life Cycle Cost Modelling of Buildings with Consideration of the Risk.* Archives of Civil Engineering, Issue 2: 149–166.

Pollak Z. 2011. *Glass in building constructions – types, requirements, research.* Świat szkła 4/2011: 6–8.

Skiba, M. & Mrówczyńska, M. & Bazan-Krzywoszańska, A. 2017. *Modeling the economic dependence between town development policy and increasing energy effectiveness with neural networks.* Case study: The town of Zielona Góra. Applied Energy, 188: 356–366.

Spisakova M. & Hyben I. & Heredos P. 2012. *Assessment of environmental risks during the construction process.* 12th International Multidisciplinary Scientific Geoconference. SGEM2012 Conference Proceedings: 75–82.

Spisakova M. & Kozlovska M. & Mesaros P. 2015. *Environmental risks in construction—case study* Environment, Energy and Applied Technology – Proceedings of the 2014 3rd International Conference on Frontier of Energy and Environment Engineering: 675–679.

Urbańska-Galewska E. & Kowalski D. 2016. *Lekka obudowa – Układy konstrukcyjne.* Builder, December 2016: 106–110. www.catalog.aluprof.eu/pl (access 01.07.2018 r.).

Advances and Trends in Engineering Sciences and Technologies III – Al Ali & Platko (Eds)
© 2019 Taylor & Francis Group, London, ISBN 978-0-367-07509-5

Usage of fly ash from biomass incineration in preparing growing media

B. Lyčková, J. Mudruňka, R. Kučerová, D. Takač, K. Ossová & I. Sobková
VŠB-TU Ostrava, Ostrava, Czech Republic

ABSTRACT: The presented paper verifies the possibility of using fly ash from biomass incineration in preparing growing media. The research aim was to determine the optimal amount of fly ash which can be added into growing media. Mixtures of materials with various fly ash contents were tested by a growing test and other analyses. Moreover, an impact of the effect of Vermesfluid solution emerging as a by-product of vermicomposting was analyzed. The experiment results prove that growing media with 5% of fly ash from biomass incineration are, due to their qualities, comparable to horticultural substrate which is normally sold on the market. In addition to this, a positive impact of Vermesfluid solution on the growth and vitality of plants has been proved.

1 INTRODUCTION

Results from various researches of using fly ash from biomass incineration were highly positive, because experiments proved a positive impact of its application in growing procedures (it promotes growth of plants because the presence of a wide scale of nutrients improves elementary balance in the ecosystem; its alkaline character provides neutralizing effect similar to that of liming; it decreases mobility and bioavailability of risk elements; it reduces toxicity of aluminium, manganese and iron for plants by decreasing exchangeable contents of their ions in acid soils; it provides biological activity and conditions for certain microorganisms; it improves texture, aeration and water capacity of soil). Therefore it might be reasonably assumed that similar usage of fly ash formed during thermal processing of biomass as a suitable supplement into various types of growing media can prove to be an appropriate way of using this—in principle otherwise non-utilizable—waste material (Arshad et al., 2012; Gómez-Rey et al., 2012; Moilanen et al., 2012; Vassilev et al., 2013). Also the legislative basis in the legal code of the Czech Republic promotes this possibility because in accordance with the Decree 131/2014 amending the Ministry of Agriculture Decree 474/2000 Coll., laying down requirements for fertilizers, as amended, and Decree 377/2013 Coll., laying down requirements for storage and using of fertilizers, as amended and in the version in force, it is possible to use ash from separate biomass incineration in agricultural areas unless limit values for risk elements and substances are exceeded. Therefore, according to the valid legislation in the Czech Republic, it is possible to apply maximum two tons of ash from the separated biomass incineration under the provision that during the same year it is not possible to use stabilized and treated sludge from wastewater treatment plant together with ash (Hanzlíček, Perná, 2011; Tlusťoš et al., 2014). The legislator took inspiration from the example of legislation in several European countries, such as Sweden, Finland, Austria or Denmark where the legislation was established specifically for determining conditions for applying ash as fertilizer, and where they declare not only the maximum acceptable concentrations of toxic or potentially hazardous elements, but also the maximum application doses of ash for arable land, as well as requirements for the minimum content of nutrients and consequently also the necessity to implement laboratory analyses of polyaromatic hydrocarbons in case the limit of 5% loss on ignition is exceeded (Ochecová, 2015).

Therefore, in this respect it is necessary to mention the possible emergence of environmental risks during applying ash from biomass incineration to arable land, as a wide range of impurities and toxic substances may occur in ash together with necessary nutrients, which mainly concerns ash from wood burning (Park et al., 2004). However, in comparison with treated sludge from a wastewater treatment plant, the content of hazardous elements in ash is relatively low and it always depends on the type and quality of incinerated biomass. Technical analyses generally emphasize that chemically treated plants or wood should not be used as input material (Obernberger, Supancic, 2009). It was proved that if high temperatures are achieved during the incineration of such materials, some metals begin to volatilize in a chamber of thermal installation and then they either may condense or they may attach to surface of very small dust particles in products of combustion or ash (Liao et al., 2007).

The aim of the research was to verify usability of the fly ash as a component of substrates for growing plants and determine its optimum volume amount, which could be processed into such products. Regarding the undesirable, yet quite possible presence of toxic heavy metals, it was also necessary to implement required tests on phytotoxicity, as the ash used originated from exhaust gas cleaning filters behind a common rotary kiln where chipped wood was incinerated. At the same time, we also examined impacts and effects of application of Vermesfluid solution at different concentrations.

2 EXPERIMENT

Within the experiment, we prepared substrates with three different concentrations of fly ash which contained vermicompost, peat and garden substrate. In case of vermicompost and garden substrate, we opted for the identical quantity of 15 and 12 volume% content of fly ash, in case of peat it was 5, 10 and 15 volume%. Thus, several variations of 10 liters (see Table 1) were formed which were seeded with perennial ryegrass (*Lolium perenne*) from commercially available grass mixtures four pitches. The seeds were sowed into the formed substrate mixtures and each mixture filled four pots at a time.

Two pots were watered with standard drinking water from public water supply and others with 3% and 10% solution of Vermesfluid (VF) which is a leachate created during vermicomposting once in 10 days. Both control samples were watered with water only. They were watered once a day in the beginning, at later stages it was once in two or three days. Half the samples were also watered with VF solution once in ten days. In case of samples marked as S5 JHA, S10 JHA and S15 JHA (see Key for Figure 1 and Figure 2), a solution of concentration only 3% was used, whereas in case of S5 JHB, S10 JHB and S15 JHB it was a solution of concentration 10%.

During the vegetation experiment, we observed and compared growth of plants in different substrates with different watering. Apart from analyzing biomass, we implemented phytotoxicity test of all the substrates.

Table 1. Material composition of substrates in volume percentages (%).

Substrates	Fly ash	Vermicompost	Peat	Garden substrate
S5	5	15	68	12
S10	10	15	63	12
S15	15	15	58	12
SK	–	–	–	100

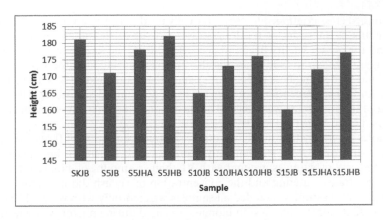

Figure 1.   Comparing heights of individual samples of perennial ryegrass (*Lolium perenne*).

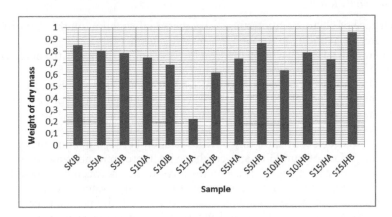

Figure 2.   The amount of harvested dry mass of perennial ryegrass (*Lolium perenne*).

3   RESULTS AND EVALUATION

3.1   *Vegetation test*

After the seeds germinated, all samples were regularly monitored. Every day we recorded vitality and height of majority number of plants in growth. Perennial ryegrass plants (*Lolium perenne*) prospered best in the variation with 10% VF solution watering in S15% substrate. Its height was over 11 cm. Also substrates S5% and control samples demonstrated high growth. S10% variation did not reach the height of other plants, but it was very similar in the speed of growths. The differences between plants without any watering VF were minimal in case of both test pots for the given mixture. In case of 3% VF watering and 10% VF watering, subjects with more concentrated solution reached greater heights, however in case of S10% substrate, the differences were minimal (Figure 1 and Figure 2).

Key for Figure 1 and 2: SKJB—garden substrate watered with water, sample B; S5 JA – 5% fly ash, watered with water, sample A; S5 JB – 5% fly ash, watered with water, sample B; S5 JHA – 5% fly ash, watered with VF 3% solution; S5 JHB – 5% fly ash, watered with VF 10% solution; S10 JA – 10% fly ash, watered with water, sample A; S10 JB – 10% fly ash, watered with water, sample B; S10 JHA – 10 fly ash, watered with VF 3% solution; S10 JHB – 5% fly ash, watered with VF 10% solution; S15 JA – 15% fly ash, watered with water, sample A; S15 JB – 15% fly ash, watered with water, sample B; S15 JHA – 15% fly ash, watered with VF 3% solution; S15 JHB – 15% fly ash, watered with 10% solution.

The examined grass continued to grow gradually in all samples, yet in many cases the grass began to wither as the result of too long straws. Colour of the plants gained darker tones of green and the overall vitality began to decrease. During the sixth week of the experiment, the growing grass began to fade. This was also apparent during its gathering because due to brittleness and high stringiness of its root system, the plants had to be cut off close above the surface and the average length of the roots was determined separately. The growth was measured and the grass mass was weighed in the obtained biomass, namely immediately after harvesting, and also after two-day drying.

The samples watered with water had, with the same concentration of fly ash, similar weight of harvested dry mass which subsequently decreases with the increasing concentration of fly ash. In the case of samples watered with VF, higher amount of biomass was harvested in more concentrated watering and the concentration of fly ash did not have any provable impact. The control samples exceeded all samples watered only with water. The established per cent representation of dry mass in biomass was used during phytotoxicity testing.

3.2 *Phytotoxicity testing*

For the purpose of phytotoxicity testing, we gathered information on the proportion of dry mass compared to humidity in the analyzed growing media (Table 2).

Next step was to take the value of pH, conductivity and turbidity of the analyzed leaches. The values of examined samples are shown in Table 3.

The values of pH were between 5.65 and 6.82. It means that the samples were slightly acidic up to neutral. The acidity was caused by relatively high addition of peat which, on the other hand, was neutralised by alcalic fly ash. From the gathered data, we were able to see that the values of pH slightly increased in substrates with higher addition of fly ash, but the results did not unequivocally show any obvious relation between pH and the type of watering. In case of values of turbidity and conductivity, there were not any signs of provable relation neither to the contents of fly ash in the mixture, nor between the types of watering.

Finally, a test of germination was carried out and the found values are shown in Table 4.

The phytotoxicity test implemented on substrates planted with perennial ryegrass (*Lolium perenne*) showed that samples S15 JB and S15 JHA belong to the category of partly mature compost with germination index 60–80%. Samples S5 JHA and S5 JHB belonged to the category > 100% (ability of stimulatory effects), as well as the control samples of garden substrate. Other substrates ranked among fully mature compost (80–100%). Table 4 shows that content of fly ash in the substrate affected germination index and with its increasing concentration its value decreased. On the other hand, the Vermesfluid solution had a posi-

Table 2. The proportion of dry mass in growing media.

Sample	Weight of backfill (g)	Weight of dry mass (g)	Proportion of dry mass (%)	Proportion of humidity (%)
SKJA	1.0032	0.5532	55.1435	44.8565
SKJB	1.0053	0.5854	58.2314	41.7686
S5 JA	1.0048	0.5659	56.3197	43.6803
S5 JB	1.0087	0.5923	58.7191	41.2809
S10 JA	1.0057	0.5819	57.8602	42.1398
S10 JB	1.0078	0.5466	54.2370	45.7630
S15 JA	1.0007	0.4057	40.5416	59.4584
S15 JB	1.0019	0.5577	55.6642	44.3358
S5 JHA	1.0030	0.4739	47.2483	52.7517
S5 JHB	1.0014	0.4589	45.8258	54.1742
S10 JHA	1.0019	0.4672	46.6314	53.3686
S10 JHB	1.0051	0.4245	42.2346	57.7654
S15 JHA	1.0089	0.4726	46.8431	53.1569
S15 JHB	1.0004	0.5091	50.8896	49.1104

Table 3.   The values of pH, turbidity and conductivity of substrates.

Substrate	Turbidity (ZF)	pH (–)	Conductivity (μS/cm)
SKJB	6.19	6.58	376
S5 JB	6.27	6.43	305
S10 JB	5.63	6.61	391
S15 JB	5.42	6.72	414
S5 JHA	6.15	5.65	502
S5 JHB	4.37	5.78	440
S10 JHA	6.16	6.69	526
S10 JHB	4.26	6.72	787
S15 JHA	4.69	6.82	472
S15 JHB	2.91	6.67	644

Table 4.   Germination index of substrates with perennial ryegrass (*Lolium perenne*).

Sample	Average germinability of a sample (%)	Average length of a radicle (mm)	Germination index (%)
Control	95.00	6.01	–
SKJB	93.50	6.47	105.95
S5 JB	83.75	6.03	88.45
S10 JB	90.00	5.09	80.23
S15 JB	87.50	4.95	75.68
S5 JHA	95.00	6.18	102.83
S5 JHB	92.50	6.44	104.33
S10 JHA	90.00	5.11	80.55
S10 JHB	90.00	5.61	80.43
S15 JHA	83.75	5.25	77.01
S15 JHB	88.75	5.75	89.38

tive influence on this parameter, specifically on a substrates with 5% and 15% of fly ash. The results of samples with 10% fly ash were approximately similar in all versions of watering.

4   DISCUSSION OF RESULTS

Observing the vegetation experiment proved the assumptions of influence of fly ash on growth of plants as well as application of VF solution. It was clearly examined that with increasing quantity of fly ash in the substrate, the amount of grown biomass at the given mixture decreases. Using the same amount of fly ash, there were better results in samples watered with VF. When comparing watering of 3% and 10% VF, then there were always better results with 10% solution.

During the experiment, there were not any significant perishing of the grown plants, except of S5 JA which was most probably caused by excessive watering which led to rotting the roots off. Ryegrass finally had to be harvested due to generally deteriorating condition of the biomass which was caused by excess length of plants due to the fact that plants were not being cut although it was necessary.

The vegetation experiment clearly showed that VF solution favourably affects growth of the observed plants and it can compensate for higher volume of fly ash from biomass incineration in growing media. Comparing the control sample, it was found out that with 5% solution, it is possible to prepare substrates comparable to conventionally sold garden substrate which contains peat for the most part. The best results can be reached using watering with 10% VF solution within 10–14 days.

This test is primarily intended for evaluating compost maturity, therefore it was only used for guidance. The eventual values of index of germination essentially confirmed trends observed during the vegetation experiment. The growing amount of fly ash in substrates decreased the index of germination. Watering with VF solution had the opposite effect. Thus, 5% substrates watered with Vermesfluid reached values over 100%, as well as the control sample which corresponds to the best category which has stimulatory effects. Most of the remaining mixtures showed germination capacity of 80–100% which ranks it among the category of mature compost.

5 CONCLUSION

The results of the experimental part show that growing media with 5% fly ash from biomass incineration are comparable with garden substrate normally sold on the market. Also the effect of Vermesfluid preparation was proved to have an impact on growth and vitality of the observed plants. These results were also proved by phytotoxicity test which proved the capacity of Vermesfluid to decrease negative impacts of higher concentration of fly ash on vegetation.

It can be therefore summed up that by applying a suitable type of ash, which is a result of biomass incineration, as a fertilizer, some hazardous elements began to immobilize in the soil which resulted in successful growth of tested plants during the vegetation experiment. Incidental phytotoxicity of increased concentration of fly ash in the substrate was then eliminated by increasing the pH value with the help of Vermesfluid preparation which comes from vermicomposting biodegradable waste. Thus it might be legitimately concluded that after properly performed and controlled fertilization of agricultural land, properties of the uppermost (humic) topsoil increase in quality.

ACKNOWLEDGEMENTS

This paper was compiled within the Project of Specific University Research (SGS) no. SP2018/27 "The possibilities of increasing contents of humus in arable land with applying products of treatment and processing biodegradable wastes."

REFERENCES

Arshad, M.A., Soon, Y.K., Azooz, R.H., Lupwayi, N.Z., Chang, S.X. 2012. Soil and crop response to wood ash and lime application in acidic soils. In *Agronomy Journal, 104 (3)*. pp. 715–721. ISSN 0002-1962.
Gómez-Rey, M.X., Madeira, M., Coutinho, J. 2012. Wood ash effects on nutrient dynamics and soil properties under Mediterranean climate. In *Annals of Forest Science, 69 (5)*. pp. 569–579. ISSN 1286-4560.
Hanzlíček, T., Perná, I. 2011. Využití popelovin ze spalování biomasy. In *Sborník přednášek XIX. Mezinárodního kongresu a výstavy Odpady—Luhačovice, 5–8. září 2011, Luhačovice.* pp. 177–181, ISBN 978-80-904356-4-3.
Liao, C., Wu, Ch., Yan, Y. 2007. The characteristics of inorganic elements in ashes from a 1 MW CFB biomass gasification power generation plant. In *Fuel Processing Technology, 8(2)*. pp. 149–156. ISSN 0378-3820.
Moilanen, M., Hytonen, J., Leppala, M. 2012. Application of wood ash accelerates soil respiration and tree growth on drained peatland. In *European Journal of Soil Science, 63(4)*. pp. 467–475. ISSN 1351-0754.
Obernberger I., Supancic K. 2009. Possibilities of ash utilisation from biomass combustion plants. In *Proceedings of the 17th European Biomass Conference & Exhibition, June/July 2009, Hamburg, ETA-Renewable Energies* (Ed.), Italy. [online]. 2009. [cit. 2018-06-17]. Dostupné z www: <https://bios-bioenergy.at/uploads/media/Paper-Obernberger-ash-utilisation-2009.pdf>.
Ochecová, P. 2015. *Popel z biomasy—významný zdroj živin.* Biom.cz. [online]. 2015-01-19. [cit. 2018-08-14]. Dostupné z www: <https://biom.cz/cz/odborne-clanky/popel-z-biomasy-vyznamny-zdroj-zivin>. ISSN: 1801-2655.
Park, B.B., Yanai, R.D., Sahm, J.M., Ballard, B.D., Abrahamson, L.P. 2004. Wood ash effects on soil solution and nutrient budgets in a willow bioenergy plantation. In *Water, Air, and Soil Pollution, 159(1)*. pp. 209–224. ISSN 1573-2932.
Tlusťoš, P., Ochecová, P., Kaplan, L., Száková, J., Habart, J. *Aplikace popelů ze spalování biomasy na zemědělskou půdu.* 1. vydání. Praha: Česká zemědělská univerzita v Praze, 2014. 26 s. ISBN 978-80-213-2514-2.
Vassilev, S.V., Baxter, D., Andersen, L.K., Vassileva, C.G. 2013. An overview of the composition and application of biomass ash. Part 1. Phase-mineral and chemical composition and classification. In *Fuel, 105.* pp. 40–76. ISSN 0016-2361.

Determination of operational criteria in the in-use certification system

D. Macek & M. Kosina
Faculty of Civil Engineering, Czech Technical University in Prague, Prague, Czech Republic

ABSTRACT: Concept behind a green building is just going with sustainable approach. This sustainable approach focuses of managing our needs with available resources without affecting the needs of future. The ideal green building would be a building project that would allow you to preserve most of the natural environment around the project site, while still being able to produce a building that is going to serve a purpose. The main principle of sustainable development is, above all, the balance between the three basic areas—the environment, social aspects and economic aspects. The aim of this paper is to define the evaluation criteria for the SBToolCZ (Czech certification system) in the area of facility management and user documentation for administrative buildings in the operational phase. Compared with foreign certificates SBToolCZ is cheaper and is kept in Czech. SBToolCZ is based on an internationally recognized method and evaluates very similar criteria to other foreign methods.

1 INTRODUCTION

The environmentally friendly building, or the so-called "Green Building", is a building characterized by high environmental friendliness and efficient use of natural resources throughout the whole life cycle of the building, with a high emphasis on user comfort. This type of building can be labeled as a "sustainable", "gentle" or "certified" building. The green building is characterized by many aspects based on the principles of sustainable development. The key principle of sustainable development is, above all, the balance between the three basic areas—the environment, social aspects and economic aspects. Green buildings are designed to achieve the most effective balance between these factors (Deka 2017).

There is no straightforward definition for the exact description of the term of "Green Building". This term could be defined through a certification system certifying the building's sustainability without which the term "Green Building" cannot be used (Jalaei & Jrade 2014). However, there is not only one certification system available, but dozens of them developed in various countries throughout the world. Each of the systems adjusts the conditions and assessment criteria to fit the characteristics relevant for the respective state or territory based on its geographical location, potential for the exploitation of existing natural resources, national legislation and codes or customs. It is, therefore, evident that a building certified according to a system used in Australia will possess some characteristics different from a building with a European certification, e.g. due to the fact that water is scarce in Australia and the local certification puts a greater emphasis on this fact (Fillipi & Sirombo 2015; Hromada 2013). Thus, certification systems differ in their conditions and assessment criteria for granting the Green Building certificate, but also in the process that leads to granting the certificate. The basic criteria and conditions are nearly identical for all certification systems, i.e. the promotion of renewable resources and their efficient use, highly economical water management, sustainability of used materials and high-quality interior environment (Kumar & Hancke 2014).

Each investor or developer can choose which certification they want to use for the design of their building. If the building is aimed to serve for their own use, they will choose the certification that best suits e.g. their business orientation. In the case that the investor intends to sell or lease the building after completion, they will be more interested in the market demand.

The certification system may also be designated by the future majority occupant who has concluded a preliminary lease agreement with the project investor on the leased office space in the administrative building. The selection of the assessment method may also be greatly affected by "fitting" the certificate onto the respective project; by its design the project can be in compliance with various ratings according to different certifications (Schneiderová Heralová 2014). Certification applies to new buildings, i.e. buildings already designed to meet the characteristics necessary for granting the respective certificate, or to already completed buildings, to reconstructed or self-contained units within interior spaces of a building, or e.g. to office spaces only (Marique & Teller 2014). At present, there is a system defining the latest sustainability rating that can be used for certifying not only buildings, but also entire urban districts, infrastructures or the landscape (Dobiáš & Macek 2014).

SBToolCZ is a Czech certification tool designed for the assessment of the quality of buildings. The certification process was officially presented in June 2010 (Kubba 2015). The benefits of the SBToolCZ certification consist in its localization directly to the Czech environment, this system is the only one fully reflecting the local climatic, construction and legislative contexts. Compared to foreign certificates, SBToolCZ is cheaper and it is kept in Czech. SBToolCZ is based on an internationally recognized method and evaluates criteria very similar to the other foreign methods (Peri et al. 2015).

2 MATERIALS AND METHODS

The objective of the study is to define the assessment criteria for the SBToolCZ certification system in the area of Facility Management and User Documentation for administrative buildings in their operational phase.

The development of the proposal for the assessment criteria structure is based on the analysis of the world's most recognized certification systems, such as the American LEED, the UK BREEAM and the German DGNB System. Each certification system is conceived and structured in a different way; therefore, a thorough comparative study of their rating systems was necessary to identify areas that might be relevant to the issue.

The analyzed outputs had to be evaluated in the context of the SBToolCZ certification system and suitably complemented with our own experience in the assessment of the quality of facility management and project documentation.

3 RESULTS AND DISCUSSION

This chapter describes the assessment criteria using the structure corresponding to the template for the formal elaboration system of handbooks on the SBToolCZ certification system assessment. It proceeds from the description of the assessment criterion for Facility Management to the assessment criterion for User Documentation.

3.1 Assessment criterion for facility management

Effective Facility Management is the necessary precondition for the economical operation of buildings ensuring the predefined quality of services. It impacts not only on the economic, but also the environmental sphere. In terms of activities related to the building's operation, it mainly covers building maintenance and renewal, cleaning and energy management in the building. The facility manager can obtain important information for optimizing the building's operation as feedback from the building users (Redmond & Pan 2015).

3.1.1 Context

Facility Management is a management sphere responsible for the operation and development of the infrastructure and related services which promote and enhance the effectiveness of the main processes of an organization. It includes the management of buildings, management of the organization's infrastructure, purchasing of support services and overall harmonization of the organization's working environment. Thus, Facility Management is responsible for the management and development of the working environment. It is closely linked to human resources management, organization management and service management. The Facility Management concept is frequently incorrectly associated with the external provision of such services only (so-called outsourcing), nevertheless, it does not matter in Facility Management whether these processes are provided via outsourcing or whether the organization sees to the respective processes or services by itself. Due to the fact that each company, each organization has some infrastructure, property or working environment, Facility Management applies to every enterprise in one way or another. The basic standard regulating Facility Management is ČSN EN 15221.

3.1.2 *Assessment description*

The assessment of the condition of a building is the necessary prerequisite for the elaboration of an effective maintenance plan for individual parts of equipment and structural components of the building. The objective is to prevent emergency situations occurring on both the building and its technical equipment. Another important aspect is to remedy all defects detected. This criterion answers the question of whether the investigation of the condition of the building and its equipment has been completed in the past 5 years and whether work has been done to remedy the deficiencies identified. The assessment score of this answer in points is presented in Table 1. To check the assessment of the criterion, records of the building's condition survey, a copy of the problem/defect elimination action plan, a proof of solving the identified problems and defects, or documents proving the age of the building if it is less than 5 years must be submitted. Energy purchasing management is related to the issue of monitoring and analyzing energy consumption and the resulting estimate of the future need for energy supplies. Table 2 lists the evaluation parameters which are added up in assessing the criterion. The assessment validation is documented by the list of software tools used, the energy consumption analyses made, the energy needs strategy document,

Table 1. Evaluation of building and condition assessment.

Requirement for building assessment and condition	K1*
Building is older than 5 years and its condition has not been assessed during the last 5 years	0
Building condition was assessed, nevertheless, no work has been done to eliminate problems/ defects detected	0
Building condition was assessed and an action plan stating when problems will be eliminated has been introduced	1
Building condition was assessed and all principal problems/defects have been remedied	2
Building condition was assessed and all principal problems/defects have been remedied and an action plan stating when all remaining problems will be eliminated has been introduced	3
Building condition was assessed and all detected problems/defects have been remedied	4
Building is not older than 5 years and its condition has not been assessed	4

*K1 credits exclude each other in the above items.

Table 2. Evaluation of energy purchasing management.

Requirement for energy purchasing management	K2*
Software tool is used for monitoring and analysing energy consumption data	1
Measurement and regulation system is used for energy consumption data collection	2
Energy needs strategy document has been elaborated	1

*K2 credits are added up in the above items. If a statement does not apply, 0 points are granted.

Table 3. Evaluation of cleaning with eco-friendly cleaning products and cleaning equipment.

Requirement for cleaning with eco-friendly cleaning products and cleaning equipment	K3*
75% of costs of cleaning products are spent on internationally or nationally certified ECO products	1
At least 40% of all driven mechanisms must comply with the above criteria. For existing devices which do not comply with the criteria, a plan for their discarding and replacement with eco-friendly products at the end of their service life must be elaborated	1

*K3 credits are added up in the above items. If a statement does not apply, 0 points are granted.

Table 4. Evaluation of feedback from building users.

Requirement for feedback from building users	K4
No regular meetings or formal communication with building users aimed at obtaining feedback on building operation are planned	0
Regular meetings or formal communication with building users aimed at obtaining feedback on building operation are planned and resulting information is reported to the management	1

the measurement system and regulation documentation. Cleaning with eco-friendly cleaning products is assessed in Table 3. The credit can be granted if at least 75% of the costs of cleaning products are spent on internationally or nationally certified ECO products. The criterion is documented by the annual audit report. For electrical devices, the following functions are required:

- protective measures, such as tubes or rubber bumpers, to avoid damage to building surfaces;
- ergonomic design to minimize vibration, noise and user fatigue, as specified in the user manual pursuant to ISO 5349-1 for shoulder vibrations, ISO 2631-1 for all body vibrations, and ISO 11201 for acoustic pressure at the operator's ear;
- vacuum programs operating at a maximum sound pressure level of 70 dB or less pursuant to ISO 11201;
- devices must be in Energy Class A or B.

The check is performed by submitting the list of used devices indicating which parameters are met and which are not met. Furthermore, a plan of discarding the existing devices which do not comply with the above criteria at the end of their service life shall be submitted. The feedback from the building users assesses whether communication with the building users on issues related to operation has been going on. The information provided must be subsequently interpreted to the management so that steps enhancing the operation can be prepared and implemented. The criterion is presented in Table 4. The compliance with the criterion can be documented e.g. by meeting minutes, feedback forms or feedback electronic forms.

The resulting number of credits K entering the benchmarks is obtained as the sum of granted $K1$, $K2$, $K3$ and $K4$ credits, thus:

$$K = K1 + K2 + K3 + K4 \tag{1}$$

3.2 Assessment criterion for user documentation

The economical operation of a building cannot be executed without perfect knowledge of the technical and structural part of the building. The necessary tool serving this purpose is the As Built documentation and operating and maintenance manuals (including crisis and emergency plans), which cover both structural components and technical equipment. The information from the above documents is intended not only for the building operator functioning as a technical report, but also for the building users allowing them to use the building and equipment installed in efficiently. For these reasons, the existence and the format of these documents must be assessed (Karásek & Pavlica 2016).

3.2.1 Context

The fundamental law regulating construction practice is Building Act No. 183/2006 Coll., as amended, which stipulates the regulations for construction and land planning activities. The Building Act, specifically Decree No. 499/2006 Coll., as amended, further specifies the content and scope of As Built documentation. This documentation is essential for the execution of property management for the reason that it is the latest documentation whose elaboration is required by the state administration body without which the building approval cannot be issued. In smaller buildings, the documentation may be replaced with a copy of validated design documentation accompanied by drawings documenting deviations, unless the clarity and comprehensibility of the documentation is affected. Once the building approval has been issued, the building use for the purpose which it was designed for can start. Property operation is an activity associated with the management of routine as well as unexpected situations. Therefore, operational documentation represents a significant partner of the facility manager, or the administrator, who is responsible for the operation in the building. Thanks to appropriately elaborated documentation, routine activities can be planned and unexpected situations minimized, if not fully avoided. The documentation has two parts: text and drawings. The rule is that if something cannot be expressed graphically, it is expressed by a verbal description. The graphical expression of information is preferred because of the unambiguous disclosure of the information. The As Built documentation captures the entire technical condition of a building in a complex way; therefore, this documentation can be used for the needs of its operation and use.

3.2.2 Assessment description

The As Built documentation of a building is the essential background material for effective management. The form of this documentation affects the speed and optimization of solving planned as well as required acts, particularly in the area of building maintenance and operation. The granting of credits for this item is presented in Table 5. The assessment is documented by a digital or paper version of the documentation. Other important documents are operating or maintenance manuals, which supply more detailed information about the building equipment and structural components and serve directly for the execution of building maintenance. The assessment is presented in Table 6. Meeting the criterion is documented by

Table 5. Evaluation of As Built documentation format.

Requirement for As Built documentation format	K1*
As Built documentation is not available	0
As Built documentation is available on paper	4
As Built documentation is available in digital format as 2D drawings	6
As Built documentation is available in digital format as a 3D model	7

*K1 credits exclude each other in the above items.

Table 6. Evaluation of completeness of operating and maintenance manuals.

Requirement for completeness of operating and maintenance manuals	K2
Complete set of manuals for operation and maintenance, which are accessible to building/ facility management, is not available	0
Complete set of manuals for operation and maintenance, which are accessible to building/ facility management, is available	2

Table 7. Evaluation of accessibility of essential information from the user manual to the building.

Requirement for accessibility of essential information from the user manual to the building	K3
Not all users have access to essential information from the user manual to the building	0
All users have access to essential information from the user manual to the building	1

the list of operating and maintenance manuals with a copy of the front page and the content of the respective documents. The building users should have access to the information and instructions of how to effectively use the building and equipment designed for them. The criterion is displayed in Table 7. Compliance with the criterion is documented by a copy of the respective parts of the user manual to the building, including details of how the information was distributed to the users.

The resulting number of credits K entering the benchmarks is obtained as the sum of granted $K1$, $K2$ and $K3$ credits, thus:

$$K = K1 + K2 + K3 \qquad (2)$$

4 CONCLUSIONS

The processing of the above criteria reflects the current trend in demands for the building management system, which relies on sophisticated facility management and high-quality and updated documentation of the building. Efforts to certify buildings and comply with the required criteria for buildings under operation lead to enhancing the efficiency and cost-effectiveness of the whole building management process, even in the context of environmental friendliness. The owners and administrators of buildings are aware of the importance of the building's operating phase seen from the perspective of its life cycle cost.

The certification system must best describe the nature of the issue in question so that the method does not turn into mere compliance with partial criteria and obtaining the necessary score for the final assessment in the certification system. The objective is to make the certified building really comfortable for its users, make it fulfill its purpose and, at the same time, become cost effective with minimal negative impacts on the environment.

The above criteria represent only a fraction of the entire SBTool certification system and need to be coordinated with the assessment of other areas where potential overlaps may occur. The presented state is the state preceding the expert assessment and, therefore, changes and potential additions can be proposed and implemented in the final version.

ACKNOWLEDGEMENTS

This work was supported by the Grant Agency of the Czech Technical University in Prague, grant No. SGS18/018/OHK1/1T/11.

REFERENCES

Deka, G.C. 2014. Cost-benefit analysis of datacenter consolidation using virtualization. IT Professional, 16 (6), art. no. 6964984, pp. 54–62. ISSN 15209202.

Dobiáš, J. & Macek, D. 2014, Leadership in Energy and Environmental Design (LEED) and its Impact on Building Operational Expenditures. Proceedings of the Creative Construction Conference 2014. Creative Construction 2014. Prague, 21.06.2014. Budapest: Diamond Congress Kft., ISBN 978-963-269-434-4.

Fillipi, M. & Sirombo, E. 2015. Green rating of existing school facilities. Energy Procedia, 78, pp. 3156–3161. ISSN 18766102.

Hromada, E. 2013. Decision-support tools and assessment methods. Central Europe towards Sustainable Building 2013. Praha, 26.06.2013–28.06.2013. Praha: Grada, pp. 669–672. ISBN 978-80-247-5018-7.

Jalaei, F. & Jrade, A. 2014. Integrating Building Information Modeling (BIM) and energy analysis tools with green building certification system to conceptually design sustainable buildings. Journal of Information Technology in Construction, 19, pp. 494–519. ISSN 14036835.

Karásek, J. & Pavlica, J. 2016. Green Investment Scheme: Experience and results in the Czech Republic. Energy Policy, 90(90), pp. 121–130. ISSN 0301-4215.

Kubba, S. 2015. LEED v4 Practices, Certification, and Accreditation Handbook: Second Edition, Elsevier Inc, pp. 1–675. ISBN 978012803900.

Kumar, A. & Hancke, G.P. 2014. An energy-efficient smart comfort sensing system based on the IEEE 1451standard for green buildings. IEEE Sensors Journal, 14 (12), art. no. 6899588, pp. 4245–4252. ISSN 1530437X.

Liu, M.M. 2014. Probabilistic prediction of green roof energy performance under parameter uncertainty. Energy, 77, pp. 667–674. ISSN: 03605442.

Marique, A.-F. & Teller, J. 2014. Towards sustainable neighbourhoods: A new handbook and its application. WIT Transactions on Ecology and the Environment, 191, pp. 177–188. ISSN 17433541.

Peri, G. et al. 2015. Design, building up and first results of three monitored green coverings over a university department building. Energy Procedia, 78, pp. 3037–3042. ISSN: 18766102.

Redmond A. & Pan, P. 2015. Defining techniques for developing an energy analysis services on a middleware for technical optimism of green buildings, IEEE Green Energy and Systems Conference, IGESC, art. no. 7359392, pp. 61–66. ISBN: 9781467372633.

Schnaiderová Heralová, R. 2014. Life Cycle Cost optimization within decision making on alternative designs of public buildings. Procedia Engineering, (85), pp. 454–463. ISSN 1877-7058.

Advances and Trends in Engineering Sciences and Technologies III – Al Ali & Platko (Eds)
© *2019 Taylor & Francis Group, London, ISBN 978-0-367-07509-5*

Innovative model of accreditation in environmental fields of university study in Slovakia

M. Majerník, N. Daneshjo & G. Sančiová
University of Economics in Bratislava, Košice, Slovakia

ABSTRACT: University education at Slovak colleges (public, state and private) and universities is carried out almost for 15 years by the European structure of fields of study. Within them, individual study programs are created and accredited. The authors as members of standing working groups of the Accreditation Commission—an advisory body of the Slovak Government—present a state of accreditation process, their experiences and lessons learned from different areas of research in the face of globalization trends and criteria for international accreditation. They thus formulate a concept of an innovative and institutional ensuring model of the accreditation process compatible with accreditation in economy of Slovakia and with rules and criteria of European accreditation.

1 INTRODUCTION

Accreditation, whether in the economic or academic field, involves an impartial, independent and expert assessment of competence of entity (company or college) by accreditation authority (commission) and an issuance of a certificate that the entity is competent to carry out the activities specified in the certificate and to fulfil the requirements of competencies permanently specified by the relevant normative documents. Its purpose is to assess conformity in a globalized environment according to internationally accepted criteria in ensuring and improving continually complex quality of products and activities. (Mikuš, Sekan, 2015; Majerník et al., 2013)

Conditions and procedures for a harmonized application of the accreditation are created in a coordinated approach by global organizations today, i.e. assessing competences of conformity assessing bodies in systems, processes and persons at national level through national accreditation bodies to meet international criteria.

The Slovak National Accreditation Service (SNAS) in economic field ensures accreditation processes in Slovakia in accordance with general criteria determined in the relevant international standards and ISO/IEC directives, European standards and under other requirements published in relevant application documents EA, IAF, etc.

Accreditation processes in the university environment are ensured in Slovakia by the Accreditation Commission (AC) as an advisory body of the government or the Ministry of Education Science Research and Sport of the Slovak republic (MESRS of SR) in particular in a form of complex accreditations (CA) in six-year cycles.

The Accreditation Commission composed of prominent experts (predominantly from the academic environment) in the exercise of its competence uses criteria that the ministry approves of its proposal and after an expression of the bodies of university representation. It is increasingly resonant in these processes with the issue of the impartiality of an AC as one of the basic criteria of accreditation processes in general as well as an incompatibility of the process in relation to activities in the economy, e.g. surveillance audits, internal audits and use of the ISO 19011 to demonstrate continuous improvement of internal quality system of colleges (Karapetovics, 2001).

Based on the above-mentioned contexts, knowledge and experience of the authors in the AC, as well as progressive global trends in accreditation, it is necessary to develop a concept of an innovative model of infrastructure ensuring of accreditation processes of colleges and universities in Slovakia. The primary objective is besides international recognition of accreditation performances of national accreditation authority.

2 CURRENT APPROACHES TO QUALITY ASSESSMENT OF UNIVERSITY EDUCATION IN SLOVAKIA

The accreditation commission as the competent authority in the process of accreditation of university education institutions in Slovakia assesses their competence according to established criteria and it issues a certificate that they are competent (and at what qualitative level)

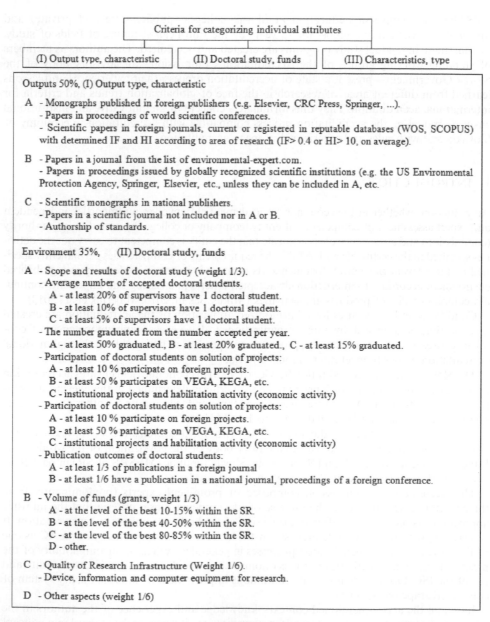

Figure 1.   Systemization of criteria, categories and attributes.

to carry out educational, scientific and research and other creative activities in the system of continuous improvement of quality of their processes, performances and outcomes.

2.1 Evaluation of research area in complex accreditation of Universities

The competence of the Accreditation Commission for complex accreditation is generally governed by the University Education Act, the Statute of the AC, and the criteria for assessing the level of research, development, artistic and other creative activities. The assessment rules represent a comparable level of quality in each of the categories (A, B, C, D) in all 24 research areas. The AC in so doing respects specificities of individual areas in quantifying and qualifying the quality levels.

In accordance with paragraphs 10 and 13 of the Criterion, a university may also submit another output, e.g. engineering work. The total number of outputs should not exceed 10.

There is shown a complex systematization of the criteria for categorization for individual attributes in Figure 1 (Efimov et al., 2015; Kehm, 2015).

2.2 Current infrastructure ensuring of accreditation of Slovak Universities

The algorithm for ensuring the accreditation of study programs at Slovak universities is used for all study fields, or within them created programs with slight modifications for some of the 24 research areas.

The bases for decision-making of the AC are elaborated Evaluation Reports by WQ, as well as other opinions of experts. The members of the WQ of the standing and temporary AC are experts from universities, from practice and from abroad, meeting the pedagogical and scientific criteria in an above standard level—level A (top international quality).

3 DEVELOPMENT TRENDS OF ACCREDITATION ACTIVITY IN SLOVAKIA

Management of accreditation activities is a process of continuous improvement in the Deming cycle P – D – C – A (Bernardo et al., 2009). Two previous complex accreditation of Slovak universities in addition to fulfilment of original primary objectives—a classification of universities into categories (research universities, colleges and professional) also pointed out several problematic areas and discrepancies in the work of the AC and its working groups in relation to globally recognized standards of conformity assessment and creation of accreditation criteria, e.g. impartiality, compatibility with abroad, transparency, etc. These discrepancies were continuously addressed through corrective and preventive measures such as personnel reconstruction of the commission and its working groups, differentiation and modification of criteria for individual research areas, formalization of top scientific teams of universities in Slovakia, etc.

3.1 Purpose, subject and objectives of formalizing top scientific teams

In connection with the process of categorizing universities, the project "Top scientific teams of universities in Slovakia" was initiated from the AC level aiming:

– To analyze and to identify workplaces and scientific teams in individual research areas with above average outputs within Slovakia, accepted internationally.
– To specify and to publish international research areas, including specific workplaces, teams and figures.
– To motivate and to encourage university students and staff to develop profiles through these excellent, European and world-recognized workplaces.
– To motivate scientific teams and workplaces at Slovak universities and top outcomes with a potential for international acceptance and benefits.
– To advise the government of the Slovak Republic and the Ministry of Education that excellent workplaces and scientific figures should have high moral, social and financial support for their research work in terms of sustainable socio-economic development.

- To formalize requirements and criteria for grant support for sustainable consumption, production and green growth through above standard support of doctoral students and postdoctoral researchers.
- To explain, that the quality-related situation is unsustainable without support of top-class workplaces and without increasing university funding the EU science's average GDP share. (Lescevica, 2015).

4 AN INNOVATIVE MODEL OF PROCESS AND INFRASTRUCTURE ENSURING THE ACCREDITATION OF UNIVERSITIES IN SLOVAKIA

The current AC is a member of three international associations dealing with quality of university education:

- INQAAHE, the International Network for Quality Assurance Agencies in University Education, the worldwide association of quality assurance agencies in University education and AC.
- Affiliate of the ENQA, European Association for Quality Assurance in University Education.
- CEENQA, Association of Germany, Austria, Czech Republic, Poland, Hungary.

Full membership in the ENQA is conditional on demonstrable independence and impartiality of the accreditation institution (Rowan 2007). In order to achieve international recognition of accreditation activities in the field of University education in Slovakia, an innovative model of process and infrastructure ensuring of accreditation is being prepared through an impartial, independent, professional institution which operation will be accepted in globalized environment—full membership in ENQA, EA, etc., Figure 2.

5 STATUS, PURPOSE AND STRUCTURE OF THE SLOVAK ACCREDITATION AGENCY FOR UNIVERSITIES

Within the framework of two previous complex accreditations, universities in Slovakia (36 in total) could not be divided into university, unclassified and professional also as a result of non-structural approach of colleges and non-system approach of the Ministry of Education to accreditation criteria, including not meeting the principle of impartiality. A new designed Slovak Accreditation Agency for universities (SAA for U) should also contribute to addressing these shortcomings. It will grant colleges a license for issuing of university education diplomas in individual study fields and within them created study programs by universities.
Slovak Accreditation Agency for universities in Slovakia will assess the internal quality system of education at universities and its continuous improvement or continuous development focusing on:

- Quality ensuring strategy.
- Quality ensuring processes.
- The relationship between education and research activity.

The Agency body will decide in particular on:

- Conformity of the internal quality system and its implementation with internationally recognized standards.
- Granting or not granting accreditation of SP with finality.
- Granting or not granting rights for HaI proceedings.
- Imposing corrective measures on non-compliance.
- Objections to bias against a proposal of PS composition, etc. and to provide advice to the MESRS of SR on:
    - Applications for state approval for establishment of university education institution.
    - Updating and adjusting quality standards.

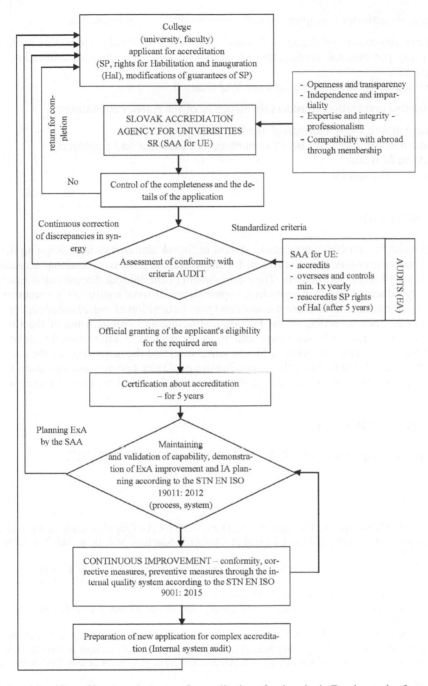

Figure 2. Algorithm of innovated process of accreditation of university in Deming cycle of continuous improvement of quality.

As in the field of "industrial accreditation", the Agency will be required to keep lists of assessors, institutions, members of the international organization of accreditation bodies in the field of quality assurance in university education, to perform surveillance audits on conformity with standards. The Agency will develop: Methodological guidelines for evaluation of standards, charging rates, trend analyses, education systems in the Slovak Republic, activity report and other documents.

Necessary activities in connection with international recognition will be:

– Regular assessments of the agency's activity by an international accreditation authority.
– Ensuring professional development of employees and, in particular, calibration of the assessors.
– Systematization of listing policies and WQ creation.

Legislative preparation considers the structure of the Agency consisting of:

– Chairman and Vice-Chairman of the Executive Board.
– A nine-member Executive Board (composed of academics and employers).
– An Appeals Board.
– A head of the agency.

6 CONCLUSION

The accreditation process management system at Slovak universities is undergoing dynamic change and continuous improvement in the accreditation process of national accreditation bodies in a globalized environment. The establishment of the "Slovak Accreditation Agency for University Education" as an independent, impartial, professional institution is a necessary step towards eliminating risks of current doubts and non-recognition of university education diplomas in study fields according to the European structure and requirements of the European socio-economic practice. Stricter requirements in the context of international standards also have to lead to the differentiation of Slovak universities and the acquisition of the status of a research university, highlighting the comprehensive quality of research and education through continuous improvement of the status in the international evaluation ranks of universities.

ACKNOWLEDGMENTS

The paper was supported by KEGA 026EU-4/2018 and VEGA 1/0251/17.

REFERENCES

Bernardo, M., Casadesus, M., Karapetrovic, S., Heras, I., 2009, An Empirical study on the integration of management system audit. [in:] Journal of Cleaner Production. Vol. 18, p. 742–750. 2009. ISSN 0959-6526.

Internal Documentation of AK, advisory body of the Government of the Slovak Republic.

Karapetrovic, S. - Willborne, W., 2001, Audit system: concepts and practices. [in:] Total Quality management. Vol. 12(1), p. 13–28. 2001. ISSN 1360-0613.

Majerník, M. a kol., 2013, Akreditácia, certifikácia, auditovanie. EUBA a UNMaS SR. 2012. 204p. ISBN 978-90-971555-0-6.

Mikuš, D., Sekan, F., 2015, Political and Economical Aspects of Scholar System in Slovak Republic. 2nd International Multidisciplinary Scientific Conference on Social Sciences & Arts, SGEM 2015. Education and Educational Research. 2015. Albena, Bulgaria, p. 729–736, ISSN 2367-5659.

Advances and Trends in Engineering Sciences and Technologies III – Al Ali & Platko (Eds)
© 2019 Taylor & Francis Group, London, ISBN 978-0-367-07509-5

Brief evaluation of moisture fighting technologies

O. Makýš, M. Hrčka & P. Šťastný
Architectural Heritage Conservation Technology Centre, Department of Building Technology
and Faculty of Civil Engineering, Slovak University of Technology in Bratislava, Slovakia

ABSTRACT: The purpose of this paper is to evaluate the ventilation effects versus the undercutting and injection technologies in fighting the unwanted moisture in a historic baroque masonry. The evaluation is based on the results from the realization of the first three levels of the operation by the Method of continuous steps (as per the ISCARS recommendations in 1999 and Charter of ICOMOS recommendations). The method was applied over a timespan of a few years. The data gathered from a regular monitoring clearly demonstrate a real effect of these controversial groups of technologies on the above mentioned baroque buildings. The data also illustrate the strong and the weak points of all mentioned technologies in a more general perspective.

1 INTRODUCTION

A capillary moisture invading traditional masonry walls of the architectural monuments is a typical wide spread problem for most of the historical buildings. Wet masonry and its plasters pose not only a technical but also a hygienic problem. Wet masonry can partially lose its strength and the wet walls are losing on their thermal insulation. On top of that, water is a medium that can in a very long term perspective dissolve all of the traditional materials of the architectural heritage building. Besides that, excess water in plasters creates a favorable environment for algae and molds. These have consequently a negative impact on human health. Stating the above, it is crucial to fight such unwanted moisture in historic masonry and its plasters.

There is a set of technologies that can be applied for fighting the unwanted moisture in historic constructions. Yet not all of them are effective. As proven by the experience, some of them have a very limited positive effect and few have no positive effect at all. Highly effective technologies include certain level of invasive and destructive interventions into the protected buildings' historic material. However, numerous international methodic documents do not like to recommend such technologies.

2 FOCUS OF THE RESEARCH

The above mentioned conflict is further supported by ambiguous statements in the internationally acclaimed methodic documents. As a result, the architectural conservation authorities (District Monument Offices in Slovakia) hesitate to accept invasive construction technologies as effective methods. The insufficient technical science education of the monument conservation officers poses yet additional challenge. On top, a problem also lies in a futile attempt to apply soft non-invasive methods. Even though considerate to the historical material, they cannot bring a desired effect. The historic construction industry was not able to use effective insulation materials and technologies. Consequently, we need to focus on solving this issue by the use of modern technologies and materials.

Therefore, our research was aimed at acquiring the relevant data that can demonstrate the effect of the three main groups of technologies fighting the moisture in the walls. During the timespan of eight years we collected the data from several different locations where moisture fighting technologies were introduced in order to get a clear image of its effects. We have decided to use three examples to demonstrate the statement towards the three different groups of moisture fighting technologies.

2.1 *The method of consecutive steps*

The method of consecutive steps is based on the repair interventions carried out in a step by step order. There needs to be a dedicated time for the impact evaluation after taking each step. The walls have some time to react to the changed situation throughout the entire process and along all the steps. There is always a possibility to adapt the following intervention as needed by the changed situation. The method was recommended by both ISCARS in 1999 and the Charter of ICOMOS in 2003.

The architectural heritage authorities have a positive approach towards this method despite its major disadvantages. Firstly, the method is time consuming which is not taken positively by the owners. On the other hand, the method is less costly and the achieved success can last for a long time. From the architectural heritage conservation view, there is also a significant advantage in a less invasive and more reversible technologies used in the first phase of its introduction. More invasive and less reversible technologies should be used only in a case when the previous group of technologies hasn't delivered a desired outcome. Nonetheless, this is the exact point of a major confusion. The so-called soft methods are usually not very effective. This is specifically true for a case such as our historic masonry is. A fully wet masonry, with over 15% of relative moisture is a very typical scenario in the architectural heritage buildings. The need to solve the problem of fully wet masonry fast leads to applying some of the invasive technologies.

On the other hand, by taking the first steps we were obliged to use the method of the sequential steps. Specifically, we took the step No.1 – Research and Exploration as well as step No.2 – Basic Arrangements (possible revitalization of the original anti-moisture arrangement, the improvement of a continuous maintenance of the buildings, the correction of a previous inappropriate intervention, etc.). Based on our experience, we highly recommend the first steps. In reality, we have very often faced the following situations: broken or deformed gutters, canalization filled with debris or waste, wildly grown green in a near proximity to the building as well as in contact with the building, closed ventilation openings, damaged plumbing, and several amateur arrangements placed on the walls.

3 MOISTURE FIGHTING TECHNOLOGIES

In our research, we have focused on the application of the three main groups of technologies that are being used in the field. The first group consists of the ventilation technologies that have a non-invasive nature and therefore are favored by the architectural heritage authorities. The injection technologies belong to the second group. Their invasive nature is limited and they are still accepted by the heritage authorities. The third group consists of the undercutting technologies that are invasive in nature and therefore not very welcomed by the heritage authorities.

3.1 *Ventilation technologies*

As a first example, we used the ventilation technologies in a crypt of a baroque church in Western Slovakia. The original ventilation openings were found and opened. As a result, the quality of air inside the crypt that was originally full of coffins improved significantly. As can be seen in the attached table, the figures clearly demonstrate a successful result: the

Table 1. Figures from the measuring of humidity in the masonry of the crypt.

Crypt of the church—measurement device: GANN UNI-1

Place/ Date of measuring		9.3. 2010	6.10. 2010	21.6. 2011	5.10. 2011	19.6. 2012	4.12. 2012	12.4. 2013	30.10. 2013	10.12. 2013	2.7. 2014
*1	*2	% mass	% mass	% mass	% mass	% mass	% mass	% mass	% mass	% mass	% mass
44	120	17,6	17,3	16,9	16,5	16,1	13,8	11,3	7,2	7,2	7,1
45	150	17,4	13,2	13,0	11,5	10,1	9,4	8,4	8,3	8,8	9,1
46	100	13,5	13,8	13,9	14,0	13,6	12,0	11,8	12,1	12,1	10,9
47	120	12,1	12,3	12,5	12,0	11,8	11,8	11,8	12,0	11,4	11,3
48	180	13,7	13,2	13,1	12,5	12,7	12,4	12,2	9,5	9,1	9,1
49	160	14,1	12,1	12,1	12,1	12,3	14,6	14,2	13,3	13,1	12,9
50	160	14,2	13,8	13,4	13,1	12,9	12,6	12,4	12,0	10,2	10,0
51	160	14,0	13,9	14,0	13,5	13,2	11,8	11,0	11,4	11,2	11,6
52	200	14,2	13,8	12,4	11,5	11,1	9,7	12,5	13,4	14,2	11,5
53	140	14,3	14,3	14,8	14,2	14,3	14,8	13,9	14,4	13,3	13,0
54	160	14,3	13,4	14,1	13,7	13,6	13,4	13,4	12,5	12,6	12,0
55	120	14,4	14,0	13,9	13,7	13,3	12,9	12,4	12,0	12,0	12,1
56	140	8,3	8,3	9,1	9,9	9,5	9,4	9,3	9,1	9,3	9,1
57	160	13,1	13,5	13,9	14,1	13,0	12,7	12,0	12,9	12,4	12,4
58	140	13,2	13,4	13,5	13,5	13,0	12,9	13,5	13,7	13,5	12,9
59	140	14,8	14,9	15,2	15,6	13,5	13,7	13,2	13,0	13,7	13,3
60	140	14,0	13,9	13,5	13,1	13,3	13,0	13,6	12,9	11,2	11,0
61	120	14,5	14,3	13,9	13,7	13,0	12,9	12,6	12,6	12,1	11,6

*1 – No. of the measuring point.
*2 – Distance of the measuring point from the floor [cm].

level of moisture in the baroque brick masonry declined. Nonetheless, we have to note that the decline in the figures was not sufficient. We have achieved a decline of approximately 5%. The minor decline was achieved by a strong air blow through the crypt. This can be seen on the picture showing the candle's flame behavior. See the measuring listed in the Table 1.

The other results point towards the common risks of unprofessional application of the ventilation technologies. Once there is no significant air flow in a ventilation channel as can be seen on the picture, there is no positive effect of the ventilation technologies (Balík – Starý, 2003).

3.2 Injection technologies

In the second example, we used the injection technologies for a moisture insulation of an altar in the baroque church in Western Slovakia. A typical baroque altar is usually constructed from the brick masonry creating a massive and thick block. The front part of the masonry's surface was partly covered by the artificial marble. The artificial marble layer created a diffusion barrier which limited the natural ventilation of the masonry. The masonry itself was affected by the capillary rise of moisture and the measurement figures were above 10% of relative moisture, representing a wet masonry. The water along with the waters' soluble salts totally destructed the artificial marble in the affected zone.

The only way to approach this problem was by using the technology of injecting the sealing and hydrophobing agent (Ashurst & Ashurst 1989). The major challenge with using this technology was linked to the fact that it was not clear what was the construction made of. In other words, it was not clear whether the construction was made of the bricks only or whether the stones were used as well.

Table 2.   Figures from the measuring of humidity in the masonry of the altar.

Interior of the church (altar) – measurement device: GANN UNI-1

Place/Date of measuring		9.3. 2010	6.10. 2010	21.6. 2011	5.10. 2011	19.6. 2012	3.12. 2012	12.4. 2013	30.10. 2013	10.12. 2013	2.7. 2014
No. of measuring point	Distance from floor [cm]	% mass	% mass	% mass	% mass	% mass	% mass	% mass	% mass	% mass	% mass
34a	20	14,0	13,6	13,2	13,0	10,9	6,7	6,1	4,7	3,8	4,6
34b	40	13,8	13,5	13,2	11,7	9,7	7,4	5,1	4,9	4,8	4,9
35a	50	10,1	9,9	9,6	9,7	9,6	5,9	4,6	4,0	4,1	3,7
35b	40	13,8	13,5	13,2	12,5	7,7	4,5	4,3	4,1	4,0	3,8
36a	50	8,0	7,6	7,9	7,0	6,1	4,9	4,7	4,6	4,1	4,2
36b	30	12,0	10,9	10,3	9,8	8,3	4,1	3,9	3,8	3,6	3,6

Note: The injection technology was applied between 19.6.2012 and 3.12.2012.

Figure 1.   Opened ventilation shaft in a crypt—visible air flow on the direction of the flame.

Figure 2.   Modern ventilation channel without any air flow—visible on the direction of the flame.

The injection resulted in a strange outcome where one part of the altar dried out quite quickly (few weeks) while the other part remained wet for several more months. Consequently, we thought that the injection method had failed and we would have to repeat it. However, following couple months the second part of the masonry also dried out in a comparable way to the first one. To conclude, we are not sure why one part dried out in a different timespan as the other. See the measuring listed in the Table 2.

3.3   Undercutting technologies

In the third example, we used the undercutting technologies for the moisture insulation in a baroque manor-house in Western Slovakia. The house was built from the bricks on a lime mortar with a partial use of stone blocks. The measurement showed the house had a serious moisture problem. We chose the undercutting as the best technology for fighting the excessive moisture in this case.

The undercutting technology is based on the cutting of the masonry with the use of a saw machine. The saw machine works with infinite steel rope, set off with individual diamonds (Lebeda et al. 1988). The diamond ensures the cut off any traditional construction material, including the stones. The cuttings are arranged in short sections (approximately 20–30 centimeters wide) that must be immediately fixed by the hard chocks. These are ensuring no movement of the above construction layers. During the cutting process the rope must be cooled with water constantly which creates a lot of dirt.

Table 3. Figures from the measuring of humidity in the masonry of the manor house.

Ground floor of the manor-house—measurement device: Hygrometer

No. of measuring	Distance from floor	% mass		Distance from floor	% mass	
Point/Date	[cm]	16.5.2017	20.6.2018	[cm]	16.5.2017	20.6.2018
M1	30	2,8	2,8	100	5,1	1,7
M2	30	2,8	3,4	100	3,8	4,1
M3	30	3,5	3,8	100	1,3	1,5
M4	30	2,5	2,2	100	7,0	2,0
M5	50	12,0	2,0	100	4,3	2,4
M6	30	6,0	3,7	120	7,4	4,6
M7	30	4,4	4,1	120	10,6	7,6
M8	30	1,7	1,9	150	13,6	6,0
M9	30	6,3	3,1	120	11,0	5,2
M10	10	9,1	3,7	180	4,4	1,9
M11	10	10,6	3,7	180	1,2	1,3
M12	10	5,2	6,5	180	2,0	2,2
M13	10	16,7	5,3	180	11,9	2,0
M14	10	11,0	1,2	180	16,4	1,5
M15	10	17,0	2,0	180	10,4	1,8
M16	10	2,9	3,8	180	1,9	3,9
M17	10	2,7	2,0	180	2,5	3,5
M18	10	4,0	4,4	180	2,2	2,8
M19	10	1,7	1,5	180	1,9	2,0
M20	10	10,0	3,1	180	8,5	2,8
M22	30	3,9	2,4	160	1,1	1,5
M23	30	11,4	2,1	150	2,0	2,0
M24	30	7,4	6,6	100	1,9	1,1
M25	30	9,4	1,6	100	5,4	3,1
M26	30	8,3	7,4	100	17,3	6,9
M27	30	17,4	8,8	100	2,9	2,5
M28	50	18,5	2,8	100	17,5	3,3
M29	30	17,5	2,6	100	5,1	2,6
M30	30	15,5	0,7	100	17,4	8,8
M32	30	8,4	8,0	100	5,0	5,4
M33	30	12,1	2,0	100	4,7	2,5
M34	30	–	1,7	150	4,0	1,9
M35	30	10,4	6,7	100	14,4	7,1
M36	30	17,4	6,7	100	6,5	4,1
M37	30	10,1	7,4	100	17,5	7,1
M38	30	17,1	11,1	100	9,5	4,0
M39	30	1,5	1,8	100	11,0	4,9
M40	30	1,6	2,0	100	13,6	4,9
M41	30	2,0	1,1	100	1,6	2,2
M42	30	3,5	2,2	100	2,0	1,7
M43	30	11,0	5,0	100	17,5	11,5
M44	30	17,5	2,8	100	17,4	14,3
M45	30	5,7	2,9	100	3,4	2,4
M46	30	3,5	1,9	100	3,7	2,7
M47	30	1,7	2,2	100	2,1	2,1
M48	30	3,4	3,0	100	3,2	1,4

Figure 3.   The implementation of the undercutting technology.

Figure 4.   The implementation of the injection technology.

The measured results of the cutting technology in the baroque house after approximately half a year showed a decline in the masonry moisture level. Given the moisture decline was measured in most of the controlled points, we can accept this technology as successful. See the measuring listed in the Table 3.

4   CONCLUSIONS

Given the above examples supported by the figures from the exact instrumental measuring, we can conclude that the effect of the soft moisture fighting technologies (ventilation technologies) is quite limited. On the other hand, there are technologies that are invasive towards the original material and are changing the original construction scheme of a historic building but are very effective.

The conclusions clearly show that the understanding of the Venice charter from 1964 calling for the protection of the original material in the historic heritage buildings cannot be taken literally. Taking the Venice charter literally would not allow for effective solving of the problems related to the unwanted moisture in the walls. As demonstrated by our research, the effective technologies ask for the intervention into the masonry.

From our perspective, it is inevitable that in order to fight the unwanted moisture in the historic masonry, the architectural heritage protection authorities will be willing to accept the invasive yet very effective technologies. It is important to realize that we are currently facing a problem that our ancestors were not able to solve. Of course, this applies to majority of the cases, not all of them. We also have a few of the historical solutions that worked very well over the centuries. Nonetheless, these are more of an exception.

It is a fact that the majority of the historic heritage buildings have an unpleasant problem with unwanted moisture. Nowadays, we possess the materials and the technologies for their application that can effectively solve this problem. With expect, with a portion of humility, that the current solutions will act minimally for more than a couple of decades.

REFERENCES

Ashurst, J. & Ashurst, N. 1989: *Practical Building Conservation. – Vol. 2. Brick, Terracotta and Earth*, Hants: English Heritage Technical Handbook, Gower Technical Press.
Balík, M. – Starý, J. 2003: *Sklepy – opravy a rekonstrukce*. Praha: Grada Publishing, 110 pp. ISBN 80-247-0221-5.
Branson, G. 2003: *Home Water and Moisture Problems: Prevention and Solutions*, Richmond Hill: Firefly Books, 144 pg. ISBN 1552978354.
Lebeda, J. et al. 1988: *Sanace zavlhlého zdiva*. Praha: SNTL.
Makýš, O. 2004: *Technologie renovace budov*, Bratislava: Jaga group, ISBN 801-8076-006-3, 262 pp.

Advances and Trends in Engineering Sciences and Technologies III – Al Ali & Platko (Eds)
© 2019 Taylor & Francis Group, London, ISBN 978-0-367-07509-5

Current state of knowledge technology in construction

P. Mesároš & T. Mandičák

Faculty of Civil Engineering, Technical University of Kosice, Košice, Slovakia

ABSTRACT: Knowledge systems and knowledge technologies are an important tool for resource efficiency in the management of construction projects. The use of knowledge technologies is essential throughout the construction period, starting with designing and ending with the building management. This paper discusses the issue of knowledge technology in construction industry. Main aim of the research was to make an overview of knowledge technologies and possibilities of using them in construction industry and defining the different levels of knowledge systems and technologies.

1 INTRODUCTION

Construction industry is a knowledge-intensive industry and an experience-based discipline (Ji-Wei et al. 2012). The traditional approach to knowledge management (KM) adopts so-called Communities of Practice (COPs) as important means of generation, sharing, exchanging, storing, and retrieving of knowledge (Wenger & Snyder 2000). Work with knowledge and tools for knowledge management will be necessary in the future. Knowledge technologies are important in every area of construction. Problem solving is related to almost all kinds of engineering services including proposal preparation, feasibility studies, architectural and engineering designs, procurement, construction supervision, and project management (Ji-Wei et al. 2012). All these areas have the potential to use knowledge for successful management of construction companies and projects. Knowledge will not bring any value unless it is used actively. A Knowledge management system is useful in actively using existing knowledge to create value. Knowledge management and knowledge technology includes process to create, secure, capture, coordinate, combine, retrieve, and distribute knowledge and tools for this purpose done (Lin et al. 2005). Construction projects are characterized by their specifics like complexity, diversity and the non-standard nature of production (Clough et al. 2000). Participants of construction project have to adapt the previous knowledge and experience quickly to face the new conditions and contents of work (Hanisch 2009). Most problems in construction project and solutions, experience and know-how are in the minds of individual engineers and experts during the construction phase of a project. An effective means of improving construction management is to share knowledge among engineers, which helps to eliminate mistakes that have already been encountered in past projects (Lin & Lee 2012).

The current trend in construction places increasing demands on technological (construction time, product quality, on-site response and so on), environmental (energy saving, potential for recycling and reuse, water consumption, waste management, CO_2 emissions, soil consumption, fuels) and social (indoor air quality, impact on the labour market, physical space, aesthetics, traffic jams) construction parameters in the context of maintaining the balance of economic efficiency and sustainability of the construction and realization of buildings. This complex and demanding process must be based on the acquisition and use of relevant multidimensional information and the availability of progressive tools and methods of information processing in the form of integrated project documentation and effective management of the planning and construction process. In the process of building preparation, the simultaneous optimization of project documentation in graphical form, time, cost and resource use is vital (Wei Feng et al. 2010).

All of the above mentioned data lead to the problem of using and enhancing a uniform methodology for the use of knowledge technologies in construction, respectively, to better understand this problem. It is important to know the current state of use of knowledge in construction and to know the level of individual options. This is also the fundamental issue of discussion and scope of exploration in this survey. The main objective of the survey is to clarify the current state and possibilities of using knowledge technology in the construction industry. Based on methodology, one step of the survey is defining the level of knowledge technology utilization. First, however, it is necessary to analyse the current state of perception of knowledge technology in construction.

2 KNOWLEDGE TECHNOLOGY IN CONSTRUCTION INDUSTRY

In spite of the construction industry is a strong, knowledge-based industry that relies heavily on knowledge input by the different participants of construction project, its nature is not conducive to effective knowledge management (Forcada et al. 2013). It is complex and heterogeneous and notorious for the level of rivalry between companies and employee migration. It operates within a dynamic and changing environment and customers are becoming more sophisticated and demand more units of construction for fewer units of expenditure (Egan, 1998). The project-based nature of construction industry also hinders effective knowledge management, since most work is carried out by one off project teams, varies between projects and is subject to time barriers so there are a few incentives to appraise performance, pass learning on and improve overall delivery (McCarthy et al. 2000).

In connection to the use of knowledge technology, the construction industry has the potential to greatly benefit from information and communication technology (ICT) tools, although only a few enterprises (mainly the largest companies) are at the cutting edge of IT use (Nitithamyong & Skibniewski 2006, Ingirige & Sexton 2007, Rivard 2004).

Knowledge technology for construction industry focuses on the obtaining and management of important knowledge and experience from previously projects and all participants of constriction project (Lin & Lee 2012). A value engineering knowledge management system was developed to make the creativity phase more systematic, more organized and more problem-focused (Zhang et al. 2009). Other groups of authors used a survey to find out how the tacit and explicit knowledge are captured, stored, shared, and used in forthcoming projects (Kivrak et al. 2008). A web-based prototype and system was developed to facilitate the live capture and reuse of project knowledge in construction (Udeaja 2008). A web-based collaborative knowledge system was introduced to solve the problem with knowledge sharing within a construction enterprise (Dave 2009). A web-based knowledge management system was introduced to support effective search of information and analysis, by enhancing communication and collaboration among researchers in the underground construction field (Forcada 2010).

The interconnectivity of global networks and the internet has presented unprecedented opportunities as well as challenges as their complexity increases. These challenges have stirred up considerable research projects in knowledge technology (Dimkovski & Deeb 2007). In spite of this, knowledge technology is still in its beginnings. Current approaches to knowledge sharing communities like recommender systems or shared ontologies often suffer from an imbalance of effort versus benefit from the individual point of view (Grather 2003). According to another study, there are general methods of knowledge sharing in construction industry. It is 22% knowledge sharing by documents. Only 22% knowledge are sharing by system usage (Lin & Lee 2012). It is very low rate. Inspite of amount approaches, researches, surveys, studies, construction industry is still unsatisfactory with the use of knowledge technologies at every stage. From a sustainable development point of view, with the support of knowledge technologies and systems, it requires rigorous identification, codification, acquisition, analysis, integration and use of existing information and knowledge as well as the development of new knowledge for the needs of building knowledge support for integrated planning buildings with an emphasis on their economic parameters. This approach covers all design dimensions, with an emphasis on optimal information inputs of economic parameters, integrating people, technology and information systems, business and economic, social, environmental and structures and processes into a process that collectively exploits the knowledge, competencies and talents of all participants maximizing resource efficiency and managing optimization at all stages of the project.

3 LEVELS OF KNOWLEDGE TECHNOLOGY IN CONSTRUCTION

Construction has its own specifics and is a unique industry in its own right. These specificities also influence to adoption of new technologies. Knowledge technologies can have different level of use or level of development (also in many literatures it stated maturity). Construction production involves a number of processes such as design, construction, facility management and more Individual processes and phases bring different possibilities and levels of use in knowledge technology.

3.1 *Web based knowledge technology in construction*

Generally, web technology is the most effective technology used in knowledge management (Wagner 2006). Web technology is based on a particular set of technologies enabling users to interact and collaborate with each other in social media (Bojars 2008). Basis for web technology using is that data can be made accessible by creating online storage of information and that it can be searched, reusedand updated as often as required (Laudon & Laudon 2000). In construction industry it is typical example for the selection of materials for specific construction works. When selecting material for the selected process (eg plastering), the knowledge system will show all the necessary materials for this construction work. It is elementary approach of knowledge technology.

3.2 *Applications and software solutions in construction*

Another approach is based on the use of applications and software solutions. Frequent interest in this approach is cost management and cost estimating in construction. It is often said about the knowledge system, which is not entirely accurate and it is possible to argue how much it can be considered as a knowledge system.

3.3 *BIM and knowledge technology*

BIM (Building Information Modelling) serves as an perfect tool for data management (Meadati & Irizarry 2010). BIMis considered to be a revolutionizing the delivery of AEC (architecture, engineering, construction) projects. BIM technology has centralized and integrated nature of design information. This can potentially provide a rich platform for capture, storage and dissemination of knowledge generated during the design and construction processes. Effective knowledge management has some requirements. Its ability in communicating and preserving knowledge effectively across several of a construction project is one of them (Dave & Koskela 2009). According to results of case study, BIM model can be effective in building knowledge system and evolving through the entire design cycle from programming through the construction stage (Deshpande et al. 2014). A lot of surveys say about pattern model of BIM and knowledge framework. It's conception of BIM technology and use of knowledge base for one purpose. But perception of BIM technology and knowledge technology it may not be just as shown in Figure 1.

The BIM approach retains knowledge in a digital format, facilitating easy updating and transfer of knowledge into the 3D BIM environment. A 3DBIM-based knowledge model is designed to be easily integrated with experience-based information and 3D objects of the model (Shu-Hui et al. 2013). But the problem is different. However, the BIM model does not have the full elements and characteristics of the knowledge system. Especially if we talk

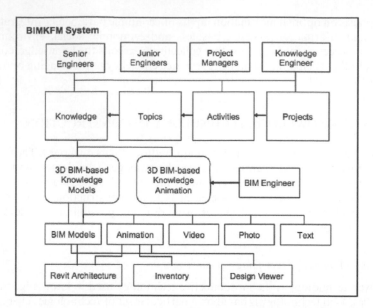

Figure 1. The concept and framework of the 3D BIM-based knowledge models (Shu-Hui et al. 2013).

Figure 2. Augmented reality in construction industry.

about 5D and consider the cost parameter too. When designing and selecting a design element, the system is still only partially linked and semi-automated to provide the necessary information (for example, in connection with the budget, etc.). The system is still not able to respond fully to any changes and automatically perform all the costs in terms of cost estimating, bill of quantities and choosing the other building elements needed for perfect design.

3.4 Augmented reality and knowledge technology

Augmented reality (AR) systems allow fast access to information and helps managers to decide on corrective actions to minimize cost and delays due to performance discrepancies (Bae & Fard 2013). Augmented reality is as the expansion of information from the physical world that allows us to perceive the virtual part that surrounds us. It enables us to distinguish the digital information flowing through the physical space. It creates a sensory interface that allows us to appreciate the virtual part that is hidden in physical space. The widespread of application of AR technology helps to the dissemination of knowledge and decision-making tool for managers (Mesárošová & Ferrer 2015).

With the development of technologies, especially smartphones and mobile applications, there is a new space for the development and use of augmented reality in this area. The use of the augmented reality technology in the construction industry is becoming a reality with innovative specific platforms and applications, for example it is AR EMS (Wang et al. 2006), UM-AR-GPS-ROVER (Behzadan & Kamat 2006), etc.

There are numerous environments of application of these technology that requires mobility of the user, needs of access to the information anytime and anywhere, in these cases it becomes necessary the use of mobile devices (Izkara et al. 2016).

3.5 *Knowledge-based approach and machine learning approach in construction industry*

According to some specialist in this field, main difference between knowledge-based approach and machine earning approach is in different way of data collection. Learning approach use more statistic tools and results shown also new trends (Bupe 2016). Machine learning has become pervasive in multiple domains, impacting a wide variety of applications, such as knowledge discovery, data mining, natural language processing, information and so on. Machine learning using in construction industry can be considered key in the coming years.

4 CONCLUSION

The issue of knowledge technologies in the construction industry is a highly topical issue. Knowledge technology and systems offer great opportunities in construction industry. These options are much greater than currently used. This survey gives an overview of the basic possibilities and levels of use of knowledge technology in construction. This paper highlighted the possibilities of using knowledge technology and the survey also points to the untapped potential in this area. Exact research findings require more extensive research in this area. The need for quantification of knowledge technology is indisputable. This survey and summarized results in the field of knowledge technology are a good first step towards extensive research in this area. This overview and individual levels of knowledge technology is a good starting point for setting further research goals, in particular the feasibility study of building pretensions in individual countries in the context of knowledge technology and the quest for quantifying the current use of these systems in individual construction markets.

ACKNOWLEDGEMENT

The paper presents a partial research results of project APVV-17-0549 "Research of knowledge-based and virtual technologies for intelligent designing and realization of building projects with emphasis on economic efficiency and sustainability" and project VEGA 1/0828/17 "Research and application of knowledge-based systems for modeling cost and economic parameters in Building Information Modeling".

REFERENCES

Bae, H., Golparvar-Fard, M., & White, J. 2013. High-precision vision-based mobile augmented reality system for context-aware architectural, engineering, construction and facility management (AEC/FM) applications. *Visualization in Engineering.* 1(1):1–8.
Behzadan, A.H., &Kamat, V.R. 2006. Animation of Construction Activities in Outdoor Augmented Reality. *Proceedings of the Joint International Conference on Computing and Decision Making in Civil and Building Engineering, American Society of Civil Engineers.* Reston.

Bojārs, U. et al. 2008. Using these mantic web for linking and reusing data across Web 2.0 communities. *Web Semantics: Science, Services and Agents on the World Wide Web*. 6:21–28.

Bupe, C. 2016. Whaz is the difference between a knowledge based approach and machine learning approach for sentiment analysis. *Quora*.

Clough, R.H., Sears, C.A. & Sears, S.K. 2000.*Construction Project Management*. 4th ed. Wiley, New York.

Dave, B. & Koskela, L. 2009. Collaborative knowledge management—a construction case study. *Journal of Automation in Construction*. 18(7):894–902.

Deshpande, A. et al. 2014. A framework for a BIM-based knowledge management system. *Procedia Engineering. Creative construction conference.*85:113–122.

Dimkovski, M. & Deeb, K. 2007. Knowledge technology through functional layered intelligence. *Future Generation Computer Systems*. 23:295–303.

Feng, CH.W. et al. 2010. Using the MD CAD model to develop the time–cost integrated schedule for construction projects. *Automation in Construction*. 19.

Forcada, N. 2013. Knowledge management perceptions in construction and design companies. *Automation in Construction*. 29:83–91.

Forcada, N. et al. 2007. A web-based system for sharing and disseminating research results: the underground construction case study. *Journal of Automation in Construction*. 19(4):458–474.

Grather, W, Klockner, K. & Kolvenbach, S. 2003. Community support and awareness enhancements for cooperative knowledge generation. *Euromicro.*165.

Hanisch, B. et al. 2009. Knowledge management in project environments. *Journal of Knowledge Management*. 13(4):148–160.

Ingirige, B. & Sexton, M. 2007. Intranets in large construction organizations: exploring advancements, capabilities and barriers. *The Electronic Journal of Information Technology in Construction*. 12: 409–427.

Izkara, J.L. et al. 2016. Mobile Augmented Reality, an Advanced Tool for the Construction Sector.

Jan, S.H.et al. 2013. Applications of Building Information Modeling (BIM) in Knowledge Sharing and Management in Construction. *International Journal of Civil and Environmental Engineering*. 7(11).

Ji-Wei, W. et al. 2012. An integrated proactive knowledge management model for enhancing engineering services. *Automation in Construction*. 24:81–88.

Kivrak, S. et al. 2008. Capturing knowledge in construction projects: knowledge platform for contractors. *Journal of Management in Engineering*. 24(2):87–95.

Laudon, K.C. & Laudon, J.P. 2000. *Management Information Systems*. New Jersey: Prentice Hall. 2000.

Lin, CH. & Lee, H.Y. 2012. Developing project communities of practice-based knowledge management system in construction. *Automation in Construction*. 22:422–432.

Lin, Y.C. et al. 2005. Enhancing knowledge & experience exchange through construction map-based knowledge management system. *Construction Research Congress 2005. Proceeding of Congress*, San Diego, 1–10.

McCarthy, T.J. et al. 2000. Knowledge Management in the Designer/Constructor Interface. *Proceedings of the 8th International Conference on Computing in Civil and Building Engineering*. Reston, VA, 279:836–843.

Mesarosova, A. & Ferrer Hernandez, M. 2015. Art behind the mind. Exploring new artforms by implementation of Electroencephalography. *Cyberworlds 2015 Conference Proceedings*. Gotland. Sweden.

Nitithamyong, P. & Skibniewski, M.J. 2006. Success/failure factors and performance measures of web-based construction project management systems: professionals' viewpoint. *Journal of Construction Engineering Management*. 132(1):80–87.

Rivard, H. et al. 2004. Case studies on the use of information technology in the Canadian construction industry. *ITcon*. 9:19–34.

Udeaja, C.E. et al. 2008. A web-based prototype for live capture and reuse of construction project knowledge. *Journal of Automation in Construction*. 17(7):839–851.

Wagner, C. 2006. Breaking the knowledge acquisition bottleneck through conversational knowledge management. *Information Resources Management Journal*. 3:70–83.

Wang, X. et al. 2006. Mixed Reality—Enhanced Operator Interface for Teleoperation Systems in Unstructured Environment. *Proceedings of the 10th Biennial ASCE Aerospace Division International Conference on Engineering, Construction and Operations in Challenging Environments, Society of Civil Engineers (ASCE)*. League City/Houston, Texas, 8.

Wenger, E.C. & Snyder, W.M.2000. Communities of practice: the organizational frontier. *Harvard Business Review*. 110:139–145.

Zhang, X. et al. 2009. Developing a knowledge management system for improved value engineering practices in the construction industry. *Journal of Automation in Construction*. 18(6):777–789.

Impact of innovation on the revenues and profit of construction companies in Slovakia

P. Mesároš, T. Mandičák & K. Krajníková
Faculty of Civil Engineering, Technical University of Kosice, Košice, Slovakia

ABSTRACT: Revenue and profit are the key performance indicators for businesses performance. The main aim of the business is to make a profit. Innovations should contribute to this aim. Innovations are also of great importance in construction industry. This paper discusses the issue of innovation in construction industry and its impact on revenues and profit of construction companies in Slovakia. Main aim of the research was to analyse the impact of innovation on the revenues and profit of construction companies in Slovakia. Main assumption of the research was set as the statement that revenues increase together with the growth of profit construction companies due to innovation. The research sample included all participants of construction project.

1 INTRODUCTION

Currently, innovation is a relatively often used term in the field of construction. Construction companies are trying to adapt to the market situation, which is constantly introducing new technologies, using new materials, and new products and services are becoming more flexible (Sičáková et al. 2017). Implementation of innovation is one of the ways to maintain a competitive market position or to improve the market position (Kotler & Trias 2015). Improving business performance is a priority for every business. Improving the performance of an enterprise by taking its indicators into account as revenue and profit is the primary objective of doing business in each area. Tracking the new trends in the construction market and continually improving is therefore very important. At present, innovation is the foundation of almost every construction company. For this reason, there is a social need to discuss this issue and examine the legality and impact of implementing innovation. An important part of the implementation of innovation is its management and continued effectiveness control. The performance in construction industry is affected by national economies (Navon 2005). However, on the other hand, there are many other factors that affect the business performance. Innovation is one of the key tools for increasing the competitiveness of businesses. Detailed examining their impact on business performance is more than desirable.

2 CURRENT STATE AND THEORETICAL BACKGROUND

2.1 *Innovation in construction industry*

The concept of innovation was introduced by American scientist Schumpeter in the late 1930s (Čimo 2010). According this scientist, innovation is a change in order to use new types of consumer goods, new production and transport means, new markets and forms of organization of production and services (Kováč 2003). In the context of construction industry, innovation can be defined as the first use of a new technology within a construction company (Tatum 1987). On innovation are known another point of view. For example, innovation can be as the translation of knowledge into physical production that is significantly affected by the level of talent at all participants of projects and holders of a venture (Arditi 1982). This view is confirmed by another description of innovation. The concept of innovation is from the Latin word "innore", which means renewing, making new. I can be say that the process of innovation is a process from idea itself to adoption, acceptance of change, realization of something new (Slaugter 1998). Even by other authors created controversy in innovation, whether it is a scientific term or art to accept the risk (Kuczmarski 1996). In any case, innovation is the use of something new for the first time, whether by a specific company or by a society. In the context of research problem, for innovation can be considered each change and implementation something new in company for purpose increasing efficiency or performance. Innovations can be divided to four groups. There are product innovation, innovation of process, marketing innovation and management innovation (Schumpeter 1911). Product innovation is in increasing qualitatively features of products (Tatum 1989). Process innovation is reflected in improving and streamlining production processes. It is a system that unifies the processes, processes and organization of a company in order to achieve the required change of business for the better. Innovation of the process can lead to a change in people's qualifications, and at the same time leads to better management of the organization (Čimo 2010). In the context of the construction industry, other author breaks down the spectrum of innovation into five types (Slaugter 1998). There are incremental, modular, architectural, system, and radical. Building innovations also concern new, progressive technologies. Technology Adoption Life Cycle is a model for understanding the acceptance of new technology (Smallwood 1973). It's been explored by Smallwood. Technology innovations play important role in construction industry. The rate of innovation in the construction equipment industry increased during the last 30 years. This increased rate of innovation can belinked to pressures generated by buyers' behaviour and to technological developments in the equipment industry as well as in other industries (Tangkar & Arditi 2000). Progressive technology includes all types, not only construction equipment. It's information and communication technology too. Especially this part from this research point of view is interesting. Increasing of information and communication technology impact on productivity levels and developing of managerial competencies of managers (Bolek et al. 2016). Innovations in this area are very important for company performance. Generally, information and communication technology for purpose of management are important part of technology innovation.

2.2 Business performance in construction industry and measurement

Increasing the performance of each business process reflects the company more competitive, better economic benefits, higher profits and a better position on the labour market. The concept of enterprise performance is important to understand as the ability of an enterprise to make the best possible return on investments that are embedded in business activities. Performance is the value of a business, and if we need to increase it, we need to increase business performance (Hudymačová & Hila 2016). From this perspective, it is therefore necessary to measure business performance. Business performance management (BPM) is an integrated set of methods, processes, metrics, and applications designed to manage the financial and operational performance of enterprise (Turban et al. 2008). Business performance management includes target management, performance management, resource management and process interface management (Balaban et al. 2011). Performance management includes receiving regular feedback on the process outputs and monitoring the actual performance by measurement dimensions set in targets. Accounting measures of performance have been the traditional abutment of quantitative approaches to business performance measurement (Otley 2003). Implementing the performance implementation plan and monitoring enhanced progress leads to continuous business process performance enhancements (Devis & Brabander 2009). Measurement of business performance is difficult process. First of all, it's necessary selection of business performance indicators.

Many researches and studies are conducted to determine so-called Key Performance Indicators (KPIs). Most of them are project specific. They focus on the performance measurement at the project level in construction. Existing research, which has been discussed for performance evaluation and comparability at the company level, is limited in the literature.

Morcover, most of the researches have been developed KPIs that are suitable for specific national features (Elshakour, et al. 2013). However, many studies point to differences in business performance measurement of business performance. According one of studies, profitability, productivity nad return on capital employed are important KPIs in construction industry for purpose of business performance measurement (Detr 2000). Schedule performance and cost performance are KPIs according another study (El-Mashaleh et al. 2007). Profitability is the most important indicator of business performance measurement according to group of authors from USA (Wang, et al, 2010). Other studies shown on indicators of growth, as revenues, profit. There are typical financial objectives of company. Classic financial performance indicators are absolute value of profit, cash flow and ratios as a measure of liquidity, activity, indebtedness, profitability, market value, efficiency and cost (Ďurkáčová & Kalafusová 2012). According to many studies, profit and revenues are basic indicators of business measurement. This was confirmed by the results of the study, where the researchers conducted a ranking of performance indicators (Elshakour, et al. 2013). Growth achieved third the biggest value in ranking. Very interesting point of view it was in this study on relationship between innovations and business performance. Measurement of business performance is possible through innovations in company. On this basis, it is also very important to measure the impact of innovation on business performance in the construction industry.

3 RESEARCH METHODOLOGY

3.1 *Research indicators, hypothesis of research and data collection*

The determination of so-called key performance indicators is an initially important step in establishing a performance measurement or impact of innovation on business performance. These indicators, when identified and implemented properly, can play important role in providing information on the performance of construction companies and innovation impact on it. Based on theoretical analysis it was set main researches indicators as profit and revenues in construction companies in Slovakia. Consequently, the basic research problem of the given issue is also conceived. That's assumption, that implementation of innovations in construction companies (as a new information and communication technology) has a positive impact on increasing of business performance indicators. But, it's so generally statement. Research problem is more specified. If revenue increases, this does not automatically mean profit growth. Therefore, if we want to prove that innovation really has a positive impact on the growth of corporate performance and above all profit, this has to be correlated. On the basis of this clarification, it is possible to establish a basic research hypothesis as follows:

H: With the growth of revenues of construction companies also increasing the profit of construction companies due to the implementation of innovations.

The data collection was done by questionnaire survey form. The questionnaire is one of the most common methods of research use. It is used for mass and faster detection of facts, attitudes, values, opinions, etc. The questionnaire contained simple and comprehensible questions about the innovation of new information and communication technologies in construction companies. The structure of the questionnaire was basic information about the respondents. Another part of the questionnaire had direct questions about the research problem. It mainly concerned the implementation of innovation in the form of new information and communication technologies in construction industry. Respondents used the Likert Scale for the quantification of impact in the main research questions (1 – low impact level on performance indicators, 2 – high impact level on performance indicators). Survey performance indicators (as revenues and profit) were quantified before and after innovation. Main aim of research was to analyse impact of innovation on revenues and profit of construction companies in Slovakia.

This research discusses in particular process innovation. The construction industry is generally more focused on process innovation. The construction process often talks about the product itself (buildings). Properly adjusting process innovations can be a good way for better products and thus for building companies generally. Research issues have also been focused on improving and implementing process innovations in construction companies trough implementation of progressive information and communication technology.

3.2 *Research sample*

The survey sample consisted of construction companies operating in Slovakia. Research sample consist participants of construction projects (investor, contractor, designer). Construction companies were selected by random selection, the sample being drawn based on the composition of the companies selected in the construction industry. Return rate was 5.25%, it means 66 construction companies, but only 52 companies did implementations for examine period. There are real research sample. Research sample so includes 9% of large companies, 18% medium sized companies and 73% of small and microenterprises.

3.3 *Data processing*

It was use a very common test called Shapiro-Wilk test to determine if the null hypothesis of normality is a reasonable assumption regarding the population distribution of a random sample. The desired significance level alpha was used 5%. The desired hypotheses:

Research sample does have normal distribution (with unspecified mean and variance) or research sample does not have normal distribution. Computation of this test was done in Matlab. It was tested differences between each measurement (Revenues before implementing minus Revenues after implementing). These results were obtained:

It also was tested differences between each measurement (Profit before implementing minus Profit after implementing). These results were obtained:

Subsequently, the Mann-Whitney test was used with these results. As it was mentioned previously, the null and two-sided research hypotheses. Based on Mann-Whitney test it can be possible confirm or not confirm statement about correlation and relationship between revenues, profit and innovation impact on this.

4 RESULTS AND DISCUSSION

The impact of innovation on construction companies may vary. On the one hand, it can be positive impact on one of business performance indicator. On the other hand, it may not be true of another. This situation is often possible in relationship between cost and revenues. In this case, the effect of innovation is not as significant as it would be. Therefore, it is also necessary to examine the relationship between revenue and profits. Based on Mann-Whitney test it was statistically significant that revenue and profit growth is related. The downgraded statistical results are in Table 3.

In both cases the result is significant on desired significance level. The mean of ranks after implementation is higher than before implementation, so revenues and also profit was increasing. Based on this facts, it can be confirm hypothesis and final statement. With the

Table 1. Results of research sample distribution (Revenues).

	Revenues before implementation	Revenues after implementation	Revenues – Differences
Mean	3.153846	3.769231	0.615385
Standard Deviation	0.871884	0.581256	0.889015
Number of respondents	52	52	52
Skewness	0.428192	0.058986	−0.88613
Kurtosis	−0.36419	−0.31134	1.480112
p value			0.0001

Table 2. Results of research sample distribution (Profit).

	Profit before implementation	Profit after implementation	Profit – Differences
Mean	2.961538	3.6922308	0.730769
Standard Deviation	0.592818	0.543715	0.717167
Number of respondents	52	52	52
Skewness	0.00695	−0.07207	−0.87343
Kurtosis	0.00325	−0.59284	0.926115
p value			0.00001

Table 3. Results of Mann-Whitney U test for Revenues and Profit.

	Revenues differences	Profit differences	Revenues BI*	Revenues AI**	Profit BI*	Profit AI**
Sum of ranks	2632	2828	2154	3306	1956	3504
Mean of ranks	50.62	54.38	41.42	63.58	37.62	67.38
Expected mean of ranks	52.5	52.5	52.5	52.5	52.5	52.5
U-value	1450	1254	1928	77	2126	578
Z-score	−0.63387		−3.74143		−5.02867	
p value	0.5287		<.05		<.00001	
Expected U-value	1352		1352		1352	
Expected sum of ranks	2730		2730		2730	

*BI – Before implementation.
**AI – After implementation.

growth of revenues of construction companies also increasing the profit of construction companies due to the implementation of innovations. New information and communication technologies in the management of construction companies have an important and positive effect. These innovations increase not only revenue but also profit. Research also showed the dependence between implementation and revenue generation. It is important to say that the companies did not make any other innovations at the time. The impact on revenue and profit was also not caused by seasonality as the same period was examined.

5 CONCLUSION

The issue of innovations in construction industry and its impact on some business performance is really important. The analysis of this issue brings new trends and information for construction industry. Based on the research sample, it was proven that increasing and implementation process innovations trough progressive information and communication tools in construction industry has positive impact on business indicators. Impact of innovation on business performance indicators like revenue and profit was statistically significant. The relationship between revenue and profit was also confirmed. Innovations have had an impact not only on revenue but also on profit, which is very important. The profitability level was more pronounced after the implementation of innovation, which was also the main motivator and prerequisite for these innovations. Research has proven the merits of implementing new information and communication technologies to the construction companies in Slovakia. Despite the statistical evidence of the survey results, it is necessary to consider whether the results of the research can be generalized across the whole construction industry. It is necessary to say that the construction industry and the studied construction market in Slovakia are relatively small. These results certainly indicate the trend in the country. It opens there is scope for exploring this region and other markets in other countries.

ACKNOWLEDGEMENT

The paper presents a partial research results of project APVV-17-0549 "Research of knowledge-based and virtual technologies for intelligent designing and realization of building projects with emphasis on economic efficiency and sustainability" and project VEGA 1/0828/17 "Research and application of knowledge-based systems for modeling cost and economic parameters in Building Information Modeling".

REFERENCES

Abd Elshakour, H. et al. 2013. Indicators for measuring performance of building construction companies in Kingdom of Saudi Arabia. Journal of King Saud University – Engineering Sciences. 25: 125–134.

Arditi, D. 1982. Diffusion of network planning in construction. Journal of Construction Engineering and Management. ASCE. 109(1): 1–13.

Balaban, N. et al. 2011. Business Process Performance Management: Theoretical and Methodological Approach and Implementation. Management Information Systems. 6: 3–9.

Bolek, V. et al. 2016. Factors affecting information security focused on SME and agricultural enterprises. Agris On-line Papers in Economics and Informatics. 8(4): 37–50.

Čimo J. 2010. Inovačný manažment. Bratislava: Ekonóm.

Department of the Environment, Transport, and the Regions (DETR). 2000. KPI Rep. for the Minister for Construction. KPI Working Group. London.

Devis, B. & Brabander, E. 2009. ARIS Design Platform – Getting Started With BPM. Berlin: Springer.

Durkáčová, M. & Kalafusová, L. 2012. Traditional and modern approaches to company performance evaluation. The 15th International Scientific Conference Trends and Innovative Approaches in Business Processes.

El-Mashaleh, M. & Minchin, R. & O'Brien, W. 2007. Management of construction firm performance using benchmarking. Journal of Management and Engineering. 23(1): 10–17.

Hudymačová, M. & Hila. M. Business performance. Internetový časopis o jakosti. 2018.

Kotler, P. & Trías, Winning At Innovation The A-to-F Model, Palgrave Macmillan UK

Kováč, M. 2003. Inovácie a technická tvorivosť. Košice: Technical university of Košice.

Kuczmarski, T.D. 1996. Innovation: Leadership Strategies for the Competitive Edge. NTC Business Books, Lincolnwood.

Mesároš, P. & Mandičák, T. 2017. Impact of ICT on performance of construction companies in Slovakia, IOP Conference Series: Materials Science and Engineering: WMCAUS 2017. Bristol: IOP Publishing, 24: 1–9

Navon, R. 2005. Automated project performance control of construction projects. Automation of Construction. 14: 467–476.

Otley, D. 2003. Management control and performance management: whence and wither? The British Accounting Review. 35: 309–326.

Schumpeter, J.A. 1911. The Theory of Development. Harvard University Press, Cambridge, MA, 1911.

Sičáková, A. et al. 2017. Long-term properties of cement-based composites incorporating natural zeolite as a feature of progressive building material. Advances in Materials Science and Engineering. 1–8.

Slaughter, E.S. 1998. Models of construction innovation. Journal of Construction Engineering and Management, ASCE. 124(3): 226–231.

Tangkar, M. & Arditi, D. 2004. Innovation in the construction industry. Dimensi Teknik Sipil, 2(2): 96–103.

Tatum, C.B. 1987. Innovation on the construction project: a process view. Project Management Journal, PMI. 18(5): 57–67.

Tatum, C.B. 1989. Organizing to increase innovation in construction firms. Journal of Construction Engineering and Management, ASCE. 115(4): 602–617.

Turban, E. et al. 2008. Information Technology for Management: Transforming Organizations in the Digital Economy.

Wang, O. & El Gafy, M. & Zha, J. 2010. Bi-level framework for measuring performance to improve productivity of construction enterprises. Construction. Res. Congr. 2: 970–979.

Advances and Trends in Engineering Sciences and Technologies III – Al Ali & Platko (Eds)
© 2019 Taylor & Francis Group, London, ISBN 978-0-367-07509-5

Differences in adhesive properties of joint sealants tested in laboratory and in-situ – effect of weathering

B. Nečasová, P. Liška & B. Kovářová
Brno University of Technology, Brno, Czech Republic

ABSTRACT: Every building and/or structure has to withstand a lot of environmental changes during the year as well as its service life. Therefore, certain attention should be given to the application of sealants in buildings where adhesive properties for the assembly of two substrates are an important indicator of the upcoming durability. A group of test samples was subjected to two normalised test procedures which may influence the resulting behaviour of the joint in the exterior. The second group of test samples was exposed to the external environment for a particular period. The obtained results show that the standardised methods are able to simulate an outdoor environment only to a certain level, since the laboratory environment is clean and dust-free, while it is not possible to ensure the same conditions in-situ. In some cases it was monitored that some sealants tested in an external environment aged rapidly compared to the ones cured and stored in the laboratory.

1   INTRODUCTION

1.1   *State of the art*

Every building and/or structure has to withstand a lot of environmental changes during the year and its service life. These changes may range between very mild to relatively extreme based on the geographical location (Chew & Zhou, 2002), (Chew, 2004), (Ding & Liu, 2006), (Ding et al., 2006). Even a minor change can have a significant effect on each component of the building if not designed and implemented properly (Yun et al., 2011), (Mirza, 2013). Certain attention should be given to the application of sealants and adhesives in constructions and buildings where adhesive properties for the assembly of two substrates are an important indicator of the upcoming service life (Pantaleo et al., 2012), (Ihari, 2014). When determining the durability and performance of joint sealants, test samples prepared in the laboratory usually have to be cured for a period which varies from 14 to 28 days before the tests are to be conducted (Chew & Zhou, 2002). This means that all samples are cured statically and uninfluenced by any form of stress, strain, or weathering. However, joint sealants installed in a real environment have to be able to meet the conditions immediately after their application, therefore, the curing procedure can be classified as dynamic (Chew, 2001) with an immediate interaction of forces and weathering effects which can accelerate the occurrence of premature failures. The conventional test methods stipulated in many national standards may be meaningless if field-installed sealants are to fail even before they are put into service (Chew, 2001). The driving force behind this case study is the necessity of stipulating a new testing method that would be able to verify curing characteristics of sealants and adhesives moulded in field, a method which will establish a suitable compromise between the existing laboratory test methods and the currently non-existent in-situ testing methods.

2   METHODOLOGY

The authors of the presented paper believe that contemporary test methods not only do not reflect unexpected weather changes, but they also do not prescribe a pertinent shape and size of test samples. Therefore, a design of testing sample of such a shape that would correspond as much as possible to the real implementation of the sealed joint was proposed (Nečasová, 2017). Subsequently put through laboratory tests, and through a set of non-standardised methods that would verify the impact of the external environment on its mechanical properties.

2.1 Materials

The cement-bonded particleboard, known as Cetris Basic, used in this study was commercially available façade board without any surface treatment supplied by CIDEM Hranice, a.s. The tested product was formulated with 63% wood chips, 25% Portland cement, 10% water and 2% hydration additives (Hranice, a.s., 2014). A thickness of 10 mm was used in the presented study. This type of cladding was selected mainly because of reported problems with adhesion.

The use of polymeric sealants for the filling of joints or cracks has been widely accepted and considered as most effective (Pattanaik, 2013). Polyurethane sealants have been generally used for sealing since they can offer a relatively long service life and excellent adhesion to a wide range of materials. However, polyurethane sealants and adhesives contain isocyanates which raise severe toxicity issues. Therefore, three different types of high-strength sealants from a group of silyl modified polymers supplied by the same manufacturer were selected and tested, see Table 1. Modified polymers are considered to be more environmentally friendly with similar strength and elongation properties as polyurethanes, and according to the manufacturer, all selected sealants are a good compromise between sealant and adhesive.

2.2 Specimen preparation

Based on previous experiences (Nečasová et al., 2014), (Nečasová et al., 2016), a new geometry of testing sample was created for this purpose, allowing the testing of a so-called 'real joint'. The real joint is a term which was already precisely described by (Nečasová et al., 2017) and, according to the opinion of the authors, suitably reflects the applied geometry of the façade cladding element as well as the procedure of their assembling. The produced Cetris boards were about 3.35 m long and 1.25 m wide. The boards were cut to obtain bonding samples of geometry 40.0 mm wide and 160.0 mm long. Each test specimen was composed of two parts. The preparation of the samples and the subsequent application of the sealant was carried out according to the principles and procedures specified by the producers of the individual system.

2.3 Conditioning

The test specimens were subjected to different methods of conditioning with the aim of influencing the resulting behaviour of the joint, see Table 2.

Table 1. General material properties of selected sealants.

Material property	Type I	Type II	Type III
Tensile stress at break	2.3 MPa	2.5 MPa	2.6 MPa
Shore A	c. 52	c. 45	c. 55
Elongation at break	c. 250%	c. 250%	c. 250%
Application Temperature	5 – 35°C		
Skinforming	approx. 15 min.	approx. 10 min.	approx. 15 min.

Table 2. Exterior air temperature changes during the monitored period.

Year	2016										2017			
Month	03	04	05	06	07	08	09	10	11	11	01	02	03	04
Min [°C]	−3	−2	1	9	8	8	3	0	−6	−11	−17	−9	−2	−2
Max [°C]	21	24	27	33	35	30	30	24	14	10	5	12	22	23

The main purpose of the laboratory testing was to provide high-quality experimental data for later comparison with data measured in the exterior. Some of the samples were subjected to conditioning based on two standardised methods that simulate weather changes. The principle of the first method (Czech-Standards, 1983) is the alternate heating of test samples with infrared lamps to a temperature of $(70 \pm 2)°C$ and their subsequent cooling with a water shower to a temperature of $(23 \pm 2)°C$. The second method (Czech-Standards, 1981) is based on the alternate freezing to a temperature of $(-22 \pm 2)°C$ and the subsequent thawing by immersion into a water bath at a temperature of $(23 \pm 2)°C$. Some of the test samples were left in the laboratory environment (at $20 \pm 2°C/ 50 \pm 5\%$ humidity) for a specific period without additional conditioning, see Table 3.

In the case of the in-situ tests, all test samples were prepared and left in the same environment for a specific period, see Table 4. These samples were 'conditioned' by real weather changes. Some of the samples were conditioned and monitored for a period of one year. The changes of air temperature were recorded to be able to compare them with conditioning methods that were performed in laboratory, see Table 2.

2.4 Test procedure

Before the testing itself, the measuring of the width of the real joint was performed in all samples. Subsequently, the test samples were placed into moulds for the testing of normal stresses and extended at a speed of approximately (5.5 ± 0.7) mm/minute. The testing was performed in the following steps:

– Extension to 100% of the original width;
– Studying of changes at 100% extension for the period of 24 hours (if possible);
– After 24 hours, extension was conducted until sample failure.

Table 3. Test samples tested under normal stress after standardised conditioning.

Tested sealant	Average max. elongation at break [mm]	Average tensibility Δ [%]	Failure mode [–]*	Percentage failure rate [%]
After 14 days – Reference Standard				
Type I	18.43	63.18	AF; CF	75; 25
Type II	21.83	89.91	CF	100
Type III	18.43	63.18	AF; CF	75; 25
After conditioning – Resistance to Temperature Variations				
Type I	16.83	46.39	AF	100
Type II	20.87	74.80	AF; CF; A/CF	33; 33; 34
Type III	13.07	42.51	AF; A/CF	33; 77
After conditioning – Frost Resistance				
Type I	15.23	29.99	AF	100
Type II	19.91	60.11	AF; CF; A/CF	33; 33; 34
Type III	12.07	32.23	AF; A/CF	33; 77
After 6 months in laboratory				
Type I	18.39	61.41	CF	100
Type II	20.07	75.84	CF	100
Type III	20.03	59.78	AF; CF; A/CF	33; 33; 34
After 12 months in laboratory				
Type I	19.46	73.02	AF; CF; A/CF	33; 33; 34
Type II	22.87	100.00	CF; FTF	75; 25
Type III	19.38	73.16	AF; CF	25; 75

*AF – Adhesive Failure; CF – Cohesive Failure; A/CF – Combination of Adhesive and Cohesive Failure; FTF – Fibre-Tear Failure.

Table 4. Test samples tested under normal stress after 6 and 12 months of curing in the exterior.

Tested sealant	Average max. elongation at break [mm]	Average tensibility Δ [%]	Failure mode [–]*	Percentage failure rate [%]
After 14 days – Reference Standard				
Type I	17.69	54.20	AF; CF	75; 25
Type II	18.42	63.21	AF; CF	75; 25
Type III	15.56	33.11	AF	100
After 6 months				
Type I	12.74	34.71	AF	100
Type II	15.14	50.18	AF	100
Type III	12.07	32.33	AF	100
After 12 months				
Type I	13.42	42.55	AF	100
Type II	17.56	53.19	AF	100
Type III	12.21	23.28	AF	100

*AF – Adhesive Failure; CF – Cohesive Failure.

Stretching was applied until 100% elongation of the original length was achieved, that is twice the width of the joint in the case of normal extension. Samples prepared in this manner were left in the testing mould for a period of 24 hours. If no failure of the sample or the sealed joint was observed during these 24 hours, a further extension was commenced until sample failure. During testing all failure modes were recorded, see Tables 3 and 4.

3 RESULTS AND DISCUSSION

The relative elongation of the tested joint was calculated to determine the maximum tensibility of the tested sealants according to Eq. 1. The relative elongation represents a ratio of the width change of the joint to its original dimension. Tensibility was expressed in a percentage of the original width of the sealed joint according to the equation (1), see Tables 2 and 3.

$$\delta = \frac{\Delta l}{l} \times 100\%$$ (1)

where δ is tensibility (%), Δl is a change of joint width in mm and l is the original joint width in mm.

One of the objectives of this research was not just the measuring of force in the application of which failure of the joint would occur, but the monitoring of the failure modes and the measuring of the maximum possible elongation of tested sealants. A scale of seven different failure modes of the real joint, based on the ČSN ISO 10365 standard (Czech-Standards, 1995) and international ASTM D 5573 standard (ASTM International, 2012), was adopted. However, as can be seen in Tables 3 and 4 only three different failure modes were monitored overall, besides one exception; see the results of the sealant Type II after 12 months in the laboratory, where a small fibre-tear failure was recorded. Such small differences in observed failure modes were quite unexpected since the Cetris cladding is composed of small particles (chips) that are only connected by cement, and, according to the previous experience with this type of material, the falling off of chips is very common. Therefore, it can be stated that even untreated and raw edges of Cetris are, from this point of view, suitable for sealing.

The presented data also show that there are considerable differences between the observed failure modes of sealants tested in the laboratory and in the exterior, even though, the same preparation and application procedures were followed. Already during the preparation of test samples in the exterior, it was observed that it is complicated to properly remove all free particles and dust. While the two parts of the test samples prepared in laboratory could be

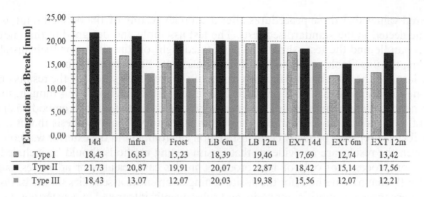

	14d	Infra	Frost	LB 6m	LB 12m	EXT 14d	EXT 6m	EXT 12m
Type I	18,43	16,83	15,23	18,39	19,46	17,69	12,74	13,42
Type II	21,73	20,87	19,91	20,07	22,87	18,42	15,14	17,56
Type III	18,43	13,07	12,07	20,03	19,38	15,56	12,07	12,21

Figure 1.   Comparison of measured data.

cleared separately to adhere to the original installation of façade panels as much as possible, the components subjected to testing in the exterior were cleaned when already installed in the testing mould, therefore, a space of only 10 mm (width of the joint) was left for cleaning.

Moreover, while samples that were cured and tested in an indoor environment, see Table 3 were usually broken by a combination of three types of failure, i.e. adhesive, cohesive and the combination of both failures, the samples prepared, cured and tested in the outdoor environment were broken by adhesive failure in c. 80% of the time. Therefore, it is evident that the tests performed in the laboratory are able to simulate the real environment only to a certain level. The conditioning methods presented in this study are compared to the real data more ruthlessly, see Figure 1 and the samples were probably subjected to a more expressive impact, however, it is said (Chew & Yi, 1997), (Lacombe, 2006), that real joint sealants have to withstand more than 140 cycles during the year which is almost six times more than what was tested here.

It was also noted from the data obtained that, after conditioning, all tested joints were contracted, in the case of Type III, more often than in other monitored cases. Also the achieved elongation was much lower compared to the other samples in which the conditioning was not conducted. The reduced elasticity of the sealant due to the temperature changes was discovered in all tested cases. These results confirmed another hypothesis, namely, that the properties of sealed joints lose their elastic properties due to aging and the impact of the environment. It is probable that a longer impact of the external environment changes could have a more significant impact on the elasticity of the materials tested.

Since all tested sealants were modified polymers and, according to the information from the manufacturer, see Table 1 should possess the same extension properties, similar results in maximum elongation were expected. However, from the presented comparison, see Figure 1 it is evident that only Type II has performed quite uniformly no matter which conditioning method was tested. On the other hand, the results obtained show that all tested sealants were not extended to even 100% of their own length, which might be a demonstration of very poor elongation properties. Moreover, the width of the tested joint was 10 mm, therefore, as was explained above, an elongation or contraction of approx. 40% of the joint width might be expected for real installations. Unfortunately, none of the tested sealants would be able to withstand such cyclical stresses. Therefore, it is possible to note that these materials are not suitable for the filling of so-called active joints and different type of sealant should be used, e.g. polyurethane, or sealant from a different manufacturer might be more suitable for the tested combination.

4   CONCLUSION

Three major accomplishments were made in this study. First was the expression of a new testing procedure for the verification of joint sealants.

The second accomplishment of the study is the installation of testing samples, which is identical to the real implementation of the sealed joints. Moreover, it follows from the measured values that these procedures offer a complex testing of the bonded joint, which is often neglected in the standardised tests. The test assembly and designed testing mould are easy to handle and the tests should be performed right on the site prior their installation, therefore, the unsuitable combination of materials could be revealed in time.

The last and the most beneficial achievement of the presented study are the results obtained from the test samples that were prepared and tested in the exterior. Even though, the steps recommended by the manufacturer were perfectly adhered to, the samples embodied diametrically different failure modes than those tested in laboratory. The hypothesis that it is almost impossible to prepare a perfectly clean installation environment in the exterior was confirmed. Moreover, the results are an evidence that in-situ testing methods should be included in the verification process, not only of new products, in this case sealants or adhesives, but more importantly that new methods for the in-situ testing of an entire assemblies should be discussed in the international scientific community and subsequently integrated in new testing standards.

REFERENCES

ASTM International, 2012. *ASTM D 5573 99(2012) – Standard Practice for Classifying Failure Modes in Fiber-Reinforced-Plastic (FRP) Joints.* West Conshohocken, PA: ASTM International.

Chew M.Y.L. & Yi L.D. 1997. Elastic Recovery of Sealants. *Build. Environ.*, 32, pp. 187–193.

Chew M.Y.L. & Zhou X. 2002. Enhanced resistance of polyurethane sealants against cohesive failure under prolonged combination of water and heat. *Polym. Test.*, 21, pp. 187–193.

Chew M.Y.L. 2001. Curing characteristics and elastic recovery of sealants. *Build. Environ.*, 36, pp. 925–929.

Chew M.Y.L. 2004. Retention of movement capability of polyurethane sealants in the tropics. *Constr. Build. Mater.*, 18, pp. 455–459.

Czech-Standards, 1981. *ČSN 73 2579 – Test for frost resistance of surface finish of building structures.* Prague: Czech Standards Institute.

Czech-Standards, 1983. *ČSN 73 2581: Test for resistance or surface finish of building structures to temperature variations.* Prague: Czech Standards Institute.

Czech-Standards, 1995. *ČSN ISO 10365 – Adhesives. Designation of main failure patterns.* Prague: Czech Standards Institute.

Ding S.H. & Liu D.Z. 2006. Durability evaluation of building sealants by accelerated weathering and thermal analysis. *Constr. Build. Mater.*, 20, pp. 878–881.

Ding S.H., Liu D.Z. & Duan L.L. 2006. Accelerated aging and aging mechanism of acrylic sealant. *Polym. Degrad. Stabil.*, 91, pp. 1010–1016.

Hranice, a.s. 2014. Basic Properties of CETRIS, Hranice, pp. 149.

Ihari T., Gustavsen A. & Jelle B.P. 2014. Sealant aging and its correlation with facade reflectance. *Constr. Build. Mater.*, 69, pp. 390–402.

Lacombe R. 2006. *Adhesion Measurement Methods: Theory and Practice.* Taylor & Francis Group, New York.

Mirza J., Bhutta M.A.R. & Tahir M.M. 2013. In situ performance of field-moulded joint sealants in dams. *Constr. Build. Mater.*, 41, pp. 889–896.

Nečasová B., Liška P. & Šlanhof J. 2017. Adhesion and Cohesion Testing of Joint Sealants after Artificial Weathering – New Test Method. *Proc. Engineering*, 109, pp. 140–147.

Nečasová B., Liška P., Šimáčková M. & Šlanhof J. 2014. Test of Adhesion and Cohesion of Silicone Sealants on Façade Cladding Materials within Extreme Weather Conditions. *Advanced Materials Research*, 1041, pp. 23–26.

Nečasová B., Liška P., Šimáčková M. & Šlanhof J. 2016. Case Study on Determination of Tensile Properties of Construction Sealants at Variable Temperatures. *Applied Mechanics and Materials*, 824, pp. 18–26.

Pantaleo A., Roma D. & Pellerano A. 2012. Influence of wood substrate on bonding joint with structural silicone sealants for wood frames applications. *Int. J. Adhes. Adhes.*, 37, pp. 121–128.

Pattanaik S. Ch. 2013. *Repair of Active Cracks of Concrete Structures with a Flexible Polyurethane Sealant for Controlled Movement.* Retrieved from http://masterbuilder.co.in, pp. 154–158.

Yun T., Lee O., Lee S.W., Kim I.T. & Cho Y. 2011. A performance evaluation method of preformed joint sealant: Slip-down failure, Constr. *Build. Mater.*, 25, pp. 1677–1684.

Traffic at rest in relation to current legislation in the Czech Republic

M. Novotný & B. Nečasová
Faculty of Civil Engineering, Institute of Technology, Mechanization and Construction Management, Brno University of Technology, Brno, Czech Republic

ABSTRACT: The article focuses on the issue of parking in towns. At present, in the Czech Republic, each new building is assessed for the state of the design even as regards the possibility of parking for users or residents of the building in relation to applicable laws and standards. However, this legislation is imperfect in terms of many people and authorities. It is perfectly fine for some objects, but it is totally inadequate or insufficient for the number of parking spaces. As part of the preparation of the article, I also addressed the building authorities with a few questions about the state of legislation that deals with parking. Questions and answers on the exact wording of building offices are not included in the article for reasons of capacity. The conclusion of the article is the assessment of suitability or inappropriateness of the legislation with regard to parking in cities, mainly related to housing construction.

1 INTRODUCTION

Parking a big problem of today's large settlements. Hand in hand with the development of economy, there has also been a development in automotive transport in the Czech Republic. There was an increase in the volume of both freight and passenger transport. Regarding means of transport, the Czech Statistical Office (hereinafter referred to as the CSO) is divided into transport by rail, road, urban mass, air and inland waterway—numbers can be found in Table 1. These data show that public transport is the most popular transport method. The second most used kind of transport is road transport, which means cars. However, the parking problem is associated with this. At present, it can be said that every family has at least one car. Of course this statement is not clear, the number of cars in families is not defined in any way, but it can be considered based on observations and counts. In fact, a large number of households own more than one car.

10,436,560 persons reside in the Czech Republic – 5,109,766 men and 5,326,794 women (data from 26 March 2011 – CSO). With this in mind, we can look at statistics on the number of registered vehicles in the Czech Republic in categories related to passenger transport and the number of newly registered vehicles—table numbers 2 and 3.

Table 1. Number of persons transported by type of transport.

	Year	2008	2010	2012	2014	2016
Personal transportation	Total transferred persons in thousands	2 882 604	2 805 938	2 748 956	2 675 430	2 798 995
	by trains	177 424	164 802	172 801	176 050	179 171
	by cars	373 395	372 548	344 988	349 515	332 763
	by plains	7 158	7 466	6 420	5 623	6 000
	public transportation	2 323 761	2 260 264	2 224 235	2 142 935	2 280 260

Table 2. Amount of vehicles for the years 2005–2016.

Year	Motorcycles	Personal cars	Buses and microbuses
2006	822 703	4 108 610	20 331
2008	892 796	4 423 370	20 375
2010	924 291	4 496 232	19 653
2012	976 911	4 706 325	19 882
2014	998 816	4 833 386	19 808
2016	1 074 880	5 307 808	20 097

Table 3. Newly registered road vehicles 2005–2013.

Year	Total summary		Personal car		Motorcycles	
	new	used	new	used	new	used
2006	238 219	210 489	120 009	181 634	19 193	13 089
2008	286 837	274 953	141 617	229 722	26 681	22 519
2010	235 575	152 126	168 372	125 859	18 277	11 566
2012	242 110	150 428	173 173	124 651	14 920	11 133
2013	235 767	154 311	164 627	127 700	16 158	11 751

It is clear from the data that the number of passenger cars in the Czech Republic steadily increases. A passenger car can be seen either as a means of transport and a working tool or as a sign of wealth. Both these views, however, combine a common factor – "where with him (by car)" at times when we are not transported, i.e. parking (transport at rest). Parking itself is the most problematic in locations where there is a lot of people and homes in a small space, especially in cities and their centers. This problem is most remarkable in reconstructions, when existing buildings change the number of dwellings and the number of people living in them. It is not always the case for such a reconstruction to build the appropriate parking capacities— whether it be a technical reason (e.g. structural static problems), a spatial impact (lack of space in the basement or the surrounding area) or the technological impact (the necessity of the foundation systems for cars). For new constructions, this problem is much better solved—the number and layout of parking areas are solved by both legislation and standards.

2 LEGISLATION AND ITS REQUIREMENTS REGARDING PARKING AND TRANSPORT SOLUTIONS AT REST

As the basic regulations, which deal with parking at rest and its execution, we can mention the Act on Spatial Planning and the Building Code (Building Act) No. 183/2006 Coll. (as amended), Decree No. 268/2009 Coll. on Technical Requirements for Buildings, The Act on Road Communications No. 13/1997 Coll. as amended, Regulation No 10/2016 laying down general land use requirements and technical requirements for buildings in the Capital City of Prague (the Prague Building Regulations) and following the Czech State Standard ČSN 736110 Design of local roads, ČSN 736056 Parking and parking areas of road vehicles, ČSN 736058 Individual, row and collective garages. These documents can be described as basic in designing and from the viewpoint of the building approval authority for parking purposes.

If we look at the Building Law, we will find data on parking, parking lots and garages (separate or built-in), especially from the point of view of building permits and management approval of their construction. The approval of the construction of parking lots and parking areas for new plans must be solved in cooperation with the Building Authority and in accordance with § 103 and 104 of the Building Act, which deals with building permits. Section 103 (5) (g) other constructions and equipment include products which perform the function of a structure, including supporting structures for them, without basement, provided they are not intended for use by persons or for the housing of animals – that is part of the parking areas could fall into of this Regulation, which would mean that they will not have to be notified or authorized. Parking places are always used by people, i.e. at least we always fall into the reporting mode. Subsequently, it is also stated that "Construction work on roads and maintenance work on them, which do not require building permission or notification, is defined by a special legal regulation – Decree No. 104/1997 Coll., Which implements the Act on roads, as amended regulations". § 104, which is called the Declaration of simple buildings, landscaping, equipment and maintenance work, then states in point (k) that reporting requires "products which perform a function of construction and which are not mentioned in Section 103 (1) g) point 5". I.e. these areas always (to a greater extent) require approval, case by special office. Of course, there is evidence of the intention of documentation according to valid legislation, which is currently Decree 405/2017 Sb. amending Decree No. 499/2006 Coll., on documentation of constructions, as amended by Decree No. 62/2013 Coll. and Decree No. 169/2016 Coll., on the determination of the scope of tender documentation for construction works and the inventory of construction works, supplies and services with a statement of merit. Subsequently, it is followed by a classic approval process.

From the perspective of the Road Act No. 13/1997 Coll., as amended, we can define a parking space on the basis of several paragraphs, e.g. § 7 A purposeful communication within the scope of paragraph 2, which states that "Useful communications are also roads in an enclosed space or object that serve the needs of the owner or operator of a closed space or object. This purposeful communication is not publicly available but in the scope and manner specified by the owner or operator of the enclosed space or facility. Concerned whether the road is an enclosed space or an object, the competent road administration decides". In addition, the Act on Communications states in Section 12 Components and Accessories item 6) that a public car park is a construction and operational area of the local or special purpose road or a separate local or special road intended for standing a motor vehicle.

From the viewpoint of Regulation 10/2016 laying down general requirements for land use and technical requirements for buildings in the Capital City of Prague – i.e. Prague Building Regulations, the following is mentioned – paragraph (2) states that "The provisions of this Regulation shall be used in the preparation of land-use planning documentation and land-use planning documents in the capital city of Prague, in particular when defining areas and determining the conditions of their use and arrangement" and paragraph (3) states "The provisions of this Regulation shall apply to the designation and placement of buildings facilities on them, changes in land use, division or jointing of land. The provisions of this Regulation shall also apply to changes to constructions or installations, to temporary construction works of a building site, to changes in the influence of the use of a building or installation on the territory, to the designation of plots of public space and to built-up building sites with buildings that are cultural monuments or monumental or historical monuments zones unless significant technical or structural reasons so exclude." Consequently, in § 2, the concepts indicate that, for the purpose of Regulation 10/2016, at point (t), the area used for parking or decommissioning of a passenger vehicle, with different types of such areas.

From the point of view of the city. of the City of Prague are probably the most important ones for the purpose of assessing the intentions of Sections 32 and 33, where the following information is given – § 32, entitled Parking Capacities, states that "for constructions, with the exception of temporary buildings for a period of not more than one year, it is necessary to establish binding and visitor standing in numbers under Regulation 10/2016". For constructions, the minimum required and b) the maximum permissible number of stalls is set. Paragraph 2 states that "the minimum required and maximum permitted number of stalls is determined by the percentage of base stables. Unless the Territorial or Regulatory Plan in accordance with Section 83 (2) stipulates otherwise, the percentage established on the basis of the center of territory and the attendance distance of public transport stations in Annex 3 to this Regulation shall be used, especially for a) (b) Covered stays for other use purposes and visitor stays for all purposes of use; for multi-zone buildings, the stack counts are determined by zone policy with a lower percentage for the required minimum. The resulting minimum required and maximum permitted number of stalls shall be rounded up to the whole station".

Details of the design of the number of parking spaces according to this regulation are set out in annexes 2 and 3 of the Annex. Annex No. 2 states "Basic number of stalls – the table specifies for individual purposes the basic number of stalls, including the percentage of coupled and visitor stalls for calculation according to § 32 For each use purpose, a basic number of stalls is defined, which is defined by the gross floor area of the purpose of use (in m^2) per 1 parking space. Percentage of fixed and visitor stays is determined. For the specific purposes of use according to item 12, the basic number of stables is determined individually according to the expected number of visitors and jobs."

From the point of view of the accuracy of the design of the parking spaces, the most up-to-date information should be from the standards. Here we can draw on three standards – ČSN 736110 Design of local roads, ČSN 736056 Parking and parking areas of road vehicles, ČSN 736058 Individual, row and collective garages. In essence, they define the nomenclature – for example, a parking bar/strip, a parking lot such as P + R, K + R and B + R and others. Next, they define the design and calculation parameters of parking lots and their accessories, such as ramps. The type of construction is also distinguished, the number of parking spaces varies according to the use of the new building – housing, production, services, etc. The entry is also the length of the parking time – long-term or short-term. In Prague and the center of Brno, so-called "resident parking" is introduced alongside this parameter. An important input, which we find here, is the assumed degree of automation, where it can be easily stated that the basic degree of the automobile is about 400 vehicles/1000 inhabitants i.e. 1: 2.5 (1 vehicle/2.5 inhabitants).

Standards also allow the number of stalls to be reduced in case of changes in constructions or renovations where it would not be possible to technically ensure the number of stalls as for a new building. As well as the standards, the rules for the placement of the resulting parking areas are amended, both in the form of garages, parking spaces, and in the form of direct communications. Part of these rules is also the attendance distance to these areas – e.g. for short-term parking of 200 m, for long-term parking of 300 m and for parking 500 m. The latter figure refers to parking lots that closely follow the urban transport and are considered part of the solution transport problems of cities such as Brno and Prague. It is a good parking lot on public transport lines – for example Park and Ride – P + R (i.e. park and continue with public transport), Bike and Ride – B + R (i.e. stop the bicycle and continue by public transport) at the edges of the central zone, the park and go – P + G (i.e., park the vehicle and go on foot), at public transportation stations then a parking Kiss and Ride – K + R, which only allows short stop to the immediate taking up and transported persons.

These and other parameters are dealt with by ČSN 736110. There are two CSNs with concrete data on parking areas, namely ČSN 736056 Parking and parking areas of road vehicles and ČSN 736058. The individual, row and collective garages, replacing older standards ČSN 736057 Individual and terraced garages – basic provisions (10.8.1987) and ČSN 736058 Massive garages – basic provisions (1.8.1988). Norms define both the basic terms, but above all the technical parameters. One of them is, for example, the basic dimension of table vehicles in individual categories, which can then be followed when all other conditions are fulfilled when designing a particular design of parking spaces. The standard further specifies the principles and parameters of the technical design. Part of the standards is also the design process, which is based on the density of the settlement, its functional use and the expected development in the planning documentation. Parking should therefore be dealt with in a broader context and not just on the basis of the current situation. More specifically, it is based on:

– the number of permanent residents in the territory,
– from the location of the area – the center – the outskirts of the city,
– a number of job opportunities,
– availability and use of public (mass) transport,
– current transport capacities at rest – public unpaid and paid and non-public,
– conducted transport surveys with a view to future development,

- parking spaces – e.g. P + R, K + R, etc.,
- composition of the user group for the given parking area.

The location and construction of areas or garages must also be addressed with regard to fire, hygiene and other regulations. A very important part of the design of parking solutions is also the aspect of noise problems or the aspect of environmental protection.

When going to ČSN 736058 Individual, row and collective garages, we can find definitions of other concepts, for example, the definition of individual, terraced and massive garages is important for our purposes. It also follows from the description in the standard that they may contain automatic loading systems of various embodiments. Here, too, we can find definitions for the relationship between garages and terrain – we have overground, underground and combined garages. From the point of view of the way vehicles are parked, then the garages with the movement of vehicles are self-contained, with an automatic parking system or a combination of both. It can also be mentioned the distribution according to the attendance of the attended and unattended operators as well as the public, non-public and the garages for special purposes – eg intervention and rescue units. Part of the design of the garages is also the design of engineering networks, in the case of permanent human service with respect to the relevant hygienic facilities and limits. In the case of public garages or public or other buildings that are used by more than one person permanently, it is usually necessary to address barrier-free access for persons with reduced mobility and their vehicles.

3 ASSESSMENT OF THE NUMBER OF PARKING SPACES IN PRACTICE FROM THE POINT OF VIEW OF THE BUILDING OFFICE

The planning and building authorities are clearly guided by their validation legislation and by the successive standards. Under the current legislation, they have no possibility to recommend or change the numbers depending on the needs of the locality or the population, but only to re-reflect the regulations – Regulation No. 10/2016 Coll. laying down general requirements for land use and technical requirements for buildings in the Capital City of Prague, Building Act No. 183/2006, as amended (most recent Act 225/2017) and Decree 268/2009 on technical requirements for construction. In close connection with them is Decree No. 398/2009 Coll. on the barrier-free use of buildings, especially in public parking facilities. Another Act to be considered is Act No. 13/1997 Coll. on land-based communications, as amended. This legislation is directly supplemented and developed by standards – ČSN 736110 Design of local roads, ČSN 736056 Parking and parking areas of vehicles and standard ČSN 736058 Individual, row and collective garages. Under the authority, the authorities only assess the submitted intentions and their compliance with the applicable legislation, and the practice often speaks of a lack of parking space. However, the planning authorities do not have the power to change the requirements for new buildings beyond the applicable regulations – which in some situations is detrimental. However, it is a good habit in the projection that the designer and the investor will advise on the plan with the building office. They can then influence the proposal itself on the safe side, or make it more "pleasant", not only to meet the requirements but to overcome it.

4 RATING FROM AN AUTHOR'S POINT OF VIEW

It is clear from the statements of the building authorities (which the author questioned about the legislation on parking at rest) that they can do nothing else but to assess the construction plan only in terms of standards and law. The authorities themselves cannot recommend increasing the number of parking spaces unless they are approached at the project preparation phase by the investor or the designer. At the stage of the completed project and the submitted application it is only possible to evaluate the presented variant, and in some localities and situations this capacity is insufficient, especially if the parking space is not part of the apartment sale. From the SÚ replies, which could not be included in the article for capacity reasons, it is also possible to make it clear that in the case of family houses in terms of transport at rest, the problem is usually not the case, unless it is a reconstruction and an associated increase in the number of dwelling units. In terms of housing construction, such as apartment buildings and complexes, it can be said that there are no problems with new buildings, where the construction of parking capacities can be anticipated, for example underground and can be technically done.

Yes, the solution to housing construction in connection with the reconstruction of buildings and the radical increase in the number of housing units and inhabitants is very problematic. There is an increasing need for parking spaces, which cannot always be solved. From the point of view of evaluation, the least problem is the construction of administrative or business centers or, where appropriate, production facilities, where the conditions of the premises and the layout of the building are very suitable and thus the problem of meeting the required number of parking spaces is not a problem. For some centers, even parking conditions are so generous that some of the sites can be allocated for long-term rentals as residential parking. Furthermore, it appears from the comments that in some respects the standard requirements are too benevolent and, on the contrary, it would be appropriate to tighten them so that certain types of constructions have to include parking lots far more. This is a question of better keying legislative requirements or, possibly, creating a more detailed methodology that takes into account all the circumstances of the assessed situation. At the same time, it would be advisable to give building authorities more authority in public buildings and city center buildings to make decisions at the design stage of the project so that the number of parking spaces can be more influenced. As you can see, it is necessary to deal with them on a continuous basis and with regard to many aspects, one of the most important being the development of automotive and the use of means of transport in the future, so that we can prepare enough parking capacity for cars.

And conclusion? How can this article be summed up? First of all, this article is essentially a big introduction to the issue of transport at rest, ie parking. This article should be followed by one to two other articles. One thing is the valid legislation and its imperfection, the second is the real situation in different cities, and the different opinions of the building authorities in the assessment of new intentions. There is also a difference between a purely urban built-up area and a possible mixed area where parking capacities are dealt with, for example, by large shopping centers. The very limited number of parties did not allow a given list to be given and a summary of the obtained opinions of the building authorities, and subsequently to make statistics of available capacities compared, for example, with Brno and Prague. In the future, we can expect this article to be expanded on these aspects. From the point of view of legislation, the members can be summed up as follows: for the construction of houses, the problem does not arise. In the case of new homes and complexes in locations with large plots also not primarily, but the big problem is the solution of parking in the reconstruction, or in the case of new buildings in a dense build-up, there is not technically sufficient capacity of the parking areas. Of course, this problem can be defined differently depending on territorial and local conditions. Regarding the possibility of improving the situation, two basic options are evident. The first is a straightforward change in legal requirements from legislation – ie the increase of quotas. The second one is more complex, it involves changes in the thinking of developers in the same direction so that they themselves realize the necessity to solve this problem in more detail, analyze more new buildings and in most cases intentionally increase the number of parking spaces. We will see what the future will bring in this respect.

REFERENCES

ČSN 736056 – *Parking and parking areas of road vehicles*. 2011. Issued in The Office for Standardization, Metrology and Testing, Prague. March 2011. 1–28.
ČSN 736058 – *Individual, row and collective garages*. 2011. Issued in The Office for Standardization, Metrology and Testing, Prague. September 2011. 1–48.
ČSN 736110 – *Design of local communications*. 2006. Issued in Czech Standards Institute, Prague. January 2006.
ČSN 736425-2 – *Bus, trolley and tram stops, hubs and habitats – Part 2: Transition nodes and habitats*. 2009. Issued in The Office for Standardization, Metrology and Testing, Prague. September 2009. 1–24.

Decree No. 169/2016 Coll. on the determination of the scope of documentation of the public works contract and the list of construction works, supplies and services with a statement of designation. 2016. In The Collection of Laws of the Czech Republic. 31 May 2016, 2824–2826. ISSN 1211-1244.

Decree No. 268/2009 Coll. on technical requirements for construction. 2009. In the Collection of Laws of the Czech Republic. Issued 26 August 2009, Volume 81, 3702–3719. ISSN 1211-1244.

Decree No. 398/2009 Coll. on general technical requirements ensuring the barrier-free use of buildings. 2009. In The Collection of Laws of the Czech Republic. 5 November 2009, 6621–6647. ISSN 1211-1244.

Decree No. 405/2017 Coll., Amending Decree No. 499/2006 Coll., On building documentation, as amended by Decree No. 62/2013 Coll., And Decree No. 169/2016 Coll., On determining the scope of documentation public works contracts and an inventory of construction works, supplies and services with a statement of merit. 2017. In The Collection of Laws of the Czech Republic. 24 November 2017, 4578–4696. ISSN 1211-1244.

Decree No. 499/2006 Coll., On building documentation. In The Collection of Laws of the Czech Republic. Issued 10 November 2006, 6872–6911. ISSN 1211-1244.

Decree No. 62/2013 Coll., Amending Decree No. 499/2006 Coll., On building documentation. 2013. In The Collection of Laws of the Czech Republic. March 14, 2013, 466–520. ISSN 1211-1244.

Law No. 13/1997 Coll. on roads. 1997. In The Collection of Laws of the Czech Republic. Issued 21 February 1997, Volume 3, 47–61.

Law No. 183/2006 Coll. on land-use planning and building regulations (Building Act). 2006. In The Collection of Laws of the Czech Republic. Issued 14 March 2006, Volume 63, 2226–2290. ISSN 1211-1244.

Law *No. 225/2017 Coll., Amending Act No. 183/2006 Coll., On Spatial Planning and Building Regulations (Building Act), as amended, and other related acts.* 2017. In The Collection of Laws of the Czech Republic. 27 June 2017, 2514–2581. ISSN 1211-1244.

Regulation No. 10 / 2016 Coll. of capital city Prague, laying down general requirements for land use and technical requirements for buildings in the Capital City of Prague (Prague Building Regulations). 2016. In regulation of the Capital City of Prague. Issued Council of the Capital City of Prague. 27 May 2016, 1–42.

Advances and Trends in Engineering Sciences and Technologies III – Al Ali & Platko (Eds)
© 2019 Taylor & Francis Group, London, ISBN 978-0-367-07509-5

Investigation of calcium and sulfur content changes in liquid medium due to bacterial attack on concrete

V. Ondrejka Harbulakova
Faculty of Civil Engineering, Institute of Environmental Engineering, Technical University of Košice, Košice, Slovakia

A. Estokova
Faculty of Civil Engineering, Institute of Material Engineering, Technical University of Košice, Košice, Slovakia

A. Luptakova
Slovak Academy of Science, Institute of Geotechnics, Bratislava, Slovakia

ABSTRACT: The trend of concrete deterioration caused by chemical or biological effects is dependent on a type of aggressive environments. Bacteria responsible for acidic corrosion of sewer systems are amply represented in waste waters. The paper brings current results of an investigation of relation between pH values and leached-out concentrations of calcium and sulfur concentrations in medium. For laboratory testing, there were preferentially selected aggressive environments with bacteria responsible for biological corrosion and pure cultivation medium without bacteria stands for the control sample. The correlation analysis confirmed that the higher was the sulfur concentration in liquid, the more massive was leaching of calcium.

1 INTRODUCTION

Sulfuric acid is present in the environment from the different sources. From the concrete corrosion point of view mostly it can be coming from water: wastewater, rainwater, groundwater (Pelikan et al. 2018, Pochwat et al. 2017) or industrial water and is often contaminated with organic and inorganic compounds (Bartoszek et al. 2015, Kida et al. 2018).

In sewer systems, the concrete corrosion is caused mainly by sulfuric acid generated by microbial sulfur oxidation because the headspace hydrogen sulfide (H_2S) and oxygen (O_2) concentrations and moisture are relatively high. It is known as biocorrosion or microbially induced concrete corrosion (MICC). Concrete corrosion is regarded as one of the most serious problems currently affecting the sewer infrastructure (Pietrucha-Urbanik 2015, Xuan Li X. et al. 2017).

Various microbial species and complex mechanisms are involved in the MICC. During the initiation stage of concrete corrosion, abiotic processes such as carbonation and H_2S acidification reduce the surface pH of concrete from ~13 to ~9 (Joseph et al. 2012). Next step after this initial abiotic acidification, biological processes, primarily sulfide oxidation, cause the direct formation of sulfuric acid which leads to lowering the pH of the concrete surface (Okabe et al. 2007). Responsible for this phenomenon is a whole family of bacteria, various strains of which thrive at different pH levels (Scrivener et al. 2013). Five species of *Acidithiobacillus* bacteria (*A. thioparus, A. novellus, A. intermedius, A. neapolitanus, A. thiooxidans*) are primarily responsible for sulfuric acid production during the biological corrosion process (Islander et al. 1992). Presence of water is essential in the deterioration processes because it acts as a solvent of an aggressive medium, then, in some cases as a constituent of the formatted reaction product (Zivica and Bajza 2001). Sewer systems represent a very aggressive environment for cementitious materials. (Scrivener et al. 2013). The production of sulfuric acid is the final step in a complex chain of processes which comprise microbially induced corrosion. The process of biogenic acid formation in sewers and subsequent attack of concrete sewer pipes is highly complex, involving a dynamic interaction between sewer environment (aqueous and gaseous), biological activity, and pipe materials (Bertron et al. 2004). The colonization of the concrete surface is progressive. The main cause of deterioration is the corrosion of concrete due to the in-situ production of sulfuric acid by bacteria.

Acid corrosion causes a gradual loss of concrete, which significantly reduces the service life of concrete structures in sewer system and may lead to failures such as a collapse threatening public safety (Wu et al. 2018). According to Neville (2004) there is a difference between a chemical and physical attack in that a chemical attack involves sulfate ions. On the other hand a physical attack involves crystallization of the salts of which one example is sulfate ion. The term sulfate attack is more often considered and limited to the chemical attack. Bertron et al. (2004) investigated the effect of pH on mineralogical and chemical modification in the cement-based matrix of ordinary Portland cement. The research done of sulfuric acid corrosion of concrete can roughly be divided in three groups: chemical tests, microbial simulation tests, and exposure tests in situ (Monteny et al. 2000). Because of a lack of standardised methods, different test methods have been used, and various parameters have been modified to evaluate the resistance of the materials.

The paper aims at the investigation of deterioration of concrete due to the bacterial attack. Experimental part was oriented on measurements of calcium and sulfur ions concentrations in leachates after bacterial effect as well as in the medium without bacterial presence (control medium). Dependency between concentrations each other and also between concentration and pH value of the liquid phase was studied by correlation analysis.

2 MATERIAL AND METHODS

2.1 *Material*

The prepared samples were divided into two sets: a set of the samples were exposed to liquid phase comprised of Waksman and Joffe cultivation medium and inoculum of suspension of *A. thiooxidans* bacterial culture.

A second set of samples (abiotic controls) served for a comparative study. The samples of the second set were being immersed only in the non-bacterial solution – solution contained cultivation medium only. All measurements presented in the text of the paper are marked with letter "C" after measured parameter (pH-C, Ca-C, S-C).

Methodology of leaching experiment was the same for both liquid media: the cut samples were brushed off, fine of impurities and then they were immersed for 2 hours in distilled water, selected and stored on filter paper and then immersed for 2 hours in ethanol. Cleaning with ethanol was due to maintaining sterile conditions. After 2 hours, the samples were selected, covered by sterile filter paper and stored in an aseptic box. The samples were dried to a constant weight at 80°C. Each sample was placed in a 700 ml glass container to which certain volume of medium was added. The volume of the liquid phase was calculated based on the determination of the volume of samples where the volume ratio of the solid and liquid phases had to be maintained at 1:10.

Concrete samples of $50 \times 50 \times 10$ mm were formed as a drilled core from concrete cube using drilling mechanism STAM. The samples were rid of impurity before the experiment and are presented in Figure 1.

The pH of the liquid medium was measured at 7 day intervals from the liquid phases of each sample were taken 5 ml of liquid needed for XRF analysis. Leaching trends of calcium and sulfur content measured in each liquid phase were studied by X-ray fluorescence analysis (XRF) using SPECTROiQ II equipment (Ametek, Germany) with SDD silicon drift detector having resolution of 145 eV at 10 000 pulses.

The primary beam was polarized by Bragg crystal and Highly Ordered Pyrolytic Graphite – HOPG target. The sample measurement was performed 300 and 180 s at the voltage of 25 kV at the applied current of 0.5 mA and 50 kV at 1.0 mA.

Table 1. Composition of concrete mixture.

Component	Amounts per 1 m³ of fresh concrete
CEM I 42.5 [kg]	360
Water (L)	162
Aggregates 0/4 [kg]	825
Aggregates 4/8 [kg]	235
Aggregates 8/16 [kg]	740

Figure 1. Simulation of corrosion conditions.

In case of sample with bacterial culture, the pH were kept at the constant values equal to 4 using 0.1 M H_2SO_4 for the reason maintaining optimal conditions for bacterial growth. The values of leachates' pH were measured by pH meter PHH – 3X Omega.

2.2 Methods

Descriptive statistics is the discipline of quantitatively describing the main features of a collection of data (Kreyszig, 2011). The leached-out masses of the calcium and sulfur from concrete matrix and pH values of liquid media were used for the subsequent mathematical evaluation using a correlation analysis. In statistics, dependence refers to any statistical relationship between two random variables or two sets of data. Correlation refers to any of a broad class of statistical relationships involving dependence. Increase of the absolute value of the correlation coefficient (R_{xy}) is proportional to linear correlation. Information about two dimensional statistical data set gives correlation coefficient R_{xy} calculated using Equation 1 (Kreyszig, 2011). R_{xy} values are from the interval <–1,1>. If $R_{xy} = 1$, the correlation is full linear, if $R_{xy} = -1$, then the correlation is inversely linear and if $R_{xy} = 0$, the pairs of values are fully independent. Than degree of the correlative closeness is: medium, if $0.3 \le |R_{xy}| < 0.5$; significant, if $0.5 \le |R_{xy}| < 0.7$; high, if $0.7 \le |R_{xy}| < 0.9$; and very high, if $0.9 \le |R_{xy}|$.

$$R_{xy} = \frac{n\sum_{i=1}^{n} x_i y_i - \left(\sum_{i=1}^{n} x_i\right)\left(\sum_{i=1}^{n} y_i\right)}{\sqrt{n\sum_{i=1}^{n} x_i^2 - \left(\sum_{i=1}^{n} x_i\right)^2 \left[n\sum_{i=1}^{n} y_i^2 - \left(\sum_{i=1}^{n} y_i\right)^2\right]}}$$

(1)

Correlation coefficient was for the purposes of our assessment obtained by the function "Pearson" in Microsoft Excel. In the paper, the correlation coefficients were calculated for dependency between leached-out masses of calcium and sulfur each other (considering type of liquid medium) and the dependency between pH and leached-out concentrations of calcium and sulfur, respectively (considering the same set of samples).

3 RESULTS AND DISCUSSION

3.1 *pH values, concentration of Ca and S ions*

The XRF measured concentrations of leached-out Ca and present S ions and pH values are reported in Figures 2 and 3.

Changes in Ca concentration and pH values in leachate for samples exposed to bacterial effects and for the samples without bacteria (control sample -C) differ. The values of Ca concentrations in control medium ranged from 57 mg/kg to 67 mg/kg (Figure 2). In the bacterial medium, the final concentration of Ca in the leachate was of 832 mg/kg, i.e. many times higher than under abiotic conditions (Figure 2). Comparing changes in Ca concentration in leachate under biotic and control conditions less significant changes in Ca concentration have been recorded during the process in case of control sample as expected. The flow of the pH curves documents a strong acidification of the leachate under the influence bacteria up to 20 days. Afterwards, due to acidic conditions, an increased trend in pH values was preserved and the concentration of Ca in the liquid phase was slightly raised until the end of experiments.

The changes in sulfur concentrations in liquids versus pH curves are shown in Figure 3. The concentrations of sulfur in control media without bacteria (S-C) ranged in the interval

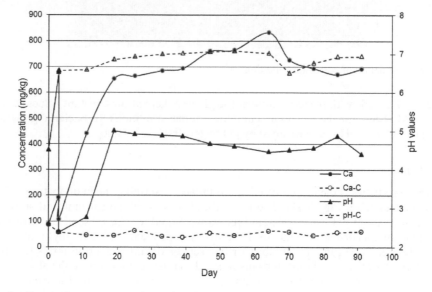

Figure 2. Leached-out concentrations of calcium and pH changes during the experiment.

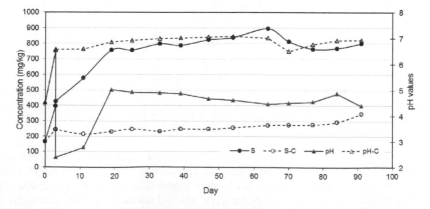

Figure 3. Leached-out concentrations of sulfur and pH changes during the experiment.

Table 2. Correlation coefficient between leached-out masses of ions and pH of media.

Correlated parameter	Ca/pH	Ca-C/pH	S/pH	S-C/pH	Ca/S	Ca-C/S-C
Correlation coefficient R_{xy}	0.21	−0.76	0.13	0.63	0.97	−0.23

from 240 mg/kg to 250 mg/kg (Figure 3) and origin mostly from the cultivation agent. The sulfur concentrations in biotic media of *A. thiooxidans* bacteria were several times higher. The bacteria have adapted to the environment after 20 days and have started their metabolic activity and oxidizing elemental sulfur (S°) found in cultivation medium to sulfate (SO_4^{2-}), resulting in formation of sulfuric acid (H_2SO_4) which subsequently acidified the leachates while acting as lye reagent for cement samples—an increase in pH concentration in the liquid phase (Figure 3).

Comparison of changes in sulfur concentration in leachates for biotic condition and in case of control set of samples (-C) points to the fact that much less significant changes in concentrations of sulfur were measured in the medium without bacteria.

3.2 *Correlation analysis*

Based on the concentrations of ions in the liquid phase, as is presented above, and pH values, the statistical evaluation was performed. Results of the calculated correlation coefficients (R_{xy}) for Ca/pH, S/pH, Ca-C/pH, S-C/pH dependences for the studied concrete samples are presented in Table 2 ("-C" stands for "control samples" – without bacteria).

The highest correlation coefficient was calculated for calcium leaching versus sulfur ($R_{xy} = 0.97$) with a correlation closeness defined as very high. This means that the higher was sulfur concentration in liquid the more massive was leaching of calcium. Correlation closeness was found as high for relation pH and calcium in control liquid medium ($R_{xy} = -0.76$) inversely; significant for pH and sulfur in control liquid medium ($R_{xy} = 0.63$). Very low or no correlation was calculated for the other pairs of evaluated parameters. These findings may show that the leaching of these elements are no mainly based on the alkali compounds.

The leached-out amounts of calcium under bacteria influence or sulfur concentrations and pH are fully independent.

4 CONCLUSION

The effect of *Acidithiobacillus thiooxidans* bacteria, which are often predominant in sewer concrete corrosion layers, was studied under laboratory conditions. Evaluation was done through measurements of leached out masses of calcium and sulfur and pH values changes in liquid phase during the experiment. The statistical analysis was used to evaluate the results of the biocorrosion process of concrete samples. The correlation analysis could help to understand and interpret the results obtained from experiment, although evaluations of more elements and more parameters are needed for more accurate conclusions. Future studies will be oriented more details studies of the environmental conditions (in situ experiments), particularly parameters related to biocorrosion processes in sewer system.

ACKNOWLEDGEMENTS

This paper has been prepared within the Project of the Scientific Grant Agency of the Ministry of Education, science, research and sport of the Slovak Republic and the Slovak Academy of Sciences (VEGA) No. 2/0145/15 and 1/0648/17.

REFERENCES

Bartoszek, L., Koszelnik, P., Gruca-Rokosz, R. & Kida, M. 2015.Assessment of agricultural use of the bottom sediments from eutrophic Rzeszów Reservoir. *RoczOchrSr* 17: 396–409.

Bertron, A., Escadeillas, G. & Duchesne, J. 2004. Cement paste alteration by liquid manure organic acids: chemical and mineralogical characterization. *Cement and Concrete Research* 34: 1823–1835.

Islander, B.R.L., Devinny, J.S., Member, A., Mansfeld, F., Postyn, A. & Shih, H. Microbial, ecology of crown corrosion insewers 1992. *J. Environ. Eng.* 117.

Joseph, A.P., Keller, J., Bustamante, H. & Bond, P.L. 2012. Surface neutralization and H_2S oxidation at early stages of sewer corrosion: influence of temperature, relative humidity and H_2S concentration. *Water Res.* 46: 4235–4245. 10.1016/j.watres.2012.05.011.

Kida, M., Ziembowicz, S. & Koszelnik, P. 2018. Removal of organochlorine pesticides (OCPs) from aqueous solutions using hydrogen peroxide, ultrasonic waves, and a hybrid process. *Separation and Purifation Technology* 192: 457–464.

Kreyszig, E. 2011. Advanced Engineering Mathematics, John Wiley and sons, 10th edition.

Monteny, J., Vincke, E., Beeldens, A., De Belie, N., Taerwe, L., Van Gemert, D. & Verstraete, V. 2000. Chemical, microbiological, and in situ test methods for biogenic sulfuric acid corrosion of concrete. *Cement and Concrete Research* 30(1): 623–634.

Neville, A. 2004. The confused world of sulfate attack on concrete. *Cement and Concrete Research* 34(8): 1275–1296.

Okabe, S., Odagiri, M., Ito, T. & Satoh, H. 2007. Succession of sulfur-oxidizing bacteria in the microbial community on corroding concrete in sewer systems. *Appl. Environ. Microbiol.* 73: 971–980. 10.1128/AEM.02054-06.

Pelikan, P., Slezingr, M. & Markova, J. The Efficiency of a Simple Stabilization Structure in a Water Reservoir. 2018. *Pol. J. Environ. Stud.* 2, 793–799.

Pietrucha-Urbanik, K. 2015. Failure Prediction in Water Supply System – Current Issues, *Theory and Engineering of Complex Systems and Dependability, Advances in Intelligent Systems and Computing* 365: 351–358.

Pochwat, K., Słyś, D. & Kordana, S. 2017. The temporal variability of a rainfall synthetic hyetograph for the dimensioning of stormwater retention tanks in small urban catchments. *Journal of Hydrology* 549: 501–511.

Scrivener, K.L. & De Biele, N. 2013. Bacterio genic sulfuric acid attack of cementitious materials in sewage system, Chapter 12, in M.G. Alexander, A. Bertron, N. De Belie (Eds.). Performance of cement-based materials in aggressive aqueous environment. University of Cape Town, South Africa: Springer. https://doi.org/10.1007/978-94-007-5413-3_12.

Wu, L., Hu, Ch. & Liu, W.V. 2018. The Sustainability of Concrete in Sewer Tunnel-A Narrative Review of Acid Corrosion in the City of Edmonton, Canada, *Sustainability* 10(2), 517.

Xuan, L.X., Guangming, J.U. & Bond P. 2017. The Ecology of Acidophilic Microorganisms in the Corroding Concrete Sewer Environment, *Front Microbiol.* 8: 683.

Zivica, V. & Bajza, A. 2001. Acidic attack of cement based materials – a review: Part 1. Principle of acidic attack. *Construction and Building Materials* 15(8): 331–340.

Research of cement mixtures with added granulated and fluidized fly ashes

R. Papesch & T. Dvorsky
Department of Environmental Engineering, Faculty of Mining and Geology, VSB-Technical University of Ostrava, Ostrava, Czech Republic

V. Vaclavik, J. Svoboda & L. Klus
Department of Environmental Engineering, Institute of Clean Technologies for Mining and Utilization of Raw Materials for Energy Use, Faculty of Mining and Geology, VSB-Technical University of Ostrava, Ostrava, Czech Republic

ABSTRACT: The positive property of fly ash to reduce the negative impact of aggressive environment on the life of the grouting mortar, together with the achieved strengths results, serve as the basis for the recommendation of their possible practical utilization. The research has been performed in laboratory of building materials and was mostly involved in the determination of the chemical or physico–mechanical parameters of the mixtures with secondary energy materials to find appropriate recipes suitable for grouting works. The results of strength and freeze tests demonstrate the possibility of replacing up to 20% of the amount of cement screed specified by the secondary fly ash in the proposed recipes. The final evaluation deals with the possible use in practice, including recommendations for next practice finish.

1 INTRODUCTION

Building materials as the main components of building structures and entire buildings play an important role as far as the overall impact on the environment is concerned (A. Estokova and Porhincak 2012). That is why there is an effort to replace the cement in construction mixtures, because its production is very energy intensive (Adriana Estokova and Palascakova 2013). The aim is to find ways how to replace this product with other materials with similar properties, such as various recycled materials (Junak and Sicakova 2017). The re-use of fly ash from coal combustion as a partial substitute of cement has been very attractive nowadays, especially in the concrete and mortar segments (Stevulova and Junak 2014; J. Junak 2016; Junak and Stevulova 2008). Specifically, granulated and fluidized fly ash from power plants could be used in the production of building materials suitable for grouting work. The effect of fluidized fly ash in concrete on its physical and mechanical properties, such as resistance to abrasion, tensibility and other properties, etc., has been studied at present (Horszczaruk and Brzozowski 2017; Chen et al., 2017). The replacement of cement with a secondary raw material in the production of mortars and concrete also brings economic benefits, which can be witnessed in the metallurgical industry as well (Jursova 2010; Jursova et al., 2016).

2 MATERIALS AND METHODS

2.1 Cement screed

Cemix C25 cement screed was used to prepare the mixtures. This mix of mineral filler with a grain size of 0–0.7 mm, Portland cement and additives improves the final processing and utility properties of the mixtures. Fine ground Portland cement is the most common inorganic hydraulic binder that becomes solid when mixed with water as a result of hydration reactions and processes. The strength and stability after hardening is also maintained under water.

2.2 Fly ash

Fly ash is the finest fraction of residues produced during the combustion of coal with a grain size of 0–1 mm that are trapped in separators. Two types of fly ash were used for the production of the mixtures: granulated fly ash from Dětmarovice power plant (granulating combustion chamber, desulphurization using wet limestone method, fly ash) and fluidized fly ash from Třinec power plant (fluid combustion chamber, fly ash).

It is a heterogeneous material with different chemical, physical, mineralogical, morphological and technological properties. These parameters depend on the quality of the combustion coal and the technology used in the combustion process. The important aspects of coal combustion include especially the content of flammable substances, ash matter and water, while in case of the used combustion process technology, it is the type of combustion equipment (wet bottom boiler, dry bottom/fluidized bed boiler), as well as the combustion temperature. An example of the chemical composition of the fly ashes arising from conventional and fluidized combustion is presented in Table 1.

2.3 Mixing water

Drinking water from the water supply system with a temperature of 18°C was used as the mixing water in the production of the mixtures. This mixing water has two basic functions. The first function is hydration (hydration of cement leads to the formation of a solid structure of cement stone), and the other one is rheological function (it makes it possible to create a fresh mixture in combination with its components).

2.4 Description of recipes and production of mixtures

The main component of the grouting mixtures was represented by cement screed, which was gradually replaced with 5, 10 and 20% of granulated and fluidized fly ash. The fly ash poses a risk, because it is a secondary material that can have variable properties. This risk was minimized using raw materials from one sample, thus assuming the same properties. A total of 7 mixtures were prepared with different input components. The composition of the individual recipes for the production of the mixtures was as follows:

- mixture no. 1 – reference sample – 100% cement screed;
- mixture no. 2–5% of granulated fly ash, Dětmarovice power plant + 95% of cement screed;
- mixture no. 3–10% of granulated fly ash, Dětmarovice power plant + 90% of cement screed;
- mixture no. 4–20% of granulated fly ash, Dětmarovice power plant + 80% of cement screed;
- mixture no. 5–5% of fluidized fly ash, Třinec power plant + 95% of cement screed;
- mixture no. 6–10% of fluidized fly ash, Třinec power plant + 90% of cement screed;
- mixture no. 7–20% of fluidized fly ash, Třinec power plant + 80% of cement screed.

The samples of fresh mortar were prepared in such a way to achieve the prescribed spill value of 230 mm, which ensured suitable consistency for the grouting work. The amount of

Table 1. Chemical composition of conventional and fluidized fly ashes.

Fly ash	SiO_2	Al_2O_3	CaO	MgO	TiO_2	Fe_2O_3	SO_3	Na_2O	K_2O
Conventional [%]	52.22	28.01	3.09	1.38	2.37	9.66	0.6	0.51	1.59
Fluidized [%]	42.34	19.44	18.21	2.49	1.55	5.79	5.26	0.37	1.41

water needed to achieve the required consistency was determined on the basis of the prepared testing mixtures.

2.5 Material properties testing methods

The input raw materials and the prepared mixtures were tested to determine their physical and mechanical properties. The tests included the determination of specific weight (ČSN EN 1097-7), particle-size analysis (ČSN EN 933-1), determination of density (ČSN EN 1015-6), determination of air content in the mixture (ČSN EN 1015-7), determination of compressive strength (ČSN EN 1015-11), frost resistance test (ČSN 72 2452), calorimetric test, determination of leachate pH, and detailed particle-size analysis of extremely fine particles (FRITSCH Laser Particle Sizer Analysette 22).

3 RESULTS AND DISCUSSION

3.1 Determination of the density of the powder substances by means of pycnometer method

The density of the powder substances was determined using the pycnometer method according to ČSN EN 1097-7 standard Testing of mechanical and physical properties of aggregate—Part 7: Determination of density of filler—Pycnometer test. The density of the cement screed was set at 2 904 kg·m^{-3}, and of granulated ash at 2 288 kg·m^{-3} and of fluidized fly ash at 2290 kg·m^{-3}.

3.2 Particle-size analyses including washing

The particle-size analysis was performed according to ČSN EN 933-1 Testing the geometrical properties of aggregate—Part 1: Particle-size determination—screen analysis. The particle-size analysis results are graphically illustrated in Figure 1.

3.3 Basic physical properties of fresh mixture

Table 2 clearly shows the physical properties of the individual mixtures, such as their density, water to solid ratio, or air content in the mixture.

3.4 Determination of leachate pH

The leachate pH was determined according to (ČSN 722071). Three samples were prepared, and the sample preparation procedure included: 150 g of dry mixture (for composition see

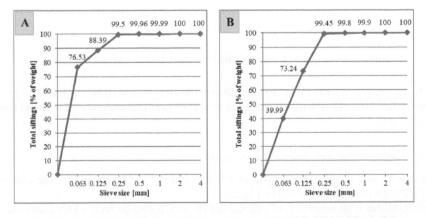

Figure 1. Particle-size curves of the used granulated fly ash (A) and fluidized fly ash (B).

Table 2.   Physical properties of fresh mixtures.

Mixture	Cement screed C25 [g]	Secondary product [g]	Water [g]	Water to solid ratio [%]	Density [kg·m⁻³]	Air content [%]
1.	3000	0	610	20.3	1792	5.7
2.	2850	150	590	19.7	1785	4.6
3.	2700	300	580	19.3	1768	3.8
4.	2400	600	580	19.3	1801	2.6
5.	2850	150	650	21.7	1859	4.0
6.	2700	300	690	23.0	1798	3.2
7.	2400	600	790	26.3	1678	2.4

Table 3.   Water leachate pH.

Sample type	pH of samples 4 h after mixing	pH of samples 3 months after mixing
Reference (100% cem. screed)	12.111	12.196
20% of granulated fly ash	11.340	11.517
20% of fluidised fly ash	11.450	11.596

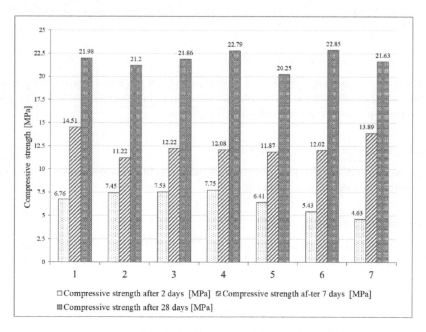

Figure 2.   Compressive strengths of the individual mixtures after 2, 7 and 28 days.

Table 3) was mixed in 1000 ml of distilled water. This was followed by the leachate pH measurement using HQ11D pH meter from Hach Company. The measurement results are presented in Table 3.

3.5 Determination of the compressive strength

The test was carried out according to (ČSN EN 1015-11). The results of the compressive strength are graphically presented in Figure 2.

In all mixtures with additions of granulated fly ash, it can be seen a faster increase in the initial strength compared to the reference sample. This property is caused by a higher content

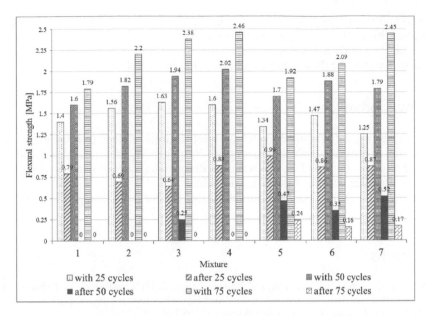

Figure 3.   Flexural strength of the individual mixtures after frost resistance test.

of the glass phase, which plays a major role in the reactivity of the fly ash with CaO. All the other samples also show higher values when comparing their long-term strengths. On the contrary, the recipes with added fluidized fly ash show lower initial strength, but after a slower beginning, the strengths after longer time (28 days) are equal to the reference mixture values. This property is caused by a higher free CaO content.

3.6   Frost resistance test

The test was performed according to the standard (ČSN 722452). Subsequently, the samples were subjected to flexural strength test and the results are presented graphically in Figure 3. The reference mixture sample and the mixtures with granulated fly ash were damaged after 50, respectively 75 cycles and that is why they were not tested for their flexural strength. The results clearly show a much higher degree of frost resistance of the mixtures containing fluidized fly ash in comparison with the mixtures containing granulated fly ash, as well as in comparison with the reference sample.

The frost resistance coefficients have been calculated according to ČSN 722452 based on the flexural bending strength results, see Figure 3. The best values of the frost resistance coefficient were achieved after 25 cycles in case of the mixture containing 5% of fluidized fly ash, namely 74%. The frost resistance coefficient after 25 cycles in case of the mixture with 20% fluidized fly ash content reached 70%. It should be noted that the above presented values of the frost resistance coefficient do not meet the condition of the frost resistance coefficient according to the standard (ČSN 722452), which is 75%. The values of the frost resistance coefficient for the other mixtures do not meet the requirement according to the standard (ČSN 722452) either.

4   CONCLUSION

The results of the strength and frost resistance tests demonstrate the possible replacement of a significant amount of cement screed with the specified secondary product in the designed recipes. This would lead to a definite reduction in costs while preserving or even improving certain parameters when compared to the reference mixture. The concerns related to the possible fluctuations in long-term strengths when using a higher share of fly ashes have not been confirmed and the recipes have been tested in the long run as well. The concerns related to the extremely high pH of the water leachate have not been confirmed either, on the contrary, thanks to the use of fly ashes, the mixtures were more stable. It is possible to deal with the potential increase of the fly ash contents in the mixtures as well.

ACKNOWLEDGMENTS

This article was written in connection with project Institute of clean technologies for mining and utilization of raw materials for energy use—Sustainability program. Identification code: LO1406. Project is supported by the National Programme for Sustainability I (2013–2020) financed by the state budget of the Czech Republic.

This research work has been supported by the student grant competition project SGS SP2018/12.

REFERENCES

Chen, Xuemei, Jianming Gao, Yun Yan, and Yuanzheng Liu. 2017. 'Investigation of Expansion Properties of Cement Paste with Circulating Fluidized Bed Fly Ash'. *Construction and Building Materials* 157 (December): 1154–62. doi:10.1016/j.conbuildmat.2017.08.159.

ČSN EN 1097-7—Tests for mechanical and physical properties of aggregates - Part 7: Determination of the particle density of filer—Pyknometer method.

ČSN EN 933-1—Tests for geometrical properties of aggregates—Part 1: Determination of particle size distribution—Sieving method.

ČSN EN 1015-6—Methods of test for mortar for masonry—Part 6: Determination of bulk density of fresh mortar.

ČSN EN 1015-7—Methods of test for mortar for masonry—Part 7: Determination of air content of fresh mortar.

ČSN EN 1015-11—Methods of test for mortar for masonry—Part 11: Determination of flexural and compressive strentgth of hardened mortar.

ČSN 72 2071—Fly ash for building industry purposes—Common provisions, requirements and test methods

ČSN 72 2452—Testing of frost resistance of mortar.

Estokova, A., and M. Porhincak. 2012. 'Reduction of Primary Energy and CO2 Emissions through Selection and Environmental Evaluation of Building Materials'. *Theoretical Foundations of Chemical Engineering* 46 (6): 704–12. doi:10.1134/S0040579512060085.

Estokova, Adriana, and Lenka Palascakova. 2013. 'Study of Natural Radioactivity of Slovak Cements'. In *Icheap-11: 11th International Conference on Chemical and Process Engineering, Pts 1-4*, edited by S. Pierucci and J.J. Klemes, 32:1675–80. Milano: Aidic Servizi Srl.

Horszczaruk, Elzbieta, and Piotr Brzozowski. 2017. 'Effects of Fluidal Fly Ash on Abrasion Resistance of Underwater Repair Concrete'. *Wear* 376 (April): 15–21. doi:10.1016/j.wear.2017.01.051.

Junak, J. 2016. 'Utilization of Crushed Glass Waste in Concrete Samples Prepared with Coal Fly Ash'. *Solid State Phenomena* 244: 102–7. doi:10.4028/www.scientific.net/SSP.244.102.

Junak, J., and N. Stevulova. 2008. 'Alkaline Modified Coal Fly Ash as an Addition to Concrete'. *Chemicke Listy* 102 (15 SPEC. ISS.): s882–83.

Junak, Jozef, and Alena Sicakova. 2017. 'Concrete Containing Recycled Concrete Aggregate with Modified Surface'. *Procedia Engineering*, International High-Performance Built Environment Conference – A Sustainable Built Environment Conference 2016 Series (SBE16), iHBE 2016, 180 (January): 1284–91. doi:10.1016/j.proeng.2017.04.290.

Jursova, S. 2010. 'Metallurgical Waste and Possibilities of Its Processing'. In, 115–20.

Jursova, S., S. Honus, and P. Pustějovská. 2016. 'Economical Evaluation of Reducibility of Compacted Metallurgical Waste with High Ratio of Fe'. In, 1838–43.

Stevulova, N., and J. Junak. 2014. 'Alkali-Activated Binder Based on Coal Fly Ash'. *Chemicke Listy* 108 (6): 620–23.

Advances and Trends in Engineering Sciences and Technologies III – Al Ali & Platko (Eds)
© 2019 Taylor & Francis Group, London, ISBN 978-0-367-07509-5

Experimental verification of mechanical properties of polystyrene for use in integral bridges design

M. Pecník, J. Panuška & V. Borzovič
Faculty of Civil Engineering, Slovak University of Technology in Bratislava, Slovak Republic

ABSTRACT: Integral bridges are progressive constructions with decreased maintenance costs. However, they are subject to cyclical horizontal movements because of thermal expansion and contraction. Expansion is typically more important in process of abutment design, as it causes passive earth pressures, which have much higher effect on the abutment than active pressures. By placing a layer of easily compressible material behind the abutment, this effect can be decreased significantly. Typical material for such layer is expanded polystyrene. This article presents the results of experimental verification of mechanical properties of this material. Specimens with 7 different densities were examined in both triaxial and compression machine, with maximal strain induced 60%. In one case, specimen was also exposed to cyclical loading.

1 PASSIVE EARTH PRESSURE DECREASE

1.1 *General principle*

Structures pushed into soil are exposed to passive earth pressures. Values of these pressures are significantly higher than those of the active earth pressures. However, only a few structures are exposed to such load. Typical example are abutments of integral bridges. Integral bridge deck expands and contracts due to temperature changes. In summer, expansion of the bridge deck is usually the greatest. Because of the rigid connection between the abutments and the bridge deck, this deformation is transferred to abutment and causes passive earth pressure.

These pressures if too high, can be lowered by placing an easily deformable layer behind bridge abutment (Figure 1). Such layer should be comprised of material that has sufficient longevity, and is relatively easily deformable in comparison to abutment. Material deformability can not be neither too stiff, but neither too yielding, to achieve load reduction. Therefore, it is necessary to work with accurate material properties. Most widespread material for construction of such layers is expanded polystyrene, because of its price, weight, and easy availability.

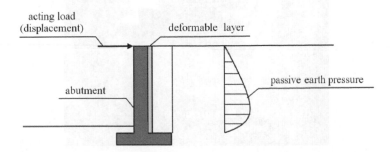

Figure 1. Setup of integral bridge abutment with compressible layer.

2 VERIFICATION OF MATERIAL PROPERTIES

2.1 *Testing procedure*

Experimental program consisted of two parts. The first part was testing in the triaxial compression machine. For this purpose, specimens of cylindrical shape, with diameter of 60 mm were needed (Figure 2a). The specimens were exposed to compression with maximal deformation 20 mm, because of equipment limitations. Deformation rate was 2 mm/min.

The second part was testing of plate specimens with dimensions 200 × 200 mm in compression machine (Figure 2b). Maximal deformation was 60 mm. Test setup is shown on Figure 3.

It should be noted that the extruded polystyrene specimen had many irregularities from manufacturing process, such as small holes, additions of different materials and visible changes in material texture.

Specimens of 7 different densities were examined. Along with expanded polystyrene, polystyrene with added graphite was also tested. Such material should have higher strength and deformation modulus. Extruded polystyrene was included in testing. The specimen's height and densities are shown in Table 1. All specimens had density above recommended 10 kg*m^{-3}, therefore fusion between bead can be considered sufficient (Horvath, J.S., 1997).

2.2 *Triaxial testing*

Triaxial testing proved to be suitable only for specimens with low strength. With high strength extruded polystyrene the cylinder specimen bended instead of compression. Results, shown

Figure 2.   Experimental specimens for a) triaxial (cylinder) and b) compression machine (prism).

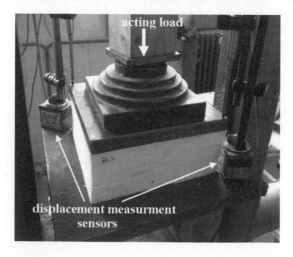

Figure 3.   Testing setup.

516

Table 1. Densities and height of the specimens.

Specimen	Measured density [kg*m^{-3}]	Height [mm]
A1	13.4	100
A2	13.5	100
B1	12.6	140
B3	13.1	140
C1	17.3	100
C2	17.3	100
D1	29.3	100
D2	29.6	100
E1	23.1	100
E2	22.1	100
F1 (added graphite)	25.3	70
F2 (added graphite)	25.0	70
G1 (extruded)	32.3	100

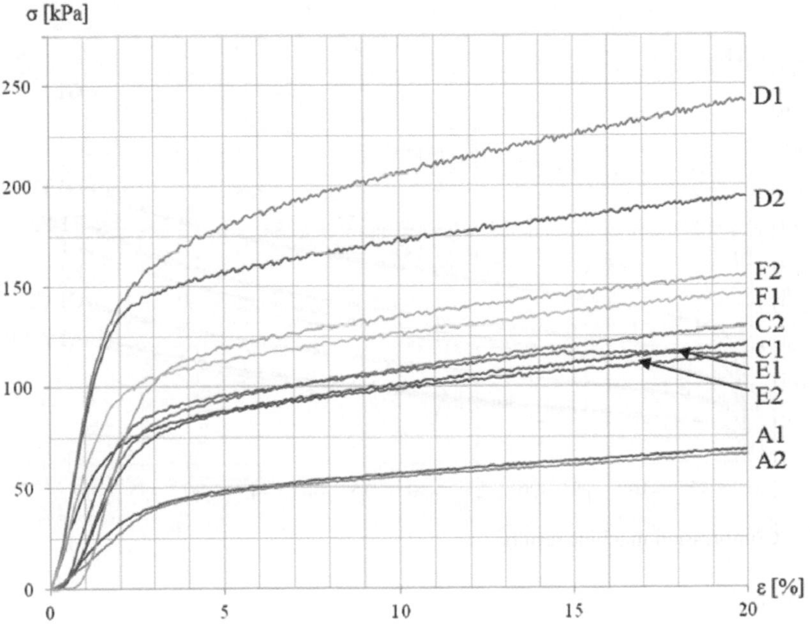

Figure 4. Triaxial testing results.

on Figure 4, shows only minor dispersion. It should be noted that in contrast with direct connection between material strength and its density, the specimen E1 and E2 was the second weakest, despite having highest density.

2.3 Triaxial testing

Triaxial testing proved to be suitable only for specimens with low strength. With high strength extruded polystyrene the cylinder specimen bended instead of compression. Results, shown on Figure 4, shows only minor dispersion. It should be noted that in contrast with direct connection between material strength and its density, the specimen E1 and E2 was the second weakest, despite having highest density.

2.4 *Compression machine testing*

Compression machine test showed almost no dispersion in results. Almost all stress-displacement relationships of same type of polystyrene overlaps. Extruded polystyrene showed highest strength, however its initial deformation modulus was very high, making it unsuitable for creation of compressive layer. Results are presented on Figure 5 and in Table 2.

Because horizontal movements of integral bridges are subject of cyclical horizontal movements, specimen B2 was examined under cyclical loading. Results are presented on Figure 6. After three cycles almost no changes in deformation modulus was observed, but plastic deformation increased.

2.5 *Poisson value*

Poisson value was not directly measured during testing, because displacement measurement sensors obstructed it. However, visible transverse contractions on polystyrene prisms appeared, which increased their size, as displacement induced by compression machine increased. This lead to conclusion that examined specimens have significant negative Poisson value, which corresponds to other authors work (Scott, E.G., et al., 2015). Detail of contractions is shown on Figure 7.

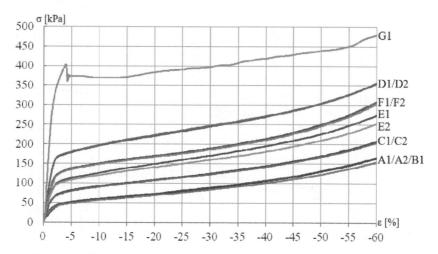

Figure 5.   Compression machine results.

Table 2.   Plastic deformations.

Specimen	Measured density [kg*m^{-3}]	Initial height [mm]	Induced deformation [mm]	Plastic deformation [mm]
A1	13.4	100	−65.3	−38.8
A2	13.5	100	−63.6	−38.0
B1	12.6	140	−96.1	−64.3
C1	17.3	100	−70.1	−45.8
C2	17.3	100	−70.4	−70.8
D1	29.3	100	−70.1	−54.1
D2	29.6	100	−70.2	−53.9
E1	23.1	100	−71.2	−50.4
E2	22.1	100	−70.3	−50.0
F1 (added graphite)	25.3	70	−50.1	−32.0
F2 (added graphite)	25.0	70	−50.1	−30.7
G1 (extruded)	32.3	100	−60.1	−30.7

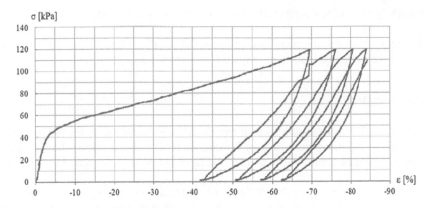

Figure 6. Cyclic loading of specimen B3.

Figure 7. Contractions on polystyrene prism under pressure.

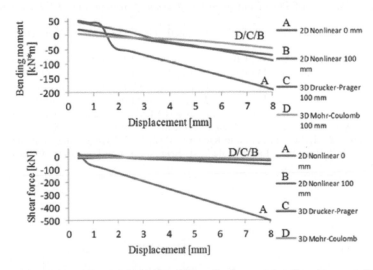

Figure 8. Internal forces with and without compressive layer a) bending moments b) shear force (Pecník, M. & Borzovič, V., 2017).

3 BENDING MOMENT AND SHEAR FORCE DECREASEMENT

Internal forces acting on abutment can decrease up to 50%. Figure 8 presents results from case study of 4 metres high abutment with 100 mm thick polystyrene layer.

4 CONCLUSIONS

Experimental testing showed that triaxial examination is suitable only for polystyrenes with low deformation modulus, because stiff specimens tend to bend rather than to compress. Compression machine results showed smaller dispersion in results, while not influenced by stiffness of the examined material.

Relationship between density and stiffness of material was clearly present, however, there was also small deviation with specimens E1 and E2, having smaller strength than expected by considering their density.

Extruded polystyrene was found to be unsuitable for purposes of deformable layer creation. Also, this material had many irregularities, such as pieces of different materials, small holes and non-uniform texture.

Cyclical loading proved that deformation modulus changes only very slightly with cycles.

All specimens, showed significant transverse contraction under pressure, leading to conclusion they have negative Poisson value.

ACKNOWLEDGEMENT

This work was supported by the Scientific Grant Agency VEGA under the contract No. VEGA 1/0810/16. This work was supported by the University Science Park (USP) of the Slovak University of Technology in Bratislava (ITMS: 26240220084).

REFERENCES

Azzam, S.A and Abdel Salam, S.S., 2015, EPS Geofoam to Reduce Lateral Earth Pressures on Rigid Walls, *International Conference on Advances in Structural and Geotechnical Engineering ICASGE'15.*

Ertugrul, O.L. and Trandafir, A.C., 2011, Reduction of Lateral Earth Forces Acting on Rigid Nonyielding Retaining Walls by EPS Geofoam Inclusion, *Journal of Materials in Civil Engineering 23 (12)*, 1711–1718.

Horvath, J.S., 1997, The Compressible Inclusion Function of EPS Geofoam, *Geotextiles and Geomembranes* 15, 77–120.

Horvath, J.S., 2000, Integral-Abutment Bridges: Problems and Innovative Solutions Using EPS Geofoam and Other Geosynthetics, Manhattan College Research Report No. CE/GE-00-2.

Horvath, J.S., 2010, Lateral Earth Pressures Reduction on Earth-Retaining Structures Using Geofoams: Correcting Some Misunderstandings, *2010 Earth Retaining Conference Proceedings*, 862–869.

Pecník, M., Borzovič, V., 2017, Comparison of various FEM approaches in analysis of passive earth pressures, *Computer Assisted methods in engineering and Science*, 24, 253–258.

Scott, E.G., Rollins, K.M., Scott, M.A., and Richards, P.W., 2015, Passive Force on Bridged Abutments and Geofoam Inclusions with Large-Scale Test, Brigham Young University, Department of Civil Engineering.

Effect of an interior green wall on the environment in the classroom

Z. Poorova, P. Kapalo & Z. Vranayova
Faculty of Civil Engineering, Kosice, Slovakia

ABSTRACT: Green infrastructure is a key strategy to provide cleaner air and water, while improving the living environment, human health and mental wellbeing. The paper is focusing on interior green walls and their qualities—on aesthetic, construction and environment. The paper describes a living wall built in one of the classrooms at Faculty of Civil Engineering, TUKE in Košice, its construction, irrigation and vegetation—the most important segments of every living wall. The research deals with the effect of a green wall on temperature in the classroom – temperature and humidity.

1   INTRODUCTION

In the winter of 2018 measurement of the indoor air quality was carried out in one of the classrooms of the Faculty of Civil Engineering. These parameters were observed: air temperature, relative humidity and CO_2 concentration. The interval of measuring data was 5 minutes. The measurements were performed between January 04, 2018 and February 02, 2018 and took place in two stages. During the first stage there was no green wall in the classroom. During the second stage, a green wall was installed in the classroom.

2   MEASURING DEVICES

2.1   *Testo 435-4*

The Testo 435-4 multifunction meter (Figure 1, Table 1) is perfect for monitoring, analyzing, and diagnosing indoor air quality, wherever you need to take important IAQ measurements. In addition, you can easily pinpoint and troubleshoot problems with HVAC systems, and immediately see the results of corrections. Commissioning, validating, benchmarking, or simply adjusting HVAC systems are all easy with the Testo 435-4.

The Testo 435-4 multifunction meter allows you to carry out all the measurements that are necessary to check and adjust ventilation and air conditioning systems, and to assess the quality of indoor air. The 435-4 accepts a wide variety of probes to measure and record

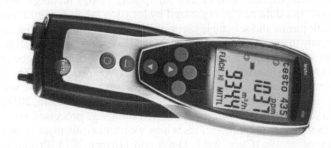

Figure 1.   Multifunction meter Testo 435-4.

Table 1.   Technical parameters of multifunction meter Testo 435-4.

Temperature—Capacitive	
Measuring range	–58° to 302°F/–50 to +150°C
Accuracy	±0.4°F (–13° to 166.8°F)/±0.2°C (25 to +74.9°C)
Resolution	0.1 °F/0.1°C
Humidity—Capacitive	
Measuring range	0 to +100 % rH
Accuracy	±2% RH (+2 do +98% rH)
Resolution	0.1% rH
Ambient CO_2	
Measuring range	0 to +10000 ppm
Accuracy	±(75 ppm ±3% z mv) (0 do +5000 ppm)
Resolution	1 ppm

Figure 2.   Testo 0632.

Figure 3.   Classroom during the first measurement stage.

airflow, humidity, temperature, and indoor air quality (IAQ) readings. With the included ComSoft Software, this data can be organized by location, analyzed, and reported.

Probe dependent menus and selectable user profiles ensure that the Testo 435-4 multifunction meter is always ready to use. Easy setup of user defined IAQ and airflow/duct profiles allows for fast measurement of many different tasks, and with a wide variety of probe choices available, the Testo 435-4 is ready to handle a vast array of HVAC/IAQ applications. The superb data handling will keep your readings organized and ready, however you need them, whether in a quick printout on-site, or for further analysis for placement into a report (Testo, 2018).

During the first evaluation phase, the measured data was processed on days when there were no people in the room (Figure 3). This is how we obtain information on air quality without the influence of people (Cuce, 2017; Davis and Hirmer, 2015; Prodanovic et al., 2017; Razzaghmanesh and Razzaghmanesh, 2017).

The Testo 0632 rapidly provides assessment of indoor climate and quality. Performs four functions: measurement of CO_2, humidity, temperature and absolute pressure. Prevents poor air quality in rooms by measuring CO_2. The Testo 0632 has a mni-DIN type connection fxed cable 1.6 m long. Absolute pressure measuring range is: 600 to 1150 hPa (240 to 460 InH2O). Absolute pressure accuracy: is +/– 10 hPA (+/– 4 InH_2O), ambient CO_2 measuring range: 0 to 10,000 ppm CO_2 Ambient CO^2 accuracy: +/–(75 ppm + 3% of mv) (0 to 5000 ppm); +/–(150 ppm + 5% of mv) (5001 to +10000 ppm). Humidity (capacitive) measuring range: 0 to 100%rH. Humidity (capacitive) accuracy: +/– 2%RH (for 2 to 98%RH range). Temperature (NTC) measuring range: 32 to 122 degrees F (0 to 50 degrees C). Temperature (NTC) accuracy: +/– 0.5 degrees F.

2.2 Testo 0632

The Testo 0632 1535 IAQ Probe (Figure 2) is used to quickly check the indoor climate, for example in offices, production areas or in warehouses. As well as temperature, humidity and absolute pressure, it also measures carbon dioxide concentration (CO_2) (Testo, 2018).

3 MEASUREMENTS

During the both stages of measurement, students were in the room during certain hours and during certain hours the class room as empty.

Figure 5 illustrates measured air quality values in the classroom. In Figure 5, we can see that the air temperature was approximately the same during the measurement. Also, the CO_2 concentration was in the same band. The biggest difference was recorded with relative air humidity.

The purpose of the measurement was to find out what effect on the air quality has the green wall (Manso and Castro-Gomes, 2015).

During the first evaluation phase, the measured data was processed on days when there were no people in the room (Figure 3). This is how we obtain information on air quality without the influence of people (Cuce, 2017; Davis and Hirmer, 2015; Prodanovic et al., 2017; Razzaghmanesh and Razzaghmanesh, 2017).

To simplify the evaluation, the arithmetic mean of the measured data was calculated for each time period, especially for green wall measurements and especially without the green wall. The resulting values of the mean values of the measured data are documented in Figure 7.

Figure 4. Classroom during the second measurement stage.

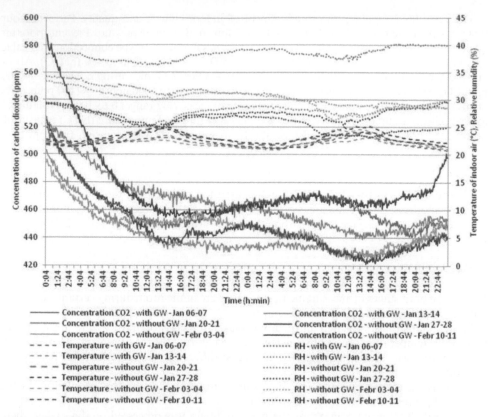

Figure 5.   Air quality in the classroom.

Figure 6.   Classroom during the second measurement stage—with the students.

From these adjusted data it can be stated that in a room with a green wall and without it:

– Air temperature difference in the classroom varies between 0 to 1.2°C. In the classroom with a green wall, the average air temperature was lower by 0.6°C.

524

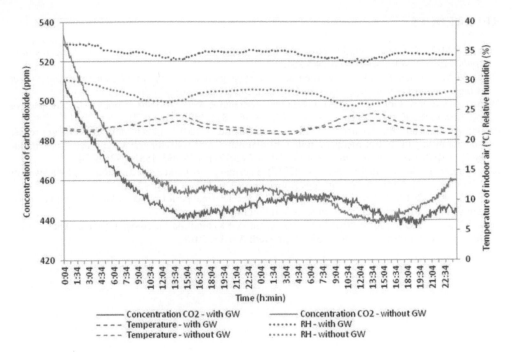

Figure 7. Mathematically modified air quality data in the classroom.

- Difference in CO_2 concentration varies between 0 ppm to 23 ppm, which can be considered a negligible difference. When staying in the room, this value is approximately 2.3%. From the measured data, we can say that in a room with a green wall there is mostly a lower concentration of CO_2 than in a room without a green wall. It can be said that the green wall has a positive effect on the persons in the room.
- The difference in relative humidity varies between 6% to 8% is significant. From the measured data we can see that the mean relative humidity in the room without the green wall is 27.7% RH, which is below the required 30% RH (STN 73 0540-3, 2012) limit. In the room with the green wall the average relative humidity was recorded from 33 to 36% RH, which is more favorable for the stay in the room.

4   RESULTS

From the preliminary assessment of indoor air quality from the measured and statistically modified data, it can be stated that the green wall favorably influenced the indoor air quality in the room. Relative air humidity increased in our case by approximately 6.9%. It is also possible to assume that the green wall achieves a CO_2 reduction of about 2.3%. As a result of increased relative air humidity, the room air temperature was reduced by 0.6°C on average. Thus, if the green wall is used in the room, it will be necessary to increase the heating medium temperature during the heating period.

ACKNOWLEDGEMENTS

This work was supported by: VEGA 1/0202/15 Sustainable and Safe Water Management in Buildings of the 3rd. Millennium and Slovak Cultural and Education Grant Agency (contract No. 073TUKE-4/2015).

REFERENCES

Cuce, E. (2017). Thermal regulation impact of green walls: *An experimental and numerical investigation.* *Applied Energy* Vol. 194 (2017) p. 247–254.

Davis, M.M., Hirmer, S. (2015). The potential for vertical gardens as evaporative coolers: *An adaptation of the 'Penman Monteith Equation'.* *Building and Environment* Vol. 92 (2015) p. 135–141.

Manso, M., Castro-Gomes, J. (2015). Green wall systems: *A review of their characteristics.* *Renewable and Sustainable Energy Reviews.* Vol. 41 (2015) p. 863–871. ISSN 1364-0321.

Prodanovic, V., Hatt, B., McCarthy, D., Zhang, K., Deletic, A. (2017). Green walls for greywater reuse: Understanding the role of media on pollutant removal. Green walls for greywater reuse: *Understanding the role of media on pollutant removal.* Vol. 102 (2017) p. 625–635. ISSN 0925-8574.

Razzaghmanesh, M., Razzaghmanesh, M. (2017). Thermal performance investigation of a living wall in a dry climate of Australia. *Building and Environment* Vol. 112 (2017) p. 45–62.

STN 73 0540-3 (2012). Thermal protection of buildings. Thermal performance of buildings and components. Part 3: Properties of environments and building products.

Testo SE & Co. KGaA, 2018, https://www.testo.com/sk-SK/sonda-iaq-pre-posudzovanie-kvality-vzduchu-v-miestnosti-meranie/p/0632-1535. Retrieved: April 5, 2018.

Advances and Trends in Engineering Sciences and Technologies III – Al Ali & Platko (Eds)
© 2019 Taylor & Francis Group, London, ISBN 978-0-367-07509-5

Case study about BIM technology and current knowledge of university students and their view on this issue

K. Prušková
Department of Civil Engineering, The Institute of Technology and Business in České Budějovice,
České Budějovice, Czech Republic

ABSTRACT: Building information modeling, further Building Information Management (BIM) technology is on the rise of use nowadays. BIM Technology's implementation for its wide use has many barriers. One of them is the low knowledge about this technology, not only among the practitioners of the AEC Industry (Architecture, Engineering and Construction Industry), but also among students, as the future practitioners. Nowadays students, as young people gaining their knowledge through the educational system, are the future for using innovative effective technologies for better and enhanced processes within wide spectrum of activities throughout the AEC industry. This article reveals the case study on the current situation of level of knowledge of the students in tertiary education, their view on the modern computer modeling and management and also on the current business operations and tracking barriers of implementing the BIM technology as a future in computer modeling. Case study is based on results from the questionnaire filled in by students studying the BIM subject. The aim of this paper is to reveal possible barriers for BIM Implementation from the view of students, future practitioners.

1 INTRODUCTION

1.1 *BIM technology implementation*

In order to reach the wide and correct use of BIM Technology, it is necessary to start with young people, students, hence future practitioners. Education is part of the processes going towards quality, time efficiency and reaching expected benefits connected to the BIM implementation. The high quality of education and change management, thus work with people, leads to effective implementation of any new software solution. Very complex spectrum of subjects and roles participating in every project and its constant individual planning for each project exists in BIM Technology. The BIM education should be in two areas: project management and system engineering (The Ministry of Industry and Trade, Czech Republic, 2017).

1.2 *BIM abbreviation and its meanings*

Speaking about BIM Technology, we can differ between three well-known explanations of this abbreviation: BIM as a Building Information Modeling, BIM as a Building Information Management and BIM as a Better Information Management. These explanations differ in area we are speaking about.

1.3 *Building information modeling*

BIM as a Building Information Modeling means the new approach to design and visualization of information. Let's start with that term. Many people understand Building Information Modeling as 3D design with realistic 3D models and complex visualization. That means, of course, the use of proper software.

Digitization is not a new topic in today's society, but its use within construction industry is at the lowest level among the other spheres of industry. The beginnings of the AEC industry digitization were many decades ago. It has started by simple computer aided design, known as CAD software in current times. This software enables creating of traditional construction drawings in a precise digital way by drawing precise lines. Unluckily this way of project documentation creation is still very widespread. The other development leads to the modeling of not just lines, but of the whole objects—the 3D modeling. This software offers more automatically generated information about the design, better visualized information. The BIM modeling is more than that. The BIM model contains next to the graphical information also the nongraphical information about modeled elements, the whole design with metadata as a time footage. It gives us the opportunity to gain precise information in the area of the amount of material, costs, environmental impacts towards sustainable building, etc. BIM Modeling also reduces human failures thanks to better visualized information, better comprehension and the limitation of human unpropitious influence via automatic creation of budgeting, appraisal, but also sub-deliveries of materials and their logistics. The BIM model is a valuable tool for visualization and calculation of future objects, variations of materials and their impacts of costs, sustainability, environment, etc. Compared to traditional project documentation, the BIM model also eliminates misunderstanding in design by collision elimination of architectural solution and engineering equipment. The BIM model contains well detailed description, makes the communication, visualization and sharing information easier and eliminates human failure factor, thus raises efficiency of the project at all its stages (Krivonogov, Zakharova, Kruglikov and Plotnikov, 2018).

1.4   *Building information management*

BIM as a Building Information Management goes further. It is connected not just to the amount of information, but also to its structure, use, further development, responsibilities and processes. As Nývlt (2018) wrote: "Managing knowledge and information seems to be the most crucial point in implementation of BIM within any country specific industry. It is obvious that we are challenged by how to transfer basic thinking about processes within building industry. It is not only about BIM in meaning of Model or Modeling, and we need to look even further that to BIM in meaning of Management. There is big potential, supported by technology development, to employ LIM (Land Information Modeling) – geographical data together with BIM-Model and to step towards "Better Information Management."

1.5   *Better information management*

The newly spread explanation of BIM is Better Information Management involving many branches. This term clearly presents the BIM principles in managing processes. Connected to better information management, very important part is knowledge transfer issues. Thanks to well-planned cooperation and data sharing processes, knowledge transfer may become very effective and benefiting factor influencing the whole building lifecycle. It is necessary to deeply understand all actions, definition of participants, responsibilities and communication streams (Ceptureanu, Florescu, 2018). Nývlt (2018) said: "Managing knowledge sharing and to change and/or improve processes within whole industry becomes challenge we are facing just now. It seems to be a way, how to deal all barriers, BIM adoption must overcome."

2   BIM TECHNOLOGY USE AND EDUCATION

2.1   *The need for education*

As Krivonogov, Zakharova, Kruglikov and Plotnikov (2018) said: "In conditions of the large deficit of qualified personnel, who possess new skills and is able to implement BIM-technologies, the government has an understanding of the need to train specialists to use the information modeling in the field of industrial and civil construction. We need educational standards and new educational programs that meet new requirements."

The BIM education can be divided into two main groups: the education of practitioners who will be involved in this revolutionary changes and the education of future practitioners—today's students who should be nowadays preparing and learning BIM at high schools and universities (The Ministry of Industry and Trade, Czech Republic, 2017). This paper focuses on the second group, university students.

2.2 *The BIM education of university students*

There is a need for implementing new professional subjects dealing with BIM issues. The goal of providing the BIM implementation is let students understand the principles, connections and needs for single processes and activities. Students also get acquainted with cases of good practice from other countries with the structure of standardization, its substance, possibilities of use and changes which brought throw the BIM Technology int the practice in AEC Industry. The important aspect is right understanding the BIM Technology and providing more complex views on this technology throughout the whole building lifecycle from the design of construction to its demolition (The Ministry of Industry and Trade, Czech, Republic, 2017).

2.3 *Possible failures within the BIM implementation into education*

As mentioned above, the BIM implementation into education is not just about adding new lectures on this issue, but also to innovate current lectures on other issues with new approach to the field of Civil Engineering with thinking in the BIM way. The BIM technology is very complicated topic overcrossing very large area not even in construction. As Matějka, Růžička, Žák, Hájek, Tomek, Kaiser and Veselka wrote (2016), there can be revealed many problems and failures within the BIM implementation into education, such as:

- "Is not deep enough and gained knowledge is only basic,
- Is too specialized (one study branch oriented) and lacks perspective (resulting even in students obtaining incorrect knowledge,
- Lacks needed complexity and integration into current subject,
- Lacks time perspective (i.e. only considers present state and not the future, which is very important for a field which develops so fast),
- Ends as BIM tool utilization or software developer presentation instead of proper BIM education."

Of course, every bigger change in study program is quite time-consuming process. That's the reason why we have already started with little changes. We added the BIM lectures as an optional lecture in current study programs. The new study programs will be modified more towards the BIM Technology by lectures on BIM itself and by evolving other lectures and implementing the new way of thinking with BIM principles.

3 EDUCATIONAL SYSTEM ON THE INSTITUTE OF TECHNOLOGY AND BUSINESS IN ČESKÉ BUDĚJOVICE

3.1 *Educational system on the Institute of Technology and Business in České Budějovice*

The teaching students about the BIM Technology on the Institute of Technology and Business in České Budějovice is new as on other universities. There is a need to get students acquainted with main principles of BIM Technology, BIM tools and connected processes.

3.2 *The BIM technology*

Speaking about principles of the BIM Technology, students have lectures on the idea of BIM, its benefits throughout the whole building lifecycle. We teach them about all participants of design process and construction. There are also lectures on project management,

Building Execution Plan and the way how to share and further develop information, facility management, intelligent buildings, etc. Students also know about BIM implementation process in the Czech Republic, planned activities and current situation.

3.3 BIM tools

As on other universities, academics from The Department of Civil Engineering have lectures on computer aided design. The goal of these lectures is to gain skills while using different software from 2D drawings, 3D modeling, visualization to BIM tools as Archi-CAD and Autodesk Revit. Students are able to create project drawings working with these tools.

3.4 BIM in Civil Engineering

Current lectures will be enhanced by adding special lectures on BIM use in individual spheres. This means that on classic lectures on many topics as static solution, sustainable development, technical equipment and others, students get acquainted with possibilities of BIM use in these spheres. Benefits of BIM Technology will be demonstrated in practical example. Professionals will be invited on lectures.

3.5 BIM best practice

The Institute of Technology and Business in České Budějovice invites many practitioners who are experienced in BIM use, BIM definition, legislation and construction. These practitioners lecture students about BIM in practice, share their knowledge and they also give students opportunity of cooperation in their joint research towards writing their qualifying thesis. Students get the opportunity for their future job, too.

3.6 Conference building defects and topic BIM

The Department of Civil Engineering organizes annual Conference Building Defects. One of conference's topic is BIM. Students, practitioners and also some stakeholders' representatives are invited to this conference. Many researchers share their knowledge, research results and experience while using BIM Technology.

4 POSSIBLE BARRIERS FOR BIM IMPLEMENTATION FROM THE POINT OF STUDENTS' VIEW

4.1 Case study

Within the optional BIM lectures performing in summer semester 2018, 47 students were responding the question about barriers which they see in the BIM implementation. Before that, they got acquainted with basic BIM principles, its benefits, the way, how to communicate and share data, basic rules and also the lectures on best practice by practitioners from the AEC industry. They also wrote their work about some topic connected to BIM they were interested in. Some of them did a little research among their colleagues from practice and discussed about the BIM implementation and their opinion on this issue. Through their work, they revealed many opinions of different practitioners, they have learned other possibilities. Students also shared their gained knowledge during lectures, talked about their own opinion and they also had a discussion on different topics in BIM field and traditional practice.

Than they responded a question, what do they see as possible barriers for BIM implementation into the practice in the AEC industry. Their responses were generalized and divided in a few groups according to their joint topics. In Figure 1, there is shown how many times they wrote the answer matching the group.

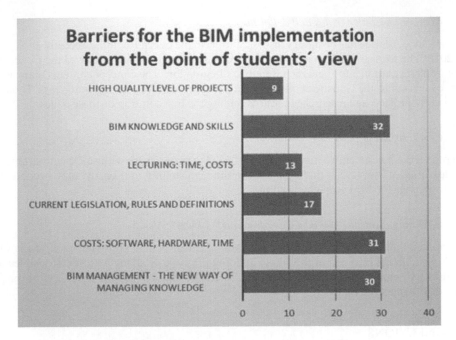

Figure 1. Barriers for the BIM implementation from the point of students' view.

4.2 *High quality level of projects*

The BIM Technology and its proper use have high demands on the input of information quality, structure and detailed solution. The traditional project documentation offers a lot of possibilities how to evade complicated elements, their connections. Next to that the BIM model needs to be provided in high quality level of all elements and details.

4.3 *BIM knowledge and skills*

When lecturing students about the basic principles of the BIM Technology, for some of them it was the first time when they get acquainted with this term. Even though the motto of The Institute of Technology in Business in České Budějovice is "Practice is the best teacher" and all of students have compulsory practice in AEC Industry companies, most of them are used to create a project documentation in 2D software, only. Most of them know BIM tools, but they used them just for 3D model and visualization. Resulting that gaining the proper knowledge about BIM means a big step for them, they still must walk through.

4.4 *Lecturing: time and costs*

The low level of BIM knowledge is not just among the young people, students, but also among the practitioners. Implementation of the BIM technology and its implementation into the company processes is a big step meaning the big change in work processes for its employees and to learn a new way, how to communicate. Companies need go through the many lecturing of their management to set the new processes and the way of communication, data sharing and enhancing current processes. The BIM project execution plan is necessary tool for well pre-pared schedule of processes, involved people and their responsibilities, goals, etc. Only well-set BIM project execution plan goes toward to the BIM project with the lowest data, time and money loses. The company employees should enhance their skills in using proper BIM software. It could be very time consuming and costly efforts for the company, because most of them maybe uses the proper BIM software tools, but not in the way of BIM Technology.

4.5 *Current legislation, rules and definition*

Task about legislation has been very discussed topic in the Czech Republic, recently. In September 2018, The Ministry of Industry and Trade of Czech Republic has approved the Concept of BIM implementation into the Czech Republic's Legislation. The concept contents many issues from education to the Construction Law which should be solved during next years. There are many early projects and researches towards finding the best solution of these issues..

4.6 *Costs: software, hardware, time*

Of course, the investment to new software, often also to hardware with sufficient parameters, and learning of employees that seems to be time-consuming and costly, may become very serious barrier for **BIM** Implementation.

4.7 *BIM management—the new way of managing knowledge*

BIM management, or Better Information Management is new approach to efficient work of all participants involved in the whole building life cycle. Starting with investor's idea and architect's design, other participants share their information and further develop project towards the construction.

5 CONLUSIONS

Education, other ways of learning and gaining knowledge is one of the most important pillars of **BIM** implementation and its effective use. Education of future practitioners in Architecture, Engineering and Construction Industry should go throw revolution in the way of **BIM** thinking. There is a long way ahead, but it is necessary to start in time, because of the complexity of the AEC industry. The AEC industry is nowadays the least digitized industry and it is time to move forward towards new innovative way of design, construction, sustainability and facility management. BIM Technology fully offers well set processes in the most effective way to reach identified goals. Students are the most opened personnel to gaining new knowledge, learning new principles and getting new experiences. To change the education system seems to be a time-consuming process, but it is needed. The aim of this paper is to summarize the situation, to try to identify the barriers from the students' view and make the procedure of **BIM** implementation into education through most effective way, which will be motivating for students' education. The paper reveals the students' overview and awareness of possible barriers in **BIM** implementation.

REFERENCES

Ceptureanu, S.I., Florescu, M. 2017. Knowledge transfer issues. *Proceedings of 29th International Business Information Management Association Conference*: 3456–3461. Vienna: Austria.

Czech Republic, Ministry of Industry and Trade. 2017. Concept of introducing the BIM method in the Czech Republic. Prague: Czech Republic.

Krivonogov, A., Zakharova, G., Kruglikov, S. & Sergey Plotnikov, S. 2018. Implementation of BIM-technologies in the educational program of the architectural university. *Proceedings of 9th International Scientific Conference Building Defects 2017, MATEC Web of Conferences 146*. Cedex: France.

Matějka, P., Růžička, J., Žák, J., Hájek, P., Tomek, A., Kaiser, J. & Veselka, J. 2016. The Implementation of Building Information Modeling into educational programs at CTU in Prague. *Proceedings of CESB2016 – Central Europe towards Sustainable Building 2016*: 853–860. Prague: Czech Republic.

Nývlt, V. 2018. The Role of Managing Knowledge and Information in BIM Implementation Processes in the Czech Republic. *Proceedings of 9th International Scientific Conference Building Defects 2017, MATEC Web of Conferences 146*. Cedex: France.

Advances and Trends in Engineering Sciences and Technologies III – Al Ali & Platko (Eds)
© 2019 Taylor & Francis Group, London, ISBN 978-0-367-07509-5

Problems in carrying out construction projects in large urban agglomerations on the example of the construction of the Varso building complex in Warsaw

E. Radziszewska-Zielina, E. Kania & G. Śladowski
Faculty of Civil Engineering, Cracow University of Technology, Cracow, Poland

ABSTRACT: The paper discusses problems in the carrying out of construction projects in large urban agglomerations on the example of a complex of buildings (including the highest skyscraper in the European Union) located in Warsaw. The location and the specifics of the work that had to be performed generated numerous technological and organisational problems. The manner of solving them, in addition to guideline proposals for similar projects are discussed in the paper.

1 INTRODUCTION

The carrying out of construction projects is a complicated process associated with many factors, especially when it is being performed in the centre of urban agglomerations. Urban agglomerations developed significantly in recent decades (Mitra, Nagar 2018). The urban agglomeration is a highly developed spatial form of integrated cities and is one of the most important conveyors of global economic development (Fang, Yu 2017).

In their earlier publications the authors pointed to problems that appear during the carrying out of construction projects in the centre of an urban agglomeration (Radziszewska-Zielina, Kania, Śladowski 2018), their characteristics (Radziszewska-Zielina, Śladowski, Kania 2018), in addition to presenting a case study (Radziszewska-Zielina, Kania 2017) on this subject.

Currently, the landscape of urban agglomerations is undergoing intensive change due to newly appearing buildings, which are to stand out from the existing ones, especially in terms of height (Craggs, 2018).

The authors of many publications, both in Poland and around the world, describe the process of designing and constructing tall buildings. The authors of (Abomoslim, Russell 2014) discussed a wide range of solutions, including innovative ones, in the design and construction of skyscrapers. Many authors highlight the problems that appear during the construction of high-rise buildings, such as: dense built-up areas and the problems associated with them (Gongalo, Gudovicheva, Gubareva, Dobrynina 2018), selection of the method of constructing the building, conducting all geotechnical works and performing deep excavation work in the centre of an urban agglomeration (Tan & Wang 2015; Ali & Moon 2007; Jasmine Nisha & Muttharam 2017), dynamic and geological impacts as well as the need to monitor neighbouring buildings (Dong, Samsonov, Yin, Ye, Cao 2014; Wójcicki, Grosel, Belostotsky, Akimov & Sidorov 2017), broadly understood problems in construction site logistics (Okolelova, Shibaeva, Trukhina 2018) and vertical transport (Koo, Hong, Yoon, Jeong 2016). Innovation and the broad application of digital technologies was pointed as currently fundamental in the carrying out of construction projects, especially complexes of high-rise buildings (Wang, 2017).

2 CHARACTERISTICS OF THE VARSO COMPLEX

The Varso building complex is being built in the centre of Warsaw, at 73 Chmielna Street. The complex consists of three buildings: Varso 1, Varso 2, Varso Tower. The beginning of the investment was December 2016 and the completion is planned for 2020. HB Reavis is the developer and general contractor.

The Varso complex (Figure 1) consists of three top-class buildings connected by a shopping and commercial service arcade on the ground floors. The underground levels are intended for an underground parking garage and technical facilities, with generally accessible areas in the case of Varso Tower, ones that will connect the building with passages leading to the Central Railway Station. The buildings consist of 4 underground floors (excavation depth reaches 19,5 meter below ground level) and, respectively: Varso 1–19 above-ground storeys (80 m high), Varso 2–21 above-ground storeys (90 m high), Varso Tower – 53 above-ground storeys (310 meter high with spire). The total usable floor area of the complex is 144 500 square meters, out of which the highest building in the European Union—Varso Tower will have feature 63 545 square meters. The supporting structure of the buildings consist of reinforced spatial concrete plate and column systems with monolithic cores ensuring rigidity (varso.com).

3 PROBLEMS ASSOCIATED WITH CARRYING OUT THE ANALYSED COMPLEX OF BUILDINGS IN WARSAW

The construction of the Varso complex in the centre of Warsaw is associated with numerous difficulties. The following are the most problematic conditions that occur during the implementation of this construction project, i.e. dense built-up area, implementation of deep excavations, collisions with existing elements either adjacent or crossing through the site and problems with logistics.

3.1 *Problem related to dense built-up area*

The Varso building complex under construction is located in the very centre of Warsaw. The surroundings of the project site include one of the main streets leading from the south to the north of the city—Aleje Jana Pawła II, the busy Chmielna Street, a railway city tunnel, underground passages, technical facilities and platforms of the Central Train Station and the adjacent residential and office buildings, all featuring underground levels. Such surroundings generate major difficulties and constraints for the work being performed and affects the manner and time of conducting them.

3.2 *Problems concerning collisions with existing elements either adjacent or crossing through the site*

Before the construction of the building complex on the site, expensive initial works had to be carried out, causing many problems and requiring a lot of additional preparation and top-level organisation. After performing a detailed survey of the site, it turned out that there were numerous collisions, particularly because of the existing and actively operating infrastructure of the Central Train Station.

Figure 1.   Visualisation of the Varso complex [varso.com].

In the southern part of the site planned for the Varso Tower building, in the immediate vicinity of the diametral tunnel, deep underground (the bottom of the tank reached the level of −18 m below the ground level) there was a caisson sewage tank of the Central Railway Station, which demolition was one of the conditions to start the exercise diaphragm walls. In order to replace an existing tank, a new, underground tank for sewage was built using diaphragm walls underneath the Varso 1 building.

The next challenge was the relocation of the transformer station which was located in the station facilities that were within the border of the site under the Varso Tower. This station provided power to the Central Tran Station and controlled train traffic throughout Poland, which caused an additional problem as the operations of the PKP could not be disrupted. The entire operation was carried out successfully and the station was moved to the buildings on the other side of Aleja Jana Pawła II.

The collision with the buildings of the Central Train Station played out in several parts. It was necessary to dismantle some of the existing facilities, especially their underground part. In order to do this, reinforced concrete walls had to be built in the building, as it altered the static system of the remaining part of the building after demolition. This conditioned the possibility of a planned demolition of a part of the building. In this building there was also a passage for passengers leading to railway platforms. Because of this, remodelling and the construction of new passages was required. The train control room had to be used for this purpose—with the necessity of moving the existing control room to a different location.

The site was crossed by elements of external infrastructure, which had to be remodelled. This affected the time and manner of constructing the buildings. On the plot for the construction of Varso 1 there was a fiber-optic bus (about 60 cables) running across the central part of the excavation. Due to there being no possibility of its relocation, it was placed in a reinforced concrete duct supported on a fragment of the bracing slab, under which a four-storey underground parking garage is being built. Ultimately, this duct will be connected to the ground floor of the Varso 1 building. All these collisions, as well as many minor ones, significantly affect the extension of the schedule of works and increase the cost of the project.

3.3 *Problems related to deep excavation*

The construction of the underground part of the buildings in a dense urban environment that features buildings several storeys high, as in the case of the Varso complex, is the most problematic and the most difficult stage of construction. The underground part of all structures is being built through the use numerous sophisticated excavation reinforcement technologies (Figure 2). The envelope is composed of diaphragm walls, the stability of which is ensured by bracing slabs, ground anchors, steel struts in corners and a ground support. Due to the location of the complex and the ground and water conditions present at the site, water level monitoring is conducted 24 hours a day.

The site was mostly free of buildings, however, it was known that before the Second World War it had featured buildings and that the discovery of non-surveyed foundations, basements, and masonry walls was likely, which had to be considered in the planning of the excavation and created additional problems after uncovering them. At the same time it was possible that unexploded bombs could be found during excavation. During the works, something very valuable was discovered, and at the same time caused a lot of additional problems and costs— a post-glacial erratic boulder measuring 12.5 meter in circumference and 2.5 meter in height, weighing 55 tons (Figure 3). A special steel platform was built to extract the boulder from the excavation. It was placed on the platform and then pulled out using a specialised crane with a very large load capacity. In order to carry out the entire operation and transport the boulder, Chmielna Street and the entire crossing route were closed for traffic. The boulder was examined by scientists from the Warsaw University of Technology and recognised as a natural monument. It will be displayed in front of the Varso Tower building after their completion.

A section of site of the Varso Tower building was built-up. It contained the technical facilities serving the Central Train Station and their pile foundation, the air intake ventilation, pump station tanks, retaining walls and ventilation ducts. Accordingly, the construction

Figure 2. Deep excavation on the Varso construction site [photo by HB Reavis].

Figure 3. Post-glacial erratic boulder found on the Varso construction site [photo by HB Reavis].

of the diaphragm wall of the building had to be adjusted after the surveying of the existing structures that would have otherwise caused a collision.

Due to the location of the excavation in a dense urban development and its depth, changes in the position of the excavation's reinforcement are constantly being monitored, in addition to the site, the adjacent buildings and the walls of the city railway tunnel and its rails.

3.4 *Problems related to logistics*

Logistics is a key element of every construction project. Particularly so in the centre of the urban agglomeration, refining all the possibilities of proper management of supply logistics and construction site is paramount. As in most cases of projects in centres of urban agglomerations, the construction of the Varso complex is affected by many problems.

In this case, these problems have accumulated specifically because of the simultaneous construction of 3 buildings without the possibility of staging. The thing that is most noticeable is the lack of space, a lack of storage yards, the need to rent additional storage yards far away from the construction site and the need to construct access roads.

Before construction work commenced, a detailed logistics plan was being developed for six months, which made it possible to greatly minimise any logistics difficulties that might have appeared.

Construction site facilities were planned in a specific manner. The lack of space and the location of external walls of parking garages (diaphragm walls) at the border of the plot made it impossible to use any part of the construction site for this purpose. Therefore, a section of the Varso 2 ground floor was made (as mentioned before, the buildings were being built using the top-down method) upon which the multi-level container facilities were placed. A number of working platforms have also been provided, based on temporary support structures that act as storage and handling yards.

The largest volume of traffic is encountered during the carrying out of earthworks and concreting large elements. Due to the lack of space and the lack of the possibility to occupy the road lanes (the construction site is adjacent to the main arteries of Warsaw) a stationary concrete pump was installed.

7 gates have been organised on the construction site through which goods can be delivered. The organisation and delivery solutions applied by the general contractor can be considered as model and well thought-through, which avoids or reduces many problems to a minimum. For the construction of the Varso complex, a mobile application has been developed that significantly increases the speed of performing work. Due to the fact that in the case of such space restrictions and the amount of daily supply deliveries, everything must be coordinated beforehand. During the first step, the engineers of subcontractor firms use the application to register a planned delivery, entering its time, gate number, handling time and book a tower crane if it is required during the unloading. Information is available to everyone, and especially to the most interested people, that is, the logistics manager that coordinates the whole and the two traffic marshals directly responsible for operating the gates, transports and access cards (thanks to coordinating the entire system in one application, data for obtaining the BREEAM certificate assessing the environmental performance of the building is automatically collected). Efficient traffic manage-

Table 1. Causes of technological and organisational problems in the carrying out of a construction project in an urban agglomerations and proposals of actions making it possible to avoid or reduce them on the basis of the Varso project. Original work.

Factors	Recommendations
Dense urban development	Developing a detailed documentation of the project during the preparatory and design phase through the comprehensive identification of soil and water conditions both underneath a planned building and in its zone of influence on adjacent structures. Determining zones of the impact of the designed building and monitoring the condition of existing structures.
The need to perform deep excavations	Selection of the most appropriate type of envelope and methods for the constructing of the underground part of the designed facility, with the possibility of rapidly responding to the need to change the technology in the event of a collision occurring during excavation. Constant monitoring of the envelope, the soil and neighbouring structures.
Logistics difficulties	The preparation of a detailed logistics plan before commencing construction work. The preparation of and strict adherence to the delivery schedule. Implementing new technologies supporting the operations of the logistics department.
Collisions with existing infrastructure.	A quick but well-thought-out reaction of the decision-making staff to the emerging difficulties, allowing to minimize the time and cost of addressing collisions.

ment and work organisation caused has made it possible for about 200 lorries of spoil to be transported off-site daily (about 4.2 m3 of spoil/car) during the excavation work.

Traffic obstructions are a very serious problem in city centres. The development project is located near large, busy streets, which is beneficial to the end users because it provides good circulation connections with the rest of the city, however, it is a significant constraint during performing construction work. The greatest logistical difficulties in this matter occur at the excavation stage—the spoil removal and the stage of building the concrete structure to be exact. Concrete mix, due to prolonged transport as a result of traffic obstruction may not be suitable for construction because it may change its consistency and should then be sent back to the plant for recycling, which generates high costs for the contractor and supplier and negatively affects the environment through air pollution from exhaust fumes produced by concrete mixers.

An additional cost incurred by the project budget in a large city is the renovation of streets adjacent to the project. The transport of large loads can damage pavements and streets, which require repair that is not included in the cost estimate. It should also be kept in particularly good order, which involves additional labour costs.

The carrying out of construction in the centre of an urban agglomeration and at such a significant height of the buildings also entails many formal and legal approvals, such as the approval of the work of tower cranes with airport management.

3.5 Summary of problems and recommendations

Table 1 shows the most common causes of problems in the carrying out of construction projects in a large city, determined on the basis of the structures that were analysed. Good practices were presented and recommendations were proposed to limit the occurrence or the effects of such problems in future during the carrying out of similar construction projects.

4 CONCLUSIONS

In the case of an investment that is still being implemented, like the Varso complex, it is difficult to assess how the solutions taken will affect the final construction result. At this point, we can consider the solutions used as good enough to implement investments in accordance with the assumed schedule.

The carrying out of construction projects within a large urban agglomeration is therefore a difficult undertaking. The space in which works are carried out, the developer's requirements and short completion time generate organisational and technological problems . In order to effectively deal with such problems, the key step in such cases is careful planning and preparation of the project and, in the course of construction, effective monitoring and quickly responding in the event of emerging problems.

ACKNOWLEDGEMENTS

The authors would like to thank the Varso construction team from HB Reavis for the information provided that is relevant to the writing of this article and the possibility of directly observing the construction of the Varso complex.

REFERENCES

Abomoslim, S., Russell, A. 2014. Screening design and construction technologies of skyscrapers. *Construction Innovation* 14(3): 307–345.

Ali, M.M., Moon, K.S. 2007. Structural Developments in Tall Buildings: Current Trends and Future Prospects. *Architectural Science Review* 50(3): 205–223.

Craggs, D. 2018. Skyscraper development and the dynamics of crisis: The new london skyline and spatial recapitalization. *Built Environment* 43(4): 500–519.

Dong, S., Samsonov, S., Yin, H., Ye, S., Cao, Y., 2014. Time-series analysis of subsidence associated with rapid urbanization in Shanghai, China measured with SBAS InSAR method. *Environmental Earth Sciences* 72(3): 677–691.

Fang, C., Yu, D. 2017. Urban agglomeration: An evolving concept of an emerging phenomenon. *Landscape and Urban Planning* 162: 126–136.

Gongalo, B., Gudovicheva, L., Gubareva, A., Dobrynina, L., 2018. High-Rise Construction in Densely Dwelled Cities: Requirements for Premises Insolation and Consequences of their Violation in Russian Law and Jurisprudence. E3S Web of Conferences, Volume 33, Article number 03069.

Jasmine Nisha, J., Muttharam, M., 2017. Deep Excavation Supported by Diaphragm Wall: A Case Study. *Indian Geotechnical Journal* 47(3): 373–383.

Koo, C., Hong, T., Yoon, J., Jeong, K., 2016. Zoning-Based Vertical Transportation Optimization for Workers at Peak Time in a Skyscraper Construction. *Computer-Aided Civil and Infrastructure Engineering* 31(11): 826–845.

Mitra, A., Nagar, J.P. 2018. City size, deprivation and other indicators of development: Evidence from India. *World Development* 106: 273–283.

Okolelova, E., Shibaeva, M., Trukhina, N., 2018. Model of investment appraisal of high-rise construction with account of cost of land resources. E3S Web of Conferences, Volume 33, Article number 03014.

Radziszewska-Zielina, E., Kania, E. 2017. Problems in Carrying Out Construction Projects in Large Urban Agglomerations on the Example of the Construction of the Axis and High5ive Office Buildings in Krakow. MATEC Web of Conferences Volume 117, Article number 00144.

Radziszewska-Zielina, E., Kania, E., Śladowski, G. 2018. Problems of the Selection of Construction Technology for Structures in the Centres of Urban Agglomerations. *Archives of Civil Engineering* 64(1): 55–71.

Radziszewska-Zielina, E., Śladowski, G., Kania, E. 2018 Structural Analysis of Conditions Determining the Selection of Construction Technology for Structures in the Centres of Urban Agglomerations. *Open Engineering* (Article in press).

Tan, Y., Wang, D. 2015. Structural behaviors of large underground earth-retaining systems in Shanghai. I: Unpropped circular diaphragm wall. *Journal of Performance of Constructed Facilities* 29(2): 04014058 varso.com.

Wang, A.J. 2017. Design and construction innovations on a skyscraper cluster in China. *Proceedings of the Institution of Civil Engineers: Civil Engineering* 171(2): 91–95.

Wójcicki, Z., Grosel, J., Belostotsky, A., Akimov, P., Sidorov, V. 2017. OMA Research of Sky Tower in Wrocław, Poland, MATEC Web of Conferences, Volume 117. Article number 00177, 26th R-S-P.

Impact of firefighting agents on diesel oil biodegradation in contaminated soil samples

J. Rakowska
The Main School for Fire Service, Warsaw, Poland

ABSTRACT: The contamination of soil by toxic, hazardous hydrophobic organic compounds is a widespread environmental problem. The solubility and bioavailability of hydrophobic pollutants can be increased by using surfactants to reduce surface tension and interfacial tension. This paper focuses on the influence of firefighting foams and degreasing agents on the biodegradation of diesel oil in the soil. This study applied three soil samples with bacterial flora, containing diesel and firefighting foam or degreasing agents. The determination of BOD was carried out using the respiratory method based on the oxygen pressure measurement through time, in a closed system. The presence of a surfactant based agent and bacterial consortium increased the effectiveness of hydrocarbon decomposition depending on the type of soil and the firefighting or degreasing agent. The best results, in terms of hydrocarbon elimination, were obtained when the soils were mixed with bacterial strains and a degreasing agent, whereas AFFF inhibited microbial activity.

1 INTRODUCTION

Soil contamination caused by petroleum and its derivatives is the most widespread problem in the environment. Fires and explosions are also an important cause of environmental pollution. All of these events related to combustion of fossil fuels often require the use of firefighting foam to extinguish the fire. However, these foams may also remain in the soil mixed with the leakage of the environmentally polluted substances or the fired hydrocarbon. These pollutants can be hazardous, as some compounds can remain in the environment for a long time. They can be absorbed by the soil particles, volatilized into air, or can contaminate surface and groundwater, which are often used for municipal supply (Mulligan, et al., 2001). Hydrocarbons are capable of blocking the gas exchange with the environment and inhibiting plant growth. For humans and animals, exposure to volatile organic compounds can lead to changes in the nervous system. Hydrocarbons can generate irritation of the respiratory pathway and digestive system and, due to their toxicity and persistently harmful influence on living organisms (Paria, 2008), should be removed from the environment.

To resolve this problem, many cleaning technologies have been developed. The bioremediation process, which uses microorganisms to degrade or transform noxious pollutants adsorbed into the soil matrix, is used because it is low-cost and highly efficient in transforming pollutants into less harmful or non-toxic compounds (Pacwa-Płociniczak, et al., 2011).

The effectiveness of hydrocarbon biodegradation in contaminated soils depends on many factors, including the type of hydrocarbon and the type of soil with different physical and chemical properties: structure, porosity, density, nutrient content, permeability, infiltration, and environmental conditions (humidity, pH and the bioavailability of contaminants to microorganisms) (Márquez-Rocha, et al., 2001), (Kopytko & Ibarra Mojica, 2009).

Surfactants can improve the mobility and removal of hydrophobic compounds from contaminated soils and, consequently, their biodegradation (Paria, 2008), (Franzetti, et al., 2009). They act to firstly mobilize hydrophobic organic compounds by reducing interfacial tension between contaminated soil and water. Therefore, surfactants can transfer hydrocarbons to the mobile phase. Secondly, they enhance the solubility of the contaminants into the core of surfactant micelles (McCray, et al., 2001), (Pacwa-Płociniczak, et al., 2011).

The effect of synthetic surfactants on increasing the cleansing of contaminated soil has been proven in many previous studies (Mohanty, et al., 2013), (Shi, et al., 2015), (Chukwura, et al., 2016). On the other hand, the addition of persistent synthetic surface active compounds could be harmful to the environment and living organisms (Mohanty, et al., 2013), (Mnif, et al., 2017).

Surfactants are basic components to obtain extinguishing agents like foam or wetting agent concentrates. Firefighting foams are used to control and extinguish fires of flammable liquids. Firefighting foams are produced from solutions containing surfactants, organic solvents, foam stabilizers and corrosion inhibitors, through the mechanical dispersion of air bubbles in liquid. Due to their chemical composition, foaming agents are classified as synthetic or protein foams. The group of synthetic agents includes concentrates based on hydrocarbon surfactants; aqueous film forming agents (Aqueous Film Forming Foam—AFFF) consist of synthetic surfactants, fluorosurfactants and polymers. Wetting agents are a kind of chemical substance that increases the spreading and moistening properties of water by lowering its surface tension in order to improve the penetration of the solution into burning materials.

The biodegradability of extinguishing agents depends on their chemical composition. The main aim of this research was to relate the course of the biodegradation process in each soil sample with both contaminants: extinguishing solutions and diesel oil. In this research the respirometric method to evaluate the impact of extinguishing foam on the remediation of soils contaminated with diesel oil was used.

2 MATERIALS AND METHODS

2.1 *Materials*

The study was carried out on three types of soils. Soil samples were prepared from horticultural products and contaminated with 4 ml/kg diesel oil. The properties of studied soils are given in Table 1.

For tests, firefighting concentrates with different chemical compositions and application scope were used. The studied extinguishing agents were marked with the symbols S, AFFF and W, and their properties at 20°C were shown in Table 2.

Table 1. Properties of studied soils.

Soil type	Grain size, mm	pH	Organic matter, %	Chemical compounds, %	
				Total nitrogen	Total phosphorus
Sandy	0.05–2	6.7	0.2	$3.6 \cdot 10^{-4}$	$0.073 \cdot 10^{-4}$
Peat	0.05–10	4.7	82.8	0.7	0.08
Chalky	0.10–5	7.3	35.9	0.12	0.003

Table 2. Properties of firefighting agents.

Firefighting agent	Surfactants	Density, g/ml	pH	Surface tension of used solution, mN/m	Biodegrability %
S	Synthetic fossil hydrocarbons	1.03	7.1	29.0	88
AFFF	Synthetic fossil hydrocarbons, fluorosurfactants, polymers	1.02	7.5	19.0	90
W	Synthetic fossil and renewable hydrocarbons	1.03	7.8	29.9	97

Furthermore, biopreparation containing *Bacillus subtilis* strain ($250 \cdot 10^6$ bacterial cells in 1 ml) and *Bacillus licheniformis* strain ($250 \cdot 10^6$ bacterial cells in 1 ml) accelerating the decomposition of post-harvest residues was used.

In all the systems, demineralized water with conductivity equal to 6 μS was used as the aqueous phase.

2.2 Biodegradation

The rate of biodegradation of diesel oil in the presence of different extinguishing agents containing surfactants under aerobic condition was measured. A respirometric test was conducted with the OxiTop®Control set. During the test, an automatic measurement of oxygen consumption and the amount of carbon dioxide released by aerobic organisms contained in polluted soil was carried out. The formed carbon dioxide was absorbed by sodium hydroxide NaOH. The measurement of oxygen consumption and recording of measured values was carried out by the measuring head of the OxiTop®Control set. The measurement data was then transferred to the controller system via the IR interface.

The soil samples weighing 50 g were placed in the bottle and sprinkled with water up to 50% of adsorbing capacity. Oil was added to the soils in concentrations of about 4 ml/kg, and then all reactors were sealed with OxiTop®Control measuring heads. Samples were tested at 23°C for 15 days. The pressure in each reactor was read every 3 hours. The OxiTop®Control device measures pressure in the bottle and the biological oxygen demand can be calculated using equation (1).

$$BOD = \frac{M(O_2)}{R \cdot T_m} \cdot \left(\frac{V_{total} - V_s}{V_s} + \alpha \frac{T_m}{T_0} \right) \cdot \Delta p(O_2) \tag{1}$$

where: BOD = biological oxygen demand of the studied chemical; M (O2) = molecular weight of oxygen (32000 mg/mol); R = gas constant (8.3144 J/(mol·K)); T_0 = Temperature (273.15 K); T_m = measuring temperature (293.15 K); V_{total} = bottle volume, mL; V_s = sample volume, mL; α = Bunsen absorption coefficient (0.03103); Δp (O₂) = difference of the partial oxygen pressure, hPa.

2.3 Content of hydrocarbons

The test of the petroleum compounds' content was performed 30 days after contamination. Before quantification, the polluted samples were stored at 25°C. The moisture of the soil was adjusted by spraying it with demineralized water.

The test of the mineral oil content in the soil was carried out using gas chromatography according to the method described in standard (EN ISO 16703:2011).

The quantitative determination in the soil of volatile aromatic hydrocarbons, halogenated hydrocarbons, and selected aliphatic ethers, using gas chromatography with GC-ECD detector, was carried out by standard method (EN ISO 22155:2016-07).

The quantification of the content of polycyclic aromatic hydrocarbons (PAH) was carried out using the method given in the standard (ISO 18287:2008).

3 RESULTS AND DISCUSSION

3.1 Biodegradation

Respirometric biodegradation tests were carried out for sandy, chalky and peaty soil samples contaminated with diesel oil. Degradation tests of petroleum substances contained in the studied soils and in soils with the addition of firefighting agents were performed.

In substances containing surfactants in soil vaccinated with a bacteria strain, biodegradation of organic compounds was observed. The highest degree of degradation was observed in the sample with the addition of a wetting agent (W). The least efficient process was observed in a sample containing fluorosurfactants (AFFF). The obtained results indicate that the presence of oil-emulsifying compounds that are less susceptible to biodegradation is less favourable than in the case of preparations with a high degree of biodegradability, substances that do not contain materials toxic to microorganisms (e.g. solvents). Surfactant toxicity is an important aspect which may adversely affect the biodegradation of oil and petroleum hydrocarbons. A high concentration of surfactants can inhibit microbial growth in soil rich in organic matter. Therefore, better results were obtained using the W agent, and worse when using the AFFF agent. The test results are shown in Figure 1.

Based on the results of the studies on hydrocarbon content 30 days after contamination in all soil samples contaminated with diesel oil, a significant loss of pollution through evaporation was found. Due to the different soil structure, organic matter content, and liquid retaining capacity in the porous structure, about 3 times more mineral oil content was observed in peaty soil than in chalky soil. The presence of substances containing surface-active compounds in contaminated soils affected slightly smaller (about 5%) diesel oil residues in the

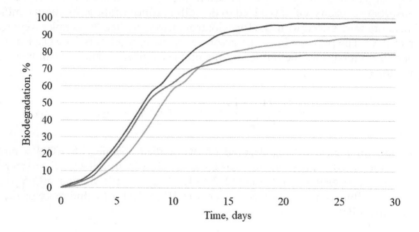

Figure 1.   Biodegradation of diesel oil in peaty soil.

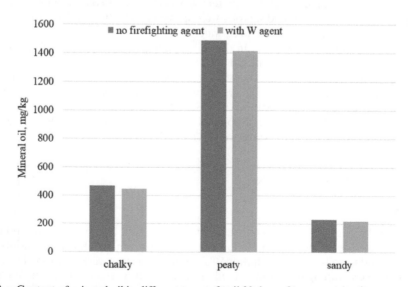

Figure 2.   Content of mineral oil in different types of soil 30 days after contamination.

tested samples. The results of the mineral oil content 30 days after contamination in different types of soil are shown in Figure 2.

3.2 *Content of petroleum derivates*

The results of the amount of individual hydrocarbon concentration in the tested soil samples make it possible to state that the amount of contamination in the chalky soil is smaller than in the sandy and peaty soils. However, when comparing changes in hydrocarbon content in the contaminated samples, faster soil remediation in the presence of surfactant-based substances was found. But in chalky soil, this effect was proven only for naphthalene, acenaphtylene and acenaphtene. Perhaps due to the presence of various types of natural organic matter in the studied soils, the proportions between the quantities of analyzed hydrocarbons were different. The results of testing the content of the PAH and BTEX present in the soils are shown in Figures 3 and 4.

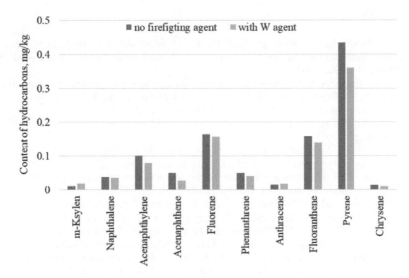

Figure 3.   Content of hydrocarbons in chalky soil 30 days after contamination.

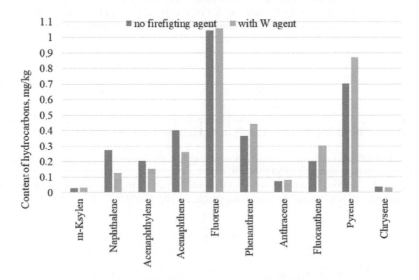

Figure 4.   Content of hydrocarbons in peaty soil 30 days after contamination.

4 CONCLUSIONS

The effectiveness of soil remediation depends on numerous factors, including the chemical composition of the firefighting agent applied to the contaminated site, and the type of hydrocarbon microorganisms present in the soil. The structure of the surfactant, its capacity for emulsification and micellar solubilization of hydrocarbons, and its toxicity to bacteria, yeast or fungi are the key parameters affecting the petroleum derivates' degradation in soil in the presence of firefighting agents. The biodegradation of oil may be negatively affected if changes in the microbial cell surface in the presence of surfactants are unfavourable.

This study proved the impact of soil parameters, particularly structure, porosity, and organic matter and nutrient content, on soil remediation contaminated with diesel oil. Thus, the natural remediation process of soil should be considered to be associated with the transfer of pollutants to other environmental components, e.g. to water and air.

REFERENCES

Carmona, N.F., Bidoia, E.D. & Montagnolli, R.N. 2017. Boidegradation of firefighting foam waste in diesel contaminated samples. In: 5th International Conference on Sustainable Solid Waste Management, Athens, 21–24 June 2017.

Chukwura, E.I., Miriam, N. & Nwankwegu, A.S. 2016. Hydrocarbon Degradation Potentials of Fungi Associated with Oil-Contaminated Soil from Selected Mechanic Workshops in Awka, Anambra State, Nigeria. Frontiers in Environmental Microbiology, 2(6): 38–44.

EN ISO 16703:2011. Soil quality. Determination of content of hydrocarbon in the range C10 to C40 by gas chromatography.

EN ISO 22155:2016-07. Soil quality. Gas chromatographic determination of volatile aromatic and halogenated hydrocarbons and selected ethers. Static headspace method.

Franzetti, A. et al., 2009. Potential applications of surface active compounds by Gordonia sp. strain BS29 in soil remediation technologies. Chemosphere, 75(6): 801–807.

ISO 18287:2008. Soil quality. Determination of the content of polycyclic aromatic hydrocarbons (PAHs). Method of gas chromatography with detection by mass spectrometry (GC-MS).

Kopytko, M. & Ibarra Mojica, D.M. 2009. Biodegradation potential of total petroleum hydrocarbons in oil industry contaminated soils. Journal of the Polish Mineral Engineering Society, 2: 31–48.

Márquez-Rocha, F.J., Hernández-Rodrí, V. & Lamela, M.T. 2001. Biodegradation of Diesel Oil in Soil by a Microbial Consortium. Water, Air, and Soil Pollution, 128(3–4): 313–320.

McCray, J.E., Bai, G., Maier, R.M. & Brusseau, M. 2001. Biosurfactant-enhanced solubilization of NAPL mixtures. Journal of Contaminant Hydrology, 48(1–2): 45–68.

Mnif, I., Sahnoun, R., Ellouz-Chaabounia, S. & Ghribi, D. 2017. Application of bacterial biosurfactants forenhanced removal and biodegradation of diesel oil in soil using a newly isolated consortium. Process Safety and Environmental Protection, 109: 72–81.

Mohanty, S., Jasmine, J. & Mukherji, S. 2013. Practical Considerations and Challenges Involved in Surfactant Enhanced Bioremediation of Oil. BioMed Research International, DOI: 10.1155/2013/328608.

Mulligan, C., Yong, R. & Gibbs, B. 2001. Surfactant-enhanced remediation of contaminated soil: a review. Engineering Geology, 60(1–4): 371–380.

Pacwa-Płociniczak, M., Płaza, G., Piotrowska-Seget, Z. & Cameotra, S. 2011. Environmental Applications of Biosurfactants: Recent Advances. International Journal of Molecular Science, 12(1): 633–654.

Paria, S. 2008. Surfactant-enhanced remediation of organic contaminated soil and water. Advances in Colloid and Interface Science, 138(1): 24–58.

Shi, Z. et al. 2015. Anionic–nonionic mixed-surfactant-enhanced remediation of PAH-contaminated soil. Environmental Science and Pollution Research, 22(16): 12769–12774.

Advances and Trends in Engineering Sciences and Technologies III – Al Ali & Platko (Eds)
© 2019 Taylor & Francis Group, London, ISBN 978-0-367-07509-5

Efficacy of protective facemasks in reducing exposure to particulate matter

J. Rakowska
The Main School for Fire Service, Warsaw, Poland

M. Tekieli
Fire Service Headquaters District Chrzanów, Poland

ABSTRACT: Most studies relate the exposure to particle matter pollution to cardiovascular illnesses, respiratory diseases, cancer, preterm birth complications and also have suggested that respiratory airborne particles can cause metabolic disease and liver failure. People exposed to toxic air pollutants at high concentrations, or for long durations may have increased risk of health problems. There has been no critical research on the effectiveness of using proper respiratory masks during periods of exposure to smog. In this study, the efficacy of wearing masks as an individual intervention attempt to reduce PM exposure in the Kraków agglomeration was examined. Among studied the personal protective equipment, all the masks can decrease the amount of particulate matter absorbed into the respiratory system. However, it was found that the masks hinder the level of airflow and breathing that is needed for people with respiratory diseases.

1 INTRODUCTION

The problem of air pollution has been widely discussed in the last ten years. The negative influence of pollution on human health and life has been scientifically confirmed (Chen, 2007), (Anderson, 2012). The atmosphere may include pollution of both anthropogenic and natural origin. Anthropogenic pollution is a result of the progress of civilization. It includes line pollution (occurring along roads due to car traffic), point pollution (occurring in the surroundings of manufacturing facilities), and surface pollution (occurring over large residential areas, mainly as a result of the individual heating of homes). In the case of pollution of natural origin; some examples are: volcanic pollution, desert sand, sea salts, flower pollen, and natural cataclysms.

Air pollution significantly influences human health, causing numerous illnesses of the cardiovascular and respiratory systems (Anderson, et al., 2012). Some studies have proved that exposure to air pollution contributes to an increased lung cancer risk (Bruce, et al., 2000). The biggest influence of air pollution on human and animal health is observed in the industrial and urbanized areas. The most vulnerable groups include: children, the elderly, and people suffering from respiratory system illnesses. Air pollution also has a negative influence on the state of ecosystems and can result in material destruction (EEA, 2017).

According to a report issued by the World Health Organization (WHO, 2016), providing measurements of air quality from about 3 thousand cities in 103 countries around the world, in 80% of these places air pollution exceeds the acceptable norms. The most polluted city in the world is Zabol in Iran, and next on the list are the Indian cities Gwalijar and Allahabad. The report shows that 33 out of the 50 most polluted cities in the European Union are located in Poland, and this list includes Kraków. This is the reason why the study was carried out in Kraków agglomeration.

Air can accumulate pollution in the form of gas, aerosol, or dust. Depending on these forms, protection measures should be adjusted accordingly. Particulate matters are a major hazard for human health because the size of the particles is so small that their rate of fall is very low, leaving them suspended in the air for a long time. Particulate matter can be divided into a respirable fraction (particles that are under 2.5 μm in size) and an inhalable fraction (particles that are under 10 μm in size). The particles of the respirable fraction can be very dangerous for the human health because their small size enables them to move into and accumulate in the lungs. The greatest air pollution with particulate matter occurs during the autumn-winter period due to the use of solid fuel in home heating stoves (Cembrzyńska, et al., 2012), (Pastuszka, et al., 2015), (EEA, 2017). Because these dusts contain solid particles, protection against them can be achieved by using appropriate barriers, e.g. protective masks, half-masks, or filters. However, the best method of protection against particulate matter is to avoid exposure, so it is advisable not to spend too much time outside, and avoid physical exertion; when smog is in the air. Air quality information systems, providing real-time information, are widely available in smartphone applications enabling residents of polluted areas to take action to protect their health. Studies conducted in India and China (Singh, et al., 2010), (Cherrie, et al., 2018) have shown that the use of filtering masks can have a positive effect on the immediate health, measured as blood pressure and heart rate (Peng, et al., 2009), (Shi, et al., 2017).

Respiratory masks retain particles from the inhaled stream of air via filtration, gravitational sedimentation, electrostatic sedimentation, and adsorption. The effectiveness of filtration depends on the size of the particles, their load, concentration and the intensity of flow through the material (Cherrie, et al., 2018).

In recent years, manufacturers have developed a variety of expensive anti-smog masks, informing people that the devices provide complete protection against air pollution, including particulate matter. The aim of this study was to compare the effectiveness of the protection provided by these expensive anti-smog masks in comparison to the cheap, alternative materials used worldwide by the inhabitants of polluted regions to protect their respiratory systems. Furthermore, the study also analyses the degree to which the masks make breathing difficult.

2 EXPERIMENTAL

Studies were carried out in the Kraków agglomeration, which included the city of Kraków and the surrounding urbanized municipalities. This area of 3231.38 km² is inhabited by 1.402 mln people. The average annual small-particle pollution with PM10, registered by air monitoring stations in 2017, was 31–55 μg/m³. The maximum value registered in a 1-hour measurement period during the study of the masks was 360 μg/m³.

2.1 *Materials*

Samples of cloth and masks with the properties described in Table 1 and presented in Figure 1 were used in the study. Samples of mask filtering materials with a multilayer structure were prepared and placed in a measurement head in the same order as they appear in the masks.

2.2 *Mask efficacy assessment*

The materials used in the protective masks were subjected to an analysis focusing on the establishment of the amount of blocked particulate matter. The samples of the six materials used in the study were weighted using analytical scales before the analysis and after exposure. The difference in mass before and after the analysis showed how much the flow of particulate matter was blocked by the material. All tests were conducted with GILIAN GilAir-3 air sampling pump and samplers for respirable and inhalable particles. The samples were placed in aspirator heads. 6 devices were working at the same time, 3 of which collected the inhaled dust fraction, and the rest the respirable dust fraction. The measurements were performed on days when the automatic measurement devices of the monitoring stations indicated a high degree of pollution with PM10 and PM2.5. The measurements for each material were about 30h long in conditions of at least medium pollution, so when the Common Air Quality Index (CAQI) exceeded the value of 50, the airflow was at 2 l/min. Each piece of the materials

Table 1. General profile of the tested mask.

The type of mask	The number of layers	Activated carbon filter	Price, €
Cotton bandana	2	–	1,00
Surgical mask	3	–	0,25
FFP2 filtering half mask	3	–	0,80
FFP3 filtering half mask	4	–	0,50
The CityMask anti-smog mask	5	+	4,00
Neoprene SmogGuard anti-smog mask	5	+	25.00 + 2.5/filter

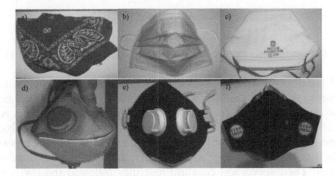

Figure 1. Tested facemask: a) Cotton bandana, b) Surgical mask, c) FFP2 filtering half mask, d) FFP3 filtering half mask, e) The CityMask anti-smog mask, f) Neoprene SmogGuard anti-smog mask.

used was measured twice, and the results show the arithmetic mean of these measurements. Differences that include the differences in filter mass before and after exposure were transferred to a bar graph, showing the effectiveness of the protection of individual materials. Moreover, the calculations of the measured background levels were made, so the amount of dust sucked in by the aspirator during measurements, were also determined.

2.3 Mask efficacy assessment

The materials used in the protective masks were subjected to an analysis focusing on the establishment of the amount of blocked particulate matter. The samples of the six materials used in the study were weighted using analytical scales before the analysis and after exposure. The difference in mass before and after the analysis showed how much the flow of particulate matter was blocked by the material. All tests were conducted with GILIAN GilAir-3 air sampling pump and samplers for respirable and inhalable particles. The samples were placed in aspirator heads. 6 devices were working at the same time, 3 of which collected the inhaled dust fraction, and the rest the respirable dust fraction. The measurements were performed on days when the automatic measurement devices of the monitoring stations indicated a high degree of pollution with PM10 and PM2.5. The measurements for each material were about 30h long in conditions of at least medium pollution, so when the Common Air Quality Index (CAQI) exceeded the value of 50, the airflow was at 2 l/min. Each piece of the materials used was measured twice, and the results show the arithmetic mean of these measurements. Differences that include the differences in filter mass before and after exposure were transferred to a bar graph, showing the effectiveness of the protection of individual materials. Moreover, the calculations of the measured background levels were made, so the amount of dust sucked in by the aspirator during measurements, were also determined.

2.4 Harvard step up test

As well as the protective efficiency of the mask, it is equally important to determine the degree to which it makes breathing difficult. The evaluation of the comfort of use of the

Figure 2.   Harvard step up test (https://bashny.net/t/en/378237).

studied masks was performed on the basis of an exertion trial using the Harvard method (Figure 2).

This allowed for the evaluation of which mask limits breathing the least. Six exercising persons took part in the study, demonstrating different degrees of physical activity and efficiency. In the study, the persons were asked to cyclically step on a dais that was 45 cm high. Based on this information, the efficiency index was determined for the subjects using each mask. Measurements were carried out in two measurement stages. Each person performed tests without the masks and, also, in selected masks that were the most successful at blocking particulate matter. Their speed was constant and they followed the rhythm of the metronome. The exertion lasted for 5 minutes each time, and the subject's heartbeat was measured three times at regular time intervals at the end of the trial.

3   RESULTS AND DISCUSSION

3.1   *Mask efficacy assessment*

The results of the effectiveness of protection of particular materials are showed in Figures 3 and 4. Figure 3 shows an average increase in filter mass for the inhaled fraction. The best result for City Mask (33.75 mg) was seven times better than the second best (bandana, 4.7 mg) and over eight times better than the professional SmogGuard mask. The efficiency of protection of anti-dust masks turned out to be small in comparison to other examples, and a surgical mask offered almost no protection at all. Figure 4 shows the average increase in filter mass for the respirable fraction. The City Mask also achieved the best results (32.15 mg) in this category as it was over eight times more effective than a bandana (3.85 mg) and the SmogGuard mask, whose protection levels were similar. The protective properties of anti-dust masks also proved to be low (less than 1.5 mg). The results of surgical masks were better than in relation to the inhaled fraction, but their effectiveness can still be considered to be insignificant (0.5 mg).

Thanks to the analysis of the measurement background, which is shown in Figures 3 and 4 as line with markers; the possibility that the high effectiveness of protection of the masks was a result of the large amounts of particulate matter present in the atmosphere during the measurement procedures can be discounted. The highest average amount of the particulate matter sucked-in by the sampling pump (PM10–277 mg, PM2.5–199 mg) occurred during the analysis of surgical masks, which proved to be the least effective; whereas an average concentration of particulate matter was obtained during the studies of bandanas (PM10–142 mg and PM2.5–92 mg) – which showed quite good results. Our results indicate that many commercially available facemasks not provide satisfactory protection.

3.2   *Harvard step up test*

The results of physical efficiency studies are shown in Figure 5. In this study, the individual masks were assigned a score depending on the results of the physical efficiency of the

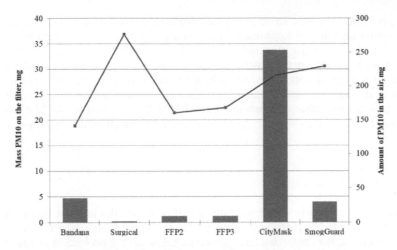

Figure 3. Comparison of the PM10 mass retained on different filter materials in relation to the amount of PM10 in the air during the experiment.

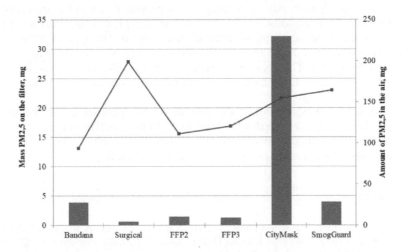

Figure 4. Comparison of the PM2,5 mass retained on different filter materials in relation to the amount of PM2,5 in the air during the experiment.

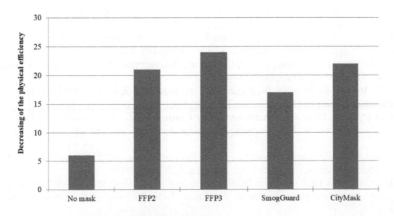

Figure 5. Effort required for breathing in facemask.

exercising persons. The best score (one point from each exercising person = 6 points) was awarded when exercising without masks, the worst (six points from the exercising person) – during a test with the mask that reduced the efficiency the most. The results confirmed the hypothesis that every mask obstructs airflow and is the reason for additional physical workload on the human organism. Even healthy and sporty adults may find uncomfortable to wear a mask because all mask increased effort required for breathing. The worst result (24 points) was achieved when using the FFP3 filtering half-mask. The SmogGuard mask achieved the best score (17 points) as it limited the airflow during physical exertion the least.

4 CONCLUSIONS

By protecting the respiratory tract with any cloth or mask, the amount of inhaled dust is reduced. Expensive and branded anti-smog masks can be less effective than their cheap counterparts, and even cotton cloth can be a good alternative for the protection of the respiratory tract against particulate matter. However, the effective management of air quality is necessary to reduce health risk to a minimum.

The results show that masks equipped with a carbon filter ensure better protection than the masks that do not have it. Also anti-smog masks provide much better protection against particulate matter than anti-dust masks. Surgical masks provide insignificant protection and there is no reason to use them. Another disadvantage of this type of protection is the possibility that pollution will slip through the narrow space between the edge of the mask and the face of the wearer. The size of this leak depends on such factors as the size and shape of the face, the presence of facial hair, the shape and material type of the respiratory mask, and the method of wearing.

Every mask that is used to protect the respiratory tract causes a load on the human organism, and leads to greater exertion for the wearer. In time, pollution particles accumulate on the filter, increasing the obstruction of the airflow. As a result of this process, the most effective anti-smog masks lead to bigger difficulties with breathing.

REFERENCES

Anderson, J.O., Thundiyil, J.G. & Stolbach, A. 2012. Clearing the Air: A Review of the Effects of Particulate Matter Air Pollution on Human Health. *Journal of Medical Toxicology*, 8(2): 166–175.

Bruce, N., Perez-Padilla, R. & Albalak, R. 2000. Indoor air pollution in developing countries: a major environmental and public health challenge. *Bulletin World Health Organization*, 78: 1078–1092.

Burton, K.A. 2015. *Efficiency of respirator filter media against diesel particulate matter*, Master of Science – Research thesis. Wollongong: School of Health and Society, University of Wollongong.

Cembrzyńska, J., Krakowiak, E. & Brewczyński, P.Z. 2012. Particulate pollution of PM10 and PM2.5 due to strong anthropopressure in Sosnowiec city. *Environmental Medicine*, 15(4): 31–38.

Cherrie, J.W. et al. 2018. Effectiveness of face masks used to protect Beijing residents against particulate air pollution. Occupational *Environmental Medicine*, 75: 446–452.

EEA, 2016. Air quality in Europe – 2016 report, Luxemburg: European Environment Agency. Publications Office of the European Union.

Pastuszka, J.S., Rogula-Kozłowska, W., Klejnowski, K. & Rogula-Kopiec, P. 2015. Optical Properties of Fine Particulate Matter in Upper Silesia, Poland. *Atmosphere*, 6(10), pp. 1521–1538.

Peng, R.D. et al. 2009. Emergency Admissions for Cardiovascular and Respiratory Diseases and the Chemical Composition of Fine Particle Air Pollution. *Environmental Health Perspectives* 117(6): 957–963.

Shi, J. et al. 2017. Cardiovascular benefits of wearing particulate- filtering respirators: a randomized crossover trial. Environmental Health Perspectives, 125(2): 175–180.

Singh, M.P. et al. 2010. Face mask application as a tool to diminish the particulate matter mediated heavy metal exposure among citizens of Lucknow, India. *Science of the Total Environment*, 408: 5723–5728.

World Health Organization, 2016. *WHO's Urban Ambient Air Pollution database – Update 2016*, Geneva.

Economical sustainability assessment for a set of façade walls with the same function equivalent – comparative analysis

D.A. Ribas & A. Curado
Instituto Politécnico de Viana do Castelo, Viana do Castelo, Portugal
CONSTRUCT LFC, Faculty of Engineering (FEUP), University of Porto, Portugal

P.B. Cachim
RISCO and Department of Civil Engineering, University of Aveiro, Portugal

ABSTRACT: The construction of new buildings as well as its retrofitting processes should be ruled by the principles of sustainable design. The applied materials and the construction processes applied in each stage that makes part of the "before use phase" of the building's life cycle, namely: "pre-construction phase", "product stage" and "construction process phase", influence its sustainability. The main objective of this paper is to compare a set of three construction solutions for opaque walls, commonly used in building envelopes, all with the same functional equivalent, i.e. with the same thermal, acoustical and fire protection performance, regarding economical sustainability. In order to evaluate the economical sustainability of the different construction solutions, the EcoSust methodology, based on the European standards framework, will be applied. According to this methodology the result of the economic performance is expressed in monetary units and the result of the sustainability performance by an index of economical sustainability (A+, A, B, C, D, E).

1 INTRODUCTION

The European Standards framework developed by CEN/TC 350 "Sustainability of construction works", proposes a system for assessing the sustainability of buildings based on life cycle analysis (LCA). The EcoSust model is based on an LCA for assessing the economic sustainability of buildings, following the European construction sustainability framework as defined by EN 15643-1:2010, EN 15643-4:2012 and EN 16627:2015. This novel approach has been developed to systematically assess the economic performance of a building within the concept of sustainability. The methodology follows the principle of modularity, where aspects and impacts that influence economic performance and building sustainability index during the "before use phase", are assigned to the categories in which they occur. The hierarchical structure of the methodology directs the flow of information relating to aspects and impacts that influence the economic performance of the indicators, modules and stages of the life cycle, based on the quantification of the 65 parameters defined for the EcoSust methodology. The parameters provide accurate information for indicators from observable or measurable data from the project or existing building.

This paper aims to present the methodology and calculation of the reference values (benchmarks) for the parameters of the hierarchical structure of EcoSust methodology, applied to a set of three construction solutions for façade walls, in order to evaluate and compare their economical sustainability. The construction solutions represent different possible scenarios for façade walls: a) masonry cavity wall filled with thermal insulation; b) a lightweight concrete wall coated with an external thermal insulation system (ETICS); and c) a double light timber frame wall filled with thermal insulation. All the three façade solutions have similar performances regarding thermal and acoustical insulation and have approximately the same fire resistance protection. It is expected to rank the three solutions concerning their economic sustainability. To apply the EcoSust methodology regarding the evaluation of the three façade solutions, a building with simple geometry and similar construction solutions for all elements, except for the façade walls, was defined. The three different façade types correspond to the three analyzed scenarios.

2 CHARACTERIZATION OF THE FAÇADE WALLS

2.1 Construction solutions

The construction solutions regarding the façades walls to be analyzed are the following (see Figure 1):

a. Solution 1 – Cavity wall with a thermal insulation layer (MW) in the air gap made with traditional units made of clay with joints plastered with cement mortar (4). The internal skin of the double wall is made with horizontal hollow bricks with 15 cm of thickness (5) and the external skin, with similar units with 22 cm (2), coated with traditional mortar rendering on the external face and with plaster on the internal face (6). Both sides are painted—the external side with a water-repellent paint layer and the internal side with a plastic paint finish (1). The thermal insulation board is 4 cm thick (3).
b. Solution 2 – Single masonry wall made with traditional lightweight concrete blocks with 25 cm of thickness (3). The blocks are vertical hollow units coated with an external thermal insulation system (ETICS) including an insulation board of MW 5 cm thick (1, 2). The external side of the wall is coated with an external cladding (1) to provide waterproofing made of a dry mortar and the internal side is coated with a plaster layer and a plastic paint finish (4).
c. Solution 3 – Double light timber frame wall, known is Portugal as "Tabique" wall. The wall is constituted of a timber frame wall panel comprising studs (2), rails and a thermal insulation layer of MW 5 cm thick (10). The wall studs are 14 cm × 4 cm to accommodate the thermal insulation layer. The internal skin is made of a gypsum fiberboard (8) fixed on an installation layer (7) which is connected to an OSB board (9). The external skin is made of a waterproofing timber wall board (5) supported by a wooden slat (1) connected to an OSB board (3). The internal skin includes a vapor barrier (6) and the external skin a waterproofing membrane (4). The total wall thickness is 23 cm.

2.2 Functional performance of the façade walls

Table 1 summarizes the functional performance of the three considered façade walls. The overall performance in terms of thermal insulation, sound insulation and fire resistance of the walls is similar. It clearly shows that the performance of the three considered façade walls is similar regarding the thermal, acoustical and fire-resistance performance. Most values

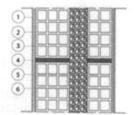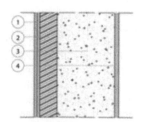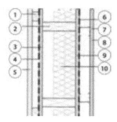

Figure 1.   Vertical section (Solutions 1 and 2) and horizontal section (Solution 3).

Table 1. Functional performance of the façade walls.

	U-value (W/m^2.°C)	Rw (dB)	Fire-resistance duration (minutes)
Solution 1	≈ 0.4	≈ 50	>90
Solution 2	≈ 0.4	≈ 50	>90
Solution 3	≈ 0.4	≈ 50	≈ 90

Table 2. Costof resources for the different solutions (€/m^2).

Resource	Solution 1 Cost (€)	Solution 2 Cost (€)	Solution 3 Cost (€)
Water	0.061	0.010	0.000
Energy	0.124	0.066	0.027
Labor	54.640	34.690	15.794
Equipment	2.037	1.647	2.681
Materials	24.410	42.460	53.899
Total	82.350	78.873	72.644

presented in Table 1 were provided by the manufacturers through laboratory tests. Therefore, the sustainability performance in comparative terms for the three solutions can be performed using EcoSust model. Calculation of the U-values for the three walls follows the method described in NP EN ISO 6946:2017, where a methodology for calculating the thermal resistance and thermal transmittance of building elements based on the electrical analogy is described.

The calculation of the weighted Sound Reduction Index (Rw) for the three walls was obtained following the method described in NP EN 20140-3:1998 and NP EN ISO 717-1:2013. This parameter is considered a fundamental parameter for laboratory comparison of different building construction solutions for walls in façades. The index considers a 1/3 octave band frequency range, between 100 Hz and 3150 Hz. The classification of the fire resistance of the three wall façades was obtained according to NP EN 13501-2:2016. This standard uses data from fire resistance and smoke leakage tests which are within the direct field of application of the relevant test method. This method is also applicable to façades and claddings that are non-loadbearing, only having a fire separating function.

2.3 Necessary resources for each constructive solution

The unitary costs for solutions 1, 2 and 3 of the resources were built based on a publication of the Portuguese National Civil Engineering Laboratory, LNEC, (Manso, et al., 2013), which is an accurate and widely used source in Portugal. In Table 2, the cost of resources for the construction of one square meter of wall are presented. The description for the materials, their quantities and unitary costs are not presented due to the article size limitations.

3 COMPARATIVE ANALYSIS

The assessment of economic performance and the level of economic sustainability of a residential building during the design phase, based on the expected behaviour for the entire building life cycle should be made for the four phases of the lifecycle of a building. Each phase of the lifecycle is divided into stages, modules, indicators and parameters. For the moment, EcoSust is developed only for the "before use phase". The object of assessment

Table 3. Stages, modules, indicators and parameters of the EcoSust Methodology (Ribas, et al., 2016).

Level 1 Stages	Level 2 Modules	Level 3 Indicators	Level 4 Parameters
Pre-construction stage (PC)	A0: Site and associated fees and counselling	A0.1: Cost of purchase and rental incurred for the site or any existing building.	P1 to P3
		A0.2: Professional fees related to the acquisition of land.	P4 to P8
Product stage (EP)	A1: Supply of raw materials	A1.1: Cost of raw materials.	P9
	A2: Transport of raw materials	A2.1: Cost of transportation of raw materials.	P10
	A3: Manufacturing	A3.1: Cost of transformation raw materials.	P11
Construction process stage (EC)	A4: Transport	A4.1: Cost of transport of materials and products from the factory gate to the building site.	P12
		A4.2: Cost of transport of construction equipment such as site accommodation, access equipment and cranes to and from the site.	P13
	A5: Construction-installation process	A5.1: Costs with exterior works and landscaping works.	P14 to P20
		A5.2: Cost of storing products including the prevision of heating, cooling, humidity etc.	P21
		A5.3: Cost of transportation of materials, products, waste and equipment within the site.	P22
		A5.4: Cost of temporary works including temporary works off-site as necessary for the construction.	P23
		A5.5: Cost on site production and transformation of a product.	P24 to P27
		A5.6: Cost of heating, cooling, ventilation, humidity control, etc. during the construction process.	P28 to P29
		A5.7: Cost of installation of the products into the building including ancillary materials.	P30 to P32
		A5.8: Cost of water used for cooling, of the construction machinery or on-site cleaning.	P33
		A5.9: Cost of waste managing processes of other wasters generated on the construction site (RCD).	P34 to P36
		A5.10: Transportation cost of waste RCD.	P37
		A5.11: Costs of commissioning and handover related costs.	P38 to P44
		A5.12: Cost for professional fees related to work on de project.	P45 to P48
		A5.13: Costs of the taxes and other costs related to the permission to build and inspection or approval of works.	P49 to P64
		A5.14: Incentives or subsidies related to the installation.	P65

is the building, including its foundations and landscaping within the building perimeter (Ribas, 2015). Table 3 shows the hierarchical structure of the method (stages, modules, indicators and parameters) that corresponds to the before use phase.

At each level, information is obtained by aggregation of information at the lower level. EcoSust assesses the performance and economic sustainability of buildings, the result of economic performance is expressed in monetary unit and the sustainability in an economic sustainability index (A+, A, B, C, D, and E), where "E" the means lowest economic sustainability and "A+" the higher economic sustainability.

3.1 Application of the methodology

The methodology was applied to a multifamily residential building located in the Porto region, on an appropriate building site located in a low density urban allotment. The lot of land destined to the construction of the building was acquired for the amount of € 723,840. In order to prepare the construction of the building it was necessary to carry out some infrastructure works related to earthmoving, road paving, peripheral containment and structures of soil support, hydraulic and electric infrastructures, public street lighting, telecommunications and gas facilities. Consequently, the cost of the building comprises the building itself plus the outside arrangements and infrastructures. The building consists of an underground floor for car parking, a ground floor and 3 more stories destined to housing apartments. The building has 29 apartments of the type T0, T1, T2 and T3 with a gross construction are of 2320 m². The structure of the building is a reinforced concrete structure resting on shallow foundations. The slabs are of prefabricated concrete. The solutions for the façades are the three previously described. The interior finishes are the following: the coverings of floors are ceramic in the kitchens and bathrooms, and totally wood surfaced in bedrooms, living rooms and halls. The common areas of the building, mainly the corridors and stairs are covered with natural stone. The partition walls of kitchens and bathrooms are covered with tiles and the remaining walls are covered with sprayed plaster painted with clear plastic paint. The façade walls are rendered and painted. For building installations conventional construction materials were applied: PVC-U for ventilation, wastewater and rainwater drainage, PPR (Polypropylene) for water piping system and copper in the natural gas piping system. The study involved a comparative analysis of the results of the assessment of economic performance and economic sustainability of the building for the three constructive solutions of walls in façades with similar performance. The original solution (Solution 1) is said to be the "basic solution" and corresponds to the described building. The façade walls have a total area of 1752 m².

3.2 Results

The results of the evaluation of the economic performance of the building in the "before use phase", for each of the three constructive solutions considered in this study are presented in Tables 4, 5 and 6, expressed in monetary units. These values correspond to the direct and indirect costs that occur within the boundary of the building, associated with all stages of the "before use phase". The normalized values (NV) and the sustainability index for the three solutions is presented in Tables 7 and 8.

Table 4. Economic performance for the level 1 attributes: before use phase (€).

| Phase: | Before use phase | | | |
	Pre-construction	Product	Construction process	Total cost
Solution 1	793711	659911	1325329	2778950
Solution 2	793711	656628	1158096	2608434
Solution 3	793711	650794	1131662	2576122

Table 5. Economic performance for the Level 2 attributes: stages (€).

Stages:	Pre-construction	Product			Construction process	
Modules:	A0	A1	A2	A3	A4	A5
Solution 1	793711	133766	178354	347791	227230	1098099
Solution 2	793711	133100	177467	346060	225460	932636
Solution 3	793711	131908	175878	342962	223145	908517

Table 6. Economic performance for the Level 3 attributes: modules (indicators A0.1 to A5.14) (€).

Modules:	A0		A1	A2	A3	A4		A5			
Indicators:	A0.1	A0.2	A1.1	A2.1	A3.1	A4.1	A4.2	A5.1	A5.2	A5.3	A5.4
Solution 1	739340	54371	133766	178354	347971	222943	4287	146995	8918	10701	156060
Solution 2	739340	54371	133100	711467	346060	221734	3626	146264	7809	9370	131991
Solution 3	739340	54371	131908	175878	342962	219848	3298	144954	6332	7598	120037

Modules:	A5									
Indicators:	A5.5	A5.6	A5.7	A5.8	A5.9	A5.10	A5.11	A5.12	A5.13	A5.14
Solution 1	346933	11147	61223	4802	8918	1338	8584	88320	244161	0
Solution 2	234741	9428	41425	4061	7010	1051	8584	87880	243021	0
Solution 3	232639	8574	41054	3693	6068	910	8584	87094	240980	0

Table 7. Stages and phases: Normalized values normalized and sustainability index.

	Solution 1		Solution 2		Solution 3	
Stages:	NV	Index	NV	Index	NV	Index
Pre-construction	0.640	B	0.640	B	0.640	B
Product	0.106	C	0.112	C	0.123	C
Construction process	0.286	C	0.384	C	0.414	B
Phase: Before use phase	0.362	C	0.418	B	0.437	B

Table 8. Modules: Normalized values normalized and sustainability index.

	Solution 1		Solution 2		Solution 3	
Modules:	NV	Index	NV	Index	NV	Index
A0: Site and associated fees and counselling	0.640	B	0.640	B	0.640	B
A1: Supply of raw materials	0.073	D	0.079	D	0.091	D
A2: Transport of raw materials	0.098	D	0.105	C	0.116	C
A3: Manufacturing	0.125	C	0.131	C	0.141	C
A4: Transport	0.327	C	0.357	C	0.377	C
A5: Construction-installation process	0.276	C	0.390	C	0.423	B

Legend: See Table 3.

4 CONCLUSIONS

The analysis of the results regarding the evaluation of the performance and the economical sustainability of the building, taken as a case study, allows the following conclusions:

- The results obtained for the assessment of the economical performance of the building in the "before use phase" shows that Solution 3 obtains the best results. In fact, Solution 3 represents a total cost of 2 576 122€, while for Solution 1 and 2 the total cost is 2 778 950€ and 2 608 434€, respectively. By converting the monetary costs for the three solutions into normalized values, the converted results are: for Solution 1, 0.362, corresponding to a sustainability index of "C", and for Solutions 2 and 3 the values of 0.418 and 0.437, respectively, which represent a sustainability index of "B" for both situations.
- Regarding Level 2, which corresponds to the "stages" level, the applied methodology attributed a score "B" for the sustainability index regarding the pre-construction stage. This stage is strongly related to the cost determined by the purchasing of the land. This cost doesn't depend on the value for the acquisition of construction materials, neither on the cost of the construction processes specified on the design of the building and applied during the construction works. The assessment of the economic performance regarding this stage determined a value of 793 711€, regardless the construction solution applied. The sustainability index "C" is calculated for the "product stage", which is associated with the material/product costs from the "raw material supply" to the "construction site gate". The normalized values obtained for Solution 1 is 0.106, for Solution 2 is 0.112, and for Solution 3 is 0.123. The "construction stage" is associated with the costs related to the processes necessary for the construction of the building – 1 325 329€, 1 158 096 € and 1 131 662€, respectively for the solutions for façades walls 1, 2 and 3. The normalized value obtained for Solution 1 is 0.286, for Solution 2 is 0.384, and for Solution 3 is 0.414, corresponding respectively to sustainability indexes "C", "C" and "B".
- By analyzing modules A2 (Transport of raw materials) and A5 (Construction-installation process) the evaluation of the building when applying Solution 3 shows an improvement in the sustainability index when compared to Solution 1—in module A2 the index improves from D to C, and on module A5 the index rises from C to B.

EcoSust methodology has proved to be a viable solution for comparison of different scenarios regarding the economic sustainability of buildings.

REFERENCES

CEN, EN 13501, 2016. Fire classification of construction products and building elements—Part 2: Classification using data from fire resistance tests, excluding ventilation services, Brussels, Belgium: CEN—ComitéEuropéen de Normalization.
CEN, NP EN 20140-3, 1998. Acústica—Medição do isolamento sonoro de edifícios e de elementos de construção—Parte 3: Medição em laboratório do isolamento sonoro a sons aéreos de elementos de construção (ISO 140-3:1995), Brussels, Belgium: CEN—Comité Européen de Normalization.
CEN. EN 15643-1, 2010. Sustainability of construction works; Sustainability assessment of buildings; Part 1: General framework, Brussels: CEN—ComitéEuropéen de Normalization.
CEN. EN 15643-4, 2012. Sustainability of construction works; Assessment of buildings; Part 4: Framework for the assessment of economic performance;, Brussels, Belgium: CEN—ComitéEuropéen de Normalization.
Delgado, J.M. et al., 2018. Indoor hygrothermal conditions and quality of life in social housing: A comparison between two neighbourhoods. Journal Sustainable Cities and Society, Volume 38, pp. 80–90.
ISO 6946, 2017. Building components and building elements—Thermal resistance and thermal transmitance—Calculation method. Genève, Switzerland: International Organization for Standardization.
ISO 717-1, 2013. Acoustics – Rating of sound insulation in buildings and of building elements – Part 1: Airborne sound insulation, Genève, Switzerland.: International Organization for Standardization.
Manso, A.C., Fonseca, M.S. & Espada, J.C., 2013. Informação sobre Custos. Fichas de rendimentos. Lisboa: Laboratório Nacional de Engenharia Civil.
Ribas, D.A., 2015. Metodologia de Avaliação da Sustentabilidade Económica de Edifícios com Base no Ciclo de Vida, Aveiro, Portugal: Universidade de Aveiro.
Ribas, D.A., Morais, M.M. & Cachim, P.B., 2016. Definition of benchmarks for the assessment of the economic performance of buildings. Proceedings of the 2nd International Conference on Engineering Sciences and Technologies, 29 June–1 July. CRC Press 2016, Taylor & Francis Group, p. 623–630.

Advances and Trends in Engineering Sciences and Technologies III – Al Ali & Platko (Eds)
© *2019 Taylor & Francis Group, London, ISBN 978-0-367-07509-5*

Evaluation of quality of sediment by infrared spectroscopy in Hornad River, Slovakia

E. Singovszka & M. Balintova
Faculty of Civil Engineering, Institute of Environmental Engineering, Technical University of Kosice, Kosice, Slovakia

ABSTRACT: Analysis of sediments provides environmentally significant information. Their chemical characterization is needed to understand the natural and anthropogenic influence on the bodies of water. In the present study, sediment collected from Hornad river in east of Slovakia is subjected to mineral analysis using FTIR technique. FTIR spectroscopy is a useful additional method for characterization of functional groups in the compounds of sediments. The paper deals with the study of sediment quality from five sampling sites from Hornad River using of XRF, XRD and FTIR analyses. It was found that analysed sediments contain mainly silicates and carbonates. On the basis of chemical analysis using the XRF method, it is possible to state that the limit values stipulated by the Slovak legislation have not been exceeded.

1 INTRODUCTION

Sediments transferred from terrestrial to aquatic ecosystems are a decisive element in identifying? the watershed dynamics, as they influence the physical characteristics of watercourses and they have a major effect on river chemistry. Consequently, sediment transfer studies are of great interest in many environmental fields, for example, the monitoring and treatment of drinking water supplies (Zimmerman et al., 2003). In addition, sediments are a major agent of transport for nutrients such as phosphorus (Sharpley, 1995), and they participate in the accumulation and transport of organic and metallic contaminants (Haag et al., 2001, Warren et al., 2003).

A major limitation of most sediment transfer studies, whether they are quantitative or qualitative, is the lack of information about the origin of the exported sediments (Collins & Walling, 2004).

Heavy metal pollution of the natural environment is a worldwide problem because of their non-biodegradable Transmission spectroscopy is the oldest and most commonly used method for identifying either organic or inorganic chemicals providing specific information on molecular structure, chemical bonding and molecular environment. It can be applied to study solids, liquids or gaseous samples being a powerful tool for qualitative and quantitative studies. From the IR absorption band or locations of the different peaks, the minerals were identified using the available literature (Ravinaskar et al., 2006; Farmer, 1974; Farmer, 1979). In addition to the band position, sharpness or diffuseness of bands helps in the identification of mineral components. The IR study on sediment samples was highly useful in identifying the various minerals in sediment. This method is non-destructive and can be used in the identification of mineralogical composition (Simonescue, 2012). These results confirmed that the applied technique is relatively quicker and more reliable in mineral analysis. XRD technique can be used to identify crystalline "fingerprints" by comparing the d-spacing (i.e., the distance between adjacent planes of atoms) of unknown samples with standard reference patterns and measurements (Chen 1977; Tankersley & Balantyne, 2010). The quantitative X-ray powder diffraction analysis of different minerals in large number of samples is based on the principles of the use of the standard powder samples and smear-oriented mounting techniques (Gibbs, 1967). Due to spatial and temporal variations in water chemistry, a monitoring program that provides a representative and reliable estimation of the quality of surface waters has become an important necessity (Angelovicova, 2013). Furthermore they facilitate the identification of the possible factors/sources influencing the system and offer not only a valuable tool for reliable management of water resources, but also provide rapid solutions to pollution problems (Simeonov, 2002).

The aim of this paper is the study of sediment quality from different sampling sites from Hornad River in Eastern Slovakia using by XRF, FTIR and XRD methods.

2 STUDY AREA

River Hornad (Figure 1) belongs to the River basin of Danube. Area of the Hornad river is 4, 414 km². In the basin is 27.6% of arable land, 15.7% of other agricultural land, 47.4% of forests, 2.7% shrubs and grasses and 6.6% is other land.

There is 165 surface water bodies while 162 are in the category of the flowing waters/rivers and 2 are in the category of standing waters/reservoirs. Ten groundwater bodies exist in the basin while 1 is in quaternary sediment, 2 is geothermal waters and 7 are in pre-quaternary rocks. The Hornad River has 11 transverse structures without fish pass in operation. Significant industrial and other pollution sources are: US Steel Kosice, Rudne bane š. p, Spišská Nová Ves, Kovohuty as, Krompachy, Solivary as Prešov). From the point of view of environmental loads, there are 11 high-risk localities which have been identified in the river basin. Diffuse pollution is from agriculture and municipalities without sewerage. The upper stretch of the Hornad River to Spišská Nová Ves is in good ecological status which gets worse to poor status or potential by pollution and hydromorphological pressures. From the Ružín Water Reservoir, the Hornad River achieves moderate ecological status. According to chemical status assessment, the Hornád River is in good status. 56 water bodies (34%) are failing

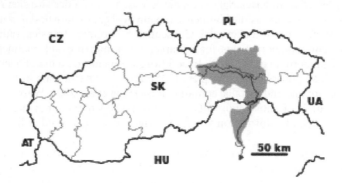

Figure 1.   The Hornad River basin.

Table 1.   GPS coordinate of sediment sampling sites.

Sample site	GPS
1	48°44′11.5″N 21°16′03.1″E
2	48°44′17.1″N 21°15′77.2″E
3	48°44′18.3″N 21°15′54.9″E
4	48°44′96.1″N 21°15′45.7″E
5	48°44′22.5″N 21°15′36.6″E

to achieve good ecological status in Hornad river basin. The water body of intergranular ground waters of quaternary alluviums of the Hornad river basin achieves poor chemical status (pollution from the point and diffuse sources) and poor quantitative status identified on the base of long-term decrease of groundwater levels. The water body of pre-quaternary rocks is in good status—quantitative and chemical (Slovak Environmental Agency, 2015). Sediment sampling sites are located at 48 south latitude and 21 east longitude. GPS coordinates of sampling sites are in Table 1.

3 MATERIAL AND METHODS

A total set of 5 sediment samples were collected in Hornad River in 2018 (Figure 1). The sediment was dried, homogenized and sieved through 0,063 mm sieve. Chemical analyses were performed by the XRF method by using SPECTRO iQ II (Ametek, Germany). Results of chemical analyses of the sediment were compared with the limited values according with

Act No. 188/2003 Coll. on application of the sewage sludge and bottom sediments into the soil (Table 2).

For infra-red spectroscopy in this study, was Alpha FT-IR Spectrometer, BRUKER OPTICS (Germany). To provide a good characterization of mineral by infrared spectroscopy, the spectrum should be recorded in the range of $4000 - 400$ cm^{-1}. Such coverage of range ensures that most of the useful vibrations active in the infrared will be included. The instrument scans the spectra 24 times in 1 minute and resolution is 5 cm^{-1}.

The crystal structure of sediments was identified with diffractometer Bruker D2 Phaser (Bruker AXS, GmbH, Germany) in Bragg-Brentano geometry (configuration Theta-2Theta), using the 1.54060 Å CuKα radiation, Ni Kβ filters and scintillation detector at a voltage of 30 kV and 10 mA current. Scan conditions were identical for all samples, recording times about 5 hours, a step size of 0.05° (2Θ) and step time of 15 s. The XRD patterns were processed using the software Diffrac.EVA v. 2.1. The ICDD PDF database (ICDD PDF – 2 Release 2009) was utilized for the phase identification.

4 RESULTS AND DISCUSSION

Based on the results in Table 2 we can state that the heavy metal concentrations according with limit values of Slovak legislation were not exceeding for all heavy metals in bottom sediments. FTIR spectra of all homogenized sediment samples showed similar features. Based on the concentration of sediment in the absorption frequencies of the peaks in the spectra of each site in wave number unit (cm^{-1}) are reported in Table 3. IR spectrum on Figure 2 it can be said that the main part of compounds are silicates (982 cm^{-1}, 825 cm^{-1}, 753 cm^{-1}, 695 cm^{-1}, 518 cm^{-1}), but carbonates (1400 cm^{-1}; 870 cm^{-1}; 870 cm^{-1}) are present, too. The Figure 2 shows thy comparison of spectrums in all sediment samples. The XRD patterns of sediments are shown in Figure 3. The spectra showed similar features and contain the phases: Q—quartz SiO$_2$ (PDF 01-075-8322), M—muscovite 2M1, C—calcit and N—nontornit.

Table 2. Heavy metal concentration in bottom sediment samples from the Hornad River.

Sample Site	Si	Fe	Ca	Al	As	Cd	Cu	Cr	Ni	Pb	Zn	Hg
		%							mg/kg			
1	18.55	1.80	1.12	4.26	<1	<5.1	36.2	12.0	39.4	<2.0	85.6	<2.0
2	17.23	1.85	2.10	3.60	<1	<5.1	29.4	28.7	40.3	2,5	179.7	<2.0
3	21.51	1.94	1.41	3.99	<1	<5.1	27.5	34.1	37.4	<2.0	55.9	<2.0
4	13.8	1.75	0.61	3.50	<1	<5.1	62.9	50.9	33.9	<2.0	17.2	<2.0
5	17.38	1.12	1.03	3.84	<1	<5.1	15.8	5.00	17.2	<2.0	21.5	<2.0
Limits	–	–	–	–	20	10	1000	1000	300	750	2500	10

Table 3. The absorption frequencies of the peaks in sediment sample of Hornad River.

Sample no.	Observed IR absorption frequencies
S1	1 634, 998, 832, 795, 777, 693, 646, 518, 456, 415
S2	1 457, 1 006, 777, 693, 518, 455
S3	1 473, 1 456, 1 418, 999, 777, 693, 518, 456
S4	1 558, 1 540, 1 521, 1 507, 1 488, 1 473, 1 456, 1 418, 999, 780, 752, 693, 517, 462
S5	1 456, 1 435, 1 163, 1 002, 879, 795, 777, 728, 693, 648, 518, 457, 417

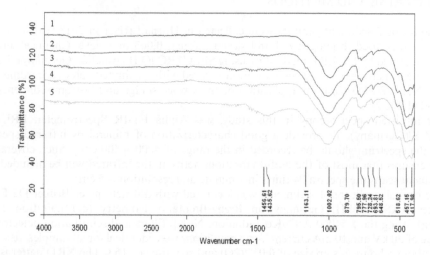

Figure 2. The FTIR analyses of sediment in for all sediments samples in Hornad River.

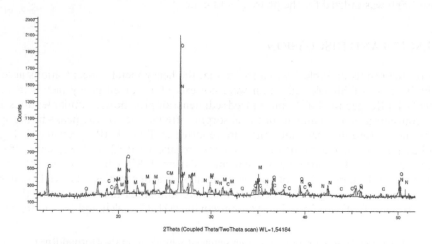

Figure 3. XRD of bottom sediments in River Hornad.

5 CONCLUSION

The analysis of bottom sediment quality is an important yet sensitive issue. The anthropological influences (i.e., urban, industrial and agricultural activities) as well as the natural processes (i.e., changes in precipitation amounts, erosion and weathering of crustal materials) degrade sediment quality and impair its use for drinking, industrial, agricultural, recreational and other purposes. Due to spatial and temporal variations in water and sediment chemistry, a monitoring program that provides a representative and reliable estimation of the quality of surface waters has become an important necessity. Analysis of sediments provides environmentally significant information. Their chemical characterization is needed to understand the natural and anthropogenic influence on the bodies of water. Consequently, comprehensive monitoring programs that include frequent sediment sampling at numerous sites and include a full analysis of a large number of physicochemical parameters designed for the proper management of water and sediment quality in surface waters.

The sediment quality in River Hornad in East of Slovakia was evaluated using XRF, FTIR and XRD analyses. Presented results by XRF analysis did not confirm exceeding limit values by Slovak legislation (Act No. 188/2003) in heavy metals concentrations. The presence of silicates and chlorates was confirmed in the sediment by FTIR method.

The use of IR microscopy provides very useful information on the spatial variation of the functional groups and organic components throughout a small section of the sediment, and demonstrates how micrometre-scale maps can help in identification of the origin of the contaminants. The presence of quartz, and iron hydroxysilicates (muscovite), calcit and nontronit was confirmed in the sediment of the Hornad River by XRD method.

ACKNOWLEDGMENT

This work has been supported by the Slovak Grant Agency for Science (Grant No. 1/0563/15).

REFERENCES

Angelovicova, L. & Fazekasova, D. 2013. The effect of heavy metal contamination to the biological and chemical soil properties in mining region of middle Spis (Slovakia). *Int J Ecosyst Ecol Sci* 3(4): 807–812.

Chen, P.Y. 1977, Table of key lines in X-ray powder diffraction patterns of minerals in clays and associated rocks. *Indiana Geol Survey Occas* Pap 21:1–67.

Clark, R.N., King, T.V.V., Kiejwa, M., Swayze, G.A. & Verge, N. 1990. High spectral resolution reflectance spectra of minerals, *Journal of Geophysics Research* 95: 126–153.

Collins, A.L. & Walling, D.E. 2004. Documenting catchment suspended sediment sources: problems, approaches and prospects. *Prog Phys Geogr* 28:159–96.

Farmer, V.C. 1974. The IR Spectra of minerals, mineralogical society, London, 42: 308–320.

Farmer, V.C. 1979. Infrared spectroscopy, Data hand book for clay materials and other non metallic minerals, Ed. Van Olphen and Fripait, 1st Ed., Pergaman press, Oxford, London. pp: 285–337.

Gibbs, R.J. 1967. Quantitative X-ray diffraction analysis using clay mineral standards extracted from the samples to be analysed. *Clay Miner* 7:79–90.

Haag, I., Kern, U. & Westrich B. 2001. Erosion investigation and sediment quality measurements for a comprehensive risk assessment of contaminated aquatic sediments. *Sci Total Environ* 266:249–57.

Lopez, M.C.B., Martinez, A. & Tascon, J.M.D. 2000 Mineral matter characterization in olivine stones by joint use of LTA, XRD, FT-IR and SEM-EDX. *App. Spect.* 54: 1712–1715.

Ravisankar, R., Rajalakshmi, A. & Manikandan 2006. Mineral characterization of soil samples in and around saltfield area, Kelambakkam, Tamil Nadu, India. *Acta Ciencia Indica* 32(3): 341–346.

Sharpley, A. 1995. Soil phosphorus dynamics: agronomic and environmental impacts. *Ecol Eng* 5: 261–275.

Simeonov, V., Einax, J.W., Stanimirova, I. & Kraft, J., 2002. Environmetric modeling and interpretation of river water monitoring data, *Analytical and Bioanalytical Chemistry* 374: 898–905.

Simonescue, C. 2012. Application of FTIR Spectroscopy in Environmental Studies, http://cdn.intechopen.com/pdfs/38543/InTechapplication_of_ftir_spectroscopy_in_environmental_studies.pdf (avaliable on internet 12.7.20018)

Slovak Environmental Agency 2015. "*1 Introduction*". Pilot Project PiP1: Hornád/Hernád, Integrated Revitalisation of the Hornád/Hernád River Valley. *TICAD*. p. 5.

Tankersley, K.B. & Balantyne, M.R. 2010. X-ray powder diffraction analysis of Late Holocene reservoir sediments. *J Archaeol Sci* 37:133–138.

Warren, N., Allan I.J., Carter, J.E., House, W.A. & Parker, A. 2003. Pesticides and other micro-organic contaminants in freshwater sedimentary environments-a review. *Appl Geochem* 18:159–94.

Zimmerman, J.K.H., Vondracek, B. & Westra, J. 2003. Agricultural land use effects on sediment loading and fish assemblages in two Minnesota (USA) watersheds. *Environ Manage* 32:93–105.

Evaluation of potential radiation hazard in a historical building in Kosice, Slovakia

E. Singovszka & A. Estokova

Faculty of Civil Engineering, Institute of Environmental Engineering, Technical University of Kosice, Kosice, Slovakia

ABSTRACT: Measurements of radioactivity in some common building materials in the second biggest town in Slovakia are reported, together with calculations of the gamma-ray exposure from walls and floors made of different materials. The building is used in calculations of the mean exposure inside concrete, brick and plaster. The monitoring of the concentration of radioactive elements in the building materials as well as the levels of radioactivity emitted by these materials in order to assess the radiation risks to human health is very important because most of the population spends about 80% of their time inside buildings. The purpose of this contribution is to evaluate the natural radioactivity in samples taken in a historical building in Košice.

1 INTRODUCTION

Naturally occurring radioactive materials are present in its crust, the floors and walls of our homes, schools, or offices and in the food we eat and drink. Building materials can contribute to ionizing radiation hazards due to their variable content of radioactive isotopes. Uranium, thorium, and potassium radioisotopes are present in various building materials due to their presence in raw materials: minerals and rocks. The main effects of ionizing radiation on living organisms are cell death, loss of reproductive capacity, or mutation (Tubiana 1990; Saha 2006; UNSECAR 1982; UNSECAR 1993). However, such effects depend on several factors, with the dose rate and the linear energy transfer (LET) of the radiation being the most important. The dose rate is the delivery of dose per time unit and the absorbed dose is typically measured in Grays, Gy, where 1 Gy = 1 J/kg, although in organisms it is also assessed in Sieverts (Sv), where 1 Sv = 1 Gy. The higher the dose rate, the greater the cell damage (Ceu 2014; Saha 2006; UNSECAR 1993). Exposure to gamma radiation in the built environment would result in low doses of radiation. Recent advances in the knowledge of the mechanisms underlying the biological effects of low doses of radiation have shown that low radiation dose effects are mechanistically different from high doses effects, with low radiation doses effects being similar to those of some chemicals in the environment (Mothersill, 2012). Thus, results under mixed exposures to radiation and chemicals may not be predictable for human health, by the consideration of single agent effects (Mothersill 2012; UNSECAR 2010; UNSECAR 2013). It has been observed that the risk of increase in cancer incidence caused by low-dose radiation is low (UNSECAR 2010), but recent epidemiological studies have indicated elevated risks of non-cancer diseases (such as perturbation of immune function or induction of inflammatory reactions with disease) at low doses below 1–2 Gy, and in some cases much lower, although the mechanisms are still unclear and the estimation of risks remains problematic (UNSECAR 2010; UNSECAR 2012). The aim of this study is monitoring the mass concentrations of radionuclides ^{40}K, ^{226}Ra and ^{232}Th in building materials taken from a 19th century building in the historical center of Kosice and determination of radiological parameters in the building.

2 MATERIAL AND METHODS

2.1 Historical building in Kosice

The object is a cultural monument no. UZKP 1188 as a house of the bourgeoisie. It is situated in the row area of one of the oldest streets of Kosice-Zamocnicka Street. This property is located in the southwestern quadrant of the medieval City Monument Reserve in Kosice. The building is situated on a narrow rectangular plot, in a row built area, on the north side of Zamocnicka Street. From the eastern side it is defined by the parcel floor of the neighboring two-storey building, on the north side of the building. On the west side, the tall object was originally tall, after which the building remained unsettled. The house is late-classicist two-storeyed, built on an older medieval building, underpinned almost entirely. The current state is from the first half of the 19th century. The house respected older parcels and adapted its size. In the 19th century, the building underwent a number of structural changes—changing of the available, vaulting systems, staircase modifications, superstructure of the floors and modification of the facade. From this period, a simpler adjustment of the street façade was preserved. The roof of the building is a countertop with a hard-baked roof. Figure 1 shows the location of the object in the historic center of Kosice.

2.2 Material samples and preparation

The sampling was carried out in May 2018 from different parts of the building and was individual depending on the nature of the material—concrete, plaster, and brick. Piece samples were obtained by random selection from several places in the basement of the historic building (Figure 2). All samples were dried at 105°C to constant weight (Figure 3). The samples were then pulled with a jaw crusher (**BRIO BCD 3**) and then ground with a planetary rotary mill (SFM-1) to a prescribed particle size of 0.5 mm and homogenized. Subsequently, the bulk samples were homogeneously dispensed into Marinelli type samples (450 ml volume), weighed and, after closure, subjected to the so-called Rn equilibrium for 36 days (Figure 4).

The mass activities of radionuclides (226Ra, 232Th and 40 K) in study materials were measured using gamma ray spectrometry. Measurements were carried out using an EMS-1 A SH (Empos, Prague, Czech Republic) detection system equipped with a NaI/Tl scintillation detection probe and a MC4 K multichannel analyzer with optimized resolution of 818 V, 4.096 channel and with 9 cm of lead shielding and internal lining of 2 mm tinned copper.

Figure 1.   Situation.

Figure 2.   Materials sampling in the study object.

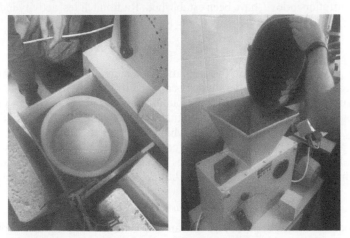

Figure 3.   Mill material.

Figure 4.   Drying materials and storage in Marinelli container.

The specific activity concentrations of ^{226}Ra, ^{232}Th and ^{40}K were determined in Bq.kg^{-1} using the count spectra. The ^{40}K radionuclide was measured directly through its gamma ray energy peak at 1461 keV, while activities of ^{226}Ra and ^{232}Th were calculated based on the mean value of their respective decay products. Activity of ^{226}Ra was measured using the 351.9 keV gamma rays from ^{214}Pb and the activity of ^{232}Th was measured using the 238.6 keV gamma rays of ^{212}Pb. The same counting time of 86,400 s (24 h) was used for all measured samples.

The chemical composition of building materials was determined by using SPECTRO iQ II (Ametek, Germany). For infra-red spectroscopy in this study, was used spectrum through 4000 cm^{-1} to 400 cm^{-1} (Alpha FT-IR Spectrometer, BRUKER OPTICS). Samples were measured in powder form (4 g) for 10 minutes.

2.3 Radiological indices

Based on the measured mass activities of the individual radionuclides A_{Ra}, A_{Th} and A_K in building materials, the following radiological parameters were determined: mass activity index, equivalent radial dose and absorbed radiation dose. Indene mass activity of radionuclides I was calculated by Within the European Union effective doses exceeding 1 mSv.y^{-1} should be taken into account from the radiation protection point of view. Since several radionuclides contribute to the overall dose, in order to assess whether the dose criterion is met, an activity concentration index I_γ have been established. Radionuclides' activity index, or shortly gamma index, I_γ was defined according to the following equation (1):

$$I\gamma = A_{Ra}/300 + A_{Th}/200 + A_K/3000 \tag{1}$$

where A_{Ra}, A_{Th}, A_K are the radium, thorium and potassium, activity concentration (Bq.kg^{-1}) in the building materials. The gamma index is derived to identify whether a dose criterion is met. The gamma index of materials used in building should not exceed limit values depending on the dose criterion ($I_\gamma \leq 1$) (UNSECAR, 2000). The gamma index should be used only as a screening tool for identifying materials which might be of concern.

Radiological index has been introduced to represent the specific radioactivity level of ^{226}Ra, ^{232}Th and ^{40}K by a common index, which takes into account the radiation hazards associated with them. This index is usually known as radium equivalent (Ra_{eq}) activity (2).

$$Ra_{eq} = a_{Ra} + 1{,}43a_{Th} + 0{,}077a_K \; [Bq/kg] \tag{2}$$

The absorbed dose rates in air in a room can be calculated by using the specific dose rates given in. The specific dose rates for radionuclides are given for different screening tool of identifying materials which might be of concern. Indoor dose rates for a model room (dimension of 4 m × 5 m × 2.8 m, thickness of 20 cm, density of 2,35 g/cm^3, and the background of 50 nGy.h^{-1}) are calculated with different structures in a building causing the irradiation as follows:

– All structures:

$$D1 = 0{,}92 \, a_{Ra} + 1{,}1 \, a_{Th} + 0{,}08 \, a_K \; [nGy/h] \tag{3}$$

– Floor and walls:

$$D2 = 0{,}67 \, a_{Ra} + 0{,}78 \, a_{Th} + 0{,}057 \, a_K \; [nGy/h] \tag{4}$$

3 RESULTS AND DISCUSSION

The results of chemical analysis of samples of building materials taken from a historic building are shown in Table 1. The representation of the basic components in the form of oxides corresponds to standard materials such as concrete, brick and lime cement plaster. Table 2 shows the content of radionuclides in well-established samples and the specified radiological parameters – mass activity index I, equivalent radium dose Ra_{eq} and absorbed radiation dose D.

Table 1. Representation of basic constituents in material samples (%).

	MgO	Al_2O_3	SiO_2	P_2O_5	SO_3	CaO	Fe_2O_3
Concrete	4,11	5,02	35,49	0,06	1,53	26,41	1,79
Brick	2,20	13,46	66,25	0,20	0,25	1,16	5,78
Plaster	1,11	1,923	10,52	0,03	6,55	46,82	0,91

Table 2. Mass activities of radionuclides and radiological parameters.

Sample	^{40}K	^{226}Ra	^{228}Th	I [-]	R_{aeq} [Bq/kg]	D_1 [nGy/h]	D_2 [nGy/h]
Concrete	178,23	11,12	18,53	0,188	51,34	44,87	32,06
Brick	617,11	14,16	61,54	0,558	149,68	130,09	92,67
Plaster	106,62	5,18	6,21	0,083	22,24	20,13	14,39

Activity index I values, calculated from the measured radionuclide mass activities, ranged from 0.083–0.558. The limit values of the mass activity index take into account the method and amount of material used in the construction industry. In all material samples evaluated, the mass activity index was calculated to be lower than the limit value. The equivalent dose values of the radio materials studied were in the range of 22.238–149.68 Bq/kg. The maximum value of the equivalent activity should be less than 370 Bq/kg, in the samples being evaluated, all values are substantially lower than the required limit (UNSEACR, 2000). The values for the absorbed radiation dose are in the range 20,125–130,089 nGy/h. The highest absorbed radiation dose of building materials that is designated abroad for the safe use of buildings is 84 nGy/h (UNSECAR 2010). In one of the studied samples (brick), the limit was exceeded (130,09 nGy/h). In the other two samples the absorbed radiation dose is lower than the world average. The limit value in the brick sample (92,67 nGy/h) was exceeded in calculating the absorbed radiation dose for floors and walls.

4 CONCLUSION

In order to better assess the radiation risk of building materials in the determination of radioactivity in the internal environment of a historical building in Kosice, radiological parameters—mass activity index, equivalent radio activity and absorbed radiation dose were determined. On the basis of the established limit values for radiological parameters it can be stated that the values of the mass activity indexes did not exceed the recommended limit. In case of the absorbed radiation dose for the brick sample, the limit value was exceeded (130,09 nGy/h> 84 nGy/h), which can lead to increased emissions of ionizing radiation to humans.

ACKNOWLEDGMENT

This work has been supported by the Slovak Grant Agency for Science (Grant No. 1/0648/17).

REFERENCES

CEU-Council of the European Union—CEU. 2014. "Council Directive 2013/59/EURATOM of 5 December 2013 laying down basic safety standards for protection against the dangers arising from exposure to ionising radiation, and repealing Directives 89/618/Euratom, 90/641/Euratom, 96/29/Euratom, 97/43/Euratom and 2003/122/Euratom. Off. J. Eur. Union L 2013, 13, 2014.

Mothersill, C.E. & Seymour, C.B. 2012. Implications for human and environmental health of low doses of radiation. In *Radiobiology and Environmental Security.* Mothersil, C.E., Korogodina, V., Seymour, C.B., Eds.; Springer: Dordrecht, The Netherlands, 2012; pp. 43–51.

Saha, G.B. 2006. *Physics and Radiobiology of Nuclear Medicine.* Springer: New York, NY, USA, ISBN 978-1-4614-4012-3.

Tubiana, M., Dutreix, J., Wambersie, A. Bewley, D.K. 1990. *Introduction to Radiobiology.* Taylor & Francis: London, UK, ISBN 978-1-4822-9277-0.

UNSCEAR (United Nations Scientific Committee on the Effects of Atomic Radiation). 1982. Ionizing Radiation: Sources and Biological Effects. United Nations Scientific Committee on the Effects of Atomic Radiation. 1982 Report to the General Assembly, with Annexes; United Nations: New York, NY, USA.

UNSCEAR (United Nations Scientific Committee on the Effects of Atomic Radiation). 1993. Ionizing Radiation: Sources and Biological Effects. United Nations Scientific Committee on the Effects of Atomic Radiation UNSCEAR 1993 Report to the General Assembly, with Annexes; United Nations: New York, NY, USA.

UNSCEAR (United Nations Scientific Committee on the Effects of Atomic Radiation). 2010. Report of the United Nations Scientific Committee on the Effects of Atomic Radiation 2010; United Nations: New York, NY, USA.

UNSCEAR (United Nations Scientific Committee on the Effects of Atomic Radiation). 2012. Biological Mechanisms of Radiation Actions at Low Doses. United Nations Scientific Committee on the Effects of Atomic Radiation UNSCEAR; United Nations: New York, NY, USA.

UNSCEAR (United Nations Scientific Committee on the Effects of Atomic Radiation). 2000. Sources, effects and risks of ionizing radiation. United Nations Scientific Committee on Effects of Atomic Radiation. Report to the General Assembly with Scientific Annexes. New York: United Nations Publication.

UNSCEAR (United Nations Scientific Committee on the Effects of Atomic Radiation). 2013. Report of the United Nations Scientific Committee on the Effects of Atomic Radiation 2013; Vols. I and II; United Nations: New York, NY, USA, 2013.

Advances and Trends in Engineering Sciences and Technologies III – Al Ali & Platko (Eds)
© *2019 Taylor & Francis Group, London, ISBN 978-0-367-07509-5*

Implementation of lean production in construction—case study

M. Spisakova & M. Kozlovska
Faculty of Civil Engineering, Technical University of Kosice, Kosice, Slovakia

ABSTRACT: Implementation of lean production in construction is the increasingly topical issue in many countries. However, the current status and extent of lean construction implementation have not been well studied. We are entering a time where "lean" will be the main method of delivering construction projects. The applying of a new "lean" philosophy in construction sector provides the condition for more competitive, construction efficient and sustainable construction. The goal of paper is to describe the possibilities of applying this lean production and its principles in the Slovak construction company which focuses on modular modern methods of construction. The selection of the analyzed construction company was targeted because the off-site construction (in production halls) represents a huge opportunity for lean philosophy implementation in construction in terms of Industry 4.0 principles. Based on a structured interview, implementation of the lean production tools, its benefits and barriers have been proposed.

1 INTRODUCTION

The construction industry is very important to the EU economy. The sector provides 18 million direct jobs and contributes to about 9% of the EU's GDP. The construction value chain includes a wide range of economic activities, going ranging from the extraction of raw materials, the manufacturing and distribution of construction products up to the design, construction, management and control of construction works, their maintenance, renovation and demolition, as well as the recycling of construction and demolition waste. It also creates new jobs, drives economic growth, and provides solutions for social, climate and energy challenges. The goal of the European Commission is to help the sector become more competitive, construction efficient and sustainable (ec.europa.eu). Inefficient management of construction resources can result in low productivity. Therefore, it is important for contractors and construction managers to be familiar with the methods leading to evaluating the productivity of the equipments and the labourers in different crafts (Shehata & El-Gohary, 2012). The McKinsey Global Institute has described a five ways to improve the construction efficiency: (i) widespread deployments and use of interoperable technology applications, also called Building Information Modelling, (ii) improved job-site efficiency through more effective interfacing of people, processes, materials, equipment, and information, (iii) greater use of prefabrication, preassembly, modularization, and off-site fabrication techniques and processes, (iv) innovative, widespread use of demonstration installations and (v) effective performance measurements to drive efficiency and support innovation. The merging element of these points is integration of new production philosophy in construction. Lean construction aims to maximize the use of materials and labour of construction, and avoid any waste and non-value-added activities (Koskela, 1993).

1.1 *Lean production in construction*

The vision of International Council for Research and Innovation in Building and Construction (CIB) for future and development of construction industry expect that the construction will be like manufacturing: very lean, very low defect, very efficient, very integrated from the materials to the final product, as the client receives. The construction processes will be more digitally modelled, simulated, controlled and maintained (Owen, 2010). Lean production in construction is a new philosophy oriented toward construction production administration. It sets productive flows in motion in order to develop control systems with the aim of reducing losses throughout the process (Issa, 2013). It was taken from lean production that can be traced to Toyota Production System (TPS), with its focus on the reduction and elimination of waste (Ohno, 1998). There are many definition of lean construction (LC) which describes the methodology and application of these issues. Lean production in construction is the development to improve the culture, organizational and managerial style of the industry to break through the hurdles, attitude, roles, relationships, actions and communications among the project stakeholders (Egan, 1998). Likewise, the culture, and the organizational and managerial styles are the crucial pillars for a continuous improvement, which implied a constant delivery of greater value and increasing mutual competitive advantages. Conversely, Lim (Lim, 2008) suggested that lean is all about achieving a balanced use of people, materials, and resources. Another authors (Green, May, 2005) claim lean production in construction are variously understood as a set of techniques, a discourse, a socio-technical paradigm or even a cultural commodity. LC is (Bertelsen, 2004) a big scale of adaptation from the Japanese manufacturing principles and the concept is implemented to the construction process. According Koskela, advantages of the new production philosophy in terms of productivity, quality, and indicators were solid enough in practice in order to enhance the rapid diffusion of the new principles. In essence, the new conceptualization implies a dual view of construction production: it consists of conversions and flows. The overall efficiency of production is attributable to both the efficiency (level of technology, skill, motivation, etc.) of the conversion activities performed, as well as the amount and efficiency of the flow activities through which the conversion activities are bound together. While all activities expend cost and consume time, only conversion activities add value to the material or piece of information being transformed to a product. Thus, the improvement of flow activities should primarily be focused their reduction or elimination, whereas conversion activities have to be made more efficient. In generally, lean production in construction is based on the following attributes (Lim, 2008):

- specification of elements which create a value for costumer (not for department or company),
- identification of all necessary steps for design, manufacture and supply of products (on the other hands, specification of activities which create no value for customer),
- focusing on the flow for creation of value without interruption, waiting and time delay,
- carry out activities only on demand of customer,
- focusing on the continuous removal of over-production sources.

1.2 *Principles of lean production in construction*

The main principle of lean philosophy is to reduce the share of non-value-adding activities. The value adding activity converts material and/or information towards that which is required by the customer. Non value-adding activity (also called waste) takes time, resources or space but does not add value. It follows that the lean production presents a technology that minimizing the wastes in the manufacturing process. In this process can be defined seven types of wastes (Ohno, 1998): (i) waste of over production (largest waste), (ii) waste of time on hand (waiting),waste of transportation, (iii) waste of processing itself, (iv) waste of stock at hand, (v) waste of movement, (vi) waste of making defective products.

Koskela in 1992set another fundament principle of lean construction:

- increase output value through systematic consideration of customer requirements—value is generated through fulfilling customer, requirements, not as an inherent merit of conversion.
- reduce variability—from the customer point of view a uniform product is better,
- reduce the cycle time—time is a more useful and universal metric than cost and quality because it can be used to drive improvements in both,

- simplify by minimizing the number of steps and parts—the very complexity of a product or process increases the costs beyond the sum of the costs of individual parts or steps,
- increase output flexibility,
- increase process transparency—lack of process transparency increases the propensity to err, reduces the visibility of errors, and diminishes motivation for improvement.
- focus control on the complete process—there must a controlling authority for the complete process,
- build continuous improvement into the process—The effort to reduce waste and to increase value is an internal, incremental, and iterative activity, that can and must be carried out continuously,
- balance flow improvement with conversion improvement,
- benchmark—is a useful stimulus to achieve breakthrough improvement through radical reconfiguration of processes.

According the authors Salem and Zimmer (2005), there are five major lean principles that are applicable in the construction industry: customer focus, culture/people, workplace standardization, waste elimination, continuous improvement/built-in quality. The goal of applying lean principles is continuous improvement based on the lean construction philosophy, not only innovation. In Table 1 are mentioned the comparison and continuous improvement. The traditional measures are most often focusing on costs, productivity or utilization rates. In lean production, measurements should support the application of the new principles (Lesniak & Zima, 2018). Thus, there are a number of requirements for measurements (according lean principles): waste reduction, adding value, variability reduction, cycle time, simplification, transparency, focus on complete process, measurements should focus on causes rather than results, e.g. costs and continuous improvement.

1.3 Drivers and barriers of lean production in construction implementation

The successful implementation process of new lean philosophy has been challenged by various impediments eversince these systems were introduced into the construction industry brought up from the manufacturing industry. Lean construction is affected by a variety of barriers that hinder its successful implementation. This effort identified the most three significant barriers to be "lack of awareness", "lack of management support and commitment", and "culture and human attitudinal issues" (Small et al, 2017). Author Kanafani (2015) identified 34 barriers to the implementation of lean thinking in the construction industry.

Table 1. Comparison of innovation and continuous improvement (Imai, 1986).

	Innovation	Continuous improvement
Focus	Efficiency of conversions	Efficiency of flow processes
Goal	Leaps in efficiency	Small steps, details, finetuning
Involvement	Company and outside specialists, champions	Everybody in the company
Time frame	Intermittent and non-incremental	Continuous and incremental
Technology relied upon	Outside technological breakthroughs, new inventions	Internal know-how, best practice
Incentive	New superior technology or need for capacity extension	Overcome constraints in variability reduction or cycle time compression
Practical requirements	Requires large investment, but little effort to maintain it	Requires little investment, but great effort to maintain it
Mode of action	Scrap and rebuild	Maintenance and improvement
Transferability	Transferable: embodied in individual equipment and related operating skill	Primarily idiosyncratic: embodied in system of equipments, skills, procedures and organization
Effort orientation	Technology	People

Table 2.   Drivers of lean production in construction implementation (Mossman et al, 2010).

Investors	Easier linking design options to business objectives
	Improved value and a higher quality product
	Greater potential for lower cost construction and operation
	Reduced energy cost of use
	Facility delivered faster with higher quality so able to begin payback sooner
Designers	Less rework, minimises iteration
	Relationships, conversations & commitments are managed
	Decisions at last responsible moment
	Easier creating excellent green buildings
	Easier design to target cost
	Reduced design documentation time
Constructors	Better integrated design—less rework, lower costs, faster completion
	More buildable, logistics considered from outset
	Relationships, conversations & commitments systematically managed
	Greater construction process reliability and cost certainty

Factors considered as first 10 barriers according to their ranking are: 1. incomplete and complicated designs, 2. cyclic nature of the construction industry (i.e. economic cycles), 3. less involvement of contractors and specialists in design, 4. lack of lean understanding, 5. poor communication among stakeholders, 6. low tender prices, 7. slow decision making process, 8. inadequate pre-planning, 9. lack of individual performance measurement & motivation, 10. lack of long-term commitment to change and innovation.

Implementation of lean construction has a huge benefit for all stakeholders. Mossman et al. (2010) defined the potential benefits lean production for the three groups of construction participants—investors, designers and constructors (Table 2).

2 MATERIAL AND METHODOLOGY

The aim of paper is to describe the possibilities of applying the lean production and its principles and tools in the Slovak construction conditions.

2.1 Object description

The analysed construction company in Slovakia has engaged in the design, production, sale and rental of residential, office, sanitary and warehouse modular construction. Modular houses and buildings are delivered either on the basis of project documentation according to customer's preference and requirements or in the form of a standardized catalogue modular house. Modern methods of construction are synonymous with off-site manufacturing and prefabrication of building components and modules in factory settings, including complete buildings. Just the off-site construction (in production halls) represents a huge opportunity for lean philosophy implementation in construction in terms of Industry 4.0 principles. The modular constructions are produced in the construction hall and then are transported on construction site

2.2 Methodology of research

Analysis of the possibilities of applying the lean production was carried out by through a structured interview with designers and constructors of company. The structured interview was consisted of four parts: (i) information about company and its production program, (ii) the current state of building production, (iii) advantages and disadvantages of lean production implementation in company, (iv) the implementation possibility of selected tools lean construction.

Table 3. Implementation of lean production tools in construction company, its goal, benefits and barriers.

Tool of lean production	Goal	Benefits	Barriers
Milk run	Usage of full transport capacity of vehicle for transportation of construction material	Usage of full transport capacity of vehicle; Cost reduction in terms of vehicle numbers; Time saving	Apply only for own transport materials
Supply Chain Management (SCM)	Simplify the transport of construction modules to the site	Transport of all modules yourself; No need to contract the carrier; No need to increase the capacity of the labour force	Need to invest in new vehicles
Vendor Managed Inventory (VMI)	Improving the information flow	Providing constant information flow; Awareness of responsibility; Prevention of changes during module manufacturing	Multiple customer communication
Lean layout	Best workspace layout	Reducing of waste transport processes; Saving time in handling; Continuity of production processes	Increase of invest costs for the reconstruction of production halls
RFID tags	Improving information about materials	Location of the material is always clear; There is no loss and looking for materials	Use only for larger materials; Selection of suitable RDFI antennas and their correct location
ITF codes	Improve information about finished modules	No rechange of modules; Provides overview of finished modules; Easier installation of modules; Save time for handling modules	Correct design of the transport and subsequent assembly of the modules; Need to mark finished modules
Just in Time (JIT)	Fluent consumption of materials and module delivery	Save space in the warehouse; Limitation of materials handling; Save time and costs associated with storage	Correct planning of materials supplies
Total Productive Maintenance (TPM)	Ensuring the correct operation of machines and equipment	Caring for machines and equipment; Extending the lifetime of machines and equipment; Prevent the downtimes	Need to control machinery and equipment before and after work; Providing the necessary repairs
Poka Yoke	Monitoring of errors by workers	Errors are immediately identified; Prevents possible errors	Continuous control of workers and processes
Kaizen	Continuous motivation of workers	Work is made better and faster; Concentration of workers; Improving working conditions	Need to invest in the workers motivation
5S	Organization of the workplace	Cleanliness of the workplace; no limited workspace; Clarity of workspace	Constantly supervising the workplace organization

3 RESULTS AND DISCUSSION

The implementation of lean construction in company is focused on the production buildings, not on design of modules or construction of module houses on the site. During the structured interview, the errors were found within the design, management and realization of modules, and communication with customers, suppliers and subscribers.

Founded errors are: (i) frequent changes in customer requirements regarding the spatial, technological and material parameters of the modules already during the modules' realization, (ii) the necessity of continuously updating the timetable for the modules production (due to possible parameters changes by the customer), (iii) damage to production materials due to its excessive handling, (iv) unnecessary handling of the materials (due to inappropriate layout of the workspace in hall), (v) no modern machinery, tools and equipment usage, (vi) inappropriate organization of workplace. The holistic survey of lean construction tools is analyzed in research of author Ansah et al (2016). The lean construction tools are divided into three groups—waste detection, waste processing and waste response. According the analysis of processes for design, management and realization of modules, in construction companies were suggested eleven lean construction tools which have to eliminate identified error in construction company according the lean philosophy. The goal, benefits and barriers of its implementation in construction company are in Table 3.

4 CONCLUSION

The journal "The Economist" (2009) describes lean production as a group of highly efficient manufacturing techniques developed (mainly by large Japanese companies) in the 1980's and early 1990's. Lean production was seen as the third step in an historical progression, which took industry from the age of the craftsman through the methods of mass production and into an era that combined the best of both. It has been described as "the most fundamental change to occur since mass production was brought to full development by Henry Ford early in the 20th century". The change of construction philosophy of all stakeholders fulfils the assumptions of the idea of Industry 4.0. The submitted paper described the possibilities of applying lean production and its principles and tools in the Slovak construction company which focuses on modular modern methods of construction. The correct implementation of lean production principles can lead to saving ½ hours of labour resources in production; reducing mistakes in production by 50%, reducing investments in equipment, machinery and tools by 50%, saving 1/3 working hours of engineers, reducing workspace by 50% and reducing of the stock of materials by 90%.

ACKNOWLEDGEMENTS

The article presents a partial research result of project VEGA – 1/0557/18.

REFERENCES

Ansah, R.H. et al. 2016. Lean construction tools. In Proceedings of the 2016 International conference on industrial engineering and operations managements, Detroit 23–25 September 2016. IEOM Society.
Bertelsen, S. 2004. Lean construction: where are we and how to proceed. Available on: http://www.kth.se, 26 June 2018.
Egan, J. 1998. Rethinking Construction: Report of the Construction Task Force. London: HMSO.
Green, S.D. & May, S. 2005. Lean construction: arenas of enactment, models of diffusion and the managing of leanness. *Building Research and Information* 33(6): 498–511.
Imai, M. 1986. *Kaizen, the key to Japan's competitive success*. Random House, New York. 259 p.
Issa, U.H. 2013. Implementation of lean construction techniques for minimizing the risks effect on project construction time. *Alexandria Engineering Journal* 52 (1): 697–704.

Koskela, L. 1992. Application of the new production philosophy to construction. *CIFE Technical report* No. 72, September 1992, Stanford University.

Koskela, L. 1993. Lean production in construction. In. *Proceedings of the first annual conference of the international group for lean construction.*

Lesniak, A. & Zima, K. 2018. Cost calculation of construction projects including sustainability factors using the Case Based Reasoning (CBR) method. *Sustainability* 10 (5): 1–14.

Lim, V.L.J. 2008. Lean construction: knowledge and barriers in implementing into Malaysia Mossman, A., Ballard, G. & Pasquire, Ch. 2010. Lean project delivery – innovation in integrated design and delivery, available on: http://bit.ly/TCB-LPD.

Ohno, T. 1998. The Toyota Production System: Beyond large scale, Productivity Press, Cambridge.

Owen, R. 2010. CIB White paper on IDDS, University of Salford.

Mossman, A., Ballard. G. & Pasquire, Ch. (2010) Lean project delivery – innovation in integrated design and delivery. *Architectural Engineering and Design Management*, p. 165–190.

Shehata, E.M. & El-Gohary, M.K. 2012. Towards improving construction labor productivity and projects' performance. *Alexandria Engineering Journal* 50 (1): 321–330.

Small, E.P., Hamouri, A.k. & Hamouri, A.H. 2017. Examination of opportunities for integration of lean principles in construction in Dubai. In *Creative construction conference 2017*, Primosten, 19–22 June 2017. Elsevier.

The Economist. 2009. Lean production. Available on: http://www.economist.com/node/14299730, 19. October 2009.

Construction site safety modelling with implementation of lean production tools

D.E. Sprengel & M.V. Petrochenko
Peter the Great St. Petersburg Polytechnic University, St. Petersburg, Russia

M.A. Romero
University of São Paulo, São Paulo, Brazil

ABSTRACT: This paper presents the methodology of construction site safety integral criterion determination with implementation of lean construction tools. For this purpose, several groups of occupational risks were defined due to classification based on its origin (general group, individual protective equipment absence, working at heights, construction machinery and mechanisms, construction site energy supply, noise and vibration, maintenance of the construction site and workplaces, materials storage), the critical boundaries of each dangerous factor are given. The mathematical dependency between the main parameters is determined. The algorithm for decision making on safety management using lean manufacturing tools is proposed depending on the value of the integral criterion. This methodology allows to clearly identify the part of the construction site that is in the greatest risk area, enabling engineers to respond quickly to deviations from baseline construction cost and schedule.

1 INTRODUCTION

Lean manufacturing or lean production is an approach to the organisation management based on principles of increasing of production quality while reducing costs.

Lean production system includes several methods, tools and approaches to organization management. The main lean production tools can be divided into three groups: analysis, improvement and involvement technologies (Kazmina, 2016).

There are more than 25 lean production tools, but not all of them can be applied in construction field. In one of the researches the main tools were considered, including: 5S-system, Bottleneck Analysis, Gemba, Kaizen, KPI, PDCA etc. (Radchenko, Petrochenko, 2016).

There are several approaches to the assessment and management of the occupational safety in the workplace. The standard of Occupational health and safety assessment series (ICS 03.100.01; 13.100) is based on methodology of "Plan – Do – Check – Act" (PDCA) and contains a model of safety management system.

Among other approaches, the Qualitative Risk Assessment Model (QRAM) can be distinguished, which reveals direct dependence of the risk on organisational shortcomings, production factors, and inverse dependence with the indicator characterising the activities implemented at the construction site in order to prevent risks (Pitroda, 2016).

One of the most common risk assessment methods is FMEA-analysis (Failure Mode and Effects Analysis). In this case, the overall construction process can be divided into subprocesses and specific types of work. After that the importance of each type of work is determined from the safety point of view, given that the assessment can be carried out in two stages—the preliminary hazard analysis and the final one. The result of the second stage is the definition of the risk category in accordance with the value of RPN-index (risk priority index) (Nezhnikova, 2014; Zeng, 2010; Song, 2007).

In accordance with the conducted analysis of other researches' works, it can be pointed out that currently there is no integrated system of risk assessment and safety management of the construction site. Therefore, the algorithm of the task solution is suggested in this work.

2 THE METHODOLOGY OF CONSTRUCTION SITE SAFETY ASSESSMENT

To create the model of the construction site safety management and risk assessment, to optimize management process, it is necessary to divide risks into groups by a certain feature. In the study case the origin of the risk is suggested as the feature for the classification. As the result, eight groups of hazardous factors (risks) were defined:

– individual protective equipment absence;
– working at heights;
– construction machinery and mechanisms, risks accuring during installation and assembly works;
– construction site energy supply;
– noise and vibration;
– maintenance of the construction site and workplaces;
– materials storage;
– general group, including risks that do not depend on precise work type or part of the construction site (such as weather conditions, concentration of harmful substance in the air, instability of support structures, sharp corners and edges etc.).

Each group includes a list of requirements that need to be taken into account during the expert assessment at the construction site; critical boundaries of the requirements are given in the special assessment form. The critical boundaries are adopted as maximum values of the factor (risk) in accordance with regulatory documents.

Safety itself, in general, is a dimensionless indicator, therefore, there is need to shift from qualitative indicator description to the quantitative one. One of the methods of quantitative description of qualitative characteristics is qualimetry method (Simankina, 2013; Savall, 2012).

Integral criterion characterizing the level of safety at the construction site can be described with the function

$$Q_{int} = f(Q_C, Q_B, Q_S) \qquad (1)$$

where Q_C = complex indicator characterising the impact of the risk consequences on the cost; Q_B = complex indicator characterising the impact of the risk consequences on the duration of the construction (building time); Q_S = complex indicator characterising the impact of the risk consequences on workers' safety.

For description of the complex indicators, characterising the impact of the risk consequences on cost, duration and safety of the construction, the following function is adopted

$$Q_{C,B,S} = f(n, q_i, \lambda_{ni}) \qquad (2)$$

or

$$Q_{C,B,S} = \frac{\sum \lambda_{ni} \cdot q_i}{n} \qquad (3)$$

where q_i = relative i-th indicator of group of factors; λ_{ni} = weight coefficient of the i-th group of factors; n = number of estimated groups of hazardous factors.

The value of the relative indicators can be determined depending on the variation of the parameters affecting the indicator. Thus, at various stages of the construction process the indicator can be determined in different ways. In the first case, at the planning stage of the project, it is necessary to take into consideration the probability of occurrence of a hazardous factor on the construction site and its impact on the cost and duration of the construction, as well as the overall impact on the health of workers

$$q_{i,plan} = P \cdot C(B,S) \tag{4}$$

where P = probability of risk occurrence (can be based on government/company's statistic data or on the expert assessment method); C = impact of the risk on construction costs; B = impact of the risk on construction duration; S = the impact of the risk on workers' safety and health.

In the second case, in addition to the main parameters, the variable D is introduced, considering the detectability of safety violations at the site as well as the difficulty of its elimination

$$q_{i,act} = P \cdot D \cdot C(B,S) \tag{5}$$

The last formula should be used for the actual assessment of the indicator at the construction site. The values of $q_{i,plan}$ and $q_{i,act}$ can be applied for project analysis and new projects planning.

The weight coefficients of each group of factors are determined by means of expert assessment, the method of points assignment (Romanovich, 2015; Fedoseev, 2012). For this purpose, several experts assess the importance (risk) of each group of factors by assigning them certain number of points on accepted scale (for e.g. from 1 to 5). The weight for the factor is calculated by formula (Gorbunov, 2010)

$$r_{ik} = \frac{h_{ik}}{\sum h_{ik}} \tag{6}$$

where r_{ik} = weight calculated for the k-factor by the i-th expert; h_{ik} = score (points) assigned by the i-th expert for the k-factor.

The final weight coefficients are determined by the formula

$$\lambda_i = \frac{r_i}{\sum r_i} \tag{7}$$

where $r_i = \Sigma r_{ji}$ – sum of weights for the k-factor by all experts.

Having the values of all variables and having calculated complex indicators and the construction site safety integral criterion, it is necessary to compare it with the standard one (benchmark). By this means, the construction site safety assessment process can be divided into 6 main stages. The visual representation of it is given in Figure 1.

The standard indicators are adopted providing that their calculation is under ideal conditions of the construction site or close to them; there are no violations at the site and, consequently, nothing affects the duration, cost of construction and safety of workers. In this case, the value of the standard indicator is taken to be equal to unity. Thus, there are three possible values of the integral criterion and safety indicator:

– $Q_{int, S} > 1$, the estimated building project has a higher safety indicator than the standard one; the works are done in compliance with occupational safety regulations and the construction technology;
– $Q_{int,S} < 1$, the estimated building site has the safety indicator which is lower than the standard one; there are violations in the occupational health and safety compliance; special measures needed to be increase the safety level;

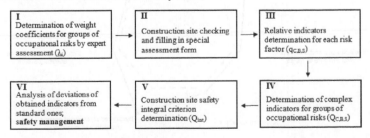

Figure 1.   Stages of construction site safety assessment.

- $Q_{int,S} = 1$, the safety of the assessed construction site is equal or close to the value of the standard indicator; there are no visible violations.

For the complex cost and duration indicators the dependence is inverse.

- $Q_{C,B} > 1$, the factor has impact on the project causing time-delays and increasing of the costs;
- $Q_{C,B} < 1$, the technology of works and safety regulations are complied, the costs and the duration either decrease or carried as planned;
- $Q_{C,B} = 1$, the cost and duration are not influenced by any factor; the works are carried out according to the baseline schedule and budget.

The results of calculations can be presented in tabular and graphical forms. The outlined methodology allows to monitor not only the overall construction site safety criterion, but also to determine the most vulnerable parts of the construction site, as well as to analyse changes in the construction duration and costs of works for selected groups of factors, identifying deviations from the approved budget and the baseline calendar plan.

Examples of diagrams reflecting the results of the construction site safety assessment based on the general group of factors, taking into account the impact on the duration and cost of construction are represented in Figures 2–3.

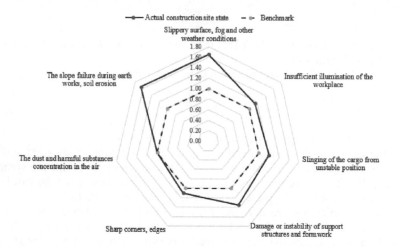

Figure 2.   The impact of the general group of hazardous factors on construction costs.

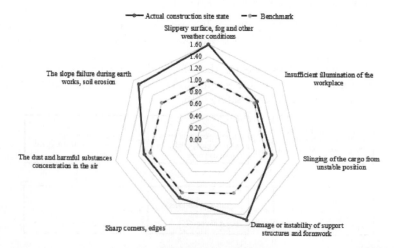

Figure 3.   The impact of the general group of hazardous factors on construction duration.

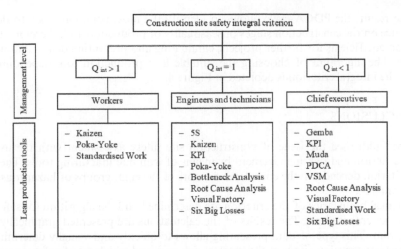

Figure 4.   Implementation of lean production tools depending on the value of the construction site safety integral criterion and management level.

The dashed line in the figure shows the benchmark—standard indicator, the solid line— the indicator, calculated as a result of assessing the current state of the construction site. If the point is inside the figure outlined with the dashed line or on it, then the safety regulations for this factor are complied and no additional measures are required; if the point is outside the figure, then this factor is critical and leads to construction costs and duration increasing.

After analysing the charts, it can be concluded that the greatest deviation from normal conditions and the impact on the cost of construction have the weather conditions as well as soil erosion and the slope failure during earth works. Violations of these factors will lead to an increase in the cost of construction and installation works and the project in general. The most crucial factors affecting the duration of the project are instability of support structures and weather conditions.

3   CONSTRUCTION SITE SEFETY MANAGEMENT WITH IMPLEMENTATION OF LEAN PRODUCTION TOOLS

Construction site management process is released at three main levels:

– workers;
– engineers and technicians;
– chief executives.

Depending on the level, different tools can be applied. For the planning stage of the project it is essential to take into consideration all internal and external factors, influencing costs and duration. At this stage the PDCA-tool with implementation of the methodology described above can be adopted. The cycle includes 4 phases:

– plan – audit and changes planning with the aim of further improvements; at this stage the $q_{i,plan}$ is defined, prognosing probable risks;
– do – implementation of the planned changes on a small scale, as a trial test at any facility, production site or in any situation; the observing the actual schedule and comparing it with the approved one;
– check – the study of the results; scheduled construction site checking, filling in the assessment forms and $q_{i,act}$ calculation;
– act – response: changed implementation or discarding, the cycle repeat with the adjustment conditions; comparing $q_{i,plan}$ and $q_{i,act}$, analysing.

As the result, the PDCA-cycle allows to identify the most common risks, to define the dependence on the construction stages or seasonality of production and, therefore, to set the correction coefficients for further projects, eliminating and preventing delays and incidental expenses. The algorithm of choosing the suitable lean production tool depending on the value of the integral criterion is depicted in Figure 4.

4 CONCLUSIONS

The paper addressed the issue of construction site safety assessment with the use of the integral criterion and risk management by means of lean manufacturing tools. The integral safety criterion depends on the complex indicators of the eight groups of hazardous production factors (risks).

The critical boundaries of the criterion are determined and the algorithm for safety violations assessments is given. The results of the calculations are presented graphically in form of a diagram. This type of data presenting allows to quickly and visually determine which part of the construction site is at the greatest risk area and which set of safety management solutions implementing lean production tools can be used.

REFERENCES

Ansah, R.H., Sorooshian, S., Mustafa, S.B., Duvvuru, G., 2016. Lean construction tools. *Proc*: 23–25.
Antipov D.A., Gonyakina Ye.N., Boychev V., Komarchev S.V., 2014. ERP-systems on Russian building sites. *Construction of Unique Buildings and Structures. Vol. 11 (26): 127*–135.
Fedoseev, V.V. 2012. Economic-mathematical methods and applied models. Moscow: YUNITY.
Gorbunov, V.M., 2010. Теория принятия решений. The theory of decision-making. *National Research Tomsk Polytechnic University* 67.
Gorelik, P.I., 2014. Lean construction as innovative construction project management method. *Construction of Unique Buildings and Structures,* (12): 40–48.
Grebneva, O.A., 2012. The theory of decision making. Irkustk: IrSTU.
Jayeshkumar Pitroda, 2016. Identification of factors affecting safety performance on construction projects. Gujarat.
Kazmina I.V, 2016. Analysis of the specifics of the introduction of lean manufacturing at domestic enterprises. *Synergy* (2).
Klochkov Y.P., 2012. Lean production: concepts, principles, mechanisms. *Engineering Journal of Don* 20 (2): 429–437.
Li, S., Wu, X., Zhou, Y., & Liu, X., 2017. A study on the evaluation of implementation level of lean construction in two Chinese firms. *Renewable and Sustainable Energy Reviews, 71*, 846–851.
Nezhnikova, E.V. & Aksenova A.A., 2014. The system of management of health and safety at work as an instrument of work safety in the construction industry. Vestnik MGSU, 7.
Pinto, A., Isabel L. Nunes, & Rita A. Ribeiro, 2011. Occupational risk assessment in construction industry-Overview and reflection. *Safety science* 49.5: 616–624.
Radchenko, A., & Petrochenko, M.V., 2016, December. Implementation of risk management theory for the development of lean thinking in a construction company. *The 2nd International Conference on Engineering Sciences and Technologies.* CRC Press Taylor & Francis Group.
Romanovich, M.A. 2015. The increasing of organizational and technological reliability of monolithic housing construction on the basis of modelling the parameters of time schedule, PhD Dissertation. Saint-Petersburg: SPbSUACE.
Savall, H., Zardet, V., Péron, M., & Bonnet, M., 2012. Possible Contributions of Qualimetrics Intervention-Research Methodology to Action Research. *International Journal of Action Research,* 8(1): 102–130.
Savall, H. & Zardet, V., 2011. The qualimetrics approach: Observing the complex object. IAP.
Simankina, T.L., & Popova, O.N., 2013. Qualimetric examination in the condition of urban areas evaluating. *Construction of Unique Buildings and Structures.* Vol. 12: 71.
Song, Ji-Won, J. Yu, & C. Kim, 2007. Construction safety management using FMEA technique: Focusing on the cases of steel frame work. *23rd Annual ARCOM Conference.*
Zeng, Sai X., Chun M. Tam & Vivian WY Tam, 2010. Integrating safety, environmental and quality risks for project management using a FMEA method. *Engineering Economics* 66.1.

A comparative case study of selected prefab façade systems

Z. Struková, R. Bašková & A. Tažiková
Faculty of Civil Engineering, Technical University of Košice, Košice, Slovakia

ABSTRACT: Recently, there has been a lot of development in prefab facade elements and systems for both renovation purposes and new buildings. Prefab systems possess many advantages; construction speed is one of the most important benefits. In the case study of a non-residential building, three variants of selected prefab façade systems are compared to each other. The Building Information Modelling (BIM) enables optimization of façade design already at the time of conceptual building designing. To compare the different façade systems in terms of construction time, construction cost, design aspect and thermal comfort, 3D BIM models of façade designs were created. The results of the comparative study are presented and discussed in the paper.

1 INTRODUCTION

The building envelope is one of the major building components. Technology driven materials, designs and construction techniques play a key role in current building facades (Paech, 2016). A variety of materials for façade cladding systems have been used in architecture. Traditional envelope materials include glass, metal, stone, timber and concrete. Over the past decades, façade technologies have undergone substantial innovations and new materials, such as glass fiber reinforced plastic and glass fiber reinforced concrete have been developed. Generally, the basic external wall materials including prefab, masonry, fixed-glass and curtain walls belong to the leading types of building façade alternatives. The building envelope prefabrication represents a serious opportunity for constructions where longer construction times may lead to the risk of incurring in unpleasant weather conditions.

Several authors have dealt with the development, analysis or evaluation of different modern façade solutions (Baldinelli 2009, Johnsen & Winther 2015). Some studies have aimed at cost efficient and sustainable solutions for refurbishment of façades. For example, Ruud et al. (2016) have conducted a cost-benefit analysis and have evaluated the energy savings of a wooden prefab façade element compared to two different conventional on-site façade refurbishment solutions, and also to compare the energy savings with other types of refurbishment options.

The selection of a building envelope is a decision characterized by multiple attributes. The success of the project is tied with assessment and selection of building envelope materials and designs that can satisfy the requirements of the stakeholders (Singhaputtangkul et al., 2013). Cost-effectiveness in application of façade is the most interesting issue for clients (Zavadskas, 2008). They mainly want to minimize the cost of the project, but they also want to achieve highest acceptable quality standards as well as satisfy technological, architectural, and comfort requirements. Sung (2016) has asserted that building envelope and façade design will inevitably be influenced using smart and low-energy systems.

Martabid & Mourques (2015) have claimed that the building façade may considerably affect the building`s sustainability performance and its selection is a complex decision. Several factors such as construction cost and time, structural behaviour and energy efficiency should be considered. They have presented an integrated approach to the envelope wall system selection problem. Their results included thermal, acoustic, and structural behaviours, cost, complexity of construction, safety and environmental impact, durability, and appearance. The most important criterion is the structural behaviour. According to the works of Brock (2005) and Gould (2005), the considerations in the assessment of the building envelope materials and combinations of their designs are associated with aesthetics, labour's skill sets, availability of manpower and equipment, building performance, durability, costs of a building project and so on. Passe and Nelson (2012) have emphasized the importance of considering the thermal behavior of the building envelope when selecting the most appropriate designs. They have claimed that the building envelope is responsible for approximately 50% of residential energy consumptions. Sighaputtangkul et al. (2014) identified the criteria for achieving sustainability and buildability for the assessment of building envelope materials and designs for residential buildings. The sustainability criteria cover environmental impacts, economic impacts and social impacts. The initial cost of the building envelope plays an important role in economic impacts. The buildability criteria involve health and safety of workers, community disturbance, simplicity of design details, material deliveries from suppliers, material handling, and ease in construction with respect to time. The thermal performance, building envelope material costs and installation costs and duration of the construction process are according to Paech (2016) involved in the most significant project specific concerns of façade systems that should be evaluated and decided during the initial stages of the design process.

The current development of Building Information Modelling (BIM) technology allows complicated building modelling to be digitally constructed with precise geometry and accurate information to support various building project stages. Lim et al. (2015) have integrated decision-making for sustainable building envelope design with BIM functionalities by considering the tropical climatic contexts and have formulated a process-driven BIM-based decision-making framework for sustainable building design in Malaysia.

2 MATERIALS AND METHODS

The case study concerns comparison of three prefab façade designs each other and selection of an optimal façade solution for a steel non-residential building. The building of Central European Training Centre with a total area of 920 m^2, which is situated 5 km from Košice city center is chosen as a reference building. The shape of the floor is rectangular with two protruding masonry extensions for garage, entry and sanitary background. The load bearing structure is designed as assembly truss structure supported by perimeter steel pillars. It comprises of fourteen transverse tracts with total length 40.6 m and trusses span 21.7 m.

To select an optimal solution of the façade for the steel building, three alternative 3D BIM models of façade design were developed. The materials for these three alternatives were found among in Europe available materials by considering the suitability for this type of building and construction complexities. The 3D BIM models of the reference building envelope provide a comprehensive view on the building envelope solution from function and aesthetic point of view. The following three feasible alternatives to carry out the façade of the steel hall building were considered suggested:

- (1) Metal modular façade system Qbiss One – as a highly cost-effective and world engineered, total wall solution with a unique rounded corner on the element. It is a complete all-in-one prefabricated wall solution replacing the whole built-up system consisting of structural wall and ventilated façade system. Its components include prefab modular wall elements, fixing and sealing material, architectural performance details and corner elements. The elements can be fixed directly on the primary steel work. It is not necessary to consider brick or concrete walls, insulation, vapour and wind protection or additional secondary sub-construction. Its manufacturer promises an efficient and rapid installation that saves construction time.
- (2) Façade system DEKMETAL based on cassettes DEKCASSETTE Special – as a system with tailor-made cassettes for every façade adapting to the originality of each architectural design in each of the works. The façade cassette is a square or rectangular bent element with an interlocking latch system, which is fastened to the supporting grid with the help of screws. The lower edge slides into the lock of another already fastened cassette while the

Alternative 1: Qbiss One   Alternative 2: DEKMETAL   Alternative 3: Equitone

Figure 1. The 3D visualization of the building façade.

upper edge is screwed to the grid. The fastening bolts are hidden within the cassette lock, which contributes to a regular façade grid of elegant joints.
- (3) Fibre cement façade material Equitone – as a high-performance façade solution with installed life expectancy of at least 50 years. The fibre cement material is through-coloured, which means the surface displays the inner texture and colour of the core eternit material. It is resistant to extreme weather conditions and temperatures. A lightweight aluminium or galvanized steel framework is fixed to the load bearing structure. The system is simple to install, robust and can be easily adjusted. The Equitone façade materials are fixed to the framework with rivets or screws and solutions with invisible fixing are available.

The methodology adopted to develop the 3D BIM models of the building façade under three alternative solutions was inspired by the study of Funtik (2012) aimed at BIM importance in a building life cycle. In building envelopes designing, the BIM offers a wonderful opportunity to improved and more effective cooperation among stakeholders (architect, contractor, client, etc.) and revision of conflicts related to façade elements fixing already in the design phase. This may save time and money in next phases of building project life cycle. The 3D visualization of the reference building façade in three studied alternatives is introduced in Figure 1.

The methodology used in the study to select an optimal façade design among the alternatives has defined three main decision-making criteria. The comparative and selection criteria involve: construction time (CT), construction cost (CC) and heat transfer coefficient (U). To assess the designed façade alternatives in terms of constructability, economy and thermal performance, the identified comparative and selection criteria were determined in all studied alternative designs. The values of the criteria were estimated based on the 3D BIM models of the building façade. Based on estimated level of efficiency for the studied alternatives of façade, an optimal solution for the reference building was recommended. The construction time covers not only duration of façade installation but also duration of façade components production and transportation of the components to construction site. The construction cost analysis was performed by estimating three major cost elements, i.e. material, transportation, and labor costs, for each alternative of façade design based on the cost data and quotes from material suppliers. To calculate the transportation costs, the distance between the building and the nearest manufacturing site for each type of façade was estimated. Thermal characteristics of façade materials clearly affect the amount of energy used during the operation phase of a building to keep its indoor temperature in a comfort range. In the study, the heat transfer coefficient was selected to compare the thermal performance of different façade designs.

3 RESULTS AND DISCUSSION

3.1 Identified comparison and selection criteria for the building façade

No one from the studied façade materials is manufactured in Slovakia. The following distances are considered in duration of the façade components transportation from factories to construction site in Košice: (1) QbiSS One: from Trebnje (Slovenia) to Košice – 786 km, (2) DEKMETAL: from Královice (Czech Rep.) to Košice – 764 km and (3) Equitone: from Neubeckum (Germany) to Košice – 1246 km. A uniform start of the façade assembly is considered in all three alternatives. The duration of assembly depends on technological procedures of different façade systems. The working group involves 8 workers in all three alternatives. The resulting values of construction time, determined based on normative labor intensity of different processes, are presented in Table 1.

To estimate the construction cost of the building façade, three major cost elements, i.e. material, transportation, and labor for each alternative of façade design was considered. In addition to main structure of the façade (i.e. load bearing structure, cladding material, thermal and diffusion insulation), the total construction cost includes scaffolding and windows. The composition of the façade from inside is identical in all three alternatives. It covers thermal insulation, permeable foil, wooden oriented strand board and plaster. The item "façade of masonry extensions" is also identical in all three alternatives. The aggregated values of the construction cost for the façade of the reference building are presented in Table 2. As can be seen, the type of the façade system may affect the associated material, labor, and transportation costs.

To compare the proposed façade solutions from thermal performance point of view, the heat transfer coefficient (U) is chosen as the comparative and selection criterion. The resulting values of this criterion in each alternative are presented in Table 3.

Table 1. The determined construction time of the building façade.

Construction phase	Construction process	CT – Construction time (day)		
		Alternative 1 Qbiss One	Alternative 2 DEKMETAL	Alternative 3 Equitone
Pre-assembly	Production of elements	32	20	30
Pre-assembly	Transport	6	3	6
Pre-assembly	Unloading	1	1	1
Assembly	Load bearing structure	3	29	16
Assembly	Thermal insulation	0	10	10
Assembly	Cladding elements	46	34	21
Assembly	Layers of façade from inside	27	27	27
Assembly	Façade of masonry extensions	9	9	9
SUM		124	133	120

Table 2. The determined construction cost of the building façade.

Cost item	CC – Construction cost (EUR)		
	Alternative 1 Qbiss One	Alternative 2 DEKMETAL	Alternative 3 Equitone
Load bearing structure of the building facade	110	52,164	39,573
Thermal insulation	0	7222	6554
Cladding material	127,319	25,659	74,345
Windows and sill boards	5430	5219	5219
Building façade from inside	36,192	36,192	36,192
Façade of masonry extensions	8015	8015	8015
Scaffolding	4645	4645	4645
SUM	181,711	139,116	174,543

Table 3. The identified heat transfer coefficient (U) of the building facade.

Alternative designs of building facade	U – Heat transfer coefficient
	(W/m^2 K) (by manufacturers' specifications)
Alternative 1: Qbiss One	0.158
Alternative 2: DEKMETAL	0.180
Alternative 3: Equitone	0.178

Table 4. The determined efficiency level of different building façade designs.

Alternative of building façade	Total utility index of façade designs				Construction cost (CC)	Efficiency level
	Construction time (CT)	Design/ aesthetic (D/A)	Heat transfer coefficient (U)	Total utility index		
	14.28%	42.86%	42.86%	100%		
Qbiss One	0.138	0.429	0.429	0.996	181,711	$0.548*10^{-5}$
DEKMETAL	0.129	0.257	0.376	0.762	139,116	$0.547*10^{-5}$
Equitone	0.143	0.343	0.381	0.867	174,543	$0.497*10^{-5}$

3.2 Selection of optimal building façade through decision-making method

After identifying and collecting all the required data, a multi-criteria decision-making problem was solved to choose an optimal solution of the building façade for the reference building. The total utility index of the façade for each alternative was estimated to determine the efficiency level of different building façade solutions. A questionnaire survey among local stakeholders (clients, project managers, contractors, architects) was conducted to set the weights of the criteria. Overall, 113 questionnaires were sent and 72 of them returned. Thus, returned/sent ratio is 0.44. Based on the results of the survey, the construction cost (CC) was determined as a crucial criterion mostly affecting efficiency level of the façade. This criterion is not included in total utility index of façade designs, but it is considerably involved in determination of efficiency level of different façade designs. In addition to construction time (CT) and heat transfer coefficient (U), the design/aesthetic (D/A) parameter was considered as one from comparison and selection criteria in the decision-making problem. The view towards beauty of a façade design usually varies from a person to another. Based on the answers of the respondents the following weights of the mentioned selecting criteria were established: construction time (CT) – 14.28%, design/aesthetic (D/A) – 42.86% and heat transfer coefficient (U) – 42.86%. The results obtained from the problem solving are shown in Table 4.

As shown, the metal modular façade system Qbiss One was found, with a slight difference, as the most suitable option for the façade of the reference building based on criteria and weights considered in this study. However, the winning option of the façade is the most expensive to procure and install, but it has rendered satisfactory results in other regarded criteria. From construction complexity point of view, its biggest advantage consists in possibility of fixing directly on the primary steel work without any additional load bearing structure. In terms of thermal performance, the system may be recognized a worthwhile investment. On the other hand, the fibre cement façade material Equitone was determined as the least appropriate option. Even though, the results of this study shouldn't be considered as a general ranking of efficiency of different prefab façade materials examined here. The achieved ranking of the studied alternatives is clearly reflective of view of stakeholders engaged in the questionnaire survey and the performance of different façade solutions may vary in dependability on the specific conditions and demands of any project.

4 CONCLUSIONS

A timely decision concerning the building's façade selection can have significant economic impact on the project. The success of a project suggests assessment and selection of building envelope materials and designs that can satisfy the demands of the stakeholders. Selecting a valid building envelope from several variants presents a crucial step in project design. The stakeholders in the design and construction process move towards using different decision criteria for selecting building materials and design. An inadequate attention to the demands may cause the selection of not entirely appropriate building envelopes. This can adversely impact the construction project phases and may cause delays, expenses increase, or even poor client satisfaction.

Three alternative designs of the prefab façade systems were compared to each other in the case study of a non-residential building. An optimal option of the building façade was then selected through multi-criteria decision-making process. The comparative and selection criteria involve construction cost, construction time, design/aesthetic parameters and heat transfer coefficient as a representative of thermal performance of the building façade. Relevant data of these criteria were identified based on three different 3D models of façades designed for the reference building. The weight of one criterion, namely design/aesthetic parameter, was estimated based on a questionnaire survey, which was conducted among local stakeholders. Based on the results of the study, the metal modular façade system Qbiss One was found, with a slight difference, as the most suitable option for the façade of the reference building. However, the achieved ranking of the studied alternatives is clearly reflective of view of the stakeholders engaged in the questionnaire survey and the performance of different façade solutions may vary in dependability on the specific conditions and demands of any project.

ACKNOWLEDGEMENT

The authors are grateful to the Cultural and Educational Grant Agency of the Ministry of Education, Science, Research and Sport of the Slovak Republic (Grant No. 059TUKE-4/2017 Supporting the skills in use of BIM technology in a building life-cycle) for financial support of this work.

REFERENCES

Baldinelli, G. 2009. Double skin facades for warm climate regions: Analysis of a solution with an integrated movable shading system. *Building and Environment* 44(6): 1107–1118.
Brock, L. 2005. *Designing the Exterior Wall: An Architectural Guide to the Vertical Envelope*. N. Jersey: John Wiley and Sons.
Funtík, T. 2012. Building information modeling from the perspective of a contractor. In *ICAMS 2012. Proceedings of the 4th International Conference on Advanced Materials and Systems., Buchurest, 27–29 September 2012*, INCDTP-ICPI, Romania.
Gould, F.E. 2005. *Managing the Construction Process: Estimating, Scheduling, and Project Control*. New Jersey: Pearson Prentice Hall.
Johnsen, K. & Winther, F.V. 2015. Dynamic facades, the smart way of meeting the energy requirements. *Energy Procedia* 78: 1568–1573.
Lim, Y.W., Shahsavari, F., Fazlenawati, N., Azli, M.N., Ossen, D.R. & Ahmad, M.H. 2015. Developing a BIM-based process-driven decision-making framework for sustainable building envelope design in the tropics. *WIT Transactions on The Built Environment* 149: 531–542.
Martabid, J.E. & Mourgues, C. 2015. Criteria Used for Selecting Envelope Wall Systems in Chilean Residential Projects. *Journal of Construction Engineering and Management* 141(12) 05015011.
Paech, Ch. 2016. Structural Membranes Used in Modern Building Facades. *Procedia Engineering* 155:61–70.
Passe, U. & Nelson, R. 2012. Constructing Energy Efficiency: Rethinking and Redesigning the Architectural Detail. *Journal of Architectural Engineering* 19(3): 193–203.
Ruud, S., Östman, L. & Orädd, P. 2016. Energy savings for a wood based modular pre-fabricated facade refurbishment system compared to other measures. Energy Procedia 96: 768–778.
Singhaputtangkul, N., Low, S.P., Teo, A.L. & Hwang, B.G. 2013. Knowledge-based Decision Support System Quality Function Deployment (KBDSS-QFD) tool for assessment of building *envelopes. Automation in Construction* 35: 314–328.
Singhaputtangkul, N., Low, S., Teo, A. & Hwang, B. 2014. Criteria for Architects and Engineers to Achieve Sustainability and Buildability in Building Envelope Designs. *Journal of Management in Engineering* 30(2): 236–245.
Sung, D. 2016. A New Look at Building Facades as Infrastructures. *Engineering* 2(1): 63–68.
Zavadskas, E.K., Kaklauskas, A., Turskis, Z. & Tamošaitiene, J. 2008. Selection of the effective dwelling house walls by applying attributes values determined at intervals. *Journal of Civil Engineering and Management* 14(2): 85–93.

Comparison of motorway costs between the Czech Republic and Germany

J. Stuchlík & J. Frková
Faculty of Civil Engineering, Czech Technical University in Prague, Prague, Czech Republic

ABSTRACT: This work compares the construction costs and the most frequent reasons of overpriced motorways between the Czech Republic (CR) and Germany (GE). The cost of 1-km (in mil. CZK) of the selected motorways was: 451 (CR) and 389 (GE). In CR, the rise in the price was incurred primarily by the land planning procedure and building permit procedure. The cost of motorways in the CR did not differ significantly from those in the GE.

1 INTRODUCTION

The growing traffic intensity is putting increasing demands on the modernization or construction of motorways. Construction of highways, such as the D8 motorway, and the ongoing, sharply monitored, modernization of the D1 motorway, raised some opinions that the projects are too expensive and lengthy. Some existing comparative studies show contradictory results: unreasonably high versus same or even lower prices of Czech highways, compared to those built in foreign countries. So where is the truth? (Oživení 2010, HospodářskéNoviny 2013, Mediafax Nova Ekonomika, 2013, Sdružení pro výstavbu silnic 2013, Echo24.cz 2018 ÚAMK 2018).

For the needs of this analysis, a sample of motorways was chosen, where construction costs of an individual construction steps are publicly known. This includes sections of the Czech motorway D8 and the German motorway A17. The Czech highway D8 is a 94 km-long highway from Prague North-West via Lovosice and Ústí nad Labem to Czech/German border (Krásný Les/Breitenau), where it freely connects to the German motorway A17. The D8 motorway starts at the northern outskirts of Prague, near Zdiby, and ends northwest of the Krásný Les village, in the Ore Mountains in the district of Ústí nad Labem. The 45 km-long German A17 motorway freely connects to the D8 motorway and leads to Dresden, towards the motorway junction with the A4 motorway (exit 77b – Dreieck Dresden-West). Both sections of the motorway—Czech and German—were built at the same time so that they could be put into operation together to connect the motorways of both countries.

2 METHODOLOGY

As part of the analysis of the motorways D8 and A17, a topology of a terrain and range of construction works were analyzed. Based on this, average prices of one kilometer of motorway in the CR and GE were calculated. This study also discusses possible reasons for non-compliance with a construction budget and failure to meet a deadline for completion of project.

Prior to the analysis of an average price of motorway construction, a development trend of a traffic intensity in the CR and GE in the period from 2005 to 2016 was analyzed. This is important before creation of new plans for modernization of motorways and a construction of a new transport network. The data for this analysis were drawn from the Road and Motorway Directorate of the CR (ŘSD ČR) and the Federal Highway Research Institute (BASt). The data from the web portals – www.dalnice-silnice.cz and www.autobahn17.de – were used to analyze the construction objects built on the D8 and A17 motorways.

The following highway sections (samples) for the traffic intensity analysis were selected: D1 (Hvězdonice – Ostředek), D5 (Rudná – Loděnice), D8 (Prague – Zdiby), D11 (HorníPočernice – Jirny). In case of GE, following sections were selected: A1 (Köln-Nord), A3 (Rosenhof (W)), A8 (Augsburg-Ost), A17 (Gompitz). Selection of these buildings was determined by their construction completeness and commissioning, including availability of their final construction costs. Another selection criterion included a similar topology of the motorway terrain and similar building technology used. For the D8 and A17 motorways, a type and a number of completed construction objects, e.g. a number of exit ramps, extensions of road lanes, bridges, tunnels, rest areas, noise barriers, etc., were identified. Terrain topology graphs of selected highways were created using the "Geocontext Profiler" application, freely available for use on the web site – http://www.geocontext.org. It is worthy of notice that the access to project documentation for these and many other public buildings is very limited for both lay and professional public.

3 RESULTS

3.1 Traffic intensity

The growth of traffic intensity recorded in the period from 2005 to 2016 in the CR and GE is presented in Figure 1 (the CR) and Figure 2 (GE).

In the CR, the most significant average daily increase in road traffic, in both directions, was recorded for motorway section "Prague – Zdiby" of the D8, i.e. increased by about 58%. This corresponds to 17,170 vehicles, which passed through the measured section in 24 hours in both directions. On the contrary, the smallest increase in the traffic intensity was recorded on the D1 motorway section "Hvězdonice – Ostředek", where the increase was by about 10% (4,195 vehicles/24 hrs). Historically, the oldest motorway in the CR is the D1 motorway, which connects Prague and Brno. The D1 is the most crowded motorway in the CR, which is reflected in its alarming state and necessity of its immediate modernization over its entire length. The D1 motorway repairs have begun in 2013 and continue to this day. Its surface is being replaced and other works are being carried out, such as extending the road for 0,75 m in both directions.

In Germany, the highest increase in traffic intensity was recorded for the A8 motorway section "Augsburg-Ost", where the increase was by about 43% (28,600 vehicles/24 hrs). On the other hand, the lowest increase was reported on the section A1 "Köln-Nord", and was by 4% (3 968 vehicles/24 hrs). The sections on the A8 motorway are controversial in terms of modernization, similarly to Czech D1 motorway, as the project has been overpriced and lengthy, with a completion deadline extended multiple times. One of the reasons for that may be the fact that the German A8 motorway has a cement-concrete cover and was extended by traffic

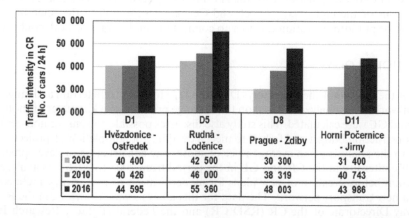

Figure 1.   Traffic intensity in the Czech Republic (CR) in years 2005–2016. References: ŘSD ČR (2005, 2010, 2016).

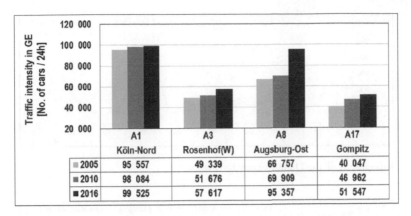

	A1 Köln-Nord	A3 Rosenhof(W)	A8 Augsburg-Ost	A17 Gompitz
2005	95 557	49 339	66 757	40 047
2010	98 084	51 676	69 909	46 962
2016	99 525	57 617	95 357	51 547

Figure 2. Traffic intensity in Germany (GE) in years 2005–2016. References: BASt (2005, 2010, 2016).

lanes, rest areas, noise barriers, etc., similarly to Czech D1 motorway. The cement-concrete cover is designed for 35–50 year-lifespan compared to asphalt cover, which has a service life of 10 to 15 years, with regular maintenance (Südwest Presse 2015).

3.2 Terrain topology of the selected highways

The topology of the terrain of the Czech D8 motorway is shown in Figure 3 and the German motorway A17 in Figure 4.

The length of the D8 motorway is 94,50 km. Half of the total length of the motorway is a sloping terrain, and the other half is a terrain with a rising tendency, i.e. the terrain is hilly. In the framework of the preparation of the D8 motorway, optimization of the Lovosice – Řehlovice highway was proposed, leading the highway through a tunnel dug through the basalt rock. An advantage of the proposed route would be its environmental-friendly character and easier winter maintenance. This proposal was abandoned for financial reasons (ŘSD ČR 2016).The highway route is eventually led by the nature of the Bohemian Central Mountains, where there are diverse geological foundations. Before the completion of the construction of the motorway section in the Central Bohemian Mountains, there was a landslide. Because of a natural disaster, it was necessary to repair the damaged railroad track, to build the motorway section and to remove the sloped ground. Costs were estimated at CZK 1 billion (Echo24 2017). The whole D8 motorway was launched on September 20, 2017.

The A17 motorway measures 44,70 km, of which more than a half (59%) is a sloping terrain, and the remaining part (41%) has a rising tendency, i.e. the terrain is classified as slightly wavy. The final costs incurred for earthworks and landscaping, including some other costs that might influence the final budget of the construction were not traceable from publicly available sources for both selected highways. However, it is clear from the topology of the terrain that the work associated with the foundation and grounding of the D8 motorway had to be more extensive compared to that associated with the German A17 motorway.

3.3 Building objects on the selected highways

The building objects built on the motorway sections of the D8 and A17 are listed in the Table 1.

Average prices of the motorway construction are influenced by building structures such as tunnels, motorway bridges, noise barriers, entrance or exit ramps, and rest areas. The most ex-pensive are motorway sections running through the agglomeration, a protected landscape area or through a mountain. In such case, noise barriers, ridges or tunnels, absorbing exhaust gases and acoustic or light emissions, must be constructed.

On the Czech D8 motorway there were 68 bridges, 4 tunnels, 13 interchanges, 32 over-passes and one ecoduct built, including some financially demanding ecological measures on sections passing the nature conservation area of the Central Bohemian Uplands. According

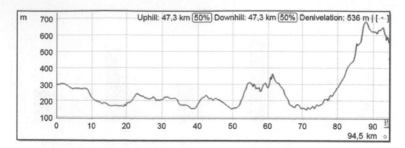

Figure 3.   Topography of the D8 motorway terrain.

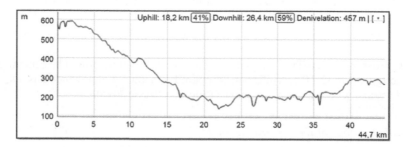

Figure 4.   Topography of the A17 motorway terrain.

to available information, final costs of construction of the D8 motorway was 42,232 million CZK. The cost of construction of selected building objects amounts to 16,477 million CZK (39% of the total cost of the D8 motorway). The most important object built on the D8 motorway was the tunnel "Panenská", which cost 5,638 million CZK, followed by the bridge "Knínice" (1 090 mil. CZK). On the German A17 motorway, Table 1, 9 bridges, 5 tunnels, 5 interchanges, and 2 noise barriers were built. Unfortunately, it was not possible to determine the final cost of the specific objects from publicly available sources.

3.4   *Average cost of construction of the selected highways*

The interpretation of the results of the average price of motorway construction should include the data obtained from previous analyses—the lengths and topologies of the terrains, the type and amount of motorway building objects, and the technology of construction (the material used). The D8 motorway is made up of 8 sections (Table 2), measuring a total of 94 km, with final cost of construction 42 232 million CZK (451 million CZK/km). The longest motorway section "Lovosice—Řehlovice", measures 16,41 km and its cost was 11 700 million CZK (713 CZK/km). It is necessary to point out that this section includes 2 highway tunnels (Prackovice, Radejčín), noise barriers, ecoduct and ecological measures over the protected area. The most expensive section of D8 was the section "Knínice—state border of the CR/GE", which is 11,48 km-long and cost 11 647 million CZK (1 015 mil. CZK/km). Two motorway tunnels, 8 highway bridges, an interchange and noise barriers were built on the section.

The A17 motorway is made up of 4 sections (Table 3) and is 45 km-long. For comparability of the final costs of construction of motorways between the CR and GE, the total construction.

costs were converted by purchasing power parity, which is set by the European Statistical Office (Eurostat). The purchasing power parity for Czech crown equaled to 17,55 in 2016 (Eurostat 2017). The final cost of construction of the A17 motorway was approximately 11,109 million CZK (248 million CZK/km). The longest section of the A17 motorway is the "Pirna – state border of CR", with length of 19,60 km and the final construction cost 2,706 million CZK, was the cheapest, with the average construction cost equal to 138 million CZK/km. Three highway bridges, one motorway tunnel or two interchanges were built on the

Table 1. Summary of building objects* implemented on the D8 and A17 motorways.

	Bridge	Inter.	Tunnel	Overpass	Ecoduct	Noise barrier	Footbridge	Retaining wall
D8	68	13	4	32	1	4	3	6
A17	9	5	5	8	0	2	0	0

*Based on data from: Dálnice – silnice.cz (2017) and Teledienstgesetz (2008); Inter. = interchange.

Table 2. The individual sections of the motorway D8 with their parameters and costs.

	Section length [km]	Final costs [mil. CZK]	Average costs of 1 km [mil. CZ]
Zdiby – Úžice	9,60	705	73
Úžice – NováVes	8,90	1 965	221
NováVes – Doksany	16,35	1 880	115
Doksany – Lovosice	14,43	2 616	181
Lovosice – Řehlovice	16,41	11 700	713
Řehlovice – Trmice	4,18	1 032	247
Trmice – Knínice	12,24	10 687	873
Knínice-state border CZ/GE	11,48	11 647	1 015
In Total	94	42 232	451

Table 3. The individual sections of the motorway A17 with their parameters and costs.

	Section length [km]	Final costs [mil. EUR]	Costs of 1 km [mil. EUR]	Final costs [mil. CZK]	Average costs of 1 km [mil. CZK]
Dresden/West – Dresden/Gorbitz	3,60	54	15	948	263
Dresden/Gorbitz – Dresden/Südvorstadt	8,85	288	33	5054	571
Dresden/Südvorstadt – Pirna	12,70	137	11	2404	189
Pirna – state border of CR	19,60	154	8	2703	138
In total	45	633	14	11109	248

section. By contrast, 8,85 km-long "Dresden-Gorbitz to Dresden-Südvorstadt" was the most expensive, built for 5053 million CZK (571 milion CZK/km). The cost per km is high due to the construction of 4 highway bridges, 3 highway tunnels, 4 overpasses and noise barriers.

4 CONCLUSIONS

So where is the truth? Are modernized or new highways too expensive in the CR, compared to other European countries? The point of this analysis was not to find the answer to this question but to point out that the problem is complex, due to many factors entering the building process that may influence the final cost of the building. The current analysis shows that the average price of construction of a 1 km-long section was 451 million CZK in CR (D8) and 248 million CZK in GE (A17), the latter being cheaper by approximately 45%.

The factors that may affect the final cost of the project are as follows: 1) different lengths of motorways—the D8 highway is longer than the A17. Not only a length massively increases the cost of an object, but this also brings higher risk of problems occurring during the construction; 2) different terrain and environment in which a motorway is built—an impact of agglomeration and an inter-connection of existing transport infrastructure; 3) an execution of different construction objects on a motorway (bridges, tunnels, ecoducts, etc.); 4) other reasons, such as problems during planning procedures and building permit process (the definition of routes), environmental activists, lengthy property settlement of land and real estate needed for construction, underestimated geological survey (landslides) or project documentation and road works, illegal buildings on the defined construction route, difficult conditions for inspection of the construction works, human failures (lack of qualification, lack of manpower, corruption), political situation (ongoing elections, changes of representatives, legal regulations).

This study drew data from a limited amount of available (relevant) sources and therefore it has not been possible to carry out an in-depth analysis of the problem. The unavailability of information on the amount of construction works and the final costs of construction of motorways is identical on the Czech and German portals and contributes to a lower degree of control and transparency of realized constructions. According to information from official independent authorities, such as the Supreme Audit Office of CR, a kilometer of the motorway is now cheaper by about 50% than five years ago (EuroZpravy.cz 2018). The CR wants to introduce the German model of motorway construction in the country. Thus, the representatives of the Czech Parliament have recently approved an amendment to the Act on Accelerating the Construction of Transport Infrastructure, which allows easier land acquisition under important buildings, which should speed up the construction of highways in the future (Poslanecká sněmovna České Republiky 2018).

REFERENCES

Bundesanstalt für Straßenwesen (BASt). 2005, 2010, 2016. Automatische Zählstellen 2005, 2010, 2016. [online] Available at: http://www.bast.de.
Bundesanstalt für Straßenwesen (BASt). 2017. Automatische Zählstellen [online] Available at: http://www.bast.de/DE/Home/home_node.html.
Dálnice—silnice. 2017. Dálnice D8. [online] Available at: http://www.dalnice-silnice.cz/D8.html.
Echo24.cz. 2018. Tragická výstavba dálnic v Česku. Do roku 2050 dálniční síť nebude. [online] Available at:https://echo24.cz/a/Sna2 U/tragicka-vystavba-dalnic-v-cesku-do-roku-2050-dalnicni-sit-nebude
Echo24. 2017. Začíná miliardový soud: kdo může za sesuv půdy na D8? [online] Available at: https://echo24.cz/a/ijfWV/zacina-miliardovy-soud-kdo-muze-za-sesuv-pudy-na-d8.
Eurostat. 2017. Purchasing power parities. [online] Available at: http://ec.europa.eu/eurostat/.
EuroZpravy.cz 2018. Kolik u nás stojí kilometr dálnice? Jsme na tom lépe než před pěti lety, ujistil prezident NKÚ. [online] Available at: Retrieved fromhttps://domaci.eurozpravy.cz/politika/.
Hospodářské noviny (HN). 2013. NKÚ: České dálnice jsou předražené. [online] Available at: https://domaci.ihned.cz/c160372380-nku-ceske-dalnice-jsou-predrazene.
Meiafax Nova Ekonomika, 2013. Propastný rozdíl: Kolik stojí dálnice v Česku a kolik v Německu či Dánsku? [online] Available at: http://www.kurzy.cz/.
Organizace Oživení. 2010. Cena dálnic. Část 5. Vývoj nákladů u vybranýchstavebdálnicD8 a D11. [online] Available at: http://www.oziveni.cz.
Poslanecká sněmovna České Republiky. 2018. [online] Available at: http://www.psp.cz/sqw/historie. sqw?o=8&t=76.
ŘSD ČR. 2005, 2010, 2016. Sčítání dopravy rok 2005, 2010, 2016. [online] Available at: http://www. rsd.cz.
ŘSD ČR. 2015. Srovnání cen dálnic se zahraničím. [online] Available at: http://www.roadmedia.cz.
ŘSD ČR. 2016. Celostátní sčítání dopravy. [online] Available at: http://scitani.rsd.cz/.
Sdružení pro výstavbu silnic Praha. 2013. Fakta a mýty o cenách českých dálnic. [online] Available at: http://denik.obce.cz/clanek.asp?id=6621709.
Südwest Presse. 2015. Freie Fahrt nach München—Ausbau der Autobahn A8 in Bayern fertig. [online] Available at: http://www.swp.de/ulm/nachrichten/politik/.
Teledienstgesetz. 2008. Autobahn A17. [online] Available at: http://www.autobahn17.de/.
ÚAMK. 2018. Kilometr dálnice stojí o půlku méně. [online] Available at: https://www.kurzy.cz/zpravy/457120-kilometr-dalnice-stoji-o-pulku-mene/.

Advances and Trends in Engineering Sciences and Technologies III – Al Ali & Platko (Eds)
© 2019 Taylor & Francis Group, London, ISBN 978-0-367-07509-5

Intensification of turbulent mixing in gases by means of active turbulence grid

N. Szaszák, P. Bencs & Sz. Szabó
Department of Fluid and Heat Engineering, University of Miskolc, Miskolc, Hungary

ABSTRACT: The objective of this work is to investigate the effect of a novel active turbulence grid on the mixing of airflow in a wind tunnel. Our grid contains flexible tubes attached onto the intersections of a square biplane grid of hollow rectangular bars. Pressurized air led to the grid bars is applied to produce random motion of the tubes in order to increase the turbulence level of the flow downstream of the grid. Upstream of the grid, an electric heater was assembled in the centerline of the wind tunnel to create an inhomogeneous temperature field. This low turbulence intensity flow was then passed through the grid. By means of an infrared camera and a porous sheet, temperature-inhomogeneity of the flow was investigated downstream of the grid in the test section while varying parameters: distance from the grid, mean flow velocity, gauge pressure of pressurized air. It was found that the grid substantially enhances the mixing.

1 INTRODUCTION

The intensive mixing of fluids (gases, liquids) with various physical and/or chemical properties is an important problem to be solved for many industrial applications. For this reason, static or dynamic mixers are applied. These instruments cause large flow disturbances, thus introduced fluids can be mixed rapidly; in a short time and/or at a short distance. Due to these flow disturbances, flow with high level of turbulence is generated, which plays the main role in the mixing process.

A novel type of active turbulence grid was developed in the recent years by our international research group Szaszák et al. (2012) but mainly for laboratory application. Based on our research results Szaszák et al. (2017) this grid can significantly increase the downstream turbulence level, therefore its area of application can certainly be broadened for intensification of mixing in gases, too.

Turbulence grids for turbulence generation in laboratory wind tunnels have been widely used both in research projects as well as in applied researches. These grids are mainly utilized to produce turbulence in the test section of the wind tunnel not far downstream of the grid Kurian & Fransson, (2009). Alternatively, these grids can be used to improve flow quality and homogeneity when the inlet flow has high turbulence intensity. Two types of these grids can be distinguished: passive Comte-Bellot, (1971), Roach, P. E., (1987) and active types Kang et al. (2003), Ling et al. (1972). Passive grids do not contain moving elements; turbulence production is mainly a result of vortex shedding from the grid itself, therefore the achievable turbulence intensity is relatively low compared to that of in case of active grids Makita et al. (1983). Active grids fundamentally contain moving elements, which enhance the vortex shedding, and also add momentum to the flow, thus much greater degree of turbulence intensity can be achieved. Our novel grid can be operated in both passive, half-active and active modes. This nature of the grid allows us to adjust the mode of operation which leads to appropriate mixing of the introduced fluids.

In this study the mixing properties of the grid were investigated by means of infrared image recording while varying flow parameters such as mean velocity of the air in the test section of the wind tunnel, operating conditions of the grid, and the downstream distance of the investigated cross-section relative to the grid.

2 APPARATUS AND INSTRUMENTATION

To determine the mixing property of the grid, inhomogeneous flows with different physical properties were produced. For this purpose, airflow with inhomogeneous temperature field was generated using a special electric heater placed in the center part of a conventional wind tunnel upstream of the grid, just upstream of the contraction section of the wind tunnel. The heater was designed and assembled especially for this purpose: it consists of square cross-sectional, low flow resistance aluminum casing which contains spirally rolled resistance heating wire inside. A toroid transformer was connected to the heater with which heating power could be controlled in the range of 0–500 W. With these helps and with controlling the flow of the wind tunnel, a flow with inhomogeneous temperature field was obtained: at the core region the temperature was a few degrees warmer than the outward part. This flow was then passed through the grid and was mixed to some extent, depending on the mode setting of the grid. Figure 1 shows a part of the wind tunnel with the test section including the heater, the grid, the experimental setup, and the location of the measurement sections, as discussed later.

During the experiments the wind tunnel was operated in open-loop mode in order to avoid temperature rise caused by the recirculated air.

The grid was placed at the beginning of the 500 × 500 mm cross sectional and 1500 mm long test section of the wind tunnel. The degree of mixing of the air flow was investigated by a VarioCAM HiRes 680 infrared camera equipped with Jenoptik IR 1.0/30 LW objective. Since the infrared radiation—and thus the temperature—of the gases cannot be seen by the infrared camera, a porous felt sheet placed in certain positions of the test section perpendicular to the mean flow was applied. Air flowing through the felt sheet warmed it up in those parts where warmer air passed through. Since the felt sheet had a relative low thermal capacity, its temperature followed quickly the temperature of the air passing through. The resulting inhomogeneous temperature field of the sheet could then be detected and captured by the IR camera, which was installed in the centerline of the test

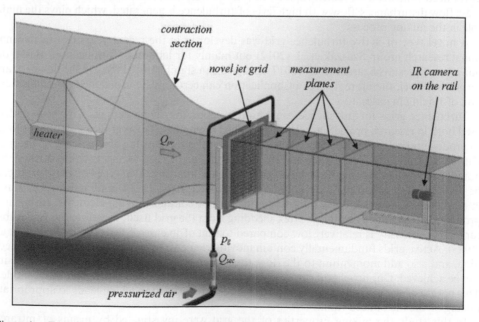

Figure 1.   Experimental setup including the heater, the novel jet grid, the IR camera, and the measurement planes in the test section discussed later in the text.

section and is movable on a mount rail so that its distance from the felt sheet can be set accurately. During our experiments the distance between the felt sheet and the camera was kept constant in order to get similar thermograms with the same aspect ratio for post processing.

The developed and investigated grid is a so called semi-active jet grid. Following numerous preliminary development, it was possible to compile such a novel semi-active grid, of which the active state is based on both the interacting secondary air jets and the chaotic movement of the colliding flexible tubes, which induce high fluctuations in the velocity field. As a result, flows with high levels of turbulence can be achieved downstream of the grid, even at relatively low incoming flow velocities.

The grid contains 20 pieces of horizontal and 20 pieces of vertical rectangular hollow brass bars with a bar thickness of 5 mm and a wall thickness of 0.45 mm, soldered perpendicular to each other. A uniform rectangular pattern biplane grid was constructed accordingly. The grid size was chosen to be $M = 25$ mm. With this setup, the solidity ratio of the grid was $\sigma = 0.36$, which is far lower than the value of 0.5, and is similar to what was used by Comte-Bellot & Corrsin (1966) ($\sigma = 0.34$ was presented in their experiment). Villermaux et al. (1991) pointed out that to avoid formation of inhomogeneous flows, solidity ratio must be kept far below the said value of 0.5.

In the active state of the grid, pressurized air was injected into the horizontal bars through both left-hand and right-hand sides of the grid, through the manifolds that were attached outside of the wind tunnel. At the grid intersections, soldered brass orifices with inner diameter of 3 mm were installed facing upstream. Flexible silicone tubes were attached to said orifices as active elements with inner diameter of 1.2 mm, wall thickness of 0.5 mm, and active length of 55 mm. The pressurized air then flowed through these flexible tubes as air-jets in counter-flow direction.

Three states of operation can be distinguished depending on the p_g gage pressure of the pressurized air (which was measured just upstream of the Y manifold applied to divide the flow to the two side manifolds, as shown in Figure 1), as follows:

a. $p_g = 0$ hPa belongs to the passive state of the grid, while the active elements were still;
b. $p_g - 250$ hPa corresponds to the conventional jet grid, in which case the flexible tubes are motionless and stiffen nearly perpendicular to the plane of the grid and thus the produced air-jets are quasi parallel;
c. $p_g = 500$ hPa corresponds to the active state of the grid, in which the active tubes move and collide randomly, hence caused by interacting jets, flow with high velocity fluctuations (i.e. high turbulence intensity) is produced.

The efficiency of the active grid was investigated not only as a function of the downstream distance from the grid, but also as a function of the mean flow velocity in the test section. To determine the role of the distance in the mixing, the temperature field was captured (i.e. the felt sheet was placed) in 4 different downstream distances from the grid, namely $x = 10$ M, 20 M, 30 M and 40 M, where M is the mesh spacing of the grid ($M = 25$ mm). In order to the influence of the mean velocity on the mixing be examined, the measurements were performed at four mean velocities with values of $U = 1.5$ m/s, 2 m/s, 2.5 m/s and 3 m/s. Since in the active cases of the grid (b and c), a secondary air flow was added to the primary airflow, here the mean velocity refers to the volume flow rate in the test section per unit cross section. The volume flow rate was determined by using Pitot-static tube in adequate distance from the grid, based on the ISO 3966:2008 standard (velocity area method). In order to ensure nearly the same temperature rise in the core region, the electric power of the heater was adjusted proportionally to the mean velocity. Thus, the power values were set to 150 W, 200 W, 250 W and 300 W, respectively. With these settings, during all measurement cases it was possible to keep the mean temperature of the flow in the range of 26–30 °C which was low enough to consider negligible heat conduction through the walls of the wind tunnel, thus the effect of the heat loss was not taken into account.

In case of each measurement setting, 10 pieces of thermogram were captured at random times with a resolution of 640×480, meaning $3 \times 4 \times 4 \times 10 = 480$ thermograms in total.

3 RESULTS

Following the measurements, it was necessary to quantify the degree of homogenization of the temperature field, since it is in relation with the mixing of the air which has been blown through the grid. For this purpose, a dedicated MATLAB script was written and then executed, which operation is the same for each measurement series, as detailed below.

Firstly, each thermogram was masked so that only the temperature field of the felt sheet was processed (since the original thermograms contain temperature values of parts of the wall of the wind tunnel). For each thermogram, the same parts have to be masked, because the distance between the felt and the camera was kept constant in all measurement cases. Therefore, during data processing only the core region (420×430 temperature values) was taken into account.

Then statistical parameters were calculated for each thermograms such as mean, standard deviation, median, mode, skewness, kurtosis, and histogram. By collecting the data obtained thermograms belong to each measurement setup it was possible to calculate the mean and the standard deviation values of the quantities mentioned previously. Different setups were compared using this information. Due to size limitations of the paper, only the trends of histograms are going to be presented.

A histogram itself shows the distribution of values of a dataset (temperatures in our case). In case of totally homogeneous temperature field, histogram is expected to be needle-shaped since only one value of temperature is present in the flow. The more inhomogeneous the temperature field is, the flatter and wider its shape of histogram is. To make the histograms of each measurement setting comparable, the temperature values were normalized by the most common value (i.e. the mode) of each thermogram. This way, the peaks (i.e. the most common value) of each thermogram coincide. The frequency values were then averaged in each measurement setup while 150 bins were set in the range of normalized temperature of 0.8–1.4.

Figure 2 shows the distribution of normalized temperature data in case of the lowest mean velocity $U = 1.5$ m/s at the furthermost investigated plane $x = 40$ $M;$ and while the highest $U = 3$ m/s velocity and the nearest plane ($x = 10$ M) were set. In both cases, results of all three states of the grid (a, b, c) are presented. Please note that the axis of frequency of values is in logarithmic scale in order to changes at less frequent values to be more emphasized.

Based on Figure 2, many statements can be drawn. If we look at the pairs of curves that belong to identical states of the grid (same shape of markers) it can be clearly seen that lower velocity and higher distance plays a significant role in the homogenization in all three states

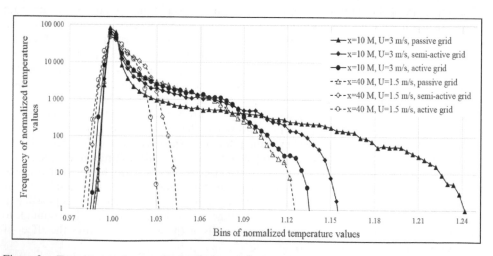

Figure 2.   Frequencies of normalized temperature data vs. histogram bins in case of six different measurement setups (see legend).

of the grid. The reason for this may be that the initially inhomogeneous flow has more time to be mixed due to swirling flow. We can state that in order to achieve better mixing at a given state of the grid, reduction of the mean velocity (evidently to a certain extent) or an increase in mixing distance is required.

Furthermore, perhaps a more important finding can be made by comparing the different states of the grid while mean velocity and distance from the grid are fixed (group of either solid lines or dashed lines). Please note that even in the worst investigated case for mixing (i.e. $U = 3$ m/s and $x = 10$ M, denoted by solid curves) activation of the grid (solid curve with filled dots) significantly enhances the homogencity of the flow downstream. This can be quantified by using the averaged standard deviation values of each settings: while in the worst case it has the value of $T_{std} = 0.692$, in the best setup it decreased to the value of 0.187 which is nearly 73% reduction in the deviation of temperature values. In this measurement setup with the activation of the grid ($p_g = 500$ hPa), nearly the same homogeneity can be achieved ($T_{std} = 0.519$) as in the best resulted case with the passive grid (dashed line with triangle markers, $T_{std} = 0.503$).

Figure 3 allows us to ascertain the effect of the distance as well as the operation mode of the grid on the flow homogeneity. For this purpose, results that belong to a constant $U = 1.5$ m/s mean velocity are presented.

It is worth noting that in both passive ($p_g = 0$ hPa, solid curves) and active cases ($p_g = 500$ hPa, dashed curves) the distance plays an important role in the homogenization. Level of mixing is increasing with it; however similarly high level of homogeneity was not achieved even at the greatest distance with passive grid as in the worst case of active grid while the lowest distance was set. This also shows the effectiveness of the activation of the grid. If we compare the cases of passive and active grids that belong to the same distance (i.e. the same shape markers), we can conclude that activation significantly reduces width of the interval of occurring values, hence the inhomogeneity reduces regardless of the distance as well.

The enhancement on mixing of a novel active grid was described and experimentally investigated in this paper. Using infrared camera, a porous felt sheet, and an electric heater the homogeneity of the temperature field of the flow downstream of the grid has been characterized. Results presented in this paper show that the developed turbulence grid can be applied to enhance mixing substantially in different gases. Activation of the grid had significant effect on the homogenization: at a given mean velocity considerably less mixing path is sufficient to reach the same level of mixing. For all of these reasons, the authors believe that the application area of this kind of turbulence grid can be extended especially for mixing.

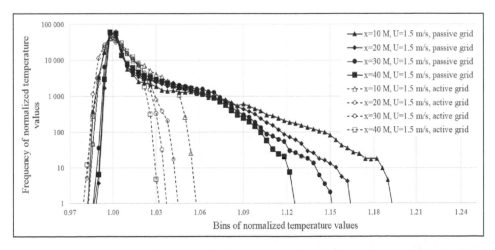

Figure 3.   Frequencies of normalized temperature data vs. histogram bins in case of measurement set-ups belong to mean velocity U = 1.5 m/s at different distances and states of the grid (see legend).

ACKNOWLEDGEMENTS

Supported BY the ÚNKP-17-3-III-ME/2 New National Excellence Program of the Ministry of Human Capacities.

The research was also supported by the EFOP-3.6.1-16-00011 "Younger and Renewing University—Innovative Knowledge City—institutional development of the University of Miskolc aiming at intelligent specialisation" project implemented in the framework of the Széchenyi 2020 program. The realization of this project is supported by the European Union, co-financed by the European Social Fund.

REFERENCES

Comte-Bellot, G., Corrsin, S. "The use of a contraction to improve the isotropy of grid-generated turbulence," *Journal of Fluid Mechanics* 25(4), 657–682 (1966).

Comte-Bellot, G. "Simple eulerian time correlation of full—and narrow-band velocity signals in grid-generated, isotropic turbulence," *Journal of Fluid Mechanics* 48(2), 273–337 (1971).

Kang, H.S., Chester, S., and Meneveau, C. "Decaying turbulence in an active-grid-generated flow and comparisons with large-eddy simulation," *Journal of Fluid Mechanics* 480, 129–160 (2003).

Kurian, T., Fransson, J.H.M. "Grid-generated turbulence revisited," *Fluid Dynamics Research* 41, 021403 (2009).

Ling, S.C., Wan, C.A. "Decay of isotropic turbulence generated by a mechanically agitated grid," *Physics of Fluids* 15(8), 1363 (1972).

Makita, H., Miyamoto, S. "Generation of high intensity turbulence and control of its structure in a low speed wind tunnel," *Proc. 2nd Asian Congress on Fluid Mechanics* Beijing, China, 101–106 (1983).

Roach, P.E. "The generation of nearly isotropic turbulence by means of grids," *International Journal of Heat and Fluid Flow* 8(2), 82–92 (1987).

Szaszák, N., Bordás, R., Mátrai, Z., Thévenin, D., and Szabó, S. "Experimental characterization of a cost-effective semi-active grid for turbulence stimulation," *Proc. Conference on Modelling Fluid Flow* (CMFF'12), Budapest, Hungary, 362–368 (2012).

Szaszák, N., Bencs, P., and Szabó, S. "Determining turbulent properties in grid-generated turbulence based on hot-wire data," *Proc. Int. Conf. on Innovative Technologies* (IN-Tech), Ljubljana, Slovenia, 73–76 (2017).

Villermaux, E., Gagne, Y., Hopfinger, E.J., and Sommeria, J. "Oscillatory instability and genesis of turbulence behind a high solidity grid," *European Journal of Mechanics*, B/Fluids, 10(4), 427–439 (1991).

Advances and Trends in Engineering Sciences and Technologies III – Al Ali & Platko (Eds)
© 2019 Taylor & Francis Group, London, ISBN 978-0-367-07509-5

Evaluation of climatic conditions affecting workers on scaffoldings

I. Szer & J. Szer
Lodz University of Technology, Lodz, Poland

B. Hoła
Wroclaw University of Science and Technology, Wroclaw, Poland

ABSTRACT: In the article a bioclimatic index—Universal Thermal Climate Index was used to present results of research on the influence of climatic conditions on people working on scaffoldings. UTCI is a measure of body's heat load calculated on the basis of thermal, radiative, wind and humidity conditions. The results have been obtained on three scaffoldings. The results show that during work on scaffoldings there is a high variability of workers' heat loads as well as the possibility of heat stress occurrence. This can have a negative influence on behaviour and health of people working on scaffoldings and may lead to fatigue and reduced concentration. As a result, there is also a higher risk of accidents. The article additionally presents co-dependencies between factors and UTCI.

1 INTRODUCTION

People working on scaffoldings at construction sites are at higher risk of having accidental injuries or participating in situations that may lead to potentially accidental events or even an accident. Since 2010, the number of people injured in construction work accidents in Poland has been decreasing; however, it remains too high. 5776 accidents arose in 2015, 84 of which were severe and 69 were lethal (Central Statistical Office. 2016). An accident is usually caused by the combination of many factors and circumstances which include technical, organisational and human factors. Research conducted in the UK, in which 62 accidents on scaffoldings were analysed, showed that 6.4% of accidents is caused by a human error (Whitaker et al. 2003). Based on the analysis of 177 accident protocols, which followed accidents related to work on scaffoldings in Poland, it was found that 27.4% of accidents was caused by a human error (Hoła et al. 2018). Human errors include lack of sufficient concentration and fatigue. People on scaffoldings do the work in dynamically changing, often uncomfortable and unfavourable conditions of the external environment, which additionally increases the risk of an accident. Too high or too low temperature, high humidity, changes in the wind speed combined with work that requires unceasing physical effort can cause changes in the organism even in people who are acclimatized. The worker can have psychophysiological disorders, which result in lower concentration, dissatisfaction, fatigue and longer reaction time (Traczyk & Trzebski 2004) which in consequence increases the number of mistakes at work. For this reason, it is significant to monitor the environment in which people on scaffoldings at construction sites work.

During past years the researchers were often focusing on examining the thermal comfort of a human being in the external environment. Evaluation of thermal comfort can be presented with a number of thermal indexes, empirical indexes and indicators based on linear equations (Coccolo et al. 2016). In the temperate climate the most frequently used indexes are, among many, OUT_SET* (Outdoor Effective Temperature) (Pickup & De Daer 1999, 2000), Physiologically Equivalent Temperature (PET) (Höppe 1999) and Universal Thermal Climate Index (UTCI) (Jendritzky et al. 2012). Thermal index OUT_SET* was first used to analyse the comfort of living in Sydney (Spagnolo & de Dear 2003). It was created through adaptation of Standard Effective Temperature (SET*) to external environment conditions by including the temperature of radiation. Thermal index PET is based on a model from Munich – a model of people's energy balance (MEMI) (Höppe 1993) and is defined as the "air temperature at which the heat balance of the human body is maintained with core and skin temperature equal to those under the conditions being assessed". UTCI index allows to define human thermal stress in different thermal conditions of external environment. UTCI was created on the basis of a multi-node model of human thermal stress (Fiala et al. 2012). Using these indicators at the construction site is frequently impeded by lack of the right software or big amount of data. For this reason, in this analysis a simplified universal indicator of thermal stress (UTCI*) was used (Błażejczyk & Kunert 2011).

2 MATERIAL AND RESEARCH METHODS

The research was conducted on three construction façade frame scaffoldings located in different cities in Poland. The scaffolding marked as L12 was in Lublin and it was examined in 2017 from April 10 to April 14. The width of the scaffolding was 45.0 m, the height – 26.36 m, and the surface was 1186.2 m^2. The Plettac 70 system frame scaffolding was located on the northern façade of the building. There was no net hanged on it. There were approximately 12 people working on it. The scaffolding marked as D17 was located in Wrocław and examined in 2017 on June 19–23. The width of the scaffolding was 9.0 m, the height – 16.12 m, and the surface was 144.9 m^2. It was produced by Layher. It was put up on the southeast façade of the building. There was a net hanged on the scaffolding. There were approximately 12 people working on it. The scaffolding W22 was located in Radzymin and it was being examined in 2017 between July 31 and August 4. The width of the scaffolding was 12.29 m, the height – 18.34 m and the surface was 223.12 m^2. The Blitz system frame scaffolding was next to the northeast façade of the building. There was a net hanged on the scaffolding. There were approximately 6 people working on it.

Selected parameters of the external environment: air temperature, relative air humidity (Szer et al. 2017), atmospheric pressure, light intensity, speed and direction of wind, sound level (Jabłoński M. et al. 2017) and dust level were examined on the scaffoldings. Additionally, technical parameters: scaffolding's perfect geometry deflections, technical condition of parts, anchoring strength, stands strength, scaffolding's vibration frequency (Błazik-Borowa et al. 2017), wind influencing the construction of the scaffolding (Jamińska-Gadomska et al. 2017), work load and physiological parameters of the workers (Czarnocki et al. 2017) were measured as well.

The measurements were carried out during one working week. On each day there were three examination rounds: first from 8 am, second from 11 am and third from 3 pm. Each examination round lasted approximately one hour and a half and depended on the number of research positions in which the measurements were carried out (Błazik-Borowa & Szer 2016). Measuring the environmental parameters was carried out in six or twelve research positions on the listed scaffoldings. The number of research positions depended on the width and height of the scaffolding (the number of scaffolding sections and work platforms) and the availability. Evaluation of working conditions was made on three work platforms. The schemes of scaffoldings with research positions marked are presented in Table 1.

The measurements were carried out with a multifunctional measuring device AMI310 which was attached to a climatic conditions module which registered air temperature and humidity and to a vane probe which measured the speed of wind. Measuring the air temperature and relative humidity was carried out in one research position within specific area of a scaffolding—in the middle of the deck at the height of a worker's face—approximately at a height of 1.5 m above the level of the deck. The duration of the measurement was 4 minutes with a sampling period of 1 second. Measuring the wind speed was carried out in a sequence in positions perpendicular and parallel to the façade, also at the height of a worker's face. Registration of the wind speed in each direction lasted 1 minute with a sampling period of 1 second. The course of the research is illustrated in Figure 1.

Table 1. Schemes of scaffoldings.

Symbol of scaffolding		
L12	D17	W22

Figure 1. The measurements on the scaffolding.

3 RESEARCH RESULTS

Table 2 presents air temperature, relative air humidity and absolute wind speed value measured in directions perpendicular and parallel to the façade. The average value was obtained from all the measurements carried out in all the research positions during one working day. In case of air temperature, the maximum value, which was 39.1°C, was observed on the D17 scaffolding on the second day of carrying out the measurements while the lowest value, which was 6.5°C, on L12 scaffolding on the fifth day. The relative humidity maximum value, which was 80.4%, was registered on L12 scaffolding on the fifth day while the lowest value, which was 21.3%, was on D17 scaffolding on the fourth day. The maximum wind speed, which was 10.1 m/s was parallel to the façade and it was registered on L12 scaffolding on the fourth day.

Simplified UTCI* index calculated according to the formula (Błażejczyk & Kunert 2011):

$$UTCI^* = 3.21 + 0.872t + 0.2459M_{rt} - 2.5078v_{10} - 0.0176RH \qquad (1)$$

where: M_{rt} – mean radiation temperature [°C], t – air temperature [°C], RH – relative air humidity [%], v_{10} – wind velocity at an altitude of 10 meters [ms^{-1}].

Table 2. Air temperature and relative humidity measured at the scaffolding.

Symbol of scaffolding		Air temperature			Relative humidity			Wind speed in the perpendicular direction		Wind speed in the parallel direction	
		Average	Minimum	Maximum	Average	Minimum	Maximum	Average	Maximum	Average	Maximum
		[°C]	[°C]	[°C]	[%]	[%]	[%]	[m/s]	[m/s]	[m/s]	[m/s]
L12	day1	18.0	13.5	23.5	40.6	29.8	51.9	1.0	1.4	1.2	5.0
	day2	11.4	9.6	14.5	50.3	39.2	64.3	0.9	1.7	1.0	6.0
	day3	9.8	7.0	12.5	57.7	51.2	70.4	1.1	2.2	2.0	5.0
	day4	10.9	9.6	13.5	63.8	51.6	73.6	0.6	1.9	1.3	10.1
	day5	9.4	6.5	11.7	63.9	51.0	80.4	0.8	2.3	1.3	5.0
D17	day1	25.8	23.7	29.5	48.3	39.9	55.3	0.5	2.7	0.6	2.3
	day2	32.8	27.2	39.1	33.2	23.2	50.8	0.5	4.0	1.2	4.4
	day3	28.3	21.2	37.6	28.2	22.5	35.5	0.6	5.0	0.7	5.0
	day4	29.6	22.6	38.0	31.1	21.3	47.1	0.4	3.8	0.8	2.7
	day5	21.1	19.1	24.1	72.6	53.9	85.6	1.0	5.0	1.3	5.0
W22	day1	32.0	28.5	35.4	49.8	40.6	71.8	0.2	1.8	0.2	1.0
	day2	32.8	28.1	38.2	49.2	38.1	65.1	0.4	2.3	0.6	2.6
	day3	30.9	26.3	38.5	49.5	37.1	71.3	0.5	5.0	0.6	3.4
	day4	29.6	26.3	32.3	56.2	47.3	70.9	0.1	0.9	0.4	2.0
	day5	31.5	29.5	35.6	50.7	41.9	58.6	0.5	4.7	0.5	2.4

A simplification has been introduced based on using the average temperature of radiation on the clothing surface, equal to the air temperature. The wind speed at a height of 10 meters in both directions was calculated according to the formula (Błażejczyk & Kunert 2011):

$$v_{10} = v_r/(h_r/h_w)^{0.2}$$ (2)

where: v_r – wind speed measured on the scaffolding [ms^{-1}], h_r – height at which the wind speed was measured [m], h_{10} – height 10 meters [m].

Then resulting wind speed was calculated:

$$v_w = \sqrt{v_{\rightarrow}^2 + v_{\uparrow}^2}$$ (3)

where: $v_{10\rightarrow}$ – wind speed parallel to the façade at 10 m height, $v_{10\uparrow}$ – wind speed perpendicular to the façade at 10 m height.

Table 3 includes human body load categorized according to the UTCI index. Respective categories were determined based on changes in physiological parameters (Błażejczyk & Kunert 2011). Table 4 presents the UTCI index calculated on the basis of an average temperature, relative air humidity and resultant wind speed at 10 metres height in each research position. The minimum and maximum values are the lowest and the highest average values from

Table 3. The scale of the estimation of heat stress of the organism according to UTCI (Błażejczyk & Kunert 2011).

UTCI [°C]	Stress category
+38.1 – + 46.0	very strong heat stress
+32.1 – + 38.0	strong heat stress
+26.1 – + 32.0	moderate heat stress
+9.1 – + 26.0	thermoneutral zone
+0.1–9.0	slight cold stress
−12.9–0.0	moderate cold stress

Table 4. UTCI.

Symbol of scaffolding		UTCI Average [°C]	Minimum [°C]	Maximum [°C]	Stress category
L12	day1	16.7	0.9	23.9	from slight cold stress to no thermal stress
	day2	9.9	−1.0	15.1	from moderate cold stress to no thermal stress
	day3	4.5	−9.5	14.0	from moderate cold stress to no thermal stress
	day4	8.6	0.2	15.4	from slight cold stress to no thermal stress
	day5	6.4	−4.9	14.4	from slight cold stress to no thermal stress
D17	day1	28.3	24.5	32.7	from no thermal stress to strong heat stress
	day2	34.5	26.3	42.0	from moderate thermal stress to very strong heat stress
	day3	31.1	23.6	43.0	from no thermal stress to very strong heat stress
	day4	32.4	25.6	40.4	from no thermal stress to very strong heat stress
	day5	19.9	10.1	26.2	from no thermal stress to moderate thermal stress
W22	day1	37.2	33.8	39.3	from strong heat stress to very strong heat stress
	day2	36.2	31.4	40.8	from moderate thermal stress to very strong heat stress
	day3	33.9	24.8	41.9	from no thermal stress to very strong heat stress
	day4	33.7	28.7	36.3	from moderate thermal stress to strong heat stress
	day5	35.0	27.1	40.6	from strong heat stress to very strong heat stress

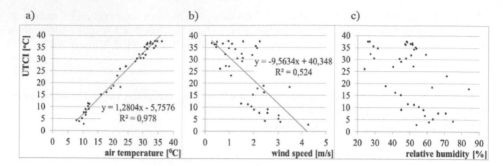

Figure 2. The dependence of the values of a) the UTCI and air temperature, b) the UTCI and relative humidity, c) the UTCI and wind speed measured at scaffolding.

all research positions. In the last column, on the basis of information from Table 4, body heat load is included. On L12 scaffolding examined in April there were the most favourable working conditions prevailing according to the index of thermal body loads. On D17 scaffolding during five days of testing and on W22 scaffolding four times, the UTCI index was at the level indicating the occurrence of very strong heat stress. Work in such environmental conditions may lead to workers' psychophysical disorders.

On the charts there is a dependency between the UTCI index and the air temperature (Fig. 2a), relative air humidity (Fig. 2b) and the resultant wind speed (Fig. 2c) measured on the scaffold presented. Since the UTCI index is calculated on the basis of temperature, relative humidity and wind speed, of course there should be a dependency between these values and UTCI. However, on the basis of the distribution of points in the graphs in Figure 2, it can be shown that the temperature has the highest impact on the value of the index and the relative humidity has the lowest impact of the air. UTCI increases linearly along with increasing air temperature.

4 SUMMARY

The performed analysis demonstrated that during the construction works on the scaffolding there was a high variability of air temperature, relative humidity of the air and wind speed. The results also showed that employees are exposed to strong or very strong heat stress. Physical effort should be periodically reduced or limited during such conditions. Unfavourable, changing environmental conditions are therefore a heavy load for people working outside. This can lead to reduced concentration, fatigue or prolonged reaction time. As a consequence, there is an increase in mistakes at work, which can cause injuries or other situations that could lead to a potentially accidental events as well as an accident. One of the solutions is to work in a brigade system. In case of extremely unfavourable conditions, it is possible to replace teams or increase the number of breaks. Therefore, monitoring the parameters of the work environment is an important element. The monitoring should be carried out by an authorised person—for example, by a person managing the construction site or occupational safety and health services. At the present stage of research we are analysing possible solutions for monitoring the parameters of the work environment on the construction site.

ACKNOWLEDGEMENTS

The paper has been prepared as a part of the project supported by the National Centre for Research and Development within Applied Research Programme (agreement No. PBS3/A2/19/2015 "Modelling of Risk Assessment of Construction Disasters. Accidents and Dangerous Incidents at Workplaces Using Scaffoldings").

REFERENCES

Błazik-Borowa, E. & Szer, J. 2016. Basic elements of the risk assessment model for the occurrence of dangerous events on scaffoldings (in Polish). *Przegląd budowlany* 10: 24–29.

Błazik-Borowa, E., Bęc, J., Robak, A., Szulej, J., Wielgos, P. & Szer, I. 2017. Technical factors affecting safety on a scaffolding. E. Fidelis & M. Behm (eds), *Towards better Safety. Health. Wellbeing. and Life in Construction*: 154–163. Bloemfointein: Department of Built Environment Central Universitty of Technology.

Błażejczyk, K. & Kunert, A. 2011. *Bioclimatic conditioning of recreation and tourism in Poland* (in Polish). Warsaw: Polish Academy of Sciences.

Central Statistical Office. 2016. *Accidents at work*. Warsaw: Information and statistical studies (in Polish).

Coccolo, S., Kämpf, J., Scartezzini, J.L. & Pearlmutter, D. 2016. Outdoor human comfort and thermal stress: A comprehensive review on models and standards. *Urban Climate* 18: 33–57.

Czarnocki, K., Błazik-Borowa, E., Czarnocka, E., Szer, J., Hoła, B., Rebelo, M. & Czarnocka K. 2017. Scaffold use risk assessment model for construction process safety. E. Fidelis & M. Behm (eds), *Towards better Safety. Health. Wellbeing. and Life in Construction*: 275–284. Bloemfointein: Department of Built Environment Central University of Technology.

De Dear, R. & Pickup, J. 2000. An outdoor thermal environment index (OUT_SET*)-Part II-Applications. *Biometeorology and urban climatology at the turn of the millennium. Selected Papers from the Conference ICB-ICUC 99.*

Fiala, D., Havenith, G., Bröde, P., Kampmann, B. & Jendritzky, G. 2012. UTCI-Fiala multi-node model of human heat transfer and temperature regulation. *International Journal of Climatology* 56: 429–441.

Hoła, A., Hoła, B., Sawicki, M. & Szóstak, M. 2018. Methodology of Classifying the Causes of Occupational Accidents Involving Construction Scaffolding Using Pareto-Lorenz Analysis. *Applied Sciences* 8(48).

Höppe, P.R. 1993. Heat balance modelling. *Experientia* 49(9): 741–746.

Höppe, P. 1999. The physiological equivalent temperature—a universal index for the biometeorological assessment of the thermal environment. *International Journal of Climatology* 43: 71–75.

Jabłoński, M., Szer, J., Szer, I. & Błazik-Borowa, E. 2017. Acoustic climate on scaffolding (in Polish). *Materiały budowlane* 8: 32–34.

Jamińska-Gadomska, P., Lipecki, T., Bęc, J. & Błazik-Borowa, E. 2017. In-situ measurements of wind action on scaffoldings. *The Proc. of European-African Conference on Wind Engineering*. Belgium.

Jendritzky, G., De Dear, R. & Havenith, G. 2012. UTCI-why another thermal index?. *Int. J. Biometeorology* 56: 421–428.

Pickup, J. & De Daer, R. 1999. An outdoor thermal comfort index (OUT_SET*) – part I—the model and his assumptions. *Biometeorology and Urban Climatology at the Turn of the Millennium* 99: 279–283.

Spagnolo, J. & De Dear, R. 2003. A field study of thermal comfort in outdoor and semi-outdoor environments in subtropical Sydney Australia. *Building Environment* 38: 721–738.

Spagnolo, J. & De Dear, R. 2003. A human thermal climatology of subtropical Sydney. *International Journal of Climatology* 23: 1383–1395.

Szer, I. Błazik-Borowa, E. & Szer, J. 2017. The influence of environmental factors on employee comfort based on an example of location temperature. *Archives of Civil Engineering* LXIII(3): 163–174.

Szer, I. Szer, J. Cyniak, P. & Błazik-Borowa, E. 2017. Influence of temperature and surroundings humidity on scaffolding work comfort. A. Bernatik, L Kocurkova & K. Jørgensen (eds), *Prevention of Accidents at Work*: 19–23 Leiden: Balkema.

Traczyk, Z. & Trzebski, A. 2004. *Human physiology with elements of clinical and applied physiology (in Polish)*. Warszawa: PZWL.

Whitaker, S.M., Graves, R.J., James, M. & McCann, P. 2003. Safety with access scaffolds: Development of a prototype decision aid based on accident analysis. *Journal of Safety Research* 34: 249–261.

Advances and Trends in Engineering Sciences and Technologies III – Al Ali & Platko (Eds)
© *2019 Taylor & Francis Group, London, ISBN 978-0-367-07509-5*

The influence of Shrinkage Reducing Agents (SRA) on the volume changes of cement pastes

Z. Štefunková & V. Gregorová
Faculty of Civil Engineering, Slovak University of Technology in Bratislava, Bratislava, Slovakia

ABSTRACT: Negative properties of concrete include the volume changes occurring during their solidification and hardening process. These deformations and cracks lead to reduction of the strength and durability of structures. The most important volume change of concrete is shrinkage resulting from the ongoing physical and chemical processes. The concrete with a high water to cement ratio is especially prone on the drying shrinkage. This article deals with the effects of Shrinkage Reducing Agents (SRA) in different volume dosages on the properties of cement pastes. The experimental study was based on the measurement of the shrinkage and the cracks of hardened cement pastes occurrence, while the dosage of SRA additives and water to cement ratios varied. In addition, the impact of SRAs on the compressive strengths at the age of 1 to 90 days was also observed. The consistency and the bulk density were monitored on fresh pastes. The achieved results have shown positive effects of SRA, but only to a certain extent.

1 INTRODUCTION

The volume changes of concrete and other cement composites occur as a result of physical and chemical processes in the cement matrix. Reduction of volume causes tensions that may result in cracks or other deformations. A volume change, where moisture loss occurs by drying, is called shrinkage. Plastic shrinkage occurs during drying of the fresh cement composites. In case of hardened concrete we are talking about drying shrinkage. We are distinguishing between two specific types of this shrinkage. First of them is the carbonation by atmospheric carbon dioxide and second is autogenous shrinkage in spontaneous drying during cement hydration.

In the case of plastic shrinking, intensive evaporation of water from the surface of fresh concrete usually results in surface cracks. These occur most frequently on large areas of concrete (slabs, walls, sidewalks). Disruption of the surface entirety reduces the durability of the concrete structure. Negative effect is increased with increasing wind speed, reducing the relative humidity and with rising temperature of the environment, etc. (Bajza et al., 2006).

The pore system of the hardened cement composite is emptied due to the evaporation of water, which results in the formation of capillary forces. These, together with the surface tension of the water, reduce the pore volume and thus the whole volume of the material. Firs the water starts to evaporate from the larger gaps and pores, later from the smaller pores. The water will keep evaporating until the balance between moisture in the composites and in the environment is reached. In composites, large tension occurs during drying, which results in the formation of cracks after exceeding the tensile strength. The character of the deformation varies depending on the humidity of the environment, i.e. the moisture content of the composite. Part of the total shrinkage of hardened cement paste, resulting from the first drying, is irreversible (Figure 1). Volume increase as a result of re-wetting and contraction in the course of the next drying are smaller (Gillinger et al., 2009). The total shrinkage value of concrete ranges from 0.3 to 0.7‰, mortar 0.7 to 1.5‰ and hardened cement paste of 3 to 5‰ (Svoboda et al., 2005).

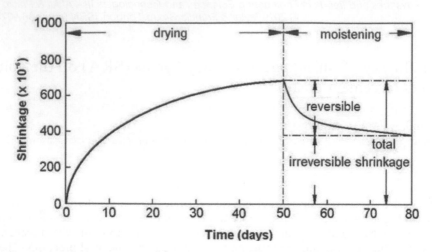

Figure 1. The behavior of concrete during drying and moistening (Bajza et al., 2006).

Composition of the cement composites significantly affects not only the drying shrinkage but also the autogenous shrinkage. Shrinkage of cement composites increases with higher water to cement ratio, higher water and fine particles content, fineness of filler granularity and a greater proportion of binder sealant. Composition of the concrete (composite) also has an impact on the type of shrinkage that occurs in concrete. Drying shrinkage is a substantial part of the regular concrete with higher water to cement ratio (above ca. 0.4). In the case of concrete with a low water to cement ratio the importance of autogenous shrinkage grows, which can form a dominant part of the total shrinkage (Neville, 1997). In practical tests it is a problem to differentiate between these two types of shrinkage. Measured values generally represent the total shrinkage of concrete at a certain time interval.

Shrinkage of the concrete can be a problem especially in the case of concrete with high water to cement ratio and finer filler such as self-leveling cement screeds. In these composites, the water to cement ratio is relatively high even with the use of quality superplasticizer agents.

The basic technological measure to reduce the shrinkage is to treat the fresh concrete. In some cases, however, the correct treatment of concrete or the reduction of the water to cement ratio by use of the additives, does not necessary guarantee the prevention of cracking (Matulová et al., 2013). One of the options for reducing the shrinkage of cement composites is using of SRAs or a various additives, such as expansive cement, which is a material based primarily on the increased formation of ettringite during hydration (Gregorová et al., 2018; Treesuwan et al., 2017). Another way is a suitable reinforcement of concrete. Different types of dispersed fibers can be used to eliminate the effects of shrinkage of self-leveling cement screeds (Ďubek et al., 2017; Ďubek et al., 2018; Štefunková et al., 2018).

Shrinkage reducing agents (SRAs) belong to the special additives based on organic compounds. They are dosed according to the exact percentage ratio of the weight of cement. SRAs greatly reduce the shrinkage process, but are known for their negative impacts too. One of them is that the hydration of cement is reduced thus the onset strength is delayed.

SRA consists of a mixture of organic compounds capable of binding water which, when mixed with water, generally reduces its surface tension. Smaller surface tension of water means a less attractive force between the walls of the capillary pores of the concrete matrix and consequently lower shrinkage due to drying. However, it is recommended to use a super plasticizing agent in the preparation of concrete with low water to cement ratio. Using the SRA may help prevent the crack formation as a result of not quite optimal concrete maturing, low relative humidity or where there are high outdoor temperatures. Generally, it is likely to be used where the water contained in the concrete will be at the stage of maturation of concrete evaporate faster (online BASF SK, 2018).

2 MATERIALS AND METHODS

The main component of cement pastes was CEM I 42.5 R produced by Cemmac, Inc. Horné Sŕnie with following properties: compressive strength after 2 days 30.0 MPa, after 28 days 53.6 MPa, initial setting time 150 min, setting time 190 min and soundness (expansion) by Le Chatelier 5.0 mm. To elimination the effect of cement pastes shrinkage, the shrinkage reducing agent (SRA) was used at the dosage 1%, 2% and 3% of the weight of cement. The SRA is made by combination of these alcohols: 2-ethylpropan-1,3-diol, prophylidyntrimethanol and 5-ethyl-1,3-dioxan-5-methanol with density of 1010 kg/m^3 and pH value 10–11 (Matulová et al., 2016). The water to cement ratio (w/c) was also changed to 0.4; 0.5 and 0.6.

Cement pastes were mixed in a normalized laboratory mixer. After that, the properties of the fresh cement pastes were determined: consistency by a flow test on the Haegermann flow table and the bulk density. The test samples to determine compressive strength were cylindrical with diameter and heights of 30 mm and were cured in the humid environment until testing time (1, 7, 28 and 90 days). The test samples to the monitoring of cracks had a triangle form (length of arms 25 mm and total length 250 mm). They were cured for 24 hours in the humid environment. After this time, the samples were kept in the laboratory environment (temperature 20 ± 5°C, relative air humidity φ = ca 50%) where they were observed for 7 days.

3 RESULTS AND DISCUSSION

The SRA proved as a plasticizing agent during the measurement of cement paste consistency (Table 1). The increasing dosage of SRA resulted in a greater spillage of cement paste. Due to the excessive flowability, the test samples with water to cement ratio of 0.6 were not produced with SRA additive. The bulk density of the fresh cement pastes was decreasing with increasing dosage of SRA. This decrease was more pronounced at a lower dosage of mixing water.

Based on the results of compressive strength it can be concluded (Table 2, Figure 5) that the positive effect of shrinkage reducing agent (SRA) was noted at the dosage 1% of the weight of cement. At this dosage, the samples of age 7, 28 and 90 days had higher values of compressive strength than samples without SRA. Test samples with 2% dosage of SRA showed similar values to the reference samples. Using the 3% dosage of SRA resulted in the lowest strength characteristics. Samples with lower water to cement ratio reached the highest values of compressive strength.

The same course was observed at the bulk density values of the hardened samples. The highest values were obtained on samples with 1% of SRA, the lowest values with 3% dosage of SRA. Also, the results of the bulk density of fresh cement pastes showed the same trend.

The incidence of cracks was significantly affected by the dosage of SRA and water to cement ratio. The most significant cracks were formed on the sample without use of the SRA and with high water to cement ratio (0.6). The cracks were deep and wide (Figure 2). In the later observed period, the incidence of cracks on samples with 1% of SRA was less than

Table 1. The properties of fresh cement pastes.

Water to cement ratio	Dosage of SRA (%)	Consistency (mm)	Bulk density (kg/m^3)
0.4	0	208	2029
	1	212	2024
	2	215	2001
	3	220	1991
0.5	0	269	1914
	1	271	1918
	2	276	1910
	3	288	1908
0.6	0	–	–

Table 2. The properties of hardened cement pastes.

Water to cement ratio	Dosage of SRA (%)	Age of samples (days)	Bulk density (kg/m³)	Compressive strength (MPa)
0.4	0	1	1970	17.1
		7	1980	38.6
		28	2023	54.3
		90	2028	68.9
	1	1	1995	15.4
		7	2000	40.8
		28	2028	59.8
		90	2039	80.2
	2	1	1970	11.5
		7	1985	36.0
		28	2001	52.2
		90	2014	75.4
	3	1	1940	8.7
		7	1972	33.7
		28	1991	48.4
		90	1998	69.1
0.5	0	1	1861	14.2
		7	1874	31.4
		28	1914	51.4
		90	1925	58.7
	1	1	1874	12.3
		7	1904	34.9
		28	1917	51.5
		90	1923	61.7
	2	1	1856	9.4
		7	1887	30.2
		28	1909	51.1
		90	1914	57.4
	3	1	1851	7.6
		7	1879	27.3
		28	1907	40.7
		90	1909	49.6

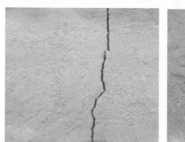  

Figure 2. The crack at w/c = 0.6 and without SRA. Figure 3. The crack at w/c = 0.5 and the dosage 2% of SRA. Figure 4. The crack at w/c = 0.4 and the dosage 1% of SRA.

on the reference samples at the same time (Figure 4). The lowest thicknesses of cracks were observed during 7 days on the samples with 2% dosage of SRA and with water to cement ratio of 0.5 (Figure 3).

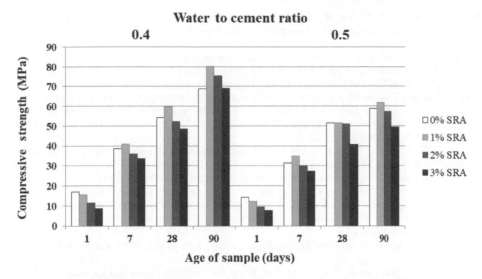

Figure 5. The compressive strength of the test samples with various dosages of SRA and water to cement ratio at the age of 1 to 90 days.

4 CONCLUSION

Based on the results can it be concluded that the positive effect of shrinkage reducing agent (SRA) on values of compressive strength was noted at the dosage 1% of the weight of cement. Test samples with 2% dosage of SRA showed similar values to the reference samples. Using the 3% dosage of SRA resulted in the lowest strength characteristics. Samples with lower water to cement ratio reached the highest values of compressive strength.

The same course was observed at the bulk density values of the hardened samples. The highest values were obtained on samples with 1% of SRA, the lowest values with 3% dosage of SRA. Also, the results of the bulk density of fresh samples showed the same trend.

The incidence of cracks was significantly affected by the dosage of SRA and water to cement ratio. The lowest thicknesses of cracks were observed during 7 days on the samples with 2% dosage of SRA and with water to cement ratio of 0.5. The most significant cracks were formed on the sample without use of the SRA and with high water to cement ratio (0.6).

ACKNOWLEDGEMENT

This article was created with the support by the Slovak Research and Development Agency under the contract No. APVV-15-0681.

REFERENCES

Bajza, A. & Rouseková, I. 2006. *Concrete technology*. Bratislava: Jaga Group, ISBN 80-8076-032-2.
BASF SK. 2018. Technical specification of MasterLife SRA 815. Available from https://assets.master-builders-solutions.basf.com/Shared%20Documents/PDF/Slovak%20(Slovakia)/TDS/masterlife-sra–815.pdf.
Ďubek, M., Makýš, P., Petro, M. & Briatka, P. 2017. Detecting Fibers in the Cross Sections of Steel Fiber-Reinforced Concrete. In: *Advances and Trends in Engineering Sciences and Technologies II – Proceedings of the 2nd International Conference on Engineering Sciences and Technologies, ESaT 2016:* 383–388.

Ďubek, M., Makýš, P., Ďubek, S., & Petro, M. 2018. The evaluation of the content of fibers in steel fiber reinforced structures and image analysis. In: *Journal of Civil Engineering and Management*, 24(3): 183–192.

Gillinger, J. et al. 2009. *Objemové zmeny betónu*. Rozborová úloha. TSUS, Košice.

Gregorova, V., Štefunková, Z., & Ledererová, M. 2018. Effects of expansive additive on cement composite properties. In: *IOP Conference Series: Materials Science and Engineering, 385: 012015.*

Matulová, K. & Unčík, S. 2013. Effect of expansive cement and waste gypsum on shrinkage of concrete. In: *ATF 2013: 2nd Conference on Acoustics, Light and Thermal Physics in Architecture and Building Structures. Book of proceedings*. Leuven, Belgium 2.-3.5.2013. – Leuven: Katholieke Universiteit Leuven, ISBN 978-90-8649-637-2. 302–307.

Matulová, K. & Unčík, S. 2016. Effect Of Shrinkage-Reducing Admixture On The Properties Of Cement Composites In: *Advances and Trends in Engineering Sciences and Technologies-Proceedings of the International Conference on Engineering Sciences and Technologies, ESaT 2015*: 329–334.

Neville, A. M. 1997, *Properties of Concrete*. Harlow: Addison Wesley Longman Limited.

Štefunková, Z., Ledererová, M., & Gregorová, V. 2018. Degradation of composites containing alkaliresistant glass fibres by ammonium nitrate solutions. In: *IOP Conference Series: Materials Science and Engineering, 385: 012054.*

Svoboda, L. et al., 2005. *Building Materials*. Bratislava: Jaga group, 2005, ISBN 80-8076-014-4 *in Czech*.

Treesuwan, S. & Maleesee, K. 2017. Effects of Shrinkage Reducing Agent and Expansive Additive on Mortar Properties. In: *Advances in Materials Science and Engineering*, 2017: art. no. 8917957.

Comparison of on-site and off-site modern methods of constructions based on wood

J. Švajlenka & M. Kozlovska
Department of Construction and Management, Faculty of Civil Engineering,
Technical University of Košice, Slovakia

ABSTRACT: In the context of traditional building materials, construction companies are increasingly offering new and modern full-bodied alternatives to housing. One of them is the modern prefabricated structural systems based on wood. Despite the undeniable benefits associated with wood-based buildings, they prevent them from widening the low awareness among investors and consumers of these structures. An important factor in deciding the majority of investors when purchasing wood-based buildings are quality parameters of construction, construction time and mainly procurement costs. For this reason, we have decided to carry out a survey aimed at examining the impact of the procurement method on existing wood buildings in the context of construction time and acquisition costs.

1   INTRODUCTION

Generally, the modern methods of construction are technologies which make use of structures or their components manufactured in factory (Lovell & Smith, 2010). The production of more or less completed components of building structures in the plants has a high potential for increasing the construction efficiency at the production stage of building components as well as in the process of their integration in the site. The MMC (Arif & Egbu, 2010) presents the technologies that provide effective procedures of construction preparation and execution, resulting in a larger volume of production with higher quality and reduced time of their procurement. The advantages of the MMC are shorter construction time, fewer errors in construction, and reduced demands on energy consumption or reducing of construction waste generation. Their ambition by Burwood and Jess (2005) is to enhance the construction efficiency through reducing of construction time, improvement of quality, sustainability and impact of the building and the building process on environment (Blismas & Wakefield, 2009). Authors Azman et al. (2012) claim that the MMC in the construction industry have a higher productivity and better quality, as well as some benefits such as reduced construction time, lower overall construction costs, better durability and better architectural appearance, increased healthy protection at work and safety, reduced materials consumption, less construction waste, fewer emissions into the environment and reduced energy and water consumption.

A range of materials is used for MMC, the most common being wood, steel and concrete. The choice of basic building materials is a vital part of each project and is usually based on professional judgment taking into consideration the importance of such criteria as economic, environmental, functional, aesthetic and health-related (Lesniak & Zima, 2015; Pošiváková et al., 2018). Responses to efficient, economic and sustainable solutions are modern methods based on wood. Regarding the modern methods of construction implementation in Slovakia, assembled buildings based on wood seem to be the most preferred construction system. This system is designed to build multi-storey buildings, apartment buildings, office buildings and houses (Thanoon et al., 2003). According to Štefko et al. (2010), they can be built as prefabricated panel constructions, framed constructions, timbered constructions, skeleton and half-timbered constructions. One of the advantages of wooden houses is the variability of structures and composition of the walls, which can be designed as a low cost, low energy and passive models. In addition, they are perceived as structures for the "healthy" housing, their main advantages are short construction time, lower environmental impact of the construction and used materials, and lower realization costs and costs of operation (Smith & Timberlake, 2011).

Despite the undeniable advantages associated with the use of modern wood-based construction systems, by Štefko (2004) prevents a wider expansion of timber structures in the Slovak Republic a low level of knowledge and information from customers and investors, as well as strong links to traditional brick technologies.

An important factor in deciding the majority of investors when purchasing wood-based buildings are quality parameters of construction, construction time and mainly procurement costs. For this reason, we decided to carry out a survey aimed at examining the impact of the procurement method on the already existing wood constructions in the context of construction time and procurement costs.

2 MATERIAL AND METHODS

This paper presents the partial results of the socio-economic exploration of modern wood-based construction methods. The results assess the impact of the procurement process on parameters construction time and the procurement costs of the wood buildings. The subject of the study was the real wood used already. A total of 80 buildings were monitored on behalf of two of the most widespread wooden construction systems realized in Slovakia (Wooden frame system, Panel construction system). The comparison parameter was subjected to a correlation analysis to determine the dependence between the analyzed parameters.

2.1 Selected construction systems buildings based on wood

The following chapters present the characteristics of the analyzed wood-based construction systems. By means of the basic design features, the differences between the comparing structural systems are clearly visible.

2.1.1 Wooden frame system

Wooden frame system originates from USA and Canada, where it is still the most widely used building system. The basic element of such a construction is supporting frame perimeter and partition walls of various timber profiles (Figure 1). Ceiling structure is composed of different profiles of timber and wood based materials. The stability is provided by the cladding of large agglomerated materials such as OSB board or gypsum board. Thermal requirements are secured by inserting thermal insulation (Figure 2). Standard construction of the walls is similar to panel construction system, but the individual elements and layers of walls are completed directly on site.

Figure 1. Wooden frame system. (Dubjel & Bobeková, 2017).

Figure 2. Inserting thermal insulation into the structure. (Dubjel & Bobeková, 2017).

Construction and assembly of wooden frame system is less demanding on a large mechanization. All layers of the structure and operation of installations are carried out on site, resulting in higher labor demands a higher proportion of the on-site works. This causes a greater probability of low quality work, including the impact of climatic conditions.

2.1.2 *Panel construction system*

Panel construction system is a main off-site construction method based on wood. Structural elements—panels (wall, ceil, roof, gable, partition wall) are produced in different stages of completion in the production hall and subsequently transported to the construction site where they are assembled to the structure. Build-up process is characterized by speed and precision. The panel generally consists of a wooden frame of profiled timber, covered on both sides with large-scale plates, filled with thermal insulation material. Instalations are prearranged in the panels during the manufacturing.

Prefabricated construction panel system fully utilizes construction, manufacturing and assembly advantages of their production to the efficiency of the entire construction process. The key moment to increase the efficiency and degree of prefabrication is panel's finalization. Panel system has enormous potential for increasing efficiency in the design, production and construction phase. Manufacturing can be automated, thus increasing the quality of production. Load bearing system of prefabricated wooden houses could be completed within a few days (Figure 3, Figure 4). Other finishing and plumbing work follows the assembly of the individual elements.

Figure 3.   Construction of panel construction system, (Haas Fertigbau, 2017).

Figure 4.   Construction of prefab panel, (Haas Fertigbau, 2017).

3 RESULTS

On the basis of the correlation analysis, we found a statistically significant dependence between the method of procurement and construction time (p = 0.5570), the method of procurement and the procurement costs for procurement of wood building (EUR) (p = –0.2776), as well as the method of procurement and the type of construction system (p = 0.3553). We also noticed the dependence between the type of construction system and the construction time (p = 0.6903).

A more detailed interpretation of the correlations between the construction system and the procurement of realization pointed out that the users of panel construction systems prefer the realization of their construction mostly through the construction company. The users of the columnar wooden constructions used the way of realization self-help in combination with the realization of the construction through the construction company. A statistically significant impact has been observed between the type of timber construction system and the construction time, which suggests that the panel timber constructions were realized in a shorter time horizon than the constructions using the column construction system.

Table 1 presents a comparison of the average construction time of the individual construction systems broken down by the method of procurement, indicating the declared construction time from woodworking producers. The table shows a breakdown according to the method of procurement, due to the fact that correlation analysis revealed statistically significant differences in terms of type of construction system and method of procurement (P = 0.3553). Declared values of construction time parameters (Table 1) and procurement costs (Table 2) from producers are determined based on the findings made on promotional materials, websites and personal interviews with representatives of companies operating in the construction sector. The findings from the mentioned sources show that the most frequently mentioned declared parameters of timber constructions can be summarized as: construction time, investment acquisition costs and energy standard, which are subsequently determined by an individual arrangement, specified and anchored in the works contract. Manufacturers also state that the construction time of the assembled dwelling completely made depends on a number of factors such as the technology used, size of the building, number of floors, the difficulty of the foundation and the construction, and, last but not least, the annual construction period. Acquisition costs as well as construction time depends on the particular technology and design. The amount of acquisition costs affected by several factors such as the energy standard of the resulting structures, diffusion of the resulting structure and quality of the building materials. Of course, such qualitative variants apply to all construction parts of the building.

From the data in Table 1 it can be stated that the shortest construction time was recorded in the panel construction system in all three ways of realization compared to the comparative construction system. The representative of the on-site construction system (wooden frame system) is largely implemented on a building site with a higher workflow and a higher demand for craftsmanship of workers, not excluding the weathering effects of the environment.

Table 1. Analysis of the construction time of comparative wood construction systems.

Construction system	Mode of procurement (number of buildings)	Average of construction time (months)	Construction time declared by suppliers (months)* (complete building)
Panel construction system	Through the supply company (40)	4.26	3–6*
	Realization by self-help (3)	7	–
	Combination (2)	10	–
Wooden frame system	Through the supply company (20)	10.47	3–6*
	Realization by self-help (13)	17.91	–
	Combination (2)	16	–

*Depending on the complexity of the project.

Table 2. Analysis procurement costs of the comparative wood construction systems.

Construction system	Mode of procurement (number of buildings)	Average of procurement costs (EUR) per m² of floor space	Procurement costs declared by suppliers (EUR)* per m² of floor space (complete building) with DPH	
		Overall without a difference in the energy standard	Low energy standard	Passive energy standard
Panel construction system	Through the supply company (40)	933	900–1200*	1400–1600*
	Realization by self-help (3)	647	–	–
	Combination (2)	1046	–	–
Wooden frame system	Through the supply company (20)	925	900–1400*	1400–1600*
	Realization by self-help (13)	635	–	–
	Combination (2)	694	–	–

*Depending on the complexity of the project.

By comparing the average construction time of the construction systems and the declared construction times by the manufacturers it can be stated that only the panel construction system has actually fulfilled the predefined parameter.

Table 2 presents a comparison of the average procurement costs of individual construction systems in terms of conversion per m² of useful area. On the basis of considerable data dissemination, there was no statistically significant effect between the procurement cost and the building energy standard, therefore we did not calculate the recalculated cost per m² of useful space in terms of the energy standards of the monitored buildings in Table 2.

By correlation analysis we recorded a statistically significant negative dependence between the method of realization and investment costs for the procurement of constructions ($p = -0.2776$), which means that if the construction was carried out by the supply company, the acquisition costs increased, whereas a decrease was made recorded when the construction was realized either alone or in combinations. These findings were inconsistent with the practice because the above-mentioned findings are in practice a standard. On the basis of average values calculated per m² in Table 2 it can be stated that in almost all methods of realization of panel and column woodwork comparable cost of acquisition per m² was recorded, except for the combined realization of the construction.

4 CONCLUSION

In the presented article, we analyzed the impact of the procurement method on the building time parameters and the procurement cost of real woodworks. Two of the most widespread wooden construction systems realized in Slovakia (Wooden frame system, Panel construction system) were analyzed. On the basis of the correlation analysis, we found a statistically significant dependence between the method of realization and the time of construction ($p = 0.5570$), the realization method and the investment costs for procurement of wood buildings (EUR) ($p = -0.2776$), and the realization method and the timber construction system ($p = 0.3553$). We also noticed the dependence between the type of construction system and the construction time ($p = 0.6903$). The conclusions of the analysis of the assessed wood construction parameters point to the fact that the timber construction based panels are the most effective in terms of construction time and are realized through a supply company. The least efficient in terms of the construction period is construction carried out by a combined construction method (a combination of realization through a supplier company and self-realization). From the point of view of procurement costs, panel and column construction system were comparable in almost all ways of realization.

ACKNOWLEDGEMENTS

The article presents a partial research result of the VEGA project-1/0557/18 "Research and development of process and product innovations of modern methods of construction in the context of the Industry 4.0 principles".

REFERENCES

Arif, M. & Egbu, C. 2010. Making a case for offsite construction in China. *ECAM* 17: 536–548.

Azman, M.N.A., Ahamad, M.S.S., Hilmi N.D. 2012. The perspective view of Malaysian industrialized building system (IBS) under IBS precast manufacturing. *The 4th International Engineering Conference – Towards engineering of 21st century.*

Blismas, N. & Wakefield, R. 2009. Concrete prefabricated housing via advances in systems Technologies, Development of a technology roadmap. *ECAM* 17: 99–110.

Burwood, S. & Jess, P. 2005. *Modern Methods of Construction Evolution or Revolution, A BURA Steering and Development forum report.* London.

Ceder. 2017. *Zrubové stavby.* Dubové: Ceder.

Dubjel, K. & Bobeková, E. 2017. *Realizácia rodinného domu drevenou stĺpikovou sústavou.* Bratislava: ASB.

Haas Fertigbau. 2017. *Montované domy.* Bratislava: Haas Fertigbau.

Kolb, J. 2008. *Dřevostavby.* Praha: Grada Publishing.

Lesniak, A. & Zima, K. 2015. Comparison of traditional and ecological wall system using the AHP method. *International Multidisciplinary Scientific GeoConference Surveying Geology and Mining Ecology Managamentt, SGEM* 3(5): 157–164.

Lovell, H. & Smith, S.J. 2010. *Agencement in housing markets, The case of the UK construction industry.* London: Geoforum.

Pošiváková, T. et al. 2018. Selected aspects of integrated environmental management. *Annals of Agricultural and Environmental Medicine* 25(3): 403–408.

Reinprecht, L. 2017. *Zrubový konštrukčný system.* Bratislava: Jaga.

Smith, R.E. & Timberlake, J. 2011. *Prefab architecture: a guide to modular design and construction.* Canada: John Wiley & Sons, Inc.

Štefko, J. & Reinprecht, L. 2004. *Dřevěné stavby – konstrukce, ochrana a údržba.* Bratislava: Jaga group.

Štefko, J. et al. 2010. *Modern wooden buildings.* Bratislava: Antar.

Thanoon, W.A.M. et al. 2003. The essential characteristics of industrialised building system. *International Conference on Industrialised Building Systems,* Malaysia.

Advances and Trends in Engineering Sciences and Technologies III – Al Ali & Platko (Eds)
© 2019 Taylor & Francis Group, London, ISBN 978-0-367-07509-5

Influence of flash floods on the drainage systems of the urbanized areas

M. Teichmann, N. Szeligova & F. Kuda
VSB-Technical University of Ostrava, Faculty of Civil Engineering, Czech Republic

ABSTRACT: The article deals with the adverse effects of flash floods on the sewer systems functionality of urbanized areas. The floods themselves have been usually advaresed and long-term impacted on urban, municipal and surrounding areas. Otherwise, this is not the case with the sewer systems that are located in these areas. The combination of sewerage systems, floods and often also the effects of other inappropriately selected local environmental conditions can cause adverse impacts. The most pervasive is the situation of transport infrastructure, where the effects of circumstances combination often result in instability in sloping areas, local dips, quarries and displacements, which are additionally intensified by underground and surface water flows, or the adverse effects of relatively powerful anthropogenic layers in combination with groundwater flow patterns. These impacts effects have a significant impact on the functionality of the transport service of the urbanized area and thus contribute to its sustainability.

1 INTRODUCTION

In recent years, larger cities especially have become significantly vulnerable to the effects of climate change, often being subject to marked changes such as high levels of precipitation, more frequent storms, extreme alternation of heat and cold, and also more frequently occurring floods and flash floods. These climatic changes also have a negative impact, of course, on the functionality of infrastructure facilities, in particular the transport and drainage systems of urbanized areas, and thus reduce the living comfort. Despite these risks, however, these issues are not addressed adequately, mainly due to a lack of understanding of the climate change and its link to strategic planning of cities and municipalities that would be able to respond to the variability and the associated disasters it causes in a timely manner. Climatic disasters, such as floods or flash floods, however, usually have markedly unfavorable and often long-lasting impacts on the territories of cities and municipalities and their surroundings. Since these are usually cases tied to specific local conditions or in combination with other effects, including the specific historical development of the territory, the examination and evaluation of these acting influences is often very specific, including a specific design procedure and the implementation of appropriate measures to eliminate the negative consequences of these effects.

The main objective of flood protection systems of towns, municipalities and their environs is not only protection against flooding of water from a watercourse, but also protection against flood waters in the protected areas of towns and municipalities. It must be realized that a combined sewer system transfers wastewater along with a portion of rainwater to wastewater treatment plants, where rainwater can be cleaned in screening chambers and discharged into waterways. Rainwater drained by storm drains is then directly diverted into watercourses. During floods, these connections of the watercourse and the sewerage network are a significant source of penetration of undesirable flood waters into protected areas of towns and municipalities, especially due to the need to maintain the capacity of the sewerage network to drain wastewater from the urbanized area, even at elevated water levels. This contradictory requirement to drain wastewater from the municipalities and at the same time to prevent flood waters from entering the protected areas of municipalities and towns is one of the basic tasks of solving the issues of flood protection of the urbanized area. (Stein 1992).

2 RAINWATER INFLOWS TO PROTECTED TERRITORIES OF TOWNS AND CITIES

Today we encounter, with increasing frequency, flash floods, mainly resulting from rainfall, and potentially occurring even outside of the developed areas of cities and municipalities, and which, despite this fact, cause flooding in urbanized areas. Flood prevention protection of watercourses as such is today at a relatively satisfactory level, but nevertheless, the actual connection of the watercourse and the sewerage network often brings with it significant problems, which may result in surrounding waters entering the urbanized territory. Often, this occurs due to the absence of closures (backflow preventers) on collecting and stem sewers where they open into the watercourse. An increase in water level on the watercourse can lead to flow of undesirable waters into the sewer system, which can then enter all the way into the lowest-lying areas in the protected urbanized territory, as shown in Figure 1, where water from the flooding watercourse flows over the road through the sewer shaft.

A similar situation then very frequently arises, for example, in basements of family homes, underground garages and other spaces situated below or at the level of public sewerage (see Figure 2), which does not manage to absorb all the torrential rainwater, whose flow profile is full and the easiest path for outflow of water is then through the floor drain or other drainage, such as a toilet, sink, or so on, precisely into the abovementioned spaces, where these wastewaters, often containing fecal matter and other impurities, can cause significant damage to property.

This type of basement flooding is most frequently caused by the either undersizing the sewerage flow profile, which is then unable to drain large amounts of wastewater blended with rainwater, or due to heavy encrustation on the walls of the pipes, which could, for one, decrease the flow profile, and secondly lead to complete blockage.

However, these two ways of spillage into protected spaces or areas can be easily eliminated by means of stoppers or backflow preventers installed in different operating sections of the sewer network. While this method is relatively easy and economically acceptable, many urbanized areas do not have it. Figure 3 shows a solid sewer closure that shuts off the entire flow profile of the sewer network in both directions, thus preventing drainage from it to the urbanized area.

This rigid seal of the sewerage system is mainly used for larger pipe profiles where there is a risk that water may flow in from watercourses. It is one of the basic elements of flood protection within the sewer network. The operation of these shutters is automatically controlled according to water levels in watercourses and water levels in the sewer network. Handling and installation

Figure 1. Overflow from watercourse into urbanized area through sewer system. Czech Republic, Ostrava 2013 (Source: Author).

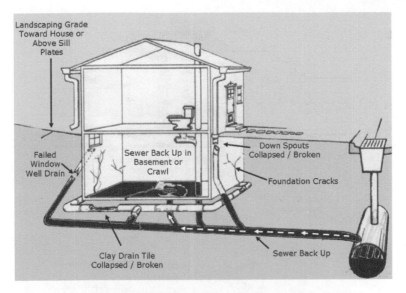

Figure 2.   Overflow of water from sewer system to basement space (Bounsall, 2017).

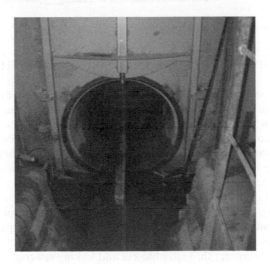

Figure 3.   Flood-preventing solid cover on sewer network. Czech Republic, Prague (Source: Author).

of such closures, however, requires good knowledge of the sewerage network and its functions. In some cases, covers with a raised overflow edge may be used, which are most often built over relief chambers, and can be used to prevent the entry of water from the watercourse into the sewer system and eventual flooding of the protected urbanized area. (Eichler, 2005).

The following Figure 4 shows a backflow preventer flap that is suitable for use on smaller flow profiles, including, in particular, sewage connections. Here, the backflow preventer is the most effective, since it prevents water from the public sewer systems from entering back into the basement areas of buildings, as was shown in Figure 2. This type of backflow prevention flap operates on the principle that wastewater is allowed to flow in only one direction, thereby preventing wastewater from the sewerage system from rising into the interior spaces of buildings. This is achieved using a flap that is open under normal circumstances, and which allows drainage of waste water to the public sewer, but when the water level increases in the public sewer system, the water damper closes and prevents water from entering.

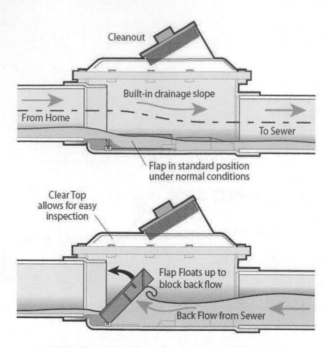

Figure 4.   Backflow preventer flap (Source: SquareOne Corporation).

3 LEAKAGE OF SEWER NETWORK IN COMBINATION WITH FLOODING

Today, it is already common for every municipality or city to have a drainage system for its territory, whereas this addresses, besides wastewater, drainage of rainwater. However, this sewerage network can very often be a source of problems in the protected areas of municipalities and towns, most often because of the poor structural-technical state of these infrastructure structures, as these networks often still fulfill their function beyond the scope of the planned lifetime. The impact of floods or flash floods, combined with the poor structural-technical status of these sewerage networks, can result not only in significant damage to property but also in a threat to the safety of the territory, the limitation of transport services and, last but not least, the disruption of the wastewater flow away from the urbanized area. Recently, such accidents have been occurring more with increased frequency. Figure 5, shows the damage to an important traffic route in the Prague—Liben district in 2014, where due to the increased flow of rainwater through obsolete and leaking sewerage, a cavern was created, which subsequently caused its collapse and damage to the surrounding engineering networks due to the dynamic and pressure loads of the road. The sinkhole caused by the collapse of the road had a size of 4×5 meters and a depth of 3 meters.

Such cave-ins do not, however, occur only due to increase flow of water through the pipes, resulting from torrential rains, but, namely, through the very poor structural-technical state of the drainage network, which is leaky. Through these leaks, the wastewater then penetrates the surrounding soil, which it drenches, gradually eroding its particles into the sewer, creating caverns around the sewer. With the onset of flooding during heavy rains, the entire sewerage profile is filled, and this wastewater creates an increased pressure on the walls of the damaged sewer pipe, which fails to withstand this pressure and is gradually destroyed, allowing the water to penetrate the surrounding soil and thus wash away soil mass, potentially creating significant ground cave-ins, virtually anywhere on the sewer route.

A similar scenario also played out in the following Figure 6 in Zaječov, Beroun, in April 2018, when a massive flow of rainwater ruptured a storm water drainage line in one of the local streets running between family homes. This happened at a time when representatives of the village and the units of the rescue fire brigade were expecting a nearby creek to be flooding, which

Figure 5. Massive collapse of road over sewer line following a flash flood. Praha – Libeň 2014; (Portal Novinky.cz).

Figure 6. Destruction of sidewalks above sewer lines during floods. Czech Repulic, Zaječov 2018 (Portal Vodarenstvi.cz, 2018).

never happened. Instead, a much worse, unforeseeable accident occurred on the storm drainage line, which leads into this creek. The surges of water literally tore apart this drainage line in two places, and due to the erosion of the surrounding soil and consequently also the construction material of the pedestrian paths over the sewerage system, this sewer line was completely blocked. All torrential rainwater thus spilled over the disrupted terrain, and due to the slope of the area, caused more damage than the threat of an overflowing stream. In this case, flooding of several cellars, destruction of fences of family houses, damage to the pedestrian walkway, collapse of part of the road into a pit and damage to a parked vehicle, occurred in its vicinity.

From these two examples, it is clear that the poor structural and technical condition of the sewerage network itself had a significant impact on these accidents. This condition is often caused by very poor and, in some cases, completely lacking management and maintenance of these sewerage networks, where often, directly prior to the emergency, various states of the surface over sewerage are visible, such as cave-ins or cracks in local roads. But objectively, it must be recognized that many other unforeseeable coincidences also have their own share, if only superfluous, in the responsibility for the accidents, as force majeure. Last but not least, the initial situation, characterized by the morphology of the terrain and geological composition of the surface layer, level of the water table and many other factors also have a significant influence on these states.

4 CONCLUSION

Floods, heavy rains and flash floods are a thorny issue these days, which is highly current and, in view of the prevailing, not entirely under control situation, in these flood-prone times, must continue being addressed. These situations that continue to threaten not only the property and technical equipment of the urbanized area, but also the safety of the residents and the quality of life. In the general approach, it is necessary that municipalities and cities manage to better react to changes in the climate and manage to resist the impending externalities on watercourses in their cadastral territory and its environs. There is a need to systematically manage and address flood risks, especially, to mitigate the extreme flow of floodwaters from extra-urban areas into the urbanized area by, for example, creating biotopes to increase the retention capacity of surrounding watercourse environments. From the point of view of the urbanized area, it is necessary to focus more on the solution of today's almost unsustainable number of impermeable surfaces, which prevent natural rainfall absorption into the subsoil, thereby producing additional loads for the sewer system. (Srytr, 2009).

One of the paths to the sustainable development of the urbanized area during floods is a clearly comprehensive approach to multi-level rainwater management in urbanized areas, especially using GIS tools, which, however, must be reinforced and shared across all stakeholders. That is, namely, representatives of municipalities and cities, administrators of individual engineering networks, rescue systems units, fire brigades, crisis management, but also citizens. It seems that an important step in this approach would be addressing the need for better management and maintenance of some infrastructure, especially of the sewerage networks, which have been referred to in this paper, because they are frequently a source of accidents, in particular, due to their poor structural and technical condition. It is therefore necessary to maintain, inspect and, if necessary, renovate such structures to avoid the negative effects that may arise during their operation.

ACKNOWLEDGEMENT

This work was supported by funds for Conceptual Development of Science, Research and Innovation for 2018 allocated to VŠB-Technical University of Ostrava by the Ministry of Education, Youth and Sports of the Czech Republic.

REFERENCES

Bounsall, D. Blog waterproof diagram whole house. HomeStars Blog, 2017. Available from: https://blog.homestars.com/basement-waterproofing/blog-waterproof-diagram-2-whole-house/.

Eichler, Ch.: Instandhaltungstechnik (VEB Verlag Technik, Berlin, 2005).

Portal Novinky.cz Petimetrovy krater v silnici Uzavel Proseckou ulici. Available from: https://www.novinky.cz/domaci/277063-petimetrovy-krater-v-silnici-proseckou-ulici-v-praze-zcela-uzavrel.html.

Portal Vodarenstvi.cz. V Zaječově roztrhala přívalová voda kanalizaci (25/05/2018). Available from: http://www.vodarenstvi.cz/2018/05/25/v-zajecove-roztrhala-privalova-voda-kanalizaci/.

Srytr, P.: Výpadky (poruchy a havárie) inženýrských sítí z hlediska udržitelného rozvoje (FSv CVUT in Prague, K126, 2009, ISBN 978-80-01-04289-2).

Stein, D., Niederehe, W.: Instandhaltung von Kanalisationen (Ernst&Sohn, Verlag für Architektur und technische Wissenschaften, Berlin, 1992).

Evaluation of operating costs in the life cycle of buildings

M. Teichmann, N. Szeligova & F. Kuda
Faculty of Civil Engineering, VSB-Technical University of Ostrava, Czech Republic

ABSTRACT: The issue of operating costs is often neglected, although their significance is reflected not only in the Czech legislation, but also in other European countries. The basic standard for the determination of costs in the lifecycle of buildings is CSN ISO 15686-5 Buildings and other structures—Life Planning—Part 5: Life Cycle Cost Assessment, which describes which areas should be included in the overall assessment. Although there are already many methodological tools, and the professional public is very interested in this issue, many owners, investors and developers are particularly interested in the costs of building acquisition. This article will offer a basic cross-section of important points that should be evaluated in the Life Cycle Costing.

1 INTRODUCTION

Building life-cycle cost analysis is an important tool for investment decision-making. Such analysis includes in its assessment not only the capital costs, but also the costs of operating, renewal and maintenance and disposal of the building. Capital costs pose the greatest financial burden on the investor during construction. However, other costs occur during the building's lifetime, associated with its use, that exceed the capital costs in their sum. For investors constructing buildings for the purpose of selling, it is understandable that they try to minimise the capital costs and disregard the other costs occurring during the building operation, since they are transferred to the future owner or user. However, these investors too should include the total building life-cycle costs in their decision-making in an effort towards sustainable development and energy consumption reduction. After all, current legislation (European Commission) prescribes the elaboration of a building energy efficiency certificate (BEEC), assessing the building in terms of its energy intensity, total energy supply to the building and breakdown into types of consumption (heating, cooling, hot water, etc.) and offering brief recommendations of measures leading to reducing the energy intensity. Both the certificate and the current legislation only deal with the issue of energy, which indeed represents a substantial portion of the building operating costs, but they are not concerned with the overall building costs. Cleaning costs, for example, may amount to hundreds of thousands Czech crowns per year in large office or residential buildings. A large proportion of the costs can be influenced by appropriate adjustments to the structures, materials or layout in the project preparation phase. Unfortunately, even public contracts fail to take into consideration the overall building costs and decisions are based mostly on the lowest bidding price for building acquisition. In public tenders for buildings, building life-cycle cost analysis should be required from all bidders. To do that, however, requires a unified methodology for building life-cycle cost analysis and definition of cost categories (Capova, 2005).

Energy costs represent a substantial portion of the operating costs and the current emphasis on assessment of these most significant operating costs had led to such great pressure to reduce energy intensity of buildings that the initial purpose of this paper has been partly accomplished by the introduction of mandatory building energy efficiency certificates and amendments to Act no. 406/2000 Coll. and Executive Decree no. 78/2013 Coll. Therefore, the main attention and contribution of the paper focus on the classification and categorisation of the various cost types and methods of identifying them and presents a methodology for identification of operating costs in the constituent phases of a building's life cycle (Langdon, 2013).

2 OPERATING COST IDENTIFICATION METHODOLOGY

At first, all the available information and data sources need to be prepared. Initial information necessary for identification of various operating costs is contained in well-developed project documentation and technical reports specifying structural, technical, material and process solutions (hereinafter referred to as PD and TR respectively). The more detailed the documentation, the better information is obtained for identifying the costs. Requirements for the level of documentation detail are increasingly less strict, and implementation documentation is often replaced with documentation for building permits, particularly for single-family houses, and thus does not include all the detailed information. The further steps are as follows:

a. identification of various operating costs;
b. identification of sources of necessary information;
c. calculation and identification of annual amounts of the various operating costs throughout the building's lifetime;
d. inclusion of operating costs among the other life-cycle costs;
e. calculation and comparison of life-cycle costs according to a chosen financial indicator, including the value of money in time (NPV, IRR, DPP, SIR,...) (European Commission, 2005).

Identification of operating costs depends on the type of the capital investment plan (hotel, hospital, school, office building, apartment building, single-family house, and others). Depending on the type of building, some operating costs may be entirely absent (greenery maintenance, guarding, etc.) or may be extended with additional specific items (clean bedding provision, periodic maintenance of specific equipment, rodent control). Another difference will be in the method of treating the investment (sale/lease). If the investment project is of a developer type with the intention to sell the property or sell its component parts one by one, the investor disregards the operating costs, since they are transferred to the future owner, thus saving both money and time by not analysing them. According ISO 15686-5, the operating costs will also differ depending on the project location and method of use of the building. The building location influences not only costs of energies and water, but also waste collection, property tax, etc. Each building has its specific features, which need to be taken into account. Provision of background information requires coordination among all the stakeholders (investor, architect, specialist technicians, project and facility manager) and cooperation of authorities and concerned public bodies (European Commission, 2007).

3 BACKGROUND INFORMATION FOR IDENTIFICATION OF OPERATING COST

Operating costs can be identified depending on the LCC analysis needs in various phases of the building life-cycle. The tables below describe the required information and its sources necessary for identification of operating costs (the breakdown of operating costs shown here is convenient for apartment and similar buildings).

The pre-investment phase concerns mainly strategic decisions whether to implement the plan or not, renovate or build new, or invest the funds in a different project (preliminary feasibility study, feasibility study). If the decision is made to build new, this is followed by comparison of various options that come into consideration; then in turn a study, draft project design and zoning proceeding documentation are made. By the European Commission, the background information for identification of the operating costs is very rough in this phase and it depends largely on expert estimates or previous experience with similar projects or information from comparable buildings operated by the investor. Up to 80% of the future costs can be influenced at the start of the project design phase (Cunningham, 2015). The information becomes more refined as the documentation is being made (documentation for building permit, project execution documentation, including tendering documentation, implementation documentation and as-built documentation) and specific structural, technical and material solutions are designed, including all equipment. According to the legislation, a building energy efficiency certificate has to be produced, informing about the energy demands of the building. Specific fixtures and fittings for the building are selected, as well as suppliers and manufacturers. The operating cost identification can make use of information from the project documentation and technical reports, as well as from the selected suppliers of materials and equipment, and necessary information can be obtained from the municipal authorities and other institutions of jurisdiction depending on the location, and price quotations and other information (cleaning services, greenery maintenance, etc.) can be requested from selected companies (Gundersen, 1998).

All the necessary information and real-world data are available after the project completion, in the use phase, but the ability to influence the operating costs is limited now. The operating cost identification can work with as-built documentation and all the technical and inspection reports, contracts with external companies providing the required services (facility management, cleaning, greenery maintenance, guarding, etc.); records of actual energy and water consumption and real-world data on expenditures are also available. All the information for identification of the actual operating costs is available before a planned renovation or cost optimisation; information for alternative options has to be identified similarly to the design phase (as mentioned in HM Treasury).

Table 1. Sources of data and information for identification of operating costs in pre-investment phase.

Operating costs (OC)		Pre-investment phase	
		Required information	Information sources
OC1	Energies		
	Energy for heating		
	Energy for hot water		
	Energy for cooling		
	Energy for ventilation		
	Energy for humidity adjustment		
	Energy for shared area lighting		
	Other energy consumption in apartment units/shared areas	Investment plan description:	
OC2	Drinking water and wastewater	– location,	– expert estimates,
	Water charges	– building type,	– data from similar projects,
	Sewerage charges	– building size,	– historical data from projects
OC3	Waste disposal	– building uses and	currently in use.
OC4	Cleaning	operations,	
OC5	Greenery maintenance	– building lifetime,	
OC6	Building guarding and security	– equipment used,	
OC7	Building and property insurance	– technical, structural and	
	Property insurance	material solutions,	
	Other insurance	– staffing.	
OC8	Administration and servicing charges		
	Property tax		
	Inspection and servicing charges		
	Other charges		

Table 2. Sources of data and information for identification of operating costs in design phase.

Operating costs (OC)		Design phase	
		Required information	Information sources
OC1	Energies: for heating, hot water, cooling, ventilation for humidity adjustment, shared area lighting, energy consumption in apartment units other energy consumption in shared areas	– energy demand for heating, – cooling, – ventilation, – humidity adjustment, – hot water, – lighting and others, – price of energy	– PD+TR, heat loss calculations, energy demand for heating, cooling and hot water, calculations of building ventilation and lighting, BEEC, – energy audit, information from material and equipment manufacturers, energy supply rates, – price trends, expert estimates, experience from similar projects, legislation and standards in force
OC2	Drinking water and wastewater Water charges Sewerage charges	– water demand, – water and sewerage charges	– PD+TR, water demand calculations, water and sewerage charges as per project location, – price trends, legislation and standards in force
OC3	Waste disposal	– waste quantities and types, – number of persons producing waste, – waste collection charges	– PD+TR, binding local municipal ordinances, waste collection companies' terms and conditions, – expert estimates, experience from similar projects, legislation and standards in force
OC4	Cleaning	– cleaning extent and frequency, – sizes of areas, – surface materials, – price of cleaning	– PD+TR, cleaning companies' quotations or wage cost estimate, expert estimates, – experience from similar projects, legislation and standards in force
OC5	Greenery maintenance	– scope of maintenance, – sizes and types of areas maintained, – price of maintenance	– PD+TR, companies' quotations or wage cost estimate, expert estimates, – experience from similar projects, legislation and standards in force
OC6	Building guarding and security	– method and scope of security arrangements, – prices	– PD+TR, companies' quotations or wage cost estimate, expert estimates, – experience from similar projects, legislation and standards in force
OC7	Building and property insurance Property insurance Other insurance	– property type and use, – roof shape, – number of floors, – total floor area, – equipment, – specific requirements	– PD+TR, insurance companies' quotations, expert estimates, – experience from similar projects, legislation and standards in force
OC8	Administration and servicing charges Property tax Inspection and servicing charges Other charges	– land and built-up area sizes, – property location, – number of floors, – property electric and gas equipment, – elevators, – boiler room type and other equipment	– PD+TR, local municipal coefficients (tax), servicing and inspection companies' quotations, – expert estimates, experience from similar projects, legislation and standards in force

Table 3. Sources of data and information for identification for operating phase.

Operating costs (OC)		Operating phase	
		Required information	Information sources
OC1	Energy for heating/cooling Energy for hot water/ventilation Energy for humidity adjustment Energy for shared area lighting Other energy consumption in apartment units/shared areas	– energy consumption, – price of energy	– records of energy consumption, – BEEC, – invoices from energy suppliers, – price trend estimates, – legislation and standards in force
OC2	Drinking water and wastewater Water charges Sewerage charges	– water consumption, – water and sewerage charges	– records of water consumption, – invoices from water suppliers (local water and sewerage utility), – price trend estimates, – legislation and standards in force
OC3	Waste disposal	– number of persons producing waste, – waste quantities, types, – waste collection charges	– records on actual waste collection expenditures, – amendments to legislation in force
OC4	Cleaning	– cleaning extent and frequency, price – sizes of areas, – surface materials,	Actual costs of cleaning (wage and material costs or contract on cleaning services with externalist)
OC5	Greenery maintenance	– scope of maintenance, – sizes and types of areas maintained, – price of maintenance	Actual costs of maintenance (wage and material costs or contract on maintenance with external company).
OC6	Building guarding and security	– method and scope of security arrangements, – prices	Actual costs of building security (wage and material costs or contract on security arrangements with external company).
OC7	Building and property insurance Property insurance Other insurance	– property type and use, – roof shape, floor area – number of floors, – equipment, – specific requirements	– PD+TR, – local survey, – current insurance policies
OC8	Administration and servicing charges Property tax Inspection and servicing charges Other charges	– land and built-up area sizes, floors – property location, – property electric and gas equipment, – elevators, – boiler room type and other equipment	– PD+TR, – current tax assesment, – contracts with servicing and inspection companies, – records on actual expenditures

In the operating phase, we can compare the operating costs planned in the design phase with the actual costs. The difference between the operating costs may be due to errors in the project documentation, wrong execution of the building or its parts by the contractor, changes in users' behaviour, legislative changes, climate change, inappropriate use of equipment, etc. Some of the causes of changes in the operating costs can be avoided (quality of building execution, errors), others have to be considered but cannot be influenced (climate change or legislation changes).

Operating costs require meticulous attention due to their amount throughout the building's life cycle. The cardinal focus on minimising operating costs should take place in the pre-investment phase of the project: the planning period, which offers the greatest room for optimisation. It is advisable to record and monitor the actual costs during the building operation, and update and optimise based on that information. For example, the insurance policy should be updated after a certain period to take account of the building's wear or increase in its value over time. An extension or heightening of the building also requires a new property tax return, resulting in a new tax assessment based on the new information. Moreover, heating sources age, which reduces their efficiency, increases the energy consumption (building energy intensity) and, most importantly, increases the operating costs.

4 CONCLUSION

Buildings have a long life cycle, which is why any improvement in techniques of assessing them when selecting the best option will significantly reduce their future environmental impacts and move them towards sustainable development. Energy costs offer the greatest saving potential. Energy-saving measures leading to reducing energy consumption are also related to the highest expenditures of capital investment, be it replacement of a heating and hot water source or replacement of windows or thermal insulation of the building envelope. Cost-saving measures may be low-cost or cost-free as well. They may not bring such significant savings as high-cost measures, but they require zero or very low investment and will result in a cost reduction. It is important to choose an appropriate combination of cost-saving measures to ensure a reasonable rate of return on investment. No general formula for calculating the relation between capital and operating costs can be provided; every project requires individual assessment, but the paper proposes effective ways of reducing operating costs.

ACKNOWLEDGEMENT

This work was supported from the funds of the Students Grant Competition of the VSB—Technical University of Ostrava. Project registration number is SP 2018/175.

REFERENCES

Čápová, D., Kremlová, L., Schneiderová, R. a Tománková, J. Metodika určování nákladů životního cyklu a stavebního objektu. In: Technické listy 2005 Díl 1: Soubor technických listů.

Cunningham, T. (2016). Cost Control during the Pre-Contract Stage of a Building Project – An Introduction. Report prepared for Dublin Institute of Technology, 2016.

European Commission. Development of horizontal standardised methods for the assessment of the integrated environmental performance of buildings, M/350 EN, Standardisation Mandate to CEN, 29 March 2004.

Gundersen, N.A. Annual Cost Analysis – Description and user's guide to the costing model. Statsbygg – Building Finance Section, 1998. [vid. 19. June 2009].

Her Majesty's Treasury (HM Treasury). The Green Book – Appraisal and Evaluation in Central Government. London, 2011. [vid. 19. June 2011].

ISO 15686-5 Buildings and constructed assets – Service life planning – Part 5: Life-cycle costing. ICS: 91.040.01, Ed. 1, June 2008. International Organization for Standardization.

Langdon, D. Life cycle costing (LCC) as a contribution to sustainable construction: a common methodology – Final Report. May 2007. Davis Langdon Management Consulting. [vid. 5. February 2013]. Available from: http://ec.europa.eu/enterprise/sectors/construction/competitiveness/life-cycle-costing/index_en.ht.

LCC Refurb: Integrated Planning for Building Refurbishment – Taking Life-Cycle-Costs into Account. European Commission, 2005. [vid. 14. June 2010]. Available from: http://www.gi-zrmk.si/euprojekti/lcc/lcc_guideline.pdf.

Office of Government Commerce. Whole-life costing and cost management – Achieving Excellence in Construction Procurement Guide. 2007. 24 s. [vid. 19. June 2011].

OPD International Total Occupancy Cost Code, © and database right Investment Property Databank Ltd. 2004.

Praha: CIDEAS-Centrum integrovaného navrhování progresivních stavebních konstrukcí, 2006, s. 3–4. ISBN 80-01-03486-0.

Task Group 4: Final report – Life Cycle Costs in Construction. 2003. European Commission, 93 s.

Advances and Trends in Engineering Sciences and Technologies III – Al Ali & Platko (Eds)
© 2019 Taylor & Francis Group, London, ISBN 978-0-367-07509-5

Use of simulation software to predict and optimize future heating demands of a family house

P. Turcsanyi & A. Sedlakova
Faculty of Civil Engineering, Institute of Architectural Engineering, Technical University of Kosice, Kosice, Slovakia

ABSTRACT: Computer programs allowing determining of the accuracy of design of buildings or their parts from the thermal engineering point of view in advance are on the rise. Finding the most suitable project design for optimizing future energy performance of a building significantly contributes to the implementation of the European Directive on Energy performance of buildings 2010/31/EU in Slovakia. Using simulation software, DesignBuilder in this case, became a very useful tool on the road to energy efficient design. In this paper, we have placed a virtual family house (which will be built in Kosice, Slovakia) to the simulation software DesignBuilder and we determined the most suitable design in case of future heating demand.

1 INTRODUCTION

The results of architectural and construction solution affect many parameters. Those parameters either can or cannot be affected by the designer; nevertheless, the first raw design usually needs to be optimized according to importance of energy efficiency of the building. Finding the optimal solution means knowing the correlation between the key parameters which are affecting the energy performance of buildings (i.e. heat demands, cooling demands, thermal comfort, etc.). After a simulation on energy performance of the building, designers are able to optimize parameters to find the best ratio between the initial costs and the long term energy save. Currently, there are more than 400 applications that can be applied to analyzing building energy and thermal simulation (Energy Efficiency and Renewable Energy. Building Energy Software Tools Directory, U.S. Department of Energy).

Building thermal simulation tools predict the thermal performance of a given building and the thermal comfort of its occupants. In general, they support the understanding of how the given building operates according to certain criteria and enable comparisons of different design alternatives. Evaluation of thermal comfort involves assessment of at least six factors: human activity levels, thermal resistance of clothing, air temperature, mean radiant temperature, air velocity and vapour pressure in ambient air (Fanger, 1970). In this paper, we are comparing two different types of external wall insulation and their effect on heat demand for heating. First is the commonly used wall thermal insulating material EPS polystyrene, part of ETICS system. On the other hand, as the age and research go further every day and new materials are being introduced into everyday construction life, insulation based on phenol insulation board is being assessed in this paper.

2 OBJECT INFORMATION

House that is being evaluated is located in Košice-Krasna, Slovak Republic. City of Košice lies at an altitude of 206 meters above sea level and covers an area of 242.77 square kilometers. It is located in eastern Slovakia, about 20 kilometers from the Hungarian borders, 80 kilometers from the Ukrainian borders, and 90 kilometers from the Polish borders (Fig. 1). It is about 400

Figure 1. Position of Kosice city to the surrounding countries.

kilometers east of Slovakia's capital Bratislava. Košice city is situated on the Hornád River in the Košice Basin, at the easternmost reaches of the Slovak Ore Mountains. More precisely it is a subdivision of the Čierna hora Mountains in the northwest and Volovské vrchy Mountains in the southwest. The basin is met on the east by the Slanské vrchy Mountains.

Two-storey family house represents a typical type of residential buildings built in this area. The floor layouts were designed for a family of two parents and three kids and the concept of a whole house was made in according to the latest energy standards. Building is based on a concrete footing foundation. The family house envelope was designed using materials with high thermal capacity. As for thermal insulation—thick layer of polystyrene (ETICS system) is being used, to minimize the heat loses during the heating period as well as to prevent the heat transfer during the summer period. Roof structure creates unconditioned spade between insulated ceiling and top of the roof. Transparent parts of building envelope were designed as wood-aluminum, triple glazed windows Internorm with outside louvers for a regulation of solar gains through the windows.

2.1 Phenolic insulation

The main use of phenolic foam in construction is as thermal insulation, to improve is thermal efficiency. The thermal efficiency of the existing building stock is of prime concern, if carbon emissions targets are to be met. For those homes with solid walls, external wall insulation can be an effective method of improving their thermal efficiency. Phenolic foam is starting to be popular material choice for external wall insulation systems owing to its low thermal conductivity and good fire performance (Densley, Tingley, Hathway, Davison, Allwood, 2017). Phenolic foam is made from three main components: phenolic resin, a blowing agent and an acid catalyst; a number of additives can also be utilised to develop specific properties within the foam. There are patents and other information available at (International Patent Application No. PCT/IE2006/000096) that describes the chemicals used to manufacture phenolic foam.

3 MATERIALS AND METHODS

The external wall insulated with polystyrene represents the most common insulation type of ETICS insulating system. On contrary, phenolic insulation materials are used quite rarely—but they are on the rise. The main reason is its price, which is extensively higher than polystyrene (or any other classical insulation material). However, phenolic insulation boards allow us to design energy efficient building while keeping the external wall construction as subtle as possible due to its low coefficient of thermal conductivity of λ = 0.020 W/m·K. With its thickness, phenolic insulation board of thickness 6 cm equals to 12 cm of polystyrene.

3.1 Technical parameters of an envelope wall type A—polystyrene insulation

Envelope wall type A is a classic type of the external envelope wall insulated with EPS polystyrene. Polystyrene insulation is the most common insulation material used in Slovak Republic. Polystyrene insulation is being put on masonry or any other wall structure, using adhesive mortar and anchors.

3.2 Technical parameters of an envelope wall type B—phenolic insulation

Envelope wall type B is a type of the external envelope wall insulated with phenolic boards, in this case Kingspan KOOLTHERM K5. The core of Kingspan Kooltherm K5 External Wall Board is a fibre-free performance rigid thermoset phenolic insulant manufactured with a blowing agent with zero Ozone Depletion Potential (ODP) and low Global Warming Potential (GWP). The core of Kooltherm K5 External Wall Board has a 90% closed cell structure. Phenolic board insulation is being put on masonry using adhesive mortar and anchors—as ETICS system.

Figure 2. Family house view—visualization.

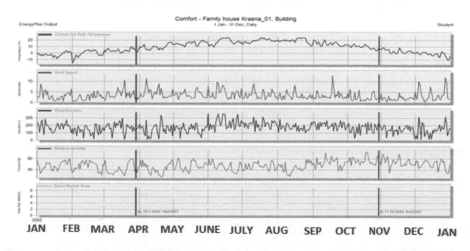

Figure 3. Weather data for the reference year Košice, Slovakia. From the top: Outside dry-bulb temperature, wind speed, wind direction, relative humidity, solar intensity.

Table 1. Physical and thermal parameters of building.

Shape factor (A/V ratio)	0.78 1/m
Volume of building space	726.18 m³
Total heat transfer surface	569.48 m²
Total floor area	222.09 m²
Wall U-Value	0.148 W/(m²·K)
Insulated ceiling U-Value	0.123 W/(m²·K)
Ground floor R-Value	4.38 (m²·K)/W
Wood-aluminum U-Value Internorm HF 350	0.60–0.95 W/(m²·K)

Table 2. Thermo-physical properties of an envelope wall type A.

	d [m]	λ [W/m·K]	c [J/kg·K]	ρ [kg/m³]	m [kg/m²]
Plaster	0.020	0.570	1000.0	1300.0	10.0
Porotherm 38 Ti Profi	0.380	0.134	1000.0	750.0	29.0
EPS insulation	0.120	0.038	1050.0	18.0	45.0
Adhesive mortar	0.005	0.800	920.0	1300.0	18.0
Silicon render	0.003	0.700	920.0	1700.0	37.0

Table 3. Thermo-physical properties of an envelope wall type B.

	d [m]	λ [W/m·K]	c [J/kg·K]	ρ [kg/m³]	m [kg/m²]
Plaster	0.020	0.570	1000.0	1300.0	10.0
Porotherm 38 Ti Profi	0.380	0.134	1000.0	750.0	29.0
Phenolic insulation	0.060	0.020	800.0	35.0	300.0
Adhesive mortar	0.005	0.800	920.0	1300.0	18.0
Silicon render	0.003	0.700	920.0	1700.0	37.0

4 SIMULATION

Using simulation software DesignBuilder, family house was put into the calculation interface a simulation of two different insulation types was done. In the first simulation, polystyrene was used as wall insulation. Second simulation shows the alternative to EPS insulation—phenolic insulation board. Weather data of the reference year for Kosice were used in this simulation. They provide inputs of dry bulb air temperature, relative humidity, solar radiation, wind speed, wind direction etc., for achieving the most accurate and realistic simulation.

5 RESULTS

Building simulations were set up with the DesignBuilder v4 software, in which building performance data were generated by the simulation engine EnergyPlus.

5.1 Envelope wall type A—polystyrene insulation

Figure 4 shows annual heating demand of a family house, with envelope wall insulated by EPS polystyrene alternative. Heating demand on square meter of floor area is 30.96 kWh/m²·a.

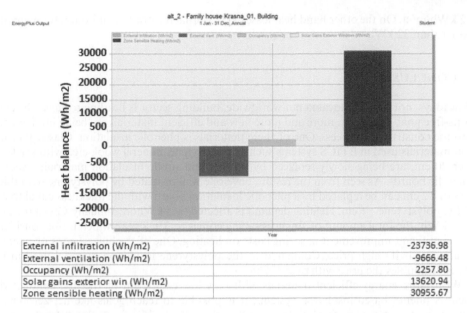

External infiltration (Wh/m2)		-23736.98
External ventilation (Wh/m2)		-9666.48
Occupancy (Wh/m2)		2257.80
Solar gains exterior win (Wh/m2)		13620.94
Zone sensible heating (Wh/m2)		30955.67

Figure 4. EnergyPlus outputs of annual heat gains, heat losses and heating demand of type A.

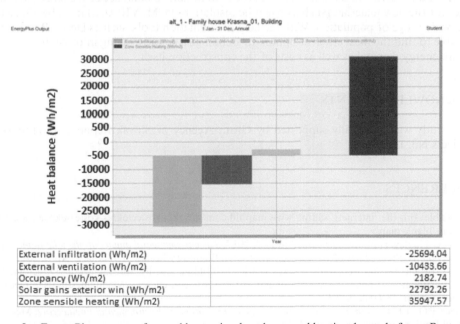

External infiltration (Wh/m2)		-25694.04
External ventilation (Wh/m2)		-10433.66
Occupancy (Wh/m2)		2182.74
Solar gains exterior win (Wh/m2)		22792.26
Zone sensible heating (Wh/m2)		35947.57

Figure 5. EnergyPlus outputs of annual heat gains, heat losses and heating demand of type B.

Solar gains through the exterior windows are 13.62 kWh/m²·a and a heat gains from the occupancy 2.2 kWh/m²·a. On the other hand heat losses via external infiltration and ventilation are little higher than 32 kWh/m²·a.

5.2 Envelope wall type B—phenolic insulation

Figure 5 shows annual heating demand of a family house, with envelope wall insulated by phenolic insulation board alternative, on square meter of floor area is 35.9 kWh/m²·a. Solar gains through the exterior windows are 13.62 kWh/m²·a and a heat gains from the occupancy

2.2 kWh/m²·a. On the other hand heat losses via external infiltration and ventilation are little higher than 32 kWh/m²·a.

6 CONCLUSION

Nowadays, not only in Slovakia but worldwide, building sector is heading towards the emissions-free goal. Every day, more and more new and efficient materials are being implemented into the construction process. One of these materials is phenolic insulation. Classical insulation materials used in ETICS systems such as polystyrene, mineral wool, etc., with thickness over 20 cm, are being slowly replaced by thinner and subtle insulation materials—such as phenolic boards. As seen from the results, envelope wall insulated by polystyrene with thickness of 12 cm can be replaced by a phenolic insulation board with the thickness equal to half of the polystyrene—6 cm. Heating demand is affected by 5 kWh/m² annually. Given the price for 1 kWh, the total amount of finances saving equals to 35€ a year, which is not considered as a significant improvement in saving costs on heating. One of the biggest cons of phenolic insulations is its high price. Compared to the polystyrene insulation material used in this paper, it doubles the price with over 17€/m².

With more energy efficient materials with very low coefficient of thermal conductivity, such as phenol based insulation systems, it is possible to reach goals like net-zero energy buildings or even energy plus buildings while making the building's envelope structure more subtle and modern. On the other hand, in order to be able to build net-zero energy buildings, price for the new materials (such as phenolic insulation, PCM, VIP, etc.,) must be affordable to a wide range of population. With the help of simulation tools, such as DesignBuilder, it is feasible to find the most efficient option for building structures, taking into account costs of future energy demand as well as costs of the construction itself.

ACKNOWLEDGEMENTS

This study was financially supported by Grant Agency of Slovak Republic to support of projects No. 1/0307/16.

REFERENCES

Accessible on the internet: <http://www.maphill.com/slovakia/kosice/okres-kosice-i/kosice/location-maps/blank-map/>.

Densley, Tingley, Hathway, Davison, Allwood, 2017, *The environmental impact of phenolic foam insulation boards,* Proceedings of the Institution of Civil Engineers, Volume 170 Issue 2, April, 2017.

Energy Efficiency and Renewable Energy (EERE). Building Energy Software Tools Directory, U.S. Department of Energy. <Available online: http://apps1.eere.energy.gov/buildings/tools_directory/> (accessed on 18 January 2017).

Fanger, P.O., 1970, *Thermal Comfort, Analysis and Applications in Environmental Engineering*; McGraw-Hill Book Company: New York, USA.

Kingspan Holdings Limited (2006) A Phenolic Foam. International Patent Application No. PCT/IE2006/000096, 8 September.

Maile, T.; Fischer, M.; Bazjanac, V., 2007, *Building Energy Performance Simulation Tools – A Life-Cycle and Interoperable Perspective*; Stanford University: Stanford, CA, USA.

STN 73 0540-2, 2012: *Thermal protection of buildings. Thermal performance of buildings and components. Part 2: Functional requirements.*

Advances and Trends in Engineering Sciences and Technologies III – Al Ali & Platko (Eds)
© 2019 Taylor & Francis Group, London, ISBN 978-0-367-07509-5

Water vapour by diffusion of PIR and mineral wool thermal insulation materials

N.I. Vatin, I.I. Pestryakov, Sh.T. Sultanov, O.T. Ogidan,
Y.A. Yarunicheva & A.P. Kiryushina
Peter the Great St. Petersburg Polytechnic University, St. Petersburg, Russian Federation

ABSTRACT: This article is aimed at studying the heat-protective properties of enclosing structures. The method models operating conditions in which thermal insulation product absorb moisture from both sides at high relative humidity of air (100%) and pressure difference of water vapour over a long period of time. The moisture absorption by diffusion of mineral wool and polyisocyanurate was obtained after 28 days of exposure to temperature and pressure drop of water vapor. Significant changes in moisture content of mineral wool were observed. From the results obtained, it can be concluded that polyisocyanurate has a lesser absorption property of water vapour, which is an important attribute in avoiding costs of upgrading the sources of thermal energy.

1 INTRODUCTION

Thermal insulation of the building's envelope is one of the effective ways to reduce energy consumption in existing buildings (Vatin et al. 2014; Borodinecs et al. 2016; Muizniece and Blumberga 2016) by implementation of a set of energy-saving measures, improving the integrity of building structures, using renewable energy sources (Gorshkov et al. 2018), (Alihodzic et al. 2014; Basok et al. 2015; V. A. Kostenko et al. 2016; Anikina et al. 2017). A typical energy-saving solution is the introduction of ventilated facades using external or internal air, for reduction of thermal loads (Nemova 2013; Petrichenko et al. 2014; Petritchenko et al. 2017; Petrichenko et al. 2018). Issues related to energy-efficiency are appropriate for both existing buildings and the ones under design. It is important to have this program correlated with the energy-saving program (V. Kostenko et al. 2016). In harsh climatic conditions, the use of thermal insulation in buildings is necessary and is gradually becoming a mandatory requirement in many countries particularly as energy becomes more precious and demand increases (Abdou and Budaiwi 2013). In this regard, the problem related to the search for a technology for energy-efficient construction has become a vital one. It is necessary to introduce not only energy-efficient designs, but also to apply meters and energy-saving technologies that allow to achieve and save on normative indicators of heat energy consumption, the corresponding class assigned to the building (Petrichenko et al. 2018).

Thermal protective properties of the fence depend on the design solutions used in the construction materials, the operating conditions of the building (Gorshkov et al. 2015; Gorshkov and Rymkevich 2015; Leshchenko and Semko 2015; Korniyenko et al. 2016; V. Kostenko et al. 2016; Muizniece and Blumberga 2016; Statsenko et al. 2016; Vasilyev et al. 2016; Tarasova and Petritchenko 2017). There are many ways to insulate buildings, the most common are insulation with fibrous structure and polymer insulation. Along with the traditional and well-developed materials in the construction industry, new thermal insulation materials are appearing on the market, the physical and mechanical characteristics of which are not fully understood. PIR also belongs to these material. In the technical characteristics of this material, indicated by the manufacturer, there is no such indicator, important from the point of view of thermos-physical properties of the material, as sorption humidity. The manufacturer indicates only the value of the thermal conductivity of the material in the dry state. Meanwhile, it is known that, depending on the level of humidity under operating conditions and the sorption properties of the thermal insulation material, its actual (operational) value of thermal conductivity can differ significantly from the experimental values in the dry state.

The thermal conductivity of insulation materials is greatly affected by their operating temperature and moisture content, yet limited information is available on the performance of insulating materials when subjected to actual climatic conditions. Many parameters should be considered when selecting thermal insulation, including cost, compression strength, water vapor absorption and transmission and, most importantly, the thermal conductivity of the material when considering thermal performance of buildings and relevant energy conservation measures (Gorshkov et al. 2018). The phase changes of vapor moisture, although not strictly an energy transfer mechanism, should also be considered in heat transfer analysis since state changes absorb and release large quantities of heat (Zach et al. 2013). This means that both vapour flow and moisture absorption is important, and they typically are more critical in insulating materials with open cell structures than with closed cell ones.

The way thermal insulating materials resist to heat flow depends on microscopic cells in which air or other gasses are trapped. Thermal insulating materials resist heat flow as a result of the countless microscopic dead air-volumes. In fact, the thermal resistance of the air entrapped within insulating materials is mainly responsible for their low thermal conductivity. Meanwhile, creating small cells or a closed cell structure within the thermal insulation across which the temperature difference is not large, reduces the radiation heat transfer mode.

Typically, air-based insulating materials do not exceed the thermal resistance of still air. However, some foam insulations such as the polyurethane encapsulate fluorocarbon gas instead of air within the insulation cells to obtain higher thermal resistance than the air. PIR plate based on polyisocyanurate, as the thermal insulation material with the lowest the indicator of heat conductivity, has been extensively used in the USA and Western Europe for a long time, more than 10 years. In North America, the roof insulation market shows that polyisocyanurate is the most widely used roof insulation, covering more than 50% of all commercial new or re-roofing applications. This is probably due to the often nominal double of thermal resistance of the polyisocyanurate when compared to fiberglass or rock wool insulation. These last products have generally a larger market share for vertical building elements and in several European countries.

Due to high performance indicators, the insulation is a considerable interest both for developers wishing to improve the energy efficiency of the constructed buildings, and for private clients interested in the most effective heat insulating material. Many benefits justify the adoption of thicker layers of thermal insulation in buildings. In fact, the use of thermal insulation in buildings helps in reducing the reliance on mechanical air-conditioning systems to realize comfortable buildings, and it allows saving energy by reducing the heat flux through the building envelope. Meanwhile, the reduced energy demand achieved by using more effective thermal insulating layers also reduces the needed HVAC equipment. The thermal insulation in building enclosure extends the periods of indoor thermal comfort, especially between seasons, and by keeping buildings with smaller temperature fluctuations, it helps in preserving the integrity of building structures, increasing their lifetime).

The basis for the preparation of polyisocyanurate is methylene diphenyl diisocyanate, which at a high temperature and in the presence of catalysts is able to react with itself, partially transforming into a tri-isocyanate-isocyanurate chemical compound. It is a rigid molecule of the ring structure, which is positively reflected on the physical properties of the final product.

This high-tech insulation is polyisocyanurate (abbreviated – PIR)—a close a relative of the well-known polyurethane foam (PUR). Polyurethane possesses exceptional properties such as high resistance to open fire (group combustibility G1) and low thermal conductivity (in the dry state) among the polymers is not more than 0.024 W/m^2. In addition, the PIR plate does not absorb moisture and is distinguished by a high resistance to compression.

At present, a number of articles have been written on the sorption humidity of insulating materials with a fibrous structure showing a change in this parameter over time during operation, which leads to its increase (V. Kostenko et al. 2016), or reflecting the efficiency of using multi-layered enclosing structures with mineral wool insulation in comparison with an unheated wall, in which the sorption characteristics are greater (V. Kostenko et al. 2016). For polymer insulation, foam polyisocyanurate, only the main characteristics characterizing the material have been studied and determined, such as: low flammability (G1), high heat-saving capacity, low density and strength (Abdou and Budaiwi 2013), (Nik et al. 2012).

Many of the works are related to the determination of the thickness of thermal insulation by calculating the temperature fields and aimed at improving the individual bearing elements of the enclosing structure (Vereecken and Roels 2016).

The purpose of this work was to determine the thermo-physical properties of slabs from foam polyisocyanurate with soft liners (PIR) with a density of between 30 and 45 kg/m³. To achieve this goal, it was necessary to solve the following tasks:

1. Using experimental methods based on National Standards of Russia GOST and GOST EN methods, to determine physical and mechanical characteristics of two types of insulation.
2. Analyzing the results and obtaining the main evidence base for the correction of normative documents in the field of heat-insulation materials. At this stage, experimental studies were carried out to determine water absorption, diffusion moisture for a long time, sorption humidity and thermal conductivity of the samples.

2 METHODS

The method simulates the operating conditions under which the samples absorb moisture both sides at high relative humidity, approximately 100% and the difference in water vapor pressure over a long period of time, from water to the form. Materials:

- Plates of mineral wool (MW) with a thickness of 50 mm and density $\rho = 130$ kg/m³;
- PIR plates 50 mm thick with double-sided lining aluminum foil 50 μm thick (PIR 1.1);
- PIR plates with a thickness of 50 mm without lining (PIR 1.2).

The experiments were carried out according to the requirements of the National Standard of Russia EN 12088-2011 "Thermal insulating products in building applications. Method for determination of long-term moisture absorption by diffusion".

Sizes of samples were measured in accordance with EN 12085. A panel of mineral wool was cut using an insulation knife, in order to obtain the required lengths and widths equal to 500 and 500 mm, respectively. Samples were weighed to the nearest 0.1 g to determine the initial mass. The sample was then placed on the frame of the container. Since the sample was lined on both sides, experiments will be proceeded with a lined surface, with either side of the sample being placed on the frame facing upwards. The lower edge of the sample is sealed around the perimeter of the frame. The width of the sealant was equal to 10 mm. A thermally insulated cooling plate is placed on the upper surface of the sample. The sample is exposed to temperature level and pressure drop of water vapor for 28 days.

The samples were conditioned for at least 6 hours at a temperature of $(23 \pm 5)°C$ before the test in a climatic chamber. In case of disagreement, the samples were kept at a temperature of $(23 \pm 2)°C$ and relative air humidity $(50 \pm 5)\%$ for the time specified in the standard, and in its absence—in the technical conditions for the product of a particular type, but not less than 6 h. The samples were then weighed to an accuracy of 0.1 g to determine the initial mass (m_0). A thermally insulated cooling plate is then placed above the upper surface of the sample to subject the sample to a lower temperature as a simulation of the winter period. On the other hand, the opposite side of the sample is placed in a thermally insulated container holding with water. Temperature in the container is controlled by a thermostat regulator at 50°C. The sample is subjected to a temperature and differential pressure of water vapor for 28 days while maintaining the water temperature $(50 \pm 1)°C$ and the temperature on the opposite side of the sample $(1 \pm 0.5)°C$. The sample is turned in the opposite direction every 7 days. After 28 days, the sample is removed from the container and water is removed from its surface with a paper or other suitable tissue. The sample is weighed and the mass after 28 days (m_D) is obtained. For each sample, the amount of absorbed moisture is estimated by mass W_{dp} in kg/m² or by volume W_{dv} in percentage. Figure 2 shows a flowchart scheme of the experiment to obtain the long-term moisture absorption by diffusion.

The thickness of the samples is equal to 50 mm for mineral wool and PIR (Table 1). In a second stage, the same samples were conditioned by setting temperature at $(23 \pm 5)°C$ and relative humidity at $(50 \pm 5)\%$ under environmental conditions for the time necessary to reach the weight stabilization, in order to obtain moist samples.

Water content (W_C) was measured using the gravimetric method by means of Equation (1):

$$W_C = (W_s - W_d)/W_d \qquad (1)$$

where W_s and W_d are the weights of the examined and of the dried samples, respectively. A precision scale with a graduation of 0.01 g was used to measure weights.

Measurements of water vapour diffusion were performed on samples of the PIR panels (Figure 1). PIR samples has been exposed to temperature and a pressure drop of water vapor for 28 days. The sample is exposed to temperature level and pressure drop of water vapor for 28 days, while maintaining the water temperature $(50 \pm 1)°C$ and the temperature on the opposite side of the sample $(1 \pm 0.5)°C$. Every 7 days the sample is turned over. After 28 days,

Figure 1.   Heating plate with water and cooling plate.

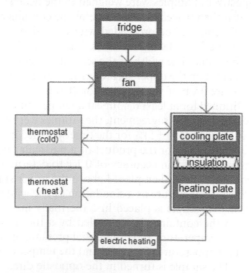

Figure 2.   Flowchart scheme of the experiment.

Figure 3. Cooling plate with mineral wool.

Figure 4. Installation for testing in accordance with GOST EN 12088-2011.

Table 1. Moisture absorption after 28 days.

| Sample | Moisture after 28 days | | | | |
	A [m^2]	d [m]	m_0 [kg]	m_d [kg]	W_{dv} [%]
PIR 1.1	0.25	0.05	0.53	0.563	0.3
PIR 1.2	0.25	0.05	0.52	0.566	0.4
MW 1	0.25	0.05	1.66	3.864	17.6
MW 2	0.25	0.05	1.69	3.622	15.5

the sample is removed from the container and the water with its surface is removed with a paper or other suitable tissue. The sample is weighed and the final mass is determined.

In view of the fact that the samples were received with a delay, to date only one PIR sample has been exposed to temperature and a pressure drop of water vapour for 28 days. Table 1 shows the amount of moisture absorbed after 28 days.

3 RESULTS AND DISCUSSION

Table 1 shows the amount of moisture absorbed after 28 days. It is also predicted that mineral wool panels will absorb more moisture than both PIR panels.

The obtained results confirm the presence of the dependence of the vapour content of thermal insulation materials of PIR and mineral wool on the relative thermal properties of the material. PIR with polymer structure of closed pores absorbs less moisture. MW with fibrous structure absorbs more moisture. Absolute values of the moisture absorption of the MW significantly exceed the analogous values for mineral wool by about 50 times.

In real operating conditions, the heat transfer regime through the outer fences always turns out to be non-stationary. However, in most practical cases, steady-state heat transfer regime is considered, characterized by constancy in time of value of the heat flow and the temperature of the enclosure envelope. By revising the steady-state heat transfer regime the solutions are greatly simplified, which makes them practical in the development of engineering calculation techniques.

4 CONCLUSION

The obtained results testify to the differences in physical and thermal properties of the materials. It is shown that the PIR is more reliable than mineral wool by this indicator. However, due to the difference in structure of this material compared with mineral wool, it is not possible to make a final conclusion about which of the materials considered is more efficient in heat-insulating structures without additional studies. For a final conclusion, it is necessary to conduct a study to determine the thermal conductivity of the PIR in the wet state and compare these values with the analogous values for mineral wool or other competing materials. This will be used with this material especially during operating conditions. The increase in the thermal insulation characteristics of the materials of the enclosing structures also makes it possible to avoid the costs of upgrading the sources of thermal energy.

REFERENCES

Abdou, A., & Budaiwi, I. 2013. The variation of thermal conductivity of fibrous insulation materials under different levels of moisture content. *Construction and Building Materials* 43: 533–544. doi: 10.1016/j.conbuildmat.2013.02.058.

Alihodzic, R., Murgul, V., Vatin, N., Aronova, E., Nikolić, V., Tanić, M., & Stanković, D. 2014. Renewable energy sources used to supply pre-school facilities with energy in different weather conditions. Applied Mechanics and Materials. Vol. 624. doi: 10.4028/www.scientific.net/AMM.624.604.

Anikina, I.D., Sergeyev, V.V., Amosov, N.T., & Luchko, M.G. 2017. Use of heat pumps in turbogenerator hydrogen cooling systems at thermal power plant. *International Journal of Hydrogen Energy* 42(1): 636–642. doi: 10.1016/j.ijhydene.2016.04.256.

Basok, B.I., Bozhko, I.K., Nedbaylo, A.N., & Lysenko, O.N. 2015. A polyvalent heating system for a passive house based on renewable energy sources. *Magazine of Civil Engineering* 58(6): 32–43. doi: 10.5862/MCE.58.4.

Borodinecs, A., Zemitis, J., Sorokins, J., Baranova, D.V., & Sovetnikov, D.O. 2016. Renovation need for apartment buildings in Latvia. *Magazine of Civil Engineering* 68(8): 58–64. doi: 10.5862/MCE.68.6.

Gorshkov, A., Vatin, N., Nemova, D., Shabaldin, A., Melnikova, L., & Kirill, P. 2015. Using life-cycle analysis to assess energy savings delivered by building insulation. *Procedia Engineering* 117(1): 1085–1094. doi: 10.1016/j.proeng.2015.08.240.

Gorshkov, A.S., & Rymkevich, P.P. 2015. A diagram method of describing the process of non-stationary heat transfer. *Magazine of Civil Engineering* 60(8): 68–82. doi: 10.5862/MCE.60.8.

Gorshkov, A.S., Vatin, N.I., Rymkevich, P.P., & Kydrevich, O.O. 2018. Payback period of investments in energy saving. *Magazine of Civil Engineering* 78(2): 65–75. doi: 10.18720/MCE.78.5.

Korniyenko, S.V., Vatin, N.I., & Gorshkov, A.S. 2016. Thermophysical field testing of residential buildings made of autoclaved aerated concrete blocks. *Magazine of Civil Engineering* 64(4): 10–25. doi: 10.5862/MCE.64.2.

Kostenko, V., Gafiyatullina, N., Zulkarneev, G., Gorshkov, A., Petrichenko, M., & Movafagh, S. 2016. Solutions to Improve the Thermal Protection of the Administrative Building. *MATEC Web of Conferences* 73: 02011. doi: 10.1051/matecconf/20167302011.

Kostenko, V.A., Gafiyatullina, N.M., Semchuk, A.A., & Kukolev, M.I. 2016. Geothermal heat pump in the passive house concept. *Magazine of Civil Engineering* 68(8): 18–25. doi: 10.5862/MCE.68.2.

Leshchenko, M.V., & Semko, V. 2015. Thermal characteristics of the external walling made of cold-formed steel studs and polystyrene concrete. *Magazine of Civil Engineering* 60(8): 44–55. doi: 10.5862/MCE.60.6.

Muizniece, I., & Blumberga, D. 2016. Thermal Conductivity of Heat Insulation Material Made from Coniferous Needles with Potato Starch Binder. *Energy Procedia* 95: 324–329. doi: 10.1016/j.egypro.2016.09.014.

Nemova, D.V. 2013. Integrated characteristics of thermogravitational convection in the air layer of ventilated facades. *Magazine of Civil Engineering* 37(2). doi: 10.5862/MCE.37.4.

Nik, V.M., Sasic Kalagasidis, A., & Kjellström, E. 2012. Assessment of hygrothermal performance and mould growth risk in ventilated attics in respect to possible climate changes in Sweden. *Building and Environment* 55: 96–109. doi: 10.1016/j.buildenv.2012.01.024.

Petrichenko, M., Vatin, N., Nemova, D., Kharkov, N., & Korsun, A. 2014. Numerical modeling of thermogravitational convection in air gap of system of rear ventilated facades. *Applied Mechanics and Materials* 672–674: 1903–1908. doi: 10.4028/www.scientific.net/AMM.672-674.1903.

Petrichenko, M.R., Nemova, D.V., Kotov, E.V., Tarasova, D.S., & Sergeev, V.V. 2018. Ventilated facade integrated with the HVAC system for cold climate. *Magazine of Civil Engineering* 77(1): 47–58. doi: 10.18720/MCE.77.5.

Petritchenko, M.R., Subbotina, S.A., Khairutdinova, F.F., Reich, E.V., Nemova, D.V., Olshevskiy, V.Y., & Sergeev, V.V. 2017. Effect of rustication joints on air mode in ventilated facade. *Magazine of Civil Engineering* 73(5): 40–48. doi: 10.18720/MCE.73.4.

Petritchenko, M.R., Kotov, E.V., Nemova, D.V., Tarasova, D.S., & Sergeev, V.V. 2018. Numerical simulation of ventilated facades under extreme climate conditions. *Magazine of Civil Engineering* 77(1): 130–140. doi: 10.18720/MCE.77.12.

Statsenko, E.A., Ostrovaia, A.F., Musorina, T.A., Kukolev, M.I., & Petritchenko, M.R. 2016. The elementary mathematical model of sustainable enclosing structure. *Magazine of Civil Engineering* 68(8): 86–91. doi: 10.5862/MCE.68.9.

Tarasova, D.S., & Petritchenko, M.R. 2017. Buildings quasi-stationary thermal behavior. *Magazine of Civil Engineering* 72(4): 28–35. doi: 10.18720/MCE.72.4.

Vasilyev, G.P., Lichman, V.A., Yurchenko, I.A., & Kolesova, M.V. 2016. Method of thermotechnical uniformity coefficient evaluation by analyzing thermograms. *Magazine of Civil Engineering* 66(6): 60–67. doi: 10.5862/MCE.66.6.

Vatin, N.I., Nemova, D.V., Kazimirova, A.S., & Gureev, K.N. 2014. Increase of energy efficiency of the building of kindergarten. *Advanced Materials Research* 953–954: 1537–1544. doi: 10.4028/www.scientific.net/AMR.953–954.1537.

Vereecken, E., & Roels, S. 2016. Capillary Active Interior Insulation Systems for Wall Retrofitting: A More Nuanced Story. *International Journal of Architectural Heritage* 10(5): 558–569. doi: 10.1080/15583058.2015.1009575.

Zach, J., Hroudová, J., Brožovský, J., Krejza, Z., & Gailius, A. 2013. Development of Thermal Insulating Materials on Natural Base for Thermal Insulation Systems. *Procedia Eng.* 57: 1288–1294. doi: 10.1016/j.proeng.2013.04.162.

Advances and Trends in Engineering Sciences and Technologies III – Al Ali & Platko (Eds)
© 2019 Taylor & Francis Group, London, ISBN 978-0-367-07509-5

BIM end-to-end training: From school to graduate school

N.I. Vatin & K.Y. Usanova
Peter the Great St. Petersburg Polytechnic University, St. Petersburg, Russian Federation

ABSTRACT: The aim of the study is to create the an end-to-end technology of already teaching BIM with the acquisition of basic skills in the first year of education at the university. Both the Moodle and MOOC technologies are used. The first-year student's task is an Autodesk Revit 3D project of a low-rise residential building. From semester to semester the designed project becomes more complicated: a restaurant, a school, a high-rise block of apartments, a shopping and recreation center. Constructional design is supplemented by FEM calculations of structures and HVAC, construction management plan and cost calculations. The training system developed by the authors can be introduced in other universities.

1 INTRODUCTION

As a powerful tool for sustainability in construction, the Building Information Modeling (BIM) is used in every sector of Architecture, Engineering and Construction, from investment plan to??. Building Information Modeling (BIM) became the norm in AEC, as well as a part of many construction project management programs (Puolitaival and Forsythe 2016). BIM technology is being actively researched on retraining in schools and universities. Key problems of university education are a the lack of teaching and learning resources for BIM; inability to find the balance between theory and practice, technology and process (Puolitaival and Forsythe 2016).

Researchers (Abdirad and Dossick 2016) suggested a framework structure of BIM curriculum, as well as a number of recommendations, which that can be used by BIM teachers as a guide for developing or evaluating of their BIM curriculum.

Student abilities in the BIM technologies aren't the only end goal. The process of mastering BIM helps students improve their cooperation and 4Cs as the most important skills required for 21st century education: critical thinking, communication, collaboration, and creativity (Zhao, McCoy, Bulbul, Fiori and Nikkhoo 2015) and achieve a higher order of learning through role play in the contexts of a real BIM project (Zhang, Wu and Li 2018).

Senior students (third year and more) are more informed in BIM, than freshmen (first and second years). This difference can even be seen in Korea (Ahn and Kim 2016). This doesn't allow use of BIM technologies for teaching a wide range of neighboring subjects, for example geodesy. To improve the course's development, it is important to pick up on student's understanding of the course related to BIM technologies (Wu, Mayo, Issa, McCuen and Smith 2017; Elliott, Glick and Valdes-Vasquez 2018).

Authors (Milyutina 2018) note the importance of development of pilot projects that should identify the ways and means of verification of the regulatory and technical base, as well as economic indicators in the transition to BIM in the construction sector.

In teaching BIM to bachelors it is recommended to use the Common Data Environment (CDE) collaboration platforms (Comiskey, McKane, Jaffrey, Wilson and Mordue 2017). However, there is a possible choice of easier media especially for the beginning of education.

Models of real buildings are too complicated to use for educational purpose (Puolitaival and Forsythe 2016). Thus creators of learning courses have a task of choosing the correct simplified model of a small house for the beginning of teaching the technology of 3D parametric modeling. However, despite the wide distribution of BIM technology learning (Matějka, Růžička, Žák, Hájek, Tomek, Kaiser and Veselka 2016), there are no specific recommendation on the structure of learned disciplines or their place in the learning plan. Tasks for the course's beginning and discipline's role in the learning plan are not clear.

Aim of this paper was the formulation of an end-to-end technology for teaching BIM from the first to the last day of a student's learning in a university while already acquiring base skills since the first year of learning.

For that, these tasks were worked on:

– development of the modeling object for freshmen BIM course in university.
– development of project summary's contents and a package of architectural drawings for starting the BIM course.
– decision on the technology and technology learning platform.
– decision on the method to grade current and final learning.
– approval of the development.

Work's results were implemented in the learning course of Institute of Civil Engineering, Peter the Great St. Petersburg Polytechnic University, Russian Federation.

2 METHODS

Main method of solving tasks was a poll of experts in field of civil engineering. The survey of experts showed the main shortcomings in the use of technology among university graduates. The experts also made recommendations on the choice of the design object in the first year of study.

Also the education process uses the possibility of new virtual learning pilot projects of St. Petersburg Polytechnic University. The core element of the project is LMS Moodle environment (Surygin, Kalmykova and Alexankov 2012; Kalmykova, Pustylnik and Razinkina 2017).

3 RESULTS AND DISCUSSION

The university create two parallel ways of BIM teaching independent from one another. Mass open self-education supports through open learning portal https://openedu.ru/ of Ministry of Education and Science, Russian Federation. Courses "Building projects. BIM" and "Basics of building structures calculation" were realized. Education oriented for university students described is this paper.

3.1 First experience of BIM implementing

The beginning stage of BIM implementing students used Allplan (in first year first semester) and AutoCAD (second semester). We described this starting phase in our earlier works (Pichugin, Usanova and Fedotova 2012; Vatin, Gamayunova, Rechinskiy and Usanova 2012). With that we organized an education principal "from easy to hard". Practice showed that for students just out of school Allplan was easier, doing modeling with objects and not primitives. Working with 3d objects like walls, ceilings columns and having an ability to see a 3D image of the object students subconsciously trained their understanding of 3D spatial relations.

Only after the full 3D model is done did we start making architectural drawings of a 2D model of the object. Things like the drawing, plan, cross section and view were taught to students only after working with a 3D model of the object. This subject was based on an already developed understanding of spatial relations. At the same time, the principle approach was "not to draw but to project". Later we started using Revit as a base graphical program.

The Building Information Model Distance Learning Course was created for University applicants. The course is a part of pre-university education and a tool of University online marketing. We use Peter the Great St.Petersburg Polytechnic University Virtual learning environments that are based on Moodle. This allows secondary school students to prepare themselves for the University's learning process and get them an outlook on the right choice of major. They use Autodesk Revit software for building and construction design. The results of the course are a three-dimensional model of a small two-storey residential building, presentation graphics, and a set of drawings in accordance with Russian construction standards. The training mission is actually a simplified version of the educational task of the first year students in Computer graphics discipline. The tasks for school students a made easier. There is no attendance control. Every student is working in their own schedule and tempo unrelated to other people in their contest. The course exists with help from Committee for Construction of St. Petersburg city Government, Russian Federation. Graduates of the course will get a certificate. Handing-out of the certificates is usually done in a celebratory form. Course graduates subsequently become successful Civil Engineering students (Usanova, Rechinsky and Vatin 2014; Usanova and Vatin 2017).

3.3 *Basics of BIM for the first year university students*

3.3.1 *Technology and technological platform of education*
We chose the technological learning platform because of an ability to educate trough the web and an ability to make joint web programs with other universities. This needed us to choose a more widespread platform for our education. And that platform is Moodle. It is free and more importantly widely used system of distance internet learning. St. Petersburg Polytechnic University actively uses Moodle and MOOC for teaching different disciplines on the first year (Bakayev, Vasilyeva, Kalmykova and Razinkina 2018).

3.3.2 *Choosing the correct simplified model of building*
Buildings and constructions are divided in assembly buildings, residence buildings and industrial buildings. Experts estimate the most common tasks for graduates of engineering universities are designing a residence building. Thus the starting task in learning on the first year is about a residence building. The modeling object is a small two storey residence building. Modeling was done in the "Basics of building design" subject in 34 hours of instructor-led lessons and 120 hours of each student's work in the semester which is equal to four ECTS. The forms of course control are a term project and a pass/fail exam. Through that already mentioned principle "not to draw but to project" was used.

3.3.3 *Contents of the project's documentation*
In opinion of polled experts many engineering university graduates are bad at imagining the contents of their project's documentation and the correlation between divisions of the project. From this opinion, we decided to make students go through all parts of modeling early on. Parts of documentation is done in full while other parts are just imitated marked on the side that this subject isn't for further development. This simple teaching maneuver allowed teaching contents of the project's documentation from the first days of education.

3.3.4 *Project from the first semester is key for all future education*
In the first semester students realized these parts of a project's documentation:

– architectural decisions,
– structural decisions.

All student's work on the project can be divided in these steps:

– structure planning work (deciding the partition-walls and placing the coordinate axis on the structure),

- creation of levels and grid plan in Revit,
- creation of first floor's 3D model (walls and partitions, foundation, ceiling, windows and doors, stairs),
- creation of second floor's 3D model by analogy with the first floor,
- creation of a roof and air vents,
- creation of structure's décor elements,
- creation of land plot layout diagram and a landscape plan,
- creation of photo-realistic images of the structure using visualization in Revit and cloud visualization from the Autodesk site,
- creation of a PowerPoint presentation,
- processing of drawings made from the 3D model (Floor plans, facades, cross section),
- printing of the drawings (perfecting the list with a frame and title block),
- compiling of the project summary.

This project can be divided in two parts: creation of structure's 3D model and work on drawings of the created 3D model. And while creating a 3D model won't cause many problems for students because the process resembles an intuitive logic game, creating drawings causes a lot more problems. First, the student doesn't know how to perfect these drawings, but already must do it by the rules. Students have to simultaneously learn building codes on drawing perfection and how to perfect them in a program.

Examples of first year student's projects done in year 2017 published in (Usanova and Sadardinova 2018; Usanova and Tarasova 2018)

3.3.5 *Usage of English*
According to Federal rule "About education in Russian Federation" Education can be done in a foreign language. In St. Petersburg Polytechnic University learning on bachelor level is in Russian and on master level in Russian or English languages depending on the students' wishes and language skills.

Learning of the English language is done in the first year and is parallel and согласовано with learning PIM-technologies (Almazova and Kogan 2014; Almazova, Kostina and Khalyapina 2016). In "Basics of building design" discipline students learn bilingual glossary on BIM technologies.

3.3.6 *Distance-learning education in university*
Distance-learning goes through the same tasks as on-campus education and at the same pace. In one same semester the same professor leads the lessons for both on-campus and distance-learning students.

Special aspects about distance-learning education associated with a different ratio of classroom (contact) hours and independent work, are these:

- distance-learning students have experience of work with project documentation,
- while doing tasks form the beginning of semester in Moodle with a distant contact with a professor they have finished and ready to be turned in projects by the consultation time at the end of the semester,
- while turning in projects a test is done to check if the student made the work themselves by making them redo part of their work with a professor present.

3.4 *Further teaching of BIM technologies in bachelor's*

In further education on later years these kids of projects with BIM-technologies are repeated every semester with higher difficulty of tasks and more documentation parts done in full. Finishing stage of bachelor's education includes a graduate qualification work which is of a big building or structure made of many parts.

As an example let's look at Arina Avdeeva's bachelor graduation thesis (Avdeeva 2017). In the graduate thesis is a business-center with a restaurant for 130 people and an underground parking lot. Building is located in St. Petersburg. This work's objective is preparing project's documentation for constructing a business-center and substitution of a steel reinforcement in the anchor plate for fiberglass reinforcement. Main tasks are: research into damage of fiberglass reinforcement by outside chemical influence, economical evaluation of steel reinforcement substitution with fiberglass, calculation of foundation weight reduction by usage of fiberglass reinforcement.

During preparation of bachelor thesis Avdeeva has published two research articles (Avdeeva, Shlykova, Antonova, Barabanschikov and Belyaeva 2016) and (Avdeeva, Shlykova, Perez, Antonova and Belyaeva 2016). These publications are a part of common students and teachers research work on this topic (Bushmanova, Barabanshchikov, Semenov, Struchkova and Manovitsky 2017), (Bushmanova, Videnkov, Semenov, Barabanshchikov, Dernakova and Korovina 2017), (Korotchenko, Ivanov, Manovitsky, Borisova, Semenov and Barabanshchikov 2017).

4 CONCLUSIONS

To form an end-to-end teaching method for **BIM** with acquiring basic skills form the first year in the university we:

- chose the modelling object for starting the BIM course in university,
- chose the contents of the project summary and drawings of the beginning of BIM course,
- chose the technologies and technological education platforms,
- chose the graduation method of current and final results of education.

The end-to-end teaching method was applied in education process in the Institute of Civil Engineering, Peter the Great St.Petersburg Polytechnic University.

REFERENCES

Abdirad, H., & Dossick, C.S. 2016. BIM curriculum design in architecture, engineering, and construction education: A systematic review. *Journal of Information Technology in Construction* 21: 250–271.
Ahn, E., & Kim, M. 2016. BIM awareness and acceptance by architecture students in Asia. *Journal of Asian Architecture and Building Engineering* 15(3): 419–424. doi: 10.3130/jaabe.15.419.
Almazova, N., & Kogan, M. 2014. Computer assisted individual approach to acquiring foreign vocabulary of students major. *Lecture Notes in Computer Science (including subseries Lecture Notes in Artificial Intelligence and Lecture Notes in Bioinformatics)*. Vol. 8524 LNCS. doi: 10.1007/978-3-319-07485-6_25.
Almazova, N.I., Kostina, E.A., & Khalyapina, L.P. 2016. The new position of foreign language as education for global citizenship. *Novosibirsk State Pedagogical University Bulletin* 6(4): 7–17. doi: 10.15293/2226-3365.1604.01.
Avdeeva, A., Shlykova, I., Perez, M., Antonova, M., & Belyaeva, S. 2016. Chemical properties of reinforcing fiberglass in aggressive media. *MATEC Web of Conferences* 53: 01004. doi: 10.1051/matecconf/20165301004.
Avdeeva, A., Shlykova, I., Antonova, M., Barabanschikov, Y., & Belyaeva, S. 2016. Reinforcement of concrete structures by fiberglass rods. *MATEC Web of Conferences* 53: 01006. doi: 10.1051/matecconf/20165301006.
Avdeeva, A.A. 2017. Business center with a monolithic foundation reinforced with fiberglass reinforcement. Peter the Great St. Petersburg Polytechnic University. doi: doi.org/10.18720/SPBPU/2/v17-3479.
Bakayev, V., Vasilyeva, V., Kalmykova, S., & Razinkina, E. 2018. Theory of physical culture- a massive open online course in educational process. *Journal of Physical Education and Sport* 18(1): 293–297. doi: 10.7752/jpes.2018.01039.
Bushmanova, A.V., Videnkov, N.V., Semenov, K.V., Barabanshchikov, Y.G., Dernakova, A.V., & Korovina, V.K. 2017. The thermo-stressed state in massive concrete structures. *Magazine of Civil Engineering* 71(3): 51–60. doi: 10.18720/MCE.71.6.

Bushmanova, A.V., Barabanshchikov, Y.G., Semenov, K.V., Struchkova, A.Y., & Manovitsky, S.S. 2017. Thermal cracking resistance in massive foundation slabs in the building period. *Magazine of Civil Engineering* 76(8): 193–200. doi: 10.18720/MCE.76.17.

Comiskey, D., McKane, M., Jaffrey, A., Wilson, P., & Mordue, S. 2017. An analysis of data sharing platforms in multidisciplinary education. *Architectural Engineering and Design Management* 13(4): 244–261. doi: 10.1080/17452007.2017.1306483.

Elliott, J.W., Glick, S., & Valdes-Vasquez, R. 2018. Student perceptions of model-based estimating. *International Journal of Construction Education and Research*: 1–18. doi: 10.1080/15578771.2018.1460642.

Kalmykova, S.V., Pustylnik, P.N., & Razinkina, E.M. 2017. Role scientometric researches' results in management of forming the educational trajectories in the electronic educational environment. *Advances in Intelligent Systems and Computing*. Vol. 545. doi: 10.1007/978-3-319-50340-0_37.

Korotchenko, I.A., Ivanov, E.N., Manovitsky, S.S., Borisova, V.A., Semenov, K.V., & Barabanshchikov, Y.G. 2017. Deformation of concrete creep in the thermal stress state calculation of massive concrete and reinforced concrete structures. *Magazine of Civil Engineering* 69(1): 56–63. doi: 10.18720/MCE.69.5.

Matějka, P., Růžička, J., Žák, J., Hájek, P., Tomek, A., Kaiser, J., & Veselka, J. 2016. The implementation of Building Information Modeling into educational programs at CTU in Prague. 853–860. CESB 2016 - Central Europe Towards Sustainable Building 2016: Innovations for Sustainable Future.

Milyutina, M.A. 2018. Introduction of Building Information Modeling (BIM) Technologies in *Construction. In Journal of Physics: Conference Series*. Vol. 1015. doi: 10.1088/1742-6596/1015/4/042038.

Pichugin, E., Usanova, K., & Fedotova, K. 2012. 3D-projects in course "Computer Graphics" made by first-year students. *Construction of Unique Buildings and Structures* 2(2): 61–112. http://unistroy.spbstu.ru/index_2012_02/10_slideshows.pdf.

Puolitaival, T., & Forsythe, P. 2016. Practical challenges of BIM education. *Structural Survey* 34(4–5): 351–366. doi: 10.1108/SS-12-2015-0053.

Surygin, A.I., Kalmykova, S.V., & Alexankov, A.M. 2012. Models of international virtual learning environment for international educational projects. 15th International Conference on Interactive Collaborative Learning, ICL 2012. doi: 10.1109/ICL.2012.6402221.

Usanova, K., & Sadardinova, L. 2018. Example of a course work on the discipline "Computer graphics" by Sadardinova L.S. *AlfaBuild* 3(1): 1–11. http://alfabuild.spbstu.ru/index_2017_3/kurs1.pdf.

Usanova, K., & Tarasova, A. 2018. Example of a course work on the discipline "Computer graphics" by Tarasova A.A. *AlfaBuild* 3(1): 12–22. http://alfabuild.spbstu.ru/index_2017_3/kurs2.pdf.

Usanova, K., & Vatin, N. 2017. University BIM distance learning course for secondary school students. 297–302. Advances and Trends in Engineering Sciences and Technologies II - Proceedings of the 2nd International Conference on Engineering Sciences and Technologies, ESaT 2016.

Usanova, K., Rechinsky, A., & Vatin, N. 2014. Academy of construction for university applicants as a tool of university online marketing. *Applied Mechanics and Materials*. 635–637: 2090–2094. doi: 10.4028/www.scientific.net/AMM.635-637.2090.

Vatin, N.I., Gamayunova, O.S., Rechinskiy, A.V., & Usanova, K.Y. 2012. Fundamental and polytechnical experience of construction education with using Moodle. *Construction of Unique Buildings and Structures* 2(2): 6–17. http://unistroy.spb.ru/index_2012_02/2_moodle.pdf.

Wu, W., Mayo, G., Issa, R.R., McCuen, T., & Smith, D. 2017. Exploring the body of knowledge for building information modeling implementation using the Delphi method. 211–219. Congress on Computing in Civil Engineering, Proceedings.

Zhang, J., Wu, W., & Li, H. 2018. Enhancing Building Information Modeling Competency among Civil Engineering and Management Students with Team-Based Learning. *Journal of Professional Issues in Engineering Education and Practice* 144(2). doi: 10.1061/(ASCE)EI.1943-5541.0000356.

Zhao, D., McCoy, A.P., Bulbul, T., Fiori, C., & Nikkhoo, P. 2015. Building Collaborative Construction Skills through BIM-integrated Learning Environment. *International Journal of Construction Education and Research* 11(2): 97–120. doi: 10.1080/15578771.2014.986251.

Advances and Trends in Engineering Sciences and Technologies III – Al Ali & Platko (Eds)
© 2019 Taylor & Francis Group, London, ISBN 978-0-367-07509-5

BIM for cost estimation

S. Vitasek & J. Žák
Faculty of Civil Engineering, Czech Technical University in Prague, Prague, Czech Republic

ABSTRACT: Building Information Modeling (BIM) is being developed in almost all construction sectors in the Czech Republic. For Cost Management, referred to as BIM 5D, this brings many opportunities and challenges associated with the need of local workflows modification in accordance to information modeling needs. This article describes how data from information models can be used today to estimate the costs using the proposed process scheme. It also indicates what needs to be done to achieve the full utilization of the BIM method in terms of cost estimation. In a case study, the current use of the information model is tested in relation to the cost estimation of a typical object.

1 INTRODUCTION

Building Information Modeling (BIM), like any innovation method, must undergo a process of verification of higher degree of applicability over the existing traditional solution in the Central European region, both with the professional public and primarily with the investors themselves. It represents a modern tool for solving both technical and economic issues related to construction projects. The up-to-date involvement of modern technologies and methods to find the most cost-effective way of determining construction costs is also underpinned by the ever-growing pressure of the society on the accuracy and transparency of published building prices in public construction projects. In the Czech Republic (CR) the information modeling is developing in almost all construction sectors and phases. (Horelica et al 2017, Ministry of Industry and Trade 2017). For cost management, which is sometimes referred to as BIM 5D, it brings many challenges in form of customizing workflows and associated processes to information modeling needs. The key area of cost management is cost estimation at the project preparation stage. The traditional way to determine it is very laborious and associated with many mistakes. The reading of the 2D documentation, which serves as a basis for the usual cost estimation of structures/construction activities, is outdated in contrast with today's modern means. Therefore, it's the right time to start using advanced tools to make estimates more realistic, faster, more reliable and flexible.

1.1 Literature review

Professional journals and publications are in limited extend describing current possibilities of BIM utilization for cost estimation or other cost management tools. Primarily due to the absence of real projects in practice and the fact that software product providers still cannot offer a comprehensive solution in this area. The specific uses of information modeling for cost estimation in relation to the local classification system is described in the paper "Utilization of BIM for Automation of Quantitative Takeoffs and Cost Estimation in Transport Infrastructure Construction Projects." (Vitasek & Matejka 2017) In this article, the authors describe real possibilities of involvement of the local price system the suitability of the traditional classification system for modern data use. The conclusions of their research call for the modernization of the classification system, as was the case in the UK. This idea to modernize or replace traditional classification systems, which are often based on the foundations from the last century, are supported by other scientific articles. (Plebankiewicz et al 2015, Ma et al

2014; Olsen & Taylor 2017) Other cost-related publications deal with the complex themes of BIM 5D. This way of thinking is completely in line with the modern concept of building monitoring throughout the life cycle. From cost estimation through cost management during construction to facility management. (Marzouk et al 2018, Röck et al 2018, Zak & Macadam 2017) Unfortunately, these articles do not describe the BIM 5D usage in such detail that they will provide desired detailed. An interesting article summarizing cost estimation requirements within the IFC standard is "Cost Estimation in Building Information Model". The authors point out: "The information requirement for the cost estimate can be summarized in 5 aspects: the building products information, the cost interim formation, the quantity information, the resource information and the price information." (Xu et al 2013) Authors agree and seek ways to implement these aspects of information modeling.

Conclusions of listed publications clearly support BIM application not only for cost estimation. A key step for the whole issue of BIM 5D is the linking of the information model of the construction with the local data standard in relation to the classification system.

2 INFORMATION MODEL REQUIREMENTS

As is clear, the quality, level of geometry and level of information of the information model is crucial aspect for successful cost estimations. The model elements must be created with all necessary associated appropriate attributes. The degree of geometry and information detail, modeling principles, data formats, coordinates and many other specifications is embedded in a contract conditions. In practice the amendment of the work contract (contract conditions) is BEP (BIM Execution Plan), that is prepared by VDC/BIM expert. Specific part of BEP is data standard. Data standard document consist specification (in form of axles. sheet, database or in any other transferable file format) of elements and associated attributes. Due to the absence of a national data standard, the level of detail specification for building structures generally refers to the Level of Development (LOD) publication (BIMForum 2017). LOD clearly defines which elements and their parts are to be represented in the model and which not. The definition of non-graphical information, i.e. the attributes (attributes), is then a separate attachment of the data standard specification. Non-graphical information is present in information models as the element's properties and attributes. For example, data such as description, volume, area, thermal resistance, strength, price, etc. The level indicating the relevant detail starts from LOD 100 (lowest detail) to LOD 500 (highest detail). Where LOD 350 appears to be optimal from the cost estimates point of view in most cases.

Such a specification is always prepared at given project phase. Choosing a level of detail and its content fundamentally influences the complexity of creating an information model. This is related to the subsequent price for these data. The problem on the Czech market is also related to the lack of experience of design companies with modeling of higher levels of detail of individual elements. Therefore, it is necessary to regularly check the delivered information model from design companies and their sup contractors, whether it complies with given data standard. Because very often it is prepared with design company libraries that do not necessarily fit the data standard requirements and are not sufficient for following cost management. Mistakes repeated by the design companies can be summarized in: lack of nongraphical data, clashes in the model, merging data that do not have the same price, elements with wrong graphical detail.

3 COST ESTIMATION

As already mentioned in the article, the key criterion for the cost estimation process is the classification system. The classification system for the valuation of building production assigns to each building structure/activity an identification code, a description and an assessment procedure in relation to a unit of measure (calculation unit). Some countries, like Czech Republic, has its own national classification system or at least the definition of construction work classifications. Unfortunately, the usability of these traditional classification systems is limited in terms of modern data processing and use of building information model. In the Czech environment, it is the Classification of Structures and Works (TSKP), which is based on price systems delivered in public tender proposals by engineering companies. These data are regularly maintained by the state that hires private company to do so. Using the software tool with available functionalities and workflows, it is almost impossible to model the information model so that it can be represented in the TSKP structure. The price systems also use types of items that have a direct connection with the construction / activity, but their modeling in the information model would be lengthy, impractical and would require designers to have cost estimation principles knowledge. Types are for example surcharges for work at heights, dusty environment, etc. That is why in many countries these classic classification systems are replaced by new ones. Such as Uniclass (UK), OmniClass (USA), CoClass (SWE), etc. These advanced classifications allow easier sorting of created elements for other data processing options.

The possibility of creating a cost estimation model from the information model is illustrated in the figure above (Figure 1). The key is a well-developed information model according

Table 1. Commercial building basic characterization.

Construction start	02/2017
Levels	7
Lent spaces	11 814 m^2
Handover to owner	11/2019
Cost estimation	10 mil. EUR
Supporting construction system	Monolithic concrete

Figure 1. Cost estimation diagram for information models.

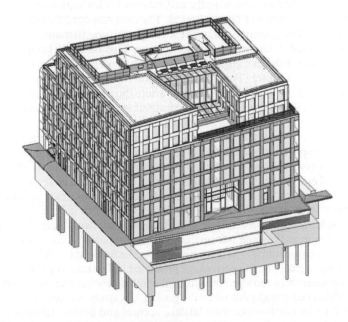

Figure 2. Information model of commercial building.

to the given data standard. Once the information model has been created, the parameters and attributes (description, volume, area, etc.) of the individual elements are transferred to the spreadsheet. Additionally, the generated data must be transformed into the classification system structure and then imported into the valuation program. Where we can already create an estimate of cost in the structure of the item budget. This way of working allows information model data to be processed when the cost estimator wants to use the conventional software tools for construction production valuation.

Other options for estimating costs using data from the information model are related to the use of specified products. A typical solution is tailor made programming through the Application Programming Interface (API), which extends the capabilities of the design software with cost estimate functionalities. These available programs and tailor-made programming is costly, especially implementation of local classification systems and hard to maintain when classification system or design software is changed. Practical use for the construction companies, which will bear the risk and cost associated with the updates of these APIs, is therefore significantly limited. Another possibility is to use the present classification system (CoClass, Uniclass, Uniformat, i.e.) directly in the creation of the information model. This avoids the subsequent transformation of the generated data. This leads to a smooth transition from the design to the valuation program, however, these classification systems do not correspond to the requirements, needs and customs of the Czech construction industry.

These two variants abroad represent the high maturity of the society and technical public for modern data work. Unprecedented phenomenon in the Czech Republic. The solution that is acceptable in our conditions is to supplement local data standards with classification systems that meet the valuation needs.

4 CASE STUDY

The application of modern technologies and procedures was carried out on the information model of the commercial (office) building. This model was prepared by construction and developing company operating both at global and Czech Republic market. The aim of the case study is to determine the current possible use of an information model to estimate the cost of a selected type of construction.

To estimate the construction cost with the utilization of information model, the scheme of the production process shown in Figure 1 is used. The data standard defines the element to be modeled in LOD 350. The quantity statement sorted according to materials used in information model was exported from the information model. This data has been transformed into a spreadsheet processor into the form of a classical statement in the corresponding TSKP (classification system) structure. The quantity takeoff was subsequently imported into the valuation software with the local pricing system, where the unit and total prices were assigned to the individual items in the takeoff. This has resulted in a targeted cost estimate and preparation of target budget broken down into budget items.

The graphical conclusions of the case study are summarized in the Figure 3 below. There are building sections (composed of specific constructions/building works—items) as defined by the TSKP classification system. Together with the percentage of estimated costs where 100% represents the traditional estimation process. That is, manual creation of a statement in the valuation software using 2D drawings. Traditional valuation of the object was made in the same valuation program and price level, making which allowed us to make both approaches direct comparison. E.g. if the building section reached 85%, it means that we can obtain 85% estimated costs through proposed process flow and information model compared to conventional cost estimations. The rest of the cost could not be determined from the model data. Additional information about methodology to calculate the percentage presented in Figure 3 is given in Tables 2 and 3. The difference between areas retrieved from the information model and manual processing was up to 4%. This difference is negligible when the goal of this study is taken in mind.

Percentage values for earthworks, foundations, vertical and horizontal constructions are at a very good level. It is primarily caused by its essential constructional role and focus of design

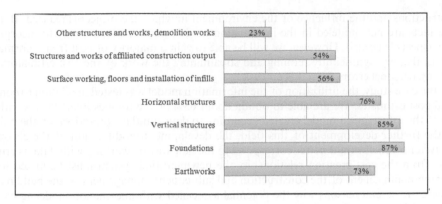

Figure 3. Comparison of cost calculated from the information model and conventional approach.

Table 2. For example, export information about doors (only 1st floor) from design software.

Symbol	Level	Count	Door width	Door height	Description	Door frame material	Door lock material
14	1st	9	700	2100	Interior doors, single-wing	Steel	Stainless steel
16	1st	17	800	2100	Interior doors, single-wing	Steel	Stainless steel
17	1st	12	900	2100	Interior doors, single-wing	Steel	Stainless steel

Table 3. Cost percentage calculation.

Building section	Building activity (item)	Quantity takeoff	Unit price [CZK]	Total price [CZK]	Percentage value from total costs for a building section
Foundations	Concrete footing C 16/20	693 m³	2,500	1,732,500	6,46
	Concrete basement slab	170 m³	2,350	339,500	1,27
	Gravel fill under basement slab TK 16/32 (grain size)	570 t	985	–	(in IM connected with another element)
	(next items)	...	...	...	...
	Sum for a building section (foundations) 26,812,656				86,93

software on these construction items. Values of the second group of construction items are affected by several factors.

The first factor is the erroneous modeling of these constructions, for example, when merging with another. Moreover, the used detail of the data standard (LOD 350) does not prescribe some elements in its specification, most notably the locksmith, plumbing and carpentry products/structures, which are listed in the building section of Associated construction production of the national classification system (TSKP). Last but not least the TSKP includes cost indirectly related to the construction items already in the construction items and this is essentially why it cannot be used in building information modeling. For example, different types of surcharges and cost for transfers of construction materials. By this the Other works cost was primarily affected. manifested in Other Works.

5 CONCLUSION

The article described how data from the information model can already be used with the help of the proposed scheme. It outlines the steps to achieve the BIM's total utilization, including the cost estimation field. It is likely that we will probably never be able to determine 100% of all construction costs during the design. Especially with regard to the variability of temporary constructions and the influence of the environment in which the project is realized. Both of these facts are not modeled in the information models and an effective way to incorporate these aspects is needed. However, we will be able to gain a majority of cost from information models, thereby significantly refining and streamlining the work of the cost estimators and omitting personal errors from this process.

In the case study, the utilization of the information model was tested in relation to conventional cost estimates. We are able to obtain about 65% of the total estimated costs with the help of the information model created according to data standard prescribed by the builder. For the further development of this field, the development/modification of the classification system, which would better correspond to the new way of working with data, is crucial. Apart from the cost-creation relationship, the new/modified system must be linked to the planning/management of the construction and subsequent management of the building.

There are several reasons why the potential associated with information modeling has not yet been fully exploited in relation to cost estimates. It is a very complex issue. Apart from the lack of significant legislative support, there is also a lack of product software or comprehensive solutions for working with information models from engineering companies operating on the market. Therefore, the activities of public organizations such as the Czech Agency for Standardization and the State Fund for Transport Infrastructure, which have the task of producing standardization documents, should be supported and will have an enormous impact in the industry.

ACKNOWLEDGEMENTS

This work was supported by the Grant Agency of the Czech Technical University in Prague, grant No. SGS18/023/OHK1/1T/11.

REFERENCES

BIMForum: Level of Development Specification – 2017 [online]. 2018 [cit. 2018-08-01]. Available from: https://bimforum.org/wp-content/uploads/2017/11/LOD-Spec-2017-Guide_2017-11-06-1.pdf.
Horelica, Z., Mertlova, O., Vykydal, I. & Zak, J. (2017). "*Plán pro rozšíření využití digitálních metod a zavedení informačního modelování staveb (Building Information Modelling – BIM) pro dopravní infrastrukturu*". State Fund for Transportation Infrastructure.
Ma, Z. & Liu, Z. (2014). "*BIM-based Intelligent Acquisition of Construction Information for Cost Estimation of Building Projects*". Paper presented at the Procedia Engineering, 85, 358–367.
Marzouk, M., Azab, S. & Metawie, M. (2018). "*BIM-based approach for optimizing life cycle costs of sustainable buildings*". Paper presented at the Procedia Engineering, 188, 217–226.
Ministry of Industry and Trade. (2017). "*Koncepce zavádění metody BIM v České republice*". Department of Construction and Building Materials.
Olsen, D. & Taylor, M. (2017). "*Quantity Take-Off Using Building Information Modeling (BIM), and Its Limiting Factors*". Paper presented at the Procedia Engineering, 196, 1098–1105.
Plebankiewicz, E., Zima K. & Skibniewski, M. (2015). "*Analysis of the First Polish BIM-Based Cost Estimation Application*". Paper presented at the Procedia Engineering, 123, 405–414.
Röck, M., Hollberg, A., Habert, G. & Passer, A. (2018). "*LCA and BIM: Visualization of environmental potentials in building construction at early design stages*". Paper presented at the Procedia Engineering, 140, 153–161.
Vitasek, S., & Matejka, P. (2017). "*Utilization of BIM for automation of quantity takeoffs and cost estimation in transport infrastructure construction projects in the Czech Republic*". Paper presented at the IOP Conference Series: Materials Science and Engineering, 236(1). DOI: 10.1088/1757-899X/236/1/012110.
Xu, S., Liu, K. & Tang, L. (2013). "*Cost Estimation in Building Information Model*". Paper presented at International Conference on Construction and Real Estate Management 2013. DOI: 10.1061/9780784413135.053.
Zak, J., & Macadam, H. (2017). "*Utilization of building information modeling in infrastructure's design and construction*". In IOP Conf. Series: Materials Science and Engineering, 1–6. DOI: 10.1088/1757-899X/236/1/012108.

Advances and Trends in Engineering Sciences and Technologies III – Al Ali & Platko (Eds)
© 2019 Taylor & Francis Group, London, ISBN 978-0-367-07509-5

Technology of rapid dewatering for excavations

D.D. Zaborova & M.R. Petrichenko
Peter the Great St. Petersburg Polytechnic University, St. Petersburg, Russia

ABSTRACT: Methods of non-stationary filtration are used in many technologies: rapid dewatering of excavation, drainage of construction sites, etc. Determination of the instantaneous groundwater level is associated with the solution of non-stationary filtration problems. Classical calculation methods of the depression curve are related to the solution of the limit problems for the Laplace equation (in domains with free boundary). This paper considers a different idea. The steady-state filtration mode is obtained as the limit of instantaneous filtering modes. Instantaneous modes are formed by the rapid lowering of the water level in the basin or in the drained well. The results of the mathematical analysis are adjusted to the calculation formulas that can be used to determine the dewatering time of excavation, drainage performance, etc.

1 INTRODUCTION

Seepage problems have major importance for power engineering and construction, with a wide range of applications extending from hydraulic engineering and melioration (Anderson, 2002; Kayode, 2018; Sainov, 2015; Loktionova, 2017) to construction technologies (rapid pumping of groundwater from the pit, drainage of construction sites) (Perazzelli, 2014; Girgidov, 2012, Titova, 2017). Seepage theory as a section of fluid mechanics includes two branches: hydromechanical and hydraulic. These branches overlap in solving stationary problems (Petrichenko, 2015).

All buildings and structures are exposed to moisture. The feeling of dampness, occurrence of cracks, wall damage, subsidence of the foundation due to leaching of the soil—all these are possible effects of groundwater on buildings and structures.

This problem is especially typical for the St. Petersburg city, Russia, where due to the low level of natural drainage of the territory a high level of groundwater ranges within 0.5–5 m.

Groundwater is extremely unstable in many factors, it affects the conditions of construction; dictates the choice of foundation and building design technologies. The further exploitation of man-made structures is also under the constant influence of the changing behavior of ground water (Sainov, 2016).

Water flow into the ground, water runoff and water evaporation always change, therefore the groundwater level does not remain constant. This level is influenced not only by the natural change in the regime of groundwater, but also by the implementation of some technical measures, such as the layout of the territory, its asphalting, drainage, storm water, etc. (Brouček, 2013).

Rise of groundwater table worsens the construction properties of soils: soil moisture increases, its "skeleton" is suspended in water, the friction and adhesion forces between the soil particles decrease, the porosity of soils increases.

If velocity of groundwater movement exceeds the critical, filtration flow washes away soil particles. Gradual removal of such particles leads to the soil loosening (Sterpi, 2003).

The study of water filtration in soils and various porous media has a great practical interest in solving numerous engineering problems in the field of hydraulic engineering, construction of water supply and sanitation facilities.

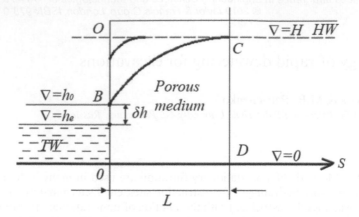

Figure 1. Depression curve TW, HW are the tailwater and the headwater; s, h are the coordinate axes (h is the seepage flow depth); ∇ is the elevation; H, h_0, h_e are the levels of the seepage flow; δh is the seepage area; L is the length of the homogeneous closing dike; BC is the depression curve; CD is the flow area; the dashed line indicates the headwater level.

2 DEPRESSION CURVE

The movement of water in the pores of the soil is called filtration. Liquid filtration can be pressure and non-pressure. In the case of pressure water flow, the filtration flow is limited to the top and bottom soils (Karamouz, 2013). The pressure in this case is excessive. The filtration flow with a free surface is called non-pressure movement. In case of non-pressure water movement in the pores of the soil the depression curve is formed as a line separating the saturated and unsaturated porous medium (Figure 1). Depression curve is the contact surface coinciding with the atmospheric pressure line in non-pressure motion (free surface).

Under the influence of the level difference, water is filtered from the headwater to the tailwater through the porous medium. If there is no pressure, then there is no filtration flow and the depression curve turns into a horizontal line (Schwartz, 2004).

The determination of the depression curve configuration and break point of the depression curve (seepage area) is necessary to evaluate the groundwater level at any time. This task is important for pumping and water—lowering technology, essentially non-stationary processes.

The purpose of this work is calculation of the seepage area and the time of formation the stationary regime. To archive this goal, the following tasks have to be solved:

– To solve the limit problem of the Boussinesq equation;
– To obtain the mathematical relation for determination of the settling time of seepage area;
– To prove, that for long closing dike the height of the seepage area is small and for such closing dikes Dupuis theory works.

3 SEEPAGE AREA

Seepage area (δh) – is a moist or wet place where water or groundwater is seeped to the atmosphere through the porous medium. Usually, water leaks to the construction pit through this point. This water interferes with construction works in it. Therefore, it is necessary to deal with groundwater by pumping or installation of watertight jumpers (Larssen sheet piles, well point and well pump) in the pit.

However, it is not always possible to define visually the water leak in the soil. This raises the challenge of mathematical definition the value of seepage area.

The starting point is the continuity condition for fluid flow:

$$\frac{\partial h}{\partial t} + \frac{\partial q}{\partial x} = 0, q = h \cdot v = -kh \frac{\partial h}{\partial x}, \tag{1}$$

where $h = h(t, s)$ is the depth of the seepage flow (t = the time, s = horizontal coordinate); k = the filtration coefficient; q = filtration flow; v = instantaneous mean velocity.

With the exception of q we obtained Boussinesq equation for the solitary flow wave:

$$\frac{\partial h}{\partial t} = \left(kh \frac{\partial h}{\partial x} \right), D(h) = (t, x : t > 0, x > 0), lm(h) = (h : 0 \leq h_e \leq h_0 \leq h \leq H) \tag{2}$$

where h_e = water level in the tailwater; $h_0 = h(t,0)$ = level of the seepage flow.

Let the boundary conditions have the form:

$$\begin{aligned} h(0,x) &= H, \\ h(t,0) &= h_0. \end{aligned} \tag{3}$$

If we start using dimensionless coordinates:

$$u : h / H; s := x / H, \tau := kt / H, u = (\tau, s)$$

Then the limit problem of Boussinesq (2), (3) takes the form:

$$\frac{\partial u}{\partial \tau} = \frac{\partial}{\partial s} \left(u \frac{\partial u}{\partial s} \right), D(u) = (\tau, s : 0 < \tau < \infty, 0 < s < \infty), lm(u) = (u : 0 \leq u_e \leq u_0 \leq u \leq 1), \tag{4}$$
$$u(0,s) - 1 = u(\tau,0) - u_0 = 0.$$

where $u_e := h_e / H$, $u_0 = h_0 / H$, $u_0 \geq u_e \geq 0$.

For the solution of (3) and (4), the method of integral relations described in (Zaborova, 2018) is used. Seepage area can be determined as:

$$\delta h = \frac{\Delta}{H} = e^{-\lambda}, \lambda = \frac{L}{H} \tag{5}$$

where L = length of the closing dike, m; H = water pressure, m; Δ = geometric height of the seepage area, m. Time of establishment of the stationary regime (setting time of seepage area):

$$T = \frac{3L^2}{kH}.$$

Projected instant length of the depression curve on the horizontal axis:

$$l = \sqrt{1/3k \cdot t \cdot H}, 0 < t < T.$$

Average fall velocity of the depression curve:

$$v = \frac{H}{T}.$$

4 PROBLEM STATEMENT AND RESULTS

Consider the breakwater that separates water area and Finland gulf as a closing dike with the length $L = 5$ m (Figure 2).

665

Water level in Finland gulf is equal 3 m. For some construction works the water area is drained and tailwater became equal zero. Water filtration occurs due to difference in the water level ($H = 3$ m). The breakwater is made of sand with filtration coefficient k = 5 m/day. Let us define the seepage area that occurs due to the water filtration.

According to formula (5), the seepage area is equal:

$$\delta h = \frac{\Delta}{H} = e^{-\lambda} = 0{,}189$$

The geometric height of the seepage area will be equal:

$$\Delta = \delta h \cdot H = 0.567 m$$

This section of the closing dike (0.57 m height) will be subject to destruction. Results for different water pressure (H) are presented in the Table 1. According to the results, we can see: the bigger water pressure (H), the higher seepage area in basin, with the same length of closing dike (L) and coefficient of filtration (k).

Let us see what happens with the seepage area if the length of closing dike changes. In Table 2 and Table 3, the length of closing dike has changed from 5 meters to 20 meters. Consider these two cases:

– Short closing dike (L = 5 m). The value $h_0 > h_e$, and seepage area is formed.
– Long closing dike (L = 20 m). The value $h_0 \to h_e + 0$, and the seepage area is small.

The first case happens if time T is small, the closing dike is short and the initial level difference $H - h_e$ is a finite value (not small). The second case happens if the closing dike is long and the initial difference of the levels $H - h_e$ is small.

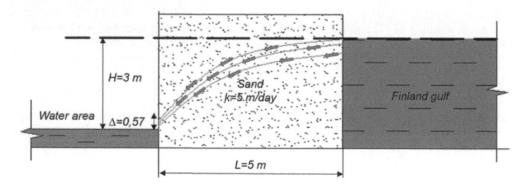

Figure 2.   Water filtration through the sand closing dike.

Table 1.   Effect of water pressure on the height of seepage area.

closing dike of length $L = 5$ M

H	λ	δh	$t = T$	l	v	H	λ	δh	$t = T$	υ
m	–	–	days	m	m/day	m	–	–	days	m/days
1	5.00	0.007	15.00	5	0.667	6	0.833	0.434	2.50	2.400
2	2.50	0.082	7.50		0.267	7	0.714	0.489	2.14	3.267
3	1.67	0.189	5.00		0.600	8	0.625	0.535	1.87	4.267
4	1.25	0.286	3.75		1.067	9	0.556	0.574	1.67	5.400
5	1.00	0.368	3.00		1.667	10	0.500	0.606	1.50	6.667

Table 2. Effect of closing dike length on the height of seepage area.

	Closing dike of length L = 5 m					Closing dike of length L = 10 m				
H	λ	δh	$t = T$	l	v	λ	δh	$t = T$	l	v
m	–	–	days	m	m/day	–	–	days	m	m/day
1	5.00	0.007	15.00	5	0.667	10.00	0.000	60	10	0.017
2	2.50	0.082	7.50		0.267	5.00	0.007	30		0.067
3	1.67	0.189	5.00		0.600	3.33	0.036	20		0.150
4	1.25	0.286	3.75		1.067	2.50	0.082	15		0.267
5	1.00	0.368	3.00		1.667	2.00	0.135	12		0.417

Table 3. Effect of closing dike length on the height of seepage area.

	Closing dike of length L = 15 m					Closing dike of length L = 20 m				
H	λ	δh	$t = T$	l	v	λ	δh	$t = T$	l	v
m	–	–	days	m	m/day	–	–	days	m	m/day
1	15.00	0.000	135.00	15	0.007	20.00	0.000	240	20	0.004
2	7.50	0.000	67.50		0.029	10.00	0.000	120		0.017
3	5.00	0.007	45.00		0.067	6.67	0.001	80		0.037
4	3.75	0.023	33.75		0.118	5.00	0.007	60		0.067
5	3.00	0.049	27.00		0.185	4.00	0.018	48		0.104

In other words, seepage area problem is actual only for short closing dikes with the high water pressure.

To optimize the drainage of the pit it is necessary to know the time of establishment of the stationary regime. For closing dike of length L = 5 m and water pressure H = 3 m, the stationary regime will be established after 5 days. Therefore, amount and power of pumps, that remove water from the pit, can be reduced.

5 CONCLUSIONS

1. Water filtration through the closing dike should be considered as a non-stationary motion in the porous medium bounded by a moving (descending and stretching) depression curve. If the motion of the depression curve stops, steady seepage develops;
2. Instantaneous height of the seepage area is determined by the length of the closing dike and differences in water levels (between tailwater and headwater). In short closing dike the seepage area is determined only by the differences in water levels (H);
3. The seepage area stabilizes during the time T (time depends on length of closing dike), so in a short closing dike this time is less than the time in long closing dike;
4. The estimation of the size of the seepage area is necessary for fixing the lower slope of the construction that will help to prevent destruction of a dam body.

REFERENCES

Anderson, M.P., Hunt, R.J., Krohelski, J.T., Chung, K. 2002. *Using high hydraulic conductivity nodes to simulate seepage lakes*. Ground Water 40(2): 117–122.
Brouček M. 2013. *Subsoil Influenced by Groundwater Flow*. Praha.

Girgidov A.D. 2012. The time of groundwater free surface lowering before foundation pit construction. *Magazine of Civil Engineering* 4(30): 52–56.

Karamouz M., Nazif S., Falahi M. 2013. *Hydrology and Hydroclimatology. Principles and Applications.* Boca Raton: CRC Press Taylor & Francis Group.

Kayode, O.T., Odukoya, A.M., Adagunodo, T.A., Adeniji, A.A. 2018. Monitoring of seepages around dams using geophysical methods: A brief review. *IOP Conference Series: Earth and Environmental Science* 173 (1): article № 012026.

Loktionova E.A., Miftakhova D.R. 2017. Fluid filtration in the clogged pressure pipelines. *Magazine of Civil Engineering* 8: 214–224.

Perazzelli P, Leone T, Anagnostou G. 2014. Tunnel face stability under seepage flow conditions. *Tunnelling and Underground Space Technology* 43: 459–469.

Petrichenko M.R., Serow D.W. 2015. Integral identities and rational approximations of the Crocco boundary problem. *Proceedings of the 3rd International Conference on Optimization and Analysis of Structures*: 86–91.

Sainov M.P. 2015. 3D performance of a seepage control wall in dam and foundation. *Magazine of Civil Engineering* 5(57): 20–33.

Sainov M.P., Anisimov O.V. 2016. Stress-strain state of seepage-control wall constructed for repairs of earth rock-fill dam. *Magazine of Civil Engineering* 8: 3–17.

Schwartz, F.W., Zhang, H. (2003). *Fundamentals of Ground Water*. New York: Wiley.

Sterpi, D. 2003. Effects of the Erosion and Transport of Fine Particles due to Seepage Flow. *International Journal of Geomechanics* 3(1): 111–122.

Titova T.S., Akhtyamov R.G., Nasyrova E.S., Longobardi A. 2017. Lifetime of earth dams. *Magazine of Civil Engineering* 1: 34–43.

Zaborova D.D., Petritchenko M.R., Musorina T.A. 2018. The Dupuis paradox and mathematical simulation of unsteady filtration in a homogeneous closing dike. *St. Petersburg Polytechnical State University Journal. Physics and Mathematics* 11 (2): 49–60.

Advances and Trends in Engineering Sciences and Technologies III – Al Ali & Platko (Eds)
© 2019 Taylor & Francis Group, London, ISBN 978-0-367-07509-5

Proposal of levels of detail LOD in building projects implementing BIM

K. Zima & E. Mitera-Kiełbasa
Institute of Construction Management, Cracow University of Technology, Cracow, Poland

ABSTRACT: BIM technology is increasingly frequently appearing on the industry market in publications, legislation and construction projects. One of the elements of this methodology is assigning a level of detail to BIM models. British Levels of Definition (LOD) are an example of this. Determining the details of elements in a design documentation is important for construction project participants. A developer who defines the expected LOD early on can prevent future disappointment. A lower level of detail at the beginning of the design process contributes to a greater motivation of the designer to introduce changes according to a developer's expectations. A greater level of detail at the design execution stage and clash detection help to prevent changes at the construction site. This article analyses Levels of Detail (LOD) at different stages of the real estate development process, including the impact of other factors.

1 INTRODUCTION

Building Information Modelling is increasingly frequently appearing on the industry market in publications, legislation, and construction investments (Howard, Björk, 2008) and can have benefits that positively impact the life cycle cost of a building (Lu, Fung, Peng, Lianf, Rowlinson, 2014). It describes tools, processes and technologies concerning a building, its performance, planning, construction and its later operation (Eastman, Björk, 2009). One of the elements of the methodology is assigning a level of detail to BIM models. British Levels of Definition (LOD) (PAS 1192-2:2013) are an example of this. They relate to the geometric data of the Level of Detail (LOD) and the non-geometric data of the Level of Information (LOI). Early reflections on LOD can avoid subsequent misunderstandings about the detail of a design for various participants in the construction process. This work analyses the levels of detail for various stages of the investment process, including the subject literature.

2 MATERIALS AND METHODS

In order to assign levels of detail to the stages of a construction project, the characteristics of these levels should first be considered. The British pattern of The AEC UK BIM Technology Protocol(AEC (UK) BIM Technology Protocol, 2015) provides a description of each level (Table 1).

To illustrate what these levels mean, the modular suspended ceiling systems section from The BIM Toolkit Project (NBS BIM Toolkit, 2018) was used as an example. It was prepared by HM UK Government's BIM Task Group and has been presented in Table 2.

The investment process is divided into stages. To determine the level of detail of a project's design model, it is necessary to specify the goals that are set for them. They have been presented in Table 3 on the basis of PAS 1192-2: 2013 and the RIBA Plan of Work 2013.

Table 1. Component Grade LOD (AEC (UK) BIM Technology Protocol, 2015).

LOD1 Symbolic	LOD2 Conceptual	LOD3 Generic	LOD4 Specific	LOD5 Construction	LOD6 As Built
– Symbolic place-holder representing an object which may not be to scale or have any dimensional values.	– Simple place-holder with absolute minimum level detail to be identifiable, e.g. as any type of chair.	– A generic model, sufficiently modelled to identify type and component materials.	– A specific object, sufficiently modelled to identify type and component materials.	– A detailed, accurate and specific object of the construction requirements and building components, including specialist sub-contract geometry and data.	– A precisely modelled representation of the constructed object.
This is particularly relevant to electrical symbols which may never exist as a 3D object.	– Superficial dimensional representation.	– Typically contains level of 2D detail suitable for the "preferred" scale.	– Accurate dimensions.	– Should include all necessary sub-components adequately represented to enable construction.	– Any construction irregularities or eccentricities should be modelled.
	– Created from consistent material: either 'Concept–White' or 'Concept–Glazing'.	– Dimensions may be approximate.	– A production, or pre-construction, "design intent" object representing the end of the design stages. – Suitable for procurement and cost analysis.	– Used only when a 3D view at a sufficient scale deems the detail necessary due to the object's proximity to the camera.	

Important! When in doubt, users should opt for less 3D geometry, rather than more, as the efficiency of the BIM is largely defined by the performance of the components contained within.

Table 2. LOD for modular suspended ceiling systems (NBS BIM Toolkit, 2018).

LOD 2	LOD 3	LOD 4

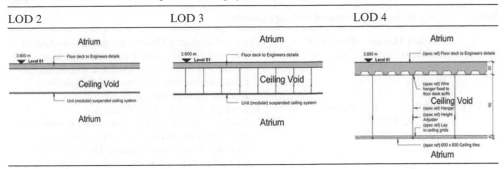

Table 3. Goals for individual stages according to PAS 1192-2:3013 and RIBA Plan of Work 2013 [Own elaboration based on (PAS 1192-2:2013), (NBS BIM Toolkit, 2018)].

Stage	According to	Name of the stage	Goals
STAGE 0	PAS 1192-2:2013	Strategy	Employer's information requirements (EIR), BIMexecution plan (BEP), Master informationdelivery plan (MIDP).
	RIBA Plan of Work 2013	Strategic Definition	Identify client's Business Case and Strategic Brief and other core project requirements.
STAGE 1	PAS 1192-2:2013	Brief	Project brief and procurement strategy. Model: Model information communicating the brief, performance requirements, performance benchmarks and site constraints.
	RIBA Plan of Work 2013	Preparation and Brief	Develop Project Objectives, including Quality Objectives and Project Outcomes, Sustainability Aspirations, Project Budget, other parameters or constraints and develop Initial Project Brief. Undertake Feasibility Studies and review of Site Information.
STAGE 2	PAS 1192-2:2013	Concept	Refined project brief and concept approval. Model: Models which communicate the initial response to the brief, aesthetic intent and outline performance requirements. The model can be used for early design development, analysis and co-ordination. Model content is not fixed and may be subject to further design development. The model can be used for co-ordination, sequencing and estimating purposes.
	RIBA Plan of Work 2013	Concept Design	Prepare Concept Design, including outline proposals for structural design, building services systems, outline specifications and preliminary Cost Information along with relevant Project Strategies in accordance with Design Programme. Agree alterations to brief and issue Final Project Brief.
STAGE 3	PAS 1192-2:2013	Definition	Approval of coordinated developed design. Model: A dimensionally correct and coordinated model which communicates the response to the brief, aesthetic intent and some performance information that can be used for analysis, design development and early contractor engagement. The model can be used for coordination, sequencing and estimating purposes including the agreement of a first stage target price.

(Continued)

671

Table 3. (Continued)

Stage	According to	Name of the stage	Goals
	RIBA Plan of Work 2013	Developed Design	Prepare Developed Design, including coordinated and updated proposals for structural design, building services systems, outline specifications, Cost Information and Project Strategies in accordance with Design Programme.
STAGE 4	PAS 1192-2:2013	Design	Integrated production information. Model: A dimensionally correct and coordinated model that can be used to verify compliance with regulatory requirements. The model can be used as the start point for the incorporation of specialist contractor design models and can include information that can be used for fabrication, coordination, sequencing and estimating purposes, including the agreement of a target price/ guaranteed maximum price.
	RIBA Plan of Work 2013	Technical Design	Prepare Technical Design in accordance with Design Responsibility Matrix and Project Strategies to include all architectural, structural and building services information, specialist subcontractor design and specifications, in accordance with Design Programme.
STAGE 5	PAS 1192-2:2013	Build and commission	Integrated production information. Complete fabrication and manufacturing details, system and element verification, operation and maintenance information Modify to represent as installed model with all associated data references. Model: An accurate model of the asset before and during construction incorporating coordinated specialist subcontract design models and associated model attributes. The model can be used for sequencing of installation and capture of as-installed information.
	RIBA Plan of Work 2013	Construction	Offsite manufacturing and onsite Construction in accordance with Construction Programme and resolution of Design Queries from site as they arise.
STAGE 6	PAS 1192-2:2013	Handover and closeout	As constructed systems, operation and maintenance information Agreed Final Account Building Log Book Information gathered as key elements are completed to feed installation. Information for the later packages.Model: An accurate record of the asset as a constructed at handover, including all information required for operation and maintenance.
	RIBA Plan of Work 2013	Handover and Close Out	Handover of building and conclusion of Building Contract.
STAGE 7	PAS 1192-2:2013	Operation	Agreed final account In use performance compared against Project Brief Project process feedback: risk, procurement, information management, Soft Landings.Model: An updated record of the asset at a fixed point in time incorporating any major changes made since handover, including performance and condition data and all information required for operation and maintenance The full content will be available in the yet to be published PAS 1192-3.
	RIBA Plan of Work 2013	In Use	Undertake In Use services in accordance with Schedule of Services.

The influence of time on the construction process in relation to the cost of design changes and the ability to affect cost and functional capabilities have been presented on Figure 2, according to (CURT, 2004).

The more detailed the model is at the design stage, the lower the probability of changes resulting from design mistakes at the construction site. According to (Plebankiewicz, Mitera, 2017) design mistakes are the most common and the most expensive source of waste produced by construction sites in Poland.

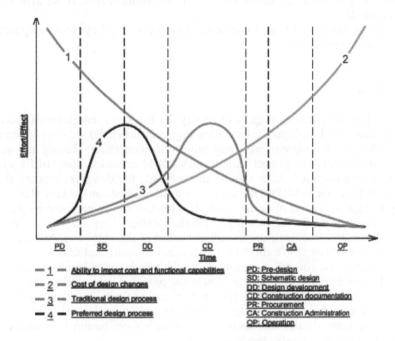

Figure 1. MacLeamy's time-effort distribution curve (CURT, 2004).

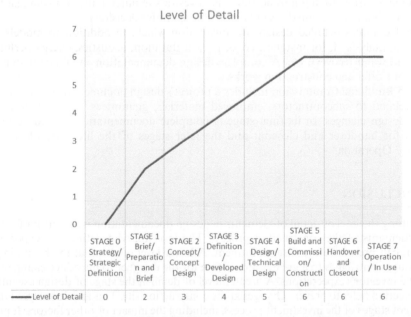

Figure 2. Proposal of levels of detail for individual stages of construction projects.

3 RESULTS

Taking into account the goals to be achieved at stages of the life cycle of a building according to PAS 1192-2: 3013 and the **RIBA** Plan of Work 2013 presented in Table 3, the impact on the cost of changes related to the stage of a project, referring to MacLeamy's curves in Figure 2 and the motivation to introduce changes by the participants of the design process, the authors proposed Levels of Detail for individual stages of a construction investment. The authors adopted the characterisation of **LOD** according to AEC (UK) **BIM** Technology Protocol (Table 1).

A proposal of Levels of Detail for individual stages of a real estate development project has been presented in Figure 2.

4 DISCUSSION

The main task of the Strategy stage is to specify the Investor's requirements. According to British guidelines, the Employer's information requirements (EIR) are a set of requirements that should be met by designers preparing project documentation. Taking into account its goals and requirements, the project team creates a **BIM** execution plan (BEP), which is an attachment to the contract. After signing the contract, the designers prepare the Master information delivery plan (MIDP) for a more detailed information delivery plan.

The Brief stage is the start of design work. It is worth having a prepared model to consult the project assumptions during meetings with the investor. They should not be advanced, because only sensitive fragments are usually consulted at this stage and their change should not be connected with excessive effort for the designer to create and adjust.

The next stage is the concept stage and it should include quite detailed data on elements important for the investor, necessary to assess the proper course of the project's path.The project should enable variant analysis, approximate cost analyses and other, e.g.the estimated duration of the construction investment, preliminary energy or acoustic analyses. It should not feature too many details, increasing the designer's mobilisation to introduce changes according to reasonable remarks by the investor.

At stage 3 Definition design documentation should be prepared, realistically reflecting the designer's vision, taking into account the investor's requirements, and being sufficiently detailed to obtain the required permits and administrative decisions.

Stage 4 contains detailed design documentation which, in addition to models, should include an analysis of, for instance, costs, project duration, acoustics, energy performance, energy and water recovery, etc. A complete design documentation should contribute to the contractor's offer and construction works.

Stage 5 Build and Commission includes a project's design documentation with additional details related to subcontractors, employed materials, guarantees, etc. and, if necessary, applied design changes. In its final stage, a complete documentation should be prepared, required for handover and closeout and the later stages of the life cycle of the facility, i.e. stage 7 Operation.

5 CONCLUSION

Determining the level of detail of elements in design documentation is important for the various participants of the construction process. The investor, who defines the expected LOD early on, can prevent future disappointment. A lower level of detail at the beginning of the design stage contributes to a greater motivation for the designer to introduce changes according to the investor's expectations. A greater level of detail at the stage of design execution and clash detection helps to prevent changes at the construction site. This article analysed the LOD at different stages of the investment process, including the impact of other factors. It proposed the following levels of detail (LOD) for the individual stages of the construction process:

Stage 0 Strategy:-, Stage 1 Brief: LOD2 (general model with details mainly allowing identification of elements), Stage 2 Concept: LOD3 (general model allowing identification of the type of the most important elements and their primary material),Stage 3 Definition: LOD4 (a more accurate model distinguishing the components of the model and the right dimensions of important elements), Stage 4 Design: LOD5 (detailed model, proper dimensions, listing the types of components and materials that they consist of), Stage 5 Build and commission: LOD6 (detailed model expanded with information from subcontractors, i.e. manufacturer, warranty, etc.), Stage 6 Handover and closeout and Stage 7 Operation: LOD 6.

REFERENCES

AEC (UK) BIM Technology Version 2.1.1. 2015.
CURT. The Construction Users Roundtable. *WP 1202 Collaboration, Integrated Information, and the Project Lifecycle in Building Design, Construction and Operation.* 2004.
Eastman C., Teicholz P., Sacks R., Liston K. *BIM handbook: a guide to building information modeling for owners, managers, designers, engineers and contractors.* New Jersey: John Wiley and Sons Inc., 2009.
Howard R., Björk B.C. "Building information modeling, expert's views on standardization and industry deployment." *Advanced Engineering Informatics*, 2008: 271–280.
Lu W., Fung A., Peng Y., Liang C., Rowlinson S. "Cost-benefit analysis of Building Information Modeling implementation in building projects through demystification of time-effort distribution curves." *Building and Environment*, 2014: 317–327.
NBS BIM Toolkit. *https://toolkit.thenbs.com/.* 26 07 2018.
PAS 1192-2:2013. *Specification for information management for the capital/delivery phase of construction projects using building information modelling.* BSI Standards Limited, 2013.
Plebankiewicz E., Mitera E. "The wastes on construction site and in production." *Acta Scientiarum Polonorum. Architectura*, 2017: 69–76.
RIBA *Plan of Work 2013.* London: RIBA, 2013.

Part C
Geodesy, surveying and mapping
Roads, bridges and geotechnics

Advances and Trends in Engineering Sciences and Technologies III – Al Ali & Platko (Eds)
© *2019 Taylor & Francis Group, London, ISBN 978-0-367-07509-5*

3D model of the historic Vltava River valley in the area of Slapy Reservoir

J. Cajthaml, P. Tobiáš & D. Kratochvílová
Department of Geomatics, Faculty of Civil Engineering, Czech Technical University in Prague, Czech Republic

ABSTRACT: The article deals with the creation of a 3D model of the historical Vltava valley from old maps within the project of the Ministry of Culture of the Czech Republic. The Vltava, probably the most popular Czech river, has been greatly influenced by the construction of dams in the middle of the 20th century. From old maps, it is possible to reconstruct the historic valley even in the areas flooded with water. In the testing area of the Slapy Reservoir, the possibility of automatic vectorization of contour lines from the State Map 1:5000-derived (SMO-5) and the creation of a 3D model in ArcGIS software was tested. Because we are working with a very interesting cultural and historical site, the goal of the project is not only to create the model, but also to present it in a web environment for broad public audience. Together with the 3D model, it is possible to present flooded areas, including lost settlements. Thanks to rich photographic documentation that can be linked to the created model, it will be possible to create a complex information system of the historic Vltava River.

1 INTRODUCTION

1.1 *Aim of our project*

The Vltava, as the longest and most famous river in the Czech Republic, is a very interesting object of its history. Because the river and its surroundings were greatly influenced by the construction of dams and water reservoirs, we decided to focus on the area flooded with water of the reservoirs in our research. In these areas, both settlements and interesting natural monuments have disappeared. The aim of our work is to reconstruct the original valley of the Vltava River before the construction of the dams. Our motivation is in making information available to the general public. Based on the 3D model of the valley, the entire information system will be created, which will allow anyone to inspect the history of the river and social and cultural activities in the river neighborhood. See more about our project in Janata & Cajthaml (2018).

1.2 *Vltava course*

Vltava is the longest river in the Czech Republic with a length of 430.2 km. It originates in Šumava, near the village Černý Kříž, as the confluence of the Teplá Vltava River and the Studená Vltava River. It flows through Český Krumlov, České Budějovice and Prague, and flows from the left into the Labe River in Mělník. Map of the course of Vltava can be seen in Figure 1. The set of hydrotechnic works on the Vltava River is called the Vltava Cascade. There are 9 dams, the first of which were built in the 1930s. The Vltava Cascade includes a dam reserving the largest volume of water from the Czech reservoirs (Orlík) as well as the largest dam in the surface area (Lipno). The Hydroelectric power plants in cascade dams produce an electrical output of up to 750 MW. The parameters of individual waterworks are clearly shown in Table 1.

Figure 1.   Course of Vltava River in the Czech Republic.

Table 1.   Overview of the Vltava Cascade.

River km	Dam name	Construction years	Altitude m	Area sq. km	Max. depth m	Volume Thousands of m³
329.540	Lipno I	1952–1959	725.6	48.7	21.5	306000
319.120	Lipno II	1952–1959	563.4	0.33	11.5	1685
210.390	Hněvkovice	1986–1992	370.1	2.68	27	21100
200.405	Kořensko*	1986–1991	353.6	–	–	–
144.700	Orlík	1954–1966	353.6	27.3	74	720000
134.730	Kamýk	1956–1962	284.6	1.95	17	12800
91.694	Slapy	1949–1955	270.6	13.92	58	270000
84.440	Štěchovice	1937–1945	219.4	1.14	22.5	11200
71.325	Vrané	1930–1936	200.1	2.51	9.7	11100

*Kořensko is part of Orlík reservoir when full.

1.3   Brief history of Vltava works

The Vltava River has long been used as an important waterway. As early as the 19th century, the largest boulders and rocks were removed from the Vltava River, new passes were built on the weirs, and waterfront walls were built. Regular boat transport has been established only in the section from České Budějovice to Týn nad Vltavou. Wooden rafts were preferred by Týn nad Vltavou. For a long time, the biggest obstacle for boats has been the Svatojánské Proudy rapids (formerly only Proudy or Slapy). Since the 18th century, the area of Svatojánské Proudy has been modified, and since the 19th century the ideas for building a water dams system with locks have begun. At the turn of the 19th and 20th centuries, the boat transport in the area of middle Vltava River was very much discussed. The topic of dams' creation was re-established after the First World War and the rise of Czechoslovakia. Many drafts of dams projects with different heights, types of dikes intended for different river sites, have been submitted. The first dam of the Vltava Cascade was built in Vrané. The construction started in 1930 and was put into operation six years later. The construction of Štěchovice dam took place predominantly during World War II and it was put into operation in 1944. The others were built between 1949–1992.

1.4   Testing area

The reservoir Slapy was chosen as the test area of our research. The dam was built in 1949–1955, less than three kilometers from the village of Slapy. Most of its plans were developed before the Second World War. Finally, the project of Libor Záruba-Pfefferman was chosen. It was based on the production of electric energy, but also on the protection of the lower parts of Vltava neighborhood against floods. The originality of the project lies in the fact that the dike's body contains the power plant's engine room, office room, and other service rooms, with a road running over the dam. Several villages and settlements were flooded after the reservoir was filled up. The construction was completed in 1955 with the power plant being put into operation. The finished dam was 68 meters high (over the foundations) and 260 meters long. Due to time and financial reasons, a vertical boat lift was not implemented, making it an unbeatable barrier for rafts and boats navigation. All boats have to be brought to the Slapy reservoir by road today.

2 METHODOLOGY

2.1 Old maps usage

The reconstruction of the flooded valley of the Vltava River is possible when using old maps. Underwater mapping with a sonar is virtually unrealistic due to financial and time costs. Unfortunately, altimetric data (meaning maps) that covers the entire Vltava River area are available since the 1950s. However, for a large part of the territory, it is sufficient because most of the dams were built after the maps were issued. For Vrané and Štěchovice dams, it will be necessary to find older maps (if there are any) in archives.

State Map 1:5000-derived (SMO-5) was produced starting in 1950 by reprocessing from existing map data. Planimetric part was derived from cadastral maps and altimetric part from existing topographic maps. The map continuously covered the entire territory of the Czech Republic. For its creation, it was used the Krovak projection with Czech national coordinate system (S-JTSK). Altitudes were reported in the Baltic altitude system after alignment. Each map sheet displayed an area of 2.5 × 2 km. On the map there were displayed settlements, transport network, water, forests, administrative boundaries, etc. in black color. The elevation was represented by contours, elevation points, if necessary by technical or topographical hatching, and signs of terrain and rocks in brown color (see Figure 2).

The accuracy of the SMO-5 is highly dependent on the accuracy and quality of the source maps used. Although the creation of SMO-5 maps was intended to be only a temporary solution, they are widely used today, despite the fact that their content does not meet the current map quality requirements. The main shortcomings are the mismatch of the position with the current state and its inhomogeneity in terms of accuracy and the fact that basic contour interval is not uniform and does not adequately represent the height mainly in the flat territories. The first issue of the SMO-5 maps was scanned by the Land Survey Office in Prague and thus the data is available in digital form (JPEG files).

Figure 2.   Sample of SMO-5 map.

Since SMO-5 has a clearly defined map sheet design in S-JTSK, it is very easy to georeference maps using four map corners and their coordinates. When using a projective transformation, the non-residual method is applied and neighboring maps are closely linked to each other. Thus, it is possible to create a seamless mosaic of SMO-5 of the whole area of interest.

2.2 Modeling of 3D valley

To create a 3D model of the valley, it is crucial to obtain vector spatial data, in our case contours, elevation points and hatching. The automatic vectorization of contours has been solved in a number of publications, but this is a complex problem. The source raster data must first be edited and cleaned. This may then be followed by automatic or semi-automatic vectorization. Finally, the appropriate attributes need to be assigned to vector elements, data should be consolidated and checked. Pacina & Havlicek (2015) describe an example of creating 3D model from digitized contours.

3 RESULTS

3.1 Cleaning map sheets

For our model territory, we have received 20 map sheets of SMO-5. The data contains a black (or rather gray) color planimetric part and a brown color altimetry. Unfortunately, the data is in JPEG format and therefore there are places in which many drawings are partially blurred. It is very difficult to separate the individual graphic elements from each other. In the first phase, it was necessary to reduce the color depth of the data. We have tested various color depth reduction settings so that the brown color can be separated as much as possible. The output of our tests was to reduce the color depth to 16 colors (4 bits). Then we chose a specific brown color out of 16, we separated it from others, and created black and white raster containing only altimetric data. Of course, some issues have also appeared in the process. On some map sheets there were separated not only altimetric features but some parts of the planimetric part as well (the gray color was due to JPEG compression approaching brown). However, we were satisfied with the results of this part. False altimetry drawing was cleared in the next phase. Black and white rasters with altimetry entered the ArcScan software, which, according to available resources, seemed to be the most appropriate. ArcScan is an extension of ArcGIS for Desktop from Esri, and work with it is quite intuitive. First, the data was cleared—where it was necessary to manually remove the false drawing. The Select Connected Cells tool was used to quickly remove some marginal notes. Morphological operations (Erosion, Dilation, Opening, Closing) provide faster map sheet cleaning. Anyway, each map sheet has different needs so that identical operations could not be used.

3.2 Vectorization

To store vectorized elements, you first need to create an empty line feature class. Then, you can start the features vectorization using Generate Features function automatically. Despite testing different input parameters, we were not fully satisfied with the results. Created features are generally divided into a number of parts and need to be merged manually. If we use semi-automatic vectorization, Vectorization Trace and Vectorization Trace Between Points functions are available. Their use in our project was omitted because of the poor graphic quality of the input data. The last option is to perform vectorization manually. Here, the user has a clear control over data creation. However, this method is most time consuming. The result of our research was a workflow in which data are automatically vectorized at first. In the next step, features are manually cleaned, merged and feature attribute tables are filled with appropriate values (see Figure 3) Following this approach, we have earned the feature class with clean contours throughout the research area. Separately, elevation points have been vectorized into point feature class, using the manual method.

Figure 3. Vectorized data (automatic method – upper left; cleaning – upper right; final merged – down).

3.3 *Issues when converting data from raster to vector*

We've identified the following issues when converting raster altimetry from old maps into vector form:

- Quality of input raster data (high compression, insufficient resolution),
- Inappropriate combination of cartographic visualizing methods of planimetry and altimetry, which leads to the inability to separate these layers automatically
- Errors in the source map (contours and elevation points mismatch, contour elevation errors),
- Rock areas expressed by hatches or drawings that do not allow to read altitude information.

3.4 *3D model creation*

After creating vector line and point elevation data, our aim was to create a 3D model of the historic Vltava valley. This model can be in vector form as TIN (triangulated irregular network), but more often it is used in GIS as a DEM (digital elevation model) in the form of a raster grid. There are a number of publications dealing with creating such model. Most GIS software includes tools for creating DEM. Algorithms work either at the point level (lines are replaced by points at vertices) and subsequent interpolation of point data into a grid, or there are more complex algorithms working with hydrologically correct models, distinguishing individual types of input data. Interpolation methods use natural neighbors, IDW (inverse distance weighted) or more complex mathematical surfaces, such as splines. See their comparison in Arun (2013). In our case, we have stayed with ArcGIS for Desktop, which includes Topo To Raster routine. Topo To Raster tool is an interpolation method specifically designed for the creation of hydrologically correct digital elevation model (DEM). It is based on the ANUDEM program developed by Michael Hutchinson. Applications of DEM to environmental modeling are discussed in Hutchinson & Gallant (2000) and Hutchinson (2008). Input data for our calculation came not only from the contours and elevation points, but also from the river shoreline and from the polygons of the rocky areas without contours. After testing the input data and output parameters settings, DEM was created with a 2-meters spatial resolution. The resulting DEM can be visualized as a 3D textured surface. In our case, it is possible to use SMO-5 map, or black and white orthophoto from the 1950s (see Figure 4).

Figure 4. 3D model of historic Vltava valley—SMO-5 as a texture.

4 CONCLUSIONS

Old maps are a valuable source of information when modeling historical landscape. In the Czech Republic there are available SMO-5 maps from 1950s, which display not only planimetric, but altimetric features as well. Using these maps it is possible to create a 3D model of historical areas, which changed completely as a result of dams' creation.

We created workflow scenario to process these maps. It includes graphic data pre-processing, automatic vectorization, cleaning the data, and creating vector feature class of contours. Along with elevation points, these contours are used for 3D modeling of the historical valley of Vltava River in the Slapy reservoir testing area. This model could be used for further modeling of lost villages and settlements in the river's neighborhood.

This work was supported by the Czech ministry of culture by the NAKI programme "Vltava—transformation of historical landscape as a result of floods, dams creation and land-use changes along with cultural and social activities in the river neighborhood" no. DG18P02OVV037.

REFERENCES

Arun, P.V. 2013. A comparative analysis of different DEM interpolation methods. *The Egyptian Journal of Remote Sensing and Space Science* 16(2): 133–139.
Hutchinson, M.F. 2008. Adding the Z-dimension. In Wilson, J.P. & Fotheringham, A.S. (eds.), *Handbook of Geographic Information Science*, Oxford: Blackwell.
Hutchinson, M.F. & Gallant, J.C. 2000. Digital elevation models and representation of terrain shape. In: J.P. Wilson and J.C. Gallant (eds.), *Terrain Analysis*. New York: Wiley.
Janata, T. & Cajthaml, J. 2018. Vltava—transformation of historical landscape along with cultural and socioeconomic activities in the river neighbourhood. In *Proceedings 13th ICA Conference Digital Approaches to Cartographic Heritage, Madrid, 18–20 April 2018.*
Pacina, J. & Havlicek, J. 2015. A vanished settlement in the Ore Mountains—the creation of 3D models. In Al Ali, M. & Platko, P. (eds.), *Advances and Trends in Engineering Sciences and Technologies, International Conference on Engineering Sciences and Technologies (ESaT), Kosice, 27–29 May 2015.*

Advances and Trends in Engineering Sciences and Technologies III – Al Ali & Platko (Eds)
© 2019 Taylor & Francis Group, London, ISBN 978-0-367-07509-5

Solar radiation analysis based on the mesh representation of point cloud data using the marching cubes algorithm

J. Faixová Chalachanová
Faculty of Civil Engineering, Slovak University of Technology in Bratislava, Slovak Republic

K. Ťapajnová
Infra Services a. s., Slovak Republic

ABSTRACT: The paper deals with the solar radiation analysis of a spruce tree stand based on its mesh representation optimized using the marching cubes algorithm. The analyzed mesh representation was obtained by the method of terrestrial laser scanning. The default method of the mesh creation was optimized using the marching cubes algorithm to specify the optimal cell size to obtain new regular data structure from the raw point cloud data. This optimized representation was used to calculate the solar exposure of the spruce tree stand. The results are the basis for detailed monitoring of the spruce tree's health and microclimate on the level of the individual tree branches as well as on the level of the whole spruce tree stand.

1   INTRODUCTION

In forestry, the aim is to obtain the most credible representation of trees in a vector structure that is usable for the varied analysis using Geographical Information System (GIS), respectively Computer Aided Design (CAD) systems. But the tree represents a complex structure where positions x, y correspond to multiple height values z. Like in the other fields, also in forestry the laser scanning method (terrestrial or airborne) is the popular method for collecting data about trees and other vegetation. The output from laser scanning is dataset with all measured points, including unwanted objects as well as the missing parts that can occur due to difficult surveying conditions and limits such as poor general visibility, bad possibilities of scanner position localization, etc. Additionally, it is not possible to use classical interpolation methods (Chen et al. 2007), which are commonly implemented in GIS and CAD systems for modeling such complicated structures with missing parts of point cloud data obtained by the laser scanning technology. In this case, the marching cubes algorithm offers much better results (Newman & Yi, 2006).

The issues of optimization of point cloud data for surface representation of objects with complex structure are discussed in (Faixová Chalachanová et al., 2017). The paper deals with the proposal of the process of cell size optimization due to raw point cloud data and computation of new regular data structure for visualization of objects using the marching cubes algorithm. In general, studying the effect of changing the cell size and shape in the marching cubes algorithm and obtaining the surfaces from all variants of the cells produced the results as the most trusted representation of the raw point cloud data using the noncubic cell $20 \times 20 \times 10$ cm. It is the optimal cell size for the purposes of detailed studying of the microclimate of the standard spruce tree in microscale. But the optimal cell size can be changed in context of the purpose of the surface usage and the tree's age. If a tree which is observed is old, so the optimal cell size is $20 \times 20 \times 20$ cm due to overhanging branches where the assumption of their vertical compactness on the level of the branch is false. And, if the observed tree is modelled in macroscale (e.g. for the purposes of predictive modeling or monitoring the regional or global changes), the generalized tree representation (simplified cover) that can be obtained by computing with the cell size $50 \times 50 \times 50$ cm is sufficient.

The impact of natural conditions on the state and development of the spruce tree is one of the bases in the field of forest protection. For studying the microclimate of a tree (e.g. in the context of the bark beetle infestation), the sunlight modelling and solar radiation analysis is essential. The sunlight penetration into the tree stand influences the temperature inside the tree, which is essential for selection of an appropriate tree by the spruce bark beetle (Jakuš et al., 2003). There are several applications and modules that can calculate and visualize the insolation. For raster data the open-source GRASS module r.sun can be used (Hofierka et al., 2017), (Huld, 2017), (Hofierka & Šúri, 2002), or the proprietary ArcGIS module Area Solar Radiation. However, those modules are not suitable for the modelling of solar radiation of such complex structures such as spruce trees because it is necessary to calculate with several values of the height in one vertical position. For vector data the specified module MIXLIGHT can be used (Stadt & Lieffers, 1998), or modules implemented in CAD systems. The continuous surface representation of a tree is the basis for the solar radiation calculation in CAD systems. This representation is usually considered in the form of triangulated mesh, but there are many restrictions due to the default implemented method of triangulation (Delaunay triangulation mostly), which is not suitable for a detailed tree structure on the level of individual branches, and it is not optimized for big point cloud data either.

Therefore, our study is based on the optimization of a tree representation created from the point cloud data obtained by the terrestrial laser scanning method and automating this process using the algorithms for big data preparation and structuring them into the form which is optimal for the marching cubes algorithm. In this article the proposed method (Faixová Chalachanová et al., 2017) was applicated in the solar radiation analysis of a concrete spruce tree stand.

2 MATERIALS AND METHODS

The method of the tree representation from the point cloud data proposed in (Faixová Chalachanová et al., 2017) was applied to data capturing from large area in the Suchá Dolina (the range was about 150×40 m). The raw point cloud data in S-JTSK (Coordinate System of Unified Trigonometric Cadastral Network) was provided by the Institute of Forest Ecology of Slovak Academy of Sciences in Zvolen. The input to the calculation of the representation of the tree stand using the proposed method was textual file with more than 4 million points. So, it was necessary to write the source code in the programming language C with dynamic allocation of the memory for restructuring the original data. The source codes were written for all variants of the cell size. Output from the calculation were files in the format VTK (Visualization ToolKit) which contain the data in the regular grid with the chosen cell size. This is the basis for the tree stand representation using the marching cubes algorithm in the open-source visualization software ParaView. The obtained representation of the tree stand was exported into the format STL (STereoLitography).

Then the calculation of solar radiation was performed in the environment of the software MicroStation. In the process of the solar radiation calculation, it is important to specify parameters such as Time Increment (minutes) – the time step of calculation, Grid Size (meters) – the grid size on which the calculations are realized, Accuracy—the accuracy of the calculation, which considerably affects the length of the calculation, Insolation Source—the source of the insolation, Output—where it is possible to select the format of the solar analysis output, Duration—the calculation period where the user-defined period can be selected (date and time of the beginning and the end of the insolation calculation) or set the standard period (day, week, month, year).

The solar radiation analysis was calculated using Solar Exposure Calculator (Figure 1) at two levels: in a microscale on the part of the tree stand obtained by the marching cubes algorithm, where the calculation was performed using a grid with cell size $20 \times 20 \times 20$ cm (grid size 0,2 m) and in a macroscale on a generalized model of the whole tree stand that was obtained by calculating the grid with cell size $50 \times 50 \times 50$ cm (grid size 0,5 m). The time step was used 10 minutes with the medium accuracy. Insolation source was defined using Weather

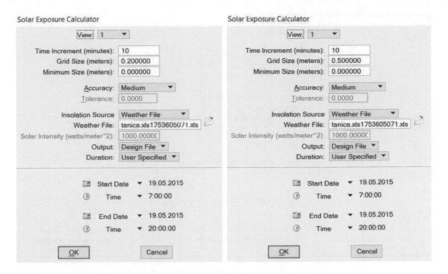

Figure 1. Parameters settings for solar radiation analysis in microscale (right) and macroscale (left).

file from the meteorological station on situ with the information about global radiation, air temperature, humidity and rainfalls.

3 RESULTS

The laser scanning method is used for collecting data in many fields of human interest. Especially in the field of forestry management it is used as a primary method for collecting the data about the tree's positions and representation of the tree's shape, which can be used for different types of environmental analysis (Yu et al., 2013); (Hyyppä et al., 2008). However, surveying conditions in the forest environment are very difficult and can be limited by a problematic visibility or inaccessibility of suitable locations of a scanner position. Due to these limits, the default methods of surface representation implemented in standard software are causing missing parts of data. First, it is necessary to applicate the methods of non-regular raw point cloud data restructuring before the marching cubes algorithm usage. The optimization of data structure was realized using the source codes written in the programming language C. The VTK files were created from an irregular structure of input point cloud data with coordinates x, y, z in TXT format using a source codes in the programming language C. Created VTK files contained data in a regular pattern in multiple variants of the 3D grid cell size with respect to the different scales of its usage. Following visualization of the spruce tree stand was done using VTK format in the open-source software ParaView, which provides the open solution with possibilities of source code development also. Diagram explaining the whole automated process of creation of the tree representation from the raw point cloud data is shown in Figure 2. The difference between the tree representation in according to scale and required level of detail is shown in (Figure 3). The issue of the implementation of created representation into the GIS, respectively CAD systems is solved using the STL exchange format that is supported in the MicroStation software also.

The result of the solar radiation analysis calculated in the MicroStation software was exported to the CSV (Comma-Separated Values) file containing data about the length of exposure (in seconds) of exposed and shadowed parts on the individual triangles representing the surface, the values of direct insolation in Wh/m^2, diffuse insolation in Wh/m^2 and the coordinates of vertices of individual triangles. The visualization of solar radiation analysis in microscale is shown in Figure 4. The visual result of solar radiation analysis in macroscale using the generalized model is shown in Figure 5. The color scale represents the period of insolation of the spruce tree stand during the day in hours.

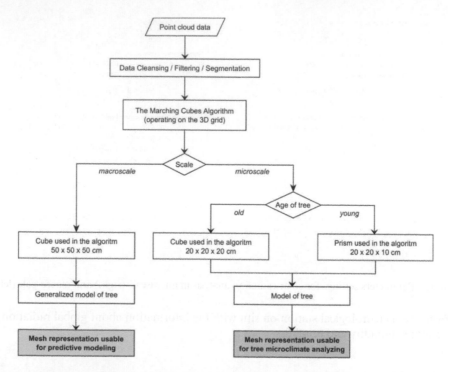

Figure 2.   Diagram of the process of the tree representation created from the point cloud data.

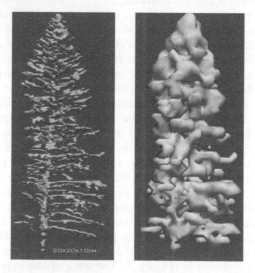

Figure 3.   Representation of the tree in microscale (left) vs macroscale (right).

4   CONCLUSION AND DISCCUSION

The result of our study is the proposal of an automated process of the tree representation created from the raw point cloud data based on the optimization by its restructuralization using the source codes written in the programming language C for several variants of the cell size. The concluding recommendation for the cell sizes of restructured regular grid depends on the following usage of the tree surface created using the marching cubes algorithm. In the microscale (if

Figure 4.   Visualization of solar radiation analysis in microscale.

Figure 5.   Visualization of solar radiation analysis in macroscale.

detailed model of the tree is needed), the recommended cell size is $20 \times 20 \times 10$ cm for a young tree and $20 \times 20 \times 20$ cm for an old tree. In the macroscale (if generalized model of the tree is needed), the recommended cell size is $50 \times 50 \times 50$ cm.

The main advantage of our approach is that it allows to optimize and to accelerate the process of the tree surface obtaining from the laser scanning point cloud data. It can be used in many types of spatial analysis, even in the solar radiation analysis. In the field of forest management, the calculation of solar radiation can be the basis for monitoring of the microclimate and health of an individual tree in microscale, and for predictive modelling in the context of potential attacks of the bark beetles in macroscale either.

ACKNOWLEDGEMENTS

This work was supported by the Grant No. 1/0682/16 of the Grant Agency of Slovak Republic VEGA.

REFERENCES

Chen, Q., Gong, P., Baldocchi, D., Xie, G. 2007. Filtering Airborne Laser Scanning Data with Morphological Methods. Photogrammetric Engineering & Remote Sensing, Number 2, American Society for Photogrammetry and Remote Sensing 2007, pp. 175–185.

Faixová Chalachanová, J., Ďuračiová, R., Ťapajnová, K. Optimization of point cloud data for surface representation of objects with complicated structure. In SGEM 2017. 17th International Multidisciplinary Scientific GeoConference. Volume 17. Informatics, Geoinformatics and Remote Sensing. Sofia: STEF 92 Technology, 2017, pp. 1021--1028. ISBN 978-619-7408-01-0.

Gerhátová, Ľ., Muňko, M., Hroššo, B., Ďuračiová, R., Faixová Chalachanová, J. The main principles of sunlight analysis based on point cloud data. In SGEM 2017. 17th International Multidisciplinary Scientific GeoConference. Volume 17. Informatics, Geoinformatics and Remote Sensing. Sofia: STEF 92 Technology, 2017, pp. 1085--1092. ISBN 978-619-7408-01-0.

Hofierka J., Šúri M., The solar radiation model for Open source GIS: implementation and applications, Open source GIS—GRASS users conference 2002, Trento, Italy 2002.

Hofierka, J., Lacko, M., Zubal, S. Parallelization of interpolation, solar radiation and water flow simulation modules in GRASS GIS using OpenMP. Computers & Geosciences, 107, 2017, pp 20–27.

Huld T. PVMAPS: Software tools and data for the estimation of solar radiation and photovoltaic module performance over large geographical areas. Solar Energy, vol. 142, 2017, pp. 171–181.

Hyyppä J., Hyyppä H., Leckie D., Gougeon F., Yu X., Maltamo M., Review of methods of small footprint airborne laser scanning for extracting forest inventory data in boreal forests, International Journal of Remote Sensing, vol. 29, Issue 5, 2008, pp. 1339–1366.

Jakuš R., Schlyter F., Zhang Q-H., Blaženec M., Vaverčák R., Grodzki W., Brutovský D., Lajzová E., Bengtsson M., Blum Z., Turcáni M., Gregoiré J-C., Overview of development of anti-attractant based technology for spruce protection against Ips typographus: from past failures to future success, Journal of Pest Science, vol. 76, 2003, pp. 89–99.

Newman, T.S. & Yi, H. A survey of the marching cubes algorithm, Computers & Graphics, vol. 30, Issue 5, 2006, pp 854–879.

Stadt K.J., Lieffers V.J., MIXLIGHT: a flexible light transmission model for mixed-species forest stands, final report 1997/98, Alberta Environmental Protection, Manning Diversified Forest Products Research Trust Fund, 1998, 44 p.

Yu X., Liang X., Hyyppä J., Kankare V., Vastaranta M., Holopainen M., Stem biomass estimation based on stem reconstruction from terrestrial laser scanning point clouds, Remote Sensing Letters, vol. 4, Issue 4, 2013, pp 344–353.

Advances and Trends in Engineering Sciences and Technologies III – Al Ali & Platko (Eds)
© 2019 Taylor & Francis Group, London, ISBN 978-0-367-07509-5

Effects of soil variability on bearing capacity of foundations

S. Harabinová, E. Panulinová, E. Kormaníková & K. Kotrasová
Faculty of Civil Engineering, TU-Košice, Košice, Slovakia

ABSTRACT: Soil is considered as a complex material produced by weathering of the solid rock. The strength of soil is a key design parameter in foundation design. The capacity of the foundation to support footing load is given by the soil's bearing capacity, which is a function of its strength parameters. Bearing capacity is the maximum pressure that the soil can support at foundation level without failure. It is the crucial parameter for foundation design. In this paper, the bearing capacity of soil foundation is numerically calculated by changing of shear strength parameters of soil foundation (angle of internal friction of soil and cohesion) and depth of square foundation.

1   INTRODUCTION

Soils consist of grains (mineral grains, rock fragments, etc.) with water and air in the voids between the grains. The water and air contents are readily changed by changes in conditions and location. Soils may be perfectly dry (without water content), fully saturated (without air content), or partly saturated (with both air and water present), respectively. Although the size and shape of the solid (granular) content rarely changes at a given point, they can vary considerably from point to point. Soil, as an engineering material, is not a coherent solid material like steel and concrete, rather it is a particulate material. It is important to understand the significance of particle size, shape and composition, and soil's internal structure or fabric. Geotechnical investigations are performed to obtain information on the physical and mechanical properties of soil and rock. Based on these data the correct way of building foundation may be suggested.

Measurements of some of the soil properties such as strength, compressibility, permeability etc. sometimes may be difficult, time consuming and expensive to obtain. In certain engineering projects, due to budget, site and other constraints, engineers and technologists are unable to carry out detailed and therefore more costly site investigations. Minimum requirements for the extent and content of geotechnical investigations, calculations and construction control check must by establish. The complexity of each geotechnical design shall be identified together with the associated risks. To establish geotechnical design requirements, three Geotechnical Categories, are introduced. The Geotechnical Category 1 should only include small and relatively simple structures with negligible risk for which it is possible to ensure the fundamental requirements to be satisfied on the basis of experience or qualitative geotechnical investigations. Simplified design procedures may be applied for these structures. Geotechnical parameters of soil or rock for design calculations shall be obtained from test results, either directly or through correlation or from other relevant data. Many geotechnical parameters are not constants—they are dependent on stress level and mode of deformation.

In Slovakia, for design structures according to Eurocode 7 (Geotechnical category 1), the geotechnical parameters may be obtained from relevant published data and local and general experience (for example: information from geological maps or available archival geological in-vestigation). Geotechnical parameters of soil are often given only as values in intervals and choosing of the correct value is difficult. The influence of changes of soil geotechnical parameters (angle of internal friction of soil and cohesion) and depth of square foundation on the bearing capacity of soil foundation is shown in this paper.

2 EFFECTS OF SOIL VARIABILITY ON THE BEARING CAPACITY

2.1 *Input parameters of the numerical experiment*

Our experimental study was focused on the city of Kosice. The town is located in the eastern part of Slovakia in the Kosice region (Figure 1), near the borders with Hungary (20 km), Ukraine (80 km) and Poland (90 km) at the crossroads of historical trade routes. The town extends on both banks of the Hornad River. The city center lies at an altitude of 208 meters above sea level. The highest point of the city is the Hradova Hill – 466.1 meters above sea level.

The geological structure is composed predominantly of quaternary sediments (namely sediments of the Hornad River). The quaternary sediments are represented principally by gravel and gravel with silt or clay. On the surface, gravel sediments are covered with silts and clays. In the case of fine soils (silts and clays), the plasticity has the biggest influence on properties. Silts and clays of low or intermediate plasticity (MI, ML or CI, CL) are in this locality. Their consistency is mostly soft to stiff. The values of the geotechnical characteristics for low plasticity clay (CL – Group F6, according to STN 72 1001) are given in the Table 1.

Characteristic values geotechnical parameters of soil using for design structures can be lower, which are less than the most probable values, or upper values, which are greater.

To calculate the bearing capacity of foundation soil was using the whole interval of stress parameters for effective angle of friction $\varphi' = (17-21)°$. Effective cohesion was considered with the values 8, 12 and 16 kPa.

Each geotechnical design shall be verified according to Eurocode 7 (EC 7). EC 7 should be used for all the problems of interaction of structures with the ground (soils and rocks), through foundations or retaining structures. The geotechnical designs of constructions (for Slovakia) are using Design Approach 2. It shall be verified that a limit state of rupture or excessive deformation will not occur with the combination (A1 "+" M1 "+" R2) of sets of partial factors, where A1 are Partial factors on actions (γ_F), M1 are Partial factors for soil parameters (γ_M) and R2 are Partial resistance factors for spread foundations (γ_R), according to EC7, Annex A.

In geotechnical design it shall be verified that no relevant limit state, is exceeded. We have two limit states: 1. Ultimate Limit States (ULS), when considering a limit state of rupture or excessive deformation of a structural element or section of the ground (STR—internal failure or excessive deformation of the structure or structural elements and GEO—failure or excessive deformation of the ground), it have to verify according to Equation (1).

Figure 1.   Map of Slovakia.

Table 1.   The geotechnical parameters of soil.

Properties	Group F6 – CL (Low Plasticity Clay)
Poisson's ratio $\upsilon(-)$	0.40
Unit weight $\gamma(kN/m^3)$	21.0
Total stress parameters – cohesion c_k (kPa)	50
Total stress parameters – angle of friction $\varphi_k(°)$	0
Effective stress parameters – cohesion c_k' (kPa)	8–16
Effective stress parameters – angle of friction φ_k' (°)	17–21

Serviceability limit states (SLS), in the ground or in a structural section, element or connection, it have to verify according to Equation (2).

$$E_d \leq R_d \quad (1)$$

$$S \leq S_{lim} \quad (2)$$

where Ed is the design values of the effects of all the actions (load from construction or ground), Rd is the design values of the corresponding resistance of the ground and s is the settlement of construction and s_{lim} is limiting settlement.

2.2 Bearing capacity

Bearing Capacity is a key design parameter for foundation design. This is the maximum pressure that the soil can support at foundation level without failure. The bearing capacity of soil was calculated according to EC 7.

The following Equation (3) to determine the general bearing capacity R_d is according to Eurocode 7 (Design Approach 2) and according to STN 73 1001. The design bearing capacity of the soil R_d(kPa) for drained conditions is determined according to STN 731001:

$$R_d = (c'_d N_C S_C d_C i_C j_C + q' N_q s_q d_q i_q j_q + \gamma' \frac{B}{2} N_\gamma S_\gamma d_\gamma i_\gamma j_\gamma) / \gamma_R \quad (3)$$

where γ_R is partial factor for a resistance, for Design Approach 2 is $\gamma_R = 1.4$, c_d' is design value of the effective cohesion $c_d' = c_k' \gamma_c$ (kPa), c_k' is characteristic value of the effective cohesion (kPa), γ_c is partial factor for the effective cohesion, for Design Approach 2 is $\gamma_c = 1.0$, q' is the design effective overburden pressure at the level of the foundation base ($q' = \gamma_l D$(kN/m²)), γ_l is the effective unit weight of soil above the base of footing level (kN/m³), D is the embedment depth (m), B is the foundation width (m), L is the foundation length (m), γ' is the design effective weight density of the soil below the foundation level (kN/m³), φ_d is design value of the effective angle ($\varphi_d = \varphi_k' \gamma_\varphi$ (°)), φ_k' is characteristic value of the effective angle (°), γ_φ is partial factor for the effective angle, for Design Approach 2 is $\gamma_\varphi = 1.0$, N_c, N_q, N_γ are the bearing capacity factors (dependent on the design value of effective angle φ_d):

$$N_C = (N_q - 1) cotg\, \varphi_d \quad for \quad \varphi_d > 0 \quad (4)$$

$$N_q = \tan^2\left(45 + \frac{\varphi_d}{2}\right) e^{(\pi \tan \varphi_d)} \quad (5)$$

$$N_\gamma = 1.5(N_q - 1) tan\varphi_d \quad N_\gamma = 2 + \pi \quad for \quad \varphi_d = 0 \quad (6)$$

The shape of foundation factors (s_c, s_q, s_γ), the depth factors for deeper shallow foundations (d_c, d_q, d_γ), the inclination factors of the vertical load (i_c, i_q, i_γ) and the inclination factors of the terrain surface (j_c, j_q, j_γ) may be calculated according to following Equations (7–10):

$$S_C = 1 + 0.2\frac{B}{L} S_q = 1 + \frac{B}{L} sin\varphi_d\, S_\gamma = 1 - 0.3\frac{B}{L} \quad (7)$$

$$d_c = 1 + 0.1\sqrt{\frac{D}{B}}, \quad d_q = 1 + 0.1\sqrt{\frac{D}{B}} \sin 2\varphi_d\, d_\gamma = 1 \quad (8)$$

$$i_c = i_q = i_r = (1 - \tan\theta)^2 \quad (9)$$

$$j_q = j_r = (1 - \tan\beta)^2\, j_c = j_q - \frac{1 - j_q}{N_c \tan \varphi_d} \quad (10)$$

where θ is the angle of deflection of the resultant force from the vertical (°). For $\theta > 30°$ is progressing individually. β is the inclination angle of the terrain from the horizontal (°).

The bearing capacity R_d is the ultimate gross bearing capacity, which is the ultimate stress value that the soil can carry at the base of the footing level. It is the ultimate bearing capacity, which is the maximum carrying stress level at the ground surface level. The bearing capacity has to be calculated on based of correct shear parameters of soil, since there may be a failure. The following text discusses the different shear strength parameters for low plasticity clay (CL), group F6 and their effect on the soil bearing capacity.

The bearing capacity was calculated for a square foundation ($B \times L = 1$ m $\times$ 1 m) which based on the cohesive soil (CL) on the depth ($D = 1.4$ m, 1.6 m to 1.8 m). The values of the geotechnical characteristics for low plasticity clay (CL), group F6 are given in the Table 1.

2.3 *Experimental results and discussion*

The design of the cohesion soils bearing capacity for drained conditions was determined according to relations (3–10). The bearing capacity was calculated for a square foundation which based on the cohesive soil with low plasticity, group F6.

Different values of cohesion $c_k' = (8, 12, 16)$ kPa and of frictional angle $\varphi_k' = (17–21)°$ in calculation of bearing capacity was used. The impact of shear strength parameters on bearing capacity for drained soil according to the embedment depth is shown in Figures 2 and 3.

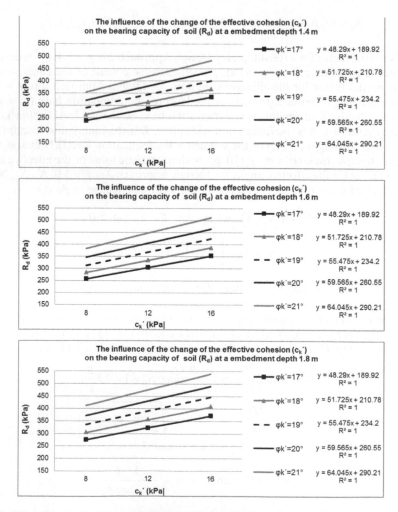

Figure 2. Bearing capacity (R_d) depending on the change of effective cohesion and the embedment depth.

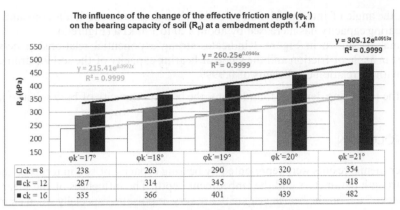

	φk'=17°	φk'=18°	φk'=19°	φk'=20°	φk'=21°
□ck = 8	238	263	290	320	354
▣ck = 12	287	314	345	380	418
■ck = 16	335	366	401	439	482

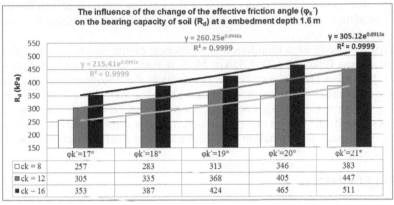

	φk'=17°	φk'=18°	φk'=19°	φk'=20°	φk'=21°
□ck = 8	257	283	313	346	383
▣ck = 12	305	335	368	405	447
■ck – 16	353	387	424	465	511

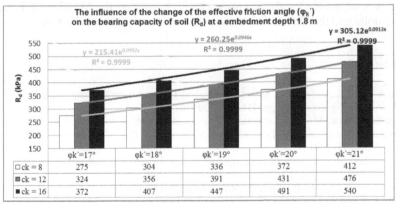

	φk'=17°	φk'=18°	φk'=19°	φk'=20°	φk'=21°
□ck = 8	275	304	336	372	412
▣ck = 12	324	356	391	431	476
■ck = 16	372	407	447	491	540

Figure 3.   Bearing capacity (R_d) depending on the change of effective frictional angle and the embedment depth.

It is seen from results, that:

– Increasing the angle of internal friction (as well as increasing cohesion) causes an increase in bearing capacity of the subsoil. The influence of shear strength parameters on bearing capacity for drained cohesive soil is shown in Figures 2 and 3.
– The change in soil cohesion for the constant value of the angle of internal friction (Figure 2) causes a slower increase in the bearing capacity of subsoil, only about 100 to 130 kPa.
– The analysis for drained conditions shown in Figure 2 indicates strong linear relationship between the change in cohesion and the value of bearing capacity of the soil—correlation coefficients R = 1.0.

- When the angle of internal friction of soil is changing and cohesion is constant, the base soil bearing capacity has increased by about 120 to 170 kPa (Figure 3).
- The analysis shown in Figure 3 indicates strong exponential relationship between the change in angle of friction and the value of bearing capacity of the soil for drained conditions with correlation coefficients of 0.9999.

With the increasing depth of foundations, the bearing capacity of base soil also increases.

3 CONCLUSION

The experimental results were represented by bearing capacity of soil in terms of the change of soil strength parameters and the embedment depth. The influences of the shear strength parameters of soil on the bearing capacity are very important, especially when changing the angle of internal friction. The correct value of the bearing capacity of soil for the optimal design of foundation without failure may be calculated after determining the correct value effective angle of friction.

The bearing capacity of a foundation is a function of the soil shear strength (Namdar, A. & Khodashenas Pelko, M., 2009, Dixit, M. S. & Patil, K. A., 2010,VandenBerghe, G. & Van Daele, M., 2006).The ultimate bearing capacity is the maximum pressure that the soil can support at foundation level without failure. The reliability of the input data is the basic prerequisite for the optimal design of foundation without failure (Kralik, J. & Simonovic, M., 1994, Kuklík, P., 2011). The most important input for durable and reliable design of the foundation is the exact determination of effective angle of internal friction of soil.

ACKNOWLEDGMENT

This work was supported by the Scientific Grant Agency of the Ministry of Education of Slovak Republic and the Slovak Academy of Sciences under Project VEGA 1/0477/15.

REFERENCES

Dixit, M.S. & Patil, K.A., Study of Effect of Different Parameter on Bearing Capacity of Soil, *Indian Geotechnical Society, GEOTID*, pp. 431–005. (2010).

Kralik, J. & Simonovic, M., Elasto-plastic analysis of deformation soil body with 3D-finite and infinite elements, *Geomechanics 93*, Proc. conference, (Editor: Z. Rakowski, Rotterdam, Brookfield, VT: Balkema), 229–232 (1994).

Kuklík, P., Preconsolidation, structural strength of soil, and its effect on subsoil upper structure interaction, *Engineering Structures* 33 (2011) 1195–1204.

Map of Slovakia, 08.07.2018. [Online]. Available: https://mapa.zoznam.sk/.

Namdar, A. & Khodashenas Pelko, M. Numerical analysis of soil bearing capacity by changing soil characteristics, *Fratturaed Integrità Strutturale*, 10, pp. 38–42. (2009).

STN 72 1001 *Classification of soils and rocks*, 2010 (in Slovak).

STN 73 1001 *Geotechnical structures. Foundation*, 2010 (in Slovak).

STN EN 1997-1, Eurocode 7, *Geotechnical design. Part 1:* General rules. 2005.

Vanden Berghe, G. & Van Daele, M., Exponentially—fitted Störmer/Verlet methods, *Journal of Numerical Analysis, Industrial and Applied Mathematics (JNAIAM)*, vol. 1, no.3, pp. 241–255. (2006).

Advances and Trends in Engineering Sciences and Technologies III – Al Ali & Platko (Eds)
© *2019 Taylor & Francis Group, London, ISBN 978-0-367-07509-5*

Possibilities of displaying the temporal component of data in Esri Story Maps

T. Janata, J. Cajthaml & J. Krejčí
Department of Geomatics, FCE, Czech Technical University in Prague, Prague, Czech Republic

ABSTRACT: The article deals with issues of displaying the temporal component of data in web map applications, specifically in Esri Story Maps. The testing of time display capabilities was performed on existing maps from the Academic Atlas of Czech History. The original data, initially intended only for the printed atlas, has been converted to Esri Story Maps templates. In addition, the individual application editing options are described to suit their purpose, for example the Time slider display, the use of bookmarks for different periods of time, or animations showing the development of the area over time. The testing was carried out with regard to the Czech Historical Atlas project of the Ministry of Culture of the Czech Republic, where the tested technology, considering results achieved, would be used for a wide range of maps.

1 INTRODUCTION

The rich history of the Czech Lands and the wider Central European region has been and continues to be the subject of scientific research by historians, art historians, individual independent researchers and the general public. Significant historical events have been regularly (and with varying degree of accuracy and rightness) captured on maps and in periodical publications where maps were often included.

In the second half of the second millennium, atlases, whose domain was foremost the twentieth century, were increasingly used as a mean of giving evidence of historical events. At the end of the twentieth century, with the development of IT and digital information recording methods, partially electronic atlases first published on digital media (CD-ROMs, etc.) began to grow as stand-alone applications. These were later becoming (due to the onset of the Internet, development of graphical formats and network tools rounded off with cloud solutions) atlases making use of the Internet, web services, JavaScript, HTML5 and other options, either supplemented by printed publications of the same (or more often with only a sample) content or being released as completely digital. This trend is still in progress at the time of release of this paper.

The publishing of electronic atlases began at the end of the 1980s (Kraak & Brown 2003) and their number was rapidly expanding. The cradle of digital atlases was represented by the USA, but in the 1990s and early 21st century a number of digital atlases were also created in most Western European countries, in Canada, Australia, Japan, Brazil and other countries.

In the Czech environment, the onset of digital technologies in the preparation of atlas works was noticeably slower and the first works originated with more than a decade of delay compared to the world cartography. Probably the last major printed historical atlas published in the Czech lands was the Academic Atlas of the Czech History (Semotanová et al. 2014), released in 2014 (Figure 1). It was most likely also one of the last extensive professional atlas publications that were published as purely printed, without an accompanying digital content. A partly successor project, which is, in printed form, mainly focusing on the Czech Lands and the Central European area in the 20th century, is an atlas being created with the draft name of the Czech Historical Atlas, which was already introduced by (Havlíček et al. 2018). A great innovation of this work is an electronic map portal, which brings a large number of maps originally intended for the Academic Atlas of the Czech History converted into

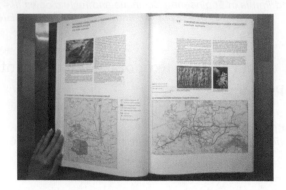

Figure 1.   The academic Atlas of the Czech History.

an electronic form, along with dozens of maps newly created based on the research of the Institute of History of the Academy of Sciences of the Czech Republic.

One of the fields where electronic maps and map applications can help better understand the topic is a map with an important temporal information—genetic, reconstruction map etc.

2   OVERVIEW OF AVAILABLE TECHNOLOGIES

2.1   *Open-source and proprietary technologies*

Mainly due to the rapid development of Web technology and expanding broadband Internet coverage there was a natural choice to create historiographical maps via web map portal, not using obsolete and outdated form of electronic stand-alone application. Especially relatively rapidly changing, emerging and fading out technologies have resulted in the fact that it is possible to relatively easy keep going applications created in the web environment thanks to substitution and updating the technology used, but this is difficult for once already designed and to the users distributed programme mapping applications.

Several technologies, which represent only a fraction of all those available for the preparation of web cartographic works, were considered in the preparation of the web map portal as a whole (Suk 2017), partly also (Kladivová 2018).

OpenLayers is an open JavaScript library, the main purpose of which is to display geographic information in a web browser environment. The main advantage is the ability to load a large amount of data formats and a wide community of users developing additional expanding content. The Leaflet is a similar JavaScript library, focused on creating interactive maps with a strong support for mobile devices. It focuses on the flawless functionality of the basic set of functions and is easier to understand for users without the necessary knowledge of programming. Like OpenLayers, Leaflet is being expanded by plug-ins prepared by a wide community of users. Unlike the previous two, MapBox is a commercial project that focuses apart from the creation of web maps also on the development of its own standards (which also find their use within OpenLayers or the Leaflet). The most important use of MapBox is in the field of vector and raster map tiles.

For fullness, the SVG data format should be mentioned in this place. It generally combines the graphical part with the content of the map or the entire web page. There are a number of web atlas projects that use the combination of SVG and XSLT transformation as a central technology, where both the maps and the entire servicing application, including graphic elements, form windows etc., are generated from the data files by appropriate transformations.

2.2   *Client-server approach*

The basis is to make digital geographic data available using Internet resources. For this purpose, the map server technology is a set of resources to provide GIS services over the network.

It can provide geographic data as well as perform various operations over it. Data access is enabled on a client-server basis where the client is meant by any computer or program (web browser) that sends server requests. Communication is not currently running directly with the map server, but via the HTTP or HTTPS protocols, when the web server connects via CGI protocol with the application running on the map server. It processes received request and returns the result. Data is often stored externally in a data store (Data server).

2.3 *Alternative Esri technologies and means of map presentation*

Whereas for maps with a significant temporal content finally the Esri Story Maps technique was chosen for creating the maps, there is a wide range of other possibilities within the Esri platform, which could be utilized.

The most powerful tool is the ArcGIS Enterprise (formerly ArcGIS Server) platform, which is a map server with additional superstructure functionality. Blažek (2018) deals in detail with this platform as well as other mentioned within this paper. It is possible to create map services according to OGC standards (e.g. WMS, WMTS), as well as native map applications, which can be used in Esri products.

3 ESRI STORY MAPS

3.1 *Time in maps*

The time aspect can be rendered within maps in many ways. As (Vít & Bláha 2016) show, sequential methods, which display basically static maps in a logical sequence that approximates the dynamics of the plot, can be used. Animation maps represent the second option. They can give an impression of a presentation, as it is known from many disciplines, can be a simple video (film), or they can allow a certain degree of interactivity—i.e. a possibility of influencing the flow of the presentation. The latest most comprehensive option are dynamic interactive maps, i.e., in principle, an application that allows users not only to change the progress of the time but also the content displayed, to query the content, to display additional elements etc.

Esri Story Maps represent a special synthetizing case, they use partial options of all the variants mentioned.

3.2 *Story maps*

The first mention of the term "Story Maps" does not have a direct connection with the form in which it is currently being used, nevertheless there is some similarity. The term Story Map was first used in 2005 in connection with the User Story Mapping technique, which aims to organize different user access stories for issued software and thus better understand the functionality of the system and avoid its failure (Patton 2018). One of the first articles in which this phrase appears in connection with Esri is (Strachan & Mitchell 2014), which deals with the perception of Story maps as an effective learning aid. Here, above all, the fact that Story Maps is used to educate, inform and visualize data and is not intended for more advanced map operations and analysis is affirmed. Esri has created a set of pre-made templates that the user simply completes with modern maps, accompanying texts and multimedia features such as charts, photos, videos, and more. The creation of an application is designed to be handled by anyone without geoinformation education.

3.3 *Temporal aspect*

There are a number of Story Maps templates, but only several selected have been found to be suitable for temporal data. These include Story Map Journal, Story Map Series and Story Map Swipe/Spyglass. Generally, Story Maps templates are conceived as an interconnection of

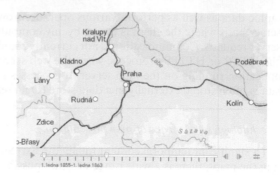

Figure 2.   Time slider when creating an application.

multimedia and maps, while a convenient way to incorporate a time component into the entire application is to use the TimeAware template within other templates designed directly for Story Maps.

The development over time can then be viewed in several ways. The most demonstrative way is probably a time slider (Figure 2). It can be configured in many ways to determine the continuity of play and the way the application responds to the time moving.

The Story Map Series and Journal are useful for displaying data genesis over time, even at irregular intervals (administrative frontiers development, route rendition, etc.). It is actually a "journeying" in time and space. Timeliness can also be expressed using bookmarks created at different time epochs.

In contrast, Story Map Swipe/Spyglass is a pair of analogous templates that allow users to browse and compare a pair of time moments at different time periods, either with a swiper or a magnifying glass—representing a hole in to the other time epoch.

4   PREPARATION OF DATA FOR USE WITHIN THE CZECH HISTORICAL ATLAS

4.1   Maps previously intended for a paper publication

A rather limiting factor in the selection was the fact that major part of the maps intended for the electronic portal had previously been created and cartographically tuned in ArcMap software. For this reason, the possibilities of transferring the current state into the web environment were explored in particular (Suk 2017).

These maps contain different items depending on the theme of the map, but in principle, three types appear:

– maps with a physico-geographical background, borders and without polygons;
– maps with a background and/or thematic content made up polygons;
– maps of selected areas—choropleth-typed.

The content of point and line elements appears practically the same in all the categories mentioned.

4.2   Conversion of maps into the story maps form

Since the maps were prepared in the form of MXD projects, it was necessary to find a way to make the publication as easy as possible, while preserving the highest possible degree of fidelity and similarity to the original map at the same time with the least number of other necessary modifications. When converting to Story Maps, the following method was followed:

– removal of unpublished layers, connected web services and redundant rasters;
– coordinate system change in geodatabase;

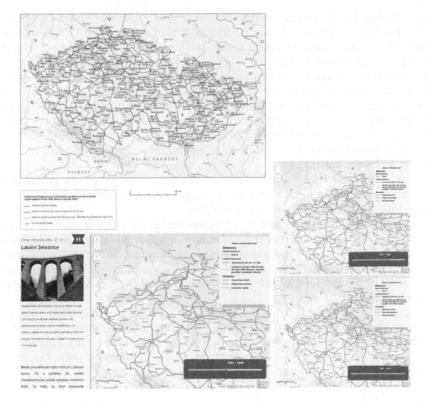

Figure 3. Original map of local railways construction (top) and its appearance after publication as a Story Map (with a slider set to limit the depiction of railways to those built before 1889 and, for comparison, also before years 1900 and 1914).

- editing original frontiers due to assembling overlapping lines;
- verifying the correct position and course of annotations (labels);
- adjusting the time information stored in attribute to fields of the type "date";
- setting the temporal options in the appropriate Properties tab for the layer;
- removal of the original time information (labels, changed symbology, etc.);
- division of the project into two: background layers and inactive items + vectors with a time component.

Other features need to be done when setting up the web map service itself. There is a need to load two services/web maps—with underlying data and with the dynamic temporal component. In the "Time settings" tab, the basic time component settings are available, such as instant data updates, epoch markers, time run controls, time units' format and others.

An example of the appearance of the Story Map test application and the original map created for the Academic Atlas of the Czech History can be seen in Figure 3.

Unfortunately, many options can not be set, but it is possible to use ArcGIS Online Assistant and customize the application by modifying the JSON files. The other option is to edit the source code of the application. These modifications require knowledge of HTML5, CSS3, JavaScript, API (Dojo Toolkit) and the appropriate JavaScript libraries: jQuery, Bootstrap.

5 CONCLUSION

The platform named Czech Historical Atlas introduces the history of the Czech Lands, including the relation to the Central European space, and brings a comprehensive portal summarizing the historical research of the Academy of Sciences of the Czech Republic. In particular, its digital part brings a missing element to the Czech and Central European historiography. It will offer relevant content both to professionals and historians as well as to lay users interested in the history of Czech Lands, while especially users actively taking advantage of the Internet or mobile devices will appreciate the electronic form.

At the beginning of the portal content preparation, a number of problems were being solved. On the one hand, how to convert the maps already practically finished and originally intended for print-only publications, which was solved by the use of Esri technology in which these maps could be used more easily, but not with a small amount of difficulty. One of the problems encountered was generally the time aspect of maps and conversion of their static symbology to dynamical one for use in the web environment. Partial results of this issue are introduced within the paper.

The Story Maps and Esri web mapping applications in general are an intermediary for making easy-to-use applications, where many things can be created automatically and for a few clicks with wizards and ready-to-use templates.

The completed maps can thus include a timeline (either the content changes or stays displayed while adding new objects) or the dynamics can be solved using sliders or other technical tools to compare two or more time levels. Since historiographic maps are very often genetic, their presentation in electronic form calls for the use of an interactive and illustrative representation of the temporal aspect of the subject, especially for events that are noticeably dynamically changing.

A great part of the intended functionality is not available in application builders, even though they often represent essential functions, for example, to limit the map window pan. These deficiencies can be resolved by editing source JavaScript files using a compiler, which means an action not for everyone, requiring advanced programming skills, but may add a welcome extension and improvement to the whole map application.

ACKNOWLEDGEMENT

The research has been financially supported by the grant of the Ministry of Culture of the Czech Republic, NAKI II programme, project No. DG16P02H010.

REFERENCES

Blažek, O. 2018. Webová mapová aplikace pro Český historický atlas (Using ESRI Story maps to visualize historical maps of landscape development over time; in Czech), diploma thesis. Prague: CTU in Prague.

Esri: Story Maps: Apps [online]. Cit. 19. 7. 2018. Available at <http://storymaps.arcgis.com/en/app-list/>.

Havlíček, J. et al. 2018. *Web-based and printed Czech Historical Atlas*. Proceedings of the 13th ICA Conference Digital Approaches to Cartographic Heritage. Greece. 134–140.

Kladivová, L. 2018. Webová mapová aplikace pro Český historický atlas (Using ESRI Story maps to visualize historical maps of landscape development over time; in Czech), diploma thesis. Prague: CTU in Prague.

Kraak, J.-M. & Brown, A. 2003. *Web cartography*. Boca Raton: CRC Press.

Patton, J. User Story Mapping [online]. Cit. 19. 7. 2018. Available at <http://jpattonassociates.com/user-story-mapping/>.

Semotanová E., Cajthaml J. et al. 2014. *Akademický atlas českých dějin* (Academic Atlas of the Czech History; in Czech), Prague: Academia.

Strachan, C. & Mitchell, J. 2014. Teachers' Perceptions of Esri Story Maps as Effective Teaching Tools. *Review of International Geographical Education Online* 4(3): 195–220.

Suk, T. 2017. Webová mapová aplikace pro Český historický atlas (Web Map Application for the Czech Historical Atlas; in Czech), diploma thesis. Prague: CTU in Prague.

Vít, L. & Bláha, J. D. 2016. A Study of the User Friendliness of Temporal Legends in Animated Maps. *AUC Geographica* [online]. 47(2): 53-61. ISSN 2336-1980.

Advances and Trends in Engineering Sciences and Technologies III – Al Ali & Platko (Eds)
© 2019 Taylor & Francis Group, London, ISBN 978-0-367-07509-5

Multicriteria analysis as a tool for estimating optimal barrier-free routes

P. Kocurová, A. Bílková & R. Zdařilová
VSB-Technical University of Ostrava, Ostrava, Czech Republic

ABSTRACT: Municipalities are in the charge of the maintenance of public spaces and areas around public amenities. In the case of planning pedestrian paths, it is necessary to count with human factor including disabled persons. The special requirements for barrier-free use of those areas are occasionally implemented incorrectly. MCA tool, especially Topsis method, is used for optimizing the barrier-free details to meet the Czech standards—decree no. 398/2009 Coll. The article explains the use of the MCA tool for pedestrian—automotive traffic intersections in case of the barrier-free solution. The implementation and solution will be explained by a case study of an area in Ostrava. Chosen area was mapped according to the decree requirements. The aim of the paper is an evaluation of the MCA method as a tool for decision-making processes of municipalities.

1 INTRODUCTION

1.1 *European strategy requirements*

According to European political strategy 10% of the population lives with a disability, varying in degree. It prevents taking part in daily life or makes it difficult to do. The number of people who are losing their autonomy is increasing due to medical and demographic reasons. The trend will continue in the coming years. It is necessary to fulfill social a physical integration like a major challenge with aiming for personal freedom. The civil engineering and a design of habitat are main factors of formation barriers in the area. A discipline of barrier-free design or removing existing barriers is obviously integrated into building and town planning.

European Union strategy also approaches six major objectives: Integrated solutions, Building for everyone (principles of universal design), Accessibility charts for existing areas, Monitoring, Integration in architectural education, and International co-operation. Accessibility charts for existing areas were described as an establishment of parameters and criteria for evaluating and potential accessibility. The requirements were implemented in the standards and legislation of the European Union. The documents exist almost in every partner country. A detailed monitoring is necessary to evaluate each solution and should be tested periodically. The plans also include developing methods, parameters, and tools for analysis and monitoring (Council of Europe 1993).

The multi-criteria analysis tool for evaluating design and realization of barrier-free spaces also follows the objectives.

1.2 *Requirements in the Czech Republic*

The legal environment of the Czech Republic has been addressing the issue of accessibility for over sixteen years. The first regulation was issued in 2001 in the form of Decree no. 369/2001 Coll., on General technical requirements enabling the use of buildings by persons with limited mobility and orientation, which featured basic definitions of terms, determination of the scope, technical parameters and dimensions for ensuring accessible use of external environment and buildings. This decree was amended in 2006 to Decree no. 492/2006 Coll. Another, and most recent amendment of regulation addressing accessibility took place in 2009 when a significant change was made to the arrangement of content, which resulted in the new and comprehensive Decree no. 398/2009 Coll. on General technical requirements ensuring accessible use of buildings, which is still in force today. In addressing the issue of proper design and implementation of transport hubs, we must also not ignore other legal regulations and Czech technical standards, especially Act no. 13/1997 Coll., on Roads, Act no. 183/2006 Coll., on Territorial planning and building regulations, CTS 73 4130 Stairs and inclined ramps, CTS 73 6110 Design of roads and CTS 73 6425-2 Bus, trolley and tram stops (part 2: Transfer hubs and stations), (Bílková et al. 2006).

These requirements are incorporated into legislation. However, they are in many cases incorrectly used in the design or a construction of the buildings. A study evaluates the implementation of Decree No. 398/2009 Coll., on general technical requirements ensuring the barrier-free use of buildings. The study is implemented for the solution of transitions and crossing points in the Ostrava, 28. Října and Opavská Street. The quality and quantitative indicators of the individual barrier elements were monitored by type crossings and the places to crossing without the zebra signs.

2 APPLICATION OF THE LEGISLATIVE REQUIREMENTS

2.1 Definition of the aims for study

In the following chapter is described the mapping of crossings and places used as crossings. The goal for research in a pre-stage was the fundamental summary of applicable legislative and technical standards related to the barrier-free design. The localities of the reconstructed areas, which should follow valid legislation, were chosen for the estimation.

2.2 Barrier-free elements

There was measured and analyzed 123 elements on the focused route (63 zebra crossings with signal lights, 55 zebra crossings without signal lights and 5 crossings without zebra signs).

Figure 1.   Traffic details of the solved area.

Table 1. Fundamental summary of barrier-free crossings.

Percentage of completion/ Type of crossing	Zebra crossing with signal lights	Zebra crossing without signal lights	Crossing without zebra signs
0%	1	6	1
10%	0	0	1
20%	2	4	0
30%	2	2	1
40%	0	0	1
50%	9	5	0
60%	2	4	0
70%	12	8	0
80%	5	1	1
90%	11	3	0
100%	19	22	0

By the requirements from the following standards were assessed those elements of pedestrian traffic. They were rated by the percentage of the parameters' accordance. The result of the measurement shows that 17% of the details achieved legislative conditions less than 50% (Table 1) and 33% by the number of details were fulfilled completely.

However, in that case, the diversity of parameters and their rate of risk is not considered. To the integrality of analysis is also necessary to complete these factors.

2.3 *Requirements*

The key for the measurement was to establish the requirements connected with the aimed traffic details. The requirements were classified by quantitative and qualitative units.

The followed requirements are expressed as:

- Zebra crossing without signal lights: transition length, signal stripes in the direction of alignment with axis, signal stripes from the guideline to the warning stripe, passage on both sides of the signal stripe, overlap of the warning stripe, warning stripe leads to the curb place, material use, visual contrast of material, location of warning stripes, minimal length of the signaling stripe, transition of the guide stripe, continuity of the transition band to the signal strip, reduction of the curb and modification in the total;
- Zebra crossing with signal lights: length of transition, signaling stripes in the direction of crossing with the axis, oblique route, possible remote activation, minimum lengths of signal strips in the islands—signaling, signaling stripes from the guideline to the warning strip, passage on both sides of the signaling strip, the belt has a prescribed overlap, a warning strip to the curb point with v. 80 mm, the standard material of the design, the acoustic signal permitting material, the visual contrast of the material, the positioning of the pillars of the acoustic signaling in the signaling stripe, the distance between the column and the edge of the signaling strip, for wheelchair crossing, detail as a whole using tactile adjustments;
- Crossing without zebra signs: corridor length, justified length extension, visual markings, tramline boundary markings, trace stripes boundary, warning strip overhang, warning strip up to 80 mm, tactile material, visual contrast of the material, warning strips located behind the curb, adherence to the length of the signage strip, reduction of the curb for the wheelchair overhang, crossing guide strip.

3 USE OF MULTI-CRITERIA ANALYSIS

3.1 *Interpretation of the use*

According to a source of the parameters and differences at their risks was followed by a method of the multicriteria analysis. The method gives the impression as available for that kind of the evaluation. The multi-criteria analysis (MCA), as the name itself indicates, deals with the evaluation of alternatives according to several criteria. The term alternative designates each of the solutions for the selected report. The criterion is an attribute that will be evaluated with the given alternative. To each criterion, such as weight, is assigned that expresses the importance of criteria regarding the others.

In case of the TOPSIS method, this is again the question of principle for the distance maximization from an ideal variant. The ideal variant means that all criteria have the best assessments. The ideal variant is mostly suppositional; the best of variants is that one which is the nearest to an ideal variant, (Dvorský et al. 2006).

3.2 Calculation

The paper describes the setting of the parameters' possibilities. Before the start of the calculation is necessary to determine the valid parameters and their weights which represents their risk significance. For the mapping and the subsequent evaluation as a tool for municipalities' decision is important the correct determination.

Then the gradual calculation ensures the basic TOPSIS rules:

- Construction of a criteria-normalized matrix
- Calculation of a criteria-weighted matrix
- Definition of the ideal variant
- Definition of the basal variant
- Calculation of the distance from the ideal variant
- Calculation of the distance from the basal variant
- Relative index of the distance
- The ranking according to the declining indicator, (Dvorský et al. 2006).

3.3 Parameters for the calculation

As an example will be used the alternatives (5 details) of the zebra crossings without traffic lights. For this type of traffic crossing are used ensuring factors in the calculation:

- F1 Transition Length (min)
- F2 Signal strip in the direction of the axis (max)
- F3 Pass route, remote activation (max)
- F4 Minimum signals' strip lengths for signaling islands (max)
- F5 Signal strips from guideline to warning strip (max)
- F6 Gap on both sides of the signaling strip (max)
- F7 The warning strip has a prescribed overlap (max)
- F8 Warning strip to curb point 80 mm (max)
- F9 Standard material (max)
- F10 Allowed acoustic signaling material (max)
- F11 Visual Contrast (max)
- F12 Location of the acoustic signaling posts in the signal strip (max)
- F13 Justification of the location of the acoustic signalization (max)
- F14 Keeping distance of the column from the edge of the signal strip (max)
- F15 Reach distance of the column from the edge of the signal strip (min)
- F16 Warning strips behind curb (max)
- F17 Containment of the signal strip length (max)
- F18 Transition Guide strip (max)
- F19 Follow-up guide strip on a signal strip (max)
- F20 Curb reduction (max)
- F21 Making detail as a whole (max)

Minimization and maximization factors, which are necessary to the optimization of parameters, are intended.

3.4 Results

In the last grade of the calculation are reached the declining indicators' values. The indicators represent the distance to the ideal alternative.

3.5 Rating of the results

The alternatives were classified according to the indicators in Table 4.

Table 2. Weights of factors and the crossings examples for MCA.

			Crossings				
Factor	Weight	Ideal	Alternative 1	Alternative 2	Alternative 3	Alternative 4	Alternative 5
F1	0.09	3000	4000	11000	10000	4000	6000
F2	0.09	1	1	0	1	1	0
F3	0.01	1	0	0	0	0	0
F4	0.13	1500	950	950	0	1600	1600
F5	0.04	1	0	0	1	1	1
F6	0.04	800	800	340	800	800	800
F7	0.02	800	800	800	800	800	800
F8	0.02	80	20	60	20	20	0
F9	0.09	1	1	1	1	1	1
F10	0.02	1	0	0	0	0	0
F11	0.01	1	1	1	1	0	1
F12	0.01	1	0	0	0	0	0
F13	0.002	1	1	1	0	0	1
F14	0.004	1200	1100	340	930	0	1100
F15	0.004	900	1100	340	930	0	1100
F16	0.07	1	1	1	1	1	1
F17	0.07	1500	1500	1500	1500	1500	1500
F18	0.04	1	0	0	1	1	1
F19	0.07	1	0	0	1	1	0
F20	0.02	1	1	1	1	1	1
F21	0.13	100	100	100	50	50	50

Table 3. The declining indicator of alternatives.

Alternatives	Indicator C_i
Ideal requirements (legislative)	0.0185
Alternative 1	0.4672
Alternative 2	0.7237
Alternative 3	0.6210
Alternative 4	0.3335
Alternative 5	0.6386

Table 4. Results.

Rating*	Alternatives
0	Ideal requirements
1	Alternative 4
2	Alternative 1
3	Alternative 3
4	Alternative 5
5	Alternative 2

*1 – best; 5 – worst.

4 CONCLUSION

The utilization of multi-criteria analysis in the evaluation is a benefit for elementary assessment. The existence of software for the calculation is also available for the key users. However, it is important to state that in case of considerable differences between the alternatives, the results may be distorted.

It can be therefore used as a simple tool for fast decision-making processes. The calculation is also conducted with the weights of parameters. That solution is more detailed than basic comparison with legislative requirements.

ACKNOWLEDGMENT

The work was supported by the Student Grant Competition VŠB-TUO. Project registration number is SP2018/156.

REFERENCES

Bílková, A., Niemiec, B., Kocurová, P., Orsáková, D. 2016. Advances and Trends in Engineering Sciences and Technologies II: proceedings of the 2nd International Conference on Engineering Sciences and Technologies. *The issue of barrier-free public places—the transport interchange Ostrava-Svinov.* London: Taylor & Francis.

Council of Europe. 1993. Accessibility: Principles and guidelines. Strasbourg: F-67075 Strasbourg Cedex.

Dvorský, J., Krejčí, P., Moldřík, P. 2006. Software MCA for computation of MCA methods. ELNET 2006 [online]. [Cited 14th July 2018]. Available from: <http://cs.vsb.cz/elnet/2006/>.

Advances and Trends in Engineering Sciences and Technologies III – Al Ali & Platko (Eds)
© *2019 Taylor & Francis Group, London, ISBN 978-0-367-07509-5*

Problems of development of bi-centric Hradecko—Pardubické residential agglomeration in Czech Republic

D. Kuta & L. Hurdalkova
Department of Urban Engineering, Faculty of Civil Engineering, VŠB-TUO, Ostrava, Czech Republic

ABSTRACT: This contribution deals with the development of centres of the unique bi-centric agglomeration in the territory of the Czech Republic, from its origin through problems of its determination to current issues related not only to its urban development. By defining individual hypotheses, this work deals with a multi-disciplinary approach using many different indicators that form jointly the input information on the agglomeration as a whole. The agglomeration as a complex requires collaboration of many scientific fields to specify in details both, the current and future nature of this unique area. Individual chapters of this contribution outline diversity and importance of this topic, which apparently requires multi-disciplinary collaboration in the development planning of such complex territory as this bi-centric Hradec—Pardubice settlement agglomeration. At the end you can find an assessment of the current state in terms of development planning of this agglomeration and possible suggestions for detail solutions of the problems found that are intended mainly for use as a basis for public authorities.

1 INTRODUCTION

The Hradec—Pardubice settlement and industrial agglomeration formed an important industrial and agricultural area in the centre of the East Bohemian region. The current extensive development of the tertiary sphere and importance of both regional towns set the pace of logistic centres development accompanied with traffic infrastructure development, but also the development of higher civil facilities. The whole organism of this agglomeration is supported by recreational function in valuable areas in terms of biological and landscape aspects. (Maier K., 2012). According to the regional development policy of the Czech Republic, this bi-centric agglomeration is included in the OB4 development area. Development areas are determined in territories with concentrated international or domestic activities, which have increased requirements for changes within the region. Urban settlements included in this bi-centric agglomeration act as localities with high concentration of the population, job opportunities, and public and commercial services. Due to their background depending on their size, they become development poles, in which significant resources are created resulting in the development not only of the agglomeration itself. Specific position of both regional towns—Hradec Králové and Pardubice—and their metropolitan areas form a very strong bipolar regional agglomeration within the CR.

2 SUBJECT MATTER OF THE RESEARCH

The subject matter of this research includes questions in the development of core municipalities of the bi-centric Hradec—Pardubice industrial and settlement agglomeration, its significance within the ranking of the developing areas of the Czech Republic, and its dependency on adjacent developing regions. Identification of development approaches in already approved but also unfinished documents outlines different looks at the agglomeration alone, and mutual collisions ma occur due to insufficient coordination of strategic objectives and their realisation. Therefore the summary of main indicators, which individual strategies often meet with on a regular basis, is not the sufficient guide for mutual coordination. Comparing approaches in the development of core municipalities and subsequent implementation of development objectives influences the future nature of this agglomeration. (SFSD, 2010). Setting principal (joint) indicators used in strategic documents not only of these core munici-palities may influence positively agglomeration development as a whole. Determining main indicators may be used for the so-called strategic framework of the bi-centric agglomeration in practice, e.g. for urban planning and development authorities not only on the local gov-ernment level. The importance of outlining rules for the conduct and decision making in the agglomeration has not been entrenched strongly in legal documents yet; the building act does not even include the term agglomeration, not to say the specific bi-centric agglomeration (CEC, 1997).

2.1 Defining the hypotheses

Based on the hypotheses presented, the overall meaning of this bi-centric agglomeration will be outlined for the development of municipalities and towns included in the agglomeration, but also its meaning for the core towns. From these hypotheses, both positive and negative impact on the development of the agglomeration but also of its individual municipalities and towns will be apparent.

2.1.1 Hypothesis 1
Determination of the bi-centric agglomeration is a time variable issue, the identification of which is based historically on the administrative structure of the area depending on the eco-nomic potential.

2.1.2 Hypothesis 2
Implementation of integrated investments strengthens pillars of the sustainable develop-ment of both core municipalities that influence the development of the whole bi-centric agglomeration.

2.1.3 Hypothesis 3
The protection of values in the area including existing limits of area utilization enable to pro-vide the current quality of facilities including its development not only in core municipalities, which strengthen the economic, demographic or commercial development within the whole agglomeration.

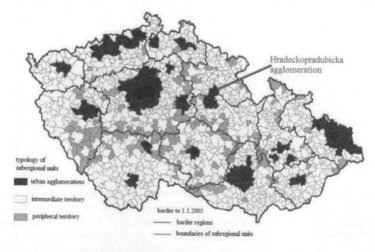

Figure 1.   Urban agglomerations in the Czech Republic. Source: cuzk.cz.

2.1.4 *Hypothesis 4*

Providing a sophisticated strategic planning not only in core municipalities but in the entire agglomeration, negative impacts of suburbanization processes can be reduced significantly.

3 DOCUMENTING THE HYPOTHESES

After the analyses completed, it is necessary to state individual documentation of input hypotheses. These hypotheses were set as a multi-disciplinary approach to issues of the bi-centric agglomeration area in question, to provide objectiveness and point out input elements that are the building stones of individual development concepts concerning markedly the development of the agglomeration itself. The summary enables to show the overall development of these issues including current problems.

3.1 *Documenting individual hypotheses*

3.1.1 *Hypothesis 1*

Determination of the bi-centric agglomeration is a time variable issue, the identification of which is based historically on the administrative structure of the area depending on the economic potential.

The changing administrative structure from the historical point of view is a time variable event, in which it is hard to rely on fixed borders of such complicated organism as the bi-centric agglomeration. Its determination in ÚPN VÚC, held strongly since 1976, is currently suppressed by several development documents. Different approach from the ZÚR point of view in both regions evokes missing coordination in preparation and decision-making approach in such considerable area. PÚR ČR has determined this agglomeration as part of the developing area of republic importance OB4; however such determined developing area in the territory of the Pardubice region (ZÚR Pk) does not connect to determination of the developing area in the Hradec Králové region (ZÚR KHK). The administrative structure of towns or municipalities is even at the present time often changed for many reasons—e.g. exchange of lands between municipalities, self-administration transfer to other municipality or town, or due to a suburbanization process assimilating small municipalities in outskirts of large municipalities where such determined agglomeration became a base for determination of a different agglomeration area concentrating on integrated investments called ITI (Epson, E.J., Edwards, M.M., 2010). Such determined agglomeration area includes incomplete territory of 2 regions and 5 districts with 145 municipalities (14 of them with the town status), which form the cores (except for those statutory towns), and it also includes 8 municipal districts with extended competence and 13 municipal districts with a delegated local authority. Determination of the agglomeration is however not reported in the documentation of regional planning or individual territorial plans. It is therefore a sort of an imaginary and purposeful determination not entrenched in legal documents. In addition, the intensity of suburbanization is mentioned here, which is divided into three zones. These zones are characterised by minimum annual housing development in 1997–2008. The suburban zone 1 is characteristic by 10 flats/1000 residents, the suburban zone 2 by 5 flats/1000 residents, and the zone 3 is below the intensity values of the annual housing development stated. These zones have been defined based on migration intensity (moving in) which is a variable indicator in time that may cause revaluation of suburban zones in the bi-centric agglomeration and subsequent determination of this agglomeration (Körner, 2005).

3.1.2 *Hypothesis 2*

Implementation of integrated investments strengthens pillars of the sustainable development of both core towns influencing the development of the whole bi-centric agglomeration.

Individual representations of administrative commissions and major stakeholders are apparent from the list of members of individual workgroups and the governing committee. The most frequent representatives in individual workgroups (partially except for PS—

horizontal) are Municipal Councils of Hradec Králové and Pardubice and their respective departments, Regional Authorities of Hradec Králové and Pardubice and their respective departments, and Regional Chambers of Commerce of Hradec Králové and Pardubice; all workgroup members stated are at the same time represented in the governing committee. We must mention here the significance of individual representations. Based on the frequency and occupancy, the principle of mutual collaboration and coordination of individual objectives is confirmed. Collaboration potential can be seen mainly in the chemistry, electronics, biomedicine, electrical engineering, mechanical engineering, and IT. Higher increase of project designs on the part of Hradec Králové that currently enter their approval process can be assumed. The most frequent projects for both centres are investments in education innovations that contribute well to the development of the tertiary sphere of both core towns and also strengthen the economic and social pillar of the sustainable development, and together with public education ensure also the development of the environmental pillar. Providing public education will enable higher development sustainability of the whole agglomeration.

3.1.3 *Hypothesis 3*

Protection of values in this area, including existing utilization limits of the area, enables to ensure the existing qualities of facilities including their development not only in core towns, which strengthen the economic, demographic or commercial development within the agglomeration as a whole.

Based on the area utilization values and limits that relate directly not only to core towns but also to the area of the whole agglomeration, it is necessary to deal wit them further in the development planning. Their limits and values are crucial for suggestions of area development, since serious natural or technical and traffic disasters may occur, if not respected (UN-Habitat, 2002). Conversely, they offer a considerable base for sustainability of these areas which needs to be further worked on to support their development enabling also sustainable development of the whole agglomeration. It is necessary to see the limits of area utilization not as a negative restriction preventing the development of both, the built-up area and the landscape component in the agglomeration territory. It results in from the values and limits stated that the natural component is greatly represented in the agglomeration territory, as well as the advanced technology of the traffic and technical infrastructure, which this area disposes of and which at the same time promotes the area to a significant area of the forth greatest agglomeration in the CR. These phenomena being part of the regional planning documentation providing coordination or partial developing areas with these limits are expanded further by partial proposals, which enable economic and demographic development, but also an increased economical potential of the whole agglomeration territory.

3.1.4 *Hypothesis 4*

Providing the sophisticated strategic planning not only in core towns but in the entire agglomeration, negative impacts of suburbanization processes can be reduced significantly.

When outlining individual characteristics resulting mainly from historical relations, matters of fellowship of both core towns show up. Regardless of significant historical milestones which both core towns had to pass through, the current situation is rather in favour of mutual collaboration. Their status of regional towns that offers their equal position, which was litigated for a long time in the past, does not play a big role. In closer examination of mutual relations and values, which both towns dispose of, a significant potential for the development was found, mainly in mutual collaboration. The synergy arising from the regional planning, where the so-called opportunities and guidelines are set for future development, makes the bicentric agglomeration quite exceptional in the whole CR that may in the future compete with agglomerations of Brno or Ostrava regions. The summary review of suggestions resulting from SWOT analyses points out possible execution of objectives for sustainable development of the bi-centric agglomeration and subsequent strengthening of its position within the CR. Considerable part of these suggestions has already been dealt with progressively in regional plans of individual municipalities and towns within the territory of this agglomeration, high readiness of these objectives for implementation is thus apparent. On the other hand, including these programs in the documentation of regional planning is not a guarantee of efficient execution of objectives that would ensure the sustainability; collaboration and coordination not only of adjacent municipalities and towns within the agglomeration but also between neighbouring regions are necessary. Under these conditions, we can consider achieving the sustainable development of the area as a whole, and strengthening individual pillars of the sustainability (Beran V., Dlask P., 2005). The time period from completion of the regional-planning documentation up to the execution is very long and represents certain threat to coordination provision; therefore it is necessary to pay attention to reducing this time period for the benefit of executed objectives that will influence positively the sustainability of specific municipality or town, but the agglomeration as a whole. Conversely, irreversible changes may occur in the area that may cause both, strengthened suburbanization processes and activities not in compliance with sustainable approaches for all pillars.

3.2 Summary of hypothesis

3.2.1 Summary of hypothesis 1
The presented documentation of this hypothesis results in historically problematic clear determination of the agglomeration area, due to continuing suburbanization processes and related changes of borders of individual administrative territories of municipalities and towns. The variability depends fully on time just as the economic development, therefore it is very problematic to determine clearly this area.

3.2.2 Summary of hypothesis 2
In spite of certain real disharmony between individual principles of the sustainable development and current development trends of the town, the conceptual tool in the form of integrated regional investments can be considered successful for the provision of coexistence and coordination of both core towns, resulting in strengthened sustainable development for the entire agglomeration area.

3.2.3 Summary of hypothesis 3
Presented documentation of the hypothesis 3 shows respecting the existing values and limits of area utilization with an effort of their collision-free protection. The agglomeration area will advance as a whole and enable the development, the purpose of which is coordination and collision-free application of partial objectives that will in consequence increase the value of the whole agglomeration territory.

3.2.4 Summary of hypothesis 4
Sophisticated strategic planning needs not necessary represent a guarantee of good coordination and the pursuit of reducing negative impacts of suburbanization. One of the significant limits is the continuously increasing time demand in the preparation not only of strategic materials but also realization of individual objectives. Reducing these limits may positively influence suburbanization processes and strengthen the pillars of sustainable development. More and more frequent legislative changes have negative impacts on these tools affecting suburbanization.

4 CONCLUSION

Based on predetermined hypotheses, the issue of the development of core towns of the bi-centric agglomeration was monitored from several points of view. As this issue is specific with its multi-disciplinary approach, these key approaches, dealing more closely with the issues and sustainability of the development of core towns and individual municipalities located within the agglomerations, need to be dealt with in more details. However, the detail list of input data and subsequent analyses and conclusions of individual fields that form a comprehensive solution for sustainable development was not the subject of this work, it is only the determination of key elements that can be applied not only to this model of agglomeration. For the purpose of this work, public resources are used primarily; the main guides and sources of information are not only the documents of regional planning (UAP), but also individual urban planning documentation predominantly of core towns, individual sub-concepts of the development from transport systems through nature protection to demographic analyses. Based on these documents, it is possible to effectively outline the current situation in relation to the agglomeration with the main and the lower centres of the settlement, the positive character of the agglomeration, as well as the outlined future development not only from the point of view of a regional plan designer. Determining individual hypotheses has led to the confirmation of multi-disciplinary issues and to pointing out the importance of each field, without which the development planning and the efforts to provide sustainability of the agglomeration would not be possible. The specificities of this agglomeration are seen in the existence of core towns, which are often stated as towns competing with each other. Pointing to a common solution of bi-centric agglomeration existence is included in the ITI strategy, which is described in details here, as the division of duties and responsibilities is outlined here, which forms the basis for the agglomeration as such.

Based on individual analyses, it must be mentioned the overall summary of individual disparities of both core towns for the final assessment, and at the same time underlined the synergy of main limits and values of both core towns. The summary confirms the overall idea of this work, which is the clarification of Hradec Králové and Pardubice coexistence in a territory based on their collaboration and mutual existential support with an objective to create a high-quality environment for economic development and living. The efforts to create a high-quality economic base, including the provision of a comprehensive system of civil facilities, will allow the future development of an economically important common space (Teslíková, 2015).

ACKNOWLEDGEMENTS

The work were supported by funds for Conceptual Development of Science, Research and Innovation for 2018 allocated to VŠB-Technical University of Ostrava by the Ministry of Education, Youth and Sports of the Czech Republic.

REFERENCES

Beran V., Dlask P., Management of sustainable development of regions, settlements, and municipalities, ACADEMIA,1, Prague 2005, ISBN 80-200-1201-X.

CEC, Commission of the European Communities, 1997, The EU Compendium of Spatial Planning Systems and Policies. Regional Development Studies, 28. Luxembourg: CEC.

Epson, E.J., Edwards, M.M., 2010: How possible is Sustainable Urban Development? An Analysis of planers' Perfections about New Urbanism, Smart Growth and the Ecological City. Planning Practice and Research 417–437 vol. 25, no.4. Taylorand Francis.

Körner M, Urbanism and Territorial Development, 06/2005, year 8, Brno, ISSN 1212-0855.

Maier K. et al., Sustainable regional development, GRADA 2012, ISBN 978-80-247-4198-7.

Strategic Framework of Sustainable Development of the CR, Ministry of the Environment, 2010,1, Prague, ISBN 978-80-7212-536-4.

Teslíková Hurdálková, Lucie, Kutá, Dagmar. Bicentric conurbation and its territorial development. In: WIT Transactions on The Built Environment: Sustainable Development and Planning VIII. Turkey: WIT Pres, Wessex institute, 2015, s. 9. ISSN 1743-3509.

Teslíková Hurdálková, L a D. Kutá. Development of the Hradec-Pardubice agglomeration from 1986 to the present. The Social Sciences. 2016, 11(19), 8. DOI: 10.3923/sscience.2016.4638.4642.

UN-Habitat (United Nations Human Settlements Programme), 2002. The Global Campaign on Urban Governance. A Concept paper. [online]. Nairobi: UN. [cit. 2016-01-27]. Available at: www.unhabitat. org/downloads/docs/2099_24326_concept_paper.doc.

Advances and Trends in Engineering Sciences and Technologies III – Al Ali & Platko (Eds)
© 2019 Taylor & Francis Group, London, ISBN 978-0-367-07509-5

Non-standardized testing methods for determination of suitability of parting agents

J. Olšová & R. Briatka
Faculty of Civil Engineering, Slovak University of Technology, Bratislava, Slovakia

P. Briatka
COLAS Slovakia, Košice, Slovakia

ABSTRACT: The parting agents (oils) are used to prevent sticking of the hot asphalt mix to the steel parts of any machinery. Their chemical properties and characteristics are tested in laboratory by standardized test methods. Are these characteristics usable in real practice? Hardly. How can we verify their efficiency by simple methods, without using costly and technically complex devices? In this article, three non-standardized simple test methods are described in order to choose the optimal oil for the asphalt paving team.

1 INTRODUCTION TO PARTING AGENTS

In daily practice, we frequently encounter requests for testing the construction materials or substances used in the construction industry which cannot make use of the standardized testing methods either for the material specificity or for the mere absence of suitable testing methods. The materials concerned include parting agents used in the construction of asphalt pavements. Parting agent is applied onto the beds of trucks carrying asphalt mixes to the work sites, onto the hoppers of pavers laying asphalt mixes and onto other auxiliary working tools, or soles of the safety footwear (Faure, M. 1998). These agents tend to be essentially mineral, or they come along as residual products in the petroleum refining processes. Accordingly, parting agents are likely to dissolve bitumen, one of the main components of asphalt mixes, and so raise a concern with the asphalt pavement contractors eager to retain the asphalt mix quality and, ultimately, pavement performance characteristics (USIRF, 2001).

1.1 *Technical and qualitative specification (TKP), Part 6: Compacted asphalt mixes*

The very TKP requires that the asphalt mixes may only be transported to the worksite on trucks with tight, smooth and clean metal beds (TKP 6, 2015). The truck bed with no parting spray applied will let the mix truckload stick. As stated by TKP the agents used to prevent the asphalt mix from sticking include a soap water solution, paraffin oil or calcic solution (optimum quantity). The use of kerosene, diesel fuel, gasoline petrol, and other similar petroleum solvents is prohibited (Bačová, et al., 2012). These substances cause dissolution of asphalt and degradation of asphalt mixes.

1.2 *Bitumen dissolution test*

An example of the efficiency of bitumen dissolution given in Figure 1 shows the formerly performed testing of dissolution of bitumen in rapeseed oil and gasoline petrol. After applied onto a composite geo-grid, the bitumen emulsion was left to wilt and then stayed submerged in solvents for 5 hours. When removed from the solvent, the geo-grid was visually inspected to prove that the rapeseed oil and gasoline petrol both managed to dissolve bitumen efficiently (Olšová, J., 2015).

Figure 1. Testing dissolution of bitumen in diesel fuel (left), rapeseed oil (right).

Table 1. Essential parameters of agents.

Agent	Base material	Yield (m²/kg)	Dilution
A	Mineral oil mixture (petroleum distillate)	27	No
B	Glycerol-based (biodegradable)	326	No
C	Mineral oil mixture (petroleum distillate)	4464	Yes (1:20 w/water)

The essential requirement for parting agents is not to dissolve bitumen so that the aforesaid degradation of asphalt mix is avoided. However, as commonly encountered, drivers tend to ignore the above solubility of bitumen and seeking savings they mistakenly pour diesel fuel or any oily substance readily available onto the truck beds rather than buying proper demoulding agents. This is a bad procedure, which eventually turns out no less expensive and may even cause unexpected potholes to form in the asphalt pavements. As often happens, the agents officially declared to be parting agents show a stronger ability to dissolve bitumen than acceptable. For these reasons, it is worth checking the parting agents for their qualitative properties to ensure that we make a good decision.

2 EXPERIMENTAL METHODS

2.1 *Materials*

This paper deals with testing of three types of parting agents by means of three non-standardized testing methods. The methods are applied with two types of bitumen as most commonly used in the asphalt plants, CA 50/70 paving grade bitumen and PmB 45/80-75 modified bitumen. The attention is drawn to the wetting power of agents (ability of agent to adhere to the steel parts and be used again), ability to dissolve bitumen, and parting ability (indicating the ability of agent to prevent bitumen adhesion to the steel parts). Table 1 lists the essential characteristic of some agents.

2.2 *Test pieces and ability to dissolve*

Six laboratory-made stainless steel strips (5 × 25 cm) are first weighed (m0), then spread with hot bitumen (3 pcs CA – 155°C and 3 pcs PmB – 180°C) for about 10 cm longwise, let cool down and later weighed again (mi).

Next, the strips are each submerged into about 100 ml of agent, each of the types, at the room temperature (23°C) and let work (cause dissolution of bitumen) for 5 hours. After that,

Figure 2. Pothole on a recently paved road.

Figure 3. Testing ability to dissolve.

the visually examined agent shows a change in colour compared to the pre-test condition. The strips are then removed from the agent, let drain and weighed again (mf) (Figure 3). The less bitumen was dissolved ((mi-m0)-(mf-m0)), the less harmful the agent is.

From practical point of view, the ability to dissolve could have an impact on quality of asphalt mix which is exposed to excess of the agent (e.g. on truck's bed) and then paved on the road. It the bitumen is dissolved (locally) it may result on formation of pothole after some time in service.

2.3 *Test pieces and wetting ability*

Wetting ability of agents is tested on 3 pieces of stainless steel strips (8 × 25 cm). The strips are each submerged into all of the agent types for 1 s. When removed from the agent, the submerged strip surface is visually examined for the percentage of agent spreading. The higher the coverage, the higher wetting ability is.

Again, from practical point of view, wetting ability has got a major impact on willingness of real users (drivers) to spray it on the trucks' beds instead of forbidden matters

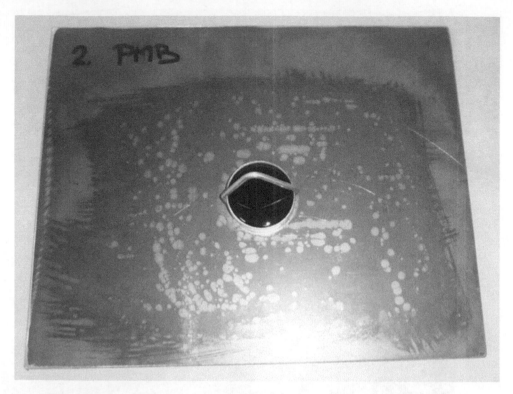

Figure 4.   Bitumen sample target on oily steel plate.

(e.g. petroleum) dissolving the asphalt, reducing adhesion and causing local, irregular and unexplainable potholes.

2.4   *Test pieces and parting ability*

Parting ability of agent is tested on steel plates making a mock-up of the truck bed surfaces. A round target, 5 cm in diameter, is placed onto an oil-spread plate. The target is filled with heated bitumen and let cool down to reach the room temperature (Figure 4). The target is then connected to a hook and dynamometer to pull the target away from the plate while applying of a direct pulling force. The maximum force achieved is recorded. The test piece spread with no parting agent (i.e. with the round target placed right onto the dry plate) serves as a reference.

Parting ability is the essential performance of parting agents which, however, is not standardized. Usually it's assessed subjectively feel- or experience-based by the drivers. As a crucial characteristic of parting agent it shall be tested and evaluated on some numerical data. In this case it is represented by maximum tensile force applied to the sample to separate the bitumen from steel plate.

3   RESULTS AND DISCUSSION

3.1   *Evaluation of ability to dissolve*

The ability of all three agents (A, B and C) to dissolve modified bitumen fails to be established. Agent A, however, managed to dissolve paving grade bitumen and cause a 78% loss on sample weight, agent changed in colour, which is not good for parting agents. Agents B and

Table 2. Test results and weights attached to parameters.

Asphalt	Agent/Sample	Wetting ability (%)	Ability to dissolve (%)	Parting ability (%)	Bitumen rating	Final rating
CA 50/70	CA A	100	–77,8	47,46	27	264
	CA B	90	0,0	93,56	558	1141
	CA C	70	0,0	99,50	568	1136
	Reference					
PmB	PmB A	100	0,0	27,42	237	
45/80-75	PmB B	90	0,0	98,61	583	
	PmB C	70	0,0	99,70	569	
	Reference					
	Weight	1	4	5		

C fail to dissolve the paving grade bitumen and cause no weight loss. Here, we must repeat that the lower the dissolving is, the more suitable the particular agent is (considering type of bitumen). Since every Hot Asphalt Mix Plant (HAMP) operates both CA and PmB. The "winner" in this characteristic must be suitable for both types of bitumen.

3.2 Evaluation of wetting ability

The best ability to spread over steel strips forming a continuous film and cause no run-off is attached to Agent A – 100% spreading. Agent B also makes a satisfactory result of 90% spreading. Agent C has a 70% ability to spread readily and uniformly over the steel surface forming a thin continuous film, which seems sufficient. Again, every Hot Asphalt Mix Plant (HAMP) operates both CA and PmB. The "winner" in this characteristic must ensure optimum wetting ability for both types of bitumen.

3.3 Evaluation of parting ability

The best parting ability, 99.6% on average, is achieved by Agent C with both of the bitumen types. Second best result is achieved by Agent B, 96.1% on average. The worst result of 37.5% is made by Agent A.

On deciding about which parting agent to choose considerations must also be given to the suitability of both bitumen types in use, CA 50/70 and PmB 45/80-75. The glycerol-based parting agents may have an insufficient effect on PmB 45/80-75. Keeping two types of agents on the asphalt plants and pavers would be uneconomical and confusion could easily arise.

The weight values attached to the parameters rate their importance. Parting ability is rated the most important parameter in making a decision on the agent suitability so the weight attached to parting ability is the highest (50%). Ability to dissolve bitumen is the second most important characteristics and it is attached the weight of 40% while the remaining 10% is attached to the wetting ability. Table 2 indicates that Agent B has the best rating. As declared by its manufacturer, biodegrability is another useful property of Agent B. Agent C is rated 5 points lower, which a negligible difference. However, pricewise Agent C is priced at 0.25 €/kg on average (price incl. transport to the asphalt plant—water diluted) much cheaper than Agent B priced at 0.85 €/kg and Agent A priced at 0.95 €/kg.

As well as good-quality and suitability of agent, the price is also important for the asphalt paving contractors to consider. This is why Agent C appears to be the most suitable parting agent.

4 CONCLUSION

Three simple non-standardized testing methods are put in place to determine the efficiency and suitability of three different parting agents used in road construction. For the asphalt pavement contractors, there is normally no informative value in the agent's chemical properties as stated by manufacturers. Accordingly, there is nothing but hope and trust in the supplier's decision about the agent. Alternatively, the agent user may perform an in-house testing as suggested above and determine the efficiency and suitability of parting agents or agree on the contract for agent supplies. The range of products available on the market is wide and price does not always have a final say when contrasted to the quality and increasingly important environmental aspects.

REFERENCES

Bačová, K., et al. 2012. Stavba ciest a diaľnic, Slovenská technická univerzita Bratislava. ISBN 978-80-227-3831-6.
Faure, M. 1998. Routes Tome 2. ENTPE ALEAS Editeur. ISBN 2-908016-90-7.
Olšová, J. 2015. Rapport de stage Master: Etude du comportement des enrobés bitumineux armés de grille de verre en vue du renforcement structurel des chaussées, Colas—document interne.
Technicko-kvalitatívne podmienky MDVRR SR, časť 6: Hutnené asfaltové zmesi, August 2015.
USIRF 2001. Les enrobés bitumineux Tome 1. RGRA Editeur. ISBN 2-913414-40-0.
US Patent: US5494502 A—William M. DeLong, Asphalt release agent.

Advances and Trends in Engineering Sciences and Technologies III – Al Ali & Platko (Eds)
© *2019 Taylor & Francis Group, London, ISBN 978-0-367-07509-5*

Mapping and modeling of the forgotten cultural heritage in areas affected by anthropogenic activity—case study for the Czech Republics

J. Pacina & M. Holá
Faculty of Environment, J.E. Purkyne University, Ústí nad Labem, the Czech Republic

ABSTRACT: The anthropogenic activity has influenced the area of the Northwest Bohemia (the Czech Republic) in many ways in the past 100 years. Many small historic landmarks were in the landscape that is now changed—some of them are still on their spot, but a lot of them are missing or forgotten. The aim of this paper is to introduce the database of stone crosses in the area of the Ore Mountains and their foothills to maintain and visualize the cultural heritage in the internet environment. The resulting comprehensive information system includes the database that contains the available historical landmarks, their history, photographs and localization. The database is connected to a web mapping application that allows visualization of the objects of interest together with processed old maps. A photogrammetric survey was used to create 3D models of the stone crosses. These 3D models are available within the database and the web mapping application.

1 INTRODUCTION

Small historical landmarks like stone crosses used to be essential parts of the landscape. They remained on their original spot for ages and they carried unique stories of their origin. Nowadays we can rarely find a stone cross on its original spot. The area of Ústí nad Labem region (districts Chomutov, Most, Teplice) was chosen for this research because of its problematic culture-historical development in the Czech-German borderland.

The Ore Mountains (in German Erzgebirge) are a 130 km long ridge in the north-west part of the Czech Republic and they create a natural border with Germany. In this region there used to be many stone crosses symbolizing character of the spot. On the crossroads there used to be crosses accompanying the wanderers on their way, or resting-stones used for resting along the way. Other crosses could be found by churches, houses, outside the villages or in forests. Such crosses can be labeled as conciliation crosses as they are placed on a spot where someone was murdered or died in an accident. The Czech-German borderland was affected by the transfer of the German speaking inhabitants after the World War 2 (WW2). The towns and villages in this region were never populated as before the WW2 and the change in the number and the origin of the inhabitants has affected the surrounding landscape structure and land use. In relation to the hate of the German-speaking inhabitants, the small historical landmarks were in many cases removed, damaged or completely destroyed shortly after the WW2. The active open-pit mining in the brown-coal basin at the foot of the Ore Mountains is another factor affecting the surrounding landscape. Many villages and towns were destroyed during the past 80 years of active coal mining—and so have disappeared the small historical landmarks.

It is very important to preserve the landscape memory as it is a part of our culture heritage. The large landscape changes, political changes or just a simple removal of a stone cross cause loss of the landscape memory and the cultural heritage that has been following this region for centuries. The new technologies of sharing spatial information (old maps, descriptive information, images, etc.) using web-maps could be a good way to preserve and offer

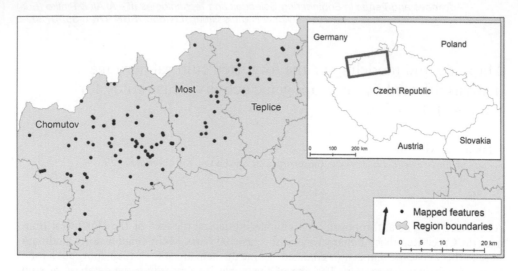

Figure 1.   Area of interest and the mapped features in regions Chomutov, Most and Teplice.

the cultural heritage to the public. The main task of this project is to perform analysis of development and current state of small historical landmarks in the region affected by anthropogenic activity. The districts Chomutov, Most and Teplice (see Figure 1) were thus depicted as they have been affected by all above mentioned factors (Brůna et al., 2014). Based on the old and current maps and archival sources a large field research was performed in this region. Part of the research was the usage of close-range photogrammetry used for 3D model data acquisition. The detailed research in the archives brought large number of useful information that has helped to create a comprehensive description of the small historical landmarks in this region. A web-page presenting information about each of the researched historical landmarks together with their 3D models was created together with a web-mapping application showing location of the depicted stone crosses.

2   DATA COLLECTION AND STORAGE

Identification of stone crosses in the landscape is in many cases difficult—thus the research in the archives, old maps and archival aerial photographs has to fore come the field survey. The maps of the 1st, 2nd and 3rd Military Survey of the Habsburg Empire, the Imperial Compulsory Imprints (Brůna et al., 2014), the Base map of the Czech Republic 1: 10 000 and the current orthophoto of the Czech Republic (CUZK, 2018) were the main data sources used for this research.

The data gathered during the background research and field survey are stored in a data structure that can be easily modified, available in the internet environment and compatible with GIS. The FileMaker Pro database was created for storing the descriptive information and photographs and made available online using the FileMaker Server.

The data were put into the database based on the archival research, consultations with historians and witnesses and our own field survey. The available literature focused on the conciliation crosses (Dreyhausen, 1940), (Procházka et al., 2001), (Brojír-Svoboda, 2001) and the knowledge of the "Society for stone crosses research, the museum in Aš" was used as well.

In the database there are stored all the information gathered during the research—historical and descriptive information, the current state of the historical landmarks, their size, reg. number and location, etc. The coordinates linking the historical landmark to their position in the field were added together with historical and current photographs if possible. The links to the online published 3D models of the stone crosses were added as well.

3   3D MODELS OF THE STONE CROSSES

The rapid development of computer technology in the past years has allowed for the creation of 3D models and orthophotos of very high spatial resolution using oblique and vertical photographs (aerial or earthbound), taken by compact digital cameras—Small Format Aerial Photography—SFAP (Aber et al., 2010). Many studies (Cardenala, et al., 2004), (Chandler et al., 2005), (Quan, 2010) have proved that the results obtained via digital compact cameras and professional surveying cameras can be comparable. The principle of 3D modelling for culture heritage preservation is used in many projects worldwide—i.e. (Alsadik et al., 2013), (Erenoglu et al. 2017) or (Qi et al., 2008).

The basic principle of creating a 3D model out of conventional digital photographs is to take pictures of the desired object from several locations in a way that the images overlap. There are several software products (free or commercial) that are able to create 3D models of conventional digital photographs. All of the images taken during this project were processed in PhotoScan by Agisoft LLC. The algorithm known as Structure From Motion (SFM) (Ullman, 1979) is used for the reconstruction of the scene, camera location and inner and outer image orientation parameters. The algorithm searches for geometrical structures within the images and traces their movement and appearance in the other images. The SFM results in a sparse point cloud that is then further processed into a dense point cloud by multiview stereo reconstruction algorithms (Seitz et al., 2006). The dense cloud is further transformed into a mesh (triangulated surface) representing the resulting DSM. The principles of 3D scene reconstruction in the PhotoScan environment is presented in many publications—i.e. (Verhoeven et al., 2012).

Terrestrial laser scanning is another option of creating 3D models of historical landmarks, statues or stone crosses. Many studies comparing the usage of laser scanning and close-range photogrammetry have been done in the past years (Barsati et al., 2013), (Doneus et al., 2011). Each of these methods has pros and cons (Kersten, 2006). With respect to the cost of the laser scanning equipment, transportation to the remote areas in the Ore Mountains and the ease of creating 3D models from digital photographs, all of the images taken during this research were taken with the Sony NEX 7 camera. The 3D model was created, using approx. 20–30 images and the SFM algorithms for each of the stone crosses identified in the field.

All of the models are available on the Sketchfab.com server and linked directly the File-Maker database and web mapping application. Example of created 3D model is on Figures 2 and 3.

Figure 2.   3D model of the stone cross created in the PhotoScan environment.

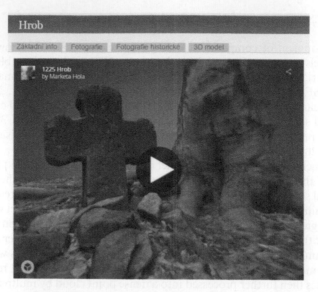

Figure 3.   Created 3D model in the web-page environment.

4   RESULTS PRESENTED ONLINE

The information system containing the resulting database of the stone crosses is available at http://zanikleobce.fzp.ujep.cz/krize/index.php. The information system consists of two parts—the textual output from the database in a form of web-page and the interactive web-map.

4.1   Web pages

The web pages present all of the gathered information and they are visualized directly from the FileMaker database using the PHP engine. This technology allows generating the web pages the in real time from the data in the database. The web page structure is simple and offers the user complete list (with overview) of the documented stone crosses and links to the web map. Each of the database record contains the above described information about the stone cross, photographs, the 3D model (if created) and hyperlink pointing to their location in the web-map.

4.2   The web-map

The interactive web-mapping application presents the locations of the researched objects together with processed old maps and archival aerial photographs that are used as basemaps. The web-map was built using the ArcGIS API for JavaScript and the WebAppBuilder environment. The web-map is available from the project homepage.

The functionality of the ArcGIS API for JavaScript is very wide (spatial data publication within ArcGIS server, geoprocessing, feature search and extraction, vector data editing, mash-ups, etc.). Within this project only basic functionality allowing the user to view the processed data, swipe in-between archival maps and aerial photographs, view attribute data for each stone cross and print of the desired information was implemented. The default base map is the orthophoto of the whole Czech Republic and the default operational layers is the point feature layer representing the researched objects. Every point has a pop-up window containing the basic information, thumbnail image and links to the database and to the 3D model (see Figure 4).

Figure 4. The web-map environment.

5 CONCLUSIONS

The current state and historical background of small historical landmarks (stone crosses) in the district Chomutov, Most and Teplice was evaluated using a background research in the archives and other archival data (old maps and aerial photographs). A comprehensive database containing all the gathered data was created using the FileMaker database system.

The actual presence of the stone crosses was evaluated by a comprehensive field survey. 134 stone crosses were filed altogether in the region and 56 were documented as missing (completely destroyed or moved to an unknown place).

The biggest contribution of this project is the creation of 3D models of the existent crosses using close range photogrammetry. Altogether 6005 images were taken and used for creation the of 74 3D models.

The 3D models were created using the Structure from Motion modelling and published as interactive models in an online 3D models gallery. Another contribution of the project is the usage of created 3D models as a tool for preservation of the cultural heritage and landscape memory. The 3D models may be used as data sources for 3D prints of the documented stone crosses in the future.

All of the processed data are available at http://zanikleobce.fzp.ujep.cz/krize/index.php.

ACKNOWLEDGEMENTS

This work was supported by the J. E. Purkyně University Student grant agency and covered by the project "Detecting landscape changes using the methods of geoinformatics".

REFERENCES

Aber, J.S., Marzolff, I., Ries, J.B. (2010). Small-Format Aerial Photography: Principles, Techniques and Geoscience Applications, Amsterdam, London: Elsevier Science.
Alsadik, B., Gerke, M., Vosselman, G. (2013). Automated camera network design for 3D modeling ofculturalheritageobjects. Journal of Cultural Heritage, 14 (6), pp. 515–526.

Barsanti, Gonizzi S., Remondino, Fabio, Visintini, Frank D. (2013). 3D surveying and modeling of archaeological sites, some critical issues, in: Grussenmeyer, P. (ed.). ISPRS Annals of the Photogrammetry, Remote Sensing and Spatial Information Sciences, Volume II-5/W1, 2013 XXIV International CIPA Symposium, 2, 6 September 2013, Strasbourg, France, s. 145, 150.

Brojír J., Svoboda F. Smírčí kříže, křížové kameny: a jiné pozoruhodné kamenné památky [Concilliation crosses, stone crosses and other stone monuments] [online]. Webarchiv,2001 [cit. 2018–01–20]. Available at: http://smircikrize.euweb.cz/Ceska_Republika/Most/Most_o.html.

Brůna, V., Pacina, J., Pacina, J., Vajsová, E. (2014). Modeling the extinct landscape and settlement for preservation of cultural heritage. Città e Storia. Vol. IX, nr. 1, pp. 131–153. ISSN: 1828-6364.

Cardenala, J., Mataa, E., Castroa, P., Delgadoa, J., Hernandeza, M.A., Pereza, J.L., Ramos, M, Torresa, M. (2004). Evaluation of a digital non metric camera (Canon D30) for the photogrammetric recording of historical buildings. In: Altan, Orhan (ed.). ISPRS Congress Istanbul. Vol. XXXV, Part B5, s 455, 460.

Chandler, Jim H., Fryer, John, G, Jack, Amanda. (2005) Metric capabilities of low-cost digital cameras for close range surface measurement. Photogrammetric Record Vol. 20, no. 109, s. 12–26.

CUZK (Czech office for surveying and cadaster), Base map of the Czech Republic at 1:10 000 and Orthophoto of the Czech Republic [cit. 2018–01–20]. Available at: http://geoportal.cuzk.cz.

Doneus, M.,Verhoeven, G., Fera, M., Briese, Ch., Kucera, M., Neubauer W. (2011). From deposit to point cloud, a study of low-cost computer vision approaches for the straightforward documentation of archaeological excavations. In Čepek, Aleš et al., (eds). International CIPA Symposium, Volume 6, s 81, 88.

Dreyhausen, Walter von. (1940). Die alten Steinkreuze in Böhmen und im Sudetengau. Reichenberg: Sudetendeutscher Verlag Franz Kraus. Beiträgezursudetendeutschen Volkskunde.

Erenoglu, R.C., Akcay, O., Erenoglu, O. (2017). An UAS-assisted multi-sensor approach for 3D modeling and reconstruction of cultural heritage site. Journal of Cultural Heritage, 26, pp. 79–90.

Kersten, Thomas P. (2006). Combination and Comparison of Digital Photogrammetry and Terrestrial Laser Scanning for the Generation of Virtual Models in Cultural Heritage Applications. In: Ioannides, M, Arnold, D., Niccolucci, F., Mania, K. (eds.). The 7th International Symposium on Virtual Reality, Archaeology and Cultural Heritage, VAST (2006), s. 207–214.

Procházka, Z., Urfus, V., Wieser, S., Karel, T., Vít, J. (2001). Kamenné kříže Čech a Moravy. [Stone crosses of Bohemia and Moravia]. Praha: Argo, 2001. ISBN 80-7203-370-0.

Qi, Y., Yang, S., Cai, S. (2008). Method for reconstruction and representation 3D models in digital museum, Journal of Computational Information Systems, vol. 4, no. 4, pp. 1721–1726.

Quan, L. (2010). Image-based Modeling. Springer, New York.

Seitz, S.M., Curless, B., Diebel, J., Scharstein, D., Szeliski, R. (2006). A comparison and Evaluation of multi-view stereo reconstruction algorithms. In: Fitzgibbon, Andrew, Taylor, Camillo, J. LeCun, Yann (eds.). 2006 IEEE Computer Society Conference on Computer Vision and Pattern Recognition (CVPR'06), vol. 1. IEEE, Washington, s. 519, 528.

Ullman, S. (1979). The interpretation of structure from motion. Proceedings of the Royal Society of London, B203, s. 405–426.

Verhoeven, G., Doneusb, M., Briese, Ch., Vermeulen F. (2012). "Mapping by matching: a computer vision-based approach to fast and accurate georeferencing of archaeological aerial photographs". Journal of Archaeological Science, Volume 39, Issue 7, July 2012, s 2060–2070.

Advances and Trends in Engineering Sciences and Technologies III – Al Ali & Platko (Eds)
© 2019 Taylor & Francis Group, London, ISBN 978-0-367-07509-5

Impact of new developed adhesion promoters for bituminous binder doping purposes on performance-based behavior of asphalt mixes

T. Valentová & J. Valentin
Department of Road Structures, Faculty of Civil Engineering, CTU in Prague, Czech Republic

J. Trejbal
Department of Mechanics, Faculty of Civil Engineering, CTU in Prague, Czech Republic

ABSTRACT: This paper is focused on the verification of functionality and overall technical applicability potential of newly developed chemical additives, which use the principle of nanotechnology techniques and poly-condensed amines. The aim of these additives is to improve adhesion between bitumen and aggregate particles whereas the later target is (not followed by the research done) to introduce such additives to the market. The verification of adhesion behavior of bitumen doped by these additives was performed on three different types of aggregates (granite porphyry, granodiorite and mixed rock) which are normally available and used by the road construction industry in the Czech Republic—they are used for asphalt mix production. These aggregate types represent different mineralogical-petrographic composition and therefore they have different level of hydrophility. The adhesion can be directly determined by various test methods, which are represented by standards and procedures worldwide.

1 INTRODUCTION

In recent years, the decreasing quality of bituminous binders has been discussed repeatedly. This lack of quality is manifested by faster progress of ageing, higher fragility and lower resistance to water immersion and/or freezing effects, lower adhesion etc. This can be partly attributed to the origin and the overall quality of the crude oil, to the processing of crude oil as well as to the particular way, how the basic paving grade bitumen is produced. Due to the continuous development of refinery processing technologies significantly more effective exploitation of the crude oil is possible mainly with respect to the yield of light and semi-dense distillates, not the crude oil residuum that forms the bitumen. Nevertheless, thanks to the research and development it is presently possible to chemically dope and modify these basic bituminous binders. Using different types of additives it is possible to set properties of these binders so that they are able to fulfil even very high requirements of quality, performance and durability, which is expected for asphalt pavements where such binder is used. This is closely connected with higher costs, which finally influences the overall price of a pavement structure. Economic markets actually offer inexhaustible variety of different types of additives, but not each of these chemical compounds really fulfills necessary requirements. Increasing attention is paid to water susceptibility resistance to UV radiation and the asphalt mix thermal susceptibility. Based on these aspects the research conducted for more than six years in the road laboratory of CTU in Prague has been focused on testing various types of adhesion promoters and their impact on asphalt mix performance-based behavior. The effort remains to select suitable additives, which would lead to a high-valuable product, which would have a secured stability for a reasonably long time.

2 INPUT MATERIALS OF ASPHALT MIXTURE

2.1 Adhesion promoters

Within experimental study presented by this paper the aim was to evaluate and verify the functionality and effective potential of newly developer types of chemical additives which are based on nanotechnology principles and poly-condensed amines based on assessments of improved adhesion between the bituminous binder and aggregate particles. The tested additives are fluids (additives with very low viscosity at normal room temperature) which are add to hot bitumen. They main function is to lower the surface energy which helps to reach a better coating of particular aggregate particles and results in stronger bonds between the binder and aggregate (the water-loving behavior/hydrophilic behavior of the aggregate is minimized). First group of adhesion promoters was based on additives marked AD and ADM. These are reactive products of unsaturated fatty acids with diethanolamine. The additives can be characterized as brown viscous fluids, which are recovered by condensation of unsaturated fatty acids at high temperature with disposal of water. The acids are dissoluble in ethanol and acetone and can be emulsified in water. In the bituminous binders, such additives are soluble at standard blending/mixing temperatures. The additive according to the producer shall be resistant to long-term heating (150 hours) at 150°C to 160°C without any changes in its activity. Second group of adhesion promoters is represented by additives TA and TA, which are based on alkylsilanes, which usually increase bitumen adhesion to acidic aggregates and improve workability and compactibility of asphalt mixtures at lower temperatures. They withstand long-term heating at 150°C to 160°C without any changes in their activity. Additives TA and TC are viscous dark fluids with various content of alkylsilanes which are soluble in polar or nonpolar organic solvents. These additives are normally dosed in bitumen by 0.2% to 0.4%. For the experimental study a regular paving grade bitumen 50/70 was used. For sake of comparison of various types of applicable additives at the same time a control sample was tested as well. In this case neat paving grade bitumen of the mentioned gradation was used.

2.2 Asphalt mix design

For impact assessment of selected types of adhesion promoters in asphalt mixture a regular ACbin 16+ mixture for binder course and AC_{surf} 11+ mixture for sur-face course were chosen and used with aggregate from the Brant quarry (granite porphyry) and Sýkořice quarry (spilite). At the same time a standard paving grade bitumen 50/70 was used in the mixtures as well. Reference mix was produced with no additive.

3 LABORATORY STUDY OF IMPACT ADHESION PROMOTERS ON ASPHALT MIXTURE CHARACTERISTICS IN SURFACE AND BINDER COURSES

3.1 Evaluation of bitumen to aggregate adhesion

The nonharmonised test of quality of adhesion between bitumen and aggregate was conducted and evaluated according to CSN 73 6161. The sample of 300 g aggregate was coated by a specific amount of bitumen (12 g). Subsequently, the sample in a glass vessel was stored for one day, and then was conditioned in water to the temperature of $60 \pm 3°C$ for 60 minutes. According to CSN 73 6161, bitumen to aggregate adhesion is evaluated using a similar system as in the case of method B given in the harmonized standard EN 12697-11.

Figure 1 presents a comparison of three different types of aggregate combined with new types of adhesion promoters on the basis of alkasilanes or unsaturated fatty acid compounds, wherein the main carrier is paving grade bitumen 50/70. The chart in Figure 1 demonstrates that only 60% of the total surface of individual aggregate particles were coated in the reference sample using just paving grade bitumen in combination with granite porphyry, and only 40% in combination with granodiorite. In contrast to that, a mineral mix achieved an

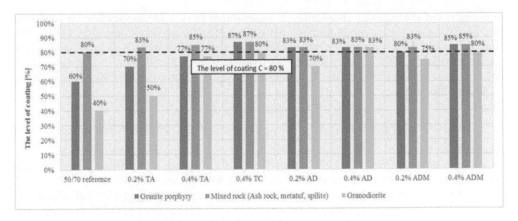

Figure 1. Results of bitumen to aggregate adhesion test according to CSN 73 6161.

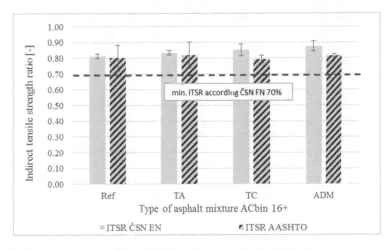

Figure 2. Evaluation of moisture susceptibility of test specimens ACbin 16+.

80% coating which is evaluated satisfactory according to the Czech standard CSN 73 6161. With this type of rock, the application of additives results in further improvement of aggregate particle coating by the added bituminous binder; even up to 87% in the case of adding 0.4% of the additive TC. Focusing on the remaining two types of rocks (minerals), the application of adhesion promoters facilitates an improved level of coating in the case of the higher dose of additive, 0.4%.

3.2 *Evaluation of resistance to moisture damage in asphalt mixtures*

The resistance of the test specimens to water immersion was conducted in compliance with the technical standard ČSN EN 12697-12. In parallel a modified test method according to the U.S. standard AASHTO T-283 was applied as well. The modification is mainly based on taking same test specimens like the one according to ČSN EN 12697-12 (compacted by impact compactor by 2 × 25 blows). It is possible to take account not only on the impact of moisture, but also on the impact of freezing effects. The result of these testing methods is the ratio of indirect tensile strength of specimens tempered in water in comparison to the indirect tensile strength of the test specimens kept at laboratory conditions dry.

Figure 2 summarises the resulting ITSR values according to both, the Czech and the US test method for ACbin 16+ mixture for binder courses with new formulations of chemical surfactants on the basis of alkasilanes or unsaturated fatty acid compounds; this mixture has according to Czech product standards a permitted threshold of just 70% ITSR. The limit was achieved in all cases; reference mixture achieved ITSR of 80%, TA, TC and ADM variants achieved ITSR of 85% on average. With respect to the ITSR results measured for the test specimens exposed to an extra freezing cycle, the value of almost 80% was achieved in all four mix variants.

3.3 Evaluation of stiffness modulus of the selected asphalt mixtures at different temperatures

The stiffness modulus characteristic was determined on cylindrical test specimens (AC_{surf} 11+) in compliance with ČSN EN 12697-26 by the IT-CY method of repetitive indirect tensile stress in a non-destructive test conducted under four different temperatures (0°C, 15°C, 27°C, 40°C). Table 1 indicates the summary of stiffness modules for unaged Marshall specimens and also for test specimens tested after long-term ageing laboratory simulation storing the test specimens for 5 days (120 hours) at 85°C. The results also show interesting finding of thermal susceptibility (ratio between stiffness at 0°C and at 40°C). The smaller the S0/S40 ratio, the less susceptible the mix is to temperature changes.

For the evaluation and/or assessment of asphalt mix strain behavior in the range of moderate temperatures there are in Czech Republic according to the existing pavement design manual determinant stiffness values which it is possible to get at the temperature of 15°C. The highest stiffness modulus at this temperature was reached by the mix variant TA (8218 MPa). In case of the variant TC and the variant ADM—the stiffness was increased in average by 15% if compared to the reference mix. Lower stiffness value variance was found for testing temperature at the range of freezing point. In comparison with control mixtures the options where different adhesion promoters were used reach more or less same stiffness values. On the other hand, at higher temperatures the stiffness variance is more visible, for variant TA the stiffness was increased by almost 30%, in contrary the variant ADM at 27°C lower values have been determined than for the control mixture. From the viewpoint of thermal susceptibility it can be stated that highest thermal susceptibility was shown by mixture containing the ADM additive, the lowest for variant with the TA additive.

During the ageing the stiffness value is usually increased. This is digestedly shown in Table 2. Due to the ageing impact the stiffness at 0°C increased by max. 10% for most variants, only for the mixture with additive ADM a slight decrease in stiffness was identified. This drop in stiffness can be caused by the low testing temperature, since the asphalt test specimens at such low temperatures are already very stiff and the ageing effect is demonstrated in limited or very limited extend. On the other hand it is necessary to take into account the allowable variance of test values during stiffness determination which can be in the range of −20% to +10% from the average. On the other hand, for higher test temperatures the specimens become softer and the tendency for stiffness increase due to bitumen ageing is more

Table 1.   Stiffness characteristics and resistance to thermal susceptibility of assessed mix variants.

Asphalt mixture	Stiffness modulus [MPa] a temperature T				Thermal susceptibility S_0/S_{40} [–]
	0°C	15°C	27°C	40°C	
AC_{surf} 11+_Ref	16 831	6 515	3 016	1 033	16.3
AC_{surf} 11+ 0.4% TA	17 341	8 218	3 687	1 336	13.0
AC_{surf} 11+ 0.4% TC	17 636	7 657	3 455	1 224	14.4
AC_{surf} 11+ 0.4% ADM	17 762	7 310	2 879	1 007	17.6
Stiffness modulus after long-term aging 5d@85°C [MPa] @ temperature T					
AC_{surf} 11+_Ref	17 461	7 309	3 400	1189	14.7
AC_{surf} 11+ 0.4% TA	19 089	9 461	4 037	1537	12.4
AC_{surf} 11+ 0.4% TC	19 039	9 104	3 554	1402	13.6
AC_{surf} 11+ 0.4% ADM	17 539	8 117	3 119	1111	15.8

Table 2. Resulting Ageing Indexes (AI) for assessed mixtures AC_{surf} 11+.

Asphalt mixture	Aging Index [AI] for comparing mixture				Thermal susceptibility [–]
	0°C	15°C	27°C	40°C	
AC_{surf} 11+_Ref	1.04	1.12	1.13	1.15	0.90
AC_{surf} 11+ 0.4% TA	1.10	1.15	1.10	1.15	0.96
AC_{surf} 11+ 0.4% TC	1.08	1.19	1.03	1.15	0.94
AC_{surf} 11+ 0.4% ADM	0.99	1.11	1.08	1.10	0.90

visible. By the impact of simulated ageing process the thermal susceptibility index was in average lowered by 5 to 10%, which shows slightly higher apparent resistance of aged test specimens to thermal changes.

4 CONCLUSION

Within this paper, the main effort was to present and discuss in a more complex way mainly the risks related to thermal susceptibility of selected adhesion promoters in regular bituminous binder. Such approach has to consider also remaining components of an asphalt mixture, which are mainly formed by particular types of aggregates. These can vary in physical-mechanical properties, which arise from their formation and genesis, mineralogy and petrographic composition, structure, texture, methamorphosis etc. These characteristics directly influence the overall behavior of the composite material—asphalt mix—as well. The main function of adhesion additives remains in securing sufficiently strong bonds between the binder and the aggregate, which are in case of a regular asphalt concrete in a wearing course of an asphalt pavement subjected to the risk of macro-texture loss, reduced resistance to climatic effects etc. Because the wearing course is directly exposed to climatic effect and the traffic, higher requirements are put on this layer with respect to its durability. Similar is the situation with pavement binder course. The only difference is that the effect of climatic impacts is indirect. The results if these effects and related stresses or strains can be significantly lower, nevertheless it is necessary to consider the overall time during which these effects act on the structure. Usually for binder courses this time is longer in comparison to wearing courses. Therefore, it was at the beginning decided to include in the experimental study two types of asphalt mixtures covering wearing as well as binder course. The comparison of the resulting effects allows by the applying of similar on both types of mixture.

Continuously for many decades attention is increasingly paid to adhesion promoters. The effort to invent new and technically more effective additives in the industry does not fade out. This can be demonstrated by vast spectrum of products, which are offered by the market. The difficult point is to specify which of these chemical compounds are durable and which disappear quickly after being in contact e.g. with hot bitumen. The paper verified potential of several new adhesion promoters using advanced procedures, which normally are not considered by the industry. Tested products were based on alkylsilanes and on products derived from unsaturated fatty acids. From the viewpoint of set bitumen—aggregate particle coating index, a simple test procedure was used as is defined and for about two decades used in the Czech Republic—subjective procedure described by the standard CSN 73 6161. Gained results show positive effect of used additives on the bitumen-aggregate adhesion for all tested aggregate types as well as with respect to increasing the content of the used doping agent. In case of the aggregate type granite porphyry, the coating level was improved by approx. 0% compared to control bitumen (without any adhesion promoter), in case of granodiorite the improvement reached even 30%. The effects of these additives were in parallel validated also on asphalt concrete mixtures—ACbin 16 mm and ACsurf 11 mm—where a hard-set aggregate type but with less suitable adhesion was used (granite porphyry). For all evaluated adhesion promoters a very small improvement in water susceptibility was found for the asphalt mixtures.

Similar trend can be seen also in case of stiffness, where in almost all cases the stiffness value was slightly increased if compared with control mix. However, a more interesting indicator might be the stiffness results after ageing and from that derived ageing index and thermal susceptibility for virgin and aged test specimens. The thermal susceptibility might be one of the suitable indicators for evaluating the effectiveness and stability of adhesion promoters. The highest resistance to thermal changes, i.e. the lowest thermal susceptibility, was found in the mixture containing the TA additive, the worst result was related to the ADM additive.

REFERENCES

Caro, S., Masad, E., Bhasin, A.: Moisture susceptibility of asphalt mixtures, Part 1: mechanisms— *International Journal of Pavement Engineering*, Vol. 9, No. 2, 2008.

Grenfell, J., Ahmad, N., Liu, Y., Apeagyei, A., Large, D., Aireyet, G.: Assessing asphalt mixture moisture susceptibility through intrinsic adhesion, bitumen stripping and mechanical damage. *Road Material Pavement* Des. 15, pp. 131–152, 2014.

Hamzah, M., Kakar, M., Quandri, S., Valentin, J.: Quantification of moisture sensitivity of warm mix asphalt using image analysis technique. *Journal of Cleaner Production.* 68 (2014) pp. 200–208, DOI: 10.1016/j.jclepro.2013.12.072. ISSN 09596526.

Hofko, B., Hospodka, M., Eberhardsteiner, L., Blab, R.: Recent Developments in the Field of Ageing of Bitumen and Asphalt Mixes. *Proceedings of Pozemní komunikace 2015 conference.* Czech Technical University in Prague, 2015.

Islam, M.R., Hossain, M.I., Tarefder, R.A. (2015). A study of asphalt aging using Indirect Tensile Strength test. *Construction and Building Materials.* DOI: 10.1016/j.conbuildmat.2015.07.159. ISSN 09500618.

Solaimanian, M., Harvey, J., Tahmoressi, M., Tandon, V.: Test Methods to Predict Moisture Sensitivity of Hot-Mix Asphalt Pavements, *Transportation Research Board*, 2003.

Valentova, T., Altman, J., Valentin, J., Impact of Asphalt Ageing on the Activity of Adhesion Promoters and the Moisture Susceptibility, *6th European Transport Research Conference*, Warsaw, Poland, 2016.

Valentova, T., Valentin, J., Alternative additives for improving the functional characteristics and performance-based behavior of asphalt mixes in the fine-grained active filler form, *Key Engineering Materials.* 731 (1–9), 2017. ISSN 1662-9795

Valentova, T., Valentin, J., Analyzing the durability of adhesion stability of asphalt mixtures, *Road and Rail Infrastructure, Proceedings of the Conference CETRA*, Zagreb, Croatia, 2018. ISBN 978-953-8168-25-3.

Advances and Trends in Engineering Sciences and Technologies III – Al Ali & Platko (Eds)
© *2019 Taylor & Francis Group, London, ISBN 978-0-367-07509-5*

Safety of visually impaired individuals in public places

R. Zdařilová, P. Kocurová & A. Bílková
VSB-Technical University of Ostrava, Ostrava, Czech Republic

ABSTRACT: Public places and their accessibility are the basic prerequisites for integrating people into the society. Even though legal regulations create conditions to remove barriers for handicapped people, one of the biggest obstacles preventing independent and safe movement of the handicapped are architectural barriers in traffic infrastructure. These are the most significant obstacles that prevent free movement of handicapped people. Public spaces must be designed based on knowledge and subsequent securing of usage conditions in terms of independent movement and navigation of people, namely visually impaired individuals.

1   INTRODUCTION

The purpose of concourses such as squares, streets, pavements and other outdoor spaces is to ensure conditions for general use by persons of any age group and with any disabilities. The concourse forms significantly life chances of their oldest people and those with disabilities. Consequences of their social utilization and architectural solution (Phillips, Siu, Cheng, 2005, Lawton 1989) and the accessibility of public facilities determine the amount of integration of persons with mobility and orientation limitations, including the old people. Accessibility of the concourse and its using is very often perceived from the perspective of seniors. Key spheres of the concourse that have influence on the quality of life of these people are as follows (Laws, 1997):

- Availability with provision of accessibility of facilities for living and civic amenities, including the buildings for employing;
- The mobility with capability of getting over the architectural barriers in traffic infrastructure to ensure barrier-free utilization of the concourse;
- Spatial segregation as a social problem and outcome of dysfunction of availability and mobility.

Secured access to utilization of environment, namely the concourse, can be understood as a chance for implementation of fundamental civil rights and integration in the society. The concourse is open to public use. It is a space where social and cultural rules of public behaviour prevail (Mitchell, Staehli, 2009). Limitation of access to concourse due to obstacles, preventing independent and free movement, including unsuitable design of municipal mobiliary, means curtailment of rights of people with mobility and orientation limitations.

Concourses must be designed providing accessible environment with an emphasis on the removal of barriers namely for people with mobility and visual limitations, including seniors. These groups of users have user specifics stemming from their handicap.

2   USER CONDITIONS OF PEOPLE WITH MOBILITY LIMITATIONS

Independent movement and orientation of every individual is given by their physical capabilities, subsequently forming the criteria for determination of technical requirements for creation of public space. The elementary problem of movement of every person with mobility limitations are different spatial and handling requirements and the possibility of good orientation resulting from the way of acquisition of information on the surrounding space or on its relation to more remote objects and targets. Each partial group of persons with mobility limitations has its own specifics and needs, different sensual and mental capabilities which we must respect when creating the unified accessible environment. It can be said in general that when creating barrier-free conditions, we must work with a bigger handling space and respect the need for clearness of the environment of each facility solved.

Minimum technical requirements on the modification of public spaces are stipulated by construction-technical regulations. Suitable solutions must be based namely on physical, sensory and mental abilities of visually impaired individuals.

2.1 *Importance of spatial orientation and independent movement of visually impaired individuals*

The basic conditions for independent movement (life) of visually impaired individuals is mobility. This refers to being able to overcome problems in the area of spatial orientation and independent movement. Excessive dependence of visually impaired individuals, due to the inability to navigate in space and move independently, significantly complicates social relations of these people, negatively affects their self-perception, and makes independent life practically impossible. This is why in the Czech Republic a lot of attention is paid to teaching spatial orientation and independent movement—an area within the field of special pedagogics; when creating public spaces, the design needs to accept the rules and conditions of this educational process. From this it also follows that technical solutions of barrier-free modification in the Czech Republic are based on this educational process and it is currently not possible to apply the provisions of foreign regulations in this legal environment.

Orientation (Jesenský, 1978) can be defined as a process for obtaining and processing information from the environment in order to literally or figuratively manipulate objects in space or in order to plan and implement relocation in space. In order to develop spatial orientation, it is necessary to have a sufficient general idea about the space, the location of orientation points in space, and its boundaries.

The main prerequisite for spatial orientation and independent movement is therefore mastering basic movement techniques, including the technique of using a long white cane, to enable blind persons to be mobile—to be able to safety and surely move in space using the learned techniques for movement and obtaining information (Wiener, 2006).

An important role in the process of spatial orientation and independent movement is played by the compensatory function of other senses. For public spaces, hearing is very important—it provides visually impaired individuals with enough information about the character of the space, about the direction of movement, and allows them to discover obstacles in their path. Another important sense is touch, which helps them obtain specific and accurate information about their immediate surroundings. In relation to public spaces, this includes namely touching with foot and using a white cane in order to use the surface structure (orientation marks) on their path to identify specific points on their path. Smell can be used to identify certain places by characteristic odors and smells, which allows the visually impaired to recognize their position on their path and also plays an important role. Another tool that can be used is the difference of temperatures, orientation using drafts in passages, etc.

From the aforementioned facts, it follows that a blind person moves in public spaces using much more than only artificially created construction modifications, such as learned techniques for independent movement and orientation. It is important to acknowledge this fact and to be aware of these techniques in order to obtain a better understanding of the principles governing modifications for blind persons. At the same time it needs to be pointed out that visually impaired individuals always move in familiar surroundings, on known paths, and all technical constructional modifications merely improve their orientation in built-up areas.

2.2 *Elements of spatial orientation and independent movement*

One of the important elements of spatial orientation of visually impaired individuals is limitation of deviations from the straight direction. When not using sight, each of us will tend to move in a spiral and in fact everyone significantly prefers to slightly deviate towards one side (Cratty, 1971). Within the teaching process, the blind person learns to walk 15 meters with a final deviation of about 500 mm. However, in order to secure suitable construction modifications we need the deviation of a blind person when walking to be as small as possible. This corresponds to walking 8 to 10 meters without any guiding element and this fact is reflected in legal requirements of construction regulations on interruption of guide lines not exceeding 8 meters.

Another, no less important element is the perception of the inclination of the path. Correct perception of the inclination is important to achieve safer and surer movement of the blind person and also to obtain more specific information about one's surroundings (for example, 78% of blind persons are able to determine an inclination of only 1°).

2.3 White cane technique

White cane is the basic compensation tool for spatial orientation and independent movement of blind people. The long cane technique (Wiener, 2006) refers to the purposeful and educated use of a white cane with a specifically determined ratio of its length to the height of the person. Using a white cane ensures safety and subjective certainty of the blind person while adhering to basic physiological and aesthetic principles of movement.

The basic techniques of using a long cane are the basic stance including grip, the sliding technique, the swinging technique and the diagonal technique. When creating public spaces, it is necessary to take these techniques into account and create the required natural guide lines and to provide a flat surface for walking without any height differences.

The sliding technique is used by beginners, namely in unfamiliar places and to obtain more specific information about the type and quality of the surface or in dangerous places (holes, areas near stairs).

The swinging technique is the most commonly used technique, whereas the bottom end of the cane moves from side to side via a low arch. It is important that the cane touches the ground at the end points of the arch (optimally, it would stop at the guide line). This is the only method that allows the visually impaired to detect holes and vertical breaks (stairs, curb). When the blind persons senses that the cane is lower (the bottom end of the canes is lower than the person themselves), they stop immediately and pull the end of the cane towards them to precisely determine the location of the vertical break. If the break exceeds 80 mm, then the blind person evaluates the place as dangerous and no additional modifications using tiles for the blind are required. On the other hand, if the break is smaller than 80 mm, then the blind person may assess it as a bump on the surface and it is necessary to mark such places as dangerous using tiles for the blind.

The diagonal technique is used mostly in familiar buildings and on stairs. When walking on stairs, feeling with legs is also used and the blind person moves perpendicularly, always on the right; the white cane is always one stair ahead of the leg. The construction of staircases is based on this technique—i.e., we use direct staircases, full and perpendicular risers, and in specialized buildings also information in Braille on the right handle of the staircase. Teaching this technique for walking up and down the stairs then eliminates the need of sensory modifications via special tiles for the blind at the beginning and end of each flight of stairs.

3 PUBLIC SPACES AND THE MOVEMENT OF VISUALLY IMPAIRED INDIVIDUALS

For visually impaired individuals, public spaces need to be understood as a set of various points, lines and marks. Points and marks represent a projection of vertical and horizontal lines at the movement level, whereas marks represent information obtained via the remaining senses. Orientation points and marks are those that can be distinguished from the surrounding space and other marks—they carry new information. Orientation points have a primary function, while orientation marks have a secondary function.

3.1 Orientation points

An orientation point is a certain place (point) that can be easily and quickly recognized when moving on the path and that does not change its place in space or shape. It must be clearly recognizable—with regards to the path and considering weather conditions. For example, orientation points include freely standing columns, small niches in facades of buildings or various obstacles that are avoided namely due to safety reasons (post boxes, phone booths). If a certain space does not have enough orientation points, a set of orientation marks can under certain circumstances be used instead.

3.2 Orientation marks

Orientation marks are phenomena that characterize the overall navigational situation of the surroundings, increase the certainty of the visually impaired individual, contribute to the creation of a better idea of the surroundings and help determine his/her position on the intended path. Depending on their perception, orientation marks can be divided into auditory (characteristic sounds of the surrounding, echolocation), tactile (feeling the sun, wind), vertical (terrain profile, ascent and descent, tilting of the path) and horizontal (changes of direction).

3.3 Guide lines

Visually impaired individuals move in urban environments always by following a guide line. To ensure their safety, it is necessary to guide their movement away from the façade of buildings, which is where most of the semi-high and high obstacles that usually reach above the waist level and cannot be recognized using a white cane are located (post boxes, balustrades, bars, etc.). On the other hand, in terms of orientation (sufficient information about the surroundings, possibility to determine one's position on the path), it is not suitable to move in the middle of the pavement. It may happen that the blind person loses the right direction (when there are more pedestrians) or is pushed by the crowd to the outer side of the pavement towards the road, where there are more obstacles (poles, traffic signs). For these reasons it is necessary for the blind person to maintain continuous contact with the guide line and the necessary distance. The optimum distance is 300–400 mm from the façade and the blind person should continuously check the guide line once every 3–5 steps by swinging the cane on the guide line. This technique clearly demonstrates the justified requirement of legal regulations to keep a clear space around the guide line of at least 900 mm. This delimits a space in which there must be no isolated obstacle that would pose a risk to the blind person. The only exception is the location of elements in the immediate vicinity to the guide line.

3.4 Complex function of orientation points, marks and lines

In the whole movement process of visually impaired individuals, orientation points, marks and guide lines must function as a complex set of senses. Consider the following situation as an example: a main road with a perpendicular secondary road and residential buildings along one side of the main road, interrupted by a secondary road. The surface of the main road is made of asphalt, while the secondary road is paved. From this it follows:

– Corners of residential buildings are orientation points
– Orientation marks are:
– Difference in surfaces: asphalt—pavement
– Noise (and therefore direction) of the traffic on the main road
– Auditory and heat changes related to the open space on the main road

When moving, a blind person intentionally looks for orientation points, while orientation marks are used subconsciously or only to check the position if the blind persons loses direction or orientation.

The above provided example demonstrates how visually impaired individuals use elements of the natural surroundings during movement and orientation and subsequently use special construction modifications. In order to make artificially created elements understandable for these users, they must be clearly identifiable by their size and surface structure.

4 MODIFICATIONS OF PUBLIC SPACES

In order to be able to move independently and safely, persons with full loss of sight need to clearly and understandably (using white cane techniques) identify sensory elements and marks; they locate their route using elements and their links determinable by touch and using auditory information. Modifications of public spaces with regards to visually impaired individuals focus namely on the following:

- Securing guide lines create mostly by elements protruding above the surface (building façade, garden curb, base wall of a fence), in justified cases (platform, surface of stops, road crossings, etc.) elements recognizable using a cane or foot, located on the walking surface (special tiles with surface that can be distinguished from its surroundings using touch).
- Identification of immediate surroundings (e.g., presence of spatial staircases, footbridges, ends of platforms, waiting rooms, etc.).
- Transfer of information about the surroundings and services, using namely auditory means.
- The following measures are intended to help persons with partial loss of sight:
- Use of non-reflective signs with a strong color contrast.
- Short, easily understandable signs in large letters.
- Location plans, instructions for purchasing tickets, etc. should be legible from very small distances and be simple and understandable.

4.1 *Touch elements for independent movement of visually impaired persons*

To allow independent movement of visually impaired persons, the elements must be recognizable by a white cane and feet via a touch contract against the surroundings and must be made of suitable materials. The touch element must always be clearly identifiable based on its size and surface.

Products for creating these elements allowing independent movement of visually impaired persons must not be used for other purposes on the structures. Namely, pavements for pedestrians must be made of materials in compliance with Government Regulations No. 163/2002 Coll., determining technical requirements on selected construction products (Annex 2, Section 12 Construction products for hygienic equipment and other special products—Products for persons with reduced mobility) and the appropriate manuals of the Technical and Testing Building Institute. These manuals define precise dimensions of not only the element itself but also the parameters of required protrusions on the tiles for warning and guiding purposes.

5 MOST COMMON MISTAKES OF BARRIER-FREE MODIFICATIONS FOR VISUALLY IMPAIRED INDIVIDUALS

As already mentioned above, barrier-free modifications for visually impaired individuals must be based on the conditions of their independent movement and orientation while meeting the valid technical requirements on barrier-free use of public buildings. It is when the user's perspective is not properly understood that serious mistakes are made in the design and implementation of barrier-free solutions of walking routes. The most common mistakes in terms of solutions for visually impaired individuals include:

- Pavement for pedestrians with isolated obstacles that does not meet the required minimum width of 900 mm from the guide line.

- The length of pedestrian crossings and places for crossing exceeds the maximum length, signal bands and guide bands of crossings do not form a direct line.
- Generally incorrect and often illogical solutions of tactile elements—signal and warning bands, artificial guide lines. Unsuitable and often dangerous shapes of tactile elements (e.g., signal bands leading outside the crossing axis, sometimes directly into the crossroad).
- Use of low pavement curbs under 80 mm (sometimes at the road level) in excessive and undesirable scope, e.g., parking lots, public spaces. Even despite a warning band, this solution is dangerous and in conflict with legal regulations.
- Incorrect material specification for tactile paving, which is in conflict with legal requirements.
- Obstacles at pavements for pedestrians and technical equipment of the road (pavement) that does not leave the minimum walking space of 1.5 m (or 0.9 m for technical equipment of the road).
- Incomplete solution of the barrier-free route in connection to important buildings/structures in the surroundings.

6 CONCLUSION

Concourses rank among the most important areas in towns and municipalities. They serve for public use and so they must meet user conditions for easy mobility and orientation of all people. Technical requirements for barrier-free use fixed by regulations must stem from these basic user conditions for people with mobility limitations. However, every structure of the traffic infrastructure, every situation of the concourse must be solved individually in connection to the particular locality and future users with their capabilities and possibilities must be kept in mind.

Accessibility of public spaces is measured by the satisfaction of visually impaired users with the degree to which they can freely move around their city. Unfortunately, the general awareness of what a barrier-free environment should look like and why it should be designed in this way still leaves a lot to be desired. Accessibility is primarily associated with physically disabled people, which is only a small group of users. However, the group is wide and not limited to only physically disabled people; it may include every one of us. A barrier-free environment is not about creating special modifications for a specific group of users, but rather about creating an environment suitable for everybody.

REFERENCES

Cratty, B.J. Movement and Spatial Awareness in Blind Children and Youth. Springfield: Charles C.Thomas—Publisher, 1971.
Jesenský, J. a kol: Studijní materiály k prostorové orientaci a samostatnému pohybu zrakově postižených. Praha: SI v ČSR, 1978.
Laws, G. 1997. "Spatiality and Age Relations." pp. 90–101 in A. Jamieson, S. Harper, C. Victor (eds.). Critical Approaches to Ageing and Later Life. Buckingham, Philadephia: Open University Press. Mitchell Don, Lynn Staeheli. 2009. "Public space." pp. 511–516 in Rob Kitchen, Nigel Thrift (eds.). International Encyclopaedia of Human Geography, vol. 8. Amsterdam: Elsevier.
Lawton, M.P. 1989. "Environmental Proactivity and Affect in Older People." pp. 135–164 in S. Spacapan, S. Oskamp (eds.). The Social Psychology of Aging. Newbury Park, CA: Sage.
Nařízení vlády č. 163/2002 Sb., kterým se stanoví technické požadavky na vybrané stavební výrobky, ve znění pozdějších předpisů.
Phillips, D.R., O.L. Siu, H.C.K. Cheng, A.G.O. Yeh. 2005. "Ageing and the Urban Environment." pp. 147–165 in G. Andrews, D.R. Phillips (eds.). Ageing and Place: Perspectives, Policy and Practice. London: Routledge.
Vyhláška č.398/2009 Sb., o obecných technických požadavcích zabezpečujících bezbariérové užívání staveb
Wiener, P. Prostorová orientace zrakově postižených. Praha: Institut rehabilitace zrakově postižených UK FHS, 2006, ISBN 80-239-6775-4.
Zákon č. 183/2006 Sb., o územním plánování a stavebním řádu (stavební zákon), ve znění pozdějších předpisů.

Advances and Trends in Engineering Sciences and Technologies III – Al Ali & Platko (Eds)
© 2019 Taylor & Francis Group, London, ISBN 978-0-367-07509-5

Laboratory design and testing of asphalt mixtures dissipating energy

J. Žák & J. Suda
Faculty of Civil Engineering, Czech Technical University, Prague, Czech Republic

ABSTRACT: Empirical methods of designing asphalt mixtures leave much room for improvement to fully utilize the potential of source materials. This approach is currently being replaced by a mechanistic-empirical approach, but only to a limited degree. In this regard, dissipating asphalt mixtures offers not only a new material, but a distinctly new approach to the design and grading of asphalt mixtures. These new dissipating asphalt mixtures have been developed for their greater resistance to the development of permanent deformations and cracks, as they dissipate the pressure from traffic and climatic influences to other forms of energy. Both conventional and innovative laboratory analyses of individual component properties are used for their development, including analyses of their mutual interactions within the asphalt mixture. This article presents the results as measured in a laboratory, including various dissipating asphalt mixtures comparison to conventionally produced mixtures.

1 INTRODUCTION

In essence, the development of an asphalt mixture is a multidisciplinary solution that must focus on the study of conditions to which this type of product is exposed (climatic impacts, traffic loading), as well as mechanical analyses considering the suitable use of asphalt mixtures dissipating energy within constructions, as well as their interaction with adjacent layers and the entire formation with the subgrade. The solution is further defined primarily by their development, composition of materials, mutual proportions, material structure and the chemical bonds between individual material-components used. Developing a material that meets these pre-defined criteria in a macro-scale and may also be effectively produced in the volumes required for the production of road infrastructure and further processed is equally important.

2 METHODOLOGY

From a certain perspective, dissipating asphalt mixtures are a new material, as well as a new approach to designing asphalt mixtures. Both conventional and innovative laboratory analyses of individual component properties, additives and admixtures, are used for the development and analysis of their mutual interactions, empirical and mechanical-physical properties (Zak et al. 2018).

2.1 *Methodology of design and evaluation*

Eight asphalt binders from different producers were selected and evaluated for the development of the dissipating asphalt mixture. Six were polymer-modified asphalt binders, along with road asphalt, and an asphalt modified by crumb rubber with increased viscosity, produced by mixing road asphalt 50/70 with approximately 15% crumb rubber granule in special mixing equipment annexed to the packing plant.

The selected binders were evaluated both in view of their properties before and during aging and with consideration of their market price. The individual binders were evaluated using both empirical and functional tests, particularly emphasizing those changes caused over the course of asphalt binder aging. Short-term aging was simulated, covering changes occurring over the course of production, application and compaction of the asphalt mixture, as well as long term aging intended to simulate the properties of the asphalt binders upon reaching the lifespan of the layer. Modified binder PmB 45/80-85 was selected from this research based on tested parameters. Comprehensive binders description can be found in (Dašek et al, 2018).

The initial analyses were executed with three conventional (reference) mixtures for base courses—ACL 16S 50/70, ACL 22S 50/70 and ACL 22S PmB 25/55-65. The choice of type of the asphalt mixture with better dissipation energy arises from the evaluation of the reference mixtures, specifically from the evaluation of functional parameters. Coarse-grain versions better resist the development of permanent deformations (Žák et al. 2017). Therefore, the ACL mixture with maximum grain D = 22 mm and qualitative class S was chosen on this basis. The aggregate structure of the individual mixtures was formed by crushed stone (granodiorite).

It is not expected in these newly designed energy-dissipating asphalt mixtures that the material be cheaper in production costs compared with conventionally used materials, but that the new material will have significantly higher usability properties and better functional characteristics. This would decrease the overall life cycle costs of transport infrastructure reconstructions. The economic benefits of using such materials primarily lie with the maintenance providers and owners of transportation-infrastructure type constructions.

New asphalt mixtures are developed with the goal of maximizing the resistance of the asphalt layer and the entire formation (focused on wearing and base courses) against the development of permanent deformations (in the form of ruts) and faults caused by fatigue cracks (in form of cracks). It is important to select suitable measuring equipment with the required parameters, as well as the appropriate measuring methodology and testing conditions:

3 RESULTS AND COMPARISON

3.1 *Optimization of asphalt mixtures*

The choice of an asphalt mixture type with improved dissipation is based on an evaluation of the reference mixtures, specifically the evaluation of their mechanical parameters. Five versions of aggregate grading were created, to correspond to the grading curves according to the Czech national standard ČSN EN 13108-1. The aggregate comes from the same source as the aggregate for reference mixtures. Basic volumetric parameters were measured for the mixtures produced, according to their respective technical specifications.

The individual asphalt mixture versions (A-E) were subsequently optimized to meet the requirements of the Czech national standard ČSN EN 13108-1. Grading curves of these mixtures were evenly distributed within requirements according to the standard.

All reference mixtures were tested for resistance against the development of permanent deformations according to Czech national standard ČSN EN 12697-22 on a Hamburg wheel tracking test in air conditioned environment. This evaluated parameter defines the susceptibility of the asphalt mixture to permanent deformation, based on the depth of ruts formed caused by repeated pressure from a wheel at a defined temperature. The asphalt mixtures intended for testing were compacted in a lamellar compaction device. Above the specified parameters in the standard, the degree of compaction was required within the range of 99.0–101.0%. The test was performed at the temperature of 50°C. The testing specimens were produced using the lamellar compaction device. The results are shown in Figure 1.

As is apparent from Figure 1, even with respect to the reference mixtures (Žák et al. 2017), the key comparison parameters of permanent deformations are not objectively better. Therefore, two additional mixtures with designation (C-D) were designed at the same time, which combine the mixtures C and D in terms of their grain and composition. Nonetheless, given the equipment and methodology used, these parameters are approaching a distinctive limit.

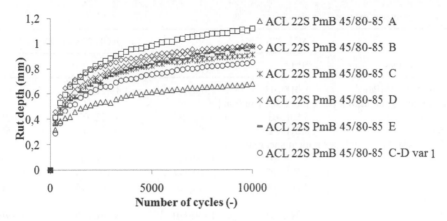

Figure 1.   Record of development of permanent deformations in the Hamburg wheel tracking test.

In essence, all mixtures show very small permanent deformation, even after 10,000 cycles. Although mixture A came out best in terms of evaluating permanent deformations, it was excluded from further evaluation due to its large content of fine particles. In this regard, the Hamburg wheel tracking test is not able to distinguish the best asphalt mixture between superior asphalt mixtures. Among the results shown Figure 1, versions C and C-D var 1 were further evaluated by monitoring their parameters of resistance against development of permanent deformations using the Uniaxial Shear Test and fatigue parameters. The 'dissipating asphalt mixture' for base courses would then be selected based on these two parameters.

3.2   Uniaxial Shear Test

The Uniaxial Shear Test (UST) is performed in standard laboratory equipment, Universal Testing Machine (UTM) with special assembly. In the case of this project, a Universal Testing Machine with a hydraulic unit was used.

Testing samples with a 150 mm diameter were tested using a method of repeated creep following rest periods at two temperatures (50°C a 60°C). The test is done ate controlled stress mode. Both the the loading curve and deformation over the course of each cycle were measured. Mechanical parameters defining the sensitivity of the asphalt mixtures to permanent deformations were obtained through the evaluation of these cycles, the phase shifts between strain and resulting deformation, as well as the size of the elastic and plastic sheer strain. Examples of measured characteristics are included with the individual asphalt mixtures. The actual equipment and testing procedure are described in publication (Zak et al. 2016). Utilization of in laboratory measured parameters and its relation to permanent deformation can be found in (Zak, Coleri, and Harvey 2018; Zak et al. 2017).

The parameters monitored that identify the resistance of asphalt mixtures against permanent deformations are those obtained from the UST. These include the shear modulus, accumulated permanent deformation regression coefficients, number of cycles to reach permanent shear strain, permanent shear strain at 5,000 and 10,000 cycles and increment of permanent shear strain. The testing bodies were produced using a gyratory compactor according to Czech national standard ČSN EN 12697-31 in the range of compaction 99.0–101.0%. The results are shown in Table 1.

3.3   Fatigue

Fatigue is defined as the consequence of disruption of the internal structure of compacted asphalt mixture by repeated strain. It becomes apparent through a gradual decrease of the complex modulus in relation to the number of strain cycles. Therefore, it measures the lifespan of the asphalt mixture defined by the number of repeated strain cycles until the damage of

Table 1. Shear parameters.

Shear parameters	ACL 22S – C		ACL 22S – C-D var 1	
	50°C	60°C	50°C	60°C
Shear modulus [MPa]	1,01E+05	7.08E+04	1.32E+05	7.76E+04
Regression of accumulated permanent deformation [–]				
parameter A	5.80E-03	5.63E-03	4.12E-03	4.62E-03
parameter B	8.21E-02	9.97E-02	9.47E-02	9.65E-02
Number of cycles to reach permanent shear strain [–]				
1% γ	3.25E+03	1.73E+02	9.45E+09	8.03E+08
3% γ	2.27E+07	2.01E+06	1.10E+14	6.15E+10
5% γ	4.26E+09	6.48E+08	8.56E+15	7.17E+12
Permanent shear strain				
at 5,000 cycles [mγ].	1.24E+01	1.49E+01	8.87E+00	1.19E+01
at 10,000 cycles) [mγ].	1.29E+01	1.61E+01	9.24E+00	1.28E+01
Increment of permanent shear strain [mγ/10^3]	9.10E-02	2.32E-01	7.49E-02	1.69E-01

the sample. In the case of a test in control strain mode, the disruption of the binding between strained particles of the material is registered as a decrease of loading resistances, where the loading resistance of the strained binding does not dissipate onto other binds.

The fatigue test was performed in the control strain mode on three levels in a manner ensuring that 50% decrease of the stiffness modulus occuring at the interval 10^4-2.10^6 cycles. The temperature is maintained at a constant value of 20 ± 1°C. The frequency of strain was 30 Hz according to Czech national standard ČSN EN 13108-20.

The testing samples were produced using a segmented compacting device with the compacting level ranging 99.0–101.0%. The specimens were subsequently cut to required testing bodies of defined sizes. The evaluation of fatigue parameters was conducted by multiple methods, specifically 50% decrease of the stiffness modulus according to Czech national standard ČSN EN 12697-24 and the dissipated energy ratio method (Hopman & Pronk) (Hopman et al. 1989, Žák et al. 2014, Boudabbous et al. 2013) and its modified method based on the proportional decrease of the complex modulus of stiffness—(Rowe) (Maggiore et al. 2014, Rowe et al. 1996, Rowe et al. 2000).

These two methods are based on the same idea of dividing the results of the fatigue tests performed in the mode of managed strain in the form of a ratio of the dissipated energy into three phases and defining the values of the cycles (N1) on the border of phase II and III as material fatigue resistance. This has been amended by simple regression analysis to obtain not subjective resistances (N1 values) of all test results (Zak, Valentin, and Mondschein 2013).

3.4 Schematic comparison

The subject of this project is not to compare conventionally produced asphalt mixtures to one another, but to study the impacts of the individual components of the mixtures on the resulting mechanical and physical properties and, on this basis, to design a new mixture with improved energy dissipation. A further goal is to broaden the options for maximizing the functional properties of asphalt mixtures, using more precise processes in evaluation and more precise testing procedures with greater distinguishing capabilities.

The parameters monitored are verified by comparison with conventionally produced mixtures. This comparison provides both information about whether the dissipating mixture was optimally designed, as well as an insight into the impact of the source material on the resulting parameters of the asphalt mixtures.

Figure 2 indicates the comparison of functional parameters of the newly designed dissipating asphalt mixture ACL 22S – C-D var 1 with reference mixtures.

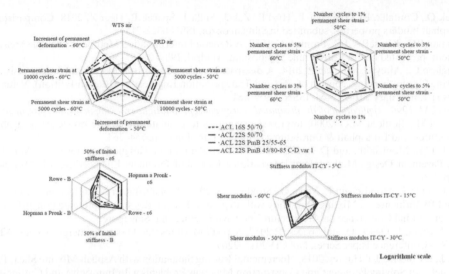

Figure 2. Schematic comparison of the designed dissipating asphalt mixture and reference mixtures.

As is clearly apparent from the above results, compared to other evaluated reference mixtures, the mixture identified as ACL 22S PmB 45/85-80 – C-D var 1 shows qualitatively very good resistance against permanent deformations and fatigue resistance. The above methodology of designing a mixture with both empirical and mechanical tests enabled designing a qualitatively higher-performance asphalt mixture from the perspective of the monitored parameters, while also maximizing parameters within the used source materials regarding permanent deformation and fatigue cycles. The aspect of comparing studied mixtures within optimization using standardized tests (Wheel tracking test and the Uniaxial shear test) must be also highlighted.

4 CONCLUSIONS

This article proves the possibilities for maximizing the functional properties of asphalt mixtures by using more precise evaluating and testing procedures with higher differentiation capabilities.

The benefits of the functional design of dissipating asphalt mixtures can be seen particularly in attaining maximum functional parameters of asphalt mixtures with regard to the source materials used—fillers, binders, additives and admixtures. This research also developed unique combination of mechanical and empirical test methods for asphalt mixture design to reach optimal designs within the parameters monitored. Using such procedures enabled the design of a mixture with a significantly greater resistance against permanent deformations while still maintaining appropriate values of other properties.

ACKNOWLEDGMENTS

This paper was elaborated within the research project No. FV10526 in the program TRIO of the Ministry of Industry and Trade of the Czech Republic.

REFERENCES

Boudabbous M., Millien A., Petit C., Neji J. 2013. Energy approach for the fatigue of thermoviscoelastic materials: Application to asphalt materials in pavement surface layers. In International Journal of Fatigue, Volume 47, Pages 308–318, ISSN 0142-1123.

Dašek O., Coufalík, P., Koudelka T., Hýzl P., Žák J., Suda J., Špaček P., Hegr Z. 2018. Comparison of asphalt binders properties, submitted in Silniční obzor, ISSN 0322-7154.

Hopman P.C., Kunst P.A.J.C., Pronk A.C. 1989. A Renewed Interpretation Method for Fatigue Measurement. presented on 4th Eurobitume Symposium, Madrid, 1989, 2, 557–561.

Maggiore C., Airey G., Marsac P. 2014. A dissipated energy comparison to evaluate fatigue resistance using 2-point bending. Journal of Traffic and Transportation Engineering (English Edition), Volume 1, Issue 1, February 2014, 49–54, ISSN 2095-7564.

Rowe G.M. 1996. Application of the dissipated energy concept to fatigue cracking in asphalt pavements.

Rowe G.M., Bouldin M.G. 2000. Improved techniques to evaluate the fatigue resistence of asphaltic mixtures, 2nd Eurasphalt & Eurobitume Congress Barcelona, pp. 754–763.

Zak J., C.L. Monismith, and D. Jarušková. 2014. Consideration of Fatigue Resistance Tests Variability in Pavement Design Methodology. International Journal of Pavement Engineering, vol. 2014, no. 8, pp. 1–6.

Zak J., Suda J., Mondschein P., Vacin O., Stoklasek S. 2015. Asphalt Binder Modification Technologies and Its Influence on Mixture Durability with Regard to Fatigue Resistance. presented on 3rd Annual international Conference on Architecture and Civil Engineering, Singapore.

Zak, J., Suda, J., Dasek, O. and Spacek, P. 2017. Asphalt Mixtures that Dissipate Energy. Athens: ATINER'S Conference Paper Series, No: CIV2017–2269.

Zak, J., E Coleri, a J.T. Harvey. 2018. "Incremental Rutting Simulation with Asphalt Mixture Shear Properties". In Solving Pavement and Construction Materials Problems with Innovative and Cutting-edge Technologies, 13–24. Hangzhou, China: Springer, Cham. https://doi.org/10.1007/978-3-319-95792-0_2.

Zak, J., E Coleri, J. Stastna, a J.T. Harvey. 2017. "Accumulated Equilibrium Compliance as a Rutting Susceptibility Parameter". Submitted to Sustainable Civil Engineering Journal.

Zak, J., C.L. Monismith, E Coleri, a J.T. Harvey. 2016. "Uniaxial Shear Tester—Test Method to Determine Shear Properties of Asphalt Mixtures". Associaltion of Asphalt Paving Technologists AAPT Journal 85.

Zak, J., J Suda, O Dasek, a P Spacek. 2018. "Asphalt Mixtures that Dissipates Energy—Comparison of Conventional and Newly Developed Mixtures". In Testing and Characterization of Asphalt Materials and Pavement Structures, 55–68. Hangzhou, China: Springer, Cham. https://doi.org/10.1007/978-3-319-95789-0_6.

Zak, J., J Valentin, a P Mondschein. 2013. "Analýza únavového chování asfaltových hutněných směsí—současné trendy a metody". Silniční obzor, Česká silniční společnost 2013 (2): 31–34.

Author index